THE INTERNATIONAL SERIES OF
MONOGRAPHS ON CHEMISTRY

THE INTERNATIONAL SERIES OF MONOGRAPHS ON CHEMISTRY

---

1. J. D. Lambert: *Vibrational and rotational relaxation in gases*
2. N. G. Parsonage and L. A. K. Staveley: *Disorder in crystals*
3. G. C. Maitland, M. Rigby, E. B. Smith, and W. A. Wakeham: *Intermolecular forces: their origin and determination*
4. W. G. Richards, H. P. Trivedi, and D. L. Cooper: *Spin-orbit coupling in molecules*
5. C. F. Cullis and M. M. Hirschler: *The combustion of organic polymers*
6. R. T. Bailey, A. M. North, and R. A. Pethrick: *Molecular motion in high polymers*
7. Atta-ur-Rahman and A. Basha: *Biosynthesis of indole alkaloids*
8. J. S. Rowlinson and B. Widom: *Molecular theory of capillarity*
9. C. G. Gray and K. E. Gubbins: *Theory of molecular fluids. Volume 1: Fundamentals*
10. C. G. Gray and K. E. Gubbins: *Theory of molecular fluids. Volume 2: Applications* (in preparation)
11. S. Wilson: *Electron correlation in molecules*
12. E. Haslam: *Metabolites and metabolism: a commentary on secondary metabolism*
13. G. R. Fleming: *Chemical applications of ultrafast spectroscopy*
14. R. R. Ernst, G. Bodenhausen, and A. Wokaun: *Principles of nuclear magnetic resonance in one and two dimensions*
15. M. Goldman: *Quantum description of high-resolution NMR in liquids*
16. R. G. Parr and W. Yang: *Density-functional theory of atoms and molecules*
17. J. C. Vickerman, A. Brown, and N. M. Reed (editors): *Secondary ion mass spectrometry: principles and applications*
18. F. R. McCourt, J. Beenakker, W. E. Köhler, and I. Kuščer: *Nonequilibrium phenomena in polyatomic gases. Volume 1: Dilute gases*
19. F. R. McCourt, J. Beenakker, W. E. Köhler, and I. Kuščer: *Nonequilibrium phenomena in polyatomic gases. Volume 2: Cross-sections, scattering, and rarefied gases*
20. T. Mukaiyama: *Challenges in synthetic organic chemistry*
21. P. Gray and S. K. Scott: *Chemical oscillations and instabilities: non-linear chemical kinetics*
22. R. F. W. Bader: *Atoms in molecules: a quantum theory*
23. J. H. Jones: *The chemical synthesis of peptides*
24. S. K. Scott: *Chemical chaos*
25. M. S. Child: *Semiclassical mechanics with molecular applications*
26. D. T. Sawyer: *Oxygen chemistry*
27. P. A. Cox: *Transition metal oxides: an introduction to their electronic structure and properties*
28. B. R. Brown: *The organic chemistry of aliphatic nitrogen compounds*

The Organic Chemistry of Aliphatic Nitrogen Compounds

B. R. BROWN

CLARENDON PRESS · OXFORD
1994

Oxford University Press, Walton Street, Oxford OX2 6DP
Oxford New York Toronto
Delhi Bombay Calcutta Madras Karachi
Kuala Lumpur Singapore Hong Kong Tokyo
Nairobi Dar es Salaam Cape Town
Melbourne Auckland Madrid
and associated companies in
Berlin Ibadan

Oxford is a trade mark of Oxford University Press

Published in the United States
by Oxford University Press, New York

A catalogue record for this book is available from the British Library

Library of Congress Cataloging in Publication Data
Brown, B. R. (Ben R.)
The organic chemistry of aliphatic nitrogen compounds / B.R. Brown.
(International series of monographs on chemistry; 28)
1. Nitrogen compounds. I. Title. II. Series.
QD305.N84B76 1994 547'.44—dc20 93-31645
ISBN 0-19-855783-3

Typeset by Joshua Associates Ltd, Oxford
Printed in Great Britain by
Bookcraft Ltd, Midsomer Norton

ACKNOWLEDGEMENTS

It is a pleasure to acknowledge the help and support I have received from a number of individuals and bodies.

For financial assistance towards the final completion of the typescript I am especially indebted to Oriel College (for the award of a Senior Research Fellowship), to the Leverhulme Trust, to the Royal Society of Chemistry, to Oxford University Chest (Walter Gordon Fund), to the Turpin Trust, and to the Tompson Trust.

For their interest and support for the project I am grateful to Sir Ewart Jones, FRS and Lord Dainton FRS.

Dr A. S. Bailey read the typescript and made valuable suggestions. Dr C. K. Prout kindly supplied data relating to bond lengths.

A team of amanuenses gave invaluable assistance with the final stages of preparation: Robert Matson, Elsa and Mark Lewis, James Kraunsoe, Duncan Crowdy, and Susanna Bahar.

My wife helped to co-ordinate these efforts, which without the dedication and patience of my nurses, Debbie, Flo, Jacqui, and Wendy, could never have reached fruition.

Northleach, Glos. B.R.B.
September 1992

POSTSCRIPT

When Ben Brown died on 28 September 1992 work on the manuscript of this book was more than two-thirds complete, the author, despite his rapidly progressing disablement caused by Motor Neurone Disease, having supervised and worked with a team of amanuenses throughout the summer. The remaining chapters were all, bar one (Azo compounds), fully worked and on disc, but needed expanding at certain points. Particular credit must go to Duncan Crowdy, the only member of the summer team still free and willing to carry through the work in the spirit and according to the principles Ben Brown had established. Without his courage and determination the final stages would have been daunting indeed. From October 1992 Duncan Crowdy was joined by Christopher Hoare and, along with Elsa Lewis, by then in Manchester, they have carried out the exacting final revisions. The overall supervision of their efforts was assumed, after Ben Brown's death, by a group of his former colleagues from the Dyson Perrins Laboratory, University of Oxford: Dr A. S. Bailey, Dr G. H. Whitham, and Dr G. D. Meakins. To all of these a debt of gratitude is owed as also to Prof. Lyn Williams (University of Durham), who gave most unstintingly and generously of his time.

Hilda M. Brown
January 1993

CONTENTS

JOURNALS USED AND THEIR ABBREVIATIONS (in bold)

Accounts of **Chem**ical **Res**earch
Acta Chemica **Scand**inavia
Acta Chemica **Scand**inavia, Series **B**
Acta Crystallographica Section **B**
Advances in **Cancer Res**earch
Advances in **Carbohydr**ate **Chem**istry
Advances in **Phys**ical **Org**anic **Chem**istry
Aldrichimica **Acta**
Angewandte **Chem**ie
Angewandte **Chem**ie **Int**ernational **Ed**itio**n** in **Engl**ish
Annalen der Chemie
Annales de **Chim**ie
Arkiv fur **Kemi**
Australian **J**ournal of **Chem**istry
Bioorganic **Chem**istry
Bulletin of the **Chem**ical **Soc**iety of **Japan**
Bulletin de la **Soc**iété **Chim**ique de **Belge**
Bulletin de la **Soc**iété **Chim**ique de **Fr**ance
Bulletin of the **Univ**ersity of **Galati**
Canadian **J**ournal of **Chem**istry
Chemical **Abs**tracts
Chemical **Pharm**acology **Bull**etin
Chemical **Soc**iety **Rev**iews
Chemische **Ber**ichte
Chemistry and **Ind**ustry (London)
Chemistry **Lett**ers
Chimia
Collection of **Czech**oslovak **Chem**ical **Commun**ications
Comptes **Rend**us Hebdomadaires des Seances de l'Acadamie des Sciences, **Series C**
Discussions of the **Faraday Soc**iety
Doklady **Chem**istry (**Engl**ish **Trans**lation)
Experimentia
Fortschritte der **Chem**ischer **Forsch**ing
Gazzetta **Chim**ica **Ital**ia
Helvetica **Chim**ica **Acta**
Heterocycles
Inorganic **Chim**ica **Acta**
Inorganic **Synth**esis
International **Cong**ress of **Chemother**apy **Proc**eedings
Journal für **Prakt**ische **Chem**ie
Journal of **Can**adian **Phys**ics
Journal of **Chem**ical **Educ**ation
Journal of **Chem**ical **Phys**ics
Journal of **Chem**ical **Res**earch **(M**iniprint**)**

Journal of **Chem**ical **Res**earch **(S**ynopses**)**
Journal of **Chem**ical and **Eng**ineering **Data**
Journal of **Heterocyclic Chem**istry
Journal of **Med**ical **Chem**istry
Journal of **Mol**ecular **Spectrosc**opy
Journal of **Org**anic **Chem**istry
Journal of **Organomet**allic **Chem**istry
Journal of **Pharm**acological **Exp**erimen**t**al **Therap**y
Journal of the **Air Pollut**ion **Control Assoc**iation
Journal of the **Am**erican **Chem**ical **Soc**iety
Journal of the **Chem**ical **Soc**iety
Journal of the **Chem**ical **Soc**iety, **Chem**ical **Commun**ications
Journal of the **Chem**ical **Soc**iety, **Perkin Trans**actions **1**: Organic and Bioorganic Chemistry
Journal of the **Chem**ical **Soc**iety, **Perkin Trans**actions **II**: Physical Organic Chemistry
Journal of the **Chem**ical **Soc**iety, Series **A**
Journal of the **Chem**ical **Soc**iety, Series **B**
Journal of the **Chem**ical **Soc**iety, Series **C**
Journal of the **Ind**ian **Chem**ical **Soc**iety
Justus Liebigs Annalen der **Chem**ie
Molecular **Phys**ics
Monatshefte für Chemie
Nature
Organic **React**ions
Organic **Synth**eses
Organic **Synth**eses, **Coll**ected **Vol**umes
Philosophical **Trans**actions
Proceedings of the **Chem**ical **Soc**iety
Proceedings of the **Nat**ional **Acad**emy of **Sci**ences of the **USA**
Propellants and Explosives
Pure and **Appl**ied **Chem**istry
Quarterly **Rev**iews **Chem**ical **Soc**iety
Recueil des **Trav**aux **Chim**ique des **Pays-Bas**
Russian **Chem**ical **Rev**iews
Science
Steroids
Synthetic **Commun**ications
Synthesis
Tetrahedron: Asymmetry
Tetrahedron Letters
Transactions of the **N**ew **Y**ork **Acad**emy of **Sci**ence
Zeitschrift **Elektrochem**ische

Note
F.&F. M. Fieser and Louis Fieser (1967–1990). *Reagents for organic synthesis*, Volumes 1–15. Wiley Interscience, New York.

CHEMICAL ABBREVIATIONS

Ac	Acetyl, CH_3CO
Acac	Acetylacetonate, $CH_3COCHCO_3$
AcOH	Acetic acid, CH_3CO_2H
AIBN	Azobisisobutyronitrile, $(CH_3)_2C(CN)N{=}NC(CN)(CH_3)_2$
Ar	Aryl
9-BBN	9-Borabicyclo [3.3.1] nonane
BMS	Borane dimethylsulphide, $BH_3.S(CH_3)_2$
Bn	Benzyl, $C_6H_5CH_2$
Boc	*t*-Butyloxycarbonyl
Bu	Butyl, C_4H_9
Bu^s	Secondary-butyl, $(CH_3)_2CHCH_2$
Bu^t	Tertiary-butyl, $C(CH_3)_3$
CAN	Ceric ammonium nitrate, $Ce(NH_4)_2(NO_3)_6$
DABCO	1,4-Diazobicyclo [2.2.2] octane
DBN	1,5-Diazabicyclo [3.4.0] nonene-5
DBU	1,5-Diazabicyclo [5.4.0] undecene-5
DCC	Dicyclohexylcarbodiimide
DDQ	Dichloro-dicyano-benzoquinone
DEAD	Diethylazodiacetate
DIBAH	Diisobutylaluminium hydride, $(iso\text{-}Bu)_2AlH$
Diglyme	Diethyleneglycoldimethylether, $CH_3OCH_2CH_2OCH_2CH_2OCH_3$
DMA	Dimethylacetamide, $CH_3CON(CH_3)_2$
DMAP	4-Dimethylaminopyridine
DME	1,2-Dimethoxyethane, $(CH_2OMe)_2$
DMF	Dimethylformamide, $HCON(CH_3)_2$
DMSO	Dimethylsulphoxide, $(CH_3)_2SO$
E, El	Electrophile
EDTA	Ethylenediaminetetraacetic acid, $(HO_2CCH_2)_2NCH_2CH_2N(CH_2CO_2H)_2$
Et	Ethyl, C_2H_5
EtOH	Ethanol, C_2H_5OH
Gly	Glycine
Hal	Halogen
HMPT, HMPA	Hexamethylphosphorictriamide, $[(CH_3)_2N]_3PO$
HOMO	Highest occupied molecular orbital
LDA	Lithium diisopropylamide, $LiN(iso\text{-}Pr)_2$
LUMO	Lowest unoccupied molecular orbital
MCPBA	*m*-Chloroperbenzoic acid, $m\text{-}ClC_6H_4CO_3H$
Me	Methyl, CH_3
MoOPH	Methylphosphoramolybdenum
NBS	*N*-Bromosuccinimide
NCS	*N*-Chlorosuccinimide
Nu	Nucleophile
Ph	Phenyl, C_6H_5
Phe	Phenylalanine
PhH	Benzene, C_6H_6
PMP	1,2,2,6,6-Pentamethylpiperidine

PPA	Polyphosphoric acid
Pr	Propyl, C_3H_7
Pr^i	Isopropyl, $(CH_3)_2CH$
PTC	Phase transfer catalyst
Py	Pyridine
R	Alkyl
Ser	Serine
TCNE	Tetracyanoethylene, $(NC)_2C{=}C(CN)_2$
TFA	Trifluoroacetic acid, CF_3CO_2H
THF	Tetrahydrofuran
TMEDA	*N*,*N*,*N′*,*N′*-Tetramethylethylenediamine
TPP	Triphenylphosphene
Ts	Tosyl, *p*-$CH_3C_6H_4SO_2$

Ben Brown, Todd Fellow of Oriel College and University Lecturer in Organic Chemistry, began his study of chemistry at Mexborough Grammar School. From there he gained a West Riding County Scholarship to Oxford, obtaining a First in Chemistry in 1947. The research carried out for his subsequent D.Phil.earned him an Exhibition of 1851 Senior Studentship, and a Research Fellowship at Trinity Hall Cambridge (1951–54). There he began work on natural products, concentrating in particular on structure elucidation of the aphin pigments.

He returned to Oxford in 1954 as Fellow and Tutor of Oriel College, where his research centred on the vegetable tannins. He studied the degradation and synthesis of the flavans and their incorporation into tannins, and later worked on heterocyclic oxygen compounds and nitrosamines.

1

AMINES[1]

(1) (2)

Structurally aliphatic amines are derived from ammonia by alkylation, i.e. the substitution of alkyl groups for the hydrogen atoms in ammonia. They are classified as primary (**1**, $R^1 = R^2 = H, R^3 =$ alkyl), secondary (**1**, $R^1 = H$, $R^2 = R^3 =$ alkyl), and tertiary (**1**, $R^1 = R^2 = R^3 =$ alkyl). Further alkylation of the nitrogen atom yields quaternary salts (**2**, $R^1 = R^2 = R^3 = R^4 =$ alkyl, $X^- =$ an anion) corresponding to the ammonium salts. The amines are pyramidal in shape (**3**) with bond angles near to those of a regular tetrahedron (109°). Rapid inversion at ordinary temperatures prevents the isolation of enantiomers[2] and results in

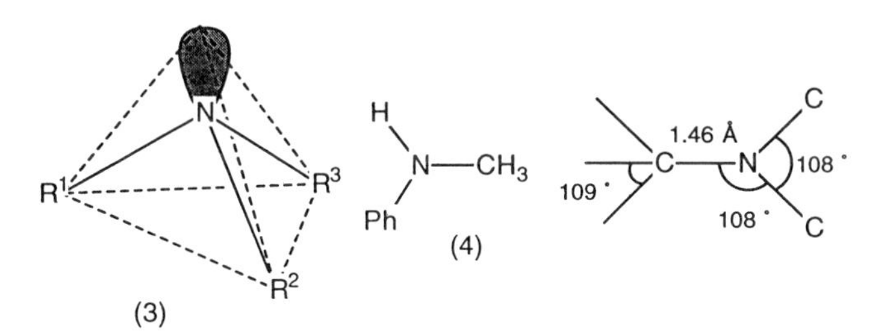

(3) (4)

no splitting of the NMR signals from the hydrogen atoms on an α-carbon atom, for example in (**4**). If the alkyl groups are incorporated into suitable cyclic structures, the inversion is prevented, for example,[3] in tetrahydro-1,3-oxazine (**5**) the hydrogen atom on the nitrogen atom is axial and its NMR signal is split with coupling constants of 13.1 and 2.9 Hz. Compounds such as Tröger's base (**6**) have been resolved.[4] The quaternary salts (**2**) have a tetrahedral nitrogen atom which can give rise to separable enantiomers.[5]

(5) (6)

As expected from their relationship with ammonia, amines are alkaline in water as shown by the equilibrium in Scheme 1.1, the position of which depends upon the structure of the amine.[2]

$$R_3N + H_2O \rightleftharpoons R_3\overset{+}{N}H + HO^- \quad (1.1)$$

Because of their basicity, amines form salts (**7**) with acids, the stability of which in solution depends upon the basic strength of the amine and the acidic strength

$$R^1R^2R^3\overset{+}{N}H\ Cl^-$$

(7)

of the acid. Amines of low molecular weight (e.g. CH_3NH_2, $(CH_3)_2NH$, and $CH_3CH_2NH_2$) are gases or volatile liquids at ordinary temperatures and pressures and most liquid amines can be distilled at atmospheric pressure without decomposition.

1.1 Preparation

1.1.1 *Nucleophilic substitution by nitrogen at a saturated carbon atom*

This is the classical method of Hofmann[6] from primary and secondary alkyl halides and ammonia or an amine in a suitable solvent (e.g. ethanol or water, or in liquid ammonia in a sealed tube). Tertiary halides undergo elimination to yield an alkene and therefore primary amines with the amino group attached to a tertiary carbon atom are better prepared from a nitrile by the Ritter reaction (see Chapter 4, p. 211). An excess of alkyl halide finally yields quaternary halides. Similarly an excess of ammonia can sometimes be used to obtain chiefly a primary amine but mixtures often result and this is a disadvantage of applying the unmodified reaction (Scheme 1.2).

$$NH_3 + R\text{-}Hal \longrightarrow R\text{-}NH_2 \xrightarrow{R\text{-}Hal} R_2NH \xrightarrow{R\text{-}Hal} R_3N \xrightarrow{R\text{-}Hal} R_4\overset{+}{N}\ Hal^- \quad (1.2)$$

However, there are several ways of making this reaction a viable preparative method.

The Hinsberg separation[7] with a sulphonyl chloride allows each class of amine (primary, secondary, and tertiary) to be isolated on hydrolysis of the sulphonamides of primary and secondary amines. Tertiary amines do not react with the chloride (Scheme 1.3).

$$R\text{-}NH_2 + ArSO_2Cl \longrightarrow R\text{-}NHSO_2Ar$$
$$(\text{soluble in alkali} \longrightarrow R\text{-}\bar{N}\text{-}SO_2Ar)$$
$$R_2NH + ArSO_2Cl \longrightarrow R_2N\text{-}SO_2Ar \quad (\text{insoluble in alkali}) \qquad (1.3)$$

A valuable variant of this preparative method by S_N mechanisms involves the use of a diacylated amide (for example, **8**) in which the nitrogen atom is protected by acyl groups and thus can only undergo monoalkylation by an alkyl halide. Subsequent hydrolysis of the alkylated diacylated amide (**9**) yields a primary amine. The original use of this procedure is the Gabriel reaction,[8] Scheme 1.4.

Gabriel

Potassium Phthalidamide (8) $\xrightarrow[\text{heat}]{\text{R-Hal}}$ (9) $\xrightarrow{\text{aq. } H_3O^+}$ $R\text{-}NH_2$

Ing and Manske

(phthalimide, N-H) $\xrightarrow[\text{heat}]{\text{R-Hal / } K_2CO_3}$ (9) $\xrightarrow{H_2N\text{-}NH_2}$ $R\text{-}NH_2$ (1.4)

For a review of the Gabriel reaction see Ref. 9.

Modifications of the Gabriel reaction

Several useful modifications of the original Gabriel procedure have been introduced. One of the first was the Ing and Manske modification;[10] shown in Scheme 1.4. Another is the method of Sheehan and Bolhofer[11] in which the alkyl halide is heated with potassium carbonate and phthalimide (**8**) and hydrazine is used for the decomposition of the amide (**9**). L. A. Carpino[12] used compound (**10**) as shown in Scheme 1.5.

$$Na^+ \; ^-N(COOCMe_3)_2 \; (10) \xrightarrow{\text{RHal}} RN(COOCMe_3)_2 \longrightarrow R\overset{+}{N}H_3 \; Cl^-$$

R = α,α'-dibromo-o-xylene
ethyl bromoacetate (1.5)

In another modification an azide was used.[13] The resulting azide was then reduced with $LiAlH_4$ or hydrogen and a catalyst. The substitution involves a Walden inversion and so stereospecificity is retained.

Phase transfer has been applied to this type of preparation (Scheme 1.6).[14]

R-X + (phthalimide) $N^- K^+$ $\xrightarrow[\text{PhCH}_3 / \text{H}_2\text{O, Phase Transfer}]{C_{16}H_{33}P^+Bu^n_3Br^-}$ (phthalimide) N—R + KX

X= Hal or SO_4R

80-95% (1.6)

Phase transfer has also been applied to the preparation of secondary amines (Scheme 1.7).[15]

R^1—NH_2 + $(EtO)_2P(H){=}O$ $\xrightarrow[\text{CHCl}_3 \text{ at } 0^\circ - 5^\circ\text{C, Phase Transfer}]{BnN^+Et_3\,Cl^- / 20\%\text{ aq. NaOH}}$ $(EtO)_2P(NHR^1){=}O$ 70-90%

$(EtO)_2P(H){=}O$ $\xrightarrow{CCl_4}$ $(EtO)_2P(Cl){=}O$ $\xrightarrow{R^1\text{-}NH_2}$ $(EtO)_2P(NHR^1){=}O$

$(EtO)_2P(NHR^1){=}O$ $\xrightarrow[\text{Phase Transfer}]{R^2Hal,\ aq.NaOH,\ Bu^n_4N^+\ HSO_4^-}$ $(EtO)_2P(NR^1R^2){=}O$ $\xrightarrow{HCl / THF}$ $R^1R^2NH_2^+\,Cl^-$

80-95% (1.7)

A further modification of the Gabriel method allows primary amines to be prepared from alcohols (Scheme 1.8).[16]

$$EtO_2C{-}N{=}N{-}CO_2Et + PPh_3 + ROH + \text{phthalimide (NH)} \longrightarrow EtO_2C{-}NH{-}NH{-}CO_2Et + O{=}PPh_3 + \text{N-R phthalimide} \xrightarrow{H_2N{-}NH_2 / EtOH} R{-}NH_2 + \text{phthalhydrazide (NH–NH)} \tag{1.8}$$

In this modification, a high retention of optical activity is achieved, with almost complete inversion occurring from optically active alcohols, such as 2-octanol (**11**), yield = 38 per cent.

$$HO{-}C(H)(C_6H_{13}){-}CH_3$$

The phthalimide method has been applied to the preparation of (*E*)-allylamines as shown in Scheme 1.9.[17]

$$\text{phthalimide } N^- \; Na^+ + CH_2{=}CH{-}P^+Bu^n{}_3 Br^- + Ph{-}CHO \xrightarrow{LiBr} \text{phthalimide } N{-}CH_2{-}C(H){=}C(H){-}Ph \xrightarrow{H_2N{-}NH_2} (H)(Ph)C{=}C(CH_2{-}NH_2)(H) \tag{1.9}$$

83%
100% E

Another variant of the Gabriel method uses a cyclic silicon amine (Scheme 1.10).[18]

$$\text{cyclic } (\text{SiMe}_2)_2\text{NH} \xrightarrow[\text{THF}]{\text{KH}} \text{cyclic } (\text{SiMe}_2)_2\text{N}^- \text{K}^+ \xrightarrow{\text{R-Br}} \text{cyclic } (\text{SiMe}_2)_2\text{N—R} \xrightarrow[\text{heat}]{\text{aq. HCl}} \text{R-NH}_2 \;\; 60\text{-}90\% \qquad (1.10)$$

Use of amide alternatives to phthalimide

Trifluoromethyl sulphonamides (triflamides) have been used with the advantage of achieving a one-pot synthesis of a primary amine (Scheme 1.11).[19]

$$\text{Ph—CH(CO}_2\text{Me)—}\overset{+}{\text{N}}\text{H}_3\ \text{Cl}^-\ (+/-) \xrightarrow[\text{Et}_3\text{N / CH}_2\text{Cl}_2]{\text{O(SO}_2\text{CF}_3)_2} \text{Ph—CH(CO}_2\text{Me)-NH—SO}_2\text{CF}_3\ (81\%) + \text{R-Br}$$

$$\xrightarrow[\text{reflux}]{\text{K}_2\text{CO}_3 \text{ in CH}_3\text{CN}} \left[\text{R—N(SO}_2\text{CF}_3\text{)—CH(Ph)—CO}_2\text{Me}\right] \xrightarrow{-\ \text{F}_3\text{C-SO}_2\text{H}} \left[\text{R—N=C(Ph)—CO}_2\text{Me}\right]$$

$$\xrightarrow{\text{aq. HCl}} \text{R—}\overset{+}{\text{N}}\text{H}_3\ \text{Cl}^- + \text{Ph—CO—CO}_2\text{Me}$$

R	Yield $R\text{—}\overset{+}{N}H_3\ Cl^-$ (%)
Bn	90
$n\text{—}C_7H_{15}$	78
$Ph(CH_2)_2$	65
$Ph(CH_2)_3$	65
$Ph\text{—}CH\text{=}CH\text{—}CH_2$	60

(1.11)

The same authors also used the sulphonamide (**11**) which gave 80 per cent $Ph(CH_2)_2NH_3^+Cl^-$ and 65 per cent $Ph(CH_2)_3NH_3^+Cl^-$.

H

$NH{-}SO_2CF_3$

(11)

J. B. Hendrickson and R. Bergeron[19,20] used phenacylsulphonamides for the conversion of primary to secondary amines but found that the conversion is only satisfactory when simple alkylating agents such as methyl and benzyl halides are used (Scheme 1.12).[21]

$R^1{-}NH_2$ + $Ph{-}CO{-}CH_2{-}SO_2Cl$ $\xrightarrow[\text{6 hours}]{CHCl_3,\ \text{pyridine}}$ $R^1{-}NH{-}SO_2{-}CH_2{-}CO{-}Ph$ (12)

(12) $\xrightarrow{R^2\text{-}Br + K_2CO_3 \text{ in acetone, 18-20 hours}}$ $R^1{-}N(R^2){-}SO_2{-}CH(R^2){-}CO{-}Ph$ (13)

(13) $\xrightarrow{\text{Zn / HOAc, trace of conc. HCl}}$ $R^1R^2\overset{+}{N}H_2\ Cl^-$ (14)

R^1	Yield (%) (**12**)
Ph	93
Cyclohexyl	91
Bu^n	92

R^1	R^2	Yield (%) (**13**)	Yield (%) (**14**)
Ph	Me	93	78
Ph	Bn	94	98
Cyclohexyl	Me	80	—
Bu^n	Me	78	75

(1.12)

A good reaction for the preparation of primary amines on a tertiary carbon atom involves the reaction of a tertiary halide with nitrogen trichloride and aluminium chloride, the reaction mixture being worked up with hydrochloric acid (Scheme 1.13).[22]

$$\mathrm{Bu^tCl} \xrightarrow[-10^\circ C]{NCl_3/AlCl_3/CH_2Cl_2} \underset{90\,\%}{\mathrm{Bu^t\text{-}NH_2}} + \underset{8\,\%}{\text{2-methylaziridine}} \quad (1.13)$$

By using sodium bis(trimethylsilyl)amide (**15**)[23] primary bromides, iodides, or tosylates can be converted in good yields into amines (Scheme 1.14).[24]

$$\mathrm{Me_3Si{-}NH{-}SiMe_3 + NaNH_2 \xrightarrow{PhH} Me_3Si{-}\overset{-}{N}(Na^+){-}SiMe_3 + NH_3} \quad (15)$$

$$\underset{X = Br,\ I\ or\ OTs}{\mathrm{R\text{-}X}} + \underset{(15)}{\mathrm{Na^+\ \overset{-}{N}(SiMe_3)_2}} \xrightarrow[\text{or } Me_3Si\text{-}SiMe_3]{\text{in HMPT}} \underset{50\text{-}80\%}{\mathrm{R\text{-}N(SiMe_3)_2}} \xrightarrow{aq.\ H_3O^+} \underset{96\text{-}100\%}{\mathrm{R\text{-}\overset{+}{N}H_3\ \overset{-}{Cl}}} + \mathrm{(Me_3Si)_2O} \quad (1.14)$$

α-Hydroxy amines result from the opening of epoxides with amines in the presence of titanum(IV) isopropoxide (Scheme 1.15).[25]

$$n\text{-}C_7H_{15}\text{-epoxide-}CO_2H \xrightarrow[HNR_2,\ 83\%]{Ti(Pr^i)_4} n\text{-}C_7H_{15}CH(N(CH_2\text{-}CH{=}CH_2)_2)CH(OH)CO_2H\ (80\%) + n\text{-}C_7H_{15}CH(OH)CH(N(CH_2\text{-}CH{=}CH_2)_2)CO_2H\ (20\%) \quad (1.15)$$

1.1.2 *Reduction of other nitrogen-containing compounds*

This is a time-honoured method which is being greatly improved by the development of more efficient and more selective reducing agents, chiefly of the complex metal hydride type.

Reduction of azides

$$\text{R-N}_3 \longrightarrow \text{R-NH}_2 \qquad (1.16)$$

Hydrogen and a catalyst or lithium aluminium hydride can be used but they are not selective in their action. However, hydrogenation of an azide over a Lindlar catalyst is selective and allows amines to be prepared in the presence of carbonyl groups and carbon–carbon double bonds,[26] e.g. Scheme 1.17.

$$\text{Ph—CH=CH-CH}_2\text{—N}_3 \xrightarrow[\text{Lindlar catalyst}]{\text{H}_2} \text{Ph—CH=CH-CH}_2\text{—NH}_2 \qquad (1.17)$$

Good yields of amines can be obtained from azides by reduction with sodium borohydride in toluene and water by phase transfer with hexadecyltributylphosphonium bromide as catalyst.[27]

A preparation from tertiary alcohols or alkenes via azides has been reported (Scheme 1.18).[28]

$$\text{Ph—C(CH}_3\text{)=CH}_2 \xrightarrow[\text{F}_3\text{C-CO}_2\text{H}]{\text{NaN}_3} \text{Ph—C(CH}_3\text{)}_2\text{—N}_3 \xrightarrow{\text{Raney Ni}} \text{Ph—C(CH}_3\text{)}_2\text{—NH}_2 \ (72\%)$$

$$\text{Ph—C(CH}_3\text{)=CH}_2 \xrightarrow{\text{H}^+} \text{Ph—}\overset{+}{\text{C}}\text{(CH}_3\text{)—CH}_3 \xrightarrow{\text{N}_3^-} \text{Ph—C(CH}_3\text{)}_2\text{—N}_3 \qquad (1.18)$$

Benzyloxy groups and carbon–carbon double bonds are unaffected by the reducing agent.

Reduction of azides with a thiol gives good yields of primary amines (Scheme 1.19).[29]

$$\text{R-N}_3 + \text{CH}_2(\text{CH}_2\text{-SH})_2 \xrightarrow[\text{MeOH}]{\text{Et}_3\text{N}} \underset{85\text{-}100\%}{\text{R-NH}_2} + \text{N}_2 + \text{CH}_2\langle{}^{\text{CH}_2\text{-S}}_{\text{CH}_2\text{-S}}\rangle \qquad (1.19)$$

The reduction is selective since double and triple carbon–carbon bonds and nitro, cyano, carboxyl, amide, and ester groups are unaffected.

The preparation of primary amines from azides by the action of triphenylphosphine and hydrolysis of the intermediate iminophosphorane was first investigated by Staudinger and Meyer in 1919.[30] Recently it has been reinvestigated and put on a satisfactory preparative basis by Vaultier *et al.* (Scheme 1.20).[31]

$$RN_3 + Ph_3P \longrightarrow R\text{-}N{=}PPh_3 \xrightarrow{H_2O} R\text{-}NH_2 + O{=}PPh_3 \quad 89\text{-}95\% \tag{1.20}$$

Reduction of amides to amines

$$R^1\text{—CO—}NR^2R^3 \longrightarrow R^1\text{—}CH_2\text{—}NR^2R^3 \tag{1.21}$$

Table 1.1 Reduction of amides to amines

Reagent	Conditions	Structure	Yield (%)	Ref.
$LiAlH_4$	In ether	$RCO.NH_2$, $R^1CO.NHR^2$, and $R^1CO.NR^2R^3$	20–80	32,35,40
$NaBH_4 + CH_3SiCl$	In THF	$Ph.CONH_2$	91	33
$NaBH_4 + CH_3SO_3H$	In DMSO	$R^1CO.NR^2R^3$, $R^1 = Ph$, $CH_3(CH_2)_4$ R^2 and $R^3 = H$ and CH_3	60–90	34
$NaBH_4 + CoCl_2$	In EtOH or ether	$R^1CO.NH_2$, $R^1CO.NHR^2$, and $R^1CO.NR^2R^3$	50	36
$NaBH_4$	In boiling pyridine	$R^1CO.NR^2R^3$	40–70	37
i, $Et_3O^+BF_4^-$ ii, $NaBH_4$/EtOH	Dry CH_2Cl_2	$R^1CO.NHR^2$, $R^1CO.NR^2R^3$ NO_2 is not reduced	80–90	38
B_2H_6	In THF	$R^1CO.NH_2$ (reflux) $R^1CO.NHR^2$ (reflux) $R^1CO.NR^2R^3$ (25 °C) Not unsaturated amides NO, COEt, and Hal are not reduced	83–90	39
AlH_3	In THF	Caproamide Benzamide *N,N*-Dimethyl caproamide *N,N*-Dimethyl benzamide	58–100	40
$LiAlH(OMe)_3$		$R^1CO.NR^2R^3$		40
B_3H_6/Me_2S	In THF (reflux)	$R^1CO.NH_2$	85	41
Li-9-BBNH*	In THF	$R^1CO.NR^2R^3$	94	42

* Li-9-BBNH = Lithium 9-boratabicyclo[3.3.1]nonane.

Sodium bis(2-methoxyethoxy)aluminium hydride (**16**),[43] has been used to reduce vinylogues of amides to γ-ketoamines, with the carbonyl group being unaffected (Scheme 1.22).[44]

$$Na + Al + 2CH_3OCH_2CH_2OH \xrightarrow[> 100^\circ C]{H_2 \text{ in PhH}} NaAlH_2(CH_3OCH_2CH_2O)_2 \ (16)$$

$$H_3C{-}C(=O){-}C(CH_3){=}CH{-}N(R^1){-}R^2 \xrightarrow{(16)} H_3C{-}C(=O){-}CH(CH_3){-}CH_2{-}N(R^1){-}R^2 \quad 10 - 55\% \qquad (1.22)$$

It is to be noted that related reducing agents convert tertiary amides to primary alcohols (Scheme 1.23).[45]

$$R^1{-}C(=O){-}NR^2R^3 \xrightarrow[THF]{\text{Li 9-BBNH}} R^1{-}CH_2{-}NR^2R^3$$

$$R^1{-}C(=O){-}NR^2R^3 \xrightarrow{\text{LiBHEt or 9- BBN}} R{-}CH_2OH \qquad (1.23)$$

Reduction of nitriles to primary amines

$$R\text{-}C{\equiv}N \xrightarrow{\text{reducing agent}} R\text{-}CH_2NH_2 \qquad (1.24)$$

Table 1.2 Reduction of nitriles to primary amines

Reagent	Conditions	Structure	Yield (%)	Ref.
H_2/5% Rh on Al_2O_3	Room temp. 2.5 atm. of H_2 + 10% NH_3 in EtOH, 2 h	$(H_3C)_2N(CH_2)_2CN$		
		1-Piperidino-$(CH_2)_2CN$		
		4-Methyl-1-piperazino$(CH_2)_2CN$		
		4-Methyl-1-piperazino$(CH_2)_3CN$		
		$C_6H_5CH_2N(CH_3)(CH_2)_2CN$		
		$C_6H_5CH_2N(CH_3)(CH_2)_3CN$		
		$(H_3C)_2N(CH_2)_4CN$	60–90	46
		$(H_3C)_2N(CH_2)_5CN$		
		$CH_3O(CH_2)_3CN$		
		Cyclopentyl-$O(CH_2)_3CN$		
		Hexahydrobenzyl-$O(CH_2)_3CN$		
		Cycloheptyl-$O(CH_2)_3CN$		
$LiAlH_4$	Ether	*o*-tolunitrile	85	
		R—CN	34–90	47

Table 1.2 (cont.)

Reagent	Conditions	Structure	Yield (%)	Ref.
$NaBH_4 + CoCl_2$	MeOH	PhCN		48
		p-$NO_2C_6H_4CH_2CN$		
		p-$HOC_6H_4CH_2CN$		
		n-$C_7H_{15}CN$		
		$C_6H_5CH(OH)CN$	35–80	49
		β-Cyanopyridine		
		α-Cyanofuran		
		$CH_2{=}CHCN$		
		Bu^nCN		
$NaBH_4 + ZrCl_4$	THF, 25 °C		91	50
Co_2B† + B_2H_6		PhCN, BnCN		
+Bu^tNH_2	MeOH	At −10 °C NO_2 is not reduced	75–85	49
($Co_2B{-}Bu^tNH_2.BH_3$)		At 40 °C NO_2 is reduced		
$B_2H_6 + Me_2S$	THF, reflux	cyclo-C_4H_7-CN		
		Capronitrile		
		Benzonitrile		
		Pivalnitrile		
		Diphenylacetonitrile		
		p-Chlorobenzonitrile	61–93	51
		o-Tolunitrile		
		p-Methoxybenzonitrile		
		Adiponitrile		
		m-Nitrobenzonitrile		
$NaBH_3(OCOCF)$‡	THF, room temp.	Benzonitrile		
		Diphenylacetonitrile		
		Phenylacetonitrile		
		p-Tolunitrile	High	52
		(CO_2R, NO_2, or Hal are not reduced)		

* Preparation:

$$NaBH_4 + 2CoCl_2 \xrightarrow{MeOH} Co_2B + H_2$$

† This is the best procedure.

‡ Preparation:

$$NaBH_4 + F_3C{-}CO_2H \xrightarrow{THF,\ 20\ °C} NaBH_3(OCOCF_3)$$

Nitriles react with diphenylphosphinodithioic acid (**17**)[53] to yield thioamides (**18**)[54] which can easily be converted into amines by the method of Borch (Scheme 1.25).[55]

$$Ph_2P(=S){-}S{-}H\ (17) + R{-}C{\equiv}N \xrightarrow[12\ \text{hours}]{40^{\circ}C} \left[R{-}C(=NH){-}S{-}P(=S)Ph_2 \quad H{-}S{-}P(=S)Ph_2 \right] \longrightarrow R{-}C(=S){-}NH_2\ (18) + Ph_2P(=S){-}S{-}P(=S)Ph_2$$

$$R{-}C(=S){-}NH_2\ (18) \xrightarrow[CH_2Cl_2]{Et_3O^+\ BF_4^-} R{-}C(SEt){=}NH_2 \xrightarrow[EtOH]{NaBH_4} R{-}NH_2 \qquad (1.25)$$

Reduction of oximes to primary amines

$$R^1R^2C{=}NOH \xrightarrow{\text{Reducing agent}} R^1R^2CH{-}NH_2 \qquad (1.26)$$

Table 1.3 Reduction of oximes to primary amines

Reagent	Conditions	Structure	Yield (%)	Ref.
H_2/Rh-Al_2O_3	Rh, 1 atm, H_2	Cycloheptanone oxime	80	56
$LiAlH_4$	In diethyl ether	$R_2C{=}NOH$	40–78	57
Ni–Al alloy	Aqueous HO^-		58	
$Na(MeOCH_2CH_2O)AlH_2$*		N–OH	90	13, 59
$NaBH_2S_3$†		CO_2R, CN, NO_2 or Hal are not reduced		60
$NaBH_4 + ZrCl_4$	THF, 25C	N–OH	85	61
$NaBH_3CN + TiCl_3$	pH 7	Acid and base sensitive groups are unaffected	59	62

* Preparation:

$$Na + Al + 2CH_3OCH_2CH_2OH \xrightarrow[H_2\ \text{under pressure}]{\text{In benzene}, > 100\ ^{\circ}C} NaAlH_2(OCH_2CH_2OCH_3)_2$$

† Preparation:

$$NaBH_4 + S \xrightarrow{THF,\ 25\ ^{\circ}C} NaBH_2S_3$$

By the reduction of tosylates or acetates of oximes with diborane, good yields of primary amines can be obtained (Scheme 1.27).[63]

$$R^1{}_2C{=}NOH \longrightarrow R^1{}_2C{=}N\text{-}OR^2 \xrightarrow[20^\circ C]{B_2H_6\,/\,THF} R^1{}_2CH\text{-}NH_2 \;\; (60\text{-}75\%)$$

$$R^1 = COCH_3 \text{ or } p\text{-}SO_2C_6H_4CH_3 \qquad (1.27)$$

In contrast, the reduction of oximes themselves with diborane yields *N*-alkylhydroxylamines, not amines (see Chapter 14, p. 553).

Reduction of sulphonamides

Moderate yields of amines result from the reduction of sulphonamides with sodium bis(2-methoxyethoxy)aluminium hydride (Scheme 1.28).[64]

$$R\text{-}NH\text{-}SO_2Ar \xrightarrow{NaAlH_2(CH_3O\text{-}CH_2CH_2O)_2} R\text{-}NH_2 \qquad (1.28)$$

Reduction of imines[65]

The reduction of imines either with sodium borohydride or by catalysis is a good way to obtain secondary amines. If borohydride is to be used, the imine is isolated and then reduced (Scheme 1.29).

$$R^1\text{-}NH_2 + R^2\text{-}CHO \longrightarrow R^1\text{-}N{=}CHR^2 \xrightarrow{NaBH_4\,/\,MeOH} R^1\text{-}NH\text{-}CH_2\text{-}R^2 \qquad (1.29)$$

Axial primary and secondary cyclohexylamines can be prepared by the reduction of suitable imines with lithium tri-*sec*-butylborohydride, $LiBu^s_3BH$ (Scheme 1.30).[66]

(1.30)

For unstable imines, catalytic reduction of a mixture of the primary amine and the aldehyde in a suitable solvent gives good yields of the secondary amine (Scheme 1.31).

$$R^1\text{-}NH_2 + R^2\text{-}CHO \xrightarrow[1\text{ atm., }20^\circ C]{H_2/Pt/EtOH} R^1\text{-}NH\text{-}CH\text{-}R^2 \quad (1.31)$$

If ammonia is used instead of a primary amine, a primary amine is produced. In a similar way tertiary amines can be obtained from a secondary amine and an aldehyde.[67] For sterically hindered secondary amines, for example those which contain Bu^t or R^1R^2CH groups, titanium tetrachloride ($TiCl_4$) and hydrogen can be used as the reducing agent.[68]

Reductive amination of aldehydes and ketones

Many methods are available for the reductive amination of aldehydes and ketones in the laboratory and some recent improved procedures are collected in Table 1.4.

Table 1.4 Reductive amination of ketones

Reducing agent	Conditions	Reactants	Yield (%)	Ref.
$NaBH_3CN$	pH 6	Cyclohexanone + Pr^nNH_2	85	69
$NaBH_3CN$ + Aliquot 33b or $Bn^n_4N^+BH_3CN^-$	In CH_2Cl_2 + HCl	Cyclohexanone, pyrrolidine (N–H)	94	70
B_2H_6	Pyridine–HOAc + CH_2Cl_2 or THF	Cyclohexanone + $PhNH_2$, p-$MeOC_6H_4NH_2$, p-$NO_2C_6H_4NH_2$, $(CH_3)_2CO$	54–93	71
NaTeH	EtOH, 20 °C	Cyclohexanone + $BuNH_2$, H_2N-cyclohexanone, $(CH_3)_2CO$, Butyraldehyde	80–95	72
$Fe(CO)_5$/KOH [i.e. $KHFe(CO)_4$] Under Argon or N_2	EtOH	$(CH_3)_2CO + Me_2CH_2NH_2$ Secondary amines can also be used	93	73
$NaBH_4$	EtOH/MeOH	Steroidal secondary amine and CH_2O or Ph–C(=O)–C(Me)(Me)–NH–CH₂CH₂–H_2N	100	74

This process of reductive amination of aldehydes and ketones has considerable importance in biochemistry, for example, the conversion of ketoglutaric acid into the amino acid glutamic acid. (Scheme 1.32).[75]

$$O{=}C(CO_2H)\text{-}CH_2\text{-}CH_2CO_2H \xrightarrow[\text{Enzyme Catalysed}]{\text{Reductive Amination}} H_2N\text{-}HC(CO_2H)\text{-}CH_2\text{-}CH_2CO_2H \quad (1.32)$$

Ketoglutaric Acid → Glutamic Acid

Leuckhart and Eschweiler–Clarke reactions

The Leuckhart reaction consists of the reductive amination of an aldehyde or a ketone with ammonium formate followed by hydrolysis of the resulting formamide to give an amine (Scheme 1.33). Several attempts have been made to

$$R^1\text{-}C(=O)\text{-}R^2 \xrightarrow[\text{heat}]{NH_4^+\ HCO_2^-} R^1R^2CH\text{-}NH\text{-}CHO \xrightarrow[\text{heat}]{\text{aqueous } H_3O^+} R^1R^2CH\text{-}NH_2 \quad (1.33)$$

postulate plausible mechanisms for the Leuckart reaction including a free radical mechanism.[76] On the basis of evidence from linear free energy relationships and a kinetic isotope effect, Awachie and Agwada[77] suggest the mechanism shown in Scheme 1.34 for the reaction between benzophenone and formamide. In this reaction the rate-determining step is C—H bond cleavage of formamide which causes reduction of the benzophenone, followed by the formation of a benzylic cation. In view of the postulation of a benzylic cation in

$$H\text{—}C(=O)\text{—}NH_2 + H_2O \longrightarrow H\text{—}C(=O)\text{—}O^-$$

$$Ph_2C{=}O \quad H\text{—}\overset{+}{N}H_3$$

$$\downarrow$$

$$Ph_2\overset{+}{C}H \quad H_2\ddot{N}\text{-}CHO \xleftarrow{H_3O^+} CO_2 + NH_3 + Ph_2CHOH$$

$$\downarrow$$

$$Ph_2CH\text{—}NH\text{—}CHO + H^+ \quad (1.34)$$

this mechanism, it seems likely that another mechanism occurs when purely aliphatic alcohols which do not easily form cations are used.

The Eschweiler–Clarke reaction is the reductive methylation of an amine with formaldehyde and formic acid (Scheme 1.35).[78] The most satisfactory

$$R_2NH + CH_2O + H\text{-}CO_2H \longrightarrow R_2\overset{+}{N}{=}CH_2 + H\text{-}C(=O)O^- + H_2O \longrightarrow R_2N\text{-}CH_3 + CO_2 \quad (1.35)$$

modification of the Eschweiler–Clarke reaction is the one introduced by Borsch and Hassid,[79] in which sodium cyanoborohydride is the reducing agent. Primary and secondary amines as well as sterically hindered amines can be methylated in good yields (Scheme 1.36).

$$R\text{-}NH_2 + \underset{(aqueous)}{CH_2O} \xrightarrow{CH_3CN} [R\text{-}N{=}CH_2] \xrightarrow{NaBH_3CN} \underset{80\text{ - }90\%}{R\text{-}NH\text{-}CH_3} \quad (1.36)$$

Reduction of enamines

Enamines can be reduced to tertiary amines by hydrogenation, either with diborane or with sodium cyanoborohydride. Hydrogenation has also been effected by reaction with mercuric acetate followed by reduction with sodium borohydride (Scheme 1.37).[80]

$$R^1_2N\text{-}CH{=}C(R^2)R^3 + AcO\text{-}Hg\text{-}O\text{-}C(=O)\text{-}CH_3 \xrightarrow[DMF]{Hg(OAc)_2} \underset{\text{Iminium ion}}{R^1_2\overset{+}{N}{=}CH\text{-}C(R^2)(HgOAc)\text{-}R^3} + AcO^- \xrightarrow{NaBH_4} \underset{50\text{ - }90\ \%}{R^1_2N\text{-}CH_2\text{-}CH(R^1)R^2} \quad (1.37)$$

Reduction of nitroalkanes and nitroalkenes

Lithium aluminium hydride in ether is a good reducing agent for the conversion of nitroalkanes to amines (Scheme 1.38).[81]

$$\text{R—NO}_2 \xrightarrow[\text{ether}]{\text{LiAlH}_4} \text{R—NH}_2 \qquad (1.38)$$

By the use of ammonium formate and palladium charcoal in methanol, transfer hydrogenation of the nitro group takes place and good yields of amines result (Scheme 1.39).[82]

$$\text{R-NO}_2 \xrightarrow[\text{Pd - C}]{\text{H-CO}_2\text{NH}_4\text{ / MeOH}} \text{R-NH}_2 \qquad (1.39)$$

Primary amines can be prepared by the reduction of nitroalkanes with titanium tetrachloride and magnesium amalgam in tetrahydrofuran and tertiary butanol (Scheme 1.40).[83] Halogen, nitrile, and ester groups are not reduced.

$$\text{C}_6\text{H}_{11}\text{-NO}_2 \xrightarrow[\text{THF + Bu}^t\text{OH, 0}^\circ\text{C}]{\text{TiCl}_4\text{ / Mg - Hg}} \text{C}_6\text{H}_{11}\text{-NH}_2 \;\; 87\% \qquad (1.40)$$

High yields (85–91 per cent) of amines are obtained by the reduction of α,β-unsaturated nitroalkenes with diborane and tetrahydrofuran in the presence of a catalytic amount of sodium borohydride (Scheme 1.41).[84]

$$\text{R-CH=CH-NO}_2 \xrightarrow[\text{NaBH}_4\text{ (catalytic amount)}]{\text{BH}_3\text{ - THF}} \text{R-CH}_2\text{CH}_2\text{NO}_2 \;\; 85 - 91\% \qquad (1.41)$$

α,β-Unsaturated nitroalkenes can also be reduced by sodium borohydride in the presence of trimethylsilyl chloride (Scheme 1.42).[85]

$$\text{R-CH=CH-NO}_2 \xrightarrow[\text{THF}]{\text{NaBH}_4\text{ / Me}_3\text{SiCl}} \text{R-CH}_2\text{CH}_2\text{NH}_2 \qquad (1.42)$$

1.1.3 *Other methods of preparation*

From alkenes

The introduction of the boron hydride reagents has enabled amines to be prepared efficiently from alkenes. Organoboranes, prepared from alkenes and diborane, react with hydroxylamine-*O*-sulphonic acid (**19**) or with chloramine ($Cl—NH_2$) to yield primary amines in yields of 50–70 per cent (Scheme 1.43).[86]

$$\text{R—CH=CH}_2 + \text{B}_2\text{H}_6 \longrightarrow (\text{R—CH}_2\text{—CH}_2)_3\text{B} \xrightarrow[(19)]{\text{H}_3\text{N}^+\text{-O-SO}_3^-} \text{R—CH}_2\cdot\text{CH}_2\text{—NH}_2 \qquad (1.43)$$

A modification has been developed in which mesityl sulphonyl hydroxylamine (**20**) is used in THF.[87]

$$\text{2,4,6-}(CH_3)_3C_6H_2\text{-}SO_2O\text{-}NH_2 \quad (20)$$

Primary amines can be prepared from alkenes by treating alkyl boranes with trimethylsilyl azide in methanol (Scheme 1.44).[88]

$$R\text{—}CH{=}CH_2 \xrightarrow{BH_3 / THF} (R\text{—}CH_2\text{—}CH_2)_3B \xrightarrow[\text{(generates } HN_3)]{Me_3SiN_3 / CH_3OH} \left[(R\text{—}CH_2\text{—}CH_2)_3\overset{-}{B}\text{—}NH\text{—}\overset{+}{N}{\equiv}N\right] \longrightarrow R\text{-}NH_2 + N_2$$

45-70%
cyclohexane 64%

(1.44)

Amines can be prepared from alkenes by the action of a sulphonium tetrafluoroborate, followed by reaction with ammonia in acetonitrile and subsequent desulphurization (Scheme 1.45).[89]

$$Me_2S + Me_3O^+ BF_4^- \,{}^{90} \longrightarrow \text{i, } MeS\text{-}\overset{+}{S}Me_2\ BF_4^-$$

$$R_2C{=}CR_2 \xrightarrow[\text{ii, aq. } NH_3 / CH_3CN]{\text{i, } MeS\text{-}\overset{+}{S}Me_2\ BF_4^-} R_2C(NH_2)\text{—}C(SMe)R_2 \longrightarrow R_2C(NH_2)\text{—}CHR_2$$

(1.45)

In a reaction analogous to the oxymercuration of a carbon–carbon double bond an alkene can be converted into a primary amine via an amide (Scheme 1.46).[91]

$$R\text{—}CH\text{=}CH_2 + Hg(O\text{—}NO_2)\text{—}O\text{—}\overset{+}{N}(O^-)\text{=}O \longrightarrow R\text{—}\overset{+}{C}H\text{-}CH_2\text{—}Hg\text{—}O\text{—}NO_2$$

$$N{\equiv}C\text{—}CH_3 \qquad {}^{-}O\text{—}NO_2$$

$$\downarrow$$

$$R\text{—}CH(NH\text{—}CO\text{-}CH_3)\text{—}CH_3 \xleftarrow{NaBH_4} R\text{—}CH(N\text{=}C(O\text{—}NO_2)\text{—}CH_3)\text{-}CH_2\text{—}Hg\text{—}O\text{—}NO_2$$

amide

$$\downarrow \text{aq. } HO^-$$

$$R\text{—}CH(NH_2)\text{-}CH_3 \qquad (1.46)$$

W. Reppe used iron pentacarbonyl, $Fe(CO)_5$, to prepare amines by the aminomethylation to alkenes (Scheme 1.47).[92] Rhodium oxide, Rh_2O_3, has been shown to be a better catalyst in this process and by the addition of a small amount of iron pentacarbonyl as a co-catalyst, yields have been boosted to *c.* 90 per cent.[93]

$$>C\text{=}C< + 3\,CO + H_2O + H\text{-}N< \xrightarrow{Fe(CO)_5} H\text{-}\overset{|}{\underset{|}{C}}\text{-}\overset{|}{\underset{|}{C}}\text{-}CH_2\text{-}N< + 2\,CO_2 \qquad (1.47)$$

From 1,3-dienes

Naphthalene sodium can be used for the preparation of tertiary amines by the amination of 1,3-dienes (Scheme 1.48). Naphthalene in dimethylether or THF plus sodium metal gives a deep green solution which contains $Na^+C_{10}H_8^-$, the colour being caused by the ion radical (**21**).[94]

(21)

$$HNEt_2 \xrightarrow[THF]{Na^+C_{10}H_8^-} {}^-NEt_2$$

$${}^-NEt_2 + CH_2{=}C(H){-}C(H){=}CH_2 \xrightarrow{H^+} Et_2N{-}CH_2{-}CH{=}C(CH_3)_2 \quad 50\text{-}95\% \qquad (1.48)$$

From allenes

Allenes can be converted into tertiary amines in 20–85 per cent yields by reaction with secondary amines in the presence of palladium or rhodium catalysts, such as $PdCl_2$, $RhCl_3$, and $(Ph_3P)_4Pd$ (Scheme 1.49).[95]

$$CH_2{=}C{=}CH_2 + HNR_2 \xrightarrow{Pd\ or\ Rh} CH_2{=}C(CH_3){-}C(CH_2 \cdot NR_2){=}CH_2 \qquad (1.49)$$

From organometallic reagents

Organometallic reagents have been used for the preparation of amines from a variety of starting compounds. The preparation of primary amines from Grignard reagents using a substituted hydroxylamine (preparation in 92 per cent yield)[96] has been reported (Scheme 1.50).[97]

$$Ph_2P(O){-}Cl + NH_2OH \longrightarrow Ph_2P(O){-}O{-}NH_2$$

Diphenylphosphinic chloride

$$Ph{-}CH_2{-}MgCl \xrightarrow[ii,\ Ph_2POONH_2]{i,\ Bu^nLi} Ph{-}CH_2{-}NH_2 \quad 70\% \qquad (1.50)$$

Amines can be prepared in yields of 60–90 per cent from Grignard reagents and a trimethylsilylated methoxymethylamine (**22**) which is prepared as shown in Scheme 1.51.[98]

$$(Me_3Si)_2NLi + Cl{-}CH_2{-}OMe \longrightarrow (Me_3Si)_2N{-}CH_2{-}OMe \ (22)$$

$$(Me_3Si)_2N{-}CH_2{-}OMe \xrightarrow{R-MgHal,\ Ether} R{-}CH_2{-}N(SiMe_3)_2 \xrightarrow{MeOH} R{-}CH_2{-}NH_2 \qquad (1.51)$$

By the action of organometallic reagents on *O*-methylhydroxylamine (**23**) primary amines can be prepared. Grignard reagents have been used (Scheme 1.52),[99] and alkyl lithium compounds similarly by reversing the polarity (Umpolung) of an alkyl halide (Scheme 1.53).[100] The methyl lithium is necessary in this preparation.

$$\text{R-Mg-Hal} + \text{MeONH}_2 \xrightarrow{-10 \text{ to } -15^\circ\text{C}} \text{R-NH}_2$$

Hal = Cl or Br (23) 45-90% (1.52)

$$\text{R-Hal} \xrightarrow[0^\circ\text{C}]{2\text{ Li}} \text{R-Li} \xrightarrow[\substack{2\text{ mole CH}_3\text{Li} \\ \text{In hexane-ether}}]{2\text{ mole MeONH}_2} \text{R-NH}_2$$

55-59% (1.53)

Secondary or tertiary amines result from the reduction of a mixture of an alcohol and a primary amine with a ruthenium compound (Scheme 1.54).[101]

$RuCl_3 + NaBH_4 + Ph_3P$ ↓

$$\text{R}^1\text{OH} + \text{R}^2\text{NH}_2 \xrightarrow{(\text{Ph}_3\text{P})_4\text{RuH}_2} \text{R}^1\text{-NH-R}^2 \text{ or } \text{R}^1{}_2\text{N-R}^2$$

50-95% (1.54)

Lithium alkyls and primary or secondary amines react with alkenes to give secondary or tertiary amines as shown in Scheme 1.55.[102]

$$\text{Pr}^n\text{—NH}_2 + \text{Et-Li} + \text{CH}_2\text{=CH}_2 \xrightarrow{130\text{-}135^\circ\text{C}} \text{Pr}^n\text{—NHEt}$$

+ tetramethylenediamine 93%

$$\text{Et}_2\text{NH} + \text{Et-Li} + \text{norbornene} \longrightarrow \text{Et}_2\text{N-norbornane}$$

+ tetramethylenediamine 17% (1.55)

Alkyl titanium compounds, prepared as shown in Scheme 1.56,[103] can be used to synthesize tertiary amines from non-enolizable aldehydes (Scheme 1.56).[104]

$$(R^1{}_2N)_3TiHal \xrightarrow{R^2Li \text{ or } R^2MgHal} (R^1{}_2N)_3TiR^2$$

$$R\text{-}C_6H_4\text{-}CHO + CH_3Ti(NEt_2)_3 \xrightarrow[-60^\circ C]{\text{ether}} H\text{—}C(R\text{-}C_6H_4)(CH_3)\text{—}NEt_2 \quad 28\text{–}48\% \qquad (1.56)$$

The reaction of Grignard reagents with aldimines constitutes a good preparation of secondary amines (see Chapter 3, p. 135), with the yields usually being in the range 50–95 per cent (Scheme 1.57).[105]

$$R^1\text{—N=CH—}R^2 + R^3\text{-MgHal} \xrightarrow{\text{ether}} R^1\text{—N(MgHal)-CHR}^2\text{-}R^3 \xrightarrow{H_2O} R^1\text{—N(H)-CHR}^2\text{-}R^3 \qquad (1.57)$$

By addition of amines to imines and ketones

Two procedures are available for the preparation of amines by the addition of other amines to imines. Primary and secondary amines react with palladium black (5 per cent by weight) at 25–200 °C (3–20 h) (Scheme 1.58).[106] The addition can be done directly to an imine (Scheme 1.59).

$$R^1\text{—}CH_2\text{—}NHR^2 \xrightarrow{Pd} H_2 + R^1\text{—CH=N—}R^2 \xrightarrow{R^1\text{-}CH_2\text{-}NHR^2} \left[R^1\text{—CH(NHR}^2\text{)—N(R}^2\text{)—}CH_2\text{—}R^1 \right] \xrightarrow{H_2} R^1\text{—}CH_2\text{—N(}R^2\text{)—}CH_2\text{—}R^1 + R^2\text{-}NH_2$$

R^2 = H or alkyl (1.58)

$$R^1\text{—CH=N—}R^2 + R^3\text{—}NHR^4 \xrightarrow{Pd} \left[R^1\text{—CH(NR}^3R^4\text{)—N(H)—}R^2 \right] \longrightarrow R^1\text{-}CH_2NR^3R^4 + R^2\text{-}NH_2 \qquad (1.59)$$

R^4 = H or alkyl

Conjugate addition of secondary amines to α,β-unsaturated ketones proceeds rapidly unless the enone is exocyclic. However, exocyclic α,β-unsaturated ketones such as (**24**) undergo Michael reactions very slowly with secondary amines unless aluminium oxide is added as a catalyst (Scheme 1.60).[107] The amine (**25**), in which the hydroxyl group is essential, is used as a catalyst for asymmetric synthesis (Scheme 1.61).[108]

O CH$_2$ (24) — $HNEt_2$, Al_2O_3, benzene → O CH$_2$-NEt$_2$ 100% (1.60)

H OH CH$_2$-NHPh Et H (25)

O + R-SH — (25) in toluene, -5°C → O H SR (26)

ee = 47-88%
yield = 75-85%
R = Ph-, *p*-MePh, *p*-ClPh, *p*-CH$_3$Ph, *p*-MeOPh (1.61)

By hydrolysis

The hydrolysis of amides (Chapter 8, p. 360), from the Beckmann rearrangement of a ketoxime, for example (Chapter 14, p. 557), and from the Ritter reaction (Chapter 4, p. 211), constitutes a route to amines as does the hydrolysis of isonitriles (Chapter 6, p. 282) and isocyanates (Chapter 7, p. 309). Another route involves the nitrosoation of aryl secondary or tertiary amines and the hydrolysis of the resulting *p*-nitroso compounds to give primary or secondary amines:

1. Secondary amines (Scheme 1.62).

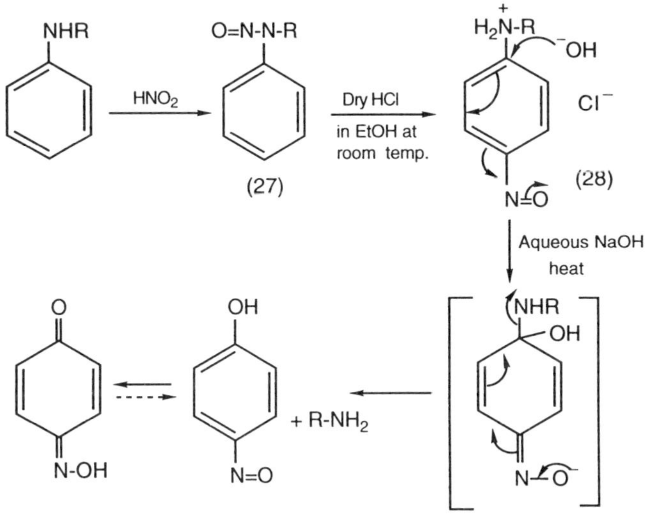

(1.62)

The conversion of the *N*-nitroso compound (**27**) into the *p*-nitroso compound (**28**) is called the Fischer–Hepp transformation.[109]

2. Tertiary amines (Scheme 1.63).

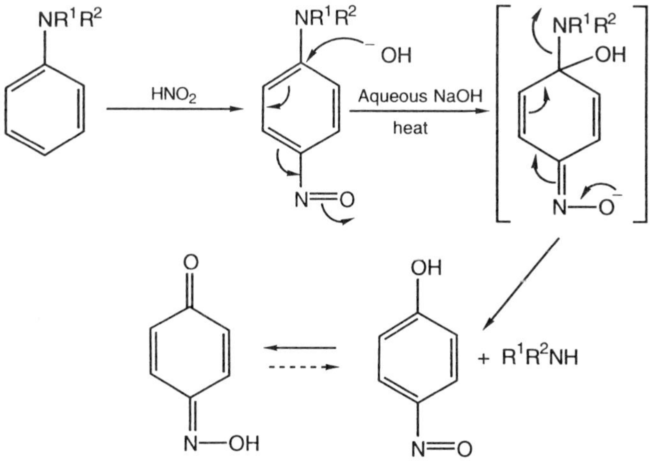

(1.63)

By rearrangement

Hofmann rearrangement The well known Hofmann rearrangement[110] occurs when an amide is treated with bromine and aqueous sodium hydroxide (Scheme 1.64). The Hofmann rearrangement gives poor yields with amides which con-

$$R{-}C({=}O){-}NH_2 \xrightarrow{Br_2/aq.NaOH} \left[R{-}C({=}O){-}N(Br){-}H\right] \xrightarrow{HO^-} \left[R{-}C({=}O){-}\bar{N}{-}Br\right] + H_2O \longrightarrow \left[O{=}C{=}N{-}R\right] \xrightarrow{H_2O} R{-}NH_2 + CO_2 \qquad (1.64)$$

tain sulphonamide groups under the usual conditions but it can be performed satisfactorily by the use of phenyliodine bis(trifluoroacetate) (Scheme 1.65).[111]

$$Ar{-}SO_2{-}NH(CH_2)_5CONH_2 \xrightarrow[CH_3CN\,/\,H_2O]{PhI(O_2C{-}CF_3)_2} \underset{38\text{-}75\%}{Ar{-}SO_2{-}NH(CH_2)_5NH_2} \qquad (1.65)$$

A Hofmann type preparation of amines by the reaction of amides with iodosyl benzene in the presence of formic acid has been reported to give good yields (Scheme 1.66).[112]

$$R{-}C({=}O){-}NH_2 \xrightarrow[H{-}CO_2H]{PhI{=}O} \underset{70\text{ - }90\%}{R{-}NH_2} \qquad (1.66)$$

Curtius rearrangement The Curtius rearrangement[113] of an acyl azide yields an isocyanate which can be hydrolysed to a primary amine (Scheme 1.67). A variant for the preparation of amines from acyl azides, obtained *in situ* under phase transfer conditions from reaction of acid chlorides and sodium azide, by a modified Curtius rearrangement has been reported (Scheme 1.68).[114]

$$R{-}C({=}O){-}N{=}\overset{+}{N}{=}\overset{-}{N} \longleftrightarrow R{-}C({=}O){-}\overset{-}{N}{-}\overset{+}{N}{\equiv}N \xrightarrow{heat} R{-}C({=}O){-}\overset{-}{N} + N_2 \longrightarrow O{=}C{=}N{-}R \xrightarrow{H_2O} R{-}NH_2 \qquad (1.67)$$

$$\text{R-COCl} + \text{NaN}_3 \xrightarrow[\text{H}_2\text{O / CH}_2\text{Cl}_2 \text{ Phase Transfer}]{\text{Bu}^n{}_4\text{N}^+ \text{Br}^-} [\text{R-CON}_3] \xrightarrow{\text{F}_3\text{C.CO}_2\text{H, CH}_2\text{Cl}_2\text{, reflux}} \text{R—NH—C(=O)—CF}_3 \xrightarrow[\text{aq. MeOH}]{\text{K}_2\text{CO}_3} \text{R-NH}_2 \tag{1.68}$$

The Schmidt rearrangement The Schmidt reaction,[115] in which a carboxylic acid reacts with hydrazoic acid in the presence of concentrated sulphuric acid, can be regarded as an extension of the Curtius reaction and therefore applied to the synthesis of amines (Scheme 1.69).

$$\text{R-C(=O)-OH} + \text{HN}_3 \xrightarrow{\text{H}_2\text{SO}_4} \left[{}^-\text{O—C(R)(OH)—}\underline{\text{N}}^-\text{—}\overset{+}{\text{N}}\equiv\overset{+}{\text{N}}\text{—H} \right] \longrightarrow \left[\text{O=C(OH)-N(H)-R} \right] + \text{N}_2 \longrightarrow \text{R-NH}_2 + \text{CO}_2 \tag{1.69}$$

From alcohols

An efficient conversion of alcohols into secondary and tertiary amines can be achieved by treatment of the alkoxide of the alcohol with an amine and *N*,*N*-methylphenyltriphenyl phosphonium iodide (**29**).[116] The reaction results in the alkylation of the amine (Scheme 1.70).[117]

$$\text{R}^1\text{-OH} \xrightarrow[\text{DMF}]{\text{NaH}} \text{R}^1\text{-O}^- \xrightarrow[\text{Ph}_3\overset{+}{\text{P}}\text{-N(CH}_3\text{)-Ph I}^- \ (29)]{\text{R}^2\text{R}^3\text{NH}} \text{R}^1\text{—N(R}^2\text{)—R}^3 \ (80-100\%) + \text{Ph}_3\text{P=O} + \text{HN(CH}_3\text{)Ph} \tag{1.70}$$

Secondary and tertiary amines can be prepared from benzyl or allyl alcohols by causing them to react with primary or secondary amines in the presence of palladium black at 80–120 °C for 6–26 h (Scheme 1.71).[118]

$$CH_2{=}CH\text{-}CH_2OH + n\text{-}C_6H_{13}NH_2 \xrightarrow{Pd} CH_2{=}CH\text{-}CH_2\text{-}NH\text{-}C_6H_{13}$$

Alcohol	Yield (%)
BnOH	98
$PhCHOHCH_3$	83
$CH_2{=}CH{-}CH_2OH$	87

(1.71)

Alkylation of nitriles with trialkyloxonium fluoroborate[119] and subsequent reduction of the nitrilium salts formed with sodium borohydride has proved a good synthesis of secondary amines (Scheme 1.72).[120] This reaction is only successful when R^2 = Me or Et but it can be extended to other alkyl groups by the use of other fluoroborates,[121] for example, see Scheme 1.73.

$$R^1\text{-}C{\equiv}N + R^2_3O^+ BF_4^- \longrightarrow R^1\text{-}C{\equiv}\overset{+}{N}\text{-}R^2\ BF_4^- \xrightarrow[EtOH]{NaBH_4} R^1\text{-}CH_2\text{-}NHR^2$$

(1.72)

$$(R^2O)_3CH \xrightarrow[-70^\circ C]{BF_3.Et_2O} \left[HC{=}OR^2(OR^2)\right]^+ BF_4^- + R^1{-}C{\equiv}N \xrightarrow[EtOH]{NaBH_4} R^1{-}C{\equiv}\overset{+}{N}{-}R^2\ BF_4^- \longrightarrow R^1\text{-}CH_2\text{-}NHR^2$$

28-94%

$R^1 = PhCH_2$

R^2 = Me, Et, $(CH_2)_2CH_3$, $CH(CH_3)_2$, $CH_2{=}CH$

(1.73)

From N-alkyl-N-benzylhydroxylamines

Primary amines have been prepared by a three step method from *N*-alkyl-*N*-benzylhydroxylamines (Scheme 1.74).[122]

$$PhCH_2NHOH + RX \longrightarrow PhCH_2N(OH)R \xrightarrow[CH_2Cl_2]{Et_3N,\ \text{2-fluoro-1-methylpyridinium } TsO^-} PhCH{=}NR + \text{1-methyl-2-pyridone}$$

$$PhCH{=}NR \xrightarrow[THF]{aq.\ HCl} PhCHO + RNH_2 \qquad (1.74)$$

From 1,1-dimethylhydrazine

Tertiary amines can be prepared by the alkylation of 1,1-dimethylhydrazine followed by the action of nitrous acid in 4 M hydrocholoric acid, which deaminates the hydrazinium salt formed (Scheme 1.75).[123]

$$Et{-}CH(CH_3)Br + (CH_3)_2N{-}NH_2 \xrightarrow{heat} Et{-}CH(CH_3){-}\overset{+}{N}(CH_3)_2{-}NH_2\ Br^- \xrightarrow[4M\ HCl]{NaNO_2} Et{-}CH(CH_3){-}N(CH_3)_2\ \ 60\% \qquad (1.75)$$

1.1.4 *Preparation of chiral amines*

Enantioselective synthesis of amines may be achieved by alkylation of chiral *S*-1-amino-2-methoxymethyl-1-pyrrolidine (SAMP) hydrazones of aldehydes using either alkyl lithium compounds,[124] or by use of organocerium compounds[125] which add stereoselectively to the C=N bond to give hydrazines. These undergo reductive cleavage with Raney Ni affording either (*R*)- or (*S*)-products and recovery of SAMP (Scheme 1.76). The substituted silylamine *N*-benzyl-*N*-methoxymethyl-*N*-trimethylsilylmethylamine (**31**) reacts with lithum fluoride to yield an azomethine (**32**) which reacts by a 1,3-dipolar addition with suitably activated double bonds, for example, as shown in Scheme 1.77 with maleic methyl ester.[126]

(−)-Norephedrine forms a complex with diborane which acts as a chiral-inducing reducing agent for prochiral oxime ethers.[127] For example, it reduces

$CeCl_3$ + CH_3-Li

i, CH_3-$CeCl_3$
ii, MeOH
iii, H_2 / Ni

Ph-CH_2-CH_2 H (30) OCH_3

NH_2 Ph-CH_2-CH_2 C H CH_3

R^1Li
73 - 98 °C

H N H Ph—CH_2—CH_2 R^1 —OCH_3

(1.76)

Me$_3$Si-CH$_2$-NH-CH$_2$-Ph $\xrightarrow[CH_3OH]{CH_2O}$ Me$_3$Si—CH$_2$—N-CH$_2$-Ph (31) CH$_2$-O-CH$_3$

LiF / CH_3CN

CH$_2$-Ph CH$_2$—N—CH$_2$ ⟷ CH$_2$═N—CH$_2$ (32)

HC═CH MeO$_2$C CO$_2$Me

CH$_2$-Ph N H$_2$C CH$_2$ MeO$_2$C CO$_2$Me

90%

(1.77)

prochiral oxime ethers to optically active amines. The enantioselectivity of the reduction is determined by the geometry of the oxime (Scheme 1.78). For R^1 = 2-naphthyl, n-C_6H_{13}, or $Ph(CH_2)_2$; R^2 = Me, then yields are 42–82 per cent and ee = 79–92 per cent.

R¹(R²)C=N~OCH₃ →[(-)-norephedrine, B_2H_6] R¹R²C(H)NH₂

(1.78)

1.1.5 *Preparation of 1,2-diamines*

In the presence of nitric oxide the cyclopentadienyl nitrosylcobalt dimer (**33**), prepared as shown in Scheme 1.79,[128] forms a *cis*- adduct with strained alkenes. These adducts may be stereoselectively reduced to 1,2-diamines with $LiAlH_4$ (Scheme 1.79).[129]

$CpCo(CO)_2$ + NO →[PhH] Cp—Co(μ-N=O)₂Co—Cp (33)

R¹(R³)C=C(R²)R⁴ →[(33), NO, 0°C] Cp–Co(N(=O)–C(R³)(R¹)–C(R²)(R⁴)–N(=O)) →[$LiAlH_4$] H_2N—C(R³)(R¹)—C(R²)(R⁴)—NH_2

50-90%

Cp =Cyclopentadienyl

(1.79)

Me(H)C=C(H)Me →[NH_2CN (Cyanamide), NBS] Me(H)C(Br)–C(H)(Me)NH-C≡N →[H_2, Pd-C] [Me(H)C(Br)–C(H)(Me)NH-CH=NH] →[NaOMe] Me, H, H, Me imidazoline (HN, N) →[50% aq. KOH] Me(H)C(NH₂)–C(H)(Me)NH₂ (+/-)

(180)

Another preparation of 1,2-diamines from alkenes in yields of 50–70 per cent has been reported (Scheme 1.80).[130]

1.1.6 *Preparation of allylamines*

Allylamines have been prepared from allyl silanes by the use of a phenyl tellurium trifluoroacetate (**34**) (Scheme 1.81).[131]

$$R^1\text{-CH=CH-CH}_2\text{SiMe}_3 \xrightarrow[BF_3.Et_2O\ /\ ClCH_2CH_2Cl]{Ph-Te(=O)-O-C(=O)-CF_3\ (34)} \left[R^1\text{-CH=CH-CH}_2\text{Te(=O)-Ph}\right] \xrightarrow[\text{Rearrangement}]{R^1\text{-NH}_2} \text{R-CH(NH-}R^1\text{)-CH=CHTe-Ph} \quad 15\text{-}86\% \qquad (1.81)$$

$R^1 = CH_3(CH_2)_5$

$E:Z = 86:14$

$R^2 = C_8H_{17}$, Ph, p-Cl-C_6H_4, p-NO-C_6H_4

By the application of the Wittig reaction to the compounds produced by the addition of primary amines to vinyltriphenylphosphonium bromide, allylamines result as mixtures of stereoisomers (Scheme 1.82).[132]

$$R^1\text{—NH}_2 + CH_2\text{=CH—}\overset{+}{P}Ph_3\ Br^- \xrightarrow{CH_3CN} R^1\text{—NH-CH}_2\text{-CH}_2\text{—}\overset{+}{P}Ph_3Br^- \xrightarrow{BuLi} R^1\text{—}\overset{-}{N}\text{—CH}_2\text{—}\overset{-}{C}H\text{—}\overset{+}{P}Ph_3 \xrightarrow[\text{Wittig Reaction}]{R^2\text{-CHO}} R^1\text{—NH-CH}_2\text{-CH=CH—}R^2 \qquad (1.82)$$

$E:Z \sim 1:1$

Allylic amines can be prepared from phenyl selenides by reaction with *N*-chlorosuccinimide (NCS), an alkyl carbamate, and triethylamine as shown in Scheme 1.83,[133] for example.

$$Me_2C\text{=CH-CH}_2\text{—SePh} + RO_2C\text{-NH}_2 \xrightarrow[\text{Et}_3\text{N, Room Temp}]{\text{NCS (excess)}} Me_2C(NH\text{—}CO_2R)(CH\text{=}CH_2) \xrightarrow{\text{Hydrolysis}} Me\text{—}C(NH_2)(Me)\text{—}CH\text{=}CH_2 \qquad (1.83)$$

R = Bu^t or Bn

Where it is relevant, the E isomer is obtained in large excess of the Z isomer (e.g.97:3)

Sharpless has developed two reagents (Scheme 1.84) which cause allylic amination of alkenes.[134] The first, generated by reaction of selenium metal with two equivalents of anhydrous chloramine-T in CH_2Cl_2 gives reagent (**35**). The second, the corresponding sulphur compound (**36**) has also been shown to effect allylic amination of alkenes and alkynes.[135]

Ts—N̄—Cl Na⁺ (chloramine-T) —Se in CH_2Cl_2→ TsN=Se=NTs (35)

Or TsN=S=NTs (36)

R^1—CH_2(R^2)C=C(R^3)(R^4) —Ts-N=Se=N-Ts→ R^1—CH(NHTs)(R^2)C=C(R^3)(R^4)

R^1	R^2	R^3	R^4	Yield (%)
$CH_3(CH_2)_3$	H	H	H	56
Cyclohexyl	H	H	H	45
H	Me	Me	Me	60

(1.84)

Vinyldiphenylphosphine oxide (**37**) is used to convert secondary amines into allylic tertiary amines (Scheme 1.85).[136]

γ-Acetoxy- or γ-chloro-allylamines can be prepared by the catalysed reaction between an allene and a secondary amine,[137] for example, see Scheme 1.86.

Hydroboration of allylic amines can be achieved if they are first protected by trimethylsilyl groups (Scheme 1.87).[138]

Allylamines can be converted into their next highest homologues by the method shown in Scheme 1.88.[139]

β,γ-Unsaturated primary amines can be prepared by a Diels–Alder reaction between 1,3-dienes and *N*-sulphinyltoluenesulphonamide (**38**) (Scheme 1.89).[140]

$Ph_2P(O)$-CH=CH$_2$ + HN(piperidine) ⟶ $Ph_2P(O)$-CH$_2$CH$_2$–N(piperidine)

(37)

i, Bu^n-Li
ii, PhCHO
80 %

Ph_2P H C CH$_2$–N(piperidine), C, HO H, Ph

\+

52 : 48

$Ph_2P(O)$ H C CH$_2$–N(piperidine), C, HO Ph, H

NaH
DMF

NaH
DMF

H, CH$_2$–N(piperidine), C=C, Ph, H

E 72 %

H, CH$_2$–N(piperidine), C=C, H, Ph

Z 6 %

(1.85)

AcO—CH$_2$—CH=CH—CH$_2$—OAc + HNEt$_2$

Pd(PPh$_3$)$_4$

AcO—CH$_2$—CH=CH-CH$_2$—NEt$_2$

70%

(1.86)

CH$_2$=CH-CH$_2$-N(SiMe$_3$)—SiMe$_3$ —[BH$_3$.SMe$_2$, 0°C]→ H$_2$B—CH$_2$·CH$_2$·CH$_2$—N(SiMe$_3$)—SiMe$_3$

MeOH

H$_2$B—CH$_2$·CH$_2$·CH$_2$—NH$_2$ —[i, HCl; ii, H$_2$O$_2$ / HO$^-$]→ HO—CH$_2$·CH$_2$·CH$_2$—NH$_2$

55%

(1.87)

CH_3 CH_2-NBu^n_2 + CO

$Pd(OAc)_2.Ph_2P$-CH_2-CH_2-CH_2-PPh_2 (catalyst)
toluene , 110° C

CH_3 O CH_2-C-NBu^n_2 77 % LiAlH_4 CH_3 CH_2-CH_2-NBu^n_2 95 % (1.88)

CH_3 CH_3 CH_3 CH_3 (Z) O S N Ts (38) $PhCH_3$ 0°C D-A CH_3 CH_3 CH_3 CH_3 $\overset{+}{S}$ O^- N Ts

CH_3 NHTs CH_3 CH_3 85% CH_3

CH_3 CH_3 CH_3 CH_3 (*E*) → 83% of CH_3 CH_3 NHTs CH_3 CH_3 (1.89)

1.1.7 *Preparation of α-amino ketones*

α-Amino ketones have been prepared by the reduction of acyl nitriles with zinc and acetic acid (Scheme 1.90).[141]

R-COCl

CuCN
CH_3CN

R—C(=O)—C≡N Zn / HOAc, AC_2O → R—C(=O)-CH_2—NHAc (1.90)

1.1.8 *Interconversions of primary, secondary, and tertiary amines and quaternary ammonium salts*

Alkylation

Primary amines to secondary and tertiary Some conversions of this kind have already been discussed: reductive amination (p. 15), the Leukart reaction (p. 16), the Eschweiler–Clarke reaction (p. 16), and methylations with reduction by sodium cyanoborohydride (p. 8). Another method for methylating a primary amine utilizes the mixed acetic–formic anhydride[142] to produce a formamide which is then reduced to the methylamine by borane:methyl sulfide (BMS) (Scheme 1.91).[143]

$$R\text{—}NH_2 \xrightarrow[\text{THF, -20}^\circ\text{C, 15 mins}]{\text{excess of } CH_3CO.O.CHO} \underset{\text{ca. 100\%}}{R\text{—}NH\text{—}CHO} \xrightarrow[65^\circ\text{C, 3 hours}]{\text{BMS / THF}} R\text{—}NH\text{—}CH_3$$

R = octyl Yield = 77% (1.91)

Primary amines have been converted into piperidines by reacting the trifluoroacetic salt of an amine with formaldehyde and an allylsilane (Scheme 1.92).[144]

$$R\text{-}NH_3^+\ F_3C\text{-}CO_2^- \xrightarrow[\text{aqueous}]{CH_2O} R\text{-}\overset{+}{N}H{=}CH_2\ F_3C\text{-}CO_2^-$$

$$\xrightarrow{Me_3Si\text{-}CH_2CH{=}CH_2} RNH\text{-}CH_2\text{-}CH_2\text{-}CH{=}CH_2$$

$$\xrightarrow[F_3C\text{-}CO_2H]{CH_2O} R\text{-}\overset{+}{N}({=}CH_2)\text{-}CH_2\text{-}CH_2\text{-}CH{=}CH_2\ \ F_3C\text{-}CO_2^-$$

$$\longrightarrow \text{4-hydroxy-}N\text{-R-piperidine (HO, NR)}$$

R^1	Yield (%)
Bn	81
2-(1-$ArSO_2$-indol-3-yl)CH_2CH_2-	63

(1.92)

Anions of amines can be generated by the action of sodium methoxide on *N*-trimethylsilylated primary amines. These can be alkylated with alkyl halides, yielding secondary amines (Scheme 1.93).[145]

$$R^1\text{—NHSiMe}_3 \xrightarrow{\text{MeONa}} R^1\text{—}\bar{N}\text{H Na}^+ + \text{MeOSiMe}_3 \xrightarrow{R^2\text{-Hal}} R^1\text{–NHR}^2 \; (45\text{-}85\%) \qquad (1.93)$$

In another conversion of primary to secondary amines, the primary amine is converted into an *N*-cyanomethyl compound which, because of the presence of the cyano group, can lose a proton to an alkyl lithium and eliminate cyanide ion. The resulting imine is alkylated by another molecule of alkyl lithium (Scheme 1.94).[146] Primary amines can be converted to secondary amines as shown in Scheme 1.95.[147]

Two derivatives of iron pentacarbonyl, (**39**) and (**40**), have been developed and have proven valuable in promoting the alkylation of primary or secondary amines by aldehydes, yielding tertiary amines (Scheme 1.96).[148] Similarly secondary amines can be alkylated by aldehydes in the presence of either of the compounds above (**39** or **40**), for example, see Scheme 1.97.[148] Primary amines

$$R^1\text{—CH}_2\text{—NH}_2 \xrightarrow[\text{or CH}_2\text{O + KCN}]{\text{Cl-CH}_2\text{CN}} R^1\text{—CH}_2\text{—NH—CH}_2\text{—C}\equiv\text{N} \xrightarrow{R^2\text{Li}} \left[R^1\text{—CH}_2\text{—N}=\text{CH}_2\right] \xrightarrow{R^2\text{Li}} R^1\text{—CH}_2\text{-NH-CH}_2\text{—R}^2 \; (50\text{-}80\%) \qquad (1.94)$$

H_3C–C(=O)–CO_2H + H_3C, NH_2, R^1, CO_2Bu^t $\xrightarrow{-H_2O}$ H_3C, CO_2CH_3, N, H, R^1, CO_2Bu^t $\xrightarrow{H_2 / Pd\text{-}C}$ H_3C, CO_2CH_3, H, NH, H, R^1, CO_2Bu^t (1.95)

$$\underset{1\ \text{mole}}{Fe(CO)_5} + \underset{3\ \text{mole}}{KOH} \longrightarrow KHFe(CO)_4 \quad (39)$$

$$Fe(CO)_5 \xrightarrow[\text{ii, } H_2O,\ 1\ \text{mole}]{\text{i, Na/Hg}} NaHFe(CO)_4 \quad (40) \qquad (1.96)$$

Pyrrolidine (NH) + Ph-CHO $\xrightarrow{(39)\text{or}(40)}$ N-CH_2-Ph (95 %) (1.97)

may be dialkylated by reaction with two equivalents of aldehyde (Scheme 1.98).[149] A molar ratio of 1 : 1 for amine : aldehyde enables monoalkylation to be achieved.

$$Bn\text{-}NH_2 + 2\ CH_3CHO \xrightarrow[\text{under CO, } 20^\circ C]{KHFe(CO)_4 \text{ in EtOH}} \underset{100\ \%}{Bn\text{-}N(CH_2CH_3)_2} \qquad (1.98)$$

Monoalkylation of primary amines can also be achieved by using organometallic reagents. This method proceeds via the *N*-alkyl-*N*-(alkylthiomethyl)-ammonium chloride (Scheme 1.99).[150]

$$R^1NH_2 \xrightarrow[\text{iii, } R^2SH]{\text{i, [CHO]; ii, HCl in dry } Et_2O} (R^1NH_2CH_2SR^2)Cl \xrightarrow{\text{iv, } 3R^3M\text{, in } Et_2O\text{; v, 1M KOH}} R^1NHCHR^3_2$$

R^1 = cyclo-C_6H_{11}, Bu^n, Bn
R_2 = Ph, Ph, Bn
R^3M = Bu^nLi, MeLi, MePhLi, AllylMgBr
Yield = 68–90% (1.99)

Secondary amines to tertiary The *N*-alkylation of secondary amines can be achieved using an ester and lithium aluminium hydride (Scheme 1.100).[151]

$$R^1COOEt + HNR^2R^3 + LiAlH_4 \xrightarrow{Et_2O \text{ or THF}} \left[R^1C(OAlH_3)(OEt)NR^2R^3 \right] \longrightarrow R^1C(=O)NR^2R^3 \xrightarrow{LiAlH_4} R^1CH_2NR^2R^3$$

R^1	R^2	R^3	Yield (%)
H	Bn	Me	72
Me	Bn	Me	55
Me		$-(C_5H_{10})-$	86
C_2H_5		$CH_3N(C_2H_4-)_2$	61
Me		$CH_3(C_2H_4-)_2$	72
C_2H_5		$PhN(C_2H_4-)_2$	62
H		$PhN(C_2H_4-)_2)$	90

(1.100)

Primary and secondary amines to quaternary salts Quaternary salts have been obtained from primary and secondary amines by addition of excess methyl iodide and a stoichiometric amount of 1,2,2,6,6,-pentamethylpiperidine (PMP) (**41**) to the amine (Scheme 1.101).[152]

$$CH_3I + R\text{-}NH_2 \text{ or } R_2NH + \text{PMP (41)} \xrightarrow[\text{20 °C, 12 hours}]{\text{Suitable solvent, e.g. DMF or } CH_3CN} R\overset{+}{N}(CH_3)_3\ I^- \text{ or } R_2\overset{+}{N}(CH_3)_2\ I^- \quad 50-95\%$$

(1.101)

Dealkylation

Quaternary salts to tertiary amines By the use of 1,4-diazabicyclo[2.2.2]octane (DABCO) (**42**) quaternary salts are converted into tertiary amines (Scheme 1.102).[153]

$$R_4\overset{+}{N}\ X^- + \text{DABCO (42)} \xrightarrow[\text{reflux}]{\text{EtOH or DMF}} \text{DABCO-}\overset{+}{N}\text{-R}\ X^- + R_3N \quad 50-90\%$$

(1.102)

Quaternary salts are also converted into tertiary amines in high yield by reduction with triethylithium borohydride (Scheme 1.103).[154]

$$R^1R^2\overset{+}{N}Me_2\ I^- \xrightarrow[\text{ii, aq. } H_3O^+]{\text{i, } LiBHEt_3 / THF / 25^\circ C} R^1R^2NMe$$

R^1	R^2	Yield (%)
Ph	Me	100
Ph	Et	96
Bn	Me	100

(1.103)

A methyl group can be selectively removed from a quaternary salt by the action of lithium *n*-propyl mercaptide (**43**) in HMPT (Scheme 1.104).[155] More vigorous conditions (50 °C and 3.4 h) are required for purely aliphatic salts.

$$R^1R^2\overset{+}{N}Me_2\ I^- \xrightarrow[\text{0° C for 30 min.}]{Pr^nS^-Li^+ \text{ (43) / HMPT}} R^1R^2NMe$$

R^1=Ph, $CH_3(CH_2)_{11}$; R^2=Et, $CH(CH_3)_2$, $(CH_3)_3C$
Yields=95–99%

(1.104)

Tertiary amines to secondary Dealkylation of tertiary amines can be achieved by use of tertiary butyl peroxide and a ruthenium complex (Scheme 1.105).[156]

$$R^1N(R^2)\text{—}R^3 \xrightarrow[C_6H_6]{Bu^tOOH,\ Ru(II)} R^1N(R^2O_2C(CH_3)_3)\text{—}R^3 \xrightarrow[93\%]{H_3O^+} R^1N(H)\text{—}R^3 + Bu^t\text{—}OH + H_2O$$

78-97%

R_1	Ph	p—CH_3Ph	p—BrPh	Ph	p—ClPh	Ph
R_2	Me	Me	Me	C_2H_5	Me	CH_2CHCH_2
R_2	Me	Me	Me	Me	$(CH_2)_2CO_2CH_3$	Me

(1.105)

von Braun[157] developed a method for converting tertiary to secondary amines by the action of cyanogen bromide.[158] This method has been applied to open chain (acyclic) tertiary amines and to cyclic tertiary amines including the degradation of alkaloids. Cyanogen bromide which is difficult to handle on account of its toxicity, volatility, and instability is prepared from bromine and

$$Br_2 + NaCN \xrightarrow[0^o - 30^o\ C,\ 2\ hours]{Aqueous} BrCN + NaBr \quad 73 - 85\ \% \qquad (1.106)$$

sodium cyanide (Scheme 1.106)[159] or from bromine and aqueous potassium cyanide.[160] The cyanogen bromide can be used neat or diluted by a solvent (e.g. by ether, chloroform, or benzene) depending upon the reactivity of the amine. The initial product of the reaction is a dialkyl cyanamide (**44**) which can be hydrolysed and decarboxylated to yield the secondary amine (Scheme 1.107[161] and 1.108[162]). The mechanism of the reaction is as shown in Scheme 1.109.

$$Pr^i_2N\text{-}CH_3 \xrightarrow{BrCN} Pr^i_2N\text{-}CN\ (44) + CH_3Br \xrightarrow{hydrolysis} [Pr^i_2N\text{-}CO_2H] \longrightarrow Pr^i_2NH + CO_2 \qquad (1.107)$$

$$\text{(octahydroindolizine)} \xrightarrow{BrCN} \text{2-}(CH_2)_3Br\text{-piperidine }N\text{-CN} \qquad (1.108)$$

$R^1R^2R^3N: \; N{\equiv}C{-}Br \longrightarrow Br^- \; R^1R^2R^3\overset{+}{N}{-}CN \longrightarrow R^2R^3N{-}CN + R^1{-}Br$

(1.109)

A more satisfactory and now more frequently used method for effecting this conversion is by the use of chloroformates.[163] Ethyl chloroformate has been used[164] for the demethylation of tropine acetate while *p*-nitrophenyl chloroformate has found use in the demethylation of a ketal of mesembrine[165] (Scheme 1.110).

$>N{-}CH_3 \xrightarrow{Cl.CO_2R} >N{-}CO_2R \xrightarrow[\text{with acid / alkali}]{\text{hydrolysis}} >N{-}H$

R	Substrate	Yield (%)
Et	Tropine acetate	70
p-Nitrophenyl	Mesembrine ketal	56

(1.110)

In the same way, methyl chloroformate has been used to demethylate morphine and codeine[166] and apomorphine;[167] phenyl chloroformate has also been used.[168]

CH_3 N O

Tropinone

$Cl_3C{-}CH_2OCOCl$

PhH, K_2CO_3

O

$C{-}OCH_2CCl_3$ N O

Zn / 90 % HOAc

25° C

NH O

(1.111)

Tropinone has been demethylated in 56 per cent yield and morphine in 75 per cent yield by the use of trichloroethyl chloroformate ($R = Cl_3CCH_2$), the hydrolysis of the intermediate carbamate being performed under mild conditions with zinc and acetic acid (Scheme 1.111).[169] For this type of conversion of tertiary to secondary amines vinyl chloroformate can be used (Scheme 1.112).[170] However, later, Olofson and his co-workers found that α-chloroethyl chloroformate is superior to vinyl chloroformate for this purpose (Scheme 1.113).[171]

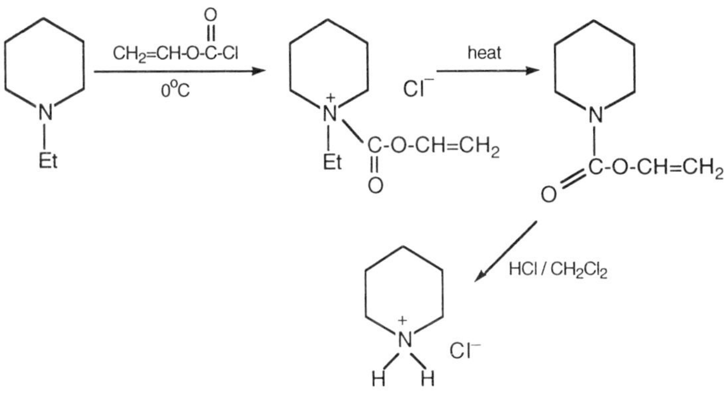

(1.112)

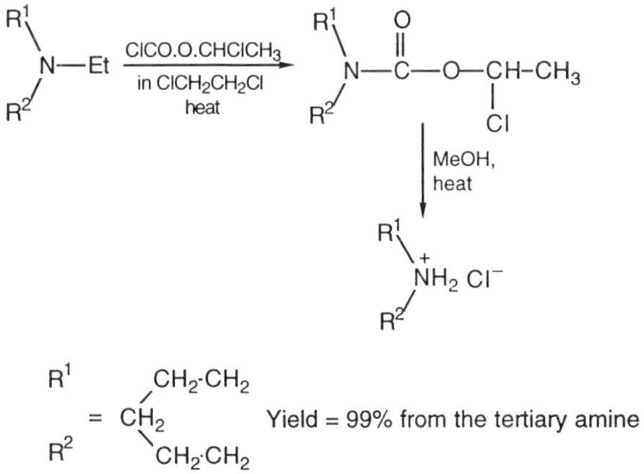

(1.113)

Tertiary amines can be converted into secondary amines and secondary amines into primary by *N*-dealkylation with diethyl azodicarboxylate (**45**) (Scheme 1.114).[172] A possible mechanism for hydrolysis is as shown in Scheme 1.115. The reaction can be carried out without isolating the intermediate amide by refluxing the amine with ester in benzene for 1 h and then adding 4 M HCl in EtOH under reflux. Tertiary amines are also dealkylated (Scheme 1.116).

$$2\,Me_2NH + EtO_2C\text{-}N{=}N\text{-}CO_2Et\ (45) \xrightarrow{0^\circ C} Me_2N\text{-}C(=O)\text{—}N{=}N\text{—}C(=O)\text{-}NMe_2\ \ 79\%$$

acid hydrolysis

$$\rightarrow 2\,Me\text{-}NH_2 + 2\,CH_2O + H_2N\text{-}NH_2 \quad (1.114)$$

H_3C, CH_3, CH_2, H_2C, NH, N^+, $CON(CH_3)_2$, $NH\text{–}NH\text{—}CON(CH_3)_2$, H_3O^+, HO_2C, CO_2H, $+\ 2\,MeNH_2$, $+\ CH_2O$

$$2\,CO_2 + H_2N\text{-}NH_2 \quad (1.115)$$

$$N(CH_3)_3 + EtO_2C\text{-}N{=}N\text{-}CO_2Et \xrightarrow{HCl} Me_2N^+H\text{-}CH_2N(CO_2Et)\text{-}NHCO_2Et\ Cl^-$$

Aqueous H_3O^+

$$\rightarrow CH_2O + EtO_2C\text{-}NHNHCO_2Et + Me_2NH \quad (1.116)$$

1.2 Protection of the amino group

The cyclic carbonate of benzoin (**46**) has been used for the protection of primary amines (Scheme 1.117).[173] Removal of the protection can be achieved either by reduction with Pd–H_2 (yields *c.* 100 per cent) or with Na–liq.NH_3

(46) + R-NH_2 —(i, DMF, 30 mins; ii, aq. HCl)→ Ph—CO—CH(Ph)—O—CO—NHR

$F_3C.CO_2H$

(1.117)

(yields 75–85 per cent) or by oxidation with excess of *m*-chloroperbenzoic acid followed by hydrolysis (Scheme 1.118). This protection has been applied to dipeptides.

For the protection of amino groups in α-amino acids[174] di-*t*-butyldicarbonate (**47**)[175] is widely used. It reacts with α-amino acid esters to give *N*-protected compounds (>N-Boc) (Scheme 1.119).[176]

Oxidation *m*-chloroperbenzoic acid

Reduction Pd / H_2 or Na liq NH_3

H_2O

R-NH_2 + CO_2 + Ph-CO-CO-Ph   R-NH_2 + CO_2 + Ph-CH_2CH_2-Ph (1.118)

$$(CH_3)_3C{-}O{-}\overset{O}{\overset{\|}{C}}{-}O{-}\overset{O}{\overset{\|}{C}}{-}O{-}C(CH_3)_3 \;(47) + R^1{-}\underset{}{\overset{NH_2}{\overset{|}{C}H}}{-}CO_2R^2 \xrightarrow[\text{reflux, 90 min.}]{\text{aq.NaHCO}_3 + \text{NaCl, CHCl}_3} R^1{-}CH(CO_2R^2){-}NH{-}\overset{O}{\overset{\|}{C}}{-}O{-}C(CH_3)_3$$

R^1	R^2	Yield (%)
H	Et	89

(1.119)

The diallyl group has been used for protection of primary amines. The protection is accomplished by reaction of the amine with allyl bromide and ethyldiisopropylamine (Scheme 1.120).[177] Deprotection is carried out using Wilkinson's Catalyst[178] in acetonitrile to isomerize the allyl groups to propenyl groups. The hydrate of rhodium trichloride ($RhCl_3.3H_2O$) has also been used to remove the allyl group as it causes isomerization of allylamines to enamines which are hydrolysed by hydrochloric acid in methanol (e.g. Scheme 1.121).[179]

$$RNH_2 + 2\,BrCH_2{-}CH{=}CH_2 \xrightarrow{EtPr^i_2N} R{-}N(CH_2{\cdot}CH{=}CH_2)_2 \xrightarrow[\text{aq. CH}_3\text{CN, reflux}]{\text{ClRh(PPh}_3)_3\text{, Wilkinson's catalyst}} R{-}NH_2 + 2\,CH_3CH_2CHO \qquad (1.120)$$

$$Ph_2N{-}CH_2{-}CH{=}CH_2 \xrightarrow{RhCl_3.3H_2O} [Ph_2N{-}CH{=}CHCH_3] \xrightarrow{\text{HCl in aq. MeOH}} Ph_2NH \qquad (1.121)$$

t-Butyldiphenylchlorosilane (**48**) in acetonitrile has been used to protect primary amines (Scheme 1.122).[180] The derivatives are stable to alkali, hydrolytic reagents, and alkylation, but can be cleaved by the action of 80 per cent acetic acid at 25 °C or by pyridine and hydrogen fluoride at 25 °C.

$$\underset{}{R\text{-}NH_2} + \underset{(48)}{Bu^tPh_2SiCl} \xrightarrow[\text{in } CH_3CN]{Et_3N} R\text{-}NH\text{-}SiBu^tPh_2 \qquad (1.122)$$

1.3 Reactivity

1.3.1 *Basicity*

The two most important factors which govern the reaction of amines with proton acids, i.e. their basicity, are their structure and the extent of solvation. There are great variations in the order of basicity of a series of amines in different solvents.[181]

The reaction of amines with Lewis acids has been investigated,[182] the particular Lewis acid, $Et_3N.BH_3$,[183] has been investigated by Matsumara and Tokura,[184] who prepared it as shown in Scheme 1.123.

$$NaBH_4 + Et_3N \xrightarrow{\text{liquid } SO_2} Et_3N.BH_3 \quad 85\,\% \qquad (1.123)$$

1.3.2 *Nucleophilicity*

Many of the well known reactions of amines stem from the fact that the nitrogen atom which they contain is nucleophilic and readily takes part in substitution reactions and in addition reactions with electrophiles. When alkyl halides react with ammonia, successive S_N2 reactions finally yield quaternary salts. In order to make this a viable preparation of amines, either the Hinsberg method with subsequent separation of the sulphonamides must be used, which is tedious, or the Gabriel method can be applied (see above, p. 3).

Amines react with acyl halides (**49**) or acid anhydrides (**50**) to produce amides (**51**), and sulphonyl chlorides (**52**) similarly give sulphonamides (**53**) (Scheme 1.124). An interesting reaction of this kind[185] is that of amines with peroxyacetyl nitrate (**54**)[186] which is safe to use in dilute solution (Scheme 1.125).

The reaction of amines with the nitrosonium ion (NO^+) is important from several points of view.[187] The reaction of nitrous acid, generated from sodium nitrite and aqueous acids, with a primary amine produces a carbonium ion whose fate depends upon its structure.[188] The product may be an alcohol, an

$$R^1R^2NH + R^3\text{-COCl} \; (49) \longrightarrow R^1R^2N\text{-CO-}R^3 \; (51)$$

$$R^1R^2NH + R^3\text{-CO.O.CO-}R^3 \; (50) \longrightarrow R^1R^2N\text{-CO-}R^3 \; (51)$$

$$R^1\text{-}NH_2 + R^2\text{-}SO_2Cl \; (52) \longrightarrow R^1\text{-NH-}SO_2R^2 \; (53) \qquad (1.124)$$

$$CH_3\text{-C(=O)-O-O-}NO_2 \; (54) + R\text{-}NH_2 \longrightarrow CH_3\text{-C(=O)-NHR} + O_2 + HNO_2 \qquad (1.125)$$

alkene, or a rearranged molecule, or a mixture of these and there is evidence that up to 10 per cent of a nitrosamine (**55**) of a secondary amine can also be produced (Scheme 1.126). Kinetic measurements[189] show that the rate is dependent upon the square of the nitrous acid concentration, and this is interpreted as involving dinitrogen trioxide (N_2O_3) as an intermediate,[190] which in reacting with the amine can lose the nitrite anion and hand on the electrophilic nitrosonium ion (Scheme 1.127).

$$R\text{-}NH_2 + 2\,HNO_2 \longrightarrow R^+ + N_2 + H_2O$$

$R^+ \longrightarrow$ Alcohol, Alkene, Rearranged product

$$R\text{-}NH_2 + R^+ \longrightarrow R_2\overset{+}{N}H_2 \longrightarrow R_2\overset{+}{N}H\text{-N=O} \longrightarrow R_2N\text{-N=O} \; (55) \qquad (1.126)$$

$$2\,HNO_2 \rightleftharpoons H_2O + N_2O_3$$

$$R\text{-}\ddot{N}H_2 + O\text{=N-O-N=O} \longrightarrow NO_2^- + R\text{-}\overset{+}{N}H_2\text{-N=O}$$

$$R\text{-}\overset{+}{N}H_2\text{-N=O} \longrightarrow R\text{-NH-N=O} \; (56) + H^+$$

$$R\text{-NH-N=O} \longrightarrow R\text{-N=N-OH} \longrightarrow R\text{-}\overset{+}{N}\equiv N + HO^-$$

$$R\text{-}\overset{+}{N}\equiv N \longrightarrow R^+ + N_2$$

$$\text{Rate} = k[\,R\text{-}NH_2\,][\,HNO_2\,]^2 \qquad (1.127)$$

The reaction of a secondary amine with nitrous acid (Scheme 1.128) follows the same route and has the same kinetic equation,[191] but the *N*-nitrosamine (**57**) (see Chapter 12, p. 495), analogous to the nitrosamine intermediate (**56**) from the primary amine, is stable at room temperature and is the product which is isolated in good yield from the reaction between a secondary amine, dilute hydrochloric acid, and sodium nitrite. For the most reactive amines (primary and secondary) the formation of N_2O_3 can be rate limiting.

$$R_2\ddot{N}H + O{=}N{-}O{-}N{=}O \longrightarrow \underset{(57)}{R_2N{-}N{=}O} + HNO_2$$

$$\text{Rate} = k[R_2NH][HNO_2]^2 \qquad (1.128)$$

It is often said that tertiary amines do not react with sodium nitrite and aqueous acid and this is the basis of a test for differentiating between primary, secondary, and tertiary amines. It should be remembered, however, that the products from secondary amines, the *N*-nitrosamines (see Chapter 12) are powerful carcinogenic substances and should only be handled by experienced chemists who take adequate precautions to prevent ingestion and inhalation. While it is true that tertiary amines do not react with dilute (2 M) hydrochloric acid and sodium nitrite under the conditions used to obtain a nitrosamine from a secondary amine, they will react with sodium nitrite and 4 M hydrochloric acid or acetic acid to yield an aldehyde and a secondary nitrosamine in a reaction whose mechanistic details are not yet fully understood.[192] Consequently the prediction of which of two alkyl groups will be split off as an aldehyde cannot be made with certainty but the importance of the reaction, in that it produces a carcinogenic *N*-nitrosamine, should not be forgotten (Scheme 1.129).

$$R^1{-}CH_2{-}NR^2R^3 \xrightarrow{NO^+} R^1{-}CH_2{-}\overset{+}{N}(N{=}O)R^2R^3 \longrightarrow R^1{-}CH{=}\overset{+}{N}R^2R^3 \xrightarrow{NO_2^-} R^1{-}CH(O{-}N{=}O){-}NR^2R^3 \longrightarrow R^1{-}CHO + R^2R^3N{-}N{=}O \qquad (1.129)$$

Primary and secondary amines react with the carbonyl group by nucleophilic addition. The reactions of primary amines with aldehydes and ketones give imines (**58**) (see Chapter 3) while secondary amines yield enamines (**59**) (see

Chapter 2). Secondary amines also react with esters to give amides (**60**) (Scheme 1.130).

$$R^1\text{-}NH_2 + R^2\text{-}CHO \longrightarrow \underset{(58)}{R^1\text{-}N{=}CHR^2} + H_2O$$

$$R^1_2NH + R^2\text{-}CO\text{-}CH_2R^3 \longrightarrow \underset{(59)}{R^2\text{-}C(=CHR^3)\text{-}NR^1_2} + H_2O$$

$$R^1\text{-}NH_2 + R^2\text{-}CO_2Et \longrightarrow \underset{(60)}{R^1\text{-}NH\text{-}CO\text{-}R^2} + EtOH \qquad (1.130)$$

The Mannich reaction[193–196]

In the Mannich reaction an aldehyde (usually formaldehyde) is condensed with ammonia, a primary amine, or a secondary amine and a compound containing an active hydrogen atom. The product of the condensation is called a Mannich base. For example when a primary amine is reacted with the enol of a ketone a β-amino ketone is formed as the Mannich base (**61**) (Scheme 1.131). The scope of the reaction is very wide[193] since ammonia can react up to three times, a primary amine twice, and a secondary amine once and almost any aldehyde can be used. Furthermore the range of compounds containing active hydrogen or a nucleophilic carbon atom (for example, phenols such as resorcinol), which can undergo the reaction is very large.

$$R^1\text{-}NH_2 + R^2\text{-}CHO + R^3\text{-}CH_2\text{-}C(=O)\text{-}R^4 \xrightarrow{\text{aq. acid}} \underset{(61)}{R^1\text{-}NH\text{-}CH(R^2)\text{-}CH(R^3)\text{-}C(=O)\text{-}R^4} \qquad (1.131)$$

The most commonly occurring mechanism involves reaction of the amino compound with the aldehyde to yield an iminium ion (**62**) which, being an electrophile, reacts with an enol or similar compound, e.g. Scheme 1.132.

Since nitroalkanes contain an activated (acidic) hydrogen atom at the α-position, they undergo a variety of Mannich reactions in good yield,[194] e.g. Scheme 1.133.

Iminium salts, for example $CH_2{=}NMe_2^+\ I^-$, can be prepared and isolated and can then be used in Mannich reactions (see Chapter 3, p. 161 for an account of these salts).

For organic synthesis many Mannich reactions are carried out with a secondary amine, often dimethylamine. Again the aldehyde is often formaldehyde. Vinyl ketones are easily prepared from a suitable Mannich base by

$$R^1R^2NH + CH_2O + HCl \longrightarrow R^1R^2\ddot{N}-CH_2-\overset{+}{O}H_2\ Cl^-$$

$$\text{(62)}\quad R^1R^2\overset{+}{N}=CH_2 + H_2O$$

$$R^3\text{-}CH_2C(=O)\text{-}R^4 \rightleftharpoons R^3\text{-}CH=C(OH)\text{-}R^4$$

$$\longrightarrow R^1R^2N\text{-}CH_2\text{-}CHR^3\text{-}C(=O)\text{-}R^4$$

$$CH_2=\overset{+}{N}R^1R^2 \quad CH_2\text{-}NR^1R^2 \quad CH_2\text{-}NR^1R^2$$

$$:OH \quad CH_2=\overset{+}{N}R^1R^2 \longrightarrow OH \quad CH_2\text{-}NR^1R^2 \quad HO \quad HO$$

$$CH_3\text{-}C\equiv CH + CH_2=\overset{+}{N}R^1R^2 \longrightarrow CH_3\text{-}C\equiv C\text{-}CH_2\text{-}NR^1R^2 \qquad (1.132)$$

elimination with alkali. Reduction of the resulting unsaturated ketone enables the homologous series of ketones to be ascended (Scheme 1.134).

Direct reduction of a Mannich base with lithium aluminium hydride in tetrahydrofuran furnishes a synthesis of β-amino alcohols,[197] e.g. Scheme 1.135.

$$Et_2NH + CH_2O \xrightarrow[18^\circ C]{aqueous} Et_2\overset{+}{N}{=}CH_2$$

$$CH_3CH_2CH_2NO_2 \longrightarrow CH_2CH_2CH{=}\overset{+}{N}(O^-)-O-H$$

$$\downarrow$$

$$Et_2N\text{-}CH_2\text{-}CH(NO_2)\text{-}CH_2CH_3 \quad 80\ \%$$

Similarly:

$$Pr^i_2NH + CH_2O + CH_3CH_2NO_2 \longrightarrow Pr^i_2N\text{-}CH_2\text{-}CH(NO_2)\text{-}CH_3 \quad 65\ \% \qquad (1.133)$$

$$CH_3\text{-}C(=O)\text{-}CH_3 + CH_2O + HNMe_2 \longrightarrow CH_3\text{-}C(=O)\text{-}CH_2\text{-}CH_2\text{-}NMe_2$$

$$\downarrow \text{hot aqueous } Na_2CO_3$$

$$CH_3\text{-}C(=O)\text{-}CH_2\text{-}CH_3 \xleftarrow{Na\ /\ liquid\ NH_3} CH_3\text{-}C(=O)\text{-}CH{=}CH_2 + HNMe_2 \qquad (1.134)$$

$$\text{2-[1-(morpholino)ethyl]cyclohexanone} \xrightarrow[THF]{LiAlH_4} \text{2-[1-(morpholino)ethyl]cyclohexanol} \quad 15\% \qquad (1.135)$$

Suitable Mannich bases react with nucleophiles by substitution for the amino group, e.g. Scheme 1.136.[198]

In biosynthesis it is known that aldehydes other than formaldehyde take part in the reaction. This happens in the biosynthesis of the isoquinoline alkaloids, papaverine and laudanosine, which occur in the opium poppy (Scheme 1.137).[198]

The range of compounds which react with iminium ions in the Mannich re-

CH_2NMe_2

NH + CH_2O + $HNMe_2$ —HOAc, 5° C→ NH

Gramine

$(EtO_2C)_2CH_2$ —EtO-N=O, EtO^-→ $(EtO_2C)_2C{=}NOH$

Zn / $H.CO_2H$

CHO

$(EtO_2C)_2CH$-NH

EtO^-

i, $(EtO_2C)_2\bar{C}$-NHCHO

ii, hydrolysis

MeI

$CH_2\overset{+}{N}Me_3$ I^-

NH

KCN / EtOH

CH_2CN

NH

CH_2-CH-CO_2H

NH_2

NH

Tryptophan

hydrolysis

CH_2-CO_2H

NH

Heteroauxin

A plant growth hormone (1.136)

action is very large. In addition to the examples shown above which yield C-Mannich bases containing carbon–carbon bonds, reactions also occur on suitable heteroatoms (e.g. N,O,S,P, and As) to yield heteroatom-Mannich bases containing heteroatom–carbon bonds.[195,196] An example in which the heteroatom is nitrogen is shown in Scheme 1.138.[199] The Mannich bases derived from benzotriazole (**63**) have been employed in reactions with enolates of ketones as nucleophiles (Scheme 1.139).[197] Other lithium enolates have also been used with yields in the range 50–85 per cent.

Reaction can occur on oxygen, as is the case with picric acid (**64**) (Scheme 1.140),[200] and on arsenic (Scheme 1.141).[201]

Esters of phosphinic acid acting as a source of active hydrogen are used with esters of α-amino acids and formaldehyde to produce after hydrolysis,

- CO_2
- H_2O

nor-Laudanosoline

Methylation

Dehydrogenation and Methylation

Laudanosine

Papaverine

(1.137)

$$\text{2,2-dimethylaziridine-type (CH}_3)_2\text{C(NH)}_2 + 2CH_2O + 2HNR_2 \longrightarrow (CH_3)_2C(N\text{-}CH_2\text{-}NR_2)_2$$

(1.138)

(63)

$R^1R^2C{=}O + R^3R^4NH$

$R^1\text{-}C(R^2)\text{-}NR^3R^4$

$OSiMe_3$

MeLi / THF

O^-

$C\text{-}NR^3R^4$, R^1, R^2

(1.139)

$$\text{Picric Acid (64)} + CH_2O + HNR^1R^2 \longrightarrow 2,4,6\text{-}(NO_2)_3C_6H_2\text{-}O\text{-}CH_2\text{-}NR^1R^2 \qquad (1.140)$$

$$R^1_2AsH + CH_2O + HNR^2_2 \longrightarrow R^1_2As\text{-}CH_2\text{-}NR^2_2 \qquad (1.141)$$

compounds which are useful as herbicides, once again employing a variation of the Mannich reaction, e.g. Scheme 1.142.[202]

$$(CH_3O)_2P(=O)\text{-}H + CH_2O + H_2N\text{-}CH_2\text{-}CO_2CH_3 \longrightarrow (CH_3O)_2P(=O)\text{-}CH_2\text{-}NH\text{-}CH_2\text{-}CO_2CH_3 \xrightarrow{\text{conc. HCl}} (CH_3O)_2P(=O)\text{-}CH_2\text{-}NH\text{-}CH_2\text{-}CO_2H \qquad (1.142)$$

Michael reaction

With α,β-unsaturated ketones, such as mesityl oxide (**65**), 1,4-addition of amines (Michael reaction) occurs (Scheme 1.143). Primary amines undergo Michael type addition reactions with alkynes. The product then tautomerizes to an imine (Scheme 1.144).[203] The product of the 1,4-addition of amines to carbon disulphide, in base, is the anion (**66**) (Scheme 1.145).

$$CH_3\text{-}C(=O)\text{-}CH{=}C(CH_3)_2\ (65) + R\text{-}\ddot{N}H_2 \longrightarrow CH_3\text{-}C(O^-){=}CH\text{-}C(CH_3)_2\text{-}\overset{+}{N}H_2R \longrightarrow CH_3\text{-}C(=O)\text{-}CH_2\text{-}C(CH_3)_2\text{-}NHR \qquad (1.143)$$

$$RC{\equiv}CR^1 + R^2NH_2 \longrightarrow RCH{=}C(NHR^2)R^1 \rightleftharpoons RCH_2{-}C({=}NR^2)R^1 \qquad (1.144)$$

$$S{=}C{=}S + RNH_2 \xrightarrow{\text{base}} RNH{-}C({=}S){-}S^- \; (66) \qquad (1.145)$$

Conversion of primary amines into other compounds via pyrylium and pyridinium salts

2,4,6-Triphenylpyrilium salts (**67**) have been applied by Katritzky and his co-workers to the conversion of primary amines into other alkyl compounds.[204]

Ph, Ph, Ph, O^+, X^-

$X = I^-, BF_4^-, SCN^-$

(67)

Primary amines to iodides Alkylamines react with (**67**) to form the pyridinium iodides (**68**) which are converted to iodides and 2,4,6-triphenyl pyridine (**69**) on heating (Scheme 1.146).[205]

$$R\text{-}NH_2 + (67)\; I^- \longrightarrow (68)\; I^- \xrightarrow{\text{heat}} (69) + R\text{-}I \qquad (1.146)$$

(67) (68) (69) 50-80%

$R = n\text{-}C_4H_9\text{-}$

Primary amines to esters Primary amines may be converted into acetates and benzoates by reaction with 2,4,6-triphenylpyrilium tetrafluoroborate (**70**),[206] which may initially be prepared from reaction with benzaldehyde as shown in Scheme 1.147.[207]

Ph-CHO + 2 Ph-COCH$_3$ $\xrightarrow{HBF_4}$ 2,4,6-triphenylpyrylium BF_4^- (70) 40 %

(70) $\xrightarrow{R^1\text{-}CH_2\text{-}NH_2}$ N-(CH_2-R^1)-2,4,6-triphenylpyridinium BF_4^-

N-(CH_2-R^1)-2,4,6-triphenylpyridinium BF_4^- $\xrightarrow[ca.\ 200^\circ C]{R^2\text{-}CO_2^-\ Na^+}$ 2,4,6-triphenylpyridine + R^2-$CO_2CH_2R^1$ 60-85%

R^1 = Me, Et, Bu^n, $PhCH_2CH_2$, 2-Picolyl, 3-picolyl, Benzyl.

R^2 = Me, Ph (1.147)

With isonitriles

Primary amines form a carbene–metal complex (**71**) on reaction with isonitriles in the presence of palladium(II) chloride. This intermediate need not be isolated but on treatment with silver oxide it may be converted into a carbodiimide (**72**) (Scheme 1.148).[208]

$$R^1\text{-}NH_2 + R^2\text{-}NC \xrightarrow{PdCl_2} PdCl_2(R^2NC)\left[:C\begin{matrix}NHR^1\\NHR^2\end{matrix}\right] \text{ (71)}$$

$$\text{(71)} \xrightarrow{Ag_2O} R^1\text{-}N{=}C{=}NR^2 \text{ (72)} \qquad (1.148)$$

1.3.3 *Oxidation*

For the use of lead tetraacetate to oxidize primary amines to nitriles, see Chapter 4 p. 187.[209]

Oxidative deamination of primary, secondary, or tertiary amines of appropriate structure takes place with potassium permanganate in a buffered aqueous solution of tertiary butanol to yield ketones (Scheme 1.149).[210]

For the oxidation of primary amines with silver(II) picolinate to nitriles along with smaller amounts of the corresponding aldehyde, see Chapter 4 p. 188.[211]

$$R^1{}_2CH\text{-}NR^2{}_2 \xrightarrow[\text{60-80°C}]{\text{KMnO}_4 \text{, aq. Bu}^t\text{OH buffered}} R^1R^1C{=}O \quad \text{9-96\%}$$

R^1 = H	H	Et	Bn	CH_3	CH_3
R^2 = H	Me	H	H	H	CH_3

(1.149)

Tertiary butyl nitrite is the nitrite of choice for the oxidative deamination of unbranched primary amines in the presence of dry cupric halides to give good yields of *gem*-dihalides (Scheme 1.150).[212] This method has been extended by the use of nitrosyls of copper halides, $(CuX_2.NO)_2$ X = Cl or Br, which may be prepared by passing nitric oxide (NO) into the cupric halide in acetonitrile.[213]

$$R\text{-}CH_2\text{-}NH_2 + 2\,CuHal_2 + Me_3C\text{-}O\text{-}N{=}O \xrightarrow[\text{1 hour at 65° C}]{CH_3CN \text{ as solvent}} R\text{-}CHHal_2 + 2\,CuHal + Me_3C\text{-}OH + N_2$$

Hal = Cl or Br 46-80 %

R = $PhCH_2$-, $Ph(CH_2)_2$-, $Ph(CH_2)_3CH_2N_2$-, $Me(CH_2)_5$-,

cyclo-C_6H_{11}-, $HO_2(CH_2)_4$-, $HO(CH_2)_{10}$-, $HOCH_2(CH_2)_8$-, (1.150)

The electro-oxidation of primary amines at a very low potential (*c.* +0.4 V) in the presence of a weak base such as 2,6-lutidine in aqueous medium yields aldehydes, whilst nitriles are produced when the medium is anhydrous acetonitrile (Scheme 1.151).[214]

$$R\text{—}NH_2 \xrightarrow[\text{aq. 2,6-lutidine}]{\text{electro-oxidation}} R\text{—}CHO$$

$$R\text{—}NH_2 \xrightarrow[\text{anhydrous } CH_3CN]{\text{electro-oxidation}} R\text{—}CN \quad (1.151)$$

Aryldialkylamines react with manganese(III) acetate in chloroform in the presence of acetic anhydride to yield amides by replacement of one of the alkyl groups by an acetyl group (Scheme 1.152).[215] Groups other than methyl are preferentially oxidized.

Primary and secondary amines which contain α-hydrogen atoms are oxidized by *p*-nitrobenzenesulphonyl peroxide $[(p\text{-}NO_2C_6H_4SO_2O)_2]$[216] to imines which can be hydrolysed to yield aldehydes (Scheme 1.153).[217]

$$\text{Ph—N(R)—CH}_3 \xrightarrow[\text{CHCl}_3,\ \text{Ac}_2\text{O},\ 25^\circ\text{C}]{\text{Mn(OAc)}_3} \text{Ph—N(COCH}_3\text{)—CH}_3 + \text{Ph—N(COCH}_3\text{)—R}$$

R	Yield (%)	
Me	61	—
Et	61	8
Bu^n	83	1

(1.152)

$$R^1\text{-CH}_2\text{-NHR}^2 + (\text{Ar-SO}_2\text{O})_2 \xrightarrow[\text{under } N_2,\ -78^\circ\text{C}]{\text{EtOAc}} [R^1\text{-CH(H)-N(O-SO}_2\text{Ar)-R}^2] \longrightarrow R^1\text{-CH=N-R}^2 \xrightarrow{H_2O} R^1\text{-CHO}$$

Ar = p-$NO_2C_6H_4$

R^1 = Bn, cyclo-C_6H_{11}, C_4H_9, i-C_3H_7
R^2 = Bn, cyclo-C_6H_{11}, C_4H_9, i-C_3H_7, Me
Yield of R^1CHO = 37–96% (1.153)

Primary amines with one α-hydrogen atom are oxidized by aqueous palladium chloride to aldehydes or ketones in variable yields (Scheme 1.154).[218] Similarly both cyclic and acyclic amines have been oxidized to ketones by 3,5-di-*t*-butyl-1,2-benzoquinone (**73**)[219] an *ortho*-quinone (Scheme 1.155).[220]

$$R^1\text{-CH(R}^2\text{)-NH}_2 \xrightarrow{\text{aqueous PdCl}_2} R^1\text{-C(=O)-R}^2 \quad 5\text{-}70\%$$

(1.154)

$$R^1R^2\text{CH-NH}_2 \xrightarrow{(73)} R^1R^2\text{C=O} \quad 83\text{-}93\%$$

(73) = 3,5-di-Bu^t-1,2-benzoquinone

(1.155)

Primary amines can be converted into aldehydes via the ditosylate which is then treated with potassium acetate and potassium iodide. Hydrolysis of the acetate formed to the alcohol and subsequent oxidation completes the conversion. This reaction can be used with all but the most hindered amines (Scheme 1.156).[221]

$$R{-}CH_2{-}NH_2 \longrightarrow R{-}CH_2{-}NTs_2 \xrightarrow[\text{HMPT}]{\text{KOAc / KI}} R{-}CH_2{-}OAc \xrightarrow{\text{i, hydrolysis; ii, oxidation}} R{-}CHO \quad (1.156)$$

Amines are oxidized to nitriles, ketones, or lactams by iodosobenzene (PhIO),[222] e.g. Scheme 1.157.

$$n\text{-}C_5H_{11}{-}CH_2NH_2 \xrightarrow[CH_2Cl_2]{2\ PhIO} n\text{-}C_5H_{11}{-}CN \quad 57\%$$

Cyclohexylamine $\xrightarrow[CH_2Cl_2]{PhIO}$ cyclohexanone 49%

Pyrrolidine $\xrightarrow[CH_2Cl_2]{PhIO}$ 2-pyrrolidinone 49% (1.157)

Tertiary amines react with acetylenic esters to give betaines (**74**) which have been found to be synthetically useful in their addition to alkoxides to yield allyloxyacrylic acids. These may undergo Claisen rearrangement to yield, γ,δ-unsaturated aldehydes (**75**) (Scheme 1.158).[223]

$$(CH_3)_3N + H{-}C{\equiv}C{-}CO_2Et \xrightarrow[0^\circ - 25^\circ C]{CH_2Cl_2 / H_2O} (CH_3)_3\overset{+}{N}{-}CH{=}CH{-}CO_2^- \quad (74)\ 76\text{-}82\%$$

(74) + HO–CH₂–CH=CH–C₆H₅ $\xrightarrow[92\%]{\text{NaH, THF}}$ $C_6H_5{-}CH{=}CH{-}CH_2{-}O{-}CH{=}CH{-}COOH$ $\xrightarrow[-CO_2]{169\text{-}180^\circ C}$ $C_6H_5{-}CH(CH{=}CH_2){-}CH_2{-}CHO$ (75) (1.158)

The oxidation of tertiary amines with chlorine oxide in the presence of sodium cyanide yields α-cyanoamines (**76**),[224] e.g. Scheme 1.159, in general, yields vary between 53 and 83 per cent.

$$(CH_3CH_2)_3N \xrightarrow{ClO_2,\ NaCN} (CH_3CH_2)_2N\text{-}CH(CN)CH_3 \ (76) \quad 69\ \% \qquad (1.159)$$

Ruthenium trichloride (hydrated) catalyses the oxidation of primary and secondary amines to a variety of compounds,[225] e.g. Scheme 1.160.

$$Ph\text{-}CH_2NH_2 + O_2 \xrightarrow[\text{in toluene at } 100^\circ C]{RuCl_3.3H_2O} Ph\text{-}C{\equiv}N\ (53\ \%) + Ph\text{-}CONH_2\ (30\ \%)$$

$$CH_3(CH_2)_3CH(NH_2)CH_3 \xrightarrow[\text{in toluene at } 100^\circ C]{RuCl_3.3H_2O} CH_3(CH_2)_3C(=O)\text{-}CH_3\ (39\ \%) + CH_3(CH_2)_3C(=NH)\text{-}CH_3\ (31\ \%) \qquad (1.160)$$

Cyclic tertiary amines are dehydrogenated by mercuric acetate in combination with ethylenediaminetetraacetic acid (EDTA) to yield lactams,[226] e.g. Scheme 1.161.

2-R-1-methylpyrrolidine $\xrightarrow[CH_2Cl_2]{Hg(OAc)_2,\ EDTA}$ 5-R-1-methylpyrrolidin-2-one (93%)

$$EDTA = (HO_2C\text{-}CH_2)_2N\text{-}CH_2\text{-}CH_2\text{-}N(CH_2\text{-}CO_2H)_2$$

R = H, β–pyridyl, Me_3C–, $Ph(CH_2)_3$– (1.161)

Amines react with acid anhydrides in the presence of di(methylcyclopentadienyl)tin (**77**) to form *N*-alkyl imides (Scheme 1.162).[227]

cyclic anhydride $(CH_2)_n$ + $R\text{-}NH_2$ $\xrightarrow[\text{Xylene reflux}]{[CH_3C_5H_4^-]_2Sn\ (77)}$ cyclic imide $(CH_2)_n$ N-R (60-90%)

n = 2 or 3 (1.162)

Primary and secondary amines are oxidized to imines by *p*-nitrobenene-sulphonyl peroxide in ethyl acetate at −78 °C under nitrogen (Scheme 1.163).[228]

$$R^1\text{—}CH_2\text{·}NHR^2 + p\text{-}NO_2C_6H_4\text{—}\overset{O}{\underset{O}{\overset{\|}{\underset{\|}{S}}}}O \xrightarrow[N_2,\ -78^\circ C]{EtOAc} R^1\text{—}CH_2\text{—}\overset{OSO_2C_6H_4NO_2\text{-}p}{\overset{|}{N}}R^2 \longrightarrow R^1\text{—}CH{=}NR^2 \xrightarrow{H_2O} \underset{6\text{-}90\%}{R^1\text{—}CHO} \qquad (1.163)$$

1.3.4 *Elimination of nitrogen*

Nitrogen acting as a nucleophile may be converted into a good leaving group. Several of the reactions described above result in the loss of nitrogen from an amine with or without attached groups. The classical way of eliminating nitrogen from an amine is by the Hofmann method of exhaustive methylation and elimination of tertiary amine from the resulting quaternary salt by the action of hot alkali to yield an alkene (Scheme 1.164).[229] While the rule for Hofmann elimination that the alkene predominantly formed is the one which contains the smallest number of alkyl groups on the double bond is clear, the reasons underlying the rule are less clear. It has been suggested that the acidity of the β-hydrogen atom is responsible.[230] However, other chemists[231] have stressed that the steric effects caused by the size of the leaving group or by the size of the base are the determining factors. For a detailed discussion of these factors see the account by Alder, Baker, and Brown.[232]

$$R\text{-}CH_2\text{-}CH_2\text{-}NH_2 \xrightarrow[MeI]{\text{excess of}} R\text{-}CH_2\text{-}CH_2\text{-}\overset{+}{N}Me_3\ \ I^- \xrightarrow[\text{aqueous } HO^-]{\text{heat}} R\text{-}CH{=}CH_2 + NMe_3 + H_2O + I^- \qquad (1.164)$$

Primary amines are converted directly or indirectly into alkenes by reaction with 5,6,8,9-tetrahydro-7-phenyldibenzo[*c*,*h*]xanthylium tetrafluoroborate (**78**) as shown in Scheme 1.165.[233]

(78) + cyclopentylamine $\xrightarrow{20^\circ C}$ cyclopentene

(1.165)

1.3.5 *Acidity of hydrogen attached to nitrogen*

The removal of a proton from the nitrogen atom of an amine transforms the amine into a better base and a better nucleophile, for example, LDA ($Li^+Pr^i{}_2N^-$).

Nitrogen Grignard reagents

A nitrogen Grignard reagent (**79**) prepared from amines as shown in Scheme 1.166 is a useful base for enolizations[234] and for aldol condensations.[235]

$$CH_3MgBr + HNPr^i{}_2 \xrightarrow[25^\circ C]{ether} BrMgNPr^i{}_2 \quad (79)$$

$$Me_2N(CH_2)_2NHCH_3 + Li \xrightarrow[\text{ultrasonic sound}]{THF} Me_2N(CH_2)_2\text{-}\bar{N}\text{-}CH_3\ Li^+ \quad (80)$$

(1.166)

Nitrogen anions with other metals

The lithium derivative (**80**) has been used for the preparation of *ortho*-hydroxyaromatic aldehydes (Scheme 1.167).[236]

Chloramines can be prepared by the action of sodium dichloroisocyanate (**81**) on a primary or a secondary amine in water or in a mixture of benzene and water (Scheme 1.168).[237] This reaction is usually applied to hindered amines. Primary and secondary amines can be used but the corresponding chloramines are relatively less stable.

Silylation of an amine renders it susceptible to oxidation. Therefore conversion of a primary or secondary silylamine to a lithium salt followed by air oxidation provides the basis of a preparation of aldehydes and ketones from amines (Scheme 1.169).[238] In the second reaction it was found that better yields could be obtained if the lithium derivative was transmetallated to the zinc analogue before oxidation.

Pyrroles can be synthesized by the action of carbon monoxide on the complex formed between lithio trimethylsilyl primary amine (**82**), a silylated amino

(80)

BuCl, Li, ultrasound or BuLi

i, [O]
ii, H_2O

[O] = molecular oxygen or MoOPH

MoOPH = methyl phosphoramolybdenum (1.167)

(81)

H_2O or H_2O and benzene

56-86% (1.168)

i, Bu^nLi, ii, O_2
- 60° to - 20° C

86 %

i,BuLi, ii, dry air

(1.169)

anion, and chlorobis(cyclopentadienyl)methylzirconium (**83**) (Scheme 1.170).[239]

The acidity of the hydrogen atom on nitrogen in a secondary amine can be used to protect that nitrogen atom so that an *N*-methyl group can be alkylated (Scheme 1.171).[240]

$$R{-}CH_2{-}\bar{N}{-}SiMe_3\ Li^+ + (Cp)_2Zr(Me)(Cl)$$

(82) (83)

THF, -78°C; 0°C, rt., THF, $R^1{-}C{\equiv}C{-}R^2$; 90 psi 80°C, 48 hours, CO, THF

R^2 = Ph, H, $CH_2CH_2CH_3$

R^1 = H

R = Ph, H, n-C_5H_{11},

(1.170)

$Bu^n{-}NH{-}CH_3$ $\xrightarrow{Bu^nLi}$ $Bu^n{-}\bar{N}{-}CH_3\ Li^+$ $\xrightarrow{CS_2}$ $Bu^n{-}N(CH_3){-}C(=S){-}S^-\ Li^+$ $\xrightarrow{Bu^nLi,\ -78\ to\ -25^{\circ}C}$ $Bu^n{-}N(\bar{C}H_2\ Li^+){-}C(=S){-}\bar{S}\ Li^+$ $\xrightarrow{Bu^nLi,\ -25^{\circ}C}$ $Bu^n{-}N(CH_2{-}Bu^n){-}C(=S){-}S^-\ Li^+$ $\xrightarrow{aq.\ H_3O^+}$ $Bu^n{-}NH{-}CH_2{-}Bu^n$

85%

(1.171)

Anions formed by the action of *t*-butyl lithium from allylic and methylallylic amines have been used for a variety of synthetic purposes,[241] e.g. Scheme 1.172.

Suitable unsaturated secondary amines can be converted into pyrrolidine (Scheme 1.173).[242]

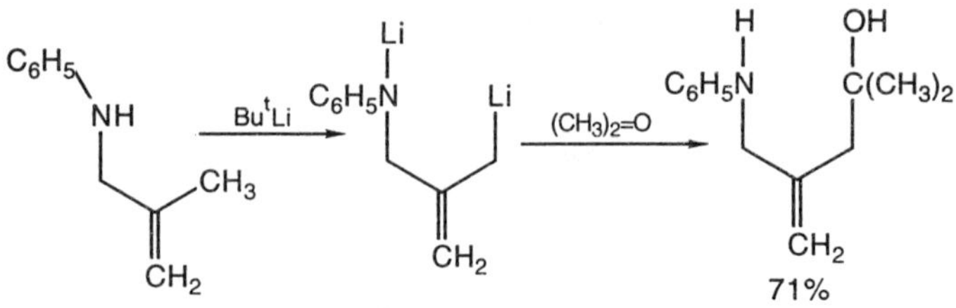

(1.172)

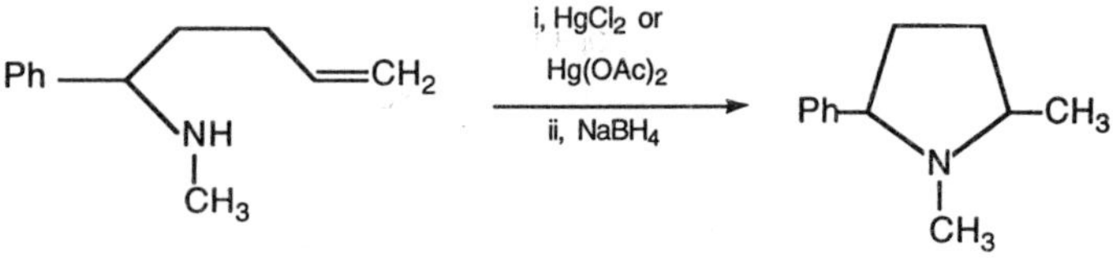

Solvent	Yield (%)	cis (%)	trans (%)
THF or CH_2Cl_2	36	90	10
THF-H_2O (1:1)	64	13	87

(1.173)

1.3.6 *α-Carbon atom of a quaternary salt as an electrophile*

Alkylation of the nitrogen atom of a tertiary amine yields a quaternary salt. The cation of the salt contains an α-carbon atom which is activated as an electrophile by the adjacent positive charge on the nitrogen atom. This activation is often taken advantage of by the alkylation of Mannich bases, an example being the Mannich base derived from indole, dimethylamine, and formaldehyde (**84**).

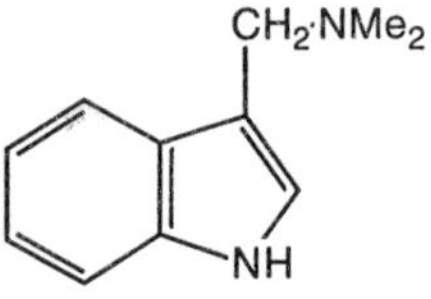

(84)

This Mannich base is the alkaloid gramine and by its reaction with dimethyl sulphate and nitromethane intermediates have been prepared for the synthesis of the tetrahydro-β-carboline nucleus which is present in many indole alkaloids (Scheme 1.174).[243]

Another important example of this procedure is the Robinson annulation reaction,[244] which has proved useful for the synthesis of alicyclic compounds such as steroids, e.g. Scheme 1.175.

CH2-NMe2 NaOMe / MeOH Me2SO4 / CH3NO2 24 hours (84) NH CH2=N-O CH2-NMe3 NH CH2—CH2 NO2 NH 86 %

(1.174)

NMe3 CH2 CH3 CH2 O=C CH3 O EtO⁻ / EtOH NMe3 CH2 CH3 CH2 O=C CH CH3 O

(1.175)

1.3.7 *α-Carbon of an amine as a nucleophile (Umpolung)*[245]

By nitrosation

The reversal of the polarity (Umpolung) of the α-carbon atom of amines so that they will then react with electrophiles is a desirable operation since it extends their synthetic utility. If a secondary amine is converted into a nitrosamine (see Chapter 12, p. 495), the desired reversal of polarity is achieved,[245] by virtue of the structure of the nitrosamine (**85**) and by removal of a proton by a base from the α-carbon atom (Scheme 1.176). Subsequent reaction with an electrophile

$$R^1\text{—}\overset{+}{N}\text{=N—}\overset{-}{O}(\text{—}CH_2\text{-}R^2)\ (85) \xrightarrow{\text{base}} R^1\text{—}\overset{+}{N}\text{=N—}\overset{-}{O}(\text{—}\overset{-}{C}H\text{-}R^2) \xrightarrow{E^+} R^1\text{—}\overset{+}{N}\text{=N—}\overset{-}{O}(\text{—}CH(E)\text{-}R^2) \xrightarrow{H^+} R^1\text{—NH—}CH(E)\text{—}R^2 + NO^+ \quad (1.176)$$

and denitrosation completes the synthetic sequence which is discussed in greater detail in Chapter 5. For tertiary amines see Ref. 246.

By butyl lithium and potassium butoxide

It has been found possible[247] to ionize the methyl group of an *N*-methylamine by the use of butyl lithium and potassium tertiary butoxide. The carbanion so formed can be used for various synthetic purposes by reaction with electrophiles as demonstrated in Scheme 1.177.

$$R^1R^2N\text{—}CH_3\ (\text{excess}) \xrightarrow[\text{KOBu}^t]{\text{BuLi}} \left[R^1R^2N\text{—}\overset{-}{C}H_2\ K^+\right] \xrightarrow{R^3Br} R^1R^2N\text{—}CH_2\text{—}R^3$$

$$\left[R^1R^2N\text{—}\overset{-}{C}H_2\ K^+\right] \xrightarrow{\text{LiBr}} \left[R^1R^2N\text{—}\overset{-}{C}H_2\ Li^+\right] \xrightarrow{R^3CHO} R^1R^2N\text{—}CH_2\text{—}CHOH\text{-}R^3$$

R^1	R^2	R^3	Yield (%) t-amine	Yield (%) β-amino alcohol
-$(CH_2)_5$-		n-C_8H_{17}	70	—
Me	Me	Ph	—	68

(1.177)

1.3.8 *N,N-Difluorination*

Conversion of primary amines to *N,N*-difluoroamines can be achieved by initial formation of an imine with benzaldehyde or the sodium salt of 4-carboxybenzaldehyde (**86**) and subsequent treatment of this with fluoroxytrifluoromethane (**87**)[248] in methanol and methylene dichloride. In the absence of methanol an alkyl fluoride is formed (Scheme 1.178).[249]

$$R\text{-}NH_2 + Ar\text{-}CHO \longrightarrow R\text{-}N{=}CHAr \xrightarrow[MeOH/CH_2Cl_2]{F_3C\text{-}OF\ (87)} R\text{-}\overset{+}{N}F_2\text{-}CH(OMe)\text{-}Ar \xrightarrow{MeOH} R\text{-}NF_2 + (MeO)_2CHAr \quad 50\text{-}75\%$$

Ar-CHO = $Na^{+}\,{}^{-}O_2C$-C₆H₄-CHO (86)

(1.178)

Benzeneselenic anhydride ($[PhSeO]_2O$) (**88**) and cyanotrimethylsilane (Me_3SiCN) (**89**) react at room temperature with a secondary amine in methylene dichloride to yield an α-cyanoamine (Scheme 1.179),[250] yields are in the range 22–36 per cent. Some examples of the amines used are shown in Scheme 1.179.

$CH_3(CH_2)_2NH(CH_2)_2CH_3$

$$R^1\text{-}NH\text{-}CH_2\text{-}R^2 + \underset{(89)}{(PhSeO)_2O} + \underset{(90)}{Me_3SiCN} \xrightarrow[20^\circ C]{CH_2Cl_2} R^1\text{-}NH\text{-}CH(CN)\text{-}R^2$$

(1.179)

1.3.9 *Arylation*

Primary and secondary amines are mono- or di-arylated by triphenylbismuth diacetate and copper (Scheme 1.180).[251]

$$\text{R-NH}_2 + \text{Ph}_3\text{Bi(OAc)}_2 \xrightarrow{\text{Cu}} \text{R-NH(Ph)} + \text{R-N(Ph)-Ph} \quad (1.180)$$

1.3.10 *β-Alkylamino-carbonyl compounds*

β-Alkylamino-carbonyl compounds have been prepared from imines, and thus originally from amines as shown in Scheme 1.181. The mechanism goes through a strain-assisted four-membered ring.[252]

$$\text{R}^1\text{-C(=O)-CH}_2\text{-CH}_2\text{Cl} + \text{R}^2\text{-NH}_2 \longrightarrow \text{R}^1\text{-C(=N-R}^2\text{)-CH}_2\text{-CH}_2\text{Cl}$$

$$\xrightarrow[\text{ii, H}_2\text{O}]{\text{i, MeO}^- \text{/ MeOH, heat}} \text{R}^1\text{-C(=O)-CH}_2\text{-CH}_2\text{-NHR}^2 \quad 60\text{-}90\% \quad (1.181)$$

1.3.11 *Quaternary ammonium salts as phase transfer catalysts*

Quaternary ammonium salts are used as phase transfer catalysts,[253] and many examples are scattered through this book.

Cetyltrimethylammonium bromide (**90**) has been used as a micellar catalyst in the Williamson ether synthesis[254] and in the oxidation of alcohols.[255]

$$n\text{-C}_{16}\text{H}_{33}\,\text{Me}_3\text{N}^+\ \text{Br}^-$$

(90)

1.3.12 *Reductive deamination to hydrocarbons*

This transformation can be achieved through a free radical reduction of an imidoyl chloride (**91**) which can be prepared from an amine (Scheme 1.182).

$$\text{R-NH}_2 \xrightarrow[\text{ii, SOCl}_2]{\text{i, PhCOCl}} \text{R-N=C(Cl)-Ph} \xrightarrow[\text{AIBN}]{\text{Bu}^n_3\text{SnH}} \text{R-H}$$

Imidoyl chloride

(91)

R^1	Yield (%)
Bn	82
Cyclohexyl	63
n-C_8H_{17}	38

(1.182)

References

1. M. S. Gibson, *The chemistry of the amino group*, Wiley-Interscience, NY, 1968.
2. H. Maskill, *The physical basis of organic chemistry*, OUP, 1985, pp. 19–23.
3. J. B. Lambert, H. F. Shirvell, L. Verbit, R. G. Cooks, and G. H. Strout, *Organic structural analysis*, Macmillan, New York/London, 1976, p. 71.
4. V. Prelog and P. Wieland, *Helv. Chim. Acta*, 1944, **27**, 1127.
5. P. Maitland and W. H. Mills, *J. Chem. Soc.*, 1936, 987.
6. A. W. Hofmann, *Philos. Trans.*, 1850, **CXL**, 93.
7. O. Hinsberg, *Ber.*, 1890, **23**, 2963; 1905, **38**, 906.
8. S. Gabriel, *Ber.*, 1887, **20**, 2224.
9. M. S. Gibson and R. W. Bradshaw, *Angew. Chem., Int. Ed. Engl.*, 1968, **7**, 919.
10. H. R. Ing and R. H. F. Manske, *J. Chem. Soc.*, 1926, 2348.
11. J. C. Sheehan and W. A. Bolhofer, *J. Am. Chem. Soc.*, 1950, **72**, 2786.
12. L. A. Carpino, *J. Org. Chem.*, 1964, **29**, 2820.
13. A. K. Bose, J. F. Kistner, and L. Farber, *J. Org. Chem.*, 1962, **27**, 2925.
14. *F. &F.*, **7**, 166; D. Landini and F. Rolla, *Synthesis*, 1976, 389.
15. A. Zwierzak, *Synthesis*, 1975, 507; *F. &F.*, **6**, 42; A. Zwierzak and J. Brylikowska-Piotrowicz, *Angew. Chem., Int. Ed. Engl.*, 1977, **16**, 107; *F. &F.*, **8**, 388.
16. *F. &F.*, **4**, 554; O. Mitsunobu, M. Wada, and T. Sano, *J. Am. Chem. Soc.*, 1972, **94**, 679.
17. *F. &F.*, **11**, 597; A. I. Meyers, J. P. Lawson, and D. R. Carver, *J. Org. Chem.*, 1981, **46**, 3119.
18. *F. &F.*, **15**, 304; preparation of the silicon amine: M. Ishikawa, M. Kumada, and H. Sakurai, *J. Organomet. Chem.*, 1970, **23**, 63; reaction: A. Hosomi, S. Kohraa, Y. Tominaga, M. Inaba, and H. Sakurai, *Chem. Pharm. Bull.*, 1988, **36**, 2342.
19. *F. &F.*, **6**, 43; J. B. Hendrickson, R. Bergeron, and D. D. Sternbach, *Tetrahedron*, 1975, **31**, 2517.
20. J. B. Hendrickson and R. Bergeron, *Tetrahedron Lett.*, 1970, 345.
21. *F. &F.*, **3**, 221.
22. *F. &F.*, **3**, 295; P. Kovacic and M. K. Lowery, *J. Org. Chem.*, 1969, **34**, 911.
23. Preparation: U. Wannagat and H. Niederprüm, *Ber.*, 1961, **96**, 2131.
24. *F. &F.*, **12**, 441; H. J. Bestmann and G. Wolfel, *Ber.*, 1984, **117**, 1250.
25. *F. &F.*, **12**, 505; J. M. Chong and K. B. Sharpless, *J. Org. Chem.*, 1985, **50**, 1560.
26. *F. &F.*, **6**, 319; E. J. Corey, K. C. Nicolaon, R. D. Balanson, and Y. Machida, *Synthesis*, 1975, 590.
27. *F. &F.*, **11**, 478; F. Rolla, *J. Org. Chem.*, 1982, **47**, 4327.
28. *F. &F.*, **8**, 433; D. Baldermann and A. Kalir, *Synthesis*, 1978, 24.
29. H. Bayley, D. N. Standring, and J. R. Knowles, *Tetrahedron Lett.*, 1978, 3633; *F. & F.*, **9**, 394.
30. H. Staudinger and J. Meyer, *Helv. Chim. Acta*, 1919, **2**, 635.
31. *F. &F.*, **12**, 529; M. Vaultier, N. Knouzi, and R. Carrié, *Tetrahedron Lett.*, 1983, **24**, 763.
32. V. Micoviz and M. L. Mihailovic, *J. Org. Chem.*, 1953, **18**, 1190.
33. *F. &F.*, **15**, 186; A. Giannis and K. Sandhoff, *Angew. Chem., Int. Ed. Engl.*, 1989, **28**, 218.
34. *F. &F.*, **11**, 478; S. R. Warren, P. T. Thorsen, and M. M. Kreevoy, *J. Org. Chem.*, 1981, **46**, 2579.

35. C. F. Lane, *Chem. Rev.*, 1976, **46**, 2579.
36. *F. &F.*, **3**, 264; T. Satoh, S. Suzuki, Y. Suzuki, Y. Miyaji, and Z. Imai, *Tetrahedron Lett.*, 1969, 61.
37. *F. &F.*, **3**, 263; Y. Kikugawa, S. Ikegani, and S. Yamada, *Chem. Pharm. Bull. Japan*, 1969, **17**, 98.
38. R. F. Borsch, *Tetrahedron Lett.*, 1968, 61.
39. *F. &F.*, **4**, 124; H. C. Brown and P. Heim, *J. Org. Chem.*, 1973, **38**, 912.
40. H. C. Brown, *J. Am. Chem. Soc.*, 1964, **86**, 3566; 1966, **88**, 1464.
41. *F. &F.*, **11**, 69; H. C. Brown, Y. M. Choin, and S. Narasimhan, *J. Org. Chem.*, 1982, **47**, 3153.
42. *F. &F.*, **12**, 276; H. C. Brown, C. P. Matthew, C. Pyun, J. C. Son, and N. M. Yoon, *J. Org. Chem.*, 1984, **49**, 3091.
43. Preparation: J. Vit, B. Cásensky, and J. Machácek, *French Patent*, 1 515 582.
44. *F. &F.*, **5**, 596; K. W. Weselowsky and A. M. Moiseenkov, *Synthesis*, 1974, 58.
45. *F. &F.*, **12**, 276; H. C. Brown, C. P. Matthew, C. Pyun, J. C. Son, and N. M. Yoon, *J. Org. Chem.*, 1984, **49**, 3091.
46. M. Freifelder, *J. Am. Chem. Soc.*, 1960, **82**, 2386.
47. R. F. Nystrom and W. G. Brown, *J. Am. Chem. Soc.*, 1948, **70**, 3738.
48. T. Satoh, S. Suzuki, Y. Suzuki, Y. Miyaji, and Z. Imai, *Tetrahedron Lett.*, 1969, 4555; *F. &F.*, **3**, 264.
49. *F. &F.*, **11**, 138; S. W. Heinzman and B. Ganem., *J. Am. Chem. Soc.*, 1982, **104**, 6801.
50. *F. &F.*, **15**, 292; S. Itsuno, Y. Sakurai, and K. Ito, *Synthesis*, 1988, 995.
51. *F. &F.*, **11**, 69; H. C. Brown, Y. M. Choin, and S. Narasimhan, *J. Org. Chem.*, 1982, **47**, 3153.
52. *F. &F.*, **7**, 342; N. Umino, T. Iwakuma, and N. Itoh, *Tetrahedron Lett.*, 1976, 2875.
53. W. Higgins, P. W. Vogel, and W. G. Craig, *J. Am. Chem. Soc.*, 1955, **77**, 1864.
54. S. A. Benner, *Tetrahedron Lett.*, 1981, **22**, 1851.
55. *F. &F.*, **10**, 172; S. Raucher and P. Klein, *Tetrahedron Lett.*, 1980, **21**, 4061.
56. M. Freifelder, W. D. Smart, and G. R. Stone, *J. Org. Chem.*, 1962, **27**, 2209.
57. E. Lanson, *Svenk Kem. Tid.*, 1949, **61**, 242; *Chem. Abstr.*, 1948, **44**, 1898.
58. S. H. Graham and A. J. S. Williams, *J. Chem. Soc. (C)*, 1966, 655.
59. M. Cerny, J. Málek, M. Capka, and V. Chvalovsky, *Czech. Commun.*, 1969, **34**, 1033; R. H. Eastman, *Org. Chem. Bull.*, 1974, **46**(1); *F. &F.*, **3**, 260.
60. *F. &F.*, **3**, 264; J. M. Lalancette and I. R. Brindle, *Can. J. Chem.*, 1968, **46**, 2754.
61. *F. &F.*, **15**, 292; S. Itsuno, Y. Sakurai, and K. Ito, *Synthesis*, 1988, 995.
62. *F. &F.*, **15**, 317; J. P. Leeds and H. A. Kirst, *Synth. Commun.*, 1988, **18**, 777.
63. A. Hassner and R. Catsoulacos, *J. Chem. Soc., Chem. Commun.*, 1967, 590.
64. E. H. Gold and E. Babad, *J. Org. Chem.*, 1972, 2208.
65. M. Freifelder, *Catalytic hydrogenation in organic synthesis*, Wiley-Interscience, New York, 1978, p. 65.
66. *F. &F.*, **12**, 287; R. O. Hutchins, W.-Y. Su, R. Sivakumar, F. Cistone, and Y. P. Stercho, *J. Org. Chem.*, 1983, **48**, 3412; R. O. Hutchins and W.-Y. Sivakumar, *Tetrahedron Lett.*, 1984, **25**, 695.
67. M. Freifelder, *Catalytic hydrogenation in organic synthesis*, Wiley-Interscience, New York, 1978, p. 10.
68. *F. &F.*, **5**, 671; J. C. Stowell and S. T. Padegimas, *Synthesis*, 1974, 127; and for optically active amines, see *F. &F.*, **5**, 682; G. Tsuchihashi, S. Iruichijima, and K. Maniwa, *Tetrahedron Lett.*, 1973, 3389.

69. *F. &F.*, **4**, 448; R. F. Borch, M. D. Bernstein, and H. D. Durst, *J. Am. Chem. Soc.*, 1971, **93**, 1793; R. F. Borsch, *Org. Synth.*, 1972, **52**, 124.
70. *F. &F.*, **11**, 499; R. O. Hutchins and M. Markowitz, *J. Org. Chem.*, 1981, **46**, 3571.
71. *F. &F.*, **12**, 65; A. Pelter, R. M. Rosser, and S. Mills, *J. Chem. Soc., Chem. Commun.*, 1984, 717.
72. *F. &F.*, **12**, 449; M. Yamashita, M. Kadokura, and R. Suemitsu, *Bull. Chem. Soc. Japan*, 1984, **57**, 3359.
73. *F. & F.*, **5**, 357; Y. Watanabe, M. Yamashita, T. Mitsudo, M. Tanaka, and Y. Takegami, *Tetrahedron Lett.*, 1974, 1879.
74. R. G. Shepherd and R. G. Wilkinson, *J. Med. Chem.*, 1962, **5**, 823; K. A. Schellenberg, *J. Org. Chem.*, 1963, **28**, 3259; C. L. Stevens, K. G. Taylor, and M. E. Munk, *J. Org. Chem.*, 1964, **29**, 3574; B. L. Sondengam, J. Hémo, and G. Charles, *Tetrahedron Lett.*, 1973, 261.
75. A. Meister, *Biochemistry of the amino acids*, Academic Press, New York, 1957, p. 157.
76. A. Lukasiewicz, *Tetrahedron*, 1963, **19**, 1789.
77. P. L. Awachie and V. C. Agwada, *Tetrahedron*, 1990, **46**, 1899.
78. W. Eschweiler, *Ber.*, 1905, **38**, 880; H. T. Clarke, H. B. Gillespie, and S. Z. Weisshaus, *J. Am. Chem. Soc.*, 1933, **38**, 880.
79. R. F. Borsch and A. I. Hassid, *J. Org. Chem.*, 1972, **37**, 1673.
80. *F. &F.*, **4**, 320; R. D. Bach and D. K. Mitra, *J. Chem. Soc., Chem. Commun.*, 1971, 1433.
81. H. C. Brown, P. M. Weissman, and N. M. Yoon, *J. Am. Chem. Soc.*, 1966, **88**, 1461.
82. *F. &F.*, **15**, 14; R. Ram and R. E. Ehrenkaufer, *Synthesis*, 1986, 133; A. G. M. Barrett and C. D. Spilling, *Tetrahedron Lett.*, 1988, **29**, 5733.
83. *F. &F.*, **12**, 503; J. George and S. Chandrasekaran, *Synth. Commun.*, 1983, **13**, 495.
84. *F. &F.*, **12**, 65; M. S. Mourad, R. S. Varma, and G. W. Kabalka, *Synth. Commun.*, 1984, **14**, 1099.
85. *F. &F.*, **15**, 186; A. Giannis and K. Sandhoff, *Angew. Chem., Int. Ed. Engl.*, 1989, **28**, 218.
86. *F. &F.*, **1**, 483; H. C. Brown, W. R. Heydkamp, E. Brewer, and W. S. Murphy, *J. Am. Chem. Soc.*, 1964, **86**, 3565; 1966, **88**, 2870.
87. Y. Tamura, J. Minamikawa, S. Fujii, and M. Ikeda, *Synthesis*, 1974, 196; *F. &F.*, **5**, 433.
88. *F. &F.*, **15**, 343; G. W. Kabalka, N. M. Goudgaon, and Y. Liang, *Synth. Commun.*, 1988, **18**, 1363.
89. *F. &F.*, **11**, 205; B. M. Trost and T. Shibata, *J. Am. Chem. Soc.*, 1982, **104**, 3225; M. C. Caserio and J. M. Kim, *J. Am. Chem. Soc.*, 1982, **104**, 3231.
90. Preparation: H. Meerwein, K. F. Zenner, and R. Gipp, *Annalen*, 1965, **688**, 67.
91. *F. &F.*, **3**, 195. H. C. Brown and J. T. Kurek, *J. Am. Chem. Soc.*, 1969, **91**, 5647.
92. W. Reppe, *Experimentia*, 1949, **5**, 93.
93. *F. &F.*, **4**, 420; A. F. M. Ibqual, *Helv. Chim. Acta*, 1971, **54**, 1440.
94. *F. &F.*, **4**, 350; T. Fujita, K. Suga, and S. Watanabe, *Chem. and Ind.*, 1973, 231.
95. *F. &F.*, **5**, 501; D. R. Coulson, *J. Org. Chem.*, 1973, **38**, 1483.
96. M. J. P. Harges, *J. Chem. Soc. Perkin Trans. I*, 1981, 3284.
97. *F. &F.*, **11**, 221; G. Boche, M. Bernheim, and W. Schrott, *Tetrahedron Lett.*, 1982, **23**, 5399.

98. *F. &F.*, **12**, 62; T. Morimoto, T. Takahashi, and M. Sekiya, *J. Chem. Soc., Chem. Commun.*, 1984, 794.
99. N. J. Sheverdina and Z. Kocheshkov, *Chem. Abstr.*, 1939, **33**, 5804; 1943, **37**, 3066; R. Brown and W. E. Jones, *J. Chem. Soc.*, 1946, 781; M. S. Kharasch and O. Reinmuth, *Grignard reactions of non-metallic substances*, Constable, London, 1954.
100. *F. &F.*, **11**, 322; P. Beak and B. J. Kokko, *J. Org. Chem.*, 1982, **47**, 2822.
101. *F. &F.*, **11**, 182; J. J. Levison and S. D. Robinson, *J. Chem. Soc. (A)*, 1970, 2947.
102. *F. &F.*, **5**, 306; H. Lehmkuhl and D. Reinehr, *J. Organomet. Chem.*, 1973, **55**, 215.
103. H. Büger and H. J. Neese, *Chimia*, 1970, **24**, 209.
104. *F. &F.*, **11**, 375; D. Seebach and M. Schiess, *Helv. Chim. Acta*, 1982, **65**, 2598.
105. M. S. Kharasch and O. Reinmuth, *Grignard reactions of non-metallic substances*, Constable, London, 1954, p. 1204; R. B. Moffett and W. H. Hoehn, *J. Am. Chem. Soc.*, 1947, **69**, 1792; C. N. Campbell, C. H. Helbing, M. P. Florkowki, and B. K. Campbell, *J. Am. Chem. Soc.*, 1948, **70**, 3868.
106. *F. &F.*, **5**, 498; N. Yoshimura, I. Moritani, T. Shimamura, and S.-I. Murahashi, *J. Am. Chem. Soc.*, 1973, **95**, 3038.
107. *F. &F.*, **10**, 9; S. W. Pelletier, A. P. Venkov, J. Finer-Moore, and N. V. Moody, *Tetrahedron Lett.*, 1980, **21**, 809.
108. *F. &F.*, **10**, 12; T. Mukaiyama, A. Ikegawa, and K. Suzuki, *Chem. Lett.*, 1981, 165.
109. O. Fischer and E. Hepp, *Ber.*, 1886, **19**, 2991.
110. A. W. Hofmann, *Ber.*, 1882, **15**, 762; E. S. Wallis and J. F. Lane, *Org. React.*, 1946, **3**, 267.
111. *F. &F.*, **15**, 257; G. M. London, A. S. Radhakrishna, M. R. Almond, J. K. Blodgett, and R. H. Boutin, *J. Org. Chem.*, 1984, **49**, 4272; V. H. Pavlidis, E. D. Chan, L. Pennington, M. McParland, M. Whitehead, and J. G. C. Coutts, *Synth. Commun.*, 1988, **18**, 1615.
112. *F. &F.*, **12**, 58; A. S. Radhakrishna, C. G. Rao, R. K. Varma, B. B. Singh, and S. P. Bhatnagar, *Synthesis*, 1983, 538.
113. T. Curtius, *J. Prakt. Chem.*, 1894, **[2]50**, 275; P. A. S. Smith, *Org. React.*, 1946, **3**, 337; *F. &F.*, **1**, 1042.
114. *F. &F.*, **12**, 529; J. R. Pfister and W. E. Wymann, *Synthesis*, 1983, 38.
115. K. F. Schmidt, *Angew. Chem.*, 1923, **36**, 511; H. Wolff, *Org. React.*, 1946, **3**, 307.
116. Preparation: H. Zimmer and G. Singh, *J. Org. Chem.*, 1963, **28**, 483.
117. *F. &F.*, **6**, 392; Y. Tinigawa, S.-I. Murahashi, and I. Moritani, *Tetrahedron Lett.*, 1975, 471.
118. *F. &F.*, **6**, 443; S.-I. Murahashi, T. Shimamura, and I. Moritani, *J. Chem. Soc., Chem. Commun.*, 1974, 931.
119. H. Meerwein, P. Laasch, R. Mersch, and J. Spille, *Ber.*, 1965, **89**, 209.
120. *F. &F.*, **3**, 303; R. F. Borch, *J. Chem. Soc., Chem. Commun.*, 1968, 442.
121. *F. &F.*, **4**, 144; R. F. Borch, *J. Org. Chem.*, 1969, **34**, 627.
122. T. Mukaiyama, T. Tsuji, and Y. Watanabe, *Chem. Lett.*, 1978, 1057.
123. *F. &F.*, **11**, 491; R. F. Smith and K. J. Coffman, *Synth. Commun.*, 1982, **12**, 801.
124. *F. &F.*, **14**, 22; D. Enders, H. Schubert, and C. Nübling, *Angew. Chem., Int. Ed. Engl.*, 1986, **25**, 1109.
125. *F. &F.*, **14**, 217; S. E. Denmark, T. Weber, and D. W. Piotrowski, *J. Am. Chem. Soc.*, 1987, **109**, 2224.
126. *F. &F.*, **14**, 30; A. Padwa and W. Dent, *J. Org. Chem.*, 1987, **52**, 235.

127. *F. &F.*, **14**, 215; Y. Sakito, Y. Yoneyoshi, and G. Suzukamo, *Tetrahedron Lett.*, 1988, **29**, 223.
128. H. Brunner, *J. Organomet. Chem.*, 1968, **12**, 517; H. Brunner and S. Loskot, *J. Organomet. Chem.*, 1973, **61**, 401.
129. *F. &F.*, **10**, 116; P. N. Becker, M. A. White, and R. G. Bergman, *J. Am. Chem. Soc.*, 1980, **102**, 5676.
130. *F. &F.*, **12**, 147; H. Kohn and S.-H. Jung, *J. Am. Chem. Soc.*, 1983, **103**, 4106; S.-H. Jung and H. Kohn, *Tetrahedron Lett.*, 1984, **25**, 399.
131. *F. &F.*, **15**, 20; N. X. Hu, Y. Aso, T. Otsubo, and F. Ogura, *Tetrahedron Lett.*, 1988, **29**, 4949.
132. *F. &F.*, **12**, 11; R. J. Linderman and A. I. Meyers, *Tetrahedron Lett.*, 1983, **23**, 3043.
133. *F. &F.*, **12**, 121; R. G. Shea, J. N. Fitzner, J. E. Fankhauser, and P. B. Hopkins, *J. Org. Chem.*, 1984, **49**, 3647.
134. *F. &F.*, **7**, 316; K. B. Sharpless, T. Hori, L. K. Truesdale, and C. O. Dietrich, *J. Am. Chem. Soc.*, 1976, **98**, 269.
135. K. B. Sharpless and T. Hori, *J. Org. Chem.*, 1976, **41**, 176; K. B. Sharpless and S. P. Singer, *J. Org. Chem.*, 1976, **41**, 2504.
136. *F. &F.*, **11**, 595; preparation: F. J. Welch and H. J. Paxton, *J. Polym. Sci.*, 1965, **A3**, 3427; D. Cavalla and S. Warren, *Tetrahedron Lett.*, 1982, **23**, 4505.
137. *F. &F.*, **12**, 472; J. P. Genêt, M. Balabane, J. E. Bächvall, and J. E. Nyström, *Tetrahedron Lett.*, 1983, **24**, 2745.
138. *F. &F.*, **14**, 53; A. Dicko, M. Motury, and M. Baboulene, *Tetrahedron Lett.*, 1987, **28**, 6041.
139. *F. &F.*, **15**, 247; S. Murahashi, Y. Imada, and K. Nishimura, *J. Chem. Soc., Chem. Commun.*, 1988, 1578.
140. *F. &F.*, **8**, 43; *F. &F.*, **12**, 455; R. S. Garigipati, J. A. Morton, and S. M. Weinreb, *Tetrahedron Lett.*, 1983, **24**, 987.
141. *F. &F.*, **12**, 568; A. Pfaltz and S. Anwar, *Tetrahedron Lett.*, 1984, **25**, 2977.
142. P. Strazzolini, A. G. Ginmmanini, and S. Cauci, *Tetrahedron*, 1990, **46**, 1097.
143. S. Krishnamurthy, *Tetrahedron Lett.*, 1982, **23**, 3315.
144. *F. &F.*, **14**, 168; S. D. Larsen, P. A. Grieco, and W. F. Forbare, *J. Am. Chem. Soc.*, 1986, **108**, 3512.
145. *F. &F.*, **11**, 489; W. Ando and H. Tsumaki, *Chem. Lett.*, 1981, 693.
146. *F. &F.*, **12**, 14; L. E. Overman and R. M. Burk, *Tetrahedron Lett.*, 1984, 1635.
147. D. Valentine and J. W. Scott, *Synthesis*, 1978, 331.
148. *F. &F.*, **6**, 305; G. Boldrini, M. Panunzio, and A. Umani-Ronchi, *Synthesis*, 1974, 733.
149. *F. & F.*, **6**, 484; Y. Watanabe, T. Mitsudo, Y. Yamashita, S. C. Shim, and Y. Takegami, *Chem. Lett.*, 1974, 1265.
150. J. Barluenga, A. M. Bayon, and G. Asensio, *J. Chem. Soc., Chem. Commun.*, 1984, 427.
151. *F. &F.*, **6**, 325; M. Khanna, V. M. Dixit, and N. Anand, *Synthesis*, 1975, 607.
152. *F. &F.*, **4**, 371; H. Z. Sommer, H. I. Lipp, and L. L. Jackson, *J. Org. Chem.*, 1971, **36**, 824; H. Z. Sommer and L. L. Jackson, *J. Org. Chem.*, 1970, **35**, 1558.
153. *F. &F.*, **4**, 119; T.-L. Ho, *Synthesis*, 1972, 702.
154. *F. &F.*, **6**, 348; M. P. Cooke, Jr and R. M. Parlman, *J. Org. Chem.*, 1975, **40**, 531.
155. *F. &F.*, **5**, 415; R. O. Hutchins and F. J. Dux, *J. Org. Chem.*, 1973, **38**, 1961.
156. *F. &F.*, **15**, 56; S.-I. Murahashi, T. Naota, and K. Yonemura, *J. Am. Chem. Soc.*, 1988, **110**, 8256.

157. J. von Braun, *Ber.*, 1907, **40**, 3914; 1909, **42**, 2219; 1911, **44**, 1252.
158. H. A. Hageman, *Org. React.*, 1953, **7**, 193.
159. W. W. Hartman and E. E. Dreger, *Org. Synth.*, **Coll. Vol. 2**, 1941, 150.
160. K. H. Slotta, *Ber.*, 1934, **67**, 1028.
161. J. von Braun, *Ber.*, 1900, **33**, 1438.
162. G. Ochai and K. Tsuda, *Ber.*, 1934, **67**, 1011.
163. Review: J. H. Cooley and E. J. Evain, *Synthesis*, 1988, 1.
164. *F. &F.*, **5**, 117; G. Kraiss and K. Nádor, *Tetrahedron Lett.*, 1971, 57.
165. P. Pfäffli and H. Hauth, *Helv. Chim. Acta*, 1973, **56**, 347.
166. *F. &F.*, **7**, 236; G. A. Brine, K. G. Boldt, C. K. Hart, and F. I. Carroll, *Org. Prep. Proc. Int.*, 1976, **8**, 103.
167. J. C. Kimm, *Org. Prep. Proc. Int.*, 1977, **9**, 1.
168. K. E. Rice, *J. Org. Chem.*, 1975, **40**, 1850.
169. *F. &F.*, **5**, 686; M. G. Reinecke and R. G. Daubert, *J. Org. Chem.*, 1973, **38**, 3281; T. A. Montzka, J. D. Matiskella, and R. A. Partyka, *Tetrahedron Lett.*, 1974, 1325.
170. *F. &F.*, **8**, 530; R. A. Olofson, R. C. Schnur, L. Bunes, and J. P. Pepe, *Tetrahedron Lett.*, 1977, 1567.
171. *F. &F.*, **12**, 113; R. A. Olofson, J. T. Martz, J. P. Senet, M. Piteau, and T. Malfroot, *J. Org. Chem.*, 1984, **49**, 2081.
172. *F. &F.*, **5**, 212; E. E. Smissman and A. Makriyannis, *J. Org. Chem.*, 1973, **38**, 1652.
173. *F. &F.*, **4**, 26; J. C. Sheehan and F. S. Guziec, Jr, *J. Am. Chem. Soc.*, 1972, **94**, 6561.
174. *F. &F.*, **4**, 128.
175. Preparation: B. M. Pope, Y. Yamamoto, and D. S. Tarbell, *Org. Synth.*, 1977, 48.
176. B. M. Pope, Y. Yamamoto, and D. S. Tarbell, *Proc. Natl. Acad. Sci. USA*, 1972, **69**, 730.
177. *F. &F.*, **10**, 98; B. C. Laguzza and B. Ganem, *Tetrahedron Lett.*, 1981, **22**, 1483.
178. *F. &F.*, **5**, 736; J. F. Young, J. A. Osborn, F. H. Jardine, and G. Wilkinson, *Chem. Commun.*, 1965, 131; *Chem. and Ind.*, 1965, 560.
179. *F. &F.*, **8**, 437; B. Moreau, S. Lavielle, and A. Marquet, *Tetrahedron Lett.*, 1977, 2591.
180. *F. &F.*, **14**, 60; L. E. Overman, M. E. Okazaki, and P. Mishra, *Tetrahedron Lett.*, 1986, **27**, 4391.
181. J. March, *Advanced organic chemistry*, Wiley-Interscience, New York, 1968, 242; I. T. Millar and H. D. Springall, *A shorter Sidgwick's organic chemistry of nitrogen*, OUP, 1969, p. 104.
182. H. C. Brown, *J. Am. Chem. Soc.*, 1946, **67**, 378; 1452.
183. *F. &F.*, **3**, 299.
184. S. Matsumara and N. Tokura, *Tetrahedron Lett.*, 1968, 4703.
185. *F. &F.*, **5**, 511; P. H. Wendschuh, H. Fuhr, J. S. Gaffney, and J. N. Pitts, Jr, *J. Chem. Soc., Chem. Commun.*, 1973, 74.
186. Preparation: E. R. Stephens, F. R. Burleson, and E. A. Cardiff, *J. Air Pollut. Control Assoc.*, 1965, **15**, 87.
187. D. L. H. Williams, *Nitrosation*, Cambridge University Press, 1988.
188. C. K. Ingold, *Structure and mechanism in organic chemistry*, G. Bell and Sons, London, 1953.
189. E. D. Hughes, C. K. Ingold, and J. H. Ridd, *J. Chem. Soc.*, 1958, **58**, 58; *Quart. Rev.*, 1961, **15**, 423.
190. L. P. Hammett, *Physical organic chemistry*, 2nd edn, McGraw-Hill, New York, 1970, pp. 53–100.

191. T. W. J. Taylor, *J. Chem. Soc.*, 1928, 1099; 1897; T. W. J. Taylor and L. S. Price, *J. Chem. Soc.*, 1929, 2052.
192. W. Lijinsky, E. Conrad, and R. Van De Bogart, No. 3, IARC Scientific Publications, N-*Nitroso compounds analysis and formation*, Lyon, 1972, p. 130.
193. F. F. Blicke, *Org. React.*, 1942, **1**, 303.
194. B. Reichert, *Die Mannich-Reaktion*, Springer Verlag, Berlin, 1959.
195. M. Tramontini, *Synthesis*, 1973, 703.
196. M. Tramontini and L. Angiolini, *Tetrahedron*, 1990, **46**, 1791.
197. A. R. Katritzky and P. H. Harris, *Tetrahedron*, 1990, **46**, 987; M. J. Brienne, C. Fouquey, and J. Jacques, *Bull. Soc. Chim. Fr.*, 1969, 2395.
198. R. Robinson, *The structural relations of natural products*, OUP, 1955.
199. O. G. Nabiev, M. A. Shakhgel, I. L. Chervin, and R. G. Kostyanovski, *Chem. Abstr.*, 1986, **105**, 133781.
200. M. A. Georgescu and M. V. Leonte, *Bull. Univ. Galati*, 1979, **2**, 69.
201. K. Kellner, B. Seidel, and A. Tzschach, *J. Organomet. Chem.*, 1978, **148**, 167.
202. G. A. Dutra, *Chem. Abstr.*, 1978, **89**, 109; D. Redmore, *J. Org. Chem.*, 1978, **43**, 992; 996; J. V. Otal Olivan and L. E. Perez Esteban, *Chem. Abstr.*, 1987, **106**, 19042.
203. C. W. Kruese and R. F. Kleinschmidt, *J. Am. Chem. Soc.*, 1961, **83**, 213.
204. *F. &F.*, **8**, 520; review: A. R. Katritzky, *Tetrahedron*, 1980, **36**, 679.
205. N. F. Eweiss. A. R. Katritzky, P.-L. Nie, and C. A. Ramsden, *Synthesis*, 1977, 634.
206. U. Gruntz, A. R. Katritzky, D. H. Kenny, M. C. Rezende, and H. Sheikh, *J. Chem. Soc., Chem. Commun.*, 1977, 701.
207. Preparation of the pyrillium tetrafluoroborate: R. Lombard and J.-P. Stephen, *Bull. Soc. Univ. Fr.*, 1958, 1458.
208. *F. &F.*, **6**, 450; Y. Ito, T. Hirao, and T. Saegusa, *J. Org. Chem.*, 1975, **40**, 2981.
209. M. L. Michailovic, A. Stojilkovic, and V. Andrejevic, *Tetrahedron Lett.*, 1965, 461; A. Stojilkovic, V. Andrejevic, and M. L. Michailovic, *Tetrahedron*, 1967, **33**, 720.
210. S. J. Rawalay and H. Shecter, *J. Org. Chem.*, 1967, **32**, 3129.
211. *F. &F.*, **5**, 20; J. B. Lee, C. Parkin, M. J. Shaw, N. A. Hampson, and K. I. MacDonald, *Tetrahedron*, 1973, **29**, 751.
212. *F. &F.*, **7**, 48; M. P. Doyle and B. Siegfried, *J. Chem. Soc., Chem. Commun.*, 1976, 433.
213. M. P. Doyle, B. Siegfried, and J. J. Hammond, *J. Am. Chem. Soc.*, 1976, **98**, 1627.
214. *F. &F.*, **12**, 480; M. F. Semmelhack and C. R. Schmidt, *J. Am. Chem. Soc.*, 1983, **105**, 6732.
215. *F. &F.*, **6**, 355; B. Rindone and C. Scolastico, *Tetrahedron Lett.*, 1974, 3379.
216. Preparation: R. L. Dannley, J. E. Gagen, and O. J. Stewart, *J. Org. Chem.*, 1970, **35**, 3076.
217. R. V. Hoffman, *J. Am. Chem. Soc.*, 1976, **98**, 6702; *F. &F.*, **7**, 251.
218. *F. &F.*, **7**, 277; M. E. Kuehne and T. C. Hall, *J. Org. Chem.*, 1976, **41**, 2742.
219. Preparation: *F. &F.*, **3**, 78; W. Flaig and H. Biergans, *Annalen.*, 1955, **597**, 196.
220. *F. &F.*, **15**, 113; R. F. X. Klein, L. M. Bargas, and V. Horak, *J. Org. Chem.*, 1988, **53**, 5994.
221. *F. &F.*, **5**, 324; N. H. Andersonn and H.-S. Uh, *Synth. Commun.*, 1972, **2**, 297.
222. *F. &F.*, **15**, 176; R. M. Moriarty, R. K. Valid, M. P. Duncan, M. Ochiai, M. Inenaga, and Y. Nagao, *Tetrahedron Lett.*, 1988, **29**, 6913; 6917.
223. *F. &F.*, **13**, 67; G. Buchi and D. E. Vogel, *J. Org. Chem.*, 1983, **48**, 5406.
224. *F. &F.*, **14**, 79; C.-K. Chen, A. G. Hortmann, and M. R. Marzabadi, *J. Am. Chem. Soc.*, 1988, **110**, 4829.

225. *F. &F.*, **8**, 437; R. Tang, S. E. Diamond, N. Neary, and F. Mares, *J. Chem. Soc., Chem. Commun.*, 1978, 562.
226. *F. &F.*, **15**, 199; E. Wenkert and E. C. Angell, *Synth. Commun.*, 1988, **19**, 1331.
227. *F. &F.*, **12**, 202; T. Mukaiyama, J. Ichikawa, M. Toba, and M. Asami, *Chem. Lett.*, 1983, 979.
228. *F. &F.*, **7**, 251; R. V. Hoffman, *J. Am. Chem. Soc.*, 1976, **98**, 6702.
229. A. W. Hofmann, *Annalen*, 1851, **78**, 253; **79**, 11.
230. C. K. Ingold, *Proc. Chem. Soc.*, 1962, 265; C. K. Ingold, *Structure and mechanism in organic chemistry*, G. Bell and Sons, London, 1953, pp. 427, 443.
231. H. C. Brown and R. L. Klimisch, *J. Am. Chem. Soc.*, 1966, **88**, 1425.
232. R. W. Alder, R. Baker, and J. M. Brown, *Mechanism in organic chemistry*, Wiley Interscience, London, 1971, p. 214.
233. *F. &F.*, **11**, 502; A. R. Katritzky and J. M. Lloyd, *J. Chem. Soc., Perkin Trans. 1*, 1980, 1895.
234. M. E. Krafft and R. A. Holton, *Tetrahedron Lett.*, 1983, **24**, 1345.
235. R. A. Holton, *J. Am. Chem. Soc.*, 1984, **106**, 5731.
236. *F. &F.*, **15**, 183; J. Einhorn, J.-L. Luche, and P. Demerseman, *J. Chem. Soc., Chem. Commun.*, 1988, 1350.
237. *F. &F.*, **15**, 292; J. Zakrzewski, *Synth. Commun.*, 1988, **18**, 2135.
238. *F. &F.*, **15**, 233; H. G. Chen and P. Knockel, *Tetrahedron Lett.*, 1988, **29**, 6701.
239. *F. &F.*, **15**, 81; S. L. Buckwald, M. W. Wannamaker, and B. T. Watson, *J. Am. Chem. Soc.*, 1989, **111**, 776.
240. *F. &F.*, **15**, 190; H. Albrecht and D. Kornetzky, *Synthesis*, 1988, 775.
241. *F. &F.*, **15**, 64; G. Stork, C. S. Shiner, C.-W. Cheng, and R. L. Polt, *J. Am. Chem. Soc.*, 1986, **108**, 304; J. Barluenga, F. J. Fafianás, F. Foubeloo, and M. Yus, *J. Chem. Soc., Chem. Commun.*, 1988, 1135.
242. M. Tokuda, Y. Yamada, and H. Suginome, *Chem. Lett.*, 1988, 1289.
243. P. H. H. Hermkens, J. H. van Maarsveen, P. L. H. M. Cobben, H. C. J. Ottenheijm, C. G. Kruse, and H. W. Scheeren, *Tetrahedron*, 1990, **46**, 833.
244. R. E. Gawley, *Synthesis*, 1976, 777; M. E. Jung *Tetrahedron*, 1976, **32**, 1.
245. D. Seebach and D. Enders, *Angew. Chem., Int. Ed. Engl.*, 1969, 639.
246. A. R. Lepley and H. G. Guimanini, *J. Org. Chem.*, 1966, **31**, 2055; 2061; 2064; *Chem. Commun.*, 1967, 1198.
247. H. Albrecht and H. Dollinger, *Tetrahedron Lett.*, 1984, **245**, 1353; *F. &F.*, **12**, 99.
248. Preparation: K. G. Kellogg and G. H. Cady, *J. Am. Chem. Soc.*, 1948, **70**, 3986.
249. *F. &F.*, **6**, 263; D. H. R. Barton, R. H. Hesse, T. R. Klose, and M. M. Pechet, *J. Chem. Soc., Chem. Commun.*, 1975, 97.
250. *F. &F.*, **13**, 28; D. H. R. Barton, A. Billion, and J. Boivin, *Tetrahedron Lett.*, 1985, **26**, 1229.
251. D. H. R. Barton, J. P. Finet, and J. Khamsi, *Tetrahedron Lett.*, 1987, **28**, 887; D. H. R. Barton, N. Yadav-Bhatnagar, J. P. Finet, and J. Khamsi, *Tetrahedron Lett.*, 1987, **28**, 3111.
252. P. Sulmon, N. De Kimpe, R. Verhé, L. De Buyck, and N. Schamp, *Synthesis*, 1986, 192; P. Sulmon, N. De Kimpe, and N. Schamp, *Tetrahedron*, 1989, **45**, 2937.
253. J. Alvarez-Builla, J. J. Vaquero, J. L. Garcia Navio, J. F. C. Carloss Sunkel, M. F. de Casa-Juana, F. Dorrego, and L. Santos, *Tetrahedron*, 1990, **46**, 967.
254. B. Jursic, *Tetrahedron*, 1988, **44**, 6677.
255. B. Jursic, *Synthesis*, 1988, 868.

2

ENAMINES AND YNAMINES

ENAMINES[1,2]

Just as the attachment of a hydroxyl group to a double bond forming an enol leads to important and unique reactions, so the attachment of an amino group directly to a carbon–carbon double bond yields molecules with interesting and synthetically useful reactions. Except in special structural circumstances the attached nitrogen atom must carry two substituents, i.e. the resulting molecule, an enamine (**1**), must be a tertiary amine; when the amino group is primary or secondary (**2**), (R^1 = H or alkyl), the tautomeric imine (**3**) is usually the stable form (Scheme 2.1). Only the C—N bond length of the enamine function has been determined.

$$R_2C{=}CH{-}NR_2 \quad (1)$$

$$\underset{(2)}{R_2C{=}CH{-}NHR^1} \xrightarrow{\text{tautomerism}} \underset{(3)}{R_2CH{-}CH{=}NR^1} \qquad (2.1)$$

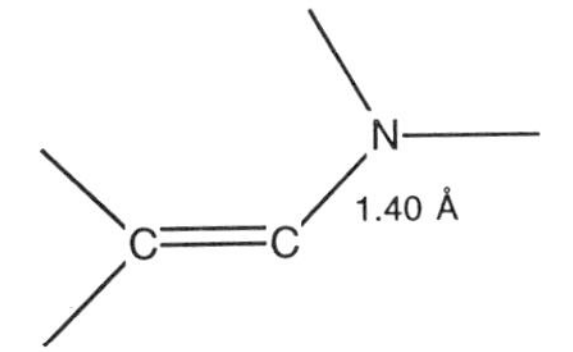

In the infra-red, enamines have a medium to strong absorption band between 1610 cm^{-1} and 1675 cm^{-1}. This can be compared with the absorption at 1640 cm^{-1} to 1690 cm^{-1} shown by iminium ions (**4**), which arise by protonation of enamines (Scheme 2.2).[3]

$$R_2C{=}CH{-}\ddot{N}R_2 \xrightarrow{H^+} \underset{(4)}{R_2CH{-}CH{=}\overset{+}{N}R_2} \qquad (2.2)$$

The olefinic protons of enamines usually give rise to an NMR signal between 4.0 and 6.0 δ dependent upon substitution.[4] The chemical shifts of these protons can be used to distinguish between the *E* and *Z* configurations (**5**) of enamines, with the *E* isomers usually giving signals at lower δ values than the *Z* isomers,[5] e.g. Scheme 2.3.

$$\underset{\substack{(Z)\\3\%\\ \tau_H = 5.87\ \delta}}{\text{Me(H)C=C(NMe}_2\text{)Et}} \rightleftharpoons \underset{\substack{(E)\\97\%\\ \tau_H = 5.46\ \delta}}{\text{Me(H)C=C(Et)NMe}_2} \quad (5) \qquad (2.3)$$

2.1 Preparation

2.1.1 *From aldehydes or ketones*

The original preparation of enamines[6] from an aldehyde or a ketone (**6**) and two equivalents of a secondary amine (**7**) in the presence of anhydrous potassium carbonate has the widest applicability and is the most useful (Scheme 2.4).

$$\underset{(6)}{R^1CH_2(R^2)C{=}O} + \underset{(7)}{HNR_2} \rightleftharpoons R^1CH_2(R^2)C(OH){-}NR_2 \rightleftharpoons R^1CH{=}C(R^2){-}NR_2 + H_2O \qquad (2.4)$$

As with imine formation (see Chapter 3, p. 125) various methods have been used to displace the equilibrium in favour of the enamine. Azeotropic distillation can be used to remove water if the reaction is carried out in a suitable solvent, e.g. in benzene, toluene, or xylene, in the presence of an acid catalyst.[7] Molecular sieves (Linde 5A) have been used to remove water providing a reasonably general, mild synthesis applicable to even medium sized ring ketones and camphor, which are both rather hindered.[8] The reaction is catalysed by Lewis acids such as titanium tetrachloride, which also functions as a water scavenger,[9] and by *p*-toluenesulphonic acid. This method has been used in the preparation of highly hindered enamines using stoichiometric amounts of ketone, secondary amine, and titanium tetrachloride, e.g. Scheme 2.5.

$$2(CH_3)_2CH{-}\overset{\overset{\displaystyle O}{\|}}{C}{-}CH(CH_3)_2 + 6\,HN(CH_3)_2 + TiCl_4 \longrightarrow 2\,(CH_3)_2C{=}\overset{\overset{\displaystyle N(CH_3)_2}{|}}{C}{-}CH(CH_3)_2 + 4\,(CH_3)_2NH_2Cl + TiO_2 \quad 72\% \tag{2.5}$$

An improvement in the use of titanium tetrachloride for the preparation of alicyclic enamines by addition of the ketone to a preformed complex of 5:1 amine:titanium tetrachloride has been reported and shown to increase the yield.[10]

For difficult condensations, e.g. of hindered ketones, tetrakis(dimethylamino)titanium (**8**), prepared from titanium tetrachloride and lithium dimethylamide (**9**), can be used,[11] e.g. Scheme 2.6.

$$TiCl_4 + 4\,\underset{(9)}{LiNMe_2} \longrightarrow \underset{(8)}{Ti(NMe_2)_4} + 4\,LiCl$$

$$CH_3{-}\overset{\overset{\displaystyle O}{\|}}{C}{-}CH(CH_3)_2 \xrightarrow{Ti(NMe_2)_4} CH_2{=}\underset{\underset{\displaystyle NMe_2}{|}}{C}{-}CH(CH_3)_2 \quad 53\% \tag{2.6}$$

Trimethylsilyl compounds (**10**), formed by the reaction of secondary amines (**11**) with trimethylsilyl chloride (**12**),[12] react with aldehydes or ketones under acid catalysis at 20 °C or by the action of heat to give high yields of enamines (Scheme 2.7).[13] This mild method of preparation is particularly useful for dimethylaminoenamines which are otherwise difficult to prepare due to the volatility of dimethylamine (b.p. 7 °C).[14]

$$\underset{(11)}{R^1R^2N{-}H} + \underset{(12)}{Me_3SiCl} \longrightarrow \underset{(10)}{R^1R^2N{-}SiMe_3}$$

$$\underset{(10)}{Me_3Si{-}NR^1R^2} + R^3{-}CH_2{-}C(R^4){=}O \longrightarrow R^1R^2N{-}\underset{\underset{\displaystyle R^4}{|}}{C}{=}CH{-}R^3 \tag{2.7}$$

Aldehydes (**13**) have been converted into enamines by the principle of Umpolung. Initially, the aldehyde is converted to an α-amino nitrile (**14**), which can then be alkylated to give another α-amino nitrile (**15**) (Scheme 2.8). On addition of a strong base this molecule eliminates hydrogen cyanide to form an

$$\text{R—CHO (13)} \xrightarrow{\text{HCN / HNR}^2_2} \text{R}^1\text{—CH(NR}^2_2\text{)—CN (14)} \xrightarrow[\text{ii, R}^3_2\text{CHHal}]{\text{i, LDA / THF}} \text{R}^3_2\text{CH—C(R}^1\text{)(NR}^2_2\text{)—CN (15) 80-90\%} \xrightarrow[\text{heat}]{\text{KOBu}^t\text{ / PhH}} \text{R}^3_2\text{C=C(NR}^2_2\text{)—R}^1 \text{ 50-90\%} \quad (2.8)$$

enamine.[15,16] This synthetic method has been applied to the preparation of dienamines[16] as well as to the relatively unstable enamines of acetone (**16**) (Scheme 2.9).

$$(CH_3)_2C{=}O \xrightarrow[\text{ii, HN(CH}_3)_2\text{; iii, KCN}]{\text{i, NaH}} \text{NC—C(CH}_3)_2\text{—N(CH}_3)_2 \xrightarrow{\text{Bu}^t\text{OK}} CH_2{=}C(CH_3)\text{—N(CH}_3)_2 \text{ (16) 64 \%} \quad (2.9)$$

The Horner–Wittig reaction, in which the amine moiety is supplied together with its end of the double bond, has been applied in various forms to prepare a variety of enamines, for example, Scheme 2.10, reactions 1[17] and 2.[18] Enamines have also been prepared by the Wittig–Horner reaction using (*N*-methyl-amino)methyldiphenylphosphine oxide (**17**), a modified Wittig–Horner reagent, which reacts with a wide variety of aldehydes and ketones forming a mixture of (*E*)- and (*Z*)-*N*-methylanilino enamines (**18**) (Scheme 2.11).[19]

1. bicyclic ketone $\xrightarrow[\text{BuLi / THF, -78}^\circ\text{C}]{(EtO)_2PO\text{-}CH_2NR_2}$ =C(H)—NR_2

2. cyclohexanone $\xrightarrow[\text{KOBu}^t\text{, pyrrolidine}]{(MeO)_2PO\text{-}CHN_2}$ cyclohexylidene=CH—N(pyrrolidine) (2.10)

$$\mathrm{Ph_2P(O){-}CH_2{-}N(Ph){-}CH_3}\ \textbf{(17)} \xrightarrow[\text{ii, } R^1\text{-CO-}R^2]{\text{i, Bu}^n\text{Li}} \mathrm{Ph_2P(O){-}CH(N(Ph)CH_3){-}C(R^1)(R^2){-}OH} \xrightarrow{\text{KOBu}^t} \mathrm{R^1R^2C{=}CH{-}N(Ph){-}CH_3}\ \textbf{(18)}\ E\ \&\ Z \qquad (2.11)$$

Enamininones (**19**) can be prepared from β-amino ketones (**20**) by the action of bis(acetonitrile)dichloropalladium in the presence of triethylamine which acts as a dehydrogenator in acetonitrile. This reaction gives general overall yields of 50–95 per cent,[20] e.g. Scheme 2.12.

$$\mathrm{CH_3COCH_2CH_2NEt_2}\ \textbf{(20)} \xrightarrow[\text{CH}_3\text{CN, Et}_3\text{N (2 mol)}]{(\text{CH}_3\text{CN})_2\text{PdCl}_2} \mathrm{CH_3COCH{=}CHNEt_2}\ \textbf{(19)}\ 70\%$$

$$\text{4,4-dimethyl-2-(dimethylaminomethyl)cyclohexanone}\ \textbf{(20)} \xrightarrow[\text{CH}_3\text{CN, Et}_3\text{N (2 mol)}]{(\text{CH}_3\text{CN})_2\text{PdCl}_2} \text{4,4-dimethyl-2-(dimethylaminomethylene)cyclohexanone}\ \textbf{(19)}\ 95\% \qquad (2.12)$$

2.1.2 *From iminium ions*

Alkylation of an imine (**21**) yields an iminium salt (**22**) which, provided that it contains an α-hydrogen atom, will react with base to give an enamine (Scheme 2.13)[21] (see Chapter 3, p. 141).

$$R^1CH_2C(R^2){=}NR^3 \ (21) \xrightarrow{MeI} R^1CH(H)C(R^2){=}N^+(R^3)Me \ I^- \ (22) \xrightarrow{Base} R^1CH{=}C(R^2){-}N(R^3)Me \quad (2.13)$$

2.1.3 *From cyanoallenes*

Addition of amines to 1-cyanoallenes (**23**) results in a fast exothermic reaction and is therefore a useful procedure for the synthesis of simple aliphatic enamines.[22] Addition may be 1,2- or 2,3- at the double bond. The mode of addition can be regulated to some extent by the amine used. Diethylamine results predominantly in 1,2-addition while piperidines give mostly 2,3-addition (Scheme 2.14).

$$R^1R^2C{=}C{=}CHCN \ (23) + R_2NH \xrightarrow{\text{1,2-Addition}} R^1R^2CH{=}C(N^+HR_2){-}\bar{C}H{-}CN \longrightarrow R^1R^2C{=}C(NR_2){-}CH_2{-}CN$$

$$R^1R^2C{=}C{=}CHCN \ (23) + R_2NH \xrightarrow{\text{2,3-Addition}} R^1R^2\bar{C}{-}C(N^+HR_2){=}CH{-}CN \longrightarrow R^1R^2CH{-}C(NR_2){=}CH{-}CN \quad (2.14)$$

2.1.4 *From alkynes and related compounds*

For unactivated terminal alkynes (**24**), mercuration–demercuration catalyses the addition of aromatic amines and the ensuing formation of an enamine,[23] e.g. Scheme 2.15.

$$Et—C{\equiv}C—H + HgHal_2 + Ph—NH—Me \xrightarrow{-HHal} [Ph—N(Me)—C(Et){=}CH—HgHal]$$

20 : 1 : 100

(24)

$$\xrightarrow{NaBH_4 / NaOH \text{ protonolysis}} [Ph—N(Me)—C(Et){=}CH_2] \xrightarrow{\text{isomerisation}} CH_3—CH{=}C(Me)—N(CH_3)—Ph$$

E and *Z* mix (2.15)

Suitably substituted aminoacetylenes (**25**) can be isomerized by the action of base to dienamines (**26**) (Scheme 2.16),[24] or, providing that the structure permits isomerism, to allenic enamines (**27**) (Scheme 2.17).[25]

$$CH_3—CH(NR_2)—C{\equiv}CH \ (25) \xrightarrow[DMSO]{KOBu^t} CH_2{=}C(NR_2)—CH{=}CH_2 \ (26) \quad 80\% \qquad (2.16)$$

$$R^1—C{\equiv}C—H + CH_2O + HNR_2 \xrightarrow{\text{Mannich reaction}} R^1—C{\equiv}C—CH_2—NR_2 \xrightarrow[\text{ii, MeOH}]{\text{i, BuLi / hexane}} R^1—CH{=}C{=}CH—NR_2 \ (27) \qquad (2.17)$$

Acetylenic amines (**28**) react with base to form allenamines (**29**) and a small amount of ynamine (**30**). The acetylenic amines may themselves be produced by reaction of dimethylamine (**31**) with an α-bromoalkylalkyne (**32**) (Scheme 2.18).[26]

$$H—C{\equiv}C—CH_2Br \ (32) + HNMe_2 \ (31) \xrightarrow{>85\%} H—C{\equiv}C—CH_2—NMe_2 \ (28)$$

$$\xrightarrow{23^\circ C, \ KOBu^t, \ -HOBu^t, \ HMPT} H_2C{=}C{=}CH—NMe_2 \ (29) + Me—C{\equiv}C—NMe_2 \ (30) \quad 96 : 4 \qquad (2.18)$$

2.1.5 *From ynamines*

Ynamines (**33**) add to various unsaturated systems to yield enamines. Ynamine addition to 1-nitrocyclopentene (**34**) gives cyclobutenamines (**35**) (Scheme 2.19).[27]

(34) + (33) → 16 hours, room temp. → (35) 40% (2.19)

Ynamines also add to 1,3-dienes (**36**) in a Diels–Alder reaction to yield six-membered cyclic enamines (**37**)[28] (see p. 118), e.g. Scheme 2.20.

(36) + $CH_3-C\equiv C-NEt_2$ → Fe / 25 °C → (37) (2.20)

Base catalysed structural isomerism may be employed to prepare enamines (**38**) from enylamines (**39**) (Scheme 2.21).[29] The intermediate butatrienylamines (**40**) may be isolated by protonation (Scheme 2.22).[29]

$$CH_2{=}CH{-}C{\equiv}C{-}NR_2 \ (39) \xrightarrow[\text{KOBu}^t,\ -100^\circ\text{C}]{\text{Bu}^n\text{Li}} HC{\equiv}C{-}CH{=}CH{-}NR_2 \ (38)\ 66\text{-}90\% \quad (2.21)$$

$$CH_2{=}C(H){-}C{\equiv}C{-}NR_2 \xrightarrow[\text{Bu}^t\text{OK}]{\text{Bu}^n\text{Li}} CH_2{=}C{=}C{=}\bar{C}{-}NR_2 \xrightarrow{\text{protonation, Bu}^t\text{OH}} CH_2{=}C{=}C{=}CHNR_2 \ (40) \quad (2.22)$$

2.1.6 *From formamides*

The reaction between a Grignard reagent (**41**) and *N*,*N*-dialkylformamides (**42**) gives addition products which eliminate spontaneously to give aldehyde enamines (**43**) (Scheme 2.23).[30]

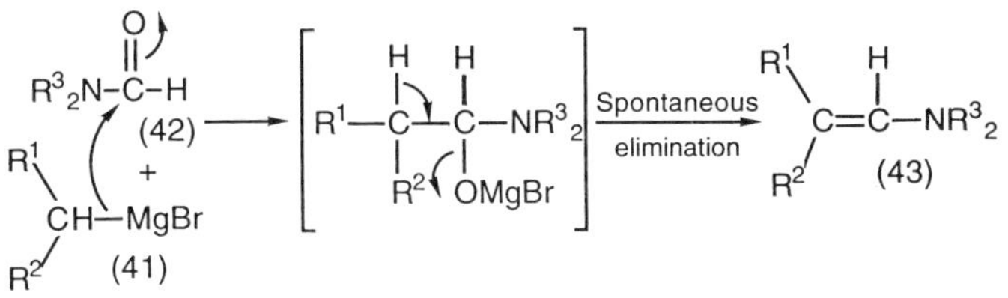

R^1	R^2	R^3	Yield (%)
Me	H	Bu^s	63
1-Pr	H	Bu^n	80
1-Pr	H	Bu^s	62
-Cyclohexyl-		Me	66
-Cyclohexyl-		Bu^n	22

(2.23)

2.1.7 *From epoxides*

Treatment of a *cis*-epoxysilane (**44**) with a secondary amine (pyrrolidine or morpholine) in the presence of alumina gives an α-amino-β-hydroxysilane (**45**) which on treatment with potassium hydride yields a pure *cis*-enamine (**46**)

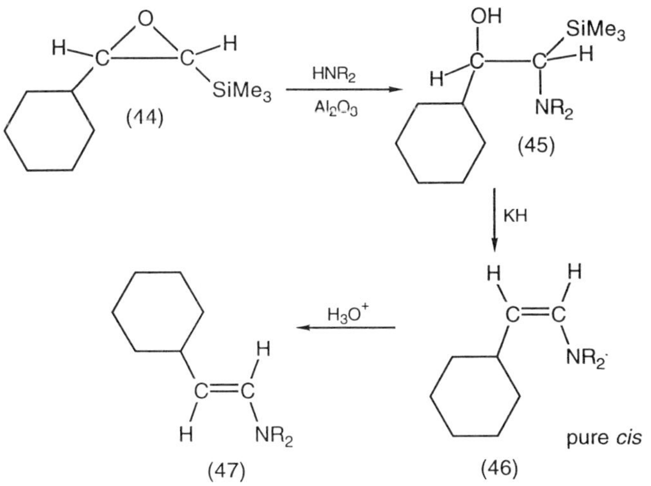

(2.24)

(Scheme 2.24). This isomer can easily be converted into the pure *trans*-enamine by the action of acid (**47**). This method provides a stereospecific synthesis.[31]

2.1.8 *From oximes*

Enamides may be prepared by a simple synthesis from ketoximes (**49**) by the use of titanium triacetate, a powerful reducing agent,[32] for example, in the preparation of 1-acetylaminocyclohexene (**48**) (Scheme 2.25).

NOH (49) — i, Ac_2O, DMF, ii, $Ti(OAc)_3$, 20°C → H, Ac, N (48) 80%

(2.25)

A further synthesis of enamides from ketoximes may be effected by refluxing with acetic anhydride in pyridine. For example, cholestanone oxime (**50**) is converted into the enimide (**51**) on refluxing for 10 h and then into the enamide (**52**) when chromatographed on alumina (Scheme 2.26).[33]

CH_3, HON, H (50) — Ac_2O, Py, Δ → CH_3, $(Ac)_2N$, H (51) 94%

Ac_2O, Py, Δ ⇅ Al_2O_3

CH_3, AcHN, H (52)

(2.26)

2.2 Reactivity

2.2.1 *Reaction with electrophiles on nitrogen or on the β-carbon atom*

Enamines can react with electrophiles either at the nitrogen atom (Scheme 2.27, equation i) or at the β-carbon atom (Scheme 2.27, equation ii). Of these two reactions the first is usually the more rapid but the product, an enammonium ion (**53**), often rearranges to the thermodynamically more stable iminium ion (**54**). A further complication in the study of enamine reactivity is that enamines are often mixtures of two isomers which affects the extent of pπ conjugation.

$$\text{(i)}\quad >C{=}C{-}\ddot{N}R_2 + RX \longrightarrow >C{=}C{-}\overset{+}{N}R_3 \ (53)$$

$$\downarrow \text{rearrangement}$$

$$\text{(ii)}\quad >C{=}C{-}\ddot{N}R_2 + RX \longrightarrow R{-}C{-}C{=}\overset{+}{N}R_2 \ (54) \qquad (2.27)$$

2.2.2 *Acid hydrolysis*

Enamines are readily hydrolysed by aqueous acid to an aldehyde or ketone and a secondary amine by reversal of their formation. Thc rate determining step is heavily pH dependent.[34] Protonation may be either at the β-carbon atom or at the nitrogen atom depending upon the nature of the protonating agent.[35] *N*-protonation is favoured by hard acids (HCl or perchloric) while *C*-protonation

$$R{-}CH{=}C(R){-}N\text{(piperidinyl)} + H{-}\overset{+}{O}H_2 \longrightarrow R{-}CH_2{-}C(R){=}\overset{+}{N}\text{(piperidinium)} \ (55) \xrightarrow{H_2\ddot{O}} R{-}CH_2{-}C(R)(OH){-}N\text{(piperidinyl)} \ (56) \xrightarrow{H{-}\overset{+}{O}H_2} R{-}CH_2{-}\overset{O}{\overset{\|}{C}}{-}R + HN\text{(piperidine)} \qquad (2.28)$$

occurs with carboxylic acids. The *N*-protonated enamine is inactive towards hydrolysis. Therefore the mechanism of hydrolysis involves the capture of a proton at the β-carbon atom, followed by attack of water at the α-carbon atom of the iminium ion (**55**) so formed, to yield a carbinolamine (**56**), which decomposes into a ketone and a secondary amine by elimination of the amine moiety (Scheme 2.28). This easy enamine formation and subsequent hydrolysis has been utilized for the protection of ketones (**57**) during syntheses,[36] e.g. Scheme 2.29.

(2.29)

2.2.3 *Hydroboration*

Hydrogenolysis of enamines by a hydroboration–protonolysis procedure has been extensively investigated as a general synthesis of olefins (**58**). The reaction involves electrophilic syn addition of diborane to produce a *trans*-β-amino organoborane which then undergoes anti-elimination of the boron group and the amine function (Scheme 2.30).[37]

Hydroboration of (*Z*)-enamines of aldehydes or ketones (**59**) with 9-borabicyclo[3.3.1]nonane (9-BBN), followed by methanolysis (retention) provides a useful conversion of aldehydes and ketones to the (*Z*)-alkene (**60**)

B_2H_6, THF, 30 mins, rt. ; CH_3CO_2H in diglyme

(58) (2.30)

(Scheme 2.31).[38] However, the same reaction with borane-dimethyl sulphide (BMS) and including a final oxidation with alkaline hydrogen peroxide produces the isomeric (*E*)-alkene (**61**) (Scheme 2.32).[38]

(*Z*) (59) — i, 9-BBN; ii, CH_3OH → 65% (*Z*) (60) (2.31)

i, BMS; ii, CH_3OH; ii, OH^-, H_2O_2 → (61) (*E*) 50% (2.32)

2.2.4 *Reaction with trichloroacetic acid*

Trichloroacetic acid reacts with cyclic enamines by protonation to form an iminium ion to which the trichloroacetic anion adds with simultaneous decarboxylation. Alkaline hydrolysis of the resulting trichloromethyl derivative yields an α,β-unsaturated carboxylic acid (**62**),[39] e.g. Scheme 2.33.

$$+ \ Cl_3C.CO_2H \longrightarrow$$

CHCl$_3$ + (62) $\xleftarrow[\text{50-81\%}]{\text{NaOH}}$ CCl$_3$ + CO$_2$

(2.33)

2.2.5 *Reduction*

Enamines can be reduced by a variety of electrophilic reducing agents to tertiary amines. Capture of a proton to give an iminium ion is usually the first stage, followed by reduction of the iminium ion. Catalytic reduction was used first by Mannich and Davidson[40] and is still used for stereoselective reduction.[41] In later work both formic acid[42] and phosphorous acid[43] have been used for the reduction of enamines with a high degree of stereoselectivity and in high yields as shown in Scheme 2.34.

A more useful and convenient reagent for use in the reduction of enamines

CH$_3$ ⟶ CH$_3$ + CH$_3$

H_3PO_3 (100°C) Yield %
75 : 25

HCO_2H Yield %
85 : 15 (2.34)

under acidic conditions is lithium or sodium cyanoborohydride. The reaction proceeds by the rapid reduction of the iminium ion (**63**) (Scheme 2.35). Using this reagent reductive amination of aldehydes and ketones, without isolation of the intermediate enamine, may be performed under mildly acidic conditions (pH ~6) (Scheme 2.36).[44] The addition of 3A molecular sieves increases the yields by removing the water formed in the first step of the reaction, thus favouring the formation of the iminium ion.

R_2N R^1 CR^2_2 H—OH_2 ⇌ $R_2\overset{+}{N}$ R^1 CHR^2_2 H—$\overset{-}{B}$ (63)

$NaBH_3CN$, pH <6, MeOH

R_2N–CH(R^1)CHR^2_2 50-85%

R	R^1	R^2	Yield (%)
Me	Me	Me	75
morpholine	-cyclohexene-		80
c-C_4H_8	-cyclohexene-		86

(2.35)

cyclohexanone + R^1R^2NH $\xrightarrow[\text{HCl / MeOH, pH 6, 25°C, 72h}]{NaBH_3CN}$ cyclohexyl-NR^1R^2

R^1	R^2	Yield (%)
H	H	61
H	CH_3	50

(2.36)

Sodium borohydride in diglyme has been used for the reduction of steroidal dienamines (**64**) in acidic medium (Scheme 2.37).[45]

Reduction with aluminium hydride (prepared *in situ* from lithium aluminium hydride and aluminium chloride (3:1) in ether) leads chiefly to hydrogenolysis

(64)

conessine series steroid

$NaBH_4,B_2H_6$ diglyme 100°C

(2.37)

of the amino group and constitutes a good method for the conversion of pyrrolidine enamines (**65**) into alkenes (**66**), e.g. Scheme 2.38. The small amount of tertiary amine also formed by hydrogenation is easily removed by an acid wash.[46] Other enamines have been reduced in a similar fashion and the yields are listed in Table 2.1. The reaction procedes by an initial electrophilic attack by aluminium hydride followed by elimination (Scheme 2.39).

$LiAlH_4$ (0.15 mole) $AlCl_3$(0.05 mole) ether, reflux, 2 days

(65) (66) 85% + 11%

(2.38)

Table 2.1 Reduction of enamines with aluminium hydride

Enamine	Alkene	Yield (%)
1-*N*-Pyrrolidinocyclohexene	Cyclohexene	80
6-Methyl-1-*N*-pyrrolidinocyclohexene	3-Methylcyclohexene	85
6-*n*-Butyl-1-*N*-pyrrolidinocyclohexene	3-*n*-Butylcyclohexene	91
3-*N*-Pyrrolidinopent-2-ene	Pent-2-ene	81
4-*N*-Pyrrolidinohept-3-ene	Hept-3-ene	85
1-*N*-Pyrrolidinoocyclopentene	Cyclopentene	92
1-*N*-Pyrrolidinocycloheptene	Cycloheptene	85

$$>C{=}C{-}N< \xrightarrow{AlH_3} -\overset{|}{\underset{|}{C}}-C{=}N^{+}< \;\; (H_2Al^{-}{-}H)$$

Tautomerism

$$H_2Al{-}N< \; + \; >C{=}C<_{H} \longleftarrow -\overset{|}{\underset{H_2Al}{C}}-\overset{|}{\underset{H}{C}}-N< \qquad (2.39)$$

2.2.6 *Reductive cyclization*

Enamines which contain a suitably placed ester group can be reductively cyclized using lithium aluminium hydride. The starting enamines (**67**) may be formed by Michael addition of electrophilic olefins to the enamines (**68**). Lithium aluminium hydride quantitatively reduces the ester to an alcohol under mild neutral conditions whilst leaving the enamine function intact. Nucleophilic addition then results in the formation of an *N,O*-acetal (**69**) which reacts under nonaqueous acidic conditions to give the vinylic ether of a pyran (**70**) and under aqueous acidic conditions to give the semiacetal of a pyran (**71**) (Scheme 2.40).[47]

4-Cyanoindole (**72**) may be synthesized by the reductive cyclization of an enamine. This is a useful synthetic reaction since the cyano group is a good precursor for many other functional groups (Scheme 2.41).[48]

2.2.7 *Alkylation*

Alkylation of enamines with simple alkyl halides usually leads to a mixture, after hydrolysis, of unalkylated (**73**), monoalkylated (**74**), and dialkylated ketones (**75**) (e.g. Scheme 2.42). Pyrrolidine enamines have generally been found to be most useful in these reactions.[49,50]

There are two ways to avoid the tedious separation of these product mixtures. One is to use highly electrophilic halides which result in almost exclusive monoalkylation,[50] for example, Scheme 2.43. A second way which enables simple alkyl compounds to be used for preponderant monoalkylation is to use sterically hindered enamines, for example, Scheme 2.44.

Potassium enolates of enamino ketones (**76**) and esters (**77**), formed by reaction with potassium hydride in THF at 0 °C may be alkylated almost exclusively in the γ-position (Scheme 2.45).[51]

The alkylation of a hindered and chiral enamine enables asymmetric monoalkylation to be performed, with very low levels of dialkylation (*c.* 5%), e.g.

(68) + $H-C(R^3)=C(R^4)-CO_2Et$ —Michael Addition→ (67) Mixture of two enamines —$LiAlH_4$→ —H_2O, HO^-→ (69) Stable in alkali

(69) —$(CO_2H)_2$ in dioxan→ (70)

(69) —aq. HCl→ (71)

n	R^1 R^2	R^3	R^4	Reaction time (h)	Yield (%) of (**70**)
2	$(CH_2)_4$	H	H	8	76
2	$(CH_2)_5$)	H	H	8	70
2	$O(CH_2CH_2)_2$	H	H	8	53
3	$(CH_2)_4$	H	H	8	70
3	$(CH_2)_5$	H	H	8	78
3	$O(CH_2CH_2)_2$	H	H	8	51
3	$(CH_2)_4$	H	Me	16	70
3	$(CH_2)_4$	Me	H	16	64
3	$(CH_2)_5$	H	Me	64	59
3	$(CH_2)_5$	Me	H	64	33

(2.40)

CN, CH_3, NO_2 — $Me_2NCH(OMe)_2$, 110°C → CN, Me, N, Me, NO_2

Fe / HOAc / EtOH ↓

CN, N, H

(72) 67%

(2.41)

N — i, MeI; ii, aq. H_3O^+ → O + O, CH_3 + CH_3, O, CH_3

(73) 16 % (74) 37 % (75) 16 %

(2.42)

N — i, $PhCH_2Br$; ii, aq. H_3O^+ → O, CH_2Ph

55%

(2.43)

CH_3, $CHCH_2CH_3$, N, $CH_2CH_2CH_2CH_3$ — i, MeI in CH_3CN at b.p.; ii, aq. NaOAc / HOAc buffer → O + CH_3, O + CH_3, O, CH_3

14% 56% 14%

i, $(MeO)_3O^+BF_4^-$ in CH_2Cl_2 at 20°C; ii, aq. buffer as above →

9% 74% 2%

(2.44)

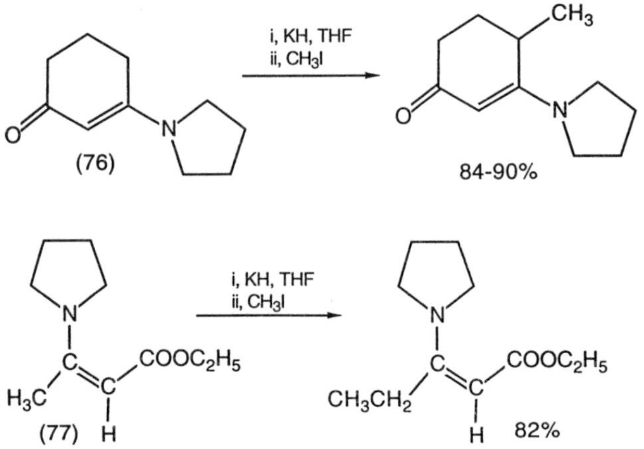

(76) i, KH, THF; ii, CH_3I → 84-90%

(77) i, KH, THF; ii, CH_3I → 82%

(2.45)

Scheme 2.46.[52] A similar procedure is described by M. Barthélémy *et al.*[53] (Scheme 2.47).

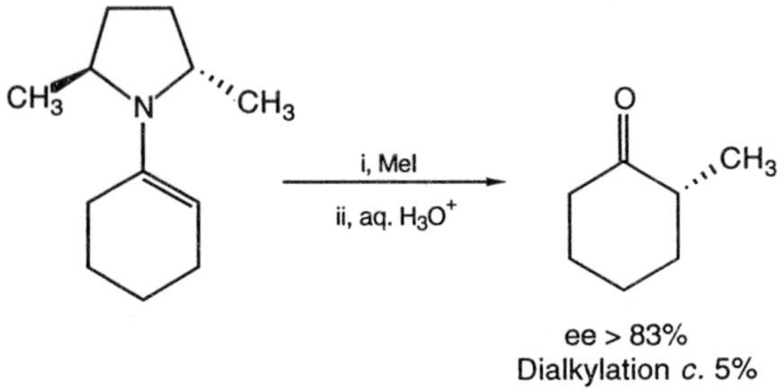

i, MeI; ii, aq. H_3O^+ → ee > 83%, Dialkylation *c.* 5%

(2.46)

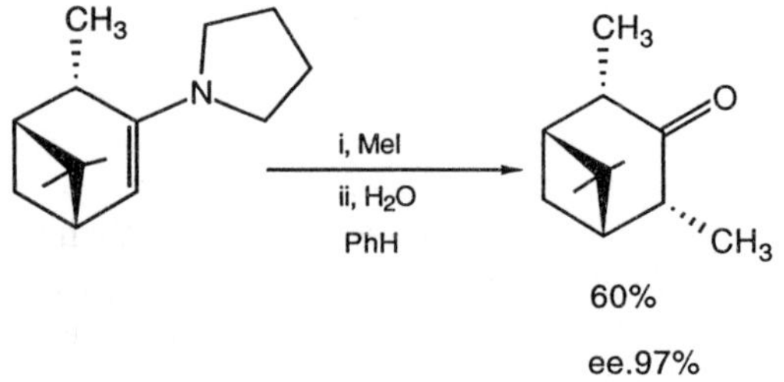

i, MeI; ii, H_2O; PhH → 60%, ee.97%

(2.47)

2.2.8 *Acylation*

β-Ketoesters (**78**) have been synthesized by acylation of pyrrolidine enamines (**79**) by reaction with chloroformic ester (**80**),[54] for example, Scheme 2.48.

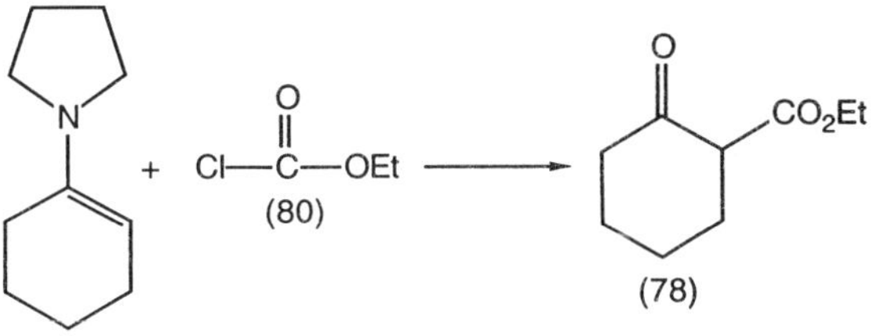
N
Cl—C—OEt
O
O
CO2Et

(80) (78)

(79) (2.48)

2.2.9 *Annulation*

The well known Robinson annulation reaction between a methylcyclohexanone (**81**) and a methyl vinyl ketone (**82**) has often been used for the synthesis of steroids (**83**) and terpenes and is regiospecific as shown in Scheme 2.49.[55]

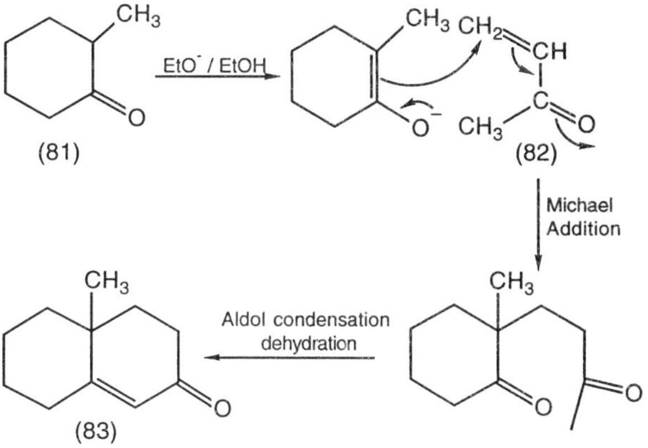
CH3
EtO⁻ / EtOH
CH3 CH2
CH
C
CH3
Michael
Addition
Aldol condensation
dehydration
CH3
CH3

(81) (82) (83)

(2.49)

However, in the Stork reaction[50,56] an enamine is used in preference to a ketone causing the regiospecificity of the annulation to change. This provides a complimentary procedure to the Robinson method (Scheme 2.50). The change in regiospecificity of the annulation (when compared with the Robinson annulation) may be attributed to the greater reactivity of the less substituted enamine isomer (**84**) compared to that of the more substituted isomer (**85**).[57]

2.2.10 *Arylation*

An organolead compound, *p*-methoxyphenyllead triacetate (**86**), reacts as an electrophile with an enamine. The arylation is very sensitive to steric effects and hence is limited in usefulness (Scheme 2.51).[58]

(85) (84)

then aq. H_3O^+

(2.50)

$CHCl_3$, rt., 5 mins

(86)

+ $Pb(OAc)_2$
+ AcO^-

H^+ / H_2O

Ar = —C6H4—OMe

83%

(2.51)

2.2.11 *Pyrrole synthesis*

Over a hundred years ago Hantzsch discovered that pyrroles (**87**) could be synthesized by the reaction of enamines with α-chloroketones (**88**) (Scheme 2.52).[59]

A more modern use of this reaction is described by Roomi and MacDonald[60] who have extended the Hantzsch synthesis generally to the production of ethyl esters of 2-alkyl and 2,4-dialkylpyrrol-3-carboxylic acids and benzyl and *t*-butyl esters when the 2-alkyl group is methyl, e.g. Scheme 2.53. Similarly

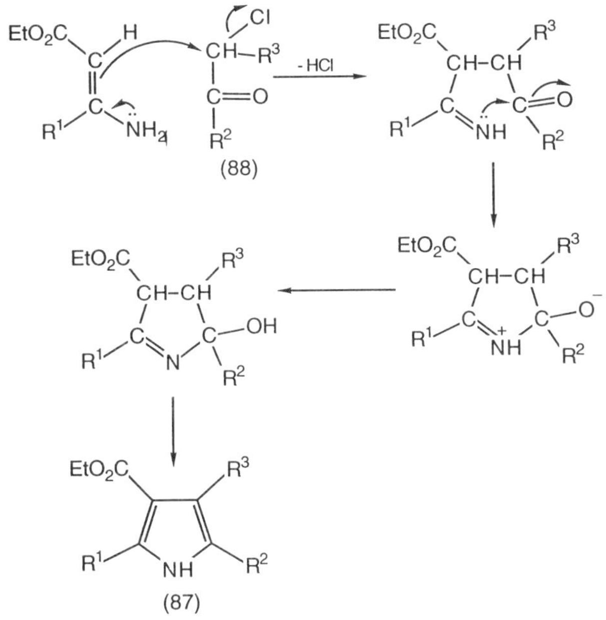

(2.52)

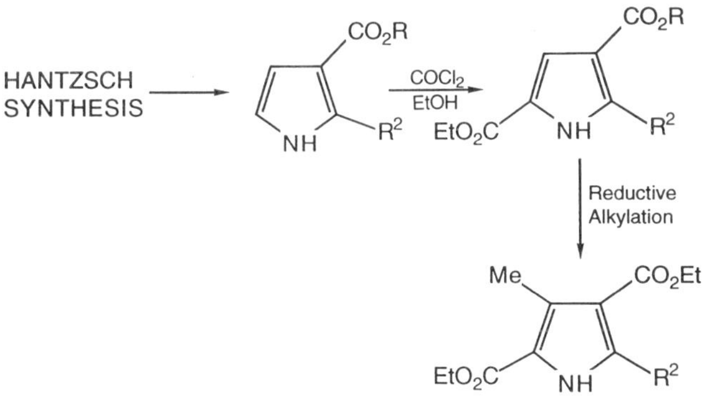

(2.53)

pyrroles result from the reaction of enamines with α-hydroxy ketones.[61] However, a more recent paper describes a pyrrole synthesis by reaction of α-cyano enamines with α-ketoaldehydes (**89**) (Scheme 2.54).[62]

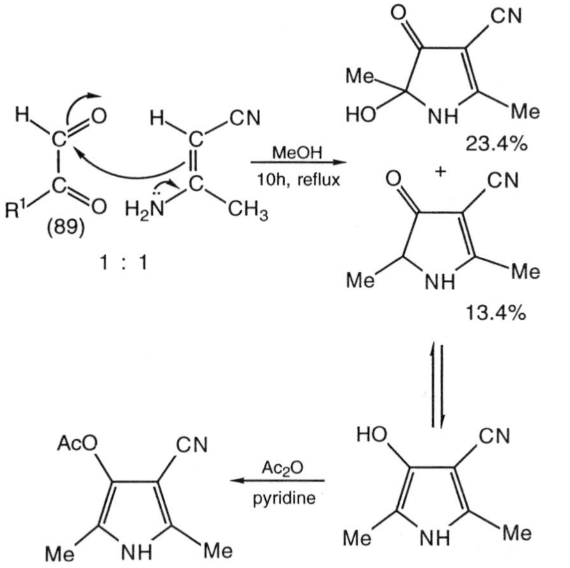

(2.54)

2.2.12 *Cycloaddition reactions*

Phenyl azide undergoes cycloaddition with an enamine to yield a *Δ*-triazoline (**90**) which, when photolysed loses nitrogen to produce a 2-amino aziridine (**91**) (Scheme 2.55).[63]

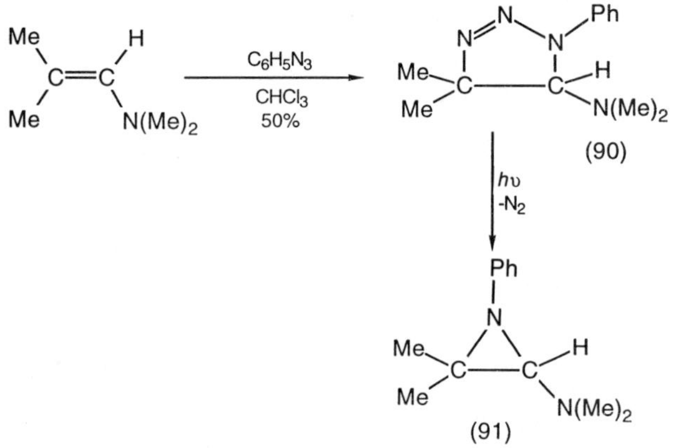

(2.55)

By a cycloaddition reaction enamines may be trapped by quinone methides (**92**) in a versatile synthesis of dihydropyrans (**93**) in which the double bond is part of the aromatic system (Scheme 2.56).[64] The quinone methides may be generated by heating an *o*-phenolic Mannich base. If the heating is performed in the presence of an enamine then the reaction ensues.

(92)

(93)

(2.56)

Enamines also undergo cycloaddition reactions with strongly polarized nitroacetylenes (**94**), with ensuing ring expansion and hydrolysis leading to the formation of α-nitro ketones (**95**),[65] e.g. Scheme 2.57.

$$Me_3C.C{\equiv}C.NO_2 \xrightarrow{Et_2O,\ 0^\circ C}$$

(94)

62%

20% H_2SO_4 40°C

(95)

(2.57)

Enamines, particularly pyrrolidine enamines (**96**), also undergo Diels–Alder reactions with reactive azadienes, for example, 1,2,4-triazine (**97**), to form substituted pyridines (**98**) (Scheme 2.58).[66]

In another example of this type of reaction, enamines undergo 1,3-dipolar cycloaddition with sulphonylaryl azides (**99**) and, in particular, with tosyl azide.

$CH_3(CH_2)_2$
C
C
C_2H_5
H
N
(96)
+
(97)
$CHCl_3$
45°C
N
N
$(CH_2)_2CH_3$
C_2H_5
$-N_2$
$CH_3(CH_2)_2$
C_2H_5
N
71%
(98)

(2.58)

This reaction furnishes a convenient preparation of diazomethane (Scheme 2.59).[67] If $R^1 = R^2 = H$, path (i) is followed and diazomethane is formed, however, when R^1, R^2, and R^3 are alkyl, path (ii), the rearrangement pathway, is followed, to form (**100**).

Ar—SO_2N_3 + $R^1R^2C{=}C(R^3)NR_2^4$
(99)

R^1 R^2 R^3 R_2^4N N N N SO_2Ar
R^1 R^2—C—N_2^+ R^3—C R_2^4N N^- SO_2Ar
R^2 R^1 C^+ R^3 R_2^4N N^- SO_2Ar

i
ii

$R^1R^2C{=}N_2$ + $R^3(R_2^4N)C{=}N—SO_2Ar$

R^1 R^2—C—C NR_2^4 R^3 N SO_2Ar
(100)

Ar = p-MeC$_6$H$_4$

(2.59)

Enamines and pyrimidines which contain an electron withdrawing group, for example, 5-nitropyrimidine (**101**), undergo interesting Diels–Alder reactions which are followed by eliminations to yield substituted pyridines (Scheme 2.60).[68] The reaction has been extended by the same authors[68] who used cyclic enamines to produce (β)-annelated pyridines (**102**), for example, Scheme 2.61.

(101)

EtOH, Reflux, 2-20 hours

-HCN

- HNR_2

R_2N = morpholino 57%
or piperidino 52%

(2.60)

2.2.13 *Oxidation*

Reaction of an aldehyde with piperidine followed by oxidation of the enamine formed by sodium dichromate in anhydrous acetic acid benzene provides a high yield method for the degradation of aldehydes to ketones, particularly in steroid systems (reverse of Darzen's reaction) (Scheme 2.62).[69]

Enamines are quickly oxidized by lead tetraacetate in benzene at room temperature to give a mixture of products, which probably results from the diacetylation of the carbon–carbon double bond, for example, Scheme 2.63.[70]

Enamines of cyclic ketones may be oxidized using lead tetraacetate in the presence of BF_3 etherate, resulting in a Favorski type rearrangement to esters of contracted cycloalkanoic acids (**103**), e.g. Scheme 2.64.[71]

Oxidation of morpholine enamines by thallium triacetate in glacial acetic acid, or chloroform, at room temperature is easier to control, and after hydrolysis gives good yields of the α-acetoxy derivatives of the parent ketone (Scheme 2.65).[72] Yields obtained are higher than those obtained by direct oxidation of the ketone.

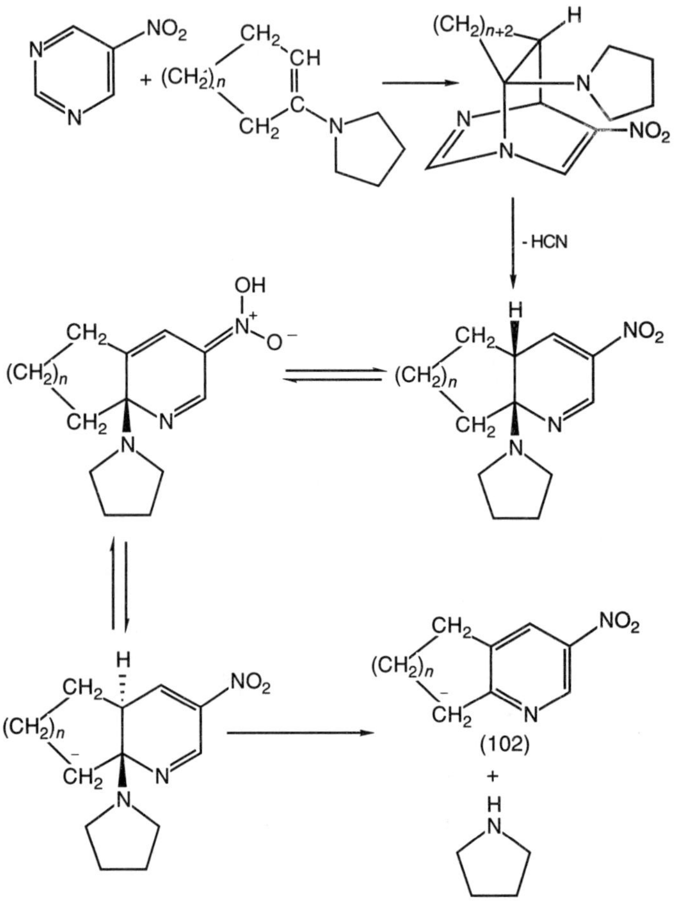

n	Yield (%)
1	60
2	80
3	75
4	76

(2.61)

2.2.14 *Michael reaction*

By causing an enamine to undergo a Michael reaction with an α,β-unsaturated ester (**104**) and oxidizing the product with oxygen and cuprous chloride, a synthesis of γ-ketoesters (**105**) has been developed (Scheme 2.66).[73]

$R_2CH\text{-}C(=O)\text{—}H$ —(piperidine; *p*-TsOH in PhH, Reflux and remove H_2O)→ $R_2C\text{=}CH\text{—}N$(piperidine) 99%

—($Na_2Cr_2O_7$, HOAc-PhH (anhydrous))→ $R_2C\text{=}O$ (2.62)

Pb(OAc)$_4$ in PhH

Room temp

10-20 minutes

OAc

OAc

OAc + OAc + O + O

7% 10% 49% <5%

\+ N OAc + OAc O

22% 10%

(2.63)

$Pb(OAc)_4$, $BF_3(C_2H_5)_2O$, C_2H_5OH, PhH, 25°C

$COOC_2H_5$

(103)

78%

(2.64)

AcOH or $CHCl_3$

+ Tl—$O_2C.CH_3$

+ $CH_3\text{-}CO_2^-$

aq. H_3O^+

70%

(2.65)

(104)

Tautomerism

O_2 / CuCl

(105)

(2.66)

2.2.15 *α-Acetoxylation of ketones*

An interesting sequence involving oximes (**106**), nitro compounds, iminium salts (**107**), enamines, and imines (**108**) resulting in the α-acetoxylation of ketones has been developed by House and Richey.[74] This demonstrates the usefulness of enamines as intermediates in synthesis (Scheme 2.67).

$CH_3CH_2CH_2C(=O)CH_2CH_2CH_3$ —(H_2NOH, Ac_2O)→ $CH_3CH_2CH_2C(=NOCOCH_3)CH_2CH_2CH_3$ (106)

↓ $(CH_3)_3O^+ BF_4^-$, CH_3NO_2

$CH_3CH_2CH_2C(=N^+(Me)OCOCH_3)CH_2CH_2CH_3$ BF_4^-
iminium salt (107)

← fluoroboric acid, Et_3N

$CH_3CH_2CH_2C(N(Me)OC(=O)Me)=CH\text{-}CH_2CH_3$
enamine

↓ Rapid Rearrangement

$CH_3CH_2CH_2C(=NMe)CH(OCOCH_3)CH_2CH_3$ (108) —(hydrolysis, H_3O^+)→ $CH_3CH_2CH_2C(=O)CH(OCOCH_3)CH_2CH_3$
48% from (106) (2.67)

2.2.16 *Oxygenation*

Photosensitized oxygenation of enamines has been shown to occur via 1,2-addition of singlet oxygen to the double bond. At low temperatures (−78 °C) the dioxetane (**109**) may be isolated,[75] but at higher temperatures the C—C bond cleaves with formation of amides and ketones (Scheme 2.68). Occasionally the C—N bond has been cleaved.[76] With cyclic enamines the dioxetane undergoes cleavage of the O—O bond by a β-elimination process to give an α-diketone (**110**),[77] e.g. Scheme 2.69.

$Me_2C{=}CHNMe_2$ —($^1O_2$)→ $Me_2C(\text{–O–O–})CHNMe_2$ (109)

↓

$Me_2C{=}O + O{=}CHNMe_2$ (2.68)

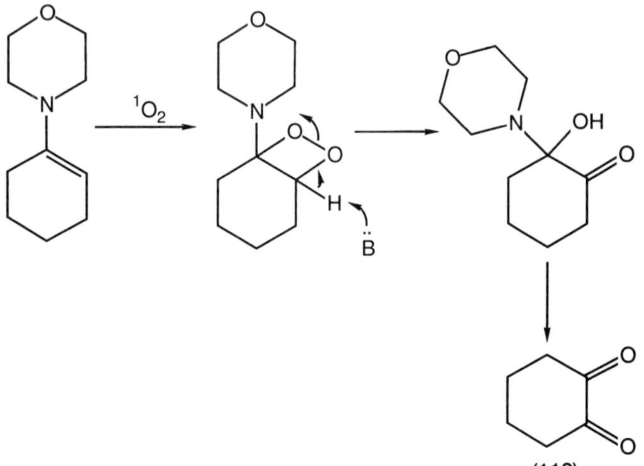

(110) (2.69)

2.2.17 *With tosyl azide*

Enamines react with tosyl azide to yield, via an unstable triazoline (**111**), a mixture of two amidines and a diazoalkane by two reaction pathways (Scheme 2.70).[78]

2.2.18 *Alkenynylamine synthesis*

Enamines react to form alkenynylamines (**114**) by a process of dehydrohalogenation accompanied by rearrangement (Scheme 2.71).[79]

1-CHLORO-*N*,*N*-2-TRIMETHYLPROPENYLAMINE

2.3 Preparation

2.3.1 *From an amide*

The enamine 1-chloro-*N*,*N*-2-trimethylpropenylamine (**115**)[80] has been developed as a useful reagent for several synthetic methods (Scheme 2.72).

(111)

(112)

(113)

Ar^2 = p-tolyl (CH_3) (p-tosyl)

Ar^1	R^1	R^2	R^3. . .R^4	Ratio	Yield (%) of (**12**) + (**13**)
Ph	H	Me	Morpholino	60:40	72
Ph	H	Me	Pyrrolidino	60:40	92
Ph	H	Me	*N*-Methylanilino	60:40	82
Ph	H	Et	Mopholino	60:40	70
Ph	Me	Me	Morpholino	65:35	65
2-Furyl	H	Me	Morpholino	65:35	70
2-Thienyl	H	Me	Morpholino	75:25	75

(2.70)

$(CH_3)_2C{=}CH{-}N(CH_3)_2 \xrightarrow[CHCl_3]{Bu^tOK}$ 1,1-dichloro-2,2-dimethyl-3-(dimethylamino)cyclopropane (Cl, Cl, CH_3, CH_3, H, NMe_2) 32 %

$\xrightarrow[DMSO]{Bu^tOK}$ $CH_2{=}C(CH_3){-}C{\equiv}C{-}N(CH_3)_2$ (114) 20-30% (2.71)

$(CH_3)_2CH{-}C(=O){-}N(Me)_2 \xrightarrow[CH_2Cl_2,\ 0\text{-}20^{\circ}C]{COCl_2} (CH_3)_2CH{-}C(Cl){=}\overset{+}{N}(Me)_2Cl^-$ (117) (2.72)

2.4 Reactivity

2.4.1 *With electron rich aromatic compounds*

The enamine may react in polar form with electron rich aromatic compounds, such as furan or pyrrole, e.g. Scheme 2.73.[81]

By addition of silver fluoroborate,[82] or less expensively, dry zinc chloride,[83] the polar form of the enamine is induced. This salt may then undergo a variety of reactions.

$(CH_3)_2C{=}C{=}\overset{+}{N}(CH_3)_2Cl^-$ + ArH $\xrightarrow{NEt_3}$ $(CH_3)_2C{=}C(Ar){-}N(CH_3)_2$

$\xrightarrow{H_3O^+}$ $(CH_3)_2CH{-}C(=O){-}Ar$ 85-95% (2.73)

2.4.2 *With alkenes and alkynes*

The salt reacts with olefins and acetylenes in [2+2] cycloadditions. The resulting imminium salts are readily hydrolysed to alkylated cyclobutanones (**116**), e.g. Scheme 2.74.[84]

$$(CH_3)_2C{=}C{=}\overset{+}{N}(CH_3)_2\ BF_4^- + (CH_3)_2C{=}C(CH_3)_2 \longrightarrow \text{2-(dimethyliminio)-1,1,4,4,3,3-hexamethylcyclobutane}\ BF_4^- \xrightarrow{OH^-} \text{(118) } 80\text{–}95\%$$

(2.74)

2.4.3 *With Schiff bases*

It also reacts with Schiff bases (**117**) to produce 2-azetidinyleneammonium chlorides which hydrolyse to give 2-azetidinones (**118**) (Scheme 2.75).[85]

$$(CH_3)_2C{=}C{=}\overset{+}{N}(CH_3)_2\ Cl^- + R^1R^2C{=}N{-}R^3\ (117) \xrightarrow[25\,^{\circ}C]{CH_2Cl_2} \left[(CH_3)_2C{=}C(N(CH_3)_2){-}\overset{+}{N}R^3{=}CR^1R^2\ Cl^-\right] \xrightarrow{47\text{–}74\%} \text{azetidinylideneammonium} \xrightarrow[42\text{–}82\%]{OH^-} (118)$$

(2.75)

2.4.4 *With magnesium*

The salt also reacts with magnesium in THF which metallates the enamine with replacement of the chlorine atom. The resulting compounds behave as equivalents of acyl derivatives (Umpolung) (Scheme 2.76).[86]

$(CH_3)_2C{=}C(Cl)N(CH_3)C_6H_6$ —Mg, THF→ $[(CH_3)_2C{=}C(MgCl)N(CH_3)C_6H_6]$ —$(CH_3CO)_2O$→ $(CH_3)_2C{=}C(COCH_3)N(CH_3)C_6H_6$ 59% + $(CH_3)_2C{=}C(H)N(CH_3)C_6H_6$ 28% (2.76)

2.4.5 *With allylic alcohols*

1-Chloro-*N*,*N*-2-trimethylpropenylamine has also proved to be an effective chlorinating reagent in the conversion of an allylic alcohol (**119**) into a chloro compound when accompanied by the Mitsunobu reagent triphenyl phosphorus isothiocyanate. The reaction takes place with inversion of configuration at the chiral centre (Scheme 2.77).[87]

(119) —$(CH_3)_2NC(Cl){=}C(CH_3)_2$, PPh_3NCS→ 59% (2.77)

YNAMINES[88]

The attachment of an amino group to a carbon–carbon triple bond yields ynamines (**120**) analogous to the enamines. Unlike enamines, ynamines are stable particularly when substituted at the β-carbon. Primary and secondary acetylenic amines are expected to be more stable in the nitrile and imine forms,

$$R-C\equiv C-NR_2$$

(120)

respectively, hence ynamines tend only to exist if the nitrogen atom is tertiary (Scheme 2.78).

$$-C\equiv C-NH_2 \longrightarrow -CH_2-CN$$

$$-C\equiv C-NH(R) \longrightarrow -CH=C=N-R \qquad (2.78)$$

In the infra-red the carbon–carbon triple bond stretching frequency is observed at 2220–2240 cm^{-1} as a strong band. Additional conjugation with C=O and C=C bonds causes a bathochromic shift to 2160–2190 cm^{-1}.

The ≡C—H stretching band in free ynamines absorbs at about 3280–3295 cm^{-1} depending on the nature of the substituents at the nitrogen atom.[89]

The ^{1}H-NMR spectrum of free ynamines is characterized by the presence of a singlet at about 2.3δ if the substituents are both alkyl.[89]

2.5 Preparation

2.5.1 *From trichloroenamines*[90]

Free ynamines may be prepared by treatment of lithiated ynamines (**121**) with mild acids such as MeOH, or $CHPh_3$.[91] The appropriate ynamines may themselves be prepared by the action of *n*-butyl lithium on trichloroenamines (**122**) (Scheme 2.79).

$$Cl_2C=CCl-NR^1R^2 \ (122) \xrightarrow{Bu^nLi} Li-C\equiv C-NR^1R^2 \ (121)$$

$$(121) \xrightarrow{MeOH \text{ or } CHPh_3} H-C\equiv C-NR^1R^2$$

$$(121) \xrightarrow{R^3X} R^3-C\equiv C-NR^1R^2$$

R^1, R^2 = Alk, Alk ; Alk, Ar
R^3 = Alk ; X= Br, OTs (2.79)

2.5.2 *From alkyne derivatives*

Chloroalkynes (**123**) give ynamines in high yields on treatment with *N*,*N*-dialkyllithium amides providing that the alkyl group is electronegative (e.g. phenyl or cyclohexen-1-yl) e.g. (Scheme 2.80).[89,92] However, if the alkyl group is *t*-butyl then high temperatures and a noncomplexing solvent are required, e.g. Scheme 2.81.

$$\text{cyclohexen-1-yl}-C{\equiv}C-Cl\ \textbf{(123)} \xrightarrow[\text{ether, }-80^{\circ}C]{LiNEt_2} \text{cyclohexen-1-yl}-C{\equiv}C-NEt_2\ \ 60\% \qquad (2.80)$$

$$(CH_3)_3C-C{\equiv}C-Cl \xrightarrow[\text{HMPT, }110^{\circ}C]{LiN(CH_3)_2} (CH_3)_3C-C{\equiv}C-N(CH_3)_2\ \ 70\% \qquad (2.81)$$

Easily accessible acetylenic ethers (**124**) have also been used in syntheses of *C*-alkylynamines (except when R = Me) by heating with lithium dialkylamides, e.g. Scheme 2.82.[93]

$$C_2H_5-C{\equiv}C-OEt\ \textbf{(124)} \xrightarrow[110\text{-}120^{\circ}C]{LiN(C_3H_7)_2} C_2H_5-C{\equiv}C-N(C_3H_7)_2\ \ 67\% \qquad (2.82)$$

2.5.3 *From haloethenes*

Instead of using fluoroalkynes or reactive chloroalkynes it is possible to use compound (**125**), which may be considered as a precursor to halogenoalkynes under the conditions of reaction employed (Scheme 2.83).[94]

$$(CH_3)_3C-CHBr-CBrF_2\ \textbf{(125)} \xrightarrow[-2HBr;\ -F]{3LiNEt_2} (CH_3)_3C-C{\equiv}C-NEt_2\ \ 48\%$$

$$FHC{=}CCl_2 \xrightarrow[\text{ether, }-80^{\circ}C]{LiNEt_2\ (3\text{ mol})} Et_2N-C{\equiv}C-NEt_2 \qquad (2.83)$$

Phenylynamines (**126**) may be prepared by the direct action of lithium dialkylamides on the appropriate dihaloethenes (**127**) (Scheme 2.84)[95] or dihaloenamines.[96] By this method *t*-butylynamines may also be prepared. In practice it has been found to be easier to use the 'dihydrobromide' of *t*-butyl fluoroalkyne (**125**), as shown above (Scheme 2.83).

$$C_6H_5{-}CH{=}CClF \;(127) \xrightarrow{2\ LiNEt_2} C_6H_5{-}C{\equiv}C{-}NEt_2 \;(126)\ 86\ \% \qquad (2.84)$$

2.5.4 *From N,N-dialkylprop-2-ynylamines*

Prototropic isomerization of *N*,*N*-dialkylprop-2-ynylamines (**128**) has been found to furnish good yields of the corresponding ynamines,[97] e.g. Scheme 2.85.

$$HC{\equiv}C{-}CH_2{-}NR_2 \;(128) \xrightarrow[PhH]{KNH_2\,/\,Al_2O_3} CH_3{-}C{\equiv}C{-}NR_2 \qquad (2.85)$$

2.5.5 *From α-haloiminium salts*

Lithium dialkylamines have been used for the dehydrohalogenation of α-haloiminium salts (**129**) to produce ynamines (Scheme 2.86).[98]

$$R^1{-}CH_2{\cdot}CCl{=}\overset{+}{N}R^2{}_2\ Cl^- \;(129) \underset{2HCl}{\overset{LiNR^3{}_2}{\rightleftharpoons}} R^1{-}C{\equiv}C{-}NR^2{}_2$$

R^1 = H, Me, Pr^n
R^2 = Et, pyrrolidino
R^3 = Me, Et, c-C_6H_{11} (2.86)

2.5.6 *From ketene-S,N-acetals*

Thiol elimination from ketene-*S*,*N*-acetals (**130**) with base has also proved successful (Scheme 2.87).[98]

$$R{-}CH{=}C(SR){-}N(CH_2CH_2)_2O \;(130) \xrightarrow[-RSH]{LiNEt_2} R{-}C{\equiv}C{-}N(CH_2CH_2)_2O \qquad (2.87)$$

2.5.7 *From amines and phenylacetylene*

Under catalysis by cupric acetate monohydrate, air oxidation of phenylacetylene and a secondary amine can be effected to produce ynamines (Scheme 2.88).[99]

$$\text{Ph—C}\equiv\text{C—H} + \text{HNMe}_2 \xrightarrow[\text{benzene}]{\text{Cu(OAc)}_2/\text{O}_2} \underset{40\%}{\text{Ph—C}\equiv\text{C—NMe}_2} \qquad (2.88)$$

2.6 Reactivity

Like enamines, ynamines are nucleophilic and therefore react with electrophiles. The readiness with which they do this varies with the substituents which they carry but is usually quite evident. However, the reactivity of $R_2N{-}C{\equiv}CH$ is very great. For a detailed discussion of the effect of substituents on the stability of ynamines see the review by Ficini.[100]

2.6.1 *Hydrolysis*

Ynamines react with aqueous acid to yield amides (**131**) (Scheme 2.89). The substituents have an effect on the rate of reaction. However, some ynamines are hydrolysed by water (e.g. *N*,*N*-diethylaminopropyne).

R¹₂N̈—C≡C—R² + H₂Ö⁺—H → R¹₂N⁺=C=C(H)—R² + H₂O: → R¹₂N—C(OH)=C(H)—R² (Enol) → (Tautomerism) → R¹₂N—C(=O)—CH₂—R² (131) (2.89)

2.6.2 *Cycloaddition reactions*

The addition of ynamines to unsaturated systems in a Diels–Alder reaction provides access to a variety of cyclic and acyclic enamine systems,[100,101] e.g. Scheme 2.90.

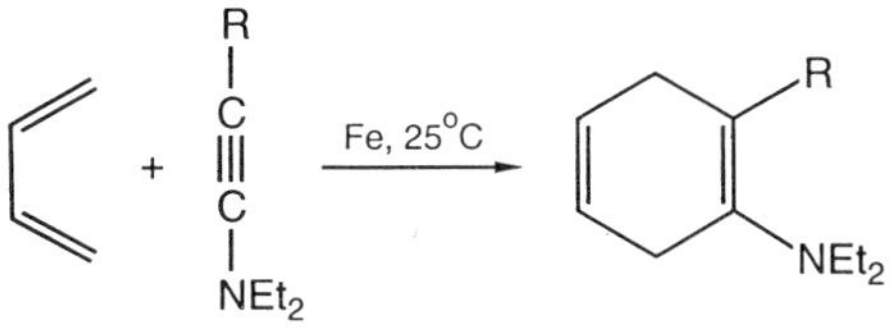

(R = Me, Pr^i, c-C_5H_{11}) (2.90)

The 1,2- and 1,3-cycloaddition reactions of *N*,*N*-diethylaminopropyne (**132**)[102] have been widely studied. For instance, it reacts stereospecifically with 5-methyl-1,2-cyclohexenone (**133**) to give a bicyclic enamine (**134**) which contains three chiral carbon atoms (Scheme 2.91).[103] A similar reaction occurs with fluoroalkenes (**135**) (Scheme 2.92).[104] Chloro- and bromo-trifluoroethylene produce only cyclobutenes (**136**); perfluorobutene-2 forms only the conjugated diene (**137**). However, in most cases a mixture of the two products is formed.

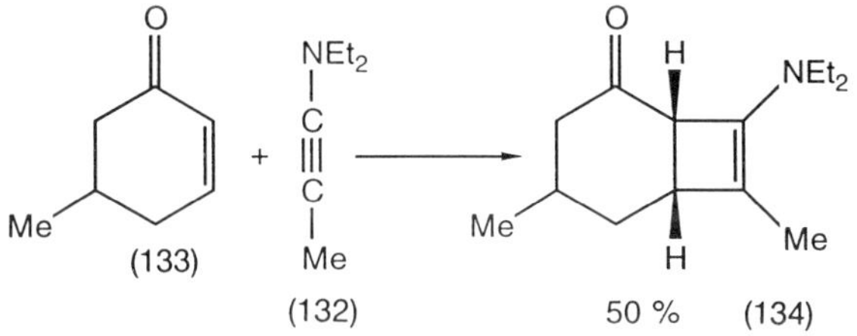

(133) (132) 50 % (134) (2.91)

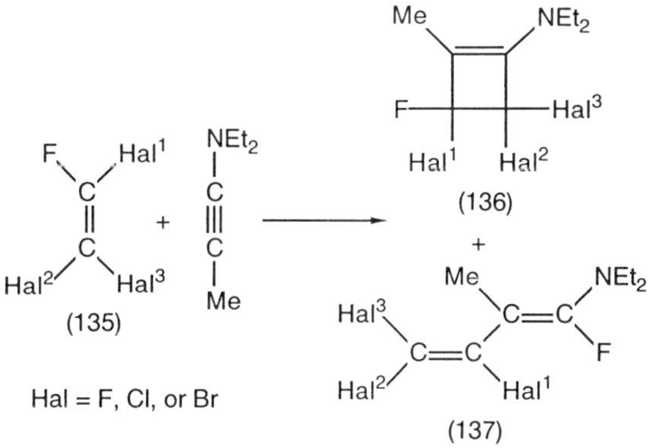

(135) Hal = F, Cl, or Br (136) + (137) (2.92)

References

1. S. F. Dykes, *The chemistry of enamines*, Cambridge University Press, 1973.
2. P. W. Hickmott, *Tetrahedron*, 1982, **38**, 1975 (Part 1); 3363 (Part 2).
3. G. Opitz, H. Hellmann, and H. W. Schubert, *Annalen*, 1959, **623**, 112; 117.
4. M. E. Kuehne and T. Garbacik, *J. Org. Chem.*, 1970, **35**, 1555; K. Nagarajan and S. Rajappa, *Tetrahedron Lett.*, 1969, 2293; P. W. Hickmott, P. J. Cox, and G. A. Sim, *J. Chem. Soc., Perkin Trans. I*, 1974, 2544.
5. R. Stradi, D. Pocar, and C. Cassio, *J. Chem. Soc., Perkin Trans. I*, 1974, 2671.
6. C. Mannich and H. Davidson, *Chem. Ber.*, 1936, **69B**, 2106.
7. G. Stork, A. Brizzolara, H. Landesman, J. Szmuszkovicz, and R. Tyrrell, *J. Am. Chem. Soc.*, 1963, **85**, 207; J. K. Whitesell and S. W. Fellman, *J. Org. Chem.*, 1977, **42**, 1663.
8. *F. &F.*, **4**, 345; K. Taguchi and F. H. Westheimer, *J. Org. Chem.*, 1971, **36**, 1570.
9. *F. &F.*, **2**, 414; W. A. White and H. Weingarten, *J. Org. Chem.*, 1967, **32**, 213; R. Carlson, R. Phan-Tan-Luu, D. Mathieu, F. S. Anouande, A. Babadjamian, and J. Metzger, *Acta Chem. Scand. B*, 1978, **32**, 335.
10. *F. &F.*, **12**, 500; R. Carlson, A. Nilsson, and M. Stromqvist, *Acta Chem. Scand.*, 1983, **37B**, 7.
11. D. C. Bradley and T. M. Thomas, *J. Chem. Soc.*, 1960, 3857.
12. *F. &F.*, **5**, 709; R. A. Pike and R. L. Schank, *J. Org. Chem.*, 1962, **27**, 2190.
13. R. Comi, R. W. Frank, M. Reitano, and M. S. Weinreb, *Tetrahedron Lett.*, 1973, 3107.
14. R. Comi, R. W. Franck, M. Reitano, and M. S. Weinreb, *Tetrahedron Lett.*, 1973, 3107.
15. *F. &F.*, **9**, 381.
16. H. Albrecht, W. Raab, and C. Vonderheid, *Synthesis*, 1979, 127.
17. S. F. Martin and R. Gompper, *J. Org. Chem.*, 1974, **39**, 2814.
18. J. C. Gilbert and U. Weerasooriya, *Tetrahedron Lett.*, 1980, 2041.
19. *F. &F.*, **12**, 334; N. L. J. M. Brockhof and A. van der Gen, *Recl. Trav.*, 1984, **103**, 305; N. L. J. M. Brockhof, P. van Elburg, D. J. Hoff, and A. van der Gen, *Recl. Trav.*, 1984, **103**, 317.
20. *F. &F.*, **12**, 50; S. I. Murahashi, T. Tsumiyama, and Y. Mitsue, *Chem. Lett.*, 1984, 1419.
21. M. E. Herr and F. W. Heyl, *J. Am. Chem. Soc.*, 1952, **74**, 3627.
22. P. M. Greaves and S. R. Landor, *J. Chem. Soc., Chem. Commun.*, 1966, 322.
23. J. Barluenga and F. Aznar, *Synthesis*, 1975, 704; J. Barluenga, F. Aznar, R. Liz, and R. Rodes, *J. Chem. Soc., Perkin Trans. I*, 1980, 2732.
24. M. L. Farmer, W. E. Billups, R. B. Greenlee, and A. N. Kurtz, *J. Org. Chem.*, 1966, **31**, 2885.
25. J. C. Craig and N. N. Ekwuribe, *Tetrahedron Lett.*, 1980, 2587.
26. *F. &F.*, **11**, 432; L. J. de Noten and L. Brandsman, *Recl. Trav.*, 1981, **100**, 244.
27. *F. &F.*, **9**, 164; A. D. DeWit, M. L. M. Pennings, W. P. Trompenaars, D. N. Reinhoudt, S. Harchema, and C. Nevestveit, *J. Chem. Soc., Chem. Commun.*, 1979, 993.
28. J. P. Genet and J. Ficini, *Tetrahedron Lett.*, 1979, 1499.
29. *F. &F.*, **11**, 103; L. Brandsman, P. E. van Rign, H. D. Verkruÿsee, and P. von R. Schleyer, *Angew. Chem., Int. Ed. Engl.*, 1982, 862.

30. *F. &F.*, **5**, 247; C. Hansson and B. Wickberg, *J. Org. Chem.*, 1973, **38**, 3074.
31. P. F. Hudlrik, A. M. Hudlrik, and A. K. Kulkarni, *Tetrahedron Lett.*, 1985, **26**, 139.
32. *F. &F.*, **6**, 587.
33. R. B. Boar, J. F. McGhie, M. Robinson, D. C. Horwell, and R. V. Stick, *J. Chem. Soc., Perkin Trans.* I, 1975, 1237.
34. W. Maas, M. J. Janssen, E. J. Stamhuis, and H. Wynberg, *J. Org. Chem.*, 1967, **32**, 1111; P. V. Sollenberger and R. B. Martin, *J. Am. Chem. Soc.*, 1970, **92**, 4261.
35. L. Alais, R. Michelot, and B. Tchoubar, *C. R. Acad. Sci. Paris C.*, 1971, **273**, 261; L. Nilsson, R. Carlson, and C. Rappe, *Acta Chem. Scand. B*, 1976, **30**, 271.
36. J. Joska, J. Fajkos, and F. Sorm, *Collect. Czech. Chem. Commun.*, 1961, 1646.
37. J. W. Lewis and A. A. Pearce, *J. Chem. Soc. (B)*, 1969, 863.
38. *F. &F.*, **15**, 43; B. Singaram, C. T. Goralski, M. V. Rangaishenui, and H. C. Brown, *J. Am. Chem. Soc.*, 1989, **111**, 384.
39. *F. &F.*, **2**, 425; G. H. Alt and A. J. Speziale, *J. Org. Chem.*, 1966, **31**, 1340.
40. C. Mannich and H. Davidson, *Ber.*, 1936, **69B**, 2106.
41. D. C. Horwell and G. H. Timms, *Synth. Commun.*, 1979, **9**, 223.
42. J. O. Madsen and P. E. Iverson, *Tetrahedron*, 1974, **30**, 3493; P. L. De Benneville and J. Macartney, *J. Am. Chem. Soc.*, 1950, **72**, 3073; N. J. Leonard and R. R. Sauers, *J. Am. Chem. Soc.*, 1957, **79**, 6210.
43. D. Redmore, *J. Org. Chem.*, 1978, **43**, 992.
44. R. F. Borch, M. D. Bernstein, and H. D. Durst, *J. Am. Chem. Soc.*, 1971, **93**, 2897.
45. J. A. Marshall and W. S. Johnson, *J. Org. Chem.*, 1963, **28**, 421; J. Schmitt, J. J. Panouse, A. Hallot, P. J. Cornu, J. A. Comoy, and H. Pluchet, *Bull. Soc. Chim. Fr.*, 1963, 798.
46. *F. &F.*, **3**, 9; J. M. Coulter, J. W. Lewis, and P. P. Lynch, *Tetrahedron*, 1968, **24**, 4489.
47. S. Carlsson and S.-O. Lawesson, *Tetrahedron*, 1980, **36**, 3585.
48. G. S. Ponticello and J. J. Baldwin, *J. Org. Chem.*, 1979, **44**, 4003.
49. T. J. Curphey, J. C. Hung, and C. C. C. Chu, *J. Org. Chem.*, 1975, **40**, 607.
50. G. Stork, A. Brizzolara, H. Landesman, I. Szmuszkovicz, and R. Tyrrell, *J. Am. Chem. Soc.*, 1963, **85**, 207.
51. *F. &F.*, **7**, 301; E. Buncel and B. Menton, *J. Chem. Soc., Chem. Commun.*, 1976, 648.
52. J. K. Whitesall and S. W. Felman, *J. Org. Chem.*, 1977, **42**, 1663.
53. M. Barthélémy, J.-P. Monthéard, and Y. Bessière-Chrétien, *Bull. Soc. Chim. Fr.*, 1969, 2725.
54. G. Stork, A. Brizzolara, H. Landesman, I. Szmuszkovicz, and R. Tyrrell, *J. Am. Chem. Soc.*, 1963, **85**, 207.
55. M. E. Jung, *Tetrahedron*, 1976, **32**, 1; B. P. Mundy, *J. Chem. Educ.*, 1973, **50**, 110.
56. A. R. Surrey, *Name reactions in organic chemistry*, Academic Press, New York, 1961, p. 231.
57. For a detailed account: P. W. Hickmott, *Tetrahedron*, 1982, **38**, 3376.
58. G. L. May and J. T. Pinhey, *Aust. J. Chem.*, 1982, **35**, 1859.
59. A. Hantzsch, *Ber.*, 1890, **23**, 1474.
60. M. W. Roomi and S. F. MacDonald, *Can. J. Chem.*, 1970, **48**, 1689.
61. F. Feist, *Ber.*, 1902, **35**, 1558; D. M. McKinnon, *Can. J. Chem.*, 1965, **43**, 2638.
62. A. S. Feliciano, A. Caballero, J. A. P. Pereira, and P. Puebla, *Tetrahedron*, 1989, **45**, 6553.

63. *F. &F.*, **5**, 513; M. De Poortere and F. C. De Schryder, *Tetrahedron Lett.*, 1970, 3949.
64. M. Von Strandtmann, M. P. Cohen, and J. Shavel, *J. Heterocycl. Chem.*, 1970, **7**, 1311.
65. V. Jäger and H. G. Viehe, *Angew. Chem., Int. Ed. Engl.*, 1970, **9**, 795.
66. *F. &F.*, **10**, 409; D. L. Boger and J. S. Panek, *J. Org. Chem.*, 1981, **46**, 2179; D. L. Boger, J. S. Panek, and M. M. Meier, *J. Org. Chem.*, 1982, **47**, 895.
67. M. Regitz and G. Himbert, *Annalen*, 1970, **734**, 70; D. Pocar and P. Trimarco, *J. Chem. Soc., Perkin Trans. I*, 1976, 622.
68. A. T. M. Marcelis and H. C. van der Plas, *Tetrahedron*, 1989, **45**, 2693.
69. *F. &F.*, **1**, 889; D. A. Shepherd, R. A. Donia, J. Allan Campbell, B. A. Johnson, R. P. Holysz, G. Slomp, Jr, J. E. Stafford, R. L. Pederson, and A. C. Ott, *J. Am. Chem. Soc.*, 1955, **77**, 1212.
70. F. Carbani, B. Rindone, and C. Scolastico, *Tetrahedron*, 1973, **29**, 3253; F. Carbani, B. Rindone, and C. Scolastico, *Tetrahedron Lett.*, 1972, 2579.
71. *F. &F.*, **10**, 228; Ž Čescović, J. Bošnjak, and M. Cvetkovic, *Tetrahedron Lett.*, 1980, **21**, 2675.
72. *F. &F.*, **2**, 406; M. E. Kuehne and T. J. Giacobbe, *J. Org. Chem.*, 1968, **33**, 3359.
73. T.-L. Ho, *Synth. Commun.*, 1974, **4**, 135.
74. *F. &F.*, **3**, 314; H. O. House and F. A. Richey, Jr, *J. Org. Chem.*, 1969, **34**, 1430.
75. C. S. Foote, A. A. Dzakapasu, and J. W. Lin, *Tetrahedron Lett.*, 1975, 1247.
76. W. Ando, T. Saiki, and T. Migita, *J. Am. Chem. Soc.*, 1975, **97**, 5028.
77. H. H. Wasserman and S. Terao, *Tetrahedron Lett.*, 1975, 1735.
78. P. D. Croce and R. Stradi, *Tetrahedron*, 1977, **33**, 865.
79. *F. &F.*, **2**, 336; T. C. Shields, W. E. Billups, and A. N. Kurtz, *Angew. Chem., Int. Ed. Engl.*, 1968, **7**, 209.
80. Preparation: B. Haveaux, A. Dekoker, M. Rens, A. R. Sidani, J. Toye, and L. Ghosez, *Org. Synth.*, 1980, **59**, 26.
81. *F. &F.*, **4**, 94; J. March-Brynaert and L. Ghosez, *J. Am. Chem. Soc.*, 1972, **94**, 2869.
82. *F. &F.*, **4**, 94; J. March-Brynaert and L. Ghosez, *J. Am. Chem. Soc.*, 1972, **94**, 2871.
83. *F. &F.*, **5**, 136; J. March-Brynaert and L. Ghosez, *Angew. Chem., Int. Ed. Engl.*, 1974, **13**, 267.
84. J. March-Brynaert and L. Ghosez, *J. Am. Chem. Soc.*, 1972, **94**, 2871; *F. &F.*, **6**, 122.
85. *F. &F.*, **5**, 136; M. De Poortere, J. March-Brynaert, and L. Ghosez, *Angew. Chem., Int. Ed. Engl.*, 1974, **13**, 267.
86. *F. &F.*, **7**, 66; C. Wiaux-Zamar, J.-P. Dejonghe, L. Ghosez, J. F. Normant, and J. Villieras, *Angew. Chem., Int. Ed. Engl.*, 1976, **15**, 371.
87. *F. &F.*, **12**, 123; A. B. Holmes, C. L. D. Jennings-White, and D. A. Kendrick, *J. Chem. Soc., Chem. Commun.*, 1983, 415; T. Fujisawa, T. Mori, K. Higuchi, and T. Sato, *Chem. Lett.*, 1983, 1791.
88. S. Patai, *Supplement F—The chemistry of amino, nitro and nitrosocompounds and their derivatives*, Part I, Wiley-Interscience, Chichester, p. 623; review of use in synthesis (contains references to preparation and properties): J. Ficini, *Tetrahedron*, 1976, **32**, 1449; other reviews for preparations: H. G. Viehe, *Angew. Chem., Int. Ed. Engl.*, 1967, **6**, 767; *Chemistry of acetylenes*, Marcel Dekker, NY, 1969, p. 861.

89. H. G. Viehe, *Chemistry of acetylenes*, Marcel Dekker, New York, 1969, Chapter 12.
90. J. Ficini and C. Barbara, *Compt. Rend.*, 1967, **265**, 1496.
91. J. Ficini and C. Barbara, *Bull. Soc. Chim. Fr.*, 1965, 2787.
92. H. G. Viehe, *Angew. Chem., Int. Ed. Engl.*, 1967, **6**, 767.
93. P. P. Montijin, E. Harryvan, and L. Brandsma, *Recl. Trav. Chim.*, 1964, **83**, 1211.
94. H. G. Viehe, *Angew. Chem.*, 1963, **75**, 638; *Angew. Chem., Int. Ed. Engl.*, 1963, **2**, 477.
95. H. G. Viehe and M. Reinstein, *Angew. Chem., Int. Ed. Engl.*, 1964, **3**, 506.
96. J. Ficini and C. Barbara, *Bull. Soc. Chim. Fr.*, 1965, 2787.
97. A. J. Hubert and H. G. Viehe, *J. Chem. Soc.*, 1968, 288.
98. R. Buijle, A. Halleux, and G. H. Viehe, *Angew. Chem., Int. Ed. Engl.*, 1966, **5**, 584.
99. *F. &F.*, **2**, 84, L. I. Peterson, *Tetrahedron Lett.*, 1968, 5357.
100. J. Ficini, *Tetrahedron*, 1976, **32**, 1449.
101. P. W. Hickmott, *Tetrahedron*, 1982, **38**, 1992; H. Naunhoffer and G. Werner, *Ann. Chem.*, 1974, 1190; J. C. Martin, *Heterocycl. Chem.*, 1980, **17**, 1111; A. T. M. Marcelis and H. C. van der Plas, *J. Org. Chem.*, 1986, **51**, 67; A. T. M. Marcelis, H. C. van der Plas, and S. Harkema, *J. Org. Chem.*, 1985, **50**, 270.
102. Preparation: *F. &F.*, **2**, 133; H. Eilingsfeld, M. Sefelder, and H. Weidinger, *Ber.*, 1963, **96**, 2671; R. Buijle, A. Halleux, and H. G. Viehe, *Angew. Chem., Int. Ed. Engl.*, 1966, **5**, 584,
103. *F. &F.*, **5**, 219; J. Ficini and A. M. Touzin, *Tetrahedron Lett.*, 1972, 2093.
104. J. C. Blazejewski, D. Cantacuzene, and C. Wakselman, *Tetrahedron Lett.*, 1974, 2055.

3

IMINES

Imines (**1**) (also called azomethines and, sometimes, Schiff's bases) and iminium ions (**2**) contain a carbon–nitrogen double bond. The C=N bond distance in

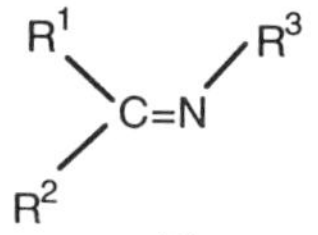

$R^1R^2C{=}NR^3$ (1) $R^1R^2C{=}N^+R^3R^4$ X^- (2)

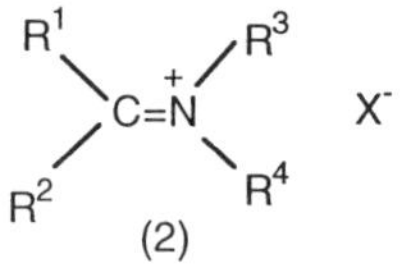

imines lies between 1.24 and 1.28 Å and the NMR signal of the hydrogen atom in imines derived from aldehydes (aldimines), RCH=NR occurs at δ_H8.1–8.7 ppm (by comparison, in aldehydes, RCHO, H is found at δ_H 9.6–9.8 ppm). The known molecular dimensions are as detailed below:

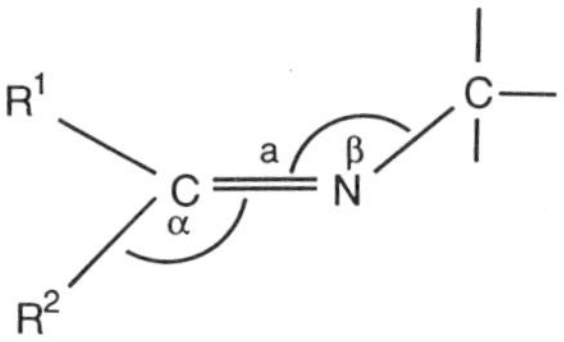

Cf. $R^1R^2C{=}O$ (1.20 Å)

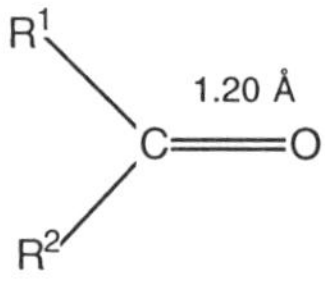

a = 1.24 - 1.28 Å

$\alpha = \beta = 120^\circ$

The imino group is polarized as shown in Scheme 3.1 and its dipole moment is 0.9 D (cf. C=O, 2.3 D). The infra-red absorption of the C=N bond is at 1660–1670 cm^{-1}. The stereochemistry is trigonal with 120° angles leading to diastereoisomerism, i.e. *E* and *Z* isomers. It is not usually possible to isolate *E* and *Z* isomers, but NMR spectroscopy shows either the more stable isomer (*E* and *anti* for aldimines) or a mixture (equilibrium) of both forms (Scheme 3.2)[1] is present under normal conditions.

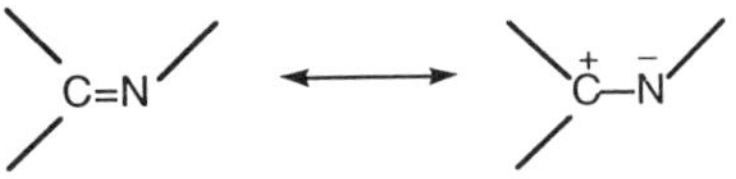

$$>C{=}N- \longleftrightarrow >\overset{+}{C}{-}\overset{-}{N}- \qquad (3.1)$$

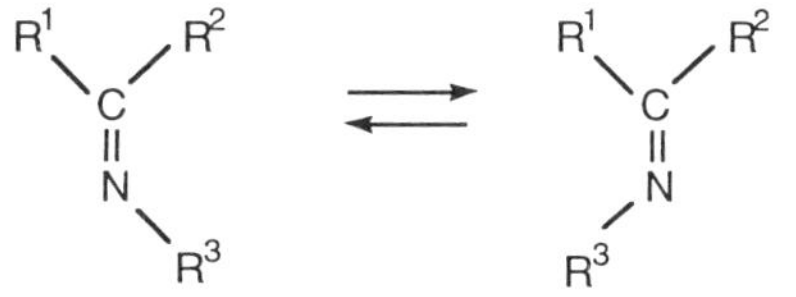

(3.2)

3.1 Preparation

3.1.1 *By condensation of aldehydes or ketones with a primary amine*

Imines may be prepared by a condensation reaction between a primary amine and either an aldehyde or a ketone, (Scheme 3.3).

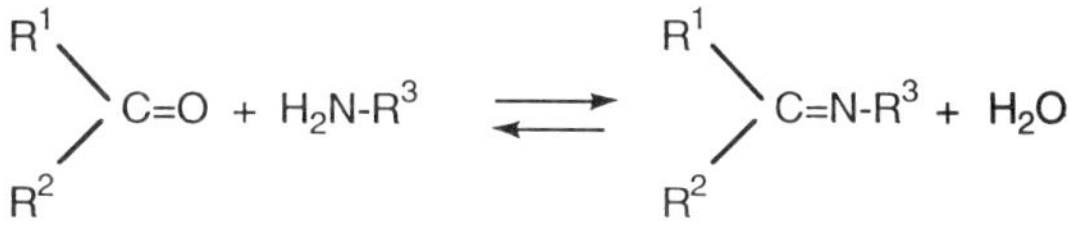

$$R^1R^2C{=}O + H_2N\text{-}R^3 \rightleftharpoons R^1R^2C{=}N\text{-}R^3 + H_2O \qquad (3.3)$$

A carbinolamine (**3**) is an intermediate in the mechanism (Scheme 3.4).

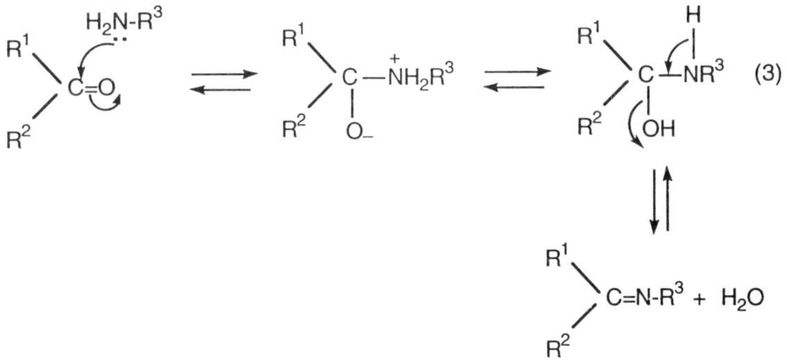

(3.4)

Effect of structure on stability

The substituent groups can have a large effect on the position of the equilibrium shown in Scheme 3.4. Imines are more stable if the substituent groups (R^1, R^2, R^3) are aryl. Aldehydes ($R^1 = H$) give more stable imines than ketones do, with the result that the equilibrium lies further to the right where aldehydes are concerned. The variable stabilities of imines stem from effects on the kinetics and on the equilibria of the steps in the mechanism of formation.

The reaction between an optically active primary amine (**4**) and an α-alkyl-substituted cyclohexanone (**5**) to yield an optically active imine, has been used to prepare optically pure *cis*-cyclohexylamines (**6**) (Scheme 3.5).[2]

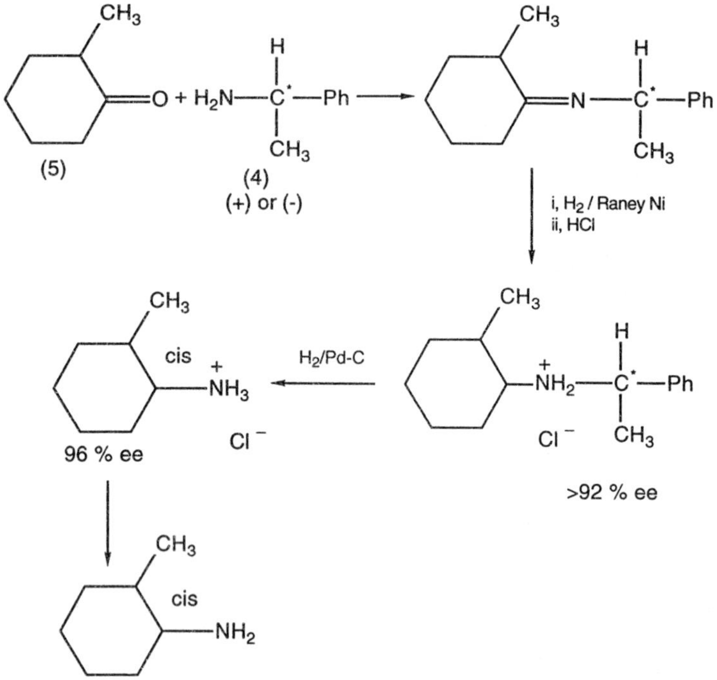

(3.5)

Procedures to aid imine formation

Removal of water moves the equilibrium to the right and this is often necessary in order to obtain a good yield of imine. Dehydration can be effected by distilling the water out of the reaction mixture, azeotropic distillation (e.g. with benzene or toluene), or by the addition of a molecular sieve of type 4A to absorb the water.[3]

Catalysts are often used to accelerate the setting up of the equilibria, for example, the formation of imines of hindered cyclic ketones (**7**), such as camphor, is facilitated by the use of titanium tetrachloride ($TiCl_4$) as catalyst (Scheme 3.6).[4]

Similarly titanium tetrachloride has been found to be useful as a catalyst for the preparation of alkyl imines of benzophenones (**8**) (Scheme 3.7).[5]

Di-*n*-butyl tin chloride (Bu_2SnCl_2) has been found to be a good neutral catalyst for the condensation of ketones with primary amines to yield imines,[6] e.g. Scheme 3.8.

A preparation of 2-aza-1,3-dienes (**9**) has been described[7] in which a trimethylsilyl substituted amine (**10**) is used to produce an intermediate imine (Scheme 3.9).

$$\text{(7)} + R^3NH_2 \xrightarrow[\text{ii, Reflux, 7 hr}]{\text{i, } TiCl_4\text{, PhMe, 0 - 5°C}} \text{2-}R^1\text{-6-}R^2\text{-cyclohexanone }N\text{-}R^3\text{ imine}$$

R^1	R^2	R^3	Yield (%)
H	H	Me	80
H	H	Pr^i	80
Me	H	Me	20
Me	Me	Me	55
Bu^t	Me	H	50
Me	H	$CH_2{=}CHCH_2$	74
Me	H	Pr^i	30
Et	H	$CH_2{=}CHCH_2$	66

(3.6)

$$Ar_2C{=}O + H_2N\text{-}R \xrightarrow[20^\circ C]{TiCl_4 / PhH} Ar_2C{=}N\text{-}R \quad (8)$$

$Ar = 4\text{-}X\text{-}C_6H_4$

R	X	Yield (%)
Me	H	97
Et	H	98
Pr^i	H	96
Bu^t	H	87
Me	Cl	98
Bu^t	Cl	75
Me	Br	95

(3.7)

$$\text{2-}(CH_2CH_2CO_2Me)\text{cyclopentanone} \xrightarrow{R\text{-}NH_2,\ Bu^n_2SnCl_2} \text{2-}(CH_2CH_2CO_2Me)\text{cyclopentanone }N\text{-}R\text{ imine}$$

RNH_2	Yield (%)
$PhCH(NH_2)Me$	82
$EtCH(CH_2OSiMe_3)NH_2$	80

(3.8)

$$R^1\text{-CHO} + H_2N\text{-CH(SiMe}_3)_2 \;(10) \longrightarrow R^1\text{-C(H)=N-CH(SiMe}_3)_2 \xrightarrow{R^2\text{-CHO} / Bu^n_4N^+ F^-} R^1\text{-CH=N-CH=CHR}^2 \;(9)$$

E : *Z* *c.* 1:1

R^1	R^2	Yield (%)
Bu^t	Bu^t	70
Bu^t	4-Me-C_6H_4	78
Bu^t	4-Cl-C_6H_4	65
Bu^t	Ph	70
Bu^t	PhCH=CH	77
Ph	4-MeC_6H_4	66

(3.9)

Acetals, ketals, enol acetates, and enol ethers which readily give carbonyl compounds all react with primary amines to give imines.

3.1.2 *Via α-cyanoamines*

Imines can be prepared indirectly from aldehydes and ketones via α-cyanoamines (**11**) by the action of dicyclohexylcarbodiimide (DCC) and 1,4-diazabicyclo[2.2.2]octane (DABCO)[8] (Scheme 3.10).[9]

$$R^1R^2C{=}O + R^3\text{-}NH_2 + HCN \longrightarrow R^1R^2C(NHR^3)(CN) \;(11) \xrightarrow{\text{DCC, DABCO}} R^1R^2C{=}N{-}R^3$$

75 - 80 %

R^1	R^2	R^3	Yield (%)
Me	Me	H	64
Me	H	Me	73
Me	Me	Me	82
Me	Me	Et	78
Me	Et	Pr^n	65
Me	Me	Pr^n	79
Pr^i	H	Et	77

(3.10)

3.1.3 *Interconversion of imines by exchange with amines*

One imine can be converted into another by exchange with an amine (Scheme 3.11).[10] By a similar exchange hydroxylamine (NH_2OH), semicarbazide ($H_2NCO.NHNH_2$), and hydrazines ($RNHNH_2$) all give good yields of carbonyl derivatives (oximes, semicarbazones, and hydrazones) with imines,[11] for example, Scheme 3.12.

$$R^1_2C{=}N\text{-}R^2 + R^3\text{-}NH_2 \longrightarrow R^1_2C{=}N\text{-}R^3 + R^2\text{-}NH_2 \qquad (3.11)$$

$$Me_2C{=}O + Cl\text{-}C_6H_4\text{-}NH_2 \longrightarrow Cl\text{-}C_6H_4\text{-}N{=}CMe_2 \xrightarrow{NH_2NHCONH_2} H_2NCONHN{=}CMe_2 \; (97\ \%) \qquad (3.12)$$

3.1.4 *Interconversion of imines by exchange with carbonyl compounds*

Interchange of an imine with a carbonyl compound also leads to the formation of a new imine (Scheme 3.13).[12] For example, ketones may be regenerated from oximes (**12**) using formaldehyde (**13**) or acetone (Scheme 3.14).[13] (see Chapter 14, p. 555).

$$R^1_2C{=}N{-}R^2 + R^3_2C{=}O \longrightarrow R^3_2C{=}N{-}R^3 + R^1_2C{=}O \quad (3.13)$$

$$\underset{(12)}{R_2C{=}N{-}OH} + \underset{(13)}{H_2C{=}O} \longrightarrow R_2C{=}O + [CH_2{=}N{-}OH] \xrightarrow{H_2O} H.CO_2H + NH_3 \quad (3.14)$$

3.1.5 *By a Wittig type reaction from an iminophosphorane and an aldehyde*

Just as alkenes can be synthesized by the Wittig reaction from ylides and carbonyl compounds, so imines result from nitrogen-containing ylides (**14**) and carbonyl compounds through a similar betaine intermediate (**15**) (Scheme 3.15).[14] The nitrogen-containing ylide, the iminophosphorane (**14**), is prepared by the reaction between an *N*-chloroamine and triphenylphosphine followed by treatment with *n*-butyllithium (Scheme 3.16).

$$\underset{(14)}{Ph_3P{=}N{-}R^1} + R^2_2C{=}O \longrightarrow \underset{\text{Betaine (15)}}{Ph_3\overset{+}{P}{-}N{-}R^1 \,/\, {}^{-}O{-}CR^2_2} \longrightarrow Ph_3P{=}O + R^1{-}N{=}CR^2_2 \quad (3.15)$$

$$R{-}NHCl + Ph_3P \longrightarrow Ph_3\overset{+}{P}{-}N(R){-}H \xrightarrow{Bu^n{-}Li} \underset{(14)}{Ph_3P{=}N{-}R} \quad (3.16)$$

3.1.6 *By oxidation of amines*

Amines may be oxidized to imines with various metal catalysts,[15] e.g. Scheme 3.17.

$$RCH_2NH_2 \xrightarrow[K_3PO_4]{\text{Cu, Chromite-nickel catalyst}} RCH{=}NH \quad (3.17)$$

The oxidation of amines is difficult to control but some specific oxidizing agents give satisfactory yields of imines from secondary amines. For example, benzylic secondary amines (**16**) can be oxidized to imines with manganese dioxide (Scheme 3.18). This can be compared with the oxidation of benzylic alcohols to aldehydes which is also possible with manganese dioxide.

$$Ar^1\text{-}CH_2\text{-}NHAr^2 \xrightarrow{MnO_2} Ar^1\text{-}CH{=}N\text{-}Ar^2$$

(16) 75-90 % (3.18)

Permanganate ions can be used for the oxidation of secondary amines (**17**) to imines; primary amines (**18**) are oxidized to the respective aldehyde (Scheme 3.19).[16] Other oxidizing agents that can be used include sulphur, amyl disulphide, or selenium.[17]

$$R_2CH\text{-}NHR \xrightarrow{MnO_4^-} R_2C{=}NR$$

(17)

$$RCH_2NH_2 \xrightarrow{MnO_4^-} RCHO$$

(18) (3.19)

Iodosyl benzene (PhIO) in the presence of a ruthenium chloride–triphenylphosphine compound ($RuCl_2\ [PPh_3]_3$) and molecular sieves has been found to oxidize certain secondary amines (**19**) to imines in good yields (Scheme 3.20). The amines must have an aromatic ring or a carbon–carbon double bond in the α-position to the C—H bond in order to be oxidized. Primary amines can be oxidized in this way and aldehydes are obtained by *in situ* hydrolysis of the intermediate imine. Two possible mechanisms for the oxidation are advanced by the authors.[18]

$$R^1\text{-}CH_2\text{-}NHR^2 \xrightarrow[\text{Molecular sieve / }20^\circ\text{ C under }N_2]{PhIO\ /\ RuCl_2(PPh_3)_3\ /\ CH_2Cl_2} R^1\text{-}CH{=}N\text{-}R^2$$

(19)

R^1	R^2	Yield (%)
Ph	Ph	77
Ph	Bn	93
PhCH=CH	Ph	86

(3.20)

The di-*t*-butyliminoxyl radical, generated from (**20**), can be used for the oxidation of primary and secondary amines (Scheme 3.21).[19]

Tertiary butyl peroxide (Bu^tOOH) can also be used with a ruthenium catalyst for this oxidation,[20] for example, Scheme 3.22.

Diphenylselenium bis(trifluoroacetate) (**21**) has been used to oxidize secondary amines to imines. The method has been used in the synthesis of large biomimetic molecules, such as 1-substituted 1,2,3,4-tetrahydroisoquinolines.[21] The reagent is prepared as shown in Scheme 3.23.

$$\mathrm{R^1\!-\!C(Ph)(H)\!-\!NHR^2} \xrightarrow[\text{In pentane, Ce catalyst}]{\mathrm{Bu^t_2C{=}NOH,\ (20)}} \mathrm{Ph(R^1)C{=}NR^2}$$

R^1	R^2	Yield (%)
Ph	H	79
Me	H	42
H	Bn	78
H	H	68

(3.21)

$$\mathrm{Ph(H)CH\text{-}NH\text{-}CH_2Ph} \xrightarrow{\text{Ru catalyst, } \mathrm{Bu^tO_2H}} \mathrm{Ph(H)C{=}N{-}CH_2Ph}\quad 80\ \%$$

(3.22)

$$\mathrm{Ph_2SeO} + \mathrm{(CF_3CO)_2O} \xrightarrow{\text{DME, 0°C}} \mathrm{Ph_2Se{:}(OCOCF_3)_2}\ (21)$$

(3.23)

Hypochlorites oxidize both primary and secondary amines to imines via an *N*-chloramine intermediate (**22**) (Scheme 3.24).[22] Hypochlorites have been shown to oxidize α-amino acids (**23**) with simultaneous decarboxylation (Scheme 3.25).

$$\mathrm{Me\text{-}CH(R)\text{-}NH_2} \xrightarrow[\mathrm{Na_2CO_3,\ dry\ ether}]{\mathrm{dry\ Bu^tOCl}} \underset{(22)}{\mathrm{Me\text{-}CH(R)\text{-}NHCl}} \xrightarrow{\mathrm{NaOEt}} \mathrm{Me\text{-}C(R){=}NH}$$

(3.24)

$$\underset{(23)}{\mathrm{R^1\text{-}C(H)(CO_2H)\text{-}NHR^2}} \xrightarrow{\mathrm{NaOCl}} \mathrm{R^1\text{-}CH(N(Cl)R^2)\text{-}C({=}O)O^-} \longrightarrow \mathrm{R^1\text{-}CH{=}N\text{-}R^2} + \mathrm{Cl^-} + \mathrm{CO_2}$$

(3.25)

Oxidation of a primary amine with sodium persulphate in aqueous alkali catalysed by silver nitrate yields an imine (Scheme 3.26). A number of primary amines can be oxidized via the imine in this manner with yields ranging between 15 and 95 per cent.[23]

$$\left[CH_3CH_2\text{-}CH{=}NH \xrightarrow{H_2O} CH_3CH_2CH{=}O \quad H_2\ddot{N}\text{-}CH_2CH_2CH_3\right] \xrightarrow{-H_2O}$$

$$CH_3CH_2CH_2\text{-}NH_2 \xrightarrow[AgNO_3\ /\ aq.\ HO^-]{Na_2S_2O_8} CH_3CH_2\text{-}CH{=}N\text{-}CH_2CH_2CH_3 \qquad (3.26)$$

The oxidation of secondary amines (**24**) with a ruthenium compound, viz. tertiary butyl hydroperoxide-dichlorotris(triphenylphosphine)ruthenium(II), yields imines (Scheme 3.27).[24]

$$\underset{(24)}{R^1\text{-}CH_2\text{-}NHR^2} \xrightarrow[\text{in benzene}]{Ru^{II}\text{ complex}} R^1\text{-}CH{=}N\text{-}R^2 \qquad (3.27)$$

3.1.7 *From nitriles and Grignard reagents*

The reaction of a Grignard reagent (**25**) with a nitrile (**26**) is one of the best methods for obtaining imines without a substituent on the nitrogen atom (**27**) (Scheme 3.28).[25]

$$\underset{(25)}{R^1\text{-}MgHal} + \underset{(26)}{R^2\text{-}C\equiv N} \longrightarrow (R^1)(R^2)C{=}N\text{-}MgHal^+ \xrightarrow[\text{ether}]{\text{dry HCl}} \underset{(27)}{(R^1)(R^2)C{=}NH} \qquad (3.28)$$

3.2 Reactivity

The polarization of the imino carbon–nitrogen double bond, C=N (Scheme 3.29), as for the C=O bond, leads to the addition of nucleophiles to carbon and of electrophiles to nitrogen and is responsible for the wide application of imines and iminium salts in organic synthesis and in living organisms. In addition, this same polarization leads to α-hydrogen atoms having acidic properties and consequently they can be removed by suitable bases.

$$>C{=}N\!-\ \longleftrightarrow\ >\overset{+}{C}\text{-}\overset{-}{N}\!- \qquad (3.29)$$

$$\ddot{B} \rightarrow H\text{-}\overset{|}{\underset{|}{C}}\text{-}\overset{|}{C}{=}N\!-$$

3.2.1 *Electrophilic reactivity of carbon*[26]

Hydrolysis

Imines are very readily hydrolysed, hence, as noted above, it is necessary to remove water in their preparation. In thin layer chromatography on silica plates, two or three spots are usually seen: sometimes there is complete hydrolysis to the amine and aldehyde, hence two spots arise. More often, the hydrolysis is partial and the imine is also seen. Very few imines are completely stable on silica plates.

The mechanism of hydrolysis of imines is the reverse of the mechanism of their formation, being a two-step mechanism through the carbinolamine (**28**) (Scheme 3.30). The rate varies with pH while the rate-determining step varies both with pH and with the structure of the imine. The rate-determining step can be attack of the hydroxyl ion on the imine (Scheme 3.31) or attack of water on the protonated imine (Scheme 3.32) or even decomposition of the carbinol-amine, the final stage of Scheme 3.32.

$$>C{=}N{-}R + H_2O \rightleftharpoons >C(OH){-}NHR \ (28) \rightleftharpoons >C{=}O + RNH_2 \quad (3.30)$$

$$>C{=}N{-}R + {}^{-}OH \underset{}{\overset{rds}{\rightleftharpoons}} >C(OH){-}\bar{N}{-}R \overset{H^+,\ fast}{\rightleftharpoons} >C(OH){-}N(H){-}R \xrightarrow{fast} >C{=}O + R{-}NH_2 \quad (3.31)$$

$$>C{=}N{-}R + H_3O^+ \overset{fast}{\rightleftharpoons} >C{=}\overset{+}{N}HR + H_2\ddot{O} \overset{rds}{\rightleftharpoons} >C(OH){-}N(H){-}R + H^+ \xrightarrow{fast} >C{=}O + R{-}NH_2 \quad (3.32)$$

Addition of hydrogen cyanide

The addition of hydrogen cyanide occurs in the same way as for aldehydes and ketones (Scheme 3.33) and the reaction has been applied to the synthesis of α-amino acids. An *N*-benzylimine (**29**) is used to facilitate the final hydrogenolysis (Scheme 3.34).[27]

$$R^1R^2C{=}N{-}R^3 \xrightarrow{HCN} R^1R^2C(C{\equiv}N){-}NHR^3 \quad (3.33)$$

$$R^1\text{-CH=N-}\underset{\displaystyle Ar}{\overset{\displaystyle R^2}{C}}\text{-H} \xrightarrow{HCN} R^1\text{-}\underset{\displaystyle C\equiv N}{CH}\text{-NH-}\underset{\displaystyle Ar}{\overset{\displaystyle R^2}{C}}\text{-H} \xrightarrow{aq.\ H_3O^+} R^1\text{-}\underset{\displaystyle CO_2H}{CH}\text{-NH-}\underset{\displaystyle Ar}{\overset{\displaystyle R^2}{C}}\text{-H} \xrightarrow{H_2 / Pd\text{-}C} R^1\text{-}\underset{\displaystyle CO_2^-}{CH}\text{-}\overset{+}{N}H_3 \quad (3.34)$$

(29)

Addition of bisulphite

The addition of bisulphite to imines occurs as for aldehydes and ketones,[28] e.g. Scheme 3.35. The resulting sodium salt of the sulphonic acid can be used for syntheses as shown in Scheme 3.35.

$$\text{Ph-CH=N-Ph} \xrightarrow{aq.\ NaHSO_3} \text{Ph-}\underset{\displaystyle SO_3^-\ \overset{+}{Na}}{CH}\text{-NHPh} \xrightarrow{aq.\ H_3O^+} \text{Ph-}\underset{\displaystyle SO_3^-}{CH}\text{-}\overset{+}{N}H_2Ph$$

An α-aminosulphonic acid

90 %

$$\text{Ph-}\underset{\displaystyle SO_3^-\ \overset{+}{Na}}{CH}\text{-NHPh} \xrightarrow{CH_2(CO_2Et)_2} \text{Ph-}\underset{\displaystyle CH(CO_2Et)_2}{CH}\text{-NHPh} \quad 85\ \% \quad (3.35)$$

With organometallic reagents

With Grignard reagents Grignard reagents (**30**) act as nucleophiles and add to imines (Scheme 3.36).[29] For aldimines this constitutes a general preparation of secondary amines and good yields are obtained when twice the ratio of the Grignard reagent to the imine is used,[30] see Chapter 1, p. 23.

$$\text{Ph-CH=N-R}^1 + \underset{(30)}{R^2\text{-MgHal}} \xrightarrow[\text{at } 0^\circ C]{\text{work-up with dil. HCl}} \text{Ph-}\underset{\displaystyle R^2}{CH}\text{-NHR}^1$$

R^1	R^2	Yield (%)
Me	Et	75
Me	Pr^n	66
Me	Pr^i	60
Et	Pr^n	40
Et	Bu^t	37
Et	Bn	74

(3.36)

With alkyllithium reagents Alkyllithiums react similarly to Grignard reagents,[31] e.g. Scheme 3.37.

$$Ph\text{-}CH{=}NBu^t \xrightarrow[H_3O^+]{CH_3Li} Ph\text{-}CH(CH_3)\text{-}NHBu^t \quad 81\%$$

(3.37)

With tin reagents Aldimines (**31**) react with allyltributyltin compounds (**32**) in the presence of Lewis acids to yield γ,δ-unsaturated secondary amines (**33**) (Scheme 3.38).[32]

$$R^1CH{=}NBn\ (31) + R^2CH{=}CH\text{-}CH_2\text{-}SnBu^n_3\ (32) \xrightarrow[\text{or } TiCl_4]{BF_3.Et_2O} R^1\text{-}CH(NHBn)\text{-}CH(R^2)\text{-}CH{=}CH_2\ (33)$$

R^1	R^2	a(BF_3) or b($TiCl_4$)	Yield (%)
cyclohexyl	H	a	48
cyclohexyl	H	b	81
cyclohexyl	Me	b	78
Ph	H	a	73
Ph	H	b	53

(3.38)

Reduction

Catalytic hydrogenation Imines may be reduced by a variety of reducing agents: catalytic reduction constitutes a good preparation of secondary amines (**34**) (Scheme 3.39). Platinum oxide (PtO_2) and palladium charcoal (Pd–C) are the best catalysts while Raney nickel is less effective,[33] for example, Scheme 3.40.

$$R^1R^2C{=}N\text{-}R^3 \xrightarrow{H_2\ /\ Ni \text{ or } Pt} R^1R^2CH\text{-}NHR^3\ (34)$$

(3.39)

$$(CH_3)_2C{=}N\text{-}C_6H_{11} \xrightarrow[5\%\ Pd\text{-}C,\ 12\ hours]{H_2\ /\ 2\ atm.\ /\ 25^\circ C} (CH_3)_2CH\text{-}NH\text{-}C_6H_{11} \quad 90\%$$

(3.40)

There is no need to isolate the intermediate imine and thus a mixture of an amine and a carbonyl compound (**35**) can be hydrogenated to give good yields, e.g. Scheme 3.41. Similarly, reductive amination of ketones can be achieved, for example, Scheme 3.42. When applied to α-keto acids (**36**) the reduction yields α-amino acids, e.g. Scheme 3.43.

$$CH_3\text{-CO-}CH_2\text{-}CH_3 + CH_3NH_2 \xrightarrow{H_2 / Pt} CH_3\text{-CH(NHCH}_3\text{)-}CH_2\text{-}CH_3 \quad (3.41)$$

(35) → 70%

Cyclohexanone $\xrightarrow{NH_3 / H_2 / Ni}$ cyclohexylamine (H, NH_2) 80 % (3.42)

$$CH_3\text{-CO-}CO_2H \xrightarrow{NH_3 / H_2 / Ni} CH_3\text{-CH(}\overset{+}{N}H_3\text{)-}CO_2^- \quad (3.43)$$

(36) → Alanine

With lithium aluminium hydride Lithium aluminium hydride has frequently been used to reduce imines to secondary amines (Scheme 3.44).[34]

$$R^1R^2C{=}NR^3 \xrightarrow{LiAlH_4} R^1R^2CH\text{-}N(R^3)H \quad (3.44)$$

R^1	R^2	R^3	Yield (%)
$-(CH_2)_5-$		Bn	65
H	PhCH=CH	cyclohexyl	70
Ph	H	cyclohexyl	80

With sodium borohydride or a trimethylamine–borane complex Imines which can be isolated and purified may be reduced by sodium borohydride, or a trimethylamine–borane complex, to secondary amines (Scheme 3.45),[35] see Chapter 1, p. 14.

$$R^1\text{-}CH{=}N\text{-}R^2 \xrightarrow{NaBH_4 / EtOH} R^1\text{-}CH_2\text{-}NHR^2$$

$$R^1\text{-}CH{=}N\text{-}R^2 \xrightarrow{Me_3N.BH_3 / HOAc} R^1\text{-}CH_2\text{-}N(COCH_3)\text{-}R^2 \xrightarrow{\text{hydrolysis}} R^1\text{-}CH_2\text{-}NHR^2$$

(3.45)

Other reducing reagents A method for the conversion of imines into amides (**37**) depends upon the combined action of carbon monoxide and organoboranes (**38**), catalysed by a cobalt carbonyl,[36] e.g. Scheme 3.46.

$$Ph\text{-}C(=O)\text{-}C(Ph){=}N\text{—}Ph + CO + Bu^n{}_3B\ (\mathbf{38}) \xrightarrow{Co_2(CO)_8,\ THF} Ph\text{-}C(=O)\text{-}CH(Ph)\text{—}N(COBu^n)\text{—}Ph\ (\mathbf{37})$$

82 %

(3.46)

Dimethylamineborane (Me_2NHBH_3) has been found to reduce imines to amines in high yield and with increased selectivity, since acids, esters, and nitro groups are not affected by this reagent.[37]

Stereoselective reduction of imines to amines can be carried out using dicyclohexylborane in acetonitrile,[38] for example, Scheme 3.47.

N-benzyl imine of 2-methylcyclohexanone (H, CH_2Ph on N; Me) $\xrightarrow[\text{in } CH_3CN]{(C_6H_{11})_2BH}$ *cis* amine + *trans* amine

79 % yield

cis : *trans* = 4 : 1

(3.47)

Formic acid, as its triethylamine salt, efficiently reduces imines to secondary amines,[39] e.g. Scheme 3.48. In this context imines are intermediate in the Leuckart and Eschweiler–Clarke preparations of amines, see Chapter 1, p. 16.

Potassium graphite (C_8K), which is prepared by melting potassium over graphite at 150–200 °C under argon, reduces imines to amines,[40] e.g. Scheme 3.49.

$$Ph\text{-}CH{=}N\text{-}Ph \xrightarrow[140-160^{\circ}C]{Et_3NH^+\ HCO_2^-} Ph\text{-}CH_2\text{-}NHPh \quad 100\ \% \qquad (3.48)$$

$$PhCH{=}NBu^t \xrightarrow{C_8K,\ THF} PhCH_2NHBu^t \quad 92\ \% \qquad (3.49)$$

Aluminium isopropoxide in isopropyl alcohol and Raney nickel reduces ketimines (**39**) to secondary amines,[41] e.g. Scheme 3.50.

$$CH_3(CH_2)_3\text{-}N{=}C_6H_{10}\ (39) \xrightarrow[\text{Raney nickel}]{Al(OPr^i)_3\ /\ Pr^iOH} CH_3(CH_2)_3\text{-}NH\text{-}CH(C_5H_{10}) \quad 66\ \% \qquad (3.50)$$

Reductive coupling Reductive coupling of imines which lack an ionizable α-hydrogen atom (**40**) occurs when they are treated with a reduced niobium chloride–tetrahydrofuran complex (**41**)[42] (Scheme 3.51).[43]

$$NbCl_5 \xrightarrow[\text{ii, THF}]{\text{i, Al in } CH_3CN} NbCl_4(THF)_2\ (41) \quad 50\ \%$$

$$Ph\text{-}CHO \xrightarrow[\text{ii, } Me_3SiCl]{\text{i, } LiN(SiMe_3)_2} Ph\text{-}CH{=}NSiMe_3\ (40)$$

$$2\ Ph\text{-}CH{=}NSiMe_3\ (40) \xrightarrow[DME]{NbCl_4(THF)_2} Ph\text{-}C({=}N\text{-}NbCl_3)\text{-}C({=}N\text{-}NbCl_3)\text{-}Ph \longrightarrow Ph\text{-}CH(NH_2)\text{-}CH(NH_2)\text{-}Ph$$

70 % yield

Racemate : Meso

19 : 1 (3.51)

Cycloaddition reactions

In Diels–Alder reactions Imines in the presence of zinc chloride act as dienophiles in Diels–Alder reactions,[44] for example, Scheme 3.52.

Formation of β-lactams Silyl imines react by a [2 + 2] cycloaddition mechanism to yield a β-lactam (**42**),[45] for example, Scheme 3.53.

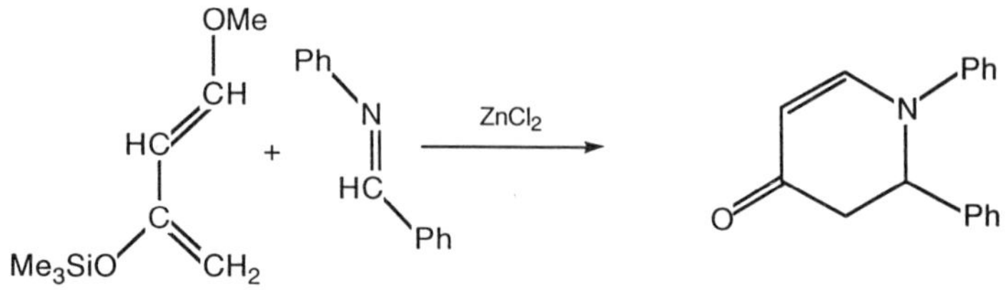

(3.52)

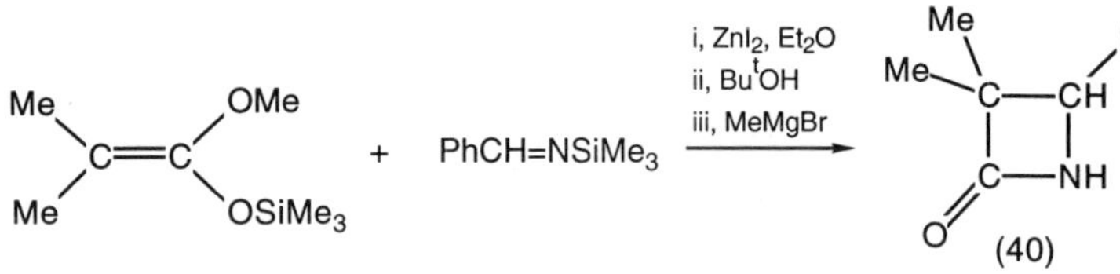

(3.53)

3.2.2 *Nucleophilic reactivity of nitrogen*

Oxidation with peroxy compounds to give oxaridines (isonitrones)

The reaction of imines with peroxy compounds is analogous to that of alkenes. Thus, at room temperature, oxaridines (isonitrones) (**43**) are produced in a reaction formally similar to the reaction of peracids with alkenes, provided anhydrous peracids are used,[46] for example, Scheme 3.54.

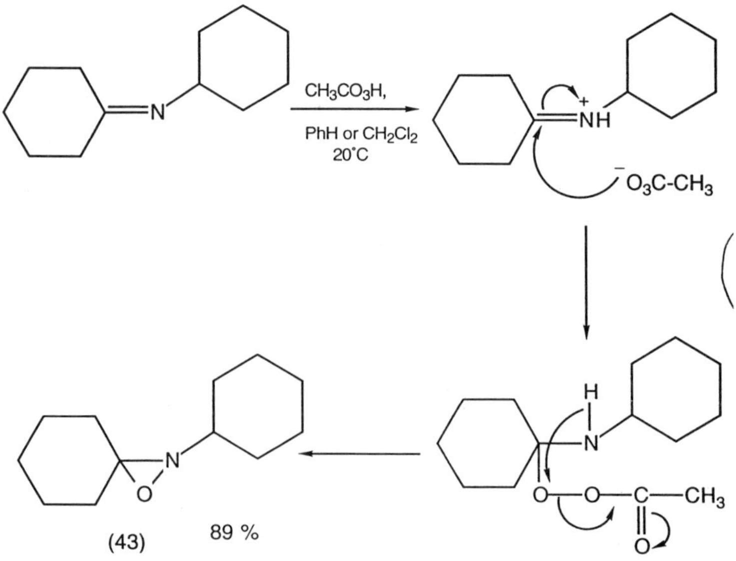

(3.54)

Oxaridines (**44**) are formed when imines react with other alkyl peroxides, such as *t*-amyl hydroperoxide ($CH_3CH_2C(OOH)(CH_3)_2$) (**45**), with molybdenum complexes in benzene (Scheme 3.55). The reaction occurs via a mechanism similar to that shown in Scheme 3.54.[47]

$$CH_3CH_2C(CH_3)_2\text{-O-}SO_2\text{-OH} \xrightarrow[0^\circ C]{\text{conc. } H_2O_2} CH_3CH_2C(CH_3)_2\text{-O-OH}$$

tertiary amyl hydrosulphate (45)

$$R^1R^2C{=}NR^3 \xrightarrow[\text{Mo(CO)}_6\text{, in PhH}]{EtCMe_2O_2H,} \text{oxaziridine (44)}$$

R^1	R^2	R^3	Yield (%)
H	Me	cyclohexyl	95
Me	Me	cyclohexyl	85
$-(CH_2)_4-$	$-(CH_2)_4-$	cyclohexyl	90
$-(CH_2)_4-$	$-(CH_2)_4-$	Ph	92
H	Me	Bn	80

(3.55)

Autoxidation

Imines autoxidize in the presence of oxygen to form oxaridines (**46**). The reaction is catalysed by cobalt diacetate,[48] for example, Scheme 3.56.

$$PhHC{=}NBu^n \xrightarrow{O_2,\ Co(OAc)_2,\ \text{catalyst}} \text{oxaziridine (46)}$$

(3.56)

With alkyl halides to give iminium salts

Alkyl halides react with imines to give iminium salts (**47**) which can be easily hydrolysed to yield secondary amines. The reaction is known as the 'Decker alkylation method',[49] e.g. Scheme 3.57.

$$\text{Ph—CH=N—CH}_2\text{-CH}_2\text{-Ph} \xrightarrow{\text{MeI}} \text{Ph—CH=}\overset{+}{\text{N}}(\text{Me})\text{—CH}_2\text{-CH}_2\text{-Ph}\ \textbf{(47)} \xrightarrow{\text{H}_2\text{O}} \text{Ph-CHO} + \text{Me—NH-CH}_2\text{-CH}_2\text{-Ph} \quad (3.57)$$

Synthesis of N-carbonyl ureas

Imines react with *N*-chlorocarbonyl isocyanate (**48**),[50] to form *N*-carbonyl ureas (**49**),[51] for example, Scheme 3.58.

$$\text{Ph-CH=N-Me} + \text{Cl-C(=O)-N=C=O}\ \textbf{(48)} \longrightarrow \text{Ph-CH(Cl)-NMe-C(=O)-N=C=O}\ \textbf{(49)}\ 66\% \quad (3.58)$$

3.2.3 *Acidity of the α-hydrogen atom*

Because the carbon–nitrogen double bond (C=N) is polarized like the carbonyl double bond, hydrogen atoms in the α-position are similarly activated and can be removed by strong bases. The resulting stabilized carbanions can then be used for the formation of new carbon–carbon bonds by reaction with electrophilic centres in other molecules.

Preparation of α,β-unsaturated aldehydes

In this way, imines have been used for the synthesis of α,β-unsaturated aldehydes (**50**) (Scheme 3.59).[52]

With organometallic reagents

Studies on the metallation and subsequent alkylation of unsymmetrical imines (**51**) have shown that butyllithium, lithium diisopropyl amine, or ethyl magnesium bromide alkylate predominantly in the less substituted (kinetically favoured) alpha position,[53] for example, Scheme 3.60.

Meyer synthesis of aldehydes, ketones, and carboxylic acids

Meyer's synthesis of aldehydes, ketones, and carboxylic acids in which a heterocyclic imine, a dihydro-oxazine (**52**) is utilized, relies on the same principle,[54] for example, Scheme 3.61.

$$CH_3\text{-}CH{=}N\text{-}Bu^t \xrightarrow{LDA} \overset{-}{C}H_2\text{-}CH{=}N\text{-}Bu^t \xrightarrow{(EtO_2)POCl} (EtO)_2P(O)\text{-}CH_2\text{-}CH{=}N\text{-}Bu^t$$

$$\xrightarrow{NaH} (EtO)_2P(O)\text{-}\overset{-}{C}H\text{-}CH{=}N\text{-}Bu^t \xrightarrow{R^1R^2CO} (EtO)_2P(O)\text{-}CH(C(O^-)R^1R^2)\text{-}CH{=}N\text{-}Bu^t$$

$$\longrightarrow \;{:}OH_2 \;\; (EtO)_2P(O)\text{-}CH(C(OH)R^1R^2)\text{-}CH{=}N\text{-}Bu^t \longrightarrow (EtO)_2P(O)\text{-}OH + Bu^tNH_2 + \underset{(50)}{R^1R^2C{=}CH\text{-}CHO}$$

R^1	R^2	Yield (%)
$PhCH_2CH_2$	H	73
PhCH=CH	H	67
Pr^n	Pr^n	53
Ph	Ph	71
Pr^i	Me	72
$-(CH_2)_5-$	$-(CH_2)_5-$	94

(3.59)

$$\underset{(51)}{CH_3\text{-}(CH_2)_4\text{-}C({=}N\text{-}C_6H_{11})\text{-}CH_3} \xrightarrow{\text{i, } Bu^iLi;\ \text{ii, BnCl};\ \text{iii, } H_2O} CH_3\text{-}(CH_2)_4\text{-}C(O)\text{-}(CH_2)_2\text{-}Ph \;\; \text{A}$$

$$+ \;\; CH_3\text{-}(CH_2)_3\text{-}CH(CH_2Ph)\text{-}C(O)\text{-}CH_3 \;\; \text{B}$$

A : B = 74 : 26

(3.60)

Me Me O Bu^nLi, THF, -78°C Me Me N CH_2-CO_2Et (52) Me O Me Me N Li CH-CO_2Et RHal

Me O $NaBH_4$ Me Me N H CH-CO_2Et R Me O Me Me N CH-CO_2Et R

H_2O, oxalic acid

R-CH-CHO CO_2Et

(3.61)

Preparation of α-silylaldehydes

The reactivity of the α-hydrogen atom of an imine has been used to prepare α-silylaldehydes (**53**) which are useful for synthetic purposes,[55] e.g. Scheme 3.62.

H_2C—CH=N— H $SiMe_2Bu^tCl$, LDA, THF, 0°C H_2C=CH—$\bar{N}$— Cl—$SiMe_2Bu^t$

Bu^tMe_2Si-CH_2-CHO 67 % (53) aq. HOAc, CH_2Cl_2 H_2C—CH=N— $SiMe_2Bu^t$

(3.62)

Preparation of primary amines

A special imine has been prepared which can be used to prepare primary amines (**54**). *N*-(Diphenylmethylene)methylamine (**55**) is metallated by lithium diisopropylamide to react with alkyl halides, as shown in Scheme 3.63.[56]

$$Et_3Al + H_2N\text{-}CH_3 \xrightarrow[55^\circ - 110^\circ C]{toluene} Et_2Al\text{-}NHCH_3 \xrightarrow{Ph_2CO} Ph_2C{=}N\text{-}CH_3\ \ 90\ \%\ (55)$$

$$Ph_2C{=}N\text{-}CH_3 \xrightarrow[-60^\circ C]{LDA / THF} Ph_2C{=}N\text{-}\underline{C}H_2 \xrightarrow{R\text{-}Br} Ph_2C{=}N\text{-}CH_2\text{-}R \xrightarrow{aq.HCl} R\text{-}CH_2\text{-}NH_2\ (54)$$

R	Yield (%)
octyl	70
cyclohexyl	51
$CH_2{=}CHCH_2$	17

(3.63)

Lithiation

Imines can be lithiated in THF at 10 °C with phenanthrene (**56**) as the hydrogen acceptor.[57] Subsequent alkylation and hydrolysis yield a ketone, for example, Scheme 3.64.

$$Me_2C{=}N\text{-}CMe_2 + Li + \text{phenanthrene (56), 1/8 mol. equiv.} \xrightarrow{THF} H_2C{=}C(Me)\text{-}N(Li)\text{-}CMe_2 + \text{(1/8 mol. equiv.)}$$

$$H_2C{=}C(Me)\text{-}N(Li)\text{-}CMe_2 \xrightarrow{Bu^nI} Me(Bu^nCH_2)C{=}N\text{-}CMe_2 \xrightarrow{H_2O} Me(Bu^nCH_2)C{=}O\ \ 52\ \%$$

(3.64)

With acyl chlorides

The reactions of acyl chlorides (**57**) with imines are interesting in that the products of the reactions vary with both the structures of the acyl chlorides and the imines. Over eighty years ago Staudinger reported that acyl chlorides and imines in the presence of an organic base (a tertiary amine) (**58**) gave β-lactams (**59**), which presumably result from the intermediate ketene reacting with the imine,[58] e.g. Scheme 3.65.

$$\underset{(57)}{\text{Ph-CH}_2\text{-COCl}} + \underset{(58)}{\text{NPh}_3} \longrightarrow \text{Ph-CH(H)-C(=O)-}\overset{+}{\text{N}}\text{Ph}_3 \longrightarrow \text{Ph(H)C=C=O} + \text{Ph}_2\text{C=N-Ph} \longrightarrow \underset{(59)}{\beta\text{-lactam}} \qquad (3.65)$$

However, acetyl chloride (**60**) reacts with triethylamine (**61**) and an imine first, to yield the ketene (**62**), which dimerizes. The dimer (**63**) under the influence of more triethylamine, undergoes ring fission to yield acetylketene (**64**). This compound undergoes a Diels–Alder reaction with the imine to form a heterocyclic compound (**65**) (Scheme 3.66).[59]

Acetyl chloride reacts in a different way giving three different products if the imine is an *N*-isopropylidene aniline (**66**) (Scheme 3.67).[60]

Phenyl malonyl chloride (**67**) has been used to convert imines into *cis*-β-lactams (**68**) (Scheme 3.68).[61]

3.2.4 *Isomerization reactions*

Reduction of α,β-unsaturated ketones

Formation of imines can be used to effect the reduction of α,β-unsaturated ketones (**69**) whilst leaving the carbonyl function intact. An imine is formed between an α,β-unsaturated ketone and a primary amine (**70**) and the resulting

$$CH_3COCl + NEt_3 \longrightarrow CH_2{-}C(=O){-}\overset{+}{N}Et_3 \longrightarrow CH_2{=}C{=}O$$

(60) (61) (62)

(63) (64) (65)

R^1	R^2	R^3	Yield (%)
Ph	H	Bn	56
Ph	H	Ph	22
Me	Me	Ph	62
Me	Me	$4\text{-}NO_2\text{-}C_6H_4$	37

(3.66)

$$Ar{-}N{=}CMe_2 + CH_3COCl \longrightarrow (Ar{-}NH)_2CMe_2 + Ar{-}N(COCH_3){-}C(CH_3){=}CH_2 + CH_3{-}C(=O){-}CH{=}C(\text{ring: }C{-}O{-}CMe_2{-}N(Ar))$$

(66)

(3.67)

$$\text{Ph-CH(COCl)}_2\ (67) + \text{H-C(R}^1\text{)=N-R}^2 \xrightarrow[\text{reflux}]{\text{PhH}} \text{azetidinone (Ph-C(COCl), C(H)R}^1\text{, C=O, N-R}^2\text{)}$$

$$\xrightarrow{\text{i, } m\text{-ClC}_6\text{H}_4\text{CO}_3\text{H / CH}_2\text{Cl}_2\text{ / 0}^\circ\text{C; ii, NEt}_3\text{; iii, NaHCO}_3\text{ aq.}}$$

A β-lactam (*cis*) (68)

(3.68)

unsaturated imine (**71**) is isomerized by the use of potassium tertiary butoxide and finally hydrolysed, thus regenerating the ketone (Scheme 3.69).

$$\underset{(70)}{\text{Ph-CH}_2\text{-NH}_2} + \underset{(69)}{\text{R}^1\text{-C(=O)-CH=CHR}^2} \longrightarrow \underset{(71)}{\text{Ph-CH}_2\text{-N=C(R}^1\text{)-CH=CH-R}^2}$$

$$\xrightarrow{\text{KOBu}^t\text{ in DMSO}} \text{Ph-CH=N-C(R}^1\text{)=CH-CH}_2\text{-R}^2 \xrightarrow{\text{dil. HOAc}} \text{Ph-CHO} + [\text{HN=C(R}^1\text{)-CH}_2\text{-CH}_2\text{-R}^2] \longrightarrow \text{R}^1\text{-C(=O)-CH}_2\text{-CH}_2\text{-R}^2$$

(3.69)

3.2.5 *Reaction as enamines*

With β-lactones

Enamines (**72**) react with β-lactones or with α,β-unsaturated carboxylic acid as shown in Scheme 3.70.[62]

or CH_2=CH-CO_2H

CH_2-CH_2-CO–N

(72)

(3.70)

N-R NHR

(73)

+ NHR NHR

NHR NHR

- H_2O - H_2O

(74) (74)

(3.71)

Because imines are known to be in equilibrium with enamines, their reaction with β-propiolactone (**73**) or with α,β-unsaturated acids to give two distinct enaminones (**74**) was investigated (Scheme 3.71).[63]

Imines from pyridoxal as enzyme models

The rearrangement of an imine, prepared from an ester of an α-amino acid (**75**), to an enamine (**76**) has been used, by reaction with pyridoxal (**77**), to synthesize imines as models for enzymes (Scheme 3.72).[64]

R^1R^2CH-CH(NH$_2$)-CO$_2$Me (75) —Bu^{t}OCl→ R^1R^2CH-CH(NHCl)-CO$_2$Me —DBU→ R^1R^2CH-C(=NH)-CO$_2$Me Imine —i, HCl ii, NH$_3$ rearrangement→ R^1R^2C=C(NH$_2$)-CO$_2$Me Enamine (76) —(77)→ R^1R^2C=C(CO$_2$Me)-N=CH-(pyridoxal ring: HO, CH$_2$OH, Me, N)

(77): CHO, HO, CH$_2$OH, Me, N

(3.72)

R—CH-NH$_2$ (CO$_2$H) (79) + OHC—(pyridine) (78) —DMF, DBU→ R—CH(CO$_2$H)-N=CH—(pyridine) ⇌ R—C(CO$_2$H)=N—CH$_2$—(pyridine) —aq. H$_3$O$^+$→ R—C(=O)—CO$_2$H

R = Me$_2$CH-CH$_2$

Yield = 85%

(3.73)

Isomerization of an imine from pyridine-4-aldehyde and an α-amino acid enables oxidative deamination of the amino acid

Pyridine-4-aldehyde (**78**) has been used for biomimetic oxidative deamination of an α-amino acid (**79**) in a reaction which goes through intermediate imines (Scheme 3.73).[65]

Similarly a 4-aldehydopyridinium (**80**) salt has been used to mimic the action of vitamin B_6 in causing oxidative deamination of a primary amine (**81**) (Scheme 3.74).[66]

i, (80) $PhSO_3^-$, CH_2Cl_2, DMF, 25°C; ii, DBU; iii, $(CO_2H)_2$; iv, H_2O

(3.74)

3.2.6 *Biochemical role of imines*

Imines play an important role in biochemistry, since the co-enzyme pyridoxal (**77**) forms imines with a variety of carbonyl groups and these imines catalyse many reactions by virtue of the fact that they are in equilibrium with enamines.[67] The following reactions are catalysed in this way:

(1) the racemization of α-amino acids;

(2) the decarboxylation of α-amino acids;

(3) the transamination of α-keto acids giving α-amino acids;
(4) the interconversion of α-amino acids;
(5) the decarboxylation of aspartic acid;
(6) the degradation of serine and of threonine;
(7) the conversion of glycine to serine.

These reactions are illustrated below.

*Racemization of α-amino acids (α-amino acid (**82**) racemase; Scheme 3.75)*

$$\text{R-C(NH}_2\text{)(H)-CO}_2\text{H (D-) (82)} \longrightarrow \text{R-C(NH}_2\text{)(H)-CO}_2\text{H (DL-)}$$

R-CH(NH$_2$)-CO$_2$H (L-) + pyridoxal phosphate (CHO, HO, Me, N, CH$_2$OPO$_3^{2-}$) ⟶ aldimine (H, R, CO$_2^-$, C, N, CH; pyridinium N$^+$–H) $\underset{+H^+}{\overset{-H^+}{\rightleftharpoons}}$ quinonoid (H$^+$, R, CO$_2^-$, C=N, CH) ⇌ H∿C(R)(CO$_2^-$)–N=CH–pyridine (HO, Me, N, CH$_2$OPO$_3^{2-}$) $\xrightarrow{H_2O}$ DL- R-CH(NH$_2$)-CO$_2$H + pyridoxal

(3.75)

Decarboxylation of α-amino acids (α-amino acid decarboxylase; Scheme 3.76)

R-CH(NH2)-CO2H → R-CH2-NH2
R-CH(NH2)-CO2H → R-CHO

O⁻
H R C=O
C
N
CH
HO CH2OPO3 2-
Me N+ H

- CO2

H R
C
N
CH
HO CH2OPO3 2−
Me N H

tautomerism

H R H
C
N
CH
HO CH2OPO3 2−
Me N

H R
C
N
CH2
HO CH2OPO3 2−
Me N

aqueous H3O+

aqueous H3O+

CHO
HO CH2OH
Me N
+ R-CH2-NH2
Pyridoxal

CH2NH2
HO CH2OH
Me N
+ R-CHO
Pyridoxamine

(3.76)

Transamination of α-keto acids to give α-amino acids

Pyridoxal catalyses the interchange between α-amino acids and α-keto acids, the overall reaction being shown in Scheme 3.77. The role of pyridoxal is as shown in Scheme 3.78.

$$R^1\text{-CH(NH}_2\text{)CO}_2\text{H} + R^2\text{-CO-CO}_2\text{H} \rightleftharpoons R^1\text{-CO-CO}_2\text{H} + R^2\text{-CH(NH}_2\text{)CO}_2\text{H} \quad (3.77)$$

H R CO_2^-
C
N
CH
HO $CH_2OPO_3^{2-}$
Me N H (+)

-H^+ / +H^+

R CO_2^-
C
N
H^+
CH
HO $CH_2OPO_3^{2-}$
Me N H

+H^+ / -H^+

R CO_2^-
C
N
CH_2
HO $CH_2OPO_3^{2-}$
Me N H (+)

H_2O

R-CH-CO_2H (with NH_2) + Pyridoxal

R-CO-CO_2^- +

CH_2NH_2
HO $CH_2OPO_3^{2-}$
Me N

Pyridoxamine phosphate

(3.78)

Decarboxylation of aspartic acid (aspartic acid decarboxylase; Scheme 3.79)

$$H_2N\text{—}CH(CH_2\text{—}CO_2H)\text{-}CO_2H \longrightarrow H_2N\text{—}CH(CH_3)\text{-}CO_2H + CO_2$$

Aspartic acid

Alanine

tautomerization

tautomerize and hydrolyse

Pyridoxal + $H_2N\text{—}CH(CH_3)\text{-}CO_2H$

Alanine

(3.79)

Interconversion of α-amino acids (Scheme 3.80)

α-Amino acids which contain a good leaving group in the position β to the carboxyl group (**83**) can be interconverted in a reaction catalysed by pyridoxal phosphate (Scheme 3.81). Nucleophiles other than the hydrosulphide ion can take part, for example, when the nucleophile is indole the amino acid, tryptophane is formed.

$$\underset{\text{Serine}}{HO\text{-}CH_2CH(NH_2)\text{-}CO_2H} \longrightarrow \underset{\text{Cysteine}}{HS\text{-}CH_2CH(NH_2)\text{-}CO_2H} \qquad (3.80)$$

$HOCH_2\text{-}CH(NH_2)\text{-}CO_2H$; O, CH, HO, $CH_2OPO_3^{2-}$, Me, N, H → $HOCH_2$, CO_2H, H, CH, N, CH, HO, $CH_2OPO_3^{2-}$, Me, N, H ⇌ $H_2\overset{+}{O}-CH_2$, CO_2H, C, N, CH, HO, $CH_2OPO_3^{2-}$, Me, N, H ↓ H_2C, CO_2H, HS^-, C, + HO^-, N, CH, HO, $CH_2OPO_3^{2-}$, Me, N, H ⇌ $HSCH_2$, CO_2H, C, N, CH, HO, $CH_2OPO_3^{2-}$, Me, N, H

tautomerize and hydrolyse

$$\underset{\text{cysteine}}{HS\text{-}CH_2\text{-}CH(NH_2)\text{-}CO_2H} + \text{pyridoxal phosphate} \qquad (3.81)$$

(83)

Biosynthesis and degradation of serine and threonine

Biosynthesis of serine (**84**) from glycine (**85**) is shown in Scheme 3.82. In a similar way, threonine (**86**) results when acetaldehyde is the aldehyde. The reverse of this reaction constitutes a degradation of serine or threonine (Scheme 3.83).

$H_2N\text{-}CH_2\text{-}CO_2H$ + CH_2O ⟶ $H_2N\text{-}CH\text{-}CO_2H$

glycine (85)

CH_2OH serine (84)

$CH_2\text{-}CO_2H$

NH_2

+

CHO

HO

$CH_2OPO_3^{2-}$

Me

N

H H CO_2^-

C

N

CH

HO

$CH_2OPO_3^{2-}$

Me

$\overset{+}{N}$

H

- H^+

+ H^+

$H_2C{=}O$

H CO_2^-

C

N

CH

HO

$CH_2OPO_3^{2-}$

Me

N

H

CH_2OH

H CO_2^-

C

N

CH

HO

$CH_2OPO_3^{2-}$

Me

$\overset{+}{N}$

H

H_2O

CH_2OH

$CH\text{-}CO_2H$

NH_2

Serine

+ pyridoxal phosphate

(3.82)

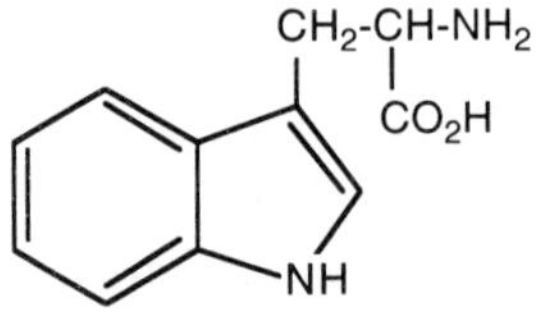

$$\text{Tryptophan (indole-3-yl)}-CH_2-CH(NH_2)-CO_2H$$

Serine transhydroxymethylase

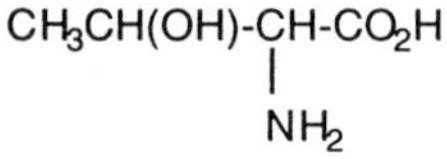

$$CH_3CH(OH)-CH(NH_2)-CO_2H$$

Threonine (86)

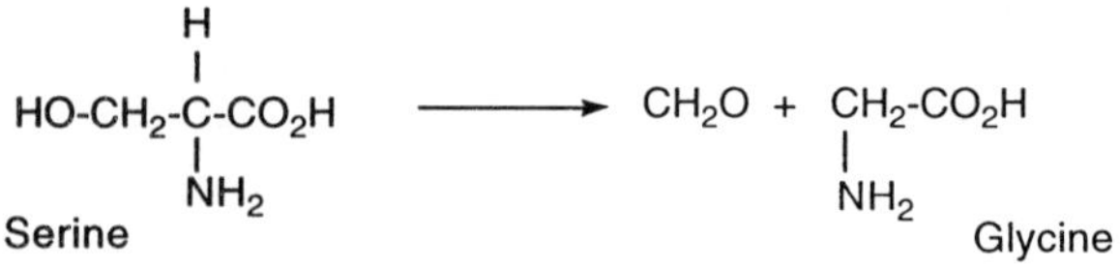

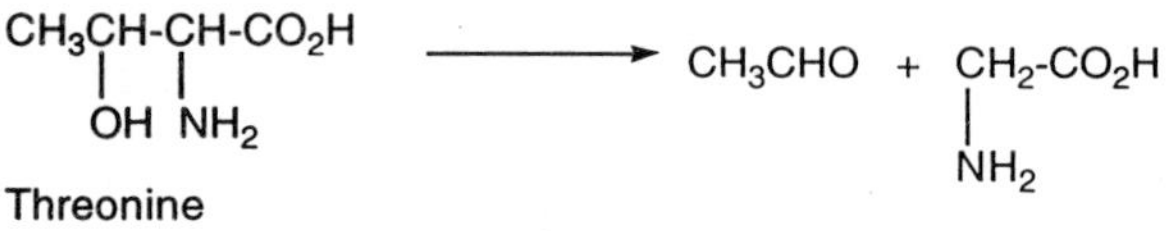

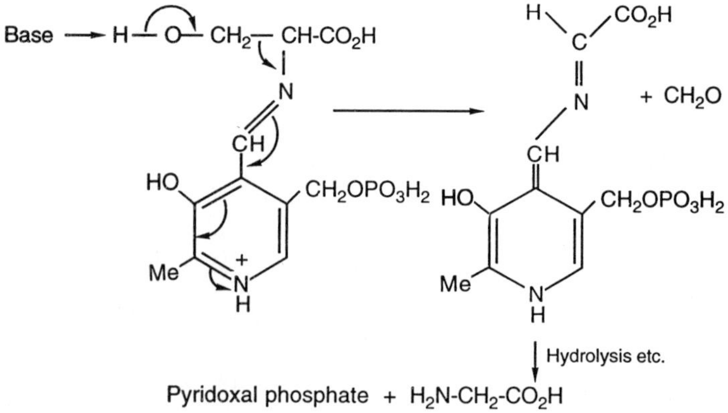

$$HO-CH_2-CH(NH_2)-CO_2H \longrightarrow CH_2O + CH_2(NH_2)-CO_2H$$

Serine Glycine

$$CH_3CH(OH)-CH(NH_2)-CO_2H \longrightarrow CH_3CHO + CH_2(NH_2)-CO_2H$$

Threonine

Base → H–O–CH$_2$–CH-CO$_2$H (N=CH– pyridinium, HO, Me, CH$_2$OPO$_3$H$_2$) → H(CO$_2$H)C=N–CH= (dihydropyridine, HO, Me, CH$_2$OPO$_3$H$_2$, NH) + CH$_2$O

Hydrolysis etc.

$$\text{Pyridoxal phosphate} + H_2N-CH_2-CO_2H \qquad (3.83)$$

Use of ylides from imines for synthetic purposes

Pyruvic acid (**87**) has been found to react with α-amino acids (**88**) to form intermediate iminium ions (**89**), as shown in Scheme 3.84.[68]

Pyruvic acid (87) + (88) ⇌ (89)

R	Yield (%)
Me	88
Bn	64
Bu^i	62
Ph	59

(3.84)

In particular, Grigg and his co-workers[69] have studied the imines formed from α-amino acids and their esters (**90**) and have demonstrated that these imines can furnish azomethine (imine) ylides of value in organic synthesis. By prototropy of an imine formed from an ester of an α-amino acid, a resonance-stabilized ylide (**91**) can be generated[70] and by decarboxylation of an imine from an aldehyde and an α-amino acid, another resonance-stabilized azomethine ylide (**92**) results (Scheme 3.85).[71]

Cycloaddition reactions of imine ylides

These azomethine ylides act as 1,3-dipoles and their addition reactions with 1,3-dipolarophiles, both intermolecular and intramolecular, have been studied by Grigg and his co-workers,[72] for example, Scheme 3.86.

Imines undergo Diels–Alder reactions with activated dienes in the presence of Lewis acids (e.g. Et_2AlCl)[73] and 1,3-dipolar additions of azomethine ylids to 1,2-dipolarophiles may be employed in the formation of pyrrolidines (Scheme 3.87).[74]

Imines in the presence of zinc chloride as a Lewis acid act as dienophiles in Diels–Alder reactions (Scheme 3.88).[75]

Diels–Alder reactions of di-imines (**93**), have been studied in depth and the highest occupied and lowest unoccupied molecular orbitals (HOMO and LUMO, respectively) have been calculated (see table on p. 161).[76]

R^1-CHO + H_2N-CH(R^2)-CO_2R^3 (90) → R^1-CH=N-CH(R^2)-CO_2R^3

R^3 = H

(91)

- CO_2

(92)

(3.85)

xylene, reflux under argon for 1 day

3:2 Mixture of two diastereoisomeric racemates

(3.86)

$H_2N\text{-}CH(CO_2Et)_2 \xrightarrow[\text{toluene}]{H_2CO} [CH_2{=}N\text{-}CH(CO_2Et)_2 \rightleftharpoons CH_2{=}\overset{+}{N}H\text{-}\overset{-}{C}(CO_2Et)_2]$; $+ EtO_2C\text{-}CH{=}CH_2 \xrightarrow{\text{heat}}$ pyrrolidine (EtO_2C, CO_2Et, CO_2Et, NH)

1.3-dipolar addition (3.87)

$Me_3SiO\text{-}C(=CH_2)\text{-}CH{=}CH\text{-}OMe + R^2\text{-}CH{=}N\text{-}R^1 \xrightarrow{ZnCl_2,\ THF}$ 2,3-dihydropyridin-4-one (R^1, H, R^2)

45 - 75 % (3.88)

$R\text{-}N{=}CH\text{-}CH{=}N\text{-}R$

(93)

R	HOMO energy (eV)	LUMO energy (eV)
Ph	−8.78	−0.79
Me	−10.52	+0.07
Pr^i	−10.44	+0.18

IMINIUM SALTS (**94**)

These electrophilic salts have wide use in synthesis and they often arise as intermediates in reaction mechanisms. The positive charge on the nitrogen atom enhances the attack of nucleophiles on the carbon atom.

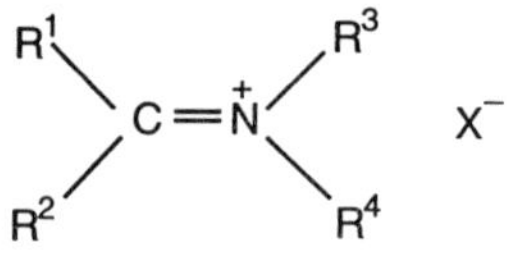

(94)

3.3 Preparation

3.3.1 *By alkylation of imines*

The alkylation of an imine produces an iminium salt (Scheme 3.89).

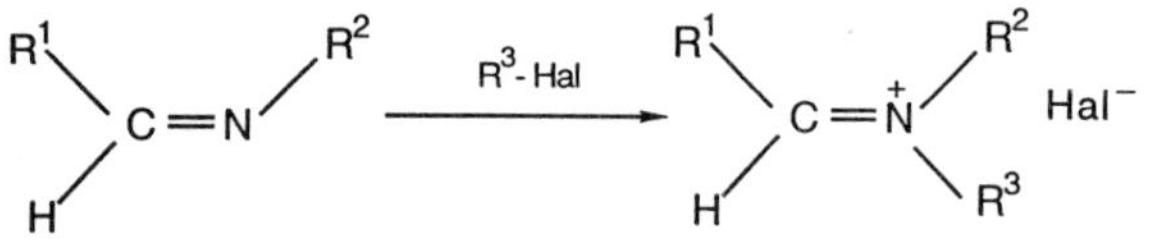

(3.89)

3.3.2 *By the reaction of mercurous acetate on tertiary amines*

By the interaction of mercuric acetate (**95**) with a tertiary amine which contains an α-hydrogen atom (**96**), an elimination occurs which furnishes an iminium salt (Scheme 3.90).[77]

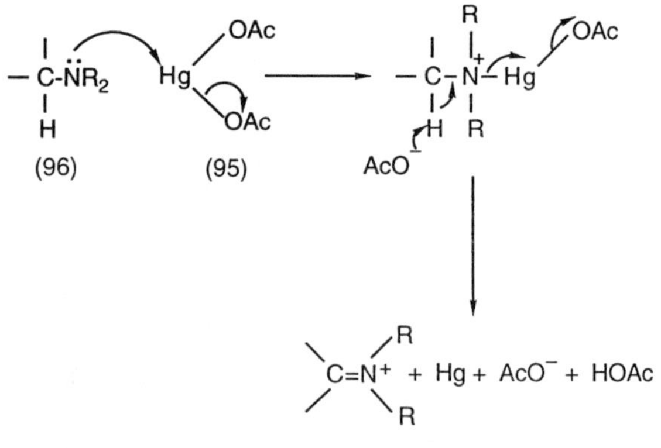

(96) (95)

(3.90)

3.3.3 *From the salt of a secondary amine and an aldehyde or a ketone*

Iminium salts (**97**) can be prepared from the salt of a secondary amine and an aldehyde or a ketone (compare imine preparations from primary amines and ketones or aldehydes) (Scheme 3.91). The iminium ions (**97**) are the electrophilic reagents in the Mannich reaction (see Chapter 1, p. 50), for example, Scheme 3.92.

$$R^1_2NH_2^+\ X^- + R^2_2C{=}O \longrightarrow R^1_2\overset{+}{N}{=}CR^2_2\ X^- + H_2O \quad (97)$$

$$X^- = F_3C\text{-}CO_2^- \text{ or } I^- \qquad (3.91)$$

$$CH_3\text{-}CO\text{-}CH_3 \rightleftharpoons CH_3\text{-}C(\text{:OH}){=}CH_2$$

$$CH_2O + H_2\overset{+}{N}Me_2\ Cl^- \longrightarrow CH_2{=}\overset{+}{N}Me_2\ (97)$$

$$\longrightarrow CH_3\text{-}CO\text{-}CH_2\text{-}CH_2\text{-}NMe_2 \text{ (Mannich base)} \qquad (3.92)$$

3.3.4 *From tertiary amines and methylene iodide—preparation of Mannich reagents*

Iminium ions have been isolated as stable salts (**98**)[78] (iodides, chlorides, or trifluoroacetates) which can then be used in Mannich reactions. For example they have been prepared from tertiary amines (**99**) and methylene iodide (**100**) (Scheme 3.93).

$$\underset{(99)}{(CH_3)_3N} + \underset{(100)}{CH_2I_2} \longrightarrow \underset{89\,\%}{(CH_3)_3\overset{+}{N}\text{-}CH_2I\ I^-} \xrightarrow[-CH_3I]{150^\circ C} \underset{(98)\ 81\,\%}{(CH_3)_2\overset{+}{N}{=}CH_2\ I^-} \qquad (3.93)$$

A more recent preparation of a Mannich reagent is shown in Scheme 3.94.[79]

$$CH_2O + HNMe_2 \xrightarrow{EtOH} EtO\text{-}CH_2\text{-}NMe_2 \xrightarrow[CH_3CN]{CH_3\text{-}SiCl_2\text{-}Cl} CH_3\text{-}SiCl_2(OEt) + H_2C{=}\overset{+}{N}Me_2\ Cl^- \quad 98\% \qquad (3.94)$$

3.4 Reactivity

3.4.1 *Use in the Mannich reaction*

Iminium salts often give better yields of Mannich bases (**101**) than the direct, classical Mannich reaction from formaldehyde, dimethylamine, and a ketone,[80] for example, Scheme 3.95.

$$Me_2CH\text{-}C(=O)\text{-}CH_3 + Me_2\overset{+}{N}{=}CH_2\ \ F_3C\text{-}CO_2^- \xrightarrow{-10\ \text{to}\ 145^\circ C} Me_2CH\text{-}C(=O)\text{-}CH_2\text{-}CH_2\text{-}NMe_2\ (101)$$

c. 50 %

The less substituted product (3.95)

3.4.2 *In Diels–Alder reactions*

These iminium ions also undergo Diels–Alder reactions with activated dienes. The product of the addition may subsequently be converted to a useful Mannich base,[81] e.g. Scheme 3.96.

OMe, H, Me_3SiO, CH_2 + I^- $\overset{+}{N}Me_2{=}CH_2$ —20°C→ OMe, $\overset{+}{N}Me_2$ I^-, Me_3SiO

aq. NaOH

OMe, H, O, CH_2, CH_2, NMe_2

95 %

(3.96)

3.4.3 *With Grignard reagents or alkyl lithiums*

Trifluoroacetate iminium ions (**102**) react with Grignard reagents or with alkyl lithiums (**103**) to give good yields of tertiary amines (Scheme 3.97)[82] (see Chapter 1, p. 23).

$$\underset{(103)}{R{-}Li} + \underset{(102)}{CH_2{=}\overset{+}{N}Me_2}\ F_3C{-}CO_2^- \longrightarrow R{-}CH_2{-}NMe_2 + Li^+\ F_3C{-}CO_2^-$$

R	Yield (%)
Ph	85
p-MeC_6H_4	72
n-C_4H_9	72

(3.97)

Coupled with the oxidation of the tertiary amine to an amine oxide (**104**) and elimination, this reaction constitutes a good synthesis of terminal alkenes (**105**) (Scheme 3.98).[83]

$$R{-}CH_2{-}MgBr + CH_2{=}\overset{+}{N}Me_2\ I^- \longrightarrow R{-}CH_2{-}CH_2{-}NMe_2 \xrightarrow{H_2O_2} \underset{(104)}{R{-}CH(H){-}CH_2{-}\overset{+}{N}Me_2{-}O^-}$$

$$\xrightarrow{160^\circ C} \underset{(105)}{R{-}CH{=}CH_2} + Me_2NOH$$

R	Yield (%) hydroxylamine	alkene
$CH_3(CH_2)_6$	91	87
Bn	85	83

(3.98)

3.4.4 *The Vilsmeier reagent*

The Vilsmeier reagent (**106**) reacts with 30 per cent hydrogen peroxide to give a peroxyiminium ion (**107**) which is useful for high yield epoxidation of alkenes,[84] for example, Scheme 3.99.

$$Me_2\overset{+}{N}{=}CH{-}Cl\ PO_2Cl_2^- \ (106) \xrightarrow{H_2O_2} Me_2\overset{+}{N}{=}CH{-}O_2H\ PO_2Cl_2^- \ (107) \xrightarrow[\text{MeOH, -20°C}]{\text{cyclohexene}} \text{cyclohexene oxide } 95\text{-}98\,\% \quad (3.99)$$

3.4.5 *Reduction by sodium cyanoborohydride*

As described earlier, iminium ions function as good carbon electrophiles and thus they are reduced by sodium cyanoborohydride to tertiary amines (Scheme 3.100)[85] (see Chapter 1, p. 14).

$$R^1R^2\overset{+}{N}{=}CR^3R^4\ X^- \xrightarrow{NaBH_3CN} R^1R^2\overset{+}{N}H{-}CHR^3R^4\ X^- \longrightarrow R^1R^2N{-}CHR^3R^4 \quad (3.100)$$

3.4.6 *With a diazoalkane—an alternative to the Wittig reaction for the conversion of a ketone into an alkene*

Being electrophiles, iminium ions readily react with diazoalkanes and the reaction with diazomethane provides an alternative to the Wittig reaction for the preparation of an alkene from a ketone (Scheme 3.101).[86]

3.4.7 *Introduction of a cyano group into activated aromatics using Viehe's salt*

Viehe's salt (**108**), $Me_2N{=}CCl_2^+Cl^-$, acts as an electrophile with activated aromatic systems and introduces a cyano group in moderate yields (Scheme 3.102).[87]

NH

70 % $HClO_4$
ether

$\overset{+}{N}H_2$ ClO_4^- $R_2C{=}O$ $\overset{+}{N}{=}CR_2$ ClO_4^-

CH_2N_2, CH_2Cl_2
- 10° C

Bu^n Li

$CH_2{=}CR_2$ + N → NH

c. 70 %

(3.101)

Cl
$C{=}\overset{+}{N}Me_2$
Cl
(108)

MeO OMe

H Cl C NMe_2 Cl
$Me\overset{+}{O}$ OMe

Me :Cl^-
$C{=}\overset{+}{N}$ Me
Cl
MeO OMe

CN
heat
- 2 MeCl
MeO OMe
28 %

(3.102)

KETENIMINES

$R^1R^2C{=}C{=}N{-}R^3$

(109)

Ketenimines (**109**) are derived from ketenes and primary amines. The bond lengths of a ketenimine are shown below.[88] The crystal determination values support the view that the resonance structure (**110**) is an important contribution to the overall polarization of the group (Scheme 3.103).[89]

R1 1.74 Å R2 1.74 Å C 1.36 Å C 1.17 Å N 1.47Å R3

$R^1R^2C{=}C{=}N{-}R^3 \rightleftharpoons R^1R^2\overset{-}{C}{-}C{\equiv}\overset{+}{N}{-}R^3 \rightleftharpoons R^1R^2C{=}C{=}N{-}R^3$

(110) (3.103)

3.5 Preparation[90]

3.5.1 *From imidoyl chlorides*

The preparation of ketenimines from imidoyl chlorides (**111**) by dehydrohalogenation,[91] for example, Scheme 3.104.

$C_6H_{11}{-}C(=O){-}NHPh \xrightarrow{PCl_5} C_6H_{11}{-}C(Cl){=}NPh \xrightarrow{Et_3N} C_6H_{10}{=}C{=}NPh$

(111) 70 % (3.104)

3.5.2 *By dehydration of N-substituted amides*

N-Substituted amides are dehydrated by phosphorus pentoxide to yield ketenimines,[93] for example, Scheme 3.105.

$$Ph_2CH\text{-}CONH\text{-}C_6H_4\text{-}Me \xrightarrow[\text{pyridine}]{P_2O_5} Ph_2C{=}C{=}N\text{-}C_6H_4\text{-}Me \quad 87\% \qquad (3.105)$$

3.5.3 *From ketenes and iminophosphoranes*

Ketenes react with iminophosphoranes ($Ph_3P{=}NR$) (**112**) to give ketenimines,[94] e.g. Scheme 3.106.

$$(Ph)(Et)C{=}C{=}O + Ph_3P{=}N\text{-}C_6H_{11}\ (112) \longrightarrow (Ph)(Et)C{=}C{=}N\text{-}C_6H_{11} \quad 58\% \qquad (3.106)$$

3.6 Reactivity[90]

3.6.1 *Hydrolysis*

Ketenimines, like ketenes, are attacked by nucleophiles on carbon and, being easily hydrolysed to amides, ketenimines function as good dehydrating agents (Scheme 3.107).

$$R^1R^2C{=}C{=}N\text{-}R^3 \xrightarrow{H_2O} \left[R^1R^2C{=}C(OH)\text{-}NHR^3\right] \longrightarrow R^1R^2CH\text{-}C({=}O)\text{-}NHR^3 \qquad (3.107)$$

3.6.2 *Cycloaddition reactions*

Ketenimines react, via cycloaddition mechanisms, with carbonyl, nitrogen–nitrogen, isocyanate, and other heteroatom double bonds,[95] for example, Scheme 3.108.

Methylketenimines (**113**) react via a [2 + 2] cycloaddition mechanism with both aromatic and aliphatic aldehydes to form 2-iminooxetanes (**114**) as intermediates in a synthesis of β-hydroxy ketones e.g. (Scheme 3.109).[96]

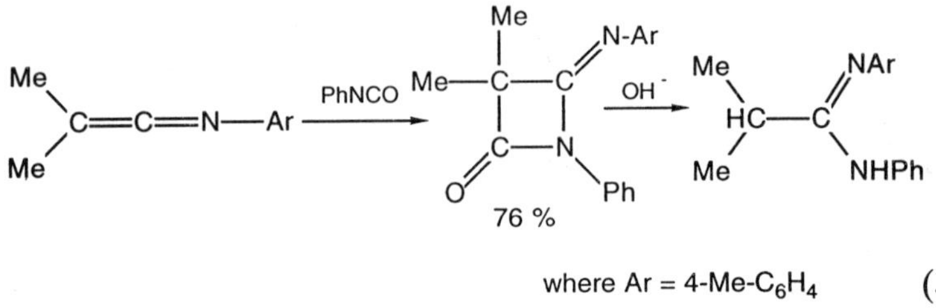

(3.108)

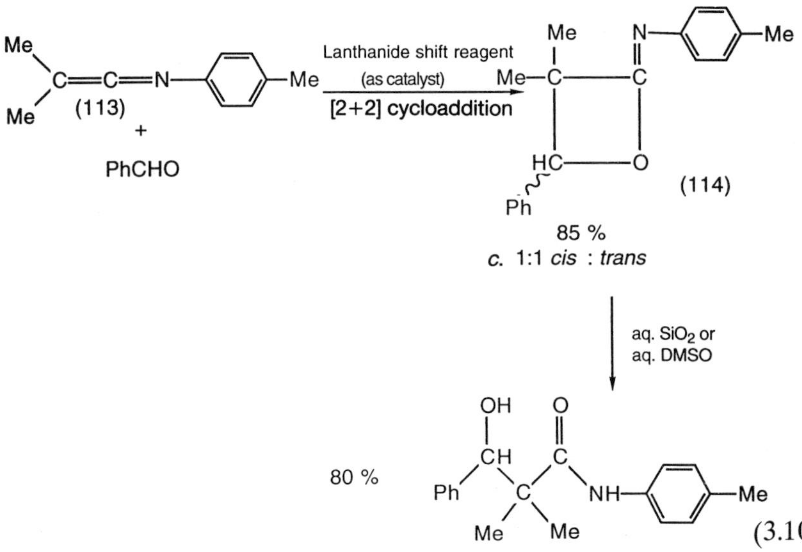

where Ar = 4-Me-C_6H_4

(3.109)

KETENIMINIUM SALTS

Tertiary amides (**115**) can be converted into ketenimine triflates (**116**) by the action of triflic anhydride followed by collidine. When these keteniminium salts are produced in the presence of alkenes or alkynes they undergo [2 + 2] cycloaddition to give cyclobutan- or cyclobuten-ones (Scheme 3.110).[97]

In the reaction with conjugated dienes, $X = BF_4^-$,[98] the product is:

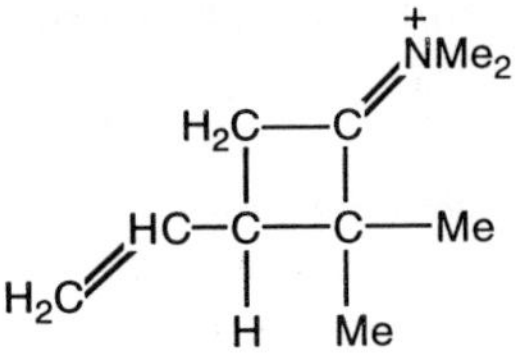

CH_3-C(=O)-NMe_2 (115) $\xrightarrow[\text{2,4,6 - collidine}]{(F_3C\text{-}SO_2)_2O}$ CH_2=C=$\overset{+}{N}Me_2$ F_3C-SO_3^- (116)

CH_2=C(OH)-NMe_2; OH, SO_2, O, S=O, O, CF_3, CF_3

R-CH=CH_2

$\overset{+}{N}Me_2$ F_3C-SO_3^-

R

H_2O, CCl_4
heat

O

R = $CH_3(CH_2)_5$, yield = *c.* 60 %

R

(3.110)

References

1. S. Patai (ed.), *The chemistry of the carbon–nitrogen double bond*, Wiley, Chichester, 1970; G. W. Buchanan and B. A. Dawson, *Can. J. Chem.*, 1977, **55**, 1437; M. Kobayashi, M. Yoshida, and H. Minato, *J. Org. Chem.*, 1976, **41**, 3322; J. Bjorgo, D. R. Boyd, C. G. Watson, and W. B. Jerina, *J. Chem. Soc., Perkin Trans. II*, 1974, 1081.
2. *F. &F.*, **11**, 411; A. W. Frahm and G. Knupp, *Tetrahedron Lett.*, 1981, **22**, 2633.
3. *F. &F.*, **3**, 206; E. P. Kyba, *Org. Prep. Proc.*, 1970, **2**, 149.
4. H. Weingarten, J. P. Chipp, and W. A. White, *J. Org. Chem.*, 1967, **32**, 3246; *F. &F.*, **2**, 414.
5. *F. &F.*, **3**, 291; I. Moretti and G. Torre, *Synthesis*, 1970, 141.
6. *F. &F.*, **11**, 161; C. Stetin, B. de Jeso, and J. C. Pommier, *Synth. Commun.*, 1982, **12**, 495.
7. *F. &F.*, **15**, 305; J. Lasarte, C. Palermo, J. P. Picard, J. Donogues, and J. M. Aizpurna, *J. Chem. Soc., Chem. Commun.*, 1989, 72.
8. *F. &F.*, **2**, 99.
9. *F. &F.*, **11**, 173; K. Findeisen, H. Heitzler, and K. Dehnicke, *Synthesis*, 1981, 702.
10. S. Dayagi and Y. Degani, *The chemistry of the carbon–nitrogen double bond* (ed. S. Patai), Wiley-Interscience, 1970, p. 81.
11. S. P. Findlay, *J. Org. Chem.*, 1956, **21**, 644.
12. R. Huls and M. Ruson, *Bull. Soc. Chim. Belg.*, 1956, **65**, 684.
13. W. P. Jenks, *J. Am. Chem. Soc.*, 1959, **81**, 475; A. Lapworth, *J. Chem. Soc.*, 1907, **91**, 1133.
14. N. P. Gambaryan, E. M. Rokhlin, Yu. V. Zeifman, C. Ching-Yu, and I. L. Knunyants, *Angew. Chem., Int. Ed. Engl.*, 1966, **5**, 947.

15. R. J. Highet and W. C. Wildman, *J. Am. Chem. Soc.*, 1955, **77**, 4399; R. E. Miller, *J. Org. Chem.*, 1960, **25**, 2126.
16. A. Skita, *Chem. Ber.*, 1915, **48**, 1685.
17. R. W. Layer, *Chem. Rev.*, 1963, **63**, 489.
18. *F. &F.*, **15**, 175; P. Müller and D. M. Gilabert, *Tetrahedron*, 1988, **44**, 7171.
19. J. J. Cornego, K. D. Larson, and G. D. Mendenhall, *J. Org. Chem.*, 1985, **50**, 5382.
20. S.-I. Murahashi, T. Naota, and H. Taki, *J. Chem. Soc., Chem. Commun.*, 1985, 613.
21. J. P. Marino and R. D. Larsen, *J. Am. Chem. Soc.*, 1981, **103**, 4642.
22. *F. &F.*, **1**, 91; W. E. Bachmann, M. P. Cava, and A. S. Dreiding, *J. Am. Chem. Soc.*, 1954, **76**, 5554; M. P. Cava and B. R. Vogt, *Tetrahedron Lett.*, 1964, 2813.
23. *F. &F.*, **3**, 267; R. G. R. Bacon and D. Stewart, *J. Chem. Soc. (C)*, 1966, 1384; 1388.
24. *F. &F.*, **13**, 54; S.-I. Murahashi, T. Naota, and H. Taki, *J. Chem. Soc., Chem. Commun.*, 1985, 613.
25. P. L. Pickard and T. L. Tolbert, *J. Org. Chem.*, 1961, **26**, 4886; E. F. Cornell, *J. Am. Chem. Soc.*, 1928, **50**, 3311.
26. S. Patai, (ed.), *Chemistry of the C—N double bond*, Wiley, Chichester, 1970, Chapter 10; W. P. Jencks and E. A. Cordes, *J. Am. Chem. Soc.*, 1963, **85**, 843.
27. K. Harada and T. Okawara, *J. Org. Chem.*, 1973, **38**, 707.
28. I. Neelakantan and W. H. Harung, *J. Org. Chem.*, 1959, **24**, 1943.
29. K. N. Campbell, C. H. Helbing, M. P. Florkowki, and B. K. Campbell, *J. Am. Chem. Soc.*, 1948, **70**, 3868.
30. M. S. Kharasch and O. Reinmuth, *Grignard reactions of non-metallic substances*, Constable and Co., London, 1954.
31. J. Harada, *The chemistry of the carbon–nitrogen double bond* (ed. S. Patai), Wiley-Interscience, Chichester, 1970, pp. 266–272; B. L. Embing, R. J. Horvath, A. J. Saraceno, E. F. Ellermeyer, L. Haile, and L. D. Hudac, *J. Org. Chem.*, 1959, **24**, 657.
32. G. E. Keck and E. J. Enholm, *J. Org. Chem.*, 1985, **50**, 146; *F. &F.*, **13**, 10.
33. M. Freifelder, *Catalytic hydrogenation in organic syntehsis. Procedures and commentary*, Wiley, New York, 1978, pp. 90 and 96.
34. M. Mousseron, R. Jacquier, M. Mousseron-Canet, and Z. Zagdoun, *Bull. Soc. Chim. Fr.*, 1952, **19**, 1042.
35. J. H. Billman and J. W. McDowell, *J. Org. Chem.*, 1962, **27**, 2640; J. H. Billman and A. C. Diesing, *J. Org. Chem.*, 1957, **22**, 1068; G. N. Walker and M. A. Moore, *J. Org. Chem.*, 1961, **26**, 432; G. N. Walker, M. A. Moore, and B. N. Weaver, *J. Org. Chem.*, 1961, **26**, 2740; E. C. Taylor, A. McKillop, and R. E. Ross, *J. Am. Chem. Soc.*, 1965, **87**, 1990.
36. *F. &F.*, **11**, 163; H. Alper and S. Armaratunga, *J. Org. Chem.*, 1982, **47**, 3593.
37. *F. &F.*, **1**, 273; H. Nöth and H. Beyer, *Ber.*, 1960, **93**, 928; J. H. Billman and J. W. McDowell, *J. Org. Chem.*, 1961, **26**, 1437.
38. *F. &F.*, **4**, 312; 450; *F. &F.*, **11**, 483; J. E. Wrobel and B. Ganem, *Tetrahedron Lett.*, 1981, **22**, 3447.
39. E. R. Alexander and R. B. Wildman, *J. Am. Chem. Soc.*, 1948, **70**, 1187.
40. D. Savoia, C. Trombini, and A. Unami-Ronchi, *Pure Appl. Chem.*, 1985, **85**, 1887.
41. M. Botta, F. De Angeli, A. Gambacorta, L. Labbiento, and R. Nicoletti, *J. Org. Chem.*, 1985, **50**, 1916.
42. Preparation: L. E. Manzer, *Inorg. Chem.*, 1977, **16**, 525.
43. E. J. Roskamp and S. F. Pederson, *J. Am. Chem. Soc.*, 1987, **109**, 3152.
44. *F. &F.*, **11**, 334; J. F. Kerwin, Jr and S. Danishefsky, *Tetrahedron Lett.*, 1982, **23**, 3739.

45. *F. &F.*, **14**, 214; E. W. Colvin and D. G. Mcgarry, *J. Chem. Soc., Chem. Commun.*, 1985, 539.
46. *F. &F.*, **1**, 788; W. D. Emmons, *J. Am. Chem. Soc.*, 1956, **78**, 6208; 1957, **79**, 5528; L. Horner and E. Jürgens, *Ber.*, 1957, **90**, 2184.
47. *F. &F.*, **4**, 20; G. A. Tolstikev, U. M. Jemilev, V. P. Jurjev, F. B. Gershanov, and S. R. Rafikof, *Tetrahedron Lett.*, 1971, 2807.
48. B. J. Auret, D. R. Boyd, and P. B. Coulter, *J. Chem. Soc., Chem. Commun.*, 1984, 463.
49. H. Decker and P. Becker, *Annalen*, 1913, **395**, 362.
50. Preparation: H. Hagemann, *Angew. Chem., Int. Ed. Engl.*, 1971, **10**, 832.
51. H. Hagemann and R. Ley, *Angew. Chem., Int. Ed. Engl.*, 1972, **11**, 1011.
52. *F. &F.*, **9**, 1; H. R. Snyder and D. S. Matterson, *J. Am. Chem. Soc.*, 1957, **79**, 2217 (preparation of the imine); A. I. Meyers, K. Tomioka, and M. P. Fleming, *J. Org. Chem.*, 1978, **43**, 3788.
53. G. Stork and S. Dowd, *J. Am. Chem. Soc.*, 1965, **85**, 2178; G. Wittig, H. D. Frommeld, and P. Suchanek, *Angew. Chem., Int. Ed. Engl.*, 1963, **2**, 683.
54. Review: J. March, *Advanced organic chemistry. Reactions, mechanisms, and structure*, 3rd Edn, McGraw Hill, New York, 1985, pp. 424–6; R. R. Schmidt, *Synthesis*, 1972, 333.
55. *F. &F.*, **11**, 88; P. F. Hudrick and A. K. Kulkarni, *J. Am. Chem. Soc.*, 1981, **103**, 6251.
56. *F. &F.*, **8**, 210; T. Kauffmann, H. Berg, E. Köppelmann, and D. Kuhlmann, *Ber.*, 1977, **110**, 2659.
57. *F. &F.*, **13**, 157; E. A. Mistryukov and I. K. Korshevetz, *Synthesis*, 1984, 947.
58. H. Staudinger, *Annalen*, 1907, **356**, 51.
59. *F. &F.*, **7**, 3; A. Maujean and J. Chuche, *Tetrahedron Lett.*, 1976, 2905.
60. H. Iwamura, M. Tsuchimoto, and N. Nishimura, *Tetrahedron Lett.*, 1975, 1405.
61. *F. &F.*, **5**, 126; A. K. Bose and J. C. Kapur, *Tetrahedron Lett.*, 1973, 1811.
62. S. O. Oleson, J. Ø. Madsen, and S.-O. Lawesson, *Bull. Soc. Chim. Belg.*, 1978, **87**, 535.
63. R. Shabana, J. B. Rasmussen, S. O. Olesen and S.-O. Lawesson, *Tetrahedron*, 1980, **36**, 3047.
64. *F. &F.*, **8**, 246; U. Schmidt and E. Öhler, *Angew. Chem., Int. Edn. Engl.*, 1977, **16**, 327; U. Schmidt and E. Prantz, *Angew. Chem., Int. Ed. Engl.*, 1977, **16**, 328.
65. *F. &F.*, **11**, 448; S. Ohta and K. Okamato, *Synthesis*, 1982, 756.
66. *F. &F.*, **11**, 244; T. F. Buckley and H. R. Rapoport, *J. Am. Chem. Soc.*, 1982, **104**, 4446.
67. A. Fersht, Enzymic catalysis: proceedings of a Royal Society discussion meeting held on 5th and 6th December, 1990, *Philos. Trans. R. Soc. London, Ser. B*, 1991, **332**, 105–84.
68. M. F. Ali, R. Grigg, S. Thiampatanagul, and V. Sridharan, *J. Chem. Soc., Perkin Trans. I*, 1988, 949; R. Grigg, D. Henderson, and A. J. Hudson, *Tetrahedron Lett.*, 1989, **30**, 2841.
69. R. G. Grigg, *Chem. Soc. Rev.*, 1987, **16**, 89.
70. R. G. Grigg and J. Kemp, *J. Chem. Soc., Chem. Commun.*, 1978, 109; M. Joucla and J. Hamelin, *Tetrahedron Lett.*, 1978, 2885.
71. R. G. Grigg, *J. Chem. Soc., Perkin Trans. 1*, 1988, 2693; R. G. Grigg, J. Idle, P. McMeckin, S. Surendrakamar, S. Thianpatanagul, and D. Vipond, *J. Chem. Soc., Perkin Trans. 1*, 1988, 2703; G. P. Rizzi, *J. Org. Chem.*, 1971, **36**, 1710.
72. R. Grigg, M. Jordan, and J. F. Malone, *Tetrahedron Lett.*, 1979, **35**, 3877; P. Arm-

strong, R. Grigg, M. S. Jordan, and J. F. Malone, *Tetrahedron Lett.*, 1985, **41**, 3547; R. Grigg, S. Thiampatanagul, and J. Kemp, *Tetrahedron Lett.*, 1988, **44**, 7283.
73. *F. &F.*, **15**, 2; M. M. Midland and J. I. McLoughlin, *Tetrahedron Lett.*, 1988, **29**, 4653.
74. *F. &F.*, **15**, 132; S. Hinec, V. Savic, and A. E. A. Porter, *Tetrahedron Lett.*, 1988, **29**, 6649.
75. *F. &F.*, **11**, 334; J. F. Kerwin, Jr and S. Danishefsky, *Tetrahedron Lett.*, 1982, **23**, 3739.
76. F. Orsini and G. Sala, *Tetrahedron Lett.*, 1989, **45**, 6531.
77. N. J. Leonard and D. F. Morrow, *J. Am. Chem. Soc.*, 1958, **80**, 371.
78. *F. &F.*, **3**, 114; 308; **4**, 135; 186.
79. *F. &F.*, **13**, 121; C. Rochin, O. Babot, J. Dunogues, and F. Doboudin, *Synthesis*, 1986, 228.
80. M. Gaudry, Y. Jasor, T. Bui Khac, and A. Marquet, *Org. Synth.*, 1980, **59**, 153; *F. & F.*, **8**, 194.
81. *F. &F.*, **7**, 131; S. Danishefsky, T. Kitahara, R. McKee, and P. F. Schuda, *J. Am. Chem. Soc.*, 1976, **98**, 6715.
82. *F. &F.*, **7**, 132; N. L. Holy, *Synth. Commun.*, 1976, **6**, 539.
83. *F. &F.*, **8**, 194; J. L. Roberts, P. S. Borromeo, and C. D. Poulter, *Tetrahedron Lett.*, 1977, 1299.
84. *F. &F.*, **11**, 256; J.-P. Dulcere and J. Rodriguez, *Tetrahedron Lett.*, 1982, **23**, 1887.
85. R. F. Borch, M. D. Bernstein, and D. Durst, *J. Am. Chem. Soc.*, 1971, **93**, 2897.
86. N. J. Leonard and J. V. Paukstellis, *J. Org. Chem.*, 1963, **28**, 3021; Y. Hata and M. Watanabe, *J. Am. Chem. Soc.*, 1973, **95**, 8450.
87. *F. &F.*, **13**, 106; J. Bergman and B. Pelcman, *Tetrahedron Lett.*, 1986, **27**, 1939.
88. J. Daly, *J. Chem. Soc.*, 1961, 2801; P. J. Wheatley, *Acta Crystallogr.*, 1954, **7**, 68; R. K. Bullough and P. J. Wheatley, *Acta Crystallogr.*, 1957, **10**, 233.
89. J. C. Jochms and F. A. L. Anet, *J. Am. Chem. Soc.*, 1970, **92**, 5524.
90. *F. &F.*, **5**, 282; G. R. Krow, *Angew. Chem., Int. Ed. Engl.*, 1971, **10**, 435.
91. *F. &F.*, **5**, 282; C. L. Stevens and J. C. French, *J. Am. Chem. Soc.*, 1954, **76**, 4398.
92. H. J. Bestman, J. Lienert, and L. Mott, *Annalen*, 1968, **718**, 24.
93. C. L. Stevens and G. H. Singal, *J. Org. Chem.*, 1964, **29**, 34.
94. H. Staudinger and J. Meyer, *Helv. Chim. Acta*, 1919, **2**, 635; 1921, **4**, 861; W. S. Wadsworth, Jr and W. D. Emmons, *J. Am. Chem. Soc.*, 1962, **84**, 1316.
95. *F. &F.*, **5**, 282; Naser-ud-din, J. Riegela, and L. Skattebøl, *J. Chem. Soc., Chem. Commun.*, 1973, 271; R. Graf, *Annalen*, 1963, **661**, 111.
96. *F. &F.*, **15**, 207; A. Battaglia, A. Dondoni, and P. Giogianni, *J. Org. Chem.*, 1982, **47**, 3998; G. Barbaro, A. Battaglia, and P. Giogianni, *J. Org. Chem.*, 1988, **53**, 5501; G. Barbaro, A. Battaglia, and P. Giogianni, *Tetrahedron Lett.*, 1987, **28**, 2995.
97. J. B. Falmagne, J. Escudero, S. Taleb-Sahraoui, and L. Ghosez, *Angew. Chem., Int. Ed. Engl.*, 1981, **20**, 879; A. Sidani, J. Marchand-Bryanaert, and L. Ghosez, *Angew. Chem., Int. Ed. Engl.*, 1974, **13**, 267; J. Marchand-Brynaert and L. Ghosez, *J. Am. Chem. Soc.*, 1972, **94**, 2869; 2891.
98. J. Marchand-Bryanaert and L. Ghosez, *Tetrahedron Lett.*, 1974, 377.

4

NITRILES (CYANIDES)

ALKYL NITRILES

$$\mathrm{H_3C\overset{1.46\,\text{Å}}{—}C\overset{1.13\,\text{Å}}{\equiv}N}$$

(1)

The carbon–carbon bond length in acetonitrile (methyl cyanide) (**1**) is 1.46 Å, which is shorter than a normal saturated carbon–carbon bond (1.54 Å) and the carbon–nitrogen bond length is 1.13 Å. Acetonitrile has a high dipole moment (3.44 D) and an infra-red absorption at 2260–2240 cm^{-1}. Thus nitriles are polarized molecules with the resonance hybrids shown in Scheme 4.1 contributing to the structure. The NMR spectrum of acetonitrile in $CDCl_3$ has one singlet: CH_3 at 2.0 δ.

$$\mathrm{H_3C{—}C{\equiv}N \longleftrightarrow H_3C{—}\overset{+}{C}{=}N^{-}} \qquad (4.1)$$

4.1 Preparation

4.1.1 *By substitution reactions*

From alkyl halides

Nitriles are obtained, usually in good yield, from alkyl halides and alkali metal cyanides (Scheme 4.2). The small amount of isocyanide (**2**) which is usually produced can easily be removed by washing with aqueous acid. However, tertiary alkyl halides (**3**) undergo elimination when they are treated with potassium cyanide. Therefore, it is advantageous to use cyanotrimethylsilane in this case[1] (Scheme 4.3).

$$\mathrm{R\text{-}CH_2\text{-}Cl \xrightarrow[\text{reflux}]{\text{KCN / EtOH}} R\text{-}CH_2\text{-}C{\equiv}N + \text{Some } R\text{-}CH_2\text{-}\overset{+}{N}{\equiv}\overset{-}{C}} \qquad (4.2)$$

(2)

$$\underset{(3)}{R_3C{-}Cl} \xrightarrow[SnCl_4\,/\,CH_2Cl_2]{Me_3SiCN} \underset{75 - 90\ \%}{R_3C{-}CN} \quad (4.3)$$

Improvements to the preparation can be made by the use of phase-transfer catalysis (e.g. $Et_4N^+CN^-$) or by the use of crown ethers. The use of cetyl quaternary ammonium salt as a surfactant has enabled good yields of nitriles to be obtained from alkyl halides (**4**), (Scheme 4.4).[2]

$$\underset{(4)}{R\text{-}Hal} + NaCN \xrightarrow[12\ hours]{C_{16}H_{33}\overset{+}{N}Me_3Br^-,\ H_2O,\ 100^\circ C} R\text{-}C{\equiv}N$$

R	Yield (%)
Bu^n	75
$n{-}C_{10}H_{21}$	85
Bn	87
$CH_2{=}CHCH_2$	84

(4.4)

Crown ethers have been used for the preparation of nitriles.[3] In one preparation potassium cyanide and 18-crown-6 dissolved in acetonitrile or chloroform were used to prepare nitriles from the corresponding benzylic chloride or bromide. This method was also used to prepare cyanotrimethylsilane, Me_3SiCN, from trimethylchlorosilane, Me_3SiCl (45%).[4] In another preparation, alkyl halides and potassium cyanide plus a catalytic amount of 18-crown-6 in acetonitrile or benzene gave good yields of nitriles (Scheme 4.5).[5]

$$R\text{-}Hal \xrightarrow{KCN,\ 18\text{-}Crown\text{-}6} R\text{-}C{\equiv}N$$

R	Hal	Yield (%)
$-CH_3CH_2-$	Cl	97
$CH_3(CH_2)_4$	Br	100
Bn	Cl	94

(4.5)

Carboxylic acids (**5**) can be converted into nitriles with chlorosulphonyl isocyanate (**6**), (Scheme 4.6).[6]

A free radical method for preparing nitriles (**7**) from alkyl halides (**8**) has recently been introduced (Scheme 4.7). The yield was 82 per cent on a steroidal halide.[7]

$$RCO_2H\ (5) + OCN\text{—}SO_2Cl\ (6) \longrightarrow [R\text{-}C(=O)\text{-}O\text{-}C(=O)\text{-}NH\text{-}SO_2Cl] \xrightarrow{-CO_2} R\text{-}C(=O)\text{-}NH\text{-}SO_2Cl \xrightarrow{DMF} R\text{—}C{\equiv}N + ClSO_3H.DMF \quad (4.6)$$

$$\underset{(8)}{R\text{-}I} \xrightarrow[\text{NaBH}_3\text{CN, AIBN}]{\text{Bu}^n{}_3\text{SnCl}} \underset{(7)}{R\text{-}C{\equiv}N} \quad (4.7)$$

Use of cyanogen/cyanogen chloride

Some Grignard reagents (**9**) react with cyanogen (**10**), or with cyanogen chloride (**11**), to yield nitriles, but the reactions have severe limitations (Scheme 4.8).[8] With cyanogen the amount of Grignard reagent must be limited to 1 mole and reverse addition must be used in order to prevent further reactions which eventually yield ketones.[9] Cyanogen chloride reacts to give nitriles but cyanogen bromide or iodide do not and even with cyanogen chloride secondary or tertiary Grignard reagents give very poor yields of nitriles. Similarly, stable carbanions (**12**) undergo substitution reactions with cyanogen chloride (**13**) to yield nitriles,[10] for example, Scheme 4.9.

$$\underset{(9)}{R\text{-}CH_2\text{-}MgHal} + \underset{(10)}{(CN)_2} \longrightarrow R\text{-}CH_2C{\equiv}N$$

$$\underset{(9)}{R\text{-}CH_2\text{-}MgHal} + \underset{(10)}{ClCN} \longrightarrow R\text{-}CH_2C{\equiv}N \quad (4.8)$$

$$\underset{(13)}{Cl\text{-}CN} + \underset{(12)}{\bar{C}H(CO_2Et)_2} \longrightarrow N{\equiv}C\text{-}CH(CO_2Et)_2 + Cl^- \quad (4.9)$$

Enamines (**14**) can be converted into β-ketonitriles (**15**) by the use of cyanogen chloride (Scheme 4.10).[11]

$$\text{(14)} \xrightarrow{\text{Cl—CN}} \text{CN-enamine} \longrightarrow \text{(15)} \quad (4.10)$$

From alcohols

Nitriles can be prepared from alcohols (**16**) by the action of sodium cyanide and trimethylsilyl chloride catalysed by sodium iodide in a mixture of dimethylformamide and acetonitrile (1:1),[12] e.g. Scheme 4.11.

$$\underset{(16)}{\text{Ph-CH}_2\text{OH}} \xrightarrow[\text{NaI (catalyst)}]{\text{NaCN / Me}_3\text{SiCl}} \underset{98\ \%}{\text{Ph-CH}_2\text{-C}\equiv\text{N}} \quad (4.11)$$

4.1.2 *By addition reactions*

For preparation of cyanohydrins

Cyanohydrins (**17**) result from the reaction of aldehydes (**18**) and ketones with hydrogen cyanide (Scheme 4.12). An extension of this reaction is the Strecker α-amino acid (**19**) synthesis in which addition of both cyanide and ammonia occurs, via an aminonitrile (**20**) (Scheme 4.13).[13]

$$\underset{(18)}{\text{R-CHO}} \xrightarrow{\text{HCN}} \underset{(17)}{\text{R-CH(OH)-C}\equiv\text{N}} \quad (4.12)$$

$$\text{R-CHO} \xrightarrow{\text{NH}_4\text{CN}} \underset{(20)}{\text{R-CH(NH}_2\text{)-C}\equiv\text{N}} \xrightarrow{\text{acid hydrolysis}} \underset{(19)}{\text{R-CH(NH}_2\text{)-CO}_2\text{H}} \quad (4.13)$$

Cyanohydrins (**21**) are best prepared from silated cyanohydrins (**22**), which result from the reaction of cyanotrimethylsilane (**23**) with an aldehyde or a ketone (**24**) in the presence of a catalytic amount of zinc iodide in methylene dichloride at 65 °C (Scheme 4.14). This is a very general reaction, and so it is possible to substitute various alkyl groups.[14] The yields are usually between 90

$$R^1\text{-}C(=O)\text{-}R^2 \;(24) + Me_3SiCN \;(23) \xrightarrow[CH_2Cl_2]{ZnI_2 \text{ at } 65^\circ C} R^1\text{-}C(OSiMe_3)(CN)\text{-}R^2 \;(22) \xrightarrow[25-45^\circ C,\ 30 \text{ min to } 23 \text{ hours}]{3M\text{-}HCl} R^1\text{-}C(OH)(CN)\text{-}R^2 \;(21) \quad (4.14)$$

and 100 per cent and the hydrolysis to the cyanohydrin is performed with aqueous acid.[15]

Cyanotrimethylsilane (**23**) can be prepared from sodium or potassium cyanide and trimethylsilylchloride (**25**) in *N*-methylpyrrolidone at *c.* 100 °C (Scheme 4.15).[16] Cyanotrimethylsilane can also be prepared from trimethylsilyl chloride, by reaction with sulphuric acid followed by addition of potassium cyanide (Scheme 4.16).[17]

$$Me_3SiCl \;(25) + KCN \xrightarrow[\text{in } N\text{-methylpyrrolidone}]{100^\circ C} Me_3SiCN \;(23) \quad c.\ 65\,\% \quad (4.15)$$

$$2\,Me_3SiCl \;(25) + H_2SO_4 \longrightarrow 2\,HCl + (Me_3SiO)_2SO_2 \xrightarrow{KCN} 2\,Me_3Si\text{-}CN \;(23) \quad (4.16)$$

Trimethylsilyl cyanohydrins can be prepared from trimethylsilyl isoselenocyanate (**26**), prepared as shown in Scheme 4.17.[18] This compound reacts with aldehydes in cyclohexane in the presence of zinc chloride to give trimethylsilyl cyanohydrins. It is selective for aldehydes, in the presence of keto groups, and for aliphatic aldehydes, in the presence of aromatic aldehydic groups.

$$Me_3Si{-}Cl + KSe{-}C{\equiv}N \xrightarrow[25\,^\circ C]{CH_3CN \text{ or } DME} Me_3Si{-}N{=}C{=}Se \;(26) \quad (4.17)$$

Chiral cyanohydrins

Enzymic methods can be used to obtain chiral cyanohydrins (**27**), for example (Scheme 4.18).[19]

More recently, an enzyme in organic solvents has been used to achieve kinetic resolution of a racemic cyanohydrin (**28**), (Scheme 4.19) (see Ref. 20 for a review of the field).

$$\text{R-CHO} \xrightarrow{\text{i}} \text{R-C(OH)(H)-CN} \quad \textbf{(27)}$$

i = mandelonitrile lyase extract in 1 M KCN/HOAc buffer, pH 5.4

R	Yield (%)	% of *R*-enantiomer
$CH_3CH_2CH_2$	98	93
$CH_2CH{=}CH$	94	95

(4.18)

$$\underset{\textbf{(28)}}{\text{NC-CH(OH)-CH}_2\text{CO}_2\text{Et}} \xrightarrow[\text{pyridine}]{\text{Ac}_2\text{O}} \text{NC-CH(OAc)-CH}_2\text{CO}_2\text{Et}$$

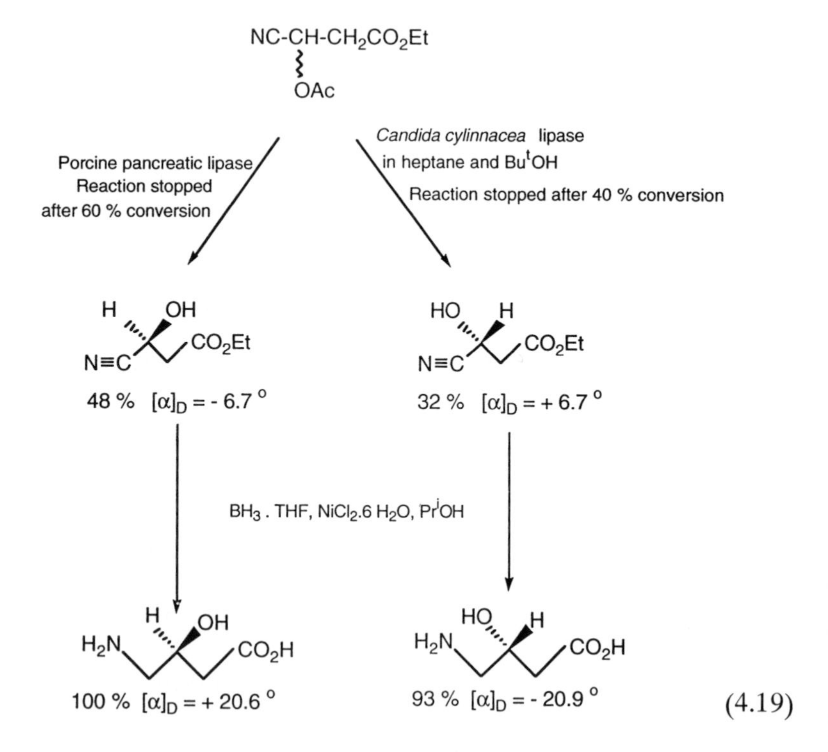

(4.19)

Michael addition

By the Michael addition of a carbanion (**29**) to acrylonitrile (**30**) (cyanoethylation), synthetically useful nitriles (**31**) can be prepared (Scheme 4.20).[21]

$$(EtO_2C)_2\bar{C}H + CH_2{=}CH{-}C{\equiv}N \longrightarrow (EtO_2C)_2CH\text{-}CH_2\text{-}CH{=}C{=}N^- \xrightarrow{H_3O^+} (EtO_2C)_2CH\text{-}CH_2\text{-}CH_2\text{-}C{\equiv}N$$

(29) (30) (31) (4.20)

α,β-Unsaturated ketones in benzene react with the 'naked' cyanide ion, produced by reaction of acetonitrile, 18-crown-6-ether, and potassium cyanide. Then acetone cyanohydrin is added and the mixture is stirred for from 3 to 20 h at 20 °C. The thermodynamically more stable product (**32**) predominates,[22] for example, from Δ^4(10)-octalin-3-one (**33**) (Scheme 4.21).

(33) $\bar{C}N$

KCN / $(CH_3)_2C(OH)CN$

18-Crown-6

PhH, 20° C, 3 hours

$H{-}O{-}C(CH_3)_2CN$

+ CN^- + $(CH_3)_2C{=}O$

(32) 3.4 : 1.0, *trans* : *cis*

Yield 85% (4.21)

Nitriles can be prepared in high yields by the Michael addition of cuprous methyltrialkylborates (**34**) (prepared from the reaction between a trialkylborane and lithium methyl (**35**), followed by reaction with cuprous bromide) and acrylonitrile (Scheme 4.22).[23]

Nitriles can also be prepared by an electrochemical method. The electrolysis of a solution of a trialkylborane (**36**) in a nitrile (**37**) containing tetraalkylammonium halide between platinum electrodes yields nitriles after oxidation with alkaline hydrogen peroxide (Scheme 4.23). This is a useful method for chain extending nitriles.[24] Yields were based on the amount of organoborane used. The authors, on good evidence, proposed the mechanism shown in Scheme 4.24.

$$R_3B + LiCH_3 \xrightarrow[\text{ether}]{\text{THF}} (R_3B\text{-}CH_3)Li \xrightarrow[0^\circ C]{\text{CuBr/THF}} (R_2B\text{-}CH_3)Cu + LiBr$$

(35) (34)

$$(R_2B\text{-}CH_3)Cu \text{ (R)} + CH_2{=}CH{-}C{\equiv}N \longrightarrow [R\text{-}CH_2\text{-}CH{=}C{=}N^-] \xrightarrow{H_2O} R\text{-}CH_2\text{-}CH_2\text{-}C{\equiv}N$$

R	Yield (%)
Pr^n	84
Bu^n	88
Bu^i	88
n-pentyl	89
n-hexyl	93

(4.22)

$$\underset{(36)}{R_3B} + \underset{(37)}{CH_3CN} \xrightarrow[\text{ii) } H_2O_2,\ \bar{O}H]{\text{i) } Et_4NI,\ Pt\text{-}Pt} RCH_2CN$$

R	Yield (%)
n-octyl	53
n-hexyl	57
n-butyl	32
sec-butyl	47

(4.23)

$$CH_3{-}CN \xrightarrow[\text{Pt-Pt, } Et_4NI]{+e} \bar{C}H_2{-}CN + 1/2\ H_2$$

$$3I^- \xrightarrow{-2e} I_3^- \xrightarrow{-e} I_2 + I^\cdot$$

$$R_3B + I^\cdot \longrightarrow RI + R_2B^\cdot$$

$$NC{-}CH_2^- + RI \longrightarrow NC{-}CH_2{-}R + I^- \quad (4.24)$$

4.1.3 *By elimination of water from amides*

A range of Lewis acids have been used to cause the dehydration of amides (**38**), the mechanism often being through the 'enol' of the amide (**39**), for example by the use of hexamethylphosphoric triamide, HMPT[27] (Scheme 4.25). Table 4.1

$$\underset{(38)}{R\text{-}C(=O)\text{-}NH_2} \rightleftharpoons \underset{(39)}{R\text{-}C(OH)=NH} \xrightarrow[220^\circ \text{ to } 240^\circ C]{O=P(NMe_2)_3} \left[R\text{-}C(\text{-}O\text{-}P(=O)(NMe_2)_2)=N\text{-}H \right] + HNMe_2 \longrightarrow HO\text{-}P(=O)(NMe_2)_2 + R\text{-}C\equiv N \qquad (4.25)$$

Table 4.1 Elimination of water from amides

$$R\text{-}CO\text{-}NH_2 \longrightarrow R\text{-}C\equiv N + H_2O \qquad (4.26)$$

Reagent	Conditions	R	Yield (%)	Ref.
$TiCl_4/Et_3N$ or *N*-methyl morpholine	In THF at 0 °C	Me, Et, Pr^n, pivalic Pr^i, *n*-pentyl	66–84	25, 26
HMPT	220–240 °C	Et, Pr^n, *n*-pentyl	49–94	26, 27
Compound (**40**)	>100 °C	Me,CH_2=CH,PhCH_2	75–100	28
1 mole of (**40**): 3 moles of amide	(e.g. reflux in PhCl)			
$(F_3C.CO)_2O$ in pyridine	<5 °C 20 °C In dioxan	Me, PhCH_2 $CH_3(CH_2)_{16}$	50–97	29
Polyphosphate ester	In $CHCl_3$	cyclohexyl, cyclopentyl, Bn, and Aryl	35–90	29
Phase transfer $BnNEt_3^+Cl^-$, $CHCl_3$, aq.NaOH	22 or 40 °C 2 or 3 h	PhCH_2, Et, PhCH=CH	45–84	28, 30
Phosphonitrilic chloride	In C_6H_5Cl, reflux for 2 or 3 h	Me, Bn, Aryl	75–100	31
Ph_3P=O (2 eq.) $(F_3C-SO_2)_2O$ (1 eq.)	25 °C	Aryl	>90	32
Methyl (carboxysulphamoyl)triethylammonium hydroxide				
Inner salt	25 °C		80–90	33
OC=N–$SOCl_2$			*c.*60	34
Ph_3PBr_2	PhH + Et_3N, 80 °C	PhCH=CH, PhCH_2	54, 64	35

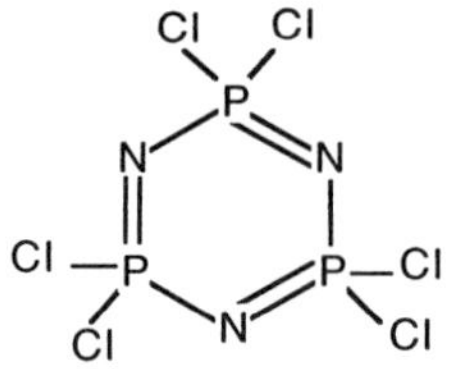

(40)

lists a variety of modern methods and reagents many of which require mild conditions (Scheme 4.26).

4.1.4 *By elimination of water from oximes*

As with the dehydration of amides, that of oximes (**41**) has been effected with a range of Lewis acids, as shown Table 4.2. The general mechanism is typified by trifluoroacetic anhydride and pyridine[40] (Scheme 4.27).

Table 4.2 Elimination of water from oximes

$(F_3C\text{-}CO)_2O$ → R-C≡N

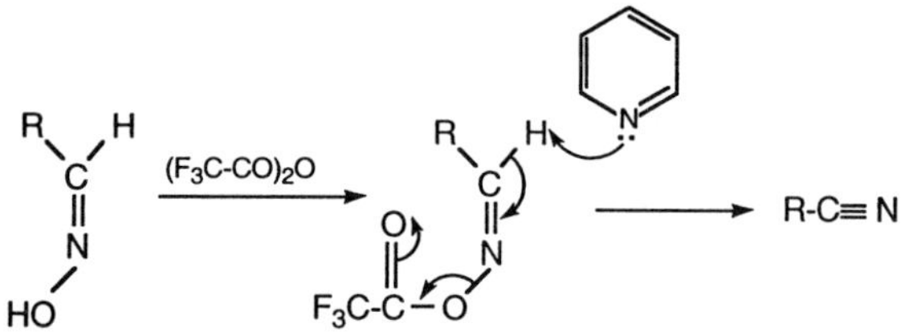

(41)

$$\text{R-CH=NOH} \longrightarrow \text{R-C}\equiv\text{N} + H_2O \qquad (4.27)$$

Reagent	Conditions	R	Yield (%)	Ref.
$TiCl_4$	22 or 80 °C, 25–50 h in THF or dioxan	Me, Et, Pr^n, Pr^i Bu^n	81–89	36
HMPT	220 °C, reflux	Pr^n, Bu^t	89, 95	37
Compound (**40**)	20 °C in ether or THF + 3 Et_3N: (1),3,7-dimethyl-2,6-octadiene	PhCH=CH (*E* = *Z*) undeca, heptyl	93–98	38
$(CF_3SO_2)_2O$	−78 to 20 °C *c.* 2 h	$PhCH_2CH_2$, $PhCH_2$ PhCH=CH, $CH_3(CH_2)_4$ in CH_2Cl_2 + Et_3N	93	39

Reagent	Conditions	R	Yield (%)	Ref.
$(F_3C.CO)_2O$	<5 then 20 °C pyridine or Et_3N 6 h polyene (2)	$CH_3(CH_2)_6$	94	40
1,1′-Dicarbonyl bi-imidazole	20 °C in CH_2Cl_2 Dehydrates chiral oximes without racemization of the α-C atom	$(CH_3)_3CCH_2$	95	45
SeO_2 in $CHCl_3$	20 °C	Pr^n, n-hexyl, n-heptyl, cyclohexyl	59–74	41
2,4,6-Trichloro-s-triazine/pyridine	20 °C 1 h	Pr^n	64	42
p-$ClC_6H_4O{-}C(Cl){=}S$*	20 °C in ether + pyridine Cl	n-hexyl	43	43
p-Chlorophenyl chlorothionoformate	pyridine (1–2 eq.) 1.5 h	$C_6H_4NO_2$	70	44
Phosphorylated imidazole	Dioxane 2 h, 50 °C	alkyl, aryl	80–97	46
$Cl_2C{=}NH_2^+Cl^-$	20 °C	CH_3, Ph, $MeCOC_6H_4$	92–98	47
$Ph_3P{=}O$ (2 eq.) + $(F_3C{-}SO_2)_2$ (1 eq.)	25 °C	4-Me—C_6H_4	>90	48
$MeN^+{=}CCl_2Cl^-$	Reflux in $CHCl_3$ for 2–4 h	Me, Aryl	>90	47
$H_3C{-}C{\equiv}CNEt^+BF_4^-$ CH_3CN $(CH_3CN + Et_3O^+BF_4^-)$	8 h at 20 °C and 30 min 80 °C	Pentyl	75–85	49, 50
PhO—SOCl[51]	Pyridine 5 °C then 25 °C	Aryl	>90	52
PhOCOCl	Heat ester at >100 °C	Aryl	85–90	53
CH_3NCO	Et_3N + DMF 110–120 °C	Aryl	65–100	54
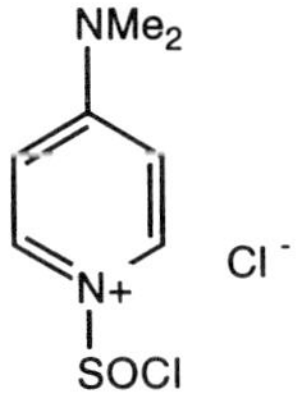 * NMe2 (pyridinium) N+ SOCl Cl^-; NMe2 (pyridine) N	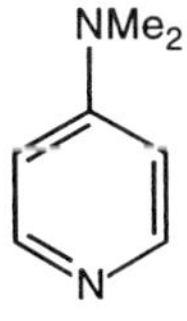 CH_2Cl_2 -10 to 10°C 10 to 30 min.	Aryl	90–100	55
$Me_2N{=}CH{-}O{-}S({=}O){-}Cl$ Cl^-	DMF, SOCl pyridine	4-Cl—C_6H_4, Ph	80–90	56
18-crown-6-ether	KBr in CH_3CN for 14 days	Ph	>90	57

Table 4.2 (*cont.*)

* preparation:

$$\text{4-Me}_2\text{N-pyridine} + SOCl_2 \xrightarrow[-10^\circ C]{CH_2Cl_2} \text{4-Me}_2\text{N-pyridinium-}N\text{-SOCl} \; Cl^-$$

Several good preparations of nitriles are known from aldehydes (**42**) via oximes (**43**). For example, the method using selenium dioxide (see Table 4.2) has been modified and the preparation can be carried out directly from the aldehyde, without isolation of the oxime. Leibscher and Hartman[58] heated aldehydes with hydroxylamine hydrochloride in DMF and obtained yields of 65–97 per cent. In another similar method,[59] the aldehyde, which is dissolved in dimethylformamide, chloroform, or ethanol and chloroform, is heated with hydroxylamine hydrochloride (**44**) for 2 to 3 h and selenium dioxide is added (Scheme 4.28). The yields are in the range 70–84 per cent for R = CH_3, Pr^i, *n*-hexyl, cyclohexyl, $HO(CH_2)_4$, $CH_3CH{=}CH$, and $CH_3CH(OH)CH_2$.

$$\underset{(42)}{\text{R-CHO}} + \underset{(44)}{H_3\overset{+}{N}OH\ Cl^-} \xrightarrow[\text{CHCl}_3\text{ - EtOH, 2 to 3 hours}]{\text{DMF or CHCl}_3\text{ or}} \underset{(43)}{[\text{R-CH=NOH}]} \xrightarrow{SeO_2} \text{R-C}\equiv\text{N} \qquad (4.28)$$

In a one-pot synthesis, dicyclohexylcarbodiimide (**45**) acts as a dehydrating agent, giving a substituted urea (**46**) (Scheme 4.29).[60]

$$\text{R-CHO} \xrightarrow[\text{Pyridine / }H_2O\text{ / 1 hour}]{H_3\overset{+}{N}OH\ \overset{-}{Cl}\text{ / }20^\circ C} [\text{R-CH=NOH}] \xrightarrow[Et_3N\text{ / DCC (45) / 2 hours}]{CuSO_4\text{ / }CH_2Cl_2} \text{R-C}\equiv\text{N}$$

DCC = Dicyclohexylcarbodiimide

$$\underset{\text{DCC (45)}}{C_6H_{11}\text{-N=C=N-}C_6H_{11}} \xrightarrow{H_2O} \underset{(46)}{C_6H_{11}\text{-NH-C(=O)-NH-}C_6H_{11}}$$

Product	Yield (%)
Ph–CH=CH-CN	93
duodecanitrile	93
cyclohexyl nitrile	88

(4.29)

In another preparation, hydrogen chloride is used to cause dehydration of the intermediate oxime (Scheme 4.30).[61] The yields were between 83 and 94 per cent, with the following starting materials: *n*-octanal, *n*-tetradecanal, *n*-hexadecanal bisulphite compound, *n*-octadecanal oxime.

$$\text{R-CHO} \xrightarrow[\text{95\% EtOH, reflux, 6 hours}]{\overset{+}{N}H_3OH\ \overset{-}{Cl}\ /\ HCl} \text{R-C}\equiv\text{N} \qquad (4.30)$$

Formic acid was first used by van Es.[62] The aldehyde was boiled under reflux with hydroxylamine hydrochloride, formic acid, and sodium formate for 1 h; the yields were low: n-$C_6H_{13}CN$ 30 per cent and Pr^nCN 42 per cent. Later[63] the method was modified by the use of hydroxylamine hydrochloride and 95–98 per cent formic acid for half an hour. The yields were considerably better, being between 77 and 95 per cent for Bu^nCN, Bu^sCN, n-$C_5H_{11}CN$, and n-$C_6H_{13}CN$.

In another method, an aldehyde (**47**) is treated with an excess of the hydroxylamine-*O*-sulphonic acid[64] (Scheme 4.31).[65]

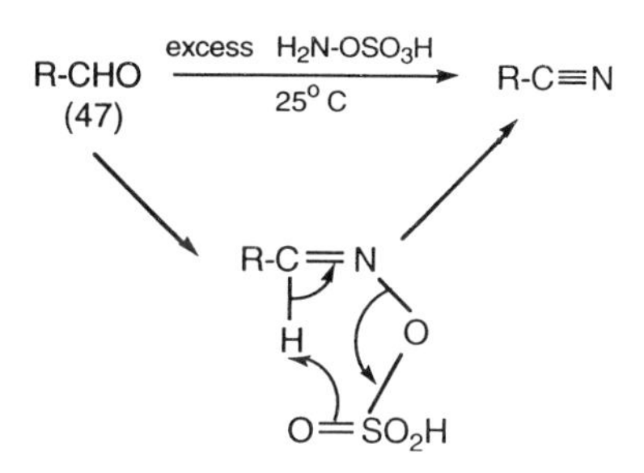

R	Yield (%)
Et	80
Pr^n	90
$HOCH_2$	60
Ph	85

(4.31)

Nitriles have been prepared directly from aldehydes (**48**) via oximes using *N*-methylpyrrolidone (**49**) as a solvent (Scheme 4.32). Satisfactory yields are recorded.[66]

4.1.5 *By oxidation of primary amines*

Primary amines (**50**) are oxidized by lead tetraacetate in benzene (Scheme 4.33).[67] As it seemed likely that imines were intermediates in these oxidations

$$\underset{(48)}{R\text{—}CHO} \xrightarrow[N\text{-methylpyrrolidone}]{NH_2OH} R\text{—}CN$$

N-methylpyrrolidone

(49) (4.32)

$$\underset{(50)}{R\text{-}CH_2\text{-}NH_2} \xrightarrow[\text{PhH, reflux}]{Pb(OAc)_4 \text{ (2 mole)}} R\text{-}C{\equiv}N$$

c. 60 % yields (4.33)

with lead tetraacetate, other workers developed a good method of preparing nitriles by oxidizing cobalt complexes of aldimines (**51**), prepared from aldehydes (**52**) and ammonia, with bromine (Scheme 4.34).[68]

$$\underset{(52)}{R\text{-}CHO} + NH_3 \xrightarrow[0^\circ C]{CH_3CN,\ Co_2(CO)_8} \underset{(51)}{(R\text{-}CH{=}NH)_n \rightarrow Co(CO)_{4-n}} \xrightarrow[0^\circ C]{Br_2} R\text{-}C{\equiv}N$$

c. 70 % yields (4.34)

The oxidation of primary amines (**53**) with silver(II) picolinate[69] yields nitriles along with smaller amounts of the corresponding aldehyde (Scheme 4.35).[70] In this synthesis an intermediate imine (**54**) is either hydrolysed to the aldehyde (**55**) or oxidized further to the nitrile (**56**) (Scheme 4.36).

$$\underset{(53)}{R\text{-}CH_2\text{-}NH_2} \xrightarrow[\text{Silver picolinate, } 20^\circ C]{\text{Aqueous}} R\text{-}C{\equiv}N + R\text{-}CHO \qquad (4.35)$$

4.1.6 *From aldehydes and ketones*

Diphenylsulphilimine (**57**)[71] has been used[72] for the conversion of aldehydes (**58**) into nitriles (**59**) (Scheme 4.37).

Aldehydes can be converted into nitriles by reaction with *N,N*-dimethylhydrazine (**61**) followed by quaternization with methyl iodide or methyl-*p*-tolylsulphonate and finally elimination in the presence of alkali (Scheme 4.38).[73]

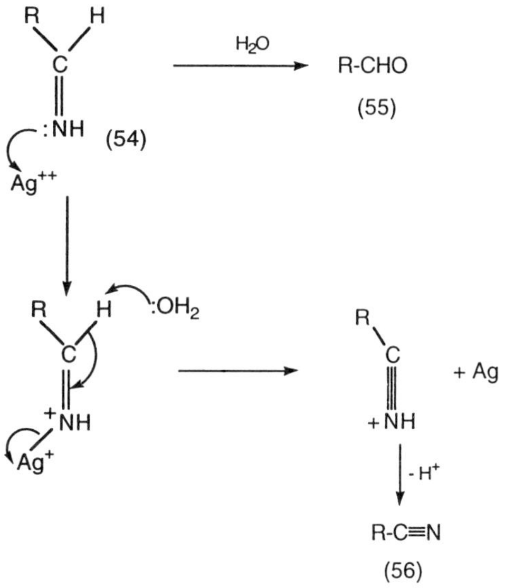

R H
C
:NH (54)
Ag^{++}

$\xrightarrow{H_2O}$ R-CHO
(55)

R H :OH_2
C
$^+$NH
Ag^+

$\longrightarrow$

R
C
+NH + Ag

$\downarrow$ - H^+

R-C≡N
(56)

R	RCN	CHO
Bu^n	77	23
n-C_5H_{11}	84	16
$CH_2CH_2CH(Et)CH_2$	90	10
$CH_3(CH_2)_3CH(Et)CH_2$	92	8
$CH_3CH(Et)CH_2CH_2$	76	24
$CH_3(CH_2)_{11}$	75	25
$CH_3(CH_2)_7$	68	32
Bn	52	48
$CH_3CH(Me)CH_2$	60	40
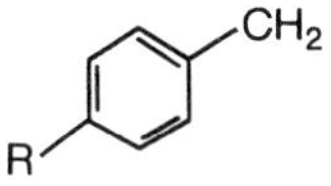 CH_2 / R (4-substituted benzyl)		
R$=$$NO_2$	12	88
R$=$$CHMe_2$	77	23
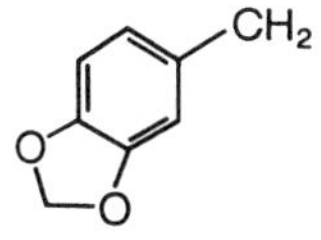 CH_2 / O, O (3,4-methylenedioxybenzyl)	84	16

(4.36)

$$\text{R-CHO (58)} + \text{HN=SPh}_2\ (57) \xrightarrow[80\text{-}85^\circ C]{\text{DMSO or PhH}} \left[\text{R-CH(O}^-\text{)-NH-}\overset{+}{\text{S}}\text{Ph}_2 \longrightarrow \text{R-CH(-O-)(-HN-)SPh}_2\right] \longrightarrow \text{Ph}_2\text{S=O} + [\text{R-CH=NH}] \longrightarrow \text{R-C}\equiv\text{N}\ (59)$$

$$\text{TsN=SPh}_2 \xrightarrow{\text{i, conc. } H_2SO_4;\ \text{ii, aq. } HO^-} \text{HN=SPh}_2\ (57)$$

R	Yield (%)
Bu^t	66
$PhCHCH_3$	40
$CH_3CH{=}CH$	43
$(CH_3)_2C{=}CH$	85

(4.37)

$$\text{R-CHO (60)} + \text{H}_2\text{N-NMe}_2\ (61) \xrightarrow[\text{reflux}]{\text{PhH}} \text{R-CH=NNMe}_2 \xrightarrow[p\text{-CH}_3\text{C}_6\text{H}_4\text{SO}_3\text{CH}_3]{\text{CH}_3\text{I or}} \text{R-CH=N-}\overset{+}{\text{N}}\text{Me}_3\ \text{I}^-\ (62) \xrightarrow[\text{NaOMe / MeOH}]{^-\text{OMe}} \text{R-C}\equiv\text{N} + \text{NMe}_3\ (63)$$

R	Yield (%) (**62**)	Yield (%) (**63**)
n-heptyl	92	62
$PhCH{=}CH$	82	83
n-valero	81	51
$PhCH_2CH_2$	85	78

(4.38)

4.1.7 *Via hydrazones*

N-Amino-4,6-diphenylpyridone (**64**)[74] is a derivative of hydrazine and reacts with aldehydes. The resulting hydrazones (**65**) decompose on being heated to yield nitriles,[75] for example, Scheme 4.39.

(4.39)

Good yields of nitriles have been obtained from aldehydes by opening epoxides (**66**) with 1,1-dimethylhydrazine (**67**) and subsequent reaction with the aldehyde to form an intermediate quaternary hydrazone (**68**) (Scheme

R^1 = H or CH_3

R^2	Yield (%)
$CH_3(CH_2)_3CHEt$	76
PhCH=CH	86

(4.40)

$$R^1R^2C{=}O + H_2N\text{-}NH\text{-}SO_2\text{-}Ar \xrightarrow[20^{\circ}\text{C, 1-2 hours}]{\text{THF or MeOH}} R^1R^2C{=}N\text{-}NH\text{-}SO_2\text{-}Ar \ (69)$$

$$\xrightarrow{\text{excess of KCN, reflux in MeOH, 1-2 hours}} R^1R^2C(CN)\text{-}N{=}NH + {}^-O_2S\text{-}Ar \longrightarrow R^1R^2CH\text{-}C{\equiv}N + N_2$$

Ar = 2,4,6-$(Me_2CH)_3C_6H_2$

Aldehyde or Ketone	Product	Yield (%)
cyclopentanone	cyclopentanecarbonitrile	47
cyclohexanone	cyclohexanecarbonitrile	71
$(CH_3)_3C\text{-}COCH_3$	$(CH_3)_3C\text{-}CH(CN)\text{-}CH_3$	60
$CH_3\text{-}CH_2\text{-}CO\text{-}(CH_2)_2\text{-}CH_3$	$CH_3\text{-}(CH_2)_2\text{-}CH(CN)\text{-}(CH_2)_2\text{-}CH_3$	74
$CH_3(CH_2)_5CHO$	$CH_3(CH_2)_6CN$	72
$CH_3(CH_2)_6CHO$	$CH_3(CH_2)_7CN$	65
$CH_2{=}CH\text{-}(CH_2)_2\text{-}CO\text{-}CH_3$	$CH_2{=}CH\text{-}(CH_2)_2\text{-}CH(CN)\text{-}CH_3$	70
17-hydroxy-5α-androstan-3-one (steroid, OH, H)	3-cyano steroid (H, CN, OH, H)	70

(4.41)

4.40). Presumably the hydrazone reacts in this regiospecific way because the inductive effect of the methyl groups outweighs their steric hindrance.[76]

Addition of potassium cyanide to a hydrazone (**69**) of aldehydes and ketones yields nitriles, as shown in Scheme 4.41.[77] The starting ketone may be unsaturated or it can contain a hydroxyl group.

Methyl carbazate (**70**), a derivative of hydrazine, reacts in the normal way with a ketone (**71**) to give a hydrazone (**72**).[78] If the hydrazone undergoes an addition reaction with hydrogen cyanide in methanol and without isolation, the intermediate (**73**) may be oxidized with bromine in alkali to give a diazene (**74**). Treatment of this with sodium methoxide in methanol yields a nitrile (**75**). Methylation of the intermediate carbanion yields an α-methylnitrile (**76**) and carboxylation gives a cyanoester (**77**) (Scheme 4.42).[79] The mechanism of the reaction with the diazine is thought to be as shown in Scheme 4.43.

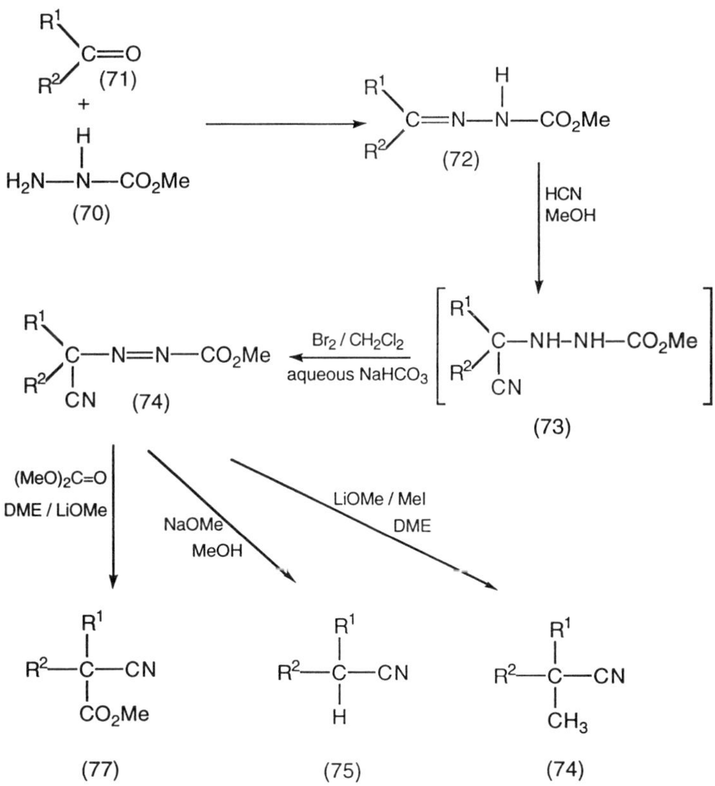

(4.42)

In a similar reaction, nitriles can be synthesized in yields of about 60 per cent by the action of potassium cyanide on tosylhydrazones of ketones (**81**), followed by thermal decomposition of the intermediate hydrogen cyanide addition product (**82**) (Scheme 4.44).[80] However, in a later paper Cacchi *et al.*[81]

R1 R2 C N=N C OMe O C N OMe

N_2 + $(MeO)_2C{=}O$ + [R1 R2 C=C=N⁻ ⟷ R1 R2 C⁻—C≡N]

aq. H_3O^+ $(MeO)_2C{=}O$ MeI

R1 R2 C—C≡N H (78)

R1 R2 C—C≡N CO_2Me (79)

R1 R2 C—C≡N CH_3 (80)

Diazene	Yield (%) (**78**)	Yield (%) (**79**)	Yield (%) (**80**)
Cyclohexyl	94	77	77
2-Methylcyclohexyl	89	—	63
4-*t*-Butylcyclohexyl	97	—	70

(4.43)

=N-NH-Ts (81) —i, KCN / MeOH / HOAc; ii, 2 M KOH→ NH-NH-Ts CN (82)

180° C

H CN + N_2 + TsH

(4.44)

pointed out that a complicating side reaction occurs by the elimination of hydrogen cyanide from the cyanohydrazide.

4.1.8 *From α-keto acids*

Good yields of nitriles result when oximes (**83**) of α-keto acids (**84**) are boiled under reflux with aqueous hydroxylamine hydrochloride.[82] The authors favour a *trans* elimination mechanism for this reaction (Scheme 4.45).

$$R\text{-}C(=O)\text{-}CO_2H \; (84) \xrightarrow[\text{reflux under } N_2]{\text{aq. } \overset{+}{N}H_3OH \; Cl^-} R\text{-}C(=N\text{-}H_2O^+)\text{-}C(=O)\text{-}O\text{-}H \; (83) \longrightarrow R\text{-}C{\equiv}N + CO_2 + H_2O$$

R	Yield (%)
CH_3	98
CH_3CH_2	55
$PhCH_2$	90
β-Indolylmethyl	62

(4.45)

Better yields result from the reaction of the oxime of an α-keto acid (**85**) with phenyl isocyanate when a double decarboxylation occurs (Scheme 4.46).[83]

$$R\text{-}C(=N\text{-}OH)\text{-}CO_2H \; (85) + Ph\text{-}N{=}C{=}O \xrightarrow[\text{under } N_2]{\text{PhH, } 20^\circ C} R\text{-}C(=N\text{-}O\text{-}C(=O)\text{-}NH\text{-}Ph)\text{-}C(=O)\text{-}O\text{-}H \longrightarrow R\text{-}C{\equiv}N + 2\,CO_2 + PhNH_2$$

R	Yield (%)
Me	91
Et	95
Pr^i	94
HO_2CCH_2	98
$HO_2C(CH_2)_2$	93

(4.46)

4.1.9 *From alcohols*

Alcohols (**86**) can be readily converted to nitriles in good yield by consecutive addition of lithium cyanide and the alcohol to a preformed complex of triphenylphosphine and diethyl azodicarboxylate (Scheme 4.47).[84] The mechanism is believed to proceed as shown in Scheme 4.48.

$$\underset{(86)}{\text{R-OH}} + \text{LiCN} \xrightarrow[\text{THF, 0 to 25°C}]{\text{Ph}_3\text{P, EtO}_2\text{C-N=N-CO}_2\text{Et}} \text{R-C}\equiv\text{N} \qquad (4.47)$$

$$\text{EtO}_2\text{C—N=N—CO}_2\text{Et} + \text{R—O—H} + \text{:PPh}_3 \longrightarrow \text{EtO}_2\text{C—}\bar{\text{N}}\text{—N(H)—CO}_2\text{Et} + \text{NC}^- \ \ \text{R—O—}\overset{+}{\text{P}}\text{Ph}_3 \longrightarrow \text{R-CN} \qquad (4.48)$$

A primary alcohol (**87**) may be converted to a nitrile by treatment with normal tri-*n*-butyl phosphine (Bu^n_3P), carbon tetrachloride, and potassium cyanide along with 18-crown-6 in acetonitrile, usually at room temperature (Scheme 4.49). In the absence of the crown ether no reaction takes place. Benzyl and primary aliphatic alcohols react smoothly but secondary and tertiary alcohols give mixtures of products.[85]

$$\underset{(87)}{\text{R-CH}_2\text{-OH}} \xrightarrow[\text{18-crown-6 in CH}_3\text{CN}]{\text{Bu}^n{}_3\text{P / CCl}_4\text{ / KCN}} \text{R-CH}_2\text{-C}\equiv\text{N}$$

R	Yield (%)
$PhCH_2$	82
$Ph(CH_2)_2$	75
n-C_8H_{17}	76
PhCH=CH-CH_2	80
[structure: geraniol-type dienol, OH]	72
[structure: cis-alkenol, OH] and *trans*	84

(4.49)

4.1.10 *From carboxylic acids*

Dicarboxylic acids (**88**) have been converted into the corresponding cyanides (**89**) by reaction with ω-dicyanides (**90**), catalysed by strong acids (e.g. H_2SO_4 or H_3PO_4) (Scheme 4.50).[86]

$HO_2C\text{-}(CH_2)_{10}\text{-}CO_2H$ (88) + 2 $NC\text{-}(CH_2)_2\text{-}CH(CH_3)\text{-}CN$ (90) ⟶ $NC\text{-}(CH_2)_{10}\text{-}CN$ (89) 97 % + 2 [3-methylglutarimide]

$NC\text{-}(CH_2)_2\text{-}CN$ or $NC\text{-}(CH_2)_3\text{-}CN$ may also be used

$HO_2C\text{-}(CH_2)_7CO_2H$ (azelaic acid) ⟶ $NC\text{-}(CH_2)_7CN$ (azelonitrile 87 %) (4.50)

Good yields of nitriles can be obtained from sodium salts of carboxylic acids (**91**) by heating them at 200–300 °C with cyanogen bromide (**92**).[87] If the carbon atom of the carboxylate group of the sodium carboxylate is labelled the resulting nitrile contains the label in the cyano group, thus showing that the alkyl–carboxyl bond in the acid is not broken.[88] Furthermore, it has been shown, using an optically active carboxylate, that there is retention of configuration in the alkyl group.[89] A possible mechanism is shown in Scheme 4.51.

$R\text{—}C^{*}(=O)\text{—}O^-$ (91) + $N\equiv C\text{—}Br$ (92) ⟶ $R\text{—}C^{*}(=O)\text{—}O\text{—}C\equiv N$

⇌ $R\text{—}C^{*+}{=}O$ + $^-N{=}C{=}O$ ⟷ $N\equiv C\text{—}O^-$

⟶ $R(O^-)C^{*}{=}\overset{+}{N}{=}C{=}O$ ⟶ $R\text{—}C^{*}\equiv\overset{+}{N}\text{—}C(=O)\text{—}O^-$

⟶ $R\text{—}C^{*}\equiv N$ + CO_2 (4.51)

Nitriles can be prepared from carboxylic acids (**93**) by the use of methanesulphonyl chloride (**94**) (Scheme 4.52).[90]

$$R{-}CO_2H\,(93) + CH_3SO_2Cl\,(94) \xrightarrow{\text{pyridine}} R{-}\overset{O}{\overset{\|}{C}}{-}O{-}SO_2CH_3 \xrightarrow{NH_3} RCONH_2 \xrightarrow{CH_3SO_2Cl} R{-}CN \quad (4.52)$$

60 - 80 % overall yields

4.1.11 *By alkylation of nitriles*

Alkylation of nitriles in a base-catalysed reaction by phase transfer catalysis is a means of production of α-alkylated nitriles (**95**) (Scheme 4.53).[91]

$$Ph{-}CH_2{-}C{\equiv}N + CH_3CH_2Br \xrightarrow[\text{aqueous NaOH, Phase Transfer}]{Et_3\overset{+}{N}Bn\ Cl^-} Ph{-}CH(CH_2CH_3){-}C{\equiv}N\ (95) \quad (4.53)$$

The yield is *c.* 80%.

2,6-di-Bu^t-phenol (OH) → 2,6-di-Bu^t-phenoxide O^-K^+ (98)

$$Et_3B\,(97) + Cl{-}CH_2{-}C{\equiv}N\,(96) \xrightarrow[\text{THF, } 0^\circ C]{(98)} CH_3CH_2CH_2{-}C{\equiv}N \quad 95\ \%$$

B-cyclohexyl-9-BBN + Cl-CH$_2$-C≡N $\xrightarrow{(98)}$ cyclohexyl-CH$_2$-C≡N 77 %

(4.54)

4.1.12 *Use of organoboranes*

Nitriles result from the reaction of α-chloronitriles (**96**) with organoboranes (**97**) in the presence of the weak, hindered base, potassium 2,6-di-*t*-butylphenoxide (**98**).[92] The use of this base is essential since no reaction occurs if potassium *t*-butoxide is used. The preparation of the hindered base (**98**) and some examples of its use are shown in Scheme 4.54. The mechanism of the reaction is as shown in Scheme 4.55. Stepwise alkylation of a perchloronitrile (**99**) is possible by this method,[93] for example, Scheme 4.56.

$$R_3B + {}^-O\text{-}C_6H_3\text{-}2,6\text{-}(Bu^t)_2 + Cl\text{-}CH_2\text{-}C\equiv N \longrightarrow R_2B\text{-}O\text{-}C_6H_3\text{-}2,6\text{-}(Bu^t)_2 + RCH_2CN + Cl^- \quad (4.55)$$

$$Cl_2CH\text{-}CN\ (99) \xrightarrow[(98)]{Et_3B} CH_3CH_2\text{-}CHCl\text{-}CN\ (89\%) \xrightarrow[R_3B]{(98)} CH_3CH_2\text{-}CH(R)\text{-}CN$$

$$Cl_2CH\text{-}CN\ (99) \xrightarrow[2\,Et_3B]{(98)} (CH_3CH_2)_2CH\text{-}CN \quad (4.56)$$

Terminal alkenes (**100**) react with dicyclohexyl borane (**101**) and the resulting organoborane (**102**) is treated with an excess of cuprous cyanide and cuprous acetate (Scheme 4.57).[94]

$$R^1R^2C{=}CH_2\ (100) + (C_6H_{11})_2B\text{-}H\ (101) \xrightarrow{THF,\ 0^\circ C} R^1R^2CH\text{-}CH_2\text{-}B(C_6H_{11})_2\ (102) \xrightarrow[Cu(acac)_2,\ \text{catalytic amount}]{CuCN\ /\ Cu(OAc)_2} R^1R^2CH\text{-}CH_2\text{-}C\equiv N \quad (4.57)$$

4.1.13 *From zinc alkyls*

Alkyl, allylic, or β,γ-acetylenic iodides (**103**) react with cyanomethylzinc bromide (**104**) to give nitriles (Scheme 4.58).[95] This reaction is particularly useful for the synthesis of γ,δ-unsaturated nitriles.

$$\underset{(103)}{R\text{-}C{\equiv}C\text{-}CH_2\text{-}I} + \underset{(104)}{BrZn\text{-}CH_2\text{-}CN} \xrightarrow{HMPT} R\text{-}C{\equiv}C\text{-}CH_2\text{-}CH_2\text{-}CN$$

R = Me_3Si, yield = 78% (4.58)

4.1.14 *Benzylic nitriles*

Trimethylsilyl nitrile and 2,3-dichloro-5,6-dicyano-1,4-benzoquinone (DDQ) react as shown in Scheme 4.59 with the benzylic compound (**105**) to yield a nitrile.[96]

$$\underset{(105)}{MeO\text{-}C_6H_4\text{-}CH_2\text{-}CH_3} \xrightarrow[CH_3CN]{Me_3SiCN,\ DDQ} MeO\text{-}C_6H_4\text{-}CH(CH_3)\text{-}C{\equiv}N \quad 69\,\% \qquad (4.59)$$

Free radical coupling of acrylonitrile (**106**) with alkyl halides (**107**) using tin reagents yields saturated nitriles (**108**) (Scheme 4.60).[97]

N-Trimethylsilyl amides (**109**) react with either aliphatic (**110**) or aromatic acid chlorides to yield nitriles (Scheme 4.61).[98]

$$\underset{(107)}{Bu^t\text{-}I} + \underset{(106)}{CH_2{=}CH\text{-}C{\equiv}N} \xrightarrow{Bu^nSnH,\ EtOH,\ NaBH_4} \underset{(108)}{Bu^t\text{-}CH_2\text{-}CH_2\text{-}C{\equiv}N} \quad 87\,\% \qquad (4.60)$$

4.1.15 *Preparation of α,β-unsaturated nitriles*

The deprotonation of cyanomethylene triphenylphosphorane (**111**) with sodium bis(trimethylsilyl)amide produces the ylide anion (**112**). Alkylation of (**112**) followed by reaction with aldehydes yields α,β-unsaturated nitriles (**113**) (Scheme 4.62).[99] The ylide anion (**112**) can also be used for the synthesis of cyclic unsaturated nitriles (**114**) (Scheme 4.63).

Vinyl halides (**115**) react only sluggishly with alkali metal cyanides but α,β-unsaturated nitriles (**116**) can be prepared from them by the use of a complex

$$R{-}C(=O){-}N(H){-}SiMe_3 + CH_3COCl \longrightarrow R{-}C(OC(=O)Me){=}N{-}SiMe_3 \;(\text{Cl}^-) \longrightarrow R{-}C{\equiv}N$$

(109) (110)

R	Yield (%)
Me	90
Bu^n	95
$CH_2{=}CMe$	95

(4.61)

$$Ph_3P{=}CHCN \xrightarrow{NaN[SiMe_3]_2} Na^+ (Ph_3P{=}C{-}CN)^- \xrightarrow{R^1X} Ph_3\overset{+}{P}{-}\overset{-}{C}(R^1){-}CN \xrightarrow{R^2CHO} R^2{-}CH{=}C(CN)R^1$$

(111) (112) (113)

(4.62)

$$(112) + Br(CH_2)_5{-}\text{(1,3-dioxolan-2-yl)} \longrightarrow Ph_3P^+{-}\overset{-}{C}(CN){-}(CH_2)_5{-}\text{(1,3-dioxolan-2-yl)} \xrightarrow{HCl} Ph_3P^+{-}CH(CN){-}(CH_2)_5{-}CHO \xrightarrow{\text{imidazole}} \text{1-cyanocycloheptene}$$

(114) 52 %

(4.63)

transition metal cyanide, potassium hexacyanodinickelate[100] in methanol (Scheme 4.64).[101]

$$\underset{(115)}{Ph(H)C{=}C(H)Br} \xrightarrow[25^{\circ}C,\ 2\ hours]{K_4Ni(CN)_6/MeOH} \underset{(116)}{Ph(H)C{=}C(H)C{\equiv}N} \qquad (4.64)$$

A one-pot synthesis of α,β-unsaturated cyanides (**117**) from ketones (**118**) can be achieved by the use of cyanotrimethylsilane (**119**)[102] (Scheme 4.65).[103]

$$\underset{(118)}{R^1\text{-}CH_2\text{-}C(=O)\text{-}R^2} + \underset{(119)}{NC\text{-}SiMe_3} \xrightarrow[PhH]{ZnI_2} \left[R^1\text{-}CH_2\text{-}C(CN)(OSiMe_3)\text{-}R^2\right] \xrightarrow[pyridine]{POCl_3} \left[R^1\text{-}CH(H)\text{-}C(CN)(O\text{-}P(=O)Cl_2)\text{-}R^2\right] \xrightarrow{H_3O^+} \underset{(117)}{R^1\text{-}CH{=}C(R^2)\text{-}C{\equiv}N}\ \ 70\text{-}95\ \% \qquad (4.65)$$

Phenyl cyanate (**120**), which can be prepared from phenol as shown in Scheme 4.66, is a useful reagent for the preparation of α,β-unsaturated nitriles (**121**) from lithium derivatives of alkenes (**122**) and alkynes.[104]

The base-catalysed addition of a nitrile to an acetylene (**123**) produces α,β-unsaturated nitriles (**124**) (Scheme 4.67).[105] The mechanism is as shown in Scheme 4.68.

$$PhOH + BrCN \xrightarrow[ether\text{ - }pentane]{Et_3N} \underset{(120)}{PhO\text{-}C{\equiv}N}\ \ 80\ \%$$

$$\underset{(120)}{PhO\text{-}C{\equiv}N} \xrightarrow{R\text{-}C{\equiv}C\text{-}Li\ \text{in ether}} R\text{-}C{\equiv}C\text{-}C{\equiv}N\ \ 85\text{-}95\%$$

$$\underset{(122)}{R(H)C{=}C(H)Li} + \underset{(120)}{PhO\text{-}C{\equiv}N} \xrightarrow[-70^{\circ}C]{ether} \underset{(121)}{R(H)C{=}C(H)C{\equiv}N} \qquad (4.66)$$

$$\text{Ar-CH(R}^1\text{)-C}\equiv\text{N} + \text{H-C}\equiv\text{C-R}^2 \ (123) \xrightarrow[\text{as catalyst}]{\text{KOH in DMSO or HMPT + BnEt}_3\text{N}^+\text{Cl}^-} \text{Ar-C(CH=CHR}^2\text{)(R}^1\text{)-C}\equiv\text{N} \ (124)$$

R^1	R^2	Yield (%)
Et	H	63
Pentyl	H	88
Pr^i	Ph	83
Ph	Bu^i	96
H	OEt	77

(4.67)

$$\text{Ar-CH(R}^1\text{)-CN} \underset{}{\overset{\text{KOH}}{\rightleftharpoons}} \text{Ar-}\bar{\text{C}}\text{(R}^1\text{)-CN} \xrightarrow{\text{H-C}\equiv\text{C-R}^2} \text{H-C(=}\bar{\text{C}}\text{-R}^2\text{)-C(Ar)(R}^1\text{)-CN} \xrightarrow{H_2O} \text{H-C(=CHR}^2\text{)-C(Ar)(R}^1\text{)-CN} \quad (4.68)$$

α,β-Unsaturated nitriles (**125**) can be prepared by the base-catalysed reaction of acetonitrile (**126**) with an aldehyde or a ketone (**127**) e.g. (Scheme 4.69). 18-Crown-6 sometimes aids the reaction.[106]

$$\text{cyclohexanone (127)} + \text{MeCN (126)} \xrightarrow{\text{KOH}} \text{cyclohexylidene=CH-C}\equiv\text{N (125)} \quad 50\ \% \quad (4.69)$$

α,β-Unsaturated nitriles (**128**) can be prepared from vinylsilanes using Lewis acid activated iodosylbenzene as shown in Scheme 4.70.[107]

Good yields of α,β-unsaturated nitriles (**129**) have been obtained by treating a saturated nitrile (**130**) with lithium *N*-isopropyl cyclohexylamide (2 mole) in tetrahydrofuran at −75 °C to remove an α-proton. The resulting anion reacts with diphenyl diselenide at 25 °C and the selenium compound (**131**) produced is isolated and oxidized with 30 per cent hydrogen peroxide solution,[108] for example, Scheme 4.71.

SiMe$_3$ | i) 1.6 $(PhIO)_n$, 1.6 BF_3-Et_2O, RT, 21 hours; ii) $NaBF_4$ → $\overset{+}{I}$-Ph, Bu^t, 80 % | $KCu(CN)_2$, DMF, 25°C, 12 hours → CN, Bu^t (128) 92 %

(4.70)

$Me{\cdot}(CH_2)_6{\cdot}CN$ (130) — i, THF, -75°C, Li^+ $\bar{N}Pr^i$-cyclohexyl; ii, PhSeSePh, 25°C → $Me{\cdot}(CH_2)_5{\cdot}CH(SePh)$–CN (131)

H_2O_2 ↓

$Me{\cdot}(CH_2)_4{\cdot}CH{=}CH{-}CN$ + $Me{\cdot}(CH_2)_4{\cdot}CH{=}CH{-}CN$

Z (129) *E*

88% yield
$Z : E = 28 : 72$

(4.71)

α,β-Unsaturated nitriles can be prepared as a mixture of *cis* and *trans* isomers from chloroacetonitrile (**132**) and aldehydes (**133**) in the presence of tributylstibine (Scheme 4.72).[109]

Cl-CH_2-C≡N (132) + R-CHO (133) —$Bu^n{}_3Sb$→ R-CH=CH-C≡N + HCl + $Bu^n{}_3Sb{=}O$

R^1	R^2	Yield (%)
pentyl	H	92
octyl	H	97
Ph	H	91
-$(CH_2)_4$-		33
-$(CH_2)_5$-		50

trans: *cis* ratio *c.* 1:1

(4.72)

α,β-Unsaturated nitriles have been prepared from chloroacetonitrile via the formation of an α-cyanosulphoxide (Scheme 4.73). The α-cyanosulphoxide may react in two ways (Scheme 4.74).[110] This preparation of α,β-unsaturated nitriles has been used for the synthesis of furans (**134**) in diisobutylaluminium hydride (DIBAH) (Scheme 4.75).[111]

$$Cl{-}CH_2{-}CN + PhSH \longrightarrow PhS{-}CH_2{-}CN \xrightarrow{NaIO_4} Ph{-}S(=O){-}CH_2{-}CN \qquad (4.73)$$

i,

$$Ph{-}S(=O){-}CH_2{-}CN \xrightarrow{R\text{-}CH_2\text{-}CHO} R{-}CH(OH){-}CH{=}CH{-}CN$$

ii,

$$Ph{-}S(=O){-}CH_2{-}CN \xrightarrow{R\text{-}CH_2\text{-}C(CH_3){=}O} (H_3C)(R{-}CH_2)C{=}C(CN)(S(=O){-}Ph)$$

MeOH: $(H_3C)(R{-}CH(OH))C{=}C(CN)(H)$

PhH, HOAc: $(HO{-}CH_2)(R{-}CH_2)C{=}C(CN)(H)$

(4.74)

$$R{-}C(=O){-}CH_3 \xrightarrow[\text{piperidine}]{PhSOCH_2CN} (HOCH_2)(R)C{=}C(H)(C{\equiv}N) \xrightarrow[\text{ii, aq. } H_3O^+]{\text{i, DIBAH}} \text{3-R-furan (134)}$$

75 % (4.75)

Trimethylsilylacetonitrile (**135**) can be prepared as shown in Scheme 4.76.[112] It has been applied as its lithium derivative to the preparation of α,β-unsaturated nitriles (**136**) from aldehydes (**137**).[113] Hexamethylphosphoric

$$Me_3SiCl + BrCH_2CN \xrightarrow{\text{activated Zn}} \underset{(135)\ 80\ \%}{Me_3Si\text{-}CH_2\text{-}CN}$$

$$\underset{(137)}{R\text{-}CHO} + Me_3Si\text{–}\underset{Li}{CH}\text{–}CN \longrightarrow \underset{(136)}{R\text{–}CH{=}CH\text{–}CN}$$

$$Me_3Si\text{–}CH_2\text{–}CN \xrightarrow{Bu^nLi} Me_3Si\text{–}CH(Li)\text{–}CN$$

R = Cyclohexyl (4.76)

triamide (HMPT) and boron isopropoxide increase the proportion of *Z*-isomers in the products:

Ratio of $Z:E$ in the presence of

Bu^nLi only	7:1
$Bu^nLi + B(OPr^i)_3$	16:1
$Bu^nLi + B(OPr^i)_3 + HMPT$	23:1

α,β-Unsaturated nitriles can be prepared from vinyl halides (**138**) by the action of potassium cyanide catalysed by a nickel complex (Scheme 4.77).[114]

$$\underset{(138)}{R^1R^2C{=}C(R)Hal} \xrightarrow[\substack{(Ph_3P)_2NiBr_2\ +\ Zn\ +\ Ph_3P \\ 1\ :\ 3\ \ \ 2}]{DMF/KCN} R^1R^2C{=}C(R)CN + R^1R^2C{=}C(CN)R$$

Halide	E:Z ratio
E-PhCH=CHBr	95:5
Z-PhCH=CHBr	12:88
E-Bu^nCH=CHCl	100:0
Z-Bu^nCH=CHCl	98:2

(4.77)

α,β-Unsaturated nitriles can be prepared from ketones using a modified Horner–Emmons reagent as shown in Scheme 4.78.[115]

α,β-Unsaturated nitriles result from the reaction of alkenes (**139**) with phenylselenocyanate (**140**) and hydrogen peroxide (Scheme 4.79).[116]

$$\mathrm{>C{=}O \xrightarrow[\text{NaH, DME}]{(EtO)_2POCH_2CN} >C{=}CH{-}CN} \qquad (4.78)$$

$$\text{(139)} \xrightarrow[\mathrm{SnCl_4,\ CH_2Cl_2}]{\text{PhSeCN}} \text{(140)} \xrightarrow{\mathrm{H_2O_2}} \qquad (4.79)$$

They can also be prepared by the action of potassium cyanide and dicyclohexyl-18-crown-6 on vinyl sulphones (**141**) in the presence of methylene blue as radical inhibitor (Scheme 4.80).[117]

$$\mathrm{R{-}CH{=}CH{-}\overset{+}{S}(O^-)(O){-}Ph}\ \text{(141)} \xrightarrow[\text{Dicyclohexyl-18 crown-6, Methylene blue}]{\mathrm{KCN,\ Bu^tOH,\ reflux,}} \mathrm{R{-}C(C{\equiv}N){=}CH_2}\ \ c\ 70\ \%$$

$$\mathrm{R{-}CH(CN){-}CH{=}S(O^-)_2{-}Ph} \longrightarrow \mathrm{R{-}C(CN)(H){-}CH_2{-}\overset{+}{S}(O^-)(O){-}Ph} \qquad (4.80)$$

Acetylenes (**142**) react with hydrogen cyanide under catalysis by tetrakis(triphenylphosphite)nickel(0) to yield α,β-unsaturated nitriles,[118] for example, Scheme 4.81. The reaction is both stereo- and regio-specific.

A ketenimine (**143**) has been used in a preparation of Z isomers of α,β-unsaturated nitriles (Scheme 4.82).[119] Yields are in the range 65–95 per cent. A

$$\mathrm{Bu^t{-}C{\equiv}C{-}Me}\ (142) + \mathrm{HCN} \xrightarrow[\mathrm{P(OPh)_3,\ PhH,\ 75^\circ C}]{\mathrm{Ni[P(OPh)_3]_4}} \mathrm{(Bu^t)(H)C{=}C(Me)(CN)} + \mathrm{(Bu^t)(NC)C{=}C(Me)(H)}$$

80 %

9 : 1 (4.81)

$$Me_3Si\text{-}CH_2CN \xrightarrow{\text{silation}} (Me_3Si)_2C{=}C{=}N\text{-}SiMe_3 \ (143)$$

$$R\text{-}CHO + (Me_3Si)_2C{=}C{=}N\text{-}SiMe_3 \xrightarrow[\text{PhH, 25°C}]{BF_3.Et_2O} R\text{—}CH(OSiMe_3)\text{-}C(SiMe_3)_2\text{—}CN \xrightarrow[-(Me_3Si)_2O]{\text{Heat}} RHC{=}C(CN)(SiMe_3) \xrightarrow{\text{NaOH, MeOH}} RHC{=}CH(CN) \quad (4.82)$$

variant of this method of preparation has more recently been reported as shown in Scheme 4.83.[120]

$$n\text{-}C_8H_{17}CHO + MeC(SiMe_3){=}C{=}N\text{-}SiMe_3 \ (\text{a ketenimine}) \xrightarrow[1:3,\ -78°C]{TiCl_4 + Ti(OPr^i)_4} n\text{-}C_8H_{17}\text{—}CH(OH)\text{-}C(CN)(SiMe_3)\text{—}Me \xrightarrow{BF_3.OEt_2} n\text{-}C_8H_{17}HC{=}C(Me)(CN)$$

80 % yield overall (4.83)

4.1.16 *Tetracyanoethylene*

Tetracyanoethylene (TCNE) (**144**) is prepared as shown in Scheme 4.84.[121] Tetracyanoethylene (**144**) is a very reactive compound, for example, in Diels-Alder reactions (Scheme 4.85).[122] Possessing an electron deficient double bond, tetracyanoethylene will react with secondary amines (**145**) to give unsaturated cyanotertiary amines (**146**),[123] for example, Scheme 4.86.

$$H_2C(CN)_2 \xrightarrow[H_2O,\ 5^\circ C]{Br_2,\ KBr,} KBr\ .\ 4\left[Br_2C(CN)_2\right] \xrightarrow{\text{Cu in benzene}} (NC)_2C{=}C(CN)_2\ (144) \quad (4.84)$$

$$(144) \xrightarrow{0^\circ C} \text{4,4,5,5-tetracyanocyclohexene}\quad 94\ \% \quad (4.85)$$

$$\underset{(145)}{R^1R^2NH} + (NC)_2C{=}C(CN)_2 \xrightarrow{-\ HCN,\ THF\ reflux} \underset{(146)}{(NC)_2C{=}C(CN)NR^1R^2}$$

R^1	R^2	Yield (%)
Bu^n	H	67
Et	Et	37
Bn	H	67
Ph	H	59
Cyclohoxyl	H	51

(4.86)

4.1.17 *Glycidic nitriles*

Glycidic nitriles (**147**) have been prepared by a reaction similar to a Darzens reaction that is phase-transfer catalysed (Scheme 4.87).[124]

$$R^1R^2C{=}O + Cl\text{-}CH_2\text{-}C{\equiv}N \xrightarrow[\text{aqueous NaOH, Phase transfer}]{BnMe_3N^+\ Cl^-} \underset{(147)}{R^1R^2C\overset{O}{—}CH\text{-}C{\equiv}N} \quad (4.87)$$

4.1.18 *Preparation of β-ketonitriles*

p-Toluenesulphonyl nitrile (**148**) is prepared by the action of cyanogen chloride (**149**) on sodium *p*-toluenesulphinate (**150**).[125] It reacts with enolates of ketones to give *β*-ketonitriles,[126] e.g. Scheme 4.88.

$$\text{Me-C}_6\text{H}_4\text{-SO}_2^-\ \text{Na}^+\ (150) + \text{Cl-CN}\ (149) \longrightarrow \text{Me-C}_6\text{H}_4\text{-SO}_2\text{CN}\ (148)$$

$$\text{2-benzylcyclohexanone} \xrightarrow[\text{ii, Ar-SO}_2\text{CN, -78°C}]{\text{i, LDA, THF, -78°C}} \text{6-benzyl-2-oxocyclohexanecarbonitrile (67 \%)} \qquad (4.88)$$

4.1.19 *Preparation of ω-cyanoaldehydes*

Cycloalkenes (**151**) can be converted into *ω*-cyanoaldehydes (**152**) by the procedure shown in Scheme 4.89.[127] Yields are in the range 45–80 per cent.

$$\text{cyclohexene}\ (151) \xrightarrow[h\nu \text{ or } H^+]{\text{NOCl}} \text{2-chlorocyclohexanone oxime} \xrightarrow[\text{ROH or THF}]{\text{RO}^-\ \text{Na}^+} \text{2-alkoxycyclohexanone O-alkyloxime} \xrightarrow{\text{Cl}_4\text{P-Cl}} \text{N}{\equiv}\text{C-(CH}_2)_4\text{-CH}{=}\text{O}^+\text{R}\ (152) \xrightarrow{\text{H}_2\text{O}} \text{N}{\equiv}\text{C-(CH}_2)_4\text{-CHO} \qquad (4.89)$$

4.1.20 *Preparation of α-cyanoethers*

α-Cyanomethyl ethers (**153**) can be prepared as shown in Scheme 4.90.[128]

$$RO{-}CH_2{-}O{-}CH_2{\cdot}CH_2{-}OMe + Et_2AlCN\ (\text{excess}) \xrightarrow[100^{\circ}C]{PhCH_3} RO{-}CH_2{-}CN\ (153)\ 65\text{-}80\% \quad (4.90)$$

4.2 Reactivity

As a result of the polarization of the carbon–nitrogen triple bond (Scheme 4.91) nitriles undergo a wide spectrum of reactions with electrophiles on nitrogen and nucleophiles on carbon, also the acidic α-hydrogen atoms can be removed by a strong base.

$$R{-}C{\equiv}N \longleftrightarrow R{-}\overset{+}{C}{=}\overset{-}{N} \quad (4.91)$$

$$H{-}\overset{|}{\underset{|}{C}}{-}C{\equiv}N$$

4.2.1 *Reaction with electrophiles on nitrogen*

The Ritter reaction

Reactions in which carbonium ions react with the nucleophilic nitrogen atom of a nitrile, are known as Ritter reactions. The carbonium ions can be generated in several ways, for example, from alcohols (**154**) (usually tertiary) which form

$$(CH_3)_3C{-}OH\ (154) \xrightarrow[RCN]{H_2SO_4} (CH_3)_3C^+\ (155)\quad \ddot{N}{\equiv}C{-}R$$

$$\xrightarrow{H_2O} (CH_3)_3C{-}\overset{+}{N}{\equiv}C{-}R \quad H_2O:$$

$$\longrightarrow (CH_3)_3C{-}N{=}\underset{OH}{\underset{|}{C}}{-}R$$

$$\longrightarrow (CH_3)_3C{-}NH{-}CO{-}R\ (156) \xrightarrow{aq.\ H_3O^+} (CH_3)_3C{-}NH_2\ (157) + R{-}CO_2H \quad (4.92)$$

stable carbonium ions (**155**)[129] (Scheme 4.92).[130] The product is an amide (**156**) which can be hydrolysed to yield the amine (**157**) derived from the alcohol. This constitutes a good method for the preparation of primary amines on a tertiary carbon atom.[131] The reaction works well with cyanohydrins (**158**) of aldehydes, the product, after oxidation and hydrolysis, being an α-keto acid (**159**) (Scheme 4.93).[132]

$$\text{R-CHO} + \text{HCN} \longrightarrow \underset{(158)}{\text{R-CH(OH)-CN}}$$

$$\text{R-CH(OH)-CN} \xrightarrow[\text{conc. } H_2SO_4]{(CH_3)_3C\text{-OH}} \text{R-CH(OH)-C(=O)-NH-C(CH}_3)_3 \xrightarrow[\text{ii, } H_2O]{\text{i, } CrO_3\text{ / HOAc}} \underset{(159)}{\text{R-C(=O)-CO}_2\text{H}}$$

R	Yield (%)
Et	55
Pr^n	61
Pr^i	73
n-C_6H_{13}	79
n-$C_{19}H_{39}$	81
Bn	78
$BnCH_2$	87
Ph_2CH	60
Bn_2CH	90

(4.93)

The reaction has also been extended to cyanamides (**160**) and enables ureas (**161**) to be synthesized (Scheme 4.94).[133]

$$\underset{(160)}{\text{R-NH-CN}} \xrightarrow[BF_3\text{.EtOH}]{Bu^tOH} \underset{(161)}{\text{R-NH-C(=O)-NH-Bu}^t}$$

R	Yield (%)
Ph	40
4-Me-C_6H_4	37
PhCO	79
2,3,4-$(MeO)_3C_6H_4$	61

(4.94)

The carbonium ion (**162**) can be generated from an alkene or from an acid (RCO_2H in which R is tertiary) and concentrated sulphuric acid.[134] A useful modification for alkenes has been introduced in which the alkene (**163**) adds to the nitrile in the presence of mercuric nitrate (**164**), followed by reduction with sodium borohydride to form an amide (Scheme 4.95).[135]

R^1	R^2	Yield (%)
Bu^n	H	92
C_8H_{17}	H	86
Bu^t	H	90
Ph	H	50

(4.95)

More recently it has been shown that by complexing benzylic or acetylenic alcohols with, for example, chromium tricarbonyl (**165**), the carbocations are stabilized and good yields are obtained (Scheme 4.96).[136] Dialkylaryl tertiary alcohols (ArR^1R^2OH) do not react. However, arylalkyl secondary alcohols (ArCHROH) react with 100 per cent retention of configuration.

Another method by which primary and secondary alcohols (**166**) can be used in the Ritter reaction involves initial reaction of the alcohol with triflic anhydride (**167**). A good leaving group is thus introduced and the formation of the carbocation (**168**) is facilitated (Scheme 4.97).[137]

$$\text{(165) } C_6H_5CH_2OH\cdot Cr(CO)_3 + CH_3CN \xrightarrow{H_2SO_4} C_6H_5CH_2\text{-NH-CO-CH}_3\cdot Cr(CO)_3 \;\; 99\,\% \xrightarrow{O_2 \text{ oxidation}} Ph\text{-}CH_2\text{-NH-CO-CH}_3$$

$$H\text{-}C\equiv C\text{-}CH_2OH\cdot Co_2(CO)_6 + CH_3CN \xrightarrow{H_2SO_4} H\text{-}C\equiv C\text{-}CH_2\text{-NH-CO-CH}_3\cdot Co_2(CO)_6 \;\; 35\,\% \xrightarrow{O_2 \text{ oxidation}} H\text{-}C\equiv C\text{-}CH_2\text{-NH-CO-CH}_3 \quad (4.96)$$

$$\underset{(166)}{R^1\text{-OH}} + \underset{(167)}{(F_3C\text{-}SO_2)_2O} \longrightarrow [R^1\text{-O-}SO_2\text{-}CF_3] \xleftarrow{} :N\equiv C-R^2 \xrightarrow{\text{in } CH_2Cl_2} \underset{(168)}{[R^1\text{-}\overset{+}{N}\equiv C\text{-}R^2]} \; HO^- \xrightarrow{\text{aq. } NaHCO_3} [R^1\text{-N=C(OH)-}R^2] \longrightarrow R^1\text{-NH-C(=O)-}R^2$$

R^1	R^2	Yield (%)
Bn	Me	87
Bn	Bu^t	75
n-C_5H_9	Me	90
Et	Ph	68
Cyclohexyl	Me	75
Pr^t	Ph	68
Bu^t	Me	50
Adamantyl	Me	98

(4.97)

Nitriles are weakly basic (pK_a ~ −11, compare R—NH_2pK_a ~ +11), but give no salts with acids in water. Salts are formed with acids and Lewis acids in dry, aprotic solvents and the nitrogen atom can be alkylated, for example, by triethyloxonium tetrafluoroborate (**169**) in dichloromethane (Scheme 4.98). This reaction has been used for the conversion of nitriles into aldehydes, the intermediate salt being reduced to an imine, which is then hydrolysed.[138]

$$\text{R-C}{\equiv}\text{N} + \underset{(169)}{\text{Et}_3\text{O}^+\,\text{BF}_4^-} \xrightarrow{\text{CH}_2\text{Cl}_2} \text{Et}_2\text{O} + \text{R-C}{\equiv}\text{NEt}^+\,\text{BF}_4^- \xrightarrow{\text{EtOH}} \text{R-C(OEt)=NEt}$$

$$\text{R-C}{\equiv}\text{NEt}^+\,\text{BF}_4^- \xrightarrow{\text{Et}_3\text{SiH}} \text{R-C(H)=N-Et} \xrightarrow{\text{H}_2\text{O}} \text{R-CHO}$$

R(RCHO)	Yield (%)
Bu^n	71
Pr^i	85
Bu^t	61
adamantyl	83
cyclopropyl	79
Bn	41
Ph	90
1-naphthyl	84

(4.98)

Acetonitrile is used as solvent and reactant in a reaction with alkenes (**170**) and nitrosonium tetrafluoroborate, which can be used for the synthesis of imidazoles (**171**) as shown in Scheme 4.99.[139]

In the presence of stannic chloride (**172**), nitriles undergo reaction with malonic esters (**173**) to give enamines, which may subsequently be converted into β-amino acids by hydrolysis, decarboxylation, and reduction (Scheme 4.100). Aromatic nitriles also undergo this reaction.[140]

By using a niobium chloride–tetrahydrofuran complex (**174**)[141] nitriles can be converted into vicinal diamines (**175**) (Scheme 4.101).[142]

The Hoesch reaction

The well known Hoesch reaction for the synthesis of phenolic ketones results from the formation of an electrophile from the nitrile and hydrogen chloride which reacts with phenols or phenolic ethers to give good yields of aromatic ketones,[143] for example, Scheme 4.102.

R^1—CH=CH—R^2 (170) —$NO^+BF_4^-$→ R^1—CH—CH(N=O)—R^2 —MeCN→ R^1—CH(N=C⁺—CH_3)—CH(R^2)(H)—N=O → cyclic intermediate (R^1, H, R^2, N=C⁺—CH_3, N—OH) → H—N⁺ imidazolium (R^1, R^2, N—OH, CH_3) —Na Al($OCH_2OMe)_2H_2$→ HN imidazole (R^1, R^2, N, CH_3) (171)

R^1	R^2	Yield (%)
Me	H	80
Me	Me	80
Ph	H	50
-($CH_2)_4$-		50

(4.99)

R-CN + $CH_2(CO_2Me)_2$ (173) —i, $SnCl_4$ / $ClCH_2CH_2Cl$; ii, Na_2CO_3 / aqueous→ R-C(-NH_2)=C($CO_2Me)_2$

R-C≡N + $SnCl_3$—Cl (172) ⇌ R-C⁺=N-$SnCl_3$ Cl⁻ —- HCl, $CH_2(CO_2Me)_2$→ R-C(=N-$SnCl_3$)-CH($CO_2Me)_2$ —H_2O→ R-C(=NH)-CH($CO_2Me)_2$ —Tautomerism→ R-C(-NH_2)=C($CO_2Me)_2$

R	Yield (%)
Et	55
Me	45
Ph	57
PhCH=CH	43
MeCH=CH	17

(4.100)

$$NbCl_5 \xrightarrow[\text{ii, THF}]{\text{i, Al in } CH_3CN} NbCl_4(THF)_2 \; (174)$$

$$RCHO \longrightarrow R{-}CH{=}N(SiMe_3)_2 \xrightarrow[\text{ii, } Bu^n{}_3SnH]{\text{i, } NbCl_4(THF)_2,\ (MeOCH_2)_2} R{-}CH(N{=}NbCl_3){-}CH(N{=}NbCl_3){-}R \xrightarrow[\text{ii, } HO^-]{\text{i, KF}} {}^{*}\,RCH(NH_2){-}CH(NH_2)R \; (175)$$

* ratio racemate to meso	R	Yield (%)
2:1	Bu^t	52
2:1	cyclopentyl	60
2:1	$CH_2{=}CH(CH_2)_3$	61
1.4:1	Bn	45
3:1	$Me_3Si(CH_2)_2$	63

(4.101)

$$MeCN + HCl \longrightarrow Me{-}C(Cl){=}NH \xrightarrow{PhOH} 4{-}HOC_6H_4{-}C(Me){=}NH.HCl \xrightarrow{H_3O^+} 4{-}HOC_6H_4{-}C(Me){=}O \; 55\% \qquad (4.102)$$

4.2.2 *Reaction with nucleophiles on carbon*

Hydrolysis

Acids or alkalis in aqueous solution hydrolyse nitriles to carboxylic acids (**176**), the reaction going through the amide (**177**) (Scheme 4.103). By taking special

$$R{-}C{\equiv}N + H_2O \longrightarrow R{-}CO{-}NH_2 \; (177) \xrightarrow{H_2O} R{-}CO_2H \; (176) \qquad (4.103)$$

precautions it is possible to stop at the amide (**177**), the yields being high for this reaction (Scheme 4.104).[144]

$$R{-}C{\equiv}N + BF_3 \xrightarrow[\text{in aq. HOAc}]{\text{Saturated solution}} R{-}\overset{+}{C}{=}N{-}\overset{-}{B}F_3 \xrightarrow{H_2O} R{-}C({=}O){-}NH_2 \quad (177) \qquad (4.104)$$

A good procedure for the hydrolysis of nitriles to amides involves the use of powdered sodium hydroxide or potassium hydroxide in tertiary butanol (Scheme 4.105).[145]

$$R{-}C{\equiv}N \xrightarrow{KOH,\ Bu^tOH} R{-}C({=}O)NH_2$$

R	Yield (%)
Bn	90
Bu^t	54
Ph	90

(4.105)

Hydrogen peroxide in dilute aqueous alkali accelerates the hydrolysis of nitriles to amides and then acids. The strongly nucleophilic peroxide ion, HOO^-, is responsible (Scheme 4.106).[146]

$$R{-}C{\equiv}N + HOO^- \xrightarrow{H_2O_2} R{-}C(OOH){=}NH \xrightarrow{OH^-\,(aq.)} RCONH_2 + O_2 + OH^- \qquad (4.106)$$

Hydrogen peroxide in the presence of methyl cyanide catalyses the epoxidation of alkenes (**178**), which is thought to be due to the formation of the peroxyacetimidic acid ($CH_3C(O_2H){=}NH$) (Scheme 4.107).[147]

Fluorine under nitrogen in acetonitrile, along with some water, reacts with alkenes (**179**) to give good yields of epoxides (**180**) (Scheme 4.108). There is some doubt about how this occurs, but again it is thought to be due to the formation of peroxyacetimidic acid.[148]

$$2\ CH_3CN + \text{cyclohexene (178)} \xrightarrow{MeOH,\ H_2O_2,\ KHCO_3} \text{cyclohexene oxide}\ \ 61\% \qquad (4.107)$$

$$R^1\text{-CH=CH-}R^2 \ (179) \xrightarrow[-15^\circ C]{F_2,\ CH_3CN,\ H_2O \text{ under } N_2} R^1\text{-CH}\overset{O}{\text{—}}\text{CH-}R^2 \ (180)$$

R^1	R^2	Yield (%)
Ph	$COCH_3$	85
Ph	Ph	80
$C_{10}H_{21}$	H	90
Ph	COEt	80

(4.108)

Halogen-substituted phthalic anhydrides have been applied to the preparation of carboxylic acids (**181**) from nitriles (Scheme 4.109).[149]

$$R\text{-C}\equiv N \xrightarrow[\text{135 to 180}^\circ\text{C for 4-6 days; Hal = F or Cl}]{\text{tetrahalophthalic anhydride}} R\text{-}CO_2H \ (181)$$

Hal	R	Yield (%)
Cl	n-C_5H_{11}	85
F	Bu^t	70
Cl	Bn	76
Cl	PhCH=CH	61
Cl	Ph	86
F	$(EtO)_2PhC$	65
F	$(EtO)_2CH(CH_2)_4$	56

(4.109)

Under different alkaline conditions with hydrogen peroxide in aqueous ethanol the hydrolysis can be stopped at the stage of the amide (Scheme 4.110).[150]

Better yields of amides are obtained by phase transfer with 30 per cent aqueous hydrogen peroxide, tetra-n-butylammonium hydrogensulphate, $Bu^n_4N^+HSO_4^-$, methylene dichloride, and 20 per cent aqueous sodium hydroxide.[151] Amides in $c.$ 90 per cent yields were obtained.

The hydration of aldehyde cyanohydrins (**182**) to α-hydroxyamides (**183**) can be achieved by treatment with aqueous borax or alkaline borates at 80 °C (Scheme 4.111).[152] Yields can be sometimes increased by the addition of potassium cyanide.

$$R{-}C{\equiv}N \xrightarrow[\text{H}_2\text{O}_2\text{ (2 mole)}, \ 40\text{-}50^{\circ}\text{C}]{\text{NaOH, aq. EtOH}} R{-}C(=O){-}NH_2 + O_2$$

50-60%

R	Yield (%)
PhCH=CH	90
Cyclohexyl	100
Cyclopentyl	70
$PhCH_2CH_2$	80
p-NO_2-C_6H_4-CMe_2	80

(4.110)

$$\underset{(182)}{R{-}CH(OH){-}C{\equiv}N} \xrightarrow[\text{or other borate, } 80^{\circ}\text{ C}]{\text{aq. Na}_2\text{B}_4\text{O}_7\text{ (borax),}} \underset{(183)}{R{-}CH(OH){-}C(=O){-}NH_2}$$

Borate	R	Yield (%)
$NaB(OH)_4$	H	39
Borax	n-hexyl	67
Borax	Me	72
$KB(OH)_4$	Ne	75
Borax	Ph	73
Borax	Pr^i	86
Borax	$MeSCH_2CH_2$	79

(4.111)

Nitriles can be hydrolysed to amides under neutral conditions by the use of a black powder produced from copper sulphate and sodium borohydride (Scheme 4.112).[153] Yields are in the range of 50–95 per cent.

$$R{-}C{\equiv}N + H_2O \xrightarrow{\text{black powder}} R{-}C(=O){-}NH_2$$

$$CuSO_4 + NaBH_4 \xrightarrow{\text{aqueous NaOH}} \text{black powder}$$

(4.112)

Nitriles are converted into carboxylic acids (**184**) by oxidative hydrolysis (Scheme 4.113).[154] The mechanism is thought to be as shown in Scheme 4.114. Yields can be increased by the presence of 18-crown-6 ether.

$$R\text{-}CH_2\text{-}C{\equiv}N \xrightarrow{O_2/KOBu^t} R\text{-}CO_2H \text{ (184)} \quad (4.113)$$

$$R\text{-}CH_2CN \longrightarrow R\text{-}\bar{C}HCN \xrightarrow{O_2} R\text{-}\dot{C}HCN \xrightarrow{O_2} R\text{-}CH(CN)\text{-}O\text{-}O^{\cdot} \longrightarrow R\text{-}CH(CN)\text{-}O\text{-}OH \longrightarrow R\text{-}CO\text{-}CN \xrightarrow{H_2O} R\text{-}CO_2H \text{ (184)}$$

R	Yield (%)
n-hexyl	89
Cyclohexyl	85
Ph	90
$CH_3(CH_2)_{14}$	86

(4.114)

Iminoethers (**185**) result from the reaction of nitriles with alcohols and Lewis acids, or an amidine (**186**) can be prepared by the action of ammonia on the intermediate salt, see Chapter 10, p. 422 (Scheme 4.115). The reaction shown in Scheme 4.115 to form an iminoether has been applied to trichloroacetonitrile (**187**) to produce a synthesis of tertiary butyl ethers (**188**) and esters (**189**) (Scheme 4.116).[155]

Another iminoether (**190**) has recently been used for the synthesis of a range of heterocyclic compounds,[156] for example, with *o*-phenylenediamine (**191**) (Scheme 4.117).

Pinner synthesis

$$R\text{-}C{\equiv}N \xrightarrow[0\text{ to }5^{\circ}C]{EtOH/\text{dry }HCl} R\text{-}C(OEt){=}NH_2^+\,Cl^- \text{ (iminoether hydrochloride)} \xrightarrow[\text{solvent}/<0^{\circ}C]{NH_3/\text{inert}} R\text{-}C(OEt){=}NH \text{ (185)}$$

$$R\text{-}C(OEt){=}NH_2^+\,Cl^- \xrightarrow{0.88\ NH_3} R\text{-}C(NH_2){=}NH \text{ Amidine (186)} \quad (4.115)$$

$Cl_3C\text{-}C{\equiv}N$ (187) $\xrightarrow[\text{ether / Bu}^tOH]{Bu^tO^- K^+}$ $Cl_3C\text{-}C(=NH)\text{-}O\text{-}Bu^t$

$\xrightarrow[BF_3 / Et_2O]{R^1OH}$ $R^1\text{-}O\text{-}C(CH_3)_3$ (188)

$\xrightarrow[BF_3 / Et_2O]{R^2CO_2H}$ $R^2\text{-}C(=O)\text{-}O\text{-}Bu^t$ (189) + $Cl_3C\text{-}C(=O)\text{-}NH_2$

R^1	Yield (ether) (%)
Bn	72
$PhCH_2CH_2$	91
$n\text{-}C_{16}H_{33}$	69
$BnOCH_2CCHOHCH_2$	67

R^2	Yield (ester) (%)
$BrCH_2$	71
$CH_2{=}CHCH_2Br$	54
PhCHMe	85
$MeCOCH_2CH_2$	83

(4.116)

$EtO_2C\text{-}C{\equiv}N + EtOH \xrightarrow{HCl} EtO\text{-}C(=\overset{+}{N}H_2)\text{-}C(=O)\text{-}OEt\ Cl^- \xrightarrow{K_2CO_3} EtO\text{-}C(=NH)\text{-}C(=O)\text{-}OEt$ (190)

(191) + (190) $\longrightarrow$ 3-amino-2(1H)-quinoxalinone, 99%

(4.117)

The reaction of nitriles with epoxides (**192**) in the presence of silicon tetrafluoride yields heterocyclic compounds (**193**) (Scheme 4.118).[157]

When a nitrile (**194**) is heated at 100 °C for a short time (<30 minutes) with thioacetamide (2 mole) in dimethylformamide saturated with hydrogen chloride it is converted into a thioamide (**195**),[158] for example, Scheme 4.119. The proposed mechanism is as shown in Scheme 4.120.

R^1	R^2	R^3 (**13**)	Yield (%) (**14**)
n-Pentyl	*n*-Pentyl	H	65
Ph	H	H	80
n-Hexyl	Me	H	51
$-(CH_2)_{11}-$		H	45
Me	$-(CH_2)_4-$		33
H	$-(CH_2)_{10}-$		38

(4.118)

(4.119)

(4.120)

An alkyl nitrile can be used to convert one β-ketoester (**196**) into another (**197**),[159] for example, Scheme 4.121. The loss of the CH_3CO group in the reaction with sodium hydroxide is by a reverse Claisen mechanism.

Nitriles react with sodium hydrazide ($NaNHNH_2$) to yield amidrazones (**198**) (Scheme 4.122). Sodium hydrazide is prepared by the reaction of hydrazine with sodium amide.[160]

Acetone cyanohydrin (**199**) can be used in the Gattermann reaction for the preparation of aromatic aldehydes (**200**) from suitably activated aromatic compounds (**201**) (Scheme 4.123).[161]

$$CH_3\text{-}CO\text{-}CH_2\text{-}CO_2Et \ (196) \xrightleftharpoons{EtCN} CH_3\text{-}C(OH){=}CH\text{-}CO_2Et \;/\; CH_3\text{-}CH_2\text{-}C{\equiv}N$$

$$\xrightarrow[\text{heat}]{SnCl_4 / PhH} CH_3\text{-}CO\text{-}CH(C(=NH)CH_2\text{-}CH_3)\text{-}CO_2Et$$

i, $NH_3OH^+ Cl^-$; ii, aq. NaOH → An isoxazole: $CH_3\text{-}CH_2$, CO_2H, N, O, CH_3

Aqueous HCl, $CHCl_3$ → $CH_3\text{-}CH_2\text{-}C(=O)\text{-}CH_2\text{-}CO_2Et$

73 % overall

(197)

(4.121)

$$NaNH_2 + H_2N\text{-}NH_2 \xrightarrow[0^\circ C \text{ under } N_2]{\text{In } Et_2O \text{ or } PhH} H_2N\text{-}NHNa + NH_3$$

$$R\text{-}C{\equiv}N + HN^{-}\text{-}NH_2 \xrightarrow[0^\circ C]{Et_2O \text{ or } Pr^i_2O} R\text{-}C(=N^-)\text{-}NH\text{-}NH_2 \xrightarrow{H_2O} R\text{-}C(=NH)\text{-}NH\text{-}NH_2 \rightleftharpoons R\text{-}C(NH_2)=N\text{-}NH_2$$

(198)

R	Yield (%)
Ph	99
Me	63
$n\text{-}C_{12}H_{25}$	85
$n\text{-}C_{17}H_{35}$	93

(4.122)

$$ArH \ (201) + (CH_3)_2C(CN)\text{-}OH \ (199) \xrightarrow[ClCH_2CH_2Cl,\ 83^\circ C]{AlCl_3} Ar\text{-}CHO \ (200)$$

Ar	Yield (%)
Ph	50
MeO-C_6H_4	50
Me-C_6H_4	40
3,4-Me_2-C_6H_3	83
3,5-Me_2-C_6H_3	73

(4.123)

4.2.3 *Reduction (see Chapter 1)*

Catalytic hydride reductions

Catalytic reduction of nitriles or reduction with lithium aluminium hydride in ether gives the primary amine in good yield (Scheme 4.124).[162]

$$R{-}C{\equiv}N \xrightarrow{LiAlH_4,\ Et_2O} R{-}CH_2{-}NH_2 \quad (4.124)$$

Catalytic reduction can be carried out efficiently on a 5 per cent rhodium–alumina catalyst. Good yields result at both room temperature and at low pressure,[163] e.g. Scheme 4.125.

CH₂-C≡N → CH₂-CH₂-NH₂ (indole, NH); H_2 (2.5 atm.); 5 % Rh on Al_2O_3; NH_3 / EtOH; 25° C / 2 hours; 78 %

(4.125)

Sodium borohydride does not usually reduce nitriles, however, in special circumstances, for example, with a dicyano-epoxide (**202**) sodium borohydride acts selectively and reduces (to a primary amine) the cyano group which is *trans* to an alkyl group,[164] for example, Scheme 4.126.

Et, H, CN, CN, O (202) → Et, H, CN, CH_2-NH_2, O 70%; $NaBH_4$; THF / H_2O

(4.126)

Modification of lithium aluminium hydride to decrease its reducing power enables an aldehyde (**203**) to be obtained from nitriles. Aliphatic and aromatic cyanides (**204**) can be reduced in this way in yields of 70–90 per cent (Scheme 4.127).[165]

Stephen reduction to the aldehyde is achieved by reduction with stannous

$$\text{LiAlH}_4 \xrightarrow[\text{in ether}]{\text{3 EtOH}} \text{LiAlH(OEt)}_3$$

$$\underset{(203)}{\text{R-C}\equiv\text{N}} \xrightarrow[\text{1 mole}]{\text{LiAlH(OEt)}_3} \text{R-CH=N}^- \xrightarrow{\text{H}_2\text{O}} \underset{(204)}{\text{R-CHO}} \quad (4.127)$$

chloride and hydrochloric acid (Scheme 4.128).[166] This reduction gives reasonable yields with aliphatic nitriles to C_6.[167]

$$\text{R-C}\equiv\text{N} \xrightarrow[\text{saturated with HCl}]{\text{SnCl}_2\text{ / ether}} [\text{R-CH=}\overset{+}{\text{N}}\text{H}_2]_2\ \text{SnCl}_6^{2-} \xrightarrow{\text{H}_2\text{O}} \text{R-CHO} \quad (4.128)$$

Reductive decyanization occurs in good yield (60–90 per cent) when a nitrile reacts with potassium and hexamethylphosphoric triamide (HMPT) in ether (Scheme 4.129).[168] More recently it has been shown that lithium, sodium, or potassium in hexamethylphosphoric triamide, with or without an alcohol at 0 °C, efficiently removes cyano groups reductively from nitriles. The best reagent is potassium and tertiary butanol. Once elimination of the cyano group from a nitrile is achieved the product is the respective alkane (**205**),[169] for example, Scheme 4.130.

$$\text{R-C}\equiv\text{N} \ \xrightarrow{2e^-\ (\text{K in HMPT})} \ \text{R}^{\frown}\underset{-}{\text{C}}\text{=N}^- \longrightarrow \text{R}^- + \text{CN}^- ;\quad \text{R}^- \xrightarrow{\text{H}_3\text{O}^+} \text{R-H} \quad (4.129)$$

$$\text{(CH}_3)_2\text{CH-CH}_2\text{CH}_2\text{-C(CH}_3)_2\text{CN} \xrightarrow{\text{K, HMPT, Bu}^t\text{OH}} \underset{(205)\ 100\%}{(\text{CH}_3)_2\text{CH-CH}_2\text{CH}_2\text{-CH(CH}_3)_2} \quad (4.130)$$

Similarly decyanization occurs in the reaction of nitriles with sodium in liquid ammonia (Scheme 4.131).[170] Other products are also possible, depending on the substrate and the conditions of the reaction,[171] for example, Scheme 4.132.

Lithium in liquid ethylamine is a less specific reagent for the decyanization of nitriles to alkenes, for example, Scheme 4.133.

Sodium in liquid ammonia can be used to decyanize α-amino nitriles, without danger of epimerization of a chiral centre. This enables biogenetic-type asymmetric syntheses of many kinds of alkaloids from optically active amino acids to be achieved.[172]

Sodium in iron triacetylacetylate (**206**) has been used to decyanize nitriles (Scheme 4.134).[173]

$$R{-}C{\equiv}N \xrightarrow{\text{2 Na, Liquid } NH_3} R{-}Na + NaCN$$

$$R{-}Na \xrightarrow{NH_3} NaNH_2 + R{-}H$$

R	Yield (%)
Me	3
Et	5
$CH_2{=}CH$	19
Bn	96

(4.131)

Na, liquid NH_3, -33°C: 5%, 95% (4.132)

Li, $EtNH_2$, 17°C: 88.5%, 11.5%

Li, $EtNH_2$, 17°C: 30%, 70% (4.133)

$$\underset{(206)}{Fe(acac)_3} \xrightarrow[\text{ii, R-CN}]{\text{i, 2 Na}} R{-}H + Fe(acac)_3^- + Na^+$$

R	Yield (%)
Me	98
n-Octyl	100
Cyclopentyl	77
Bn	56

(4.134)

4.2.4 *Reaction with organometallic reagents*

With Grignard reagents

The addition of a Grignard reagent (**207**) to a nitrile gives an intermediate compound (**208**) from which the imine (**209**) can be obtained by the action of ammonia or the ketone (**210**) results from hydrolysis (Scheme 4.135).[174] This is a slower reaction than Grignard addition to a carbonyl group. Higher yields of the imine are possible if anhydrous methanol is used instead of ammonia.[175]

$$\underset{(207)}{R^1MgHal} + R^2{-}C{\equiv}N \longrightarrow \underset{(208)}{R^1{-}C({=}NMgHal){-}R^2} \xrightarrow{H_2O} \underset{(210)}{R^1R^2C{=}O}$$

$$\underset{(208)}{R^1{-}C({=}NMgHal){-}R^2} \xrightarrow{NH_3} \underset{(209)}{R^1R^2C{=}NH} \qquad (4.135)$$

Addition of a cuprous salt greatly improves the reaction of nitriles with Grignard reagents (**211**) which is capable of yielding a variety of products depending upon the method of work up,[176] for example, Scheme 4.136.

$$Bu^t{-}C{\equiv}N + \underset{(211)}{C_6H_{11}{-}MgCl} \xrightarrow[\text{heat}]{\text{CuBr / THF}} Bu^t{-}C({=}N{-}MgCl){-}C_6H_{11}$$

$$\xrightarrow{\text{aq. } H_3O^+} Bu^t{-}C({=}O){-}C_6H_{11}\ (64\%) \qquad \xrightarrow{NH_3} Bu^t{-}C({=}NH){-}C_6H_{11}\ (4\%) \qquad \xrightarrow{\text{Li / liquid } NH_3} Bu^t{-}CH(NH_2){-}C_6H_{11}\ (64\%) \qquad (4.136)$$

O-Trimethylsilyl cyanohydrins (**212**) react with Grignard reagents to yield acyloins (**213**) after aqueous work up (Scheme 4.137).[177]

With triethylaluminium

The reaction of one mole of a nitrile with two moles of triethylaluminium followed by an aqueous work up provides a good synthesis of ethyl ketones (**214**) (Scheme 4.138).[178]

$$R\text{-}CH(=O) \xrightarrow{SiMe_3CN} R\text{-}CH(OSiMe_3)\text{-}C{\equiv}N \ (212) \xrightarrow{ArCH_2MgCl} R\text{-}CH(O\text{-}SiMe_3)\text{-}C(=N\text{-}MgCl)\text{-}CH_2\text{-}Ar \xrightarrow{aq.\ H_3O^+} R\text{-}CH(OH)\text{-}C(=O)\text{-}CH_2\text{-}Ar \ (213)$$

R	Ar	Yield (%)
Et	Ph	78
Pr^n	4-MeO-C_6H_4	78
H	Ph	79

(4.137)

$$R\text{-}C{\equiv}N \xrightarrow{AlEt_3} [\text{cyclic complex: } R\text{-}C{\equiv}N \cdots AlEt_2,\ Me\text{-}H_2C \cdots Et_2Al\text{-}\text{-}CH_2Me] \xrightarrow{H_2O} R\text{-}C(=O)\text{-}Et \ (214)$$

R	Yield (%)
n-$C_{12}H_{25}$	89
Bn	90

(4.138)

Blaise's reaction

Blaise's method[179] for the preparation of β-ketoesters (**215**) from nitriles and α-bromoesters (**216**), which is analogous to the Reformatsky reaction, has recently been greatly improved (Scheme 4.139).[180]

4.2.5 *Reactivity of α-hydrogen*

The Thorpe–Ziegler reaction

The Thorpe–Ziegler synthesis of cyclic ketones (**217**) from dinitriles (**218**) is one of the best known reactions which depends upon the ionization of a

$$R^1O_2C{-}CH(Br){-}R^2 \; (216) \; (\text{excess added slowly}) + R^3{-}C{\equiv}N \xrightarrow{\text{Zn, THF}} (R^2)(R^1O_2C)C{=}C(R^3)(NH_2) \; 70\text{-}95\% \xrightarrow{\text{aq. } H_3O^+} R^1O_2C{-}CH(R^2){-}C(=O){-}R^3 \; (215) \; 80\text{-}85\% \tag{4.139}$$

hydrogen α to a cyano group. Thorpe discovered that high yields of small ring ketones could be synthesized by the action of ethoxide ions in ethanol on ω-dinitriles,[181] for example, Scheme 4.140. When the method was applied to the synthesis of rings of size greater than six, polymers resulted.

$$NC{-}(CH_2)_4{-}CN \; (218) \xrightarrow{EtO^-,\ EtOH} \text{(carbanion } \overset{-}{C}H(CN) \text{ attacking } C{\equiv}N) \xrightarrow{EtOH} \text{2-cyanocyclopentanimine (C=NH)} \xrightarrow{\text{cold dil. aq. } H_3O^+} \text{2-cyanocyclopentanone} \xrightarrow{\text{hot dil. aq. } H_3O^+} \text{cyclopentanone } (217) \tag{4.140}$$

Some years later, Ziegler introduced the base lithium diethylamide for the alkylation of nitriles (Scheme 1.141).[182] In the following year, Ziegler used the same base with his high dilution technique in which small amounts of the dinitrile (**219**) are added to the solvent containing the base over a period of time, to synthesize large ring ketones (**220**) in about 50 per cent yield,[183] for example, Scheme 1.142.

Shortly afterwards, Ziegler introduced the sodium salt of *N*-methylaniline ($Na^{+-}NPhMe$) as the base,[184] and this base has been used in subsequent research.[185]

$$R^1{-}CH_2CN \xrightarrow{LiNEt_2,\ Et_2O} R^1{-}\overset{-}{C}HCN \xrightarrow{R_2Hal} R^1CH(R^2)CN \tag{4.141}$$

$$NC\text{-}(CH_2)_{16}\text{-}CN \ (219) \longrightarrow \text{cyclo-}[(CH_2)_{16}\text{-}C(=O)] \ (220) \qquad (4.142)$$

The polar cyanide group enables a proton to be removed from the α-position of a nitrile by a strong base to yield a stabilized anion (**221**) which can be used for various syntheses (Scheme 4.143).[186]

$$R\text{-}Li + Pr^i_2NH \longrightarrow Pr^i_2N^- Li^+ + R\text{-}H$$

$$R^1\text{-}CH(H)\text{-}C\equiv N \xrightarrow[-75^\circ C]{Pr^i_2N^- Li^+ / THF} \left[R^1\text{-}\bar{C}H\text{-}C\equiv N \longleftrightarrow R^1\text{-}CH{=}C{=}N^- \right] \ (221)$$

$$\xrightarrow{Cl\text{-}CO_2Et} R^1\text{-}CH(CO_2Et)\text{-}C\equiv N; \quad \xrightarrow{CCl_4} R^1\text{-}CH(Cl)\text{-}C\equiv N; \quad \xrightarrow{R^2\text{-}I} R^1\text{-}CH(R^2)\text{-}C\equiv N \qquad (4.143)$$

Isovaleronitrile [$(CH_3)_2CHCH_2CNH$) has been used in this way for the synthesis of a variety of compounds,[187] and lithium derivatives of other nitriles have been used for synthetic purposes, for example, the lithium derivative of acetonitrile, $Li^{+-}CH_2CN$,[188] and the derivative of isopropylnitrile, $Li^{+-}CMe_2CN$.[189]

Synthesis of alkenes

Nitriles with an α-hydrogen atom can be converted into alkenes (**222**) as shown in Scheme 4.144.[190]

$$Me\text{-}CH(CN)\text{-}(CH_2)_{11}\text{-}CH_3 \xrightarrow[\text{ii, } Bu^t_3SnCH_2I \text{; iii, LiMe}]{\text{i, LDA}} Me\text{-}C(NC)(CH_2\text{-}SnBu^t_3)\text{-}(CH_2)_{11}\text{-}CH_3 \ (Li\text{-}CH_3)$$

$$\xrightarrow{CH_3Li / LiBr} CH_2{=}C(Me)\text{-}(CH_2)_{11}\text{-}CH_3 \ (222) \qquad (4.144)$$

Synthesis of N-heterocyclic compounds

The removal of an α-hydrogen atom with lithium diethylamide has been used for the synthesis of nitrogen heterocyclic compounds (**223**) (Scheme 4.145).[191] Yields are increased when $n = 3$ if the iodide is used instead of the chloride.

R^1,R^2	n	Yield (%)
$-(CH_2)_5-$	2	81
$-(CH_2)_4-$	2	85
$-(CH_2)_5-$	3	54
$-(CH_2)_5-$	3	63

(4.145)

Synthesis of aldehydes

Nitriles which contain an α-hydrogen atom (**224**) can be converted into aldehydes (**225**) by successive treatment with diisobutyl aluminium hydride, lithium diisopropylamide in hexamethylphosphortriamide (HMPT), and an alkyl halide, without isolation of any of the intermediates (Scheme 4.146).[192]

A synthesis of β-amino ketones

An aminonitrile (**226**) is produced by an addition reaction between an aldehyde (**227**), an amine (**228**), and potassium cyanide. This may be followed by removal of an activated α-proton and reaction of the resulting carbanion with an aldehyde. Thermal elimination of hydrogen cyanide then yields a β-amino ketone (**229**) via the enol (Scheme 4.147).[193]

With potassium–graphite

Potassium–graphite (KC_n, $n = 8, 24, 36$, etc.) (prepared by heating metallic potassium with the calculated amount of graphite at 70 °C in the absence of air) in tetrahydrofuran at −60 °C, deprotonates nitriles which possess an α-hydrogen atom. The resulting anion can then be alkylated by an alkyl halide,[194] for example, Scheme 4.148.

R^1	R^2	R^3Hal	Yield (%)
Bu^n	H	n-$C_5H_{11}Br$	85
Bu^n	H	$CH_2{=}CH(CH_2)_2Br$	72
Bu^n	H	Me_2SO_4	71
Bu^n	H	Pr^iBr	55
Me	Me	Bu^nBr	74
Me	Me	BnBr	65

(4.146)

R^1	R^2	R^3	Yield (%)
Me	Me	Cyclohexyl	72
Et	Me	Bu^n	67
Pr^n	Me	Pr^n	60
Me	Et	Pr^n	75
Et	Et	Et	55
H	Pr^i	Pr^n	44
Me	Et	PhCH=CH	79

(4.147)

$$CH_3\text{-}C\equiv N \xrightarrow[-60^\circ C]{\text{i, }KC_8\text{ / THF}} \left[\overset{-}{C}H_2\text{-}C\equiv N\right] \xrightarrow{\text{ii, Ph-}CH_2Cl} \text{Ph-}CH_2\text{-}CH_2\text{-}C\equiv N \quad 42\ \% \qquad (4.148)$$

A synthesis of cis-diols

Cyanoacetic acid has been used by Corey[195] in an elegant transformation of *trans*-bromohydrins (**230**) into *cis*-diols (**231**), for example, Scheme 4.149. This method has steric advantages and consequences different from the action of osmium tetroxide, which adds to the less hindered face of a complicated alkene.

NC-CH_2-CO_2H, TsCl
CH_2Cl_2, pyridine
Br
OH
(230)
Br
O-CO-CH_2CN
85 %
NaH
0-23°C
Br
O
O—C—CH—CN
O
O
C═CH–CN
98 %
HCl, THF
O-CO-CH_2-CN
OH
100 %
OH
OH
(231)
95 %

(4.149)

With disulphides

Solid potassium hydroxide, suspended in tetrahydrofuran, has also been used to remove an α-hydrogen atom from a nitrile and the resulting anion has been caused to react with disulphides (**232**). Good yields of thioethers (**233**) result,[196] for example, Scheme 4.150. Alkylation by an alkyl halide can be carried out simultaneously with the formation of the thioether, for example, Scheme 4.151.

$$Ph\text{-}CH(C{\equiv}N)\text{-}CH_3 \xrightarrow[KOH]{THF} Ph\text{-}\overset{-}{C}(C{\equiv}N)\text{-}CH_3 + CH_3S\text{-}SCH_3 \; (232) \longrightarrow Ph\text{-}C(SCH_3)(C{\equiv}N)\text{-}CH_3 \; (233) + CH_3S^- \qquad (4.150)$$

$$Ph\text{-}C(CH_3)(CN)\text{-}CH_2\text{-}CN \xrightarrow[KOH / THF]{CH_3\text{-}S\text{-}S\text{-}CH_3 / CH_3I} Ph\text{-}C(CH_3)(CN)\text{—}C(CH_3)(CN)\text{-}S\text{-}CH_3 + SMe_2 \quad 90\% \qquad (4.151)$$

With phenylsulphenyl chloride

Aryl alkyl ketones (**234**) can be obtained from the correspondingly substituted nitriles (**235**) by the use of phenylsulphenyl chloride (**236**) which can be obtained from the reaction between benzenethiol (**237**) and sulphuryl chloride (**238**) (Scheme 4.152).[197]

$$C_6H_5SH \; (237) + SO_2Cl_2 \; (238) \longrightarrow C_6H_5SCl \; (236)$$

$$Ar\text{—}CH(R)\text{-}CN \; (235) \xrightarrow{LDA} \left[Ar\text{—}\overset{-}{C}(R)\text{—}CN \longleftrightarrow Ar\text{—}C(R){=}C{=}N^- \right]$$

$$\left[Ar\text{—}\overset{-}{C}(R)\text{—}CN \right] \xrightarrow{PhSCl \; (236), \; 70 - 90\%} Ar\text{—}C(SPh)(R)\text{—}CN$$

$$\left[Ar\text{—}C(R){=}C{=}N^- \right] \xrightarrow{ClSiMe_2\text{-}CMe_3} Ar\text{—}C(R){=}C{=}N\text{—}SiMe_2\text{-}CMe_3 \xrightarrow{PhSCl, \; (236), \; 85 - 100\%} Ar\text{—}C(SPh)(R)\text{—}CN$$

$$Ar\text{—}C(SPh)(R)\text{—}CN \xrightarrow{NaOH, \; NBS, \; MeCN} Ar(R)C{=}O \; (234) \qquad (4.152)$$

N,N-Diethylaminoacetonitrile

N,N-Diethylaminoacetonitrile (**239**) can be prepared as shown in Scheme 4.153.[198] The nitrile (**239**) reacts with lithium diethylamide to give an anion (**240**), which can be alkylated and subsequent hydrolysis yields an aldehyde (monoalkylation) or a ketone (dialkylation),[199] for example, Scheme 4.154.

Et_2NH + CH_2O ⇌ CH_2-OH / NEt_2

NaHSO$_3$

CH_2=$\overset{+}{N}Et_2$ $\overline{C}N$

NaCN

CH_2-CN / NEt_2 (239) 90 %

(4.153)

Et_2N-CH_2CN (239) —LiNEt$_2$ / THF→ Et_2N-$\overline{C}$HCN (240) —BrCH$_2$CH$_2$CH(CH$_2$O)$_2$→ (CH$_2$)$_2$-CH(OCH$_2$CH$_2$O) / Et_2N-CH-CN

—H$_2$O→ (CH$_2$)$_2$-CH(OCH$_2$CH$_2$O) / CHO 83%

—i, LiNEt$_2$ / THF; ii, BrCH$_2$CH=CHEt→ (CH$_2$)$_2$-CH(OCH$_2$CH$_2$O) / Et_2N-C-CN / CH$_2$CH=CHEt

—H$_2$O→ (CH$_2$)$_2$-CH(OCH$_2$CH$_2$O) / O=C / CH$_2$CH=CHEt 72%

(4.154)

Dicyanomethane

Dicyanomethane, NC—CH$_2$—CN, undergoes a Knoevanagel reaction with aldehydes and ketones in the presence of the catalyst alumina, Al_2O_3, (Scheme 4.155).[200]

α-Silylation

The α-silylation of nitriles can be achieved by the use of trimethylsilyl trifluoromethyl sulphonate (**241**)[201] in the presence of triethylamine,[203] for example, Scheme 4.156.

$$R^1R^2C{=}O + NC\text{-}CH_2\text{-}X \longrightarrow R^1R^2C{=}C(CN)X + H_2O$$

R^1	R^2	X	Yield (%)
Me	H	CN	75
Me	H	CO_2Me	54
Pr^i	H	CN	88
Me	Me	CN	98
Et	Et	CN	81
Me	Me	CO_2Me	53

(4.155)

$$F_3C\text{-}SO_2\text{-}O\text{-}SiMe_3\ (241) + :NEt_3 \longrightarrow F_3C\text{-}SO_3^- + Et_3\overset{+}{N}\text{-}SiMe_3$$

$$CH_3\text{-}CH_2\text{-}C{\equiv}N \xrightarrow[\text{ether / }5^\circ C]{Et_3N} CH_3\text{-}\overset{-}{C}H\text{-}C{\equiv}N \quad (+\ Et_3\overset{+}{N}\text{-}SiMe_3) \longrightarrow CH_3\text{-}CH(SiMe_3)\text{-}C{\equiv}N \xrightarrow{\text{again}} CH_3\text{-}C(SiMe_3)_2\text{-}C{\equiv}N \quad 33\ \%$$

and

$$NC\text{-}CH_2\text{-}CN + F_3C\text{-}SO_2\text{-}O\text{-}SiMe_3 \xrightarrow{Et_3N\text{ in ether}} NC\text{-}CH(SiMe_3)\text{-}CN \quad 80\ \%$$

(4.156)

Umpolung

When applied to cyanohydrins (**242**) in which the hydroxy group is protected, removal of the α-proton enables reverse polarization (Umpolung) of an aldehyde to be achieved. Ensuing alkylation yields a ketone (**243**) (Scheme 4.157).

R^1-CHO —HCN→ R^1-C(H)(OH)-CN —(dihydropyran) Protection→ R^1-C(H)(CN)-O-(tetrahydropyranyl)

—LDA→ R^1-$\bar{C}$(CN)-O-(tetrahydropyranyl) —R^2 Hal→ R^1R^2C(CN)-O-(tetrahydropyranyl) —i, H_3O^+; ii, aq. HO^-→ R^1R^2C=O (243)

(4.157)

A variation of this procedure utilizes the ethyl ether of the enol of acetaldehyde to react with the hydroxyl group of a cyanohydrin to protect it as an acetal (Scheme 4.158).[203] The reaction has been extended to the synthesis of cyclopropanone and cyclobutanone.[204]

R^1-CHO —HCN→ R^1-CHOH(CN) —CH_2=CH-OEt→ R^1-CH(CN)-O-CH(OEt)-CH_3

—LDA in HMPT→ R^1-$\bar{C}$(CN)-O-CH(OEt)-CH_3 —R^2-Br→ R^1R^2C(CN)-O-CH(OEt)-CH_3 —i, aq. H_3O^+; ii, aq. HO^-→ R^1R^2C=O

R^1	R^2	Yield (%)
Me	Bu^n	80–85
Me	*n*-hexyl	80–85
Me	*n*-decyl	80–85
Me	Pr^i	80
Me	Cyclopentyl	80
Me	Cyclohexyl	41
Me	$PhCH_2CH_2$	84
$CH_2{=}CH$	*n*-hexyl	75

(4.158)

α-Arylation of malononitriles

α-Arylation of malononitriles (**244**) can be achieved by removal of an α-proton and reaction of the resulting anion with an aromatic iodide in the presence of cyclopentadienylpalladium chloride and triphenylphosphine (Scheme 4.159).[205]

$$\underset{(244)}{\text{R-CH(CN)}_2} \xrightarrow{\text{NaH}} \text{R-}\bar{\text{C}}\text{(CN)}_2 \xrightarrow[\text{(cyclopentadiene)PdCl}_2]{\text{ArI / Ph}_3\text{P / THF / N}_2\text{ / heat}} \text{Ar-C(R)(CN)}_2 \qquad (4.159)$$

α,β-Unsaturated ketone synthesis

Anions from 2-aminonitriles of morpholine (**245**) react with acetylene to yield unsaturated compounds (**246**) which are easily transformed into α,β-unsaturated ketones (**247**),[206] for example, Scheme 4.160.

$$\text{Morpholino-CH(Ph)-CN} \xrightarrow[\text{Bu}^n{}_4\text{N}^+\text{Br}^-\text{ (catalyst), HC}\equiv\text{CH}]{\text{NaOH / DMSO}} \underset{(246)\ 77\,\%}{\text{Morpholino-C(Ph)(CN)-CH=CH}_2} \xrightarrow[\text{Heptane / H}_2\text{O}]{\text{CuSO}_4} \underset{(247)\ 63\,\%}{\text{Ph-C(=O)-CH=CH}_2}$$

$$\text{(via: } \xrightarrow{\text{HO}^-} \underset{(245)}{\text{Morpholino-}\bar{\text{C}}\text{(Ph)-CN}} + \text{HC}\equiv\text{CH} \xrightarrow{\text{H}^+} (246)\text{)} \qquad (4.160)$$

γ-Diketone synthesis

The more general reaction with 2-aminonitriles of amines other than morpholine is as shown in Scheme 4.161, to yield enamines (**248**) which can be hydrolysed to give γ-diketones (**249**).

4.2.6 *Cycloaddition reactions*

With β-naphthol

Acrylonitrile (**250**) undergoes a photochemical reaction with β-naphthol (**251**) to yield a cyclobutyl nitrile (**252**) (Scheme 4.162).[207] Irradiation of (**252**) results in ring expansion via the hypoiodite reaction.

With adiponitrile

Nitriles react with adiponitrile (**253**) in a reaction catalysed by iron carbonyl to yield heterocyclic compounds (**254**) (Scheme 4.163).[208]

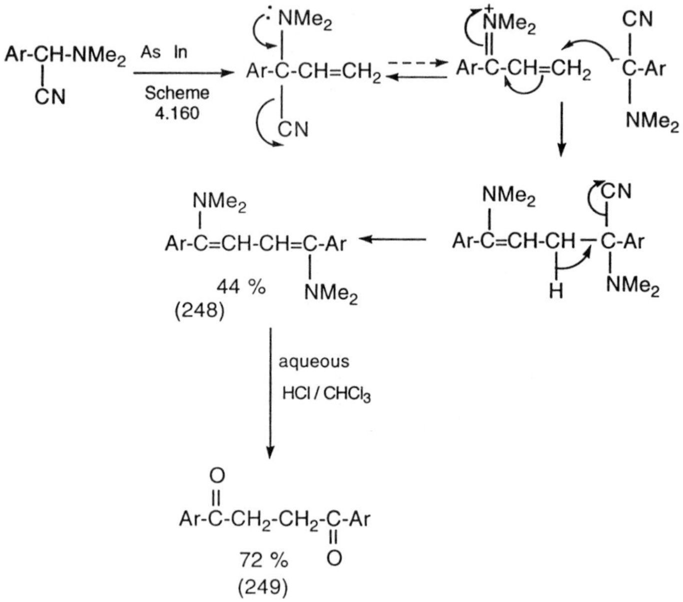

(4.161)

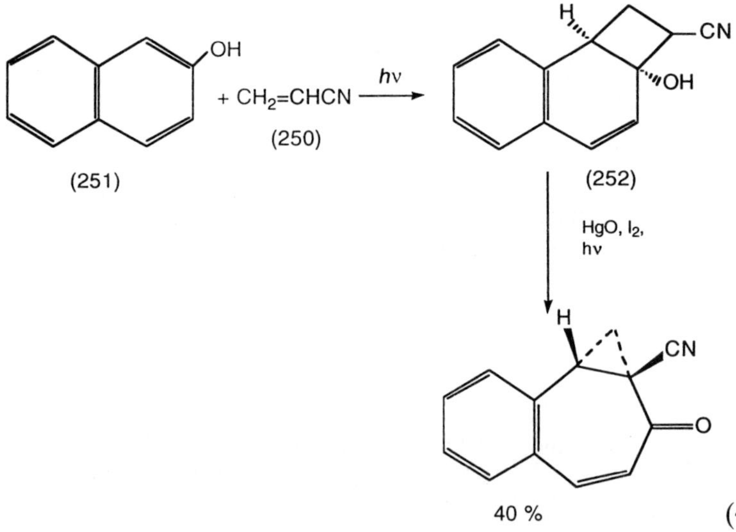

(4.162)

With acetylenic allylic ethers

[2+2+2] Addition of acetonitrile (**255**) to acetylenic allylic ethers (**256**), in the presence of dicyclopentadienyl cobalt, yields pyridines (**257**) (Scheme 4.164). This reaction has been used as a route to pyridoxine.[209]

CH_2-CN
H_2C
H_2C
CH_2-CN
(253)
+
N≡C–R
$Fe_2(CO)_9$
220° C
N N N R
(254) 42-71 % (4.163)

CH_2-C≡C-$SiMe_3$
O
CH_2-C≡C-$SiMe_3$
(256)
+ $CH_3C\equiv N$
(255)
Cp_2Co or
$Cp(CO)_2Co$
Me_3Si
O
CH_3
N
$SiMe_3$
(257) 73 % (4.164)

Reductive coupling

Nitriles can couple reductively via a [2+2] cycloaddition to Ti–C double bonds of titanacyclobutanes (C_5H_5, the cyclopentadienyl ligand is abbreviated Cp),[210] for example, Scheme 4.165.

Cp_2Ti CH_2 CH–Bu^t CH_2 + Bu^t–C≡N
Bu^t-CH=CH_2
Cp_2Ti CH_2 C–Bu^t N
Bu^t-CN
Cp_2Ti N= Bu^t N= Bu^t
Cp_2Ti N= Bu^t N H Bu^t
i, HCl
ii, H_3O^+
O= Bu^t H N H Bu^t
+ Cp_2TiCl_2

(4.165)

Heterocyclic syntheses

Nitriles participate in heterocyclic syntheses (**258**), catalysed by sulphuric acid,[211] for example, Scheme 4.166. The mechanism probably involves a carbonium ion and is analogous to the mechanism of the Ritter reaction (see p. 00).

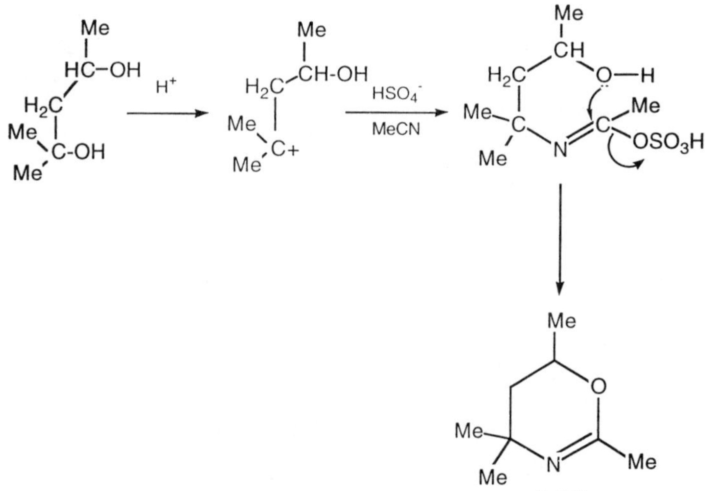

(258) (4.166)

ACYL NITRILES

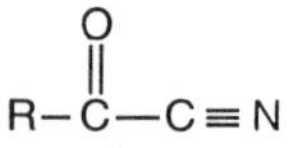

$$R-\overset{O}{\overset{\|}{C}}-C\equiv N$$

(259)

Acyl nitriles (**259**) are reactive compounds. They are potent acylating agents, for example, the methyl compound, CH_3COCN, is rapidly hydrolysed by water giving acetic acid and it reacts very quickly with methanol yielding methyl acetate.[212] They polymerize readily in the presence of alkali. In the infra-red they produce the following absorption bands: $\nu_{C=O}$ 1710–1730 cm^{-1} and ν_{CN} 2220–2225 cm^{-1}. The methyl compound (CH_3COCN) has an ultraviolet absorption in 2,2,4-trimethylpentane at 309 $m\mu$ compared to CH_3CN which absorbs at 304 $m\mu$.[212]

4.3 Preparation

4.3.1 *From acyl halides*

Acyl nitriles (**259**) can be prepared by the action of metal cyanides (**260**) on acyl halides (**261**) (Scheme 4.167).[213]

$$R{-}C(=O)Cl \;(261) + CuCN \;(260) \xrightarrow{CH_3CN} R{-}C(=O)CN \;(259) + CuCl$$

R	Yield (%)
Heptyl	78
PhCH=CH	53
Pr^i	27
Bn	49

(4.167)

By the action of cuprous cyanide on an acyl chloride in the presence of lithium iodide in acetonitrile or phenyl cyanide, good yields of acyl nitriles result (Scheme 4.168).[214]

$$R{-}\overset{O}{\overset{\|}{C}}{-}Cl \xrightarrow[\text{CH}_3\text{CN or PhCN, reflux}]{\text{CuCN / LiI}} R{-}\overset{O}{\overset{\|}{C}}{-}C{\equiv}N$$

R	Yield (%)
Me	50
Et	49
Pr^n	51
Bu^n	62
Bu^t	16

(4.168)

Aliphatic acyl nitriles can be prepared as dimers by the reaction of acyl chlorides (**262**) with thallium cyanide (**263**), which is conveniently prepared from thallium phenate (**264**) and hydrogen cyanide in ether (Scheme 4.169).[215]

$$\underset{(264)}{\text{TlOPh}} \xrightarrow[\text{ether}]{\text{HCN}} \underset{(263)}{\text{TlCN (ppt.)}}$$

$$\underset{(262)}{2\ \text{R-C(=O)-Cl}} + \underset{(263)}{2\,\text{TlCN}} \xrightarrow[\text{30 to 60 min.}]{\text{ether}} (\text{R-CO-CN})_2 + 2\ \text{TlCl}$$

R	Yield (%)
Me	75
Et	48
Bu^t	60

(4.169)

4.3.2 *From α,β-unsaturated cyanohydrins*

α,β-Unsaturated cyanohydrins (**265**) are oxidized to *α,β*-unsaturated acyl nitriles (**266**), by tertiary butylhydroperoxide in the presence of dichlorotris(triphenylphosphine)ruthenium(II) e.g. (Scheme 4.170).[216]

$$\underset{(265)}{CH_3(CH_2)_2\text{-}CH{=}CH\text{-}CH(OH)\text{-}CN} \xrightarrow[\text{in benzene}]{Bu^tOOH\ /\ Ru^{II}} \underset{\substack{81\ \%\\(266)}}{CH_3(CH_2)_2\text{-}CH{=}CH\text{-}C(=O)\text{-}CN} \quad (4.170)$$

4.4 Reactivity

4.4.1 *With alkenes*

α,β-Unsaturated acyl nitriles (**267**) react with alkenes in the presence of aluminium chloride to yield dihydropyrans (**268**) (Scheme 4.171).[217]

$$CH_2{=}C(Me)_2 + \underset{(267)}{Me_2C{=}CH\text{-}C(=O)\text{-}CN} \xrightarrow[0\text{-}5^{\circ}\,C]{AlCl_3\ /\ PhH} \underset{\substack{(268)\\35\ \%}}{\text{2,2,4,4-tetramethyl-3,4-dihydro-2H-pyran-6-carbonitrile}} \quad (4.171)$$

4.4.2 *With dienes*

α,β-Unsaturated acyl nitriles undergo Diels–Alder reactions with reactive dienes (**269**) (Scheme 4.172).

(269) + $CMe_2{=}CH{-}C(=O){-}CN$ $\xrightarrow[\text{PhH, 25° C}]{Et_2AlCl}$ Exo + Endo

Yield of both isomers = 90 %

(4.172)

4.4.3 *With silyl enol ethers*

Acetyl cyanide (CH_3COCN) reacts at −78 °C in the presence of titanium tetrachloride with the silyl enol ether of a ketone (**270**) to give, in good yield, the cyanohydrin of a β-diketone (**271**), which can be hydrolysed to the β-diketone (**272**),[218] for example, Scheme 4.173.

(270) $\xrightarrow[\text{- 78° C}]{CH_3COCN,\ TiCl_4/CH_2Cl_2}$ (271) $\xrightarrow{\text{aq. }HO^-}$ (272)

(4.173)

CYANOTRIMETHYLSILANE

4.5 Preparation

4.5.1 *From trimethylsilane*

Cyanotrimethylsilane (**273**), can be prepared from sodium or potassium cyanide and trimethylsilyl chloride (**274**) in *N*-methylpyrrolidine at *c.* 100 °C (Scheme 4.174).[219]

$$Me_3SiCl + NaCN \text{ or } KCN \xrightarrow[100^\circ C]{\text{N-methylpyrrolidine}} Me_3SiCN + NaCl \text{ or } KCl$$

(274) (273)

(4.174)

Another preparation from trimethylsilyl chloride (**274**) and potassium cyanide is carried out under different conditions (Scheme 4.175).[220]

$$Me_3SiCl + KCN \xrightarrow[\text{NaI (catalytic amount)}]{\text{MeCN, Pyridine}} Me_3SiCN + KCl$$

(274)

(4.175)

4.6 Reactivity

Below is a selection of preparations of various nitrogen compounds. Cyanotrimethylsilane is used for the synthesis of many other classes of compounds.

4.6.1 *With acyl halides*

Cyanotrimethylsilane (**273**) reacts readily with acyl chlorides or bromides (**275**) to yield acyl nitriles (**276**) (Scheme 4.176).[221]

$$R{-}C(=O){-}Hal + Me_3SiCN \xrightarrow[50-70^\circ C]{1.5-4.0 \text{ hours}} R{-}C(=O){-}CN + Me_3SiHal$$

(275) (273) (276)

R	Yield (%)
Et	40
Br^n	33
Bu^t	56
$BrCH_2$	66
CH_3CHBr	81

(4.176)

There is an induction period in the reaction and the authors on this account suggest the mechanism shown in Scheme 4.177 in which a small amount of trimethylsilyl halide is the initiator. The reaction proceeds via nucleophilic attack of the nitrogen lone pair on the silicon atom (i) to produce the cation (**277**) and a halide ion. The halide ion now reattacks one of the two silicon atoms, resulting either in reversal of the first step (ii) or production of the trimethylsilyl isonitrile (**278**) and a trimethylsilyl halide (iii) in a reversible reaction (iv). Compound (**278**) now reacts with an acyl chloride (**275**) as shown to produce the acyl nitrile (**276**).

$$Me_3Si{-}C{\equiv}N: + Me_3Si{-}Hal \underset{ii}{\overset{i}{\rightleftharpoons}} Me_3Si{-}C{\equiv}\overset{+}{N}{-}SiMe_3\ (277) + Hal^-$$

$$Hal^- + Me_3Si{-}C{\equiv}\overset{+}{N}{-}SiMe_3\ (277) \underset{iv}{\overset{iii}{\rightleftharpoons}} Me_3Si{-}\overset{+}{N}{\equiv}\overset{-}{C}\ (278) + Me_3Si{-}Hal$$

$$Me_3Si{-}\overset{+}{N}{\equiv}\overset{-}{C}\ (278) + R{-}C({=}O){-}Hal\ (275) \longrightarrow R{-}C(Hal)(O^-){-}C{\equiv}\overset{+}{N}{-}SiMe_3 \longrightarrow R{-}C(Hal)(O{-}SiMe_3){-}C{\equiv}N \longrightarrow R{-}C({=}O){-}C{\equiv}N\ (276) \quad (4.177)$$

4.6.2 *β-Amino alcohol synthesis*

In addition to its use for the preparation of acyl nitriles, cyanotrimethylsilane has wide applications in synthesis.[222] β-Amino alcohols (**279**) can be prepared by the hydrolysis of β-hydroxy isonitriles (**280**), which result from the reaction of cyanotrimethylsilane with epoxides (**281**) in the presence of zinc iodide (Scheme 4.178).[223]

However, when catalysed by diethyl aluminium chloride, epoxide (**281**) ring opening with cyanotrimethylsilane, yields hydroxynitriles (**283**), for example, Scheme 4.179. The ring opening of an epoxide group by cyanotrimethylsilane is thus dependent on the nature of the Lewis acid catalyst.

4.6.3 *Application to the conversion of unsaturated aldehydes to unsaturated ketones*

Cyanotrimethylsilane reacts with aldehydes and ketones to give *O*-trimethylsilylcyanohydrins (**284**)[225] which can be alkylated and in this way unsaturated aldehydes (**285**) have been converted into unsaturated ketones (**286**) (Scheme 4.180).[226]

(281) $\xrightarrow{Me_3SiCN,\ ZnI_2}$ $R^2R^1C(OSiMe_3)–CR^3R^4(NC)$ $\xrightarrow{KF}$ $R^2R^1C(OH)–CR^3R^4(NC)$ (280) → $R^2R^1C(OH)–CR^3R^4(NH_3^+\ Cl^-)$ (279); (280) → (282)

R^1	R^2	R^3	R^4	Yield (**279**) (%)	Yield (**282**)
H	cyclohexyl		H or Me	80–88	
H	Et	Et	H		82

(4.178)

$$H_2C(O)CH–CH_3 + Me_3SiCN \xrightarrow[20^\circ C]{Et_2AlCl} N{\equiv}C–CH_2–CH(CH_3)–OSiMe_3$$

(281) (283) 83% (4.179)

$R^1R^2C{=}CR^3–CHO$ (285) $\xrightarrow[ZnI_2]{Me_3SiCN}$ $R^1R^2C{=}CR^3–CH(CN)OSiMe_3$ (284) $\xrightarrow{i,\ LDA;\ ii,\ R^4\text{-}Br}$ $R^1R^2C{=}CR^3–CR^4(CN)OSiMe_3$ $\xrightarrow{aq.\ H_3O^+\ or\ aq.\ F^-}$ $R^1R^2C{=}CR^3–C(=O)R^4$ (286)

60 - 80 %

(4.180)

4.6.4 *With ketals and orthoesters*

Cyanotrimethylsilane reacts with ketals (**287**) to yield ethers of cyanohydrins (**288**) and with orthoesters (**289**) to give cyanoketals (**290**) (Scheme 4.181).[227] The products are useful in synthesis, since by the action of lithium diisopropylamide they can be alkylated and then hydrolysed.

$$\underset{(287)}{R^1CH(OR^2)_2} + Me_3SiCN \xrightarrow[25^\circ C]{BF_3.OEt_2 \text{ or } SnCl_2} \underset{(288)}{R^1-C(OR^2)(CN)-H}$$

$$\underset{(289)}{HC(OR^1)_3} + Me_3SiCN \xrightarrow[25^\circ C]{BF_3.OEt_2 \text{ or } SnCl_2} \underset{(290)}{H-C(OR^1)(CN)-OR^1} \xrightarrow{\text{i, LDA, THF, HMPT; ii, } R^2\text{-Ha}} R^2-C(OR^1)(CN)-OR^1 \xrightarrow{\text{TsOH in hot aq. acetone}} R^2-C(=O)-OR^1 \quad (4.181)$$

4.6.5 *Phenylacetic acid preparation*

Phenylacetic acids (**291**) may be prepared from ketones via silylcyanohydrins (**292**) (Scheme 4.182).[228]

$$Ar-C(=O)-R \xrightarrow[ZnI_2]{Me_3SiCN} \underset{(292)}{Ar-C(OSiMe_3)(CN)-R} \xrightarrow[\text{iii, HCl and heat}]{\text{i, } SnCl_2\text{, ii, HOAc}} \underset{(291)}{Ar-CH(CO_2H)-R} \quad (4.182)$$

4.6.6 *α-Chloronitrile synthesis*

Cyanotrimethylsilane is used in the preparation of α-chloronitriles (**293**) (Scheme 4.183).[230]

$$R^1C(=O)R^2 + Me_3Si\text{-}CN \xrightarrow[\text{ii, } H_2O]{\text{i, } TiCl_4,\ CH_2Cl_2} R^1\text{—}C(Cl)(CN)\text{—}R^2$$

(293)

R^1	R^2	Yield (%)
Ph	H	85
Ph	Me	65
Ph	Ph	61
4-MeO-C_6H_4	H	77
4-MeO-C_6H_4	Me	65
4-Me-C_6H_4	Me	48
4-Me-C_6H_4	H	86

(4.183)

4.6.7 *α-Alkylthionitrile synthesis*

Cyanotrimethylsilane is also used in the preparation of α-alkylthionitriles (**294**) (Scheme 4.184).[231]

$$R^1R^2C(SEt)_2 + Me_3SiCN \xrightarrow{SiCl_4,\ CH_2Cl_2} R^1R^2C(SEt)(CN)$$

(294)

R^1	R^2	Yield (%)
Pr^i	H	93
cyclohexyl	H	85
Bu^t	H	81
4-Me-C_6H_4	Me	70
-$(CH_2)_5$-		41

(4.184)

USE OF OTHER NITRILES IN SYNTHESIS

Other nitriles have been used for synthetic purposes, some examples of which are discussed below.

4.7 Di-iminosuccinonitrile

Di-iminosuccinonitrile (**295**)[232] can be prepared in 96 per cent yield by the reaction between hydrogen cyanide and cyanogen (**296**) at low temperatures (Scheme 4.185).[233] As expected from its imino structure, di-iminosuccinonitrile is a very reactive compound: it reacts with water releasing hydrogen cyanide. Di-iminosuccinonitrile (**295**) has been applied to the synthesis of heterocyclic compounds (**297**),[234] for example, Scheme 4.186.

$$2\,HCN + (CN)_2 \xrightarrow{-40^\circ C} HN{=}C(CN){-}C(CN){=}NH$$

(296) → (295) (4.185)

$$NC{-}C({=}NH){-}C({=}NH){-}CN + SCl_2 \longrightarrow \text{3,4-dicyano-1,2,5-thiadiazole} \quad 93\ \%$$

(295) → (297) (4.186)

4.8 Trichloromethylnitrile

Trichloromethylnitrile (**298**) has been utilized as shown in Scheme 4.187 for the production of benzylethers of sugar alcohols.[235] The intermediate compound (**299**) is prepared from trichloromethylnitrile and benzyl alcohol in the presence of potassium carbonate.[236]

$$Cl_3C{-}C{\equiv}N + Ph{-}CH_2OH \xrightarrow{K_2CO_3} Cl_3C{-}C(OCH_2Ph){=}NH \quad 91\ \%$$

(298) → (299)

$$\text{Sugar-OH} + Cl_3C{-}C(OCH_2Ph){=}NH \longrightarrow \text{Sugar-O-}CH_2Ph$$

(Sugar alcohol) (4.187)

References

1. M. T. Reetz and I. Chatziiosifidis, *Angew. Chem., Int. Ed. Engl.*, 1981, **20**, 1017.
2. *F. &F.*, **15**, 253; B. Juršić, *J. Chem. Res. (S)*, 1988, 336.
3. *F. &F.*, **6**, 135.
4. J. W. Zubrick, B. I. Dunbar, and H. D. Durst, *Tetrahedron Lett.*, 1975, 71; C. L. Liotta and E. E. Grisdale, *Tetrahedron Lett.*, 1975, 4205.
5. F. L. Cook, C. W. Bowers, and C. L. Liotta, *J. Org. Chem.*, 1974, **39**, 3416.
6. R. Graf, *Angew. Chem.*, 1968, **80**, 183.
7. *F. &F.*, **15**, 328; A. R. Daniewski, M. M. Kabat, M. Manyk, J. Wieha, W. Wojciechowska, and H. Duddock, *J. Org. Chem.*, 1988, **53**, 4855.
8. M. S. Kharasch and O. Reinmuth, *Grignard reactions of non-metallic substances*, Constable and Co., London, 1954, p. 786.
9. V. Grignard, E. Bellet, and C. Courtet, *Ann. Chim.*, 1920, **12**, 364.
10. A. Haller, *Compts. Rend.*, 1882, **95**, 142.
11. M. E. Kuehne, *J. Am. Chem. Soc.*, 1959, **81**, 5400.
12. *F. &F.*, **11**, 481; R. Davis and K. G. Untch, *J. Org. Chem.*, 1981, **46**, 2985.
13. A. Strecker, *Annalen*, 1850, **75**, 27.
14. W. Lidy and W. Sindermeyer, *Chem. Ber.*, 1973, **106**, 587; D. A. Evans, G. L. Carroll, and L. K. Truesdale, *J. Org. Chem.*, 1974, **39**, 914; J. K. Rasmussen and S. M. Heilmann, *Synthesis*, 1978, 219.
15. P. G. Gassman and J. T. Talley, *Tetrahedron Lett.*, 1978, 3773.
16. *F. &F.*, **9**, 127; S. Hünig and G. Whehner, *Synthesis*, 1979, 522; J. K. Rasmussen and S. M. Heilmann, *Synthesis*, 1979, 523.
17. *F. &F.*, **14**, 107.
18. *F. &F.*, **15**, 346; K. Sukata, *J. Org. Chem.*, 1989, **54**, 2015.
19. J. Brussee, W. T. Loos, C. G. Kruse, and A. Van Der Gen, *Tetrahedron*, 1990, **46**, 979.
20. Y. Lu, C. Miet, N. Kunessch, and J. Poisson, *Tetrahedron Asymmetry*, 1990, **1**, 707.
21. H. A. Bruson, *Org. React.*, 1949, **5**, 79.
22. C. L. Liotta, A. M. Dabdoub, and L. H. Zalkow, *Tetrahedron Lett.*, 1977, 1117.
23. *F. &F.*, **7**, 4; N. Miyaura, M. Itoh, and A. Suzuki, *Tetrahedron Lett.*, 1976, 255.
24. Y. Takahashi, M. Tokuda, M. Itoh, and A. Suzuki, *Chem. Lett.*, 1975, **1**, 523.
25. F. Campagna, A. Carotti, and G. Casini, *Tetrahedron Lett.*, 1977, 1813; *F. &F.*, **4**, 386.
26. W. Lehnert, *Tetrahedron Lett.*, 1971, 1501.
27. *F. &F.*, **4**, R. S. Monson and D. N. Prist, *Can. J. Chem.*, 1971, **49**, 2897.
28. J. C. Graham and D. H. Marr, *Can. J. Chem.*, 1972, **50**, 3857; *F. &F.*, **4**, 386.
29. *F. &F.*, **3**, 231; Y. Kanaoka, T. Kuga, and K. Tanizawa, *Chem. Pharm. Bull. Japan*, 1970, **18**, 397.
30. T. Sarraie, T. Ishiguro, K. Kawashima, and K. Morita, *Tetrahedron Lett.*, 1973, 2121.
31. J. C. Graham and D. H. Marr, *Can. J. Chem.*, 1972, **50**, 3857.
32. *F. &F.*, **14**, 337; J. B. Hendrickson and M. S. Hussoin, *J. Org. Chem.*, 1987, **52**, 4137.
33. *F. &F.*, **14**, 208; D. A. Claremon and B. T. Phillips, *Tetrahedron Lett.*, 1988, **29**, 2155.
34. *F. &F.*, **11**, 125; T. Tanaka and T. Miyadera, *Synthesis*, 1982, 1497.

35. L. Horner, H. Oediger, and H. Hoffmann, *Annalen*, 1959, **626**, 26.
36. W. Lehnert, *Tetrahedron Lett.*, 1971, 559.
37. N. O. Vesterager, E. B. Pedersen, and S. O. Lawesson, *Tetrahedron*, 1974, **30**, 2509; *F. &F.*, **5**, 323.
38. G. Rosini, G. Baccolini, and S. Cacchi, *J. Org. Chem.*, 1973, **33**, 1060; *F. &F.*, **4**, 387.
39. J. B. Hendrickson, K. W. Blair, and P. M. Keehn, *Tetrahedron Lett.*, 1976, 603; *F. & F.*, **7**, 390.
40. A. Carotti, F. Campagna, and R. Ballini, *Synthesis*, 1979, 56; *F. &F.*, **9**, 484.
41. G. Sosnovsky and J. A. Krogh, *Synthesis*, 1978, 703; G. Sosnovsky, J. A. Krogh, and S. G. Umhoefer, *Synthesis*, 1979, 722; *F. &F.*, **9**, 409.
42. *F. & F.*, **4**, 523; J. K. Chakrabarti and T. M. Hotten, *J. Chem. Soc., Chem. Commun.*, 1972, 1226.
43. D. L. J. Clive, *J. Chem. Soc., Chem. Commun.*, 1970, 1014; *F. &F.*, **3**, 50; D. L. Garmaise, A. Uchiyama, and A. F. McKay, *J. Org. Chem.*, 1962, **27**, 4509.
44. *F. &F.*, **3**, 50; D. L. J. Clive, *J. Chem. Soc., Chem. Commun.*, 1970, 1014.
45. H. G. Foley and D. R. Dalton, *J. Chem. Soc., Chem. Commun.*, 1973, 628.
46. G. Sosvnosky and J. A. Krogh, *Z. Naturforsch., Teil B*, 1978, **32**, 1179.
47. V. P. Kukhar and V. I. Pasternak, *Synthesis*, 1974, 563.
48. *F. &F.*, **14**, 337; J. B. Hindrickson and M. S. Hussoin, *J. Org. Chem.*, 1987, **52**, 4137.
49. Preparation: H. Meerwein, P. Laasch, R. Mersch, and J. Spille, *Ber.*, 1956, **89**, 209.
50. T.-L. Ho, *Synthesis*, 1975, 401.
51. W. E. Bissinger and F. E. Kung, *J. Am. Chem. Soc.*, 1948, **70**, 2664. Preparation from phenol and thionyl chloride, reflux.
52. J. G. Kraus and S. Shaikh, *Synthesis*, 1975, 502.
53. J. M. Prokipcak and P. A. Forte, *Can. J. Chem.*, 1971, **49**, 1321; *F. &F.*, **6**, 456.
54. J. A. Albright and M. L. Alexander, *Org. Prep. Proc. Int.*, 1972, **4**, 215; *F. &F.*, **4**, 341.
55. *F. &F.*, **12**, 122; A. Arrieta and C. Palomo, *Synthesis*, 1983, 472.
56. *F. &F.*, **12**, 204; A. Arrieta, J. M. Aizopura, and C. Palomo, *Tetrahedron Lett.*, 1984, **25**, 3365.
57. J. K. Rasmussen, *Chem. Lett.*, 1977, 1295.
58. J. Liebscher and H. Hartman, *Z. Chem.*, 1975, **15**, 302.
59. G. Sosvnosky, J. A. Krogh, and S. G. Umhoefer, *Synthesis*, 1979, 722.
60. *F. &F.*, **5**, 206; E. Vowinkel and J. Bartel, *Ber.*, 1974, **107**, 1221.
61. *F. &F.*, **2**, 217; J. A. Findlay and C. S. Tang, *Can. J. Chem.*, 1967, **45**, 1014.
62. T. van Es, *J. Chem. Soc.*, 1965, 1564.
63. G. D. Olah and T. Keumi, *Synthesis*, 1979, 12; *F. &F.*, **9**, 245.
64. Preparation: *F. &F.*, **1**, 481; H. J. Matsugama and L. F. Audrieth, *Inorg. Synth.*, 1957, **5**, 122.
65. *F. &F.*, **6**, 290; C. Fizet and J. Streith, *Tetrahedron Lett.*, 1974, 3187.
66. *F. &F.*, **11**, 346; P. Andoye, A. Gaset, and J. P. Gorrichon, *Chimia*, 1982, **36**, 4.
67. A. Stojilkovic, V. Andrejevic, and M. L. Mihailovic, *Tetrahedron*, 1967, **23**, 721; *Quart. Rev.*, 1971, **25**, 407; G. Tennant, *Comprehensive organic chemistry* (vol. ed. I. O. Sutherland, series ed. D. H. R. Barton and W. D. Ollis), Pergamon Press, Oxford, 1979, Vol. 2, p. 535.
68. I. Rhee, M. Ryang, and S. Tsutsumi, *Tetrahedron Lett.*, 1970, 3419.
69. Preparation: T. G. Clarke, N. A. Hampson, J. B. Lee, J. R. Morley, and B. Scanlon, *Can. J. Chem.*, 1969, **47**, 3729.

70. *F. &F.*, **5**, 20; J. B. Lee, C. Parkin, M. J. Shaw, N. A. Hampson, and K. I. MacDonald, *Tetrahedron*, 1973, **29**, 751.
71. Preparation: T. Yoshimura, T. Omata, N. Furukawa, and S. Oae, *J. Org. Chem.*, 1976, **41**, 1728.
72. *F. &F.*, **10**, 174; Y. Gelas-Mialhe and R. Vessière, *Synthesis*, 1980, 1005.
73. *F. &F.*, **1**, 289; R. F. Smith and L. E. Walker, *J. Org. Chem.*, 1962, **27**, 4372.
74. Preparation: I. El-S. El- Kholy, F. K. Rafla, and M. M. Mishrikey, *J. Chem. Soc., (C)*, 1970, 1578.
75. *F. &F.*, **7**, 10; J. B. Bapat, R. J. Blade, A. J. Boulton, J. Epsztajn, A. R. Katritzky, J. Lewis, P. Molina-Buendia, P.-L. Nie, and C. A. Ramsden, *Tetrahedron Lett.*, 1976, 2691.
76. I. Ikeda, Y. Machii, and M. Okahara, *Synthesis*, 1978, 301.
77. D. M. Orere and C. B. Reese, *J. Chem. Soc., Chem. Commun.*, 1977, 280; *F. &F.*, **8**, 409.
78. M. C. Chaco and N. Rabjohn, *J. Org. Chem.*, 1962, **27**, 2765; *F. &F.*, **7**, 235, cf. **4**, 333; 512.
79. F. E. Ziegler and P. A. Wender, *J. Am. Chem. Soc.*, 1971, **93**, 4318; P. A. Wender, M. A. Eissenstat, N. Sapuppo, and F. E. Ziegler, *Org. Synth.*, 1978, **58**, 101.
80. *F. &F.*, **4**, 512; S. Cacchi, L. Caglioti, and G. Paolucci, *Chem. &Ind.*, 1972, 213.
81. S. Cacchi, L. Caglioti, and G. Paolucci, *Synthesis*, 1975, 120.
82. *F. &F.*, **1**, 480; A. Ahmad and I. D. Spencer, *Can. J. Chem.*, 1961, **39**, 1340.
83. *F. &F.*, **7**, 284; A. Ahmad, *Synthesis*, 1976, 418.
84. *F. &F.*, **13**, 332; S. Manna, J. R. Falck, and C. Mioskowski, *Synth. Commun.*, 1985, **15**, 663.
85. *F. &F.*, **10**, 324; A. Mizuno, Y. Hamada, and T. Shioiri, *Synthesis*, 1980, 1007.
86. *F. &F.*, **4**, 241; D. A. Klein, *J. Org. Chem.*, 1971, **36**, 3050.
87. *F. &F.*, **1**, 176; E. V. Zappi and O. Bonso, *Anales Asoc. Quim. Argentina*, 1947, **35**, 137.
88. D. E. Douglas and A. M. Burnett, *Can. J. Chem.*, 1953, **31**, 1127.
89. J. A. Barltrop, A. C. Day, and D. B. Bigley, *J. Chem. Soc.*, 1961, 3185.
90. *F. &F.*, **11**, 322; A. D. Dunn, M. T. Mills, and W. Henry, *Org. Prep. Proc. Int.*, 1982, **14**, 396.
91. M. Makosza and A. Jonczyc, *Org. Synth.*, 1976, **55**, 91.
92. *F. &F.*, **3**, 236; H. C. Brown, H. Nambu, and M. M. Rogic, *J. Am. Chem. Soc.*, 1969, **91**, 6854.
93. H. Nambu and H. C. Brown, *J. Am. Chem. Soc.*, 1970, **92**, 5790.
94. *F. & F.*, **15**, 130; Y. Masuda, M. Hoshi, and A. Arase, *J. Chem. Soc., Chem. Commun.*, 1989, 266.
95. *F. &F.*, **13**, 221; F. Orsini, *Synthesis*, 1985, 500.
96. *F. &F.*, **15**, 126; M. Lemaire, D. Doussot, and A. Guy, *Chem. Lett.*, 1988, 1581.
97. *F. &F.*, **12**, 518; B. Giese, J. A. González-Gomez, and T. Witzel, *Angew. Chem., Int. Ed. Engl.*, 1984, **23**, 69; B. Giese and J. Dupuis, *Angew. Chem. Int., Ed. Engl.*, 1983, **22**, 753.
98. M. L. Hallensleben, *Tetrahedron Lett.*, 1972, 2057.
99. *F. &F.*, **14**, 106; H. J. Bestmann and M. Schmidt, *Angew. Chem., Int. Ed. Engl.*, 1987, **26**, 79.
100. W. M. Burgess and J. W. Eastes, *Inorg. Synth.*, 1957, **5**, 197.
101. I. Hashamito, N. Tsuruta, M. Ryang, and S. Tsutsumi, *J. Org. Chem.*, 1970, **35**, 3748; *F. &F.*, **3**, 212.
102. W. C. Groutas and D. Felker, *Synthesis*, 1980, 861.

103. *F. &F.*, **10**, 112; M. Oda, A. Yamamuro, and T. Watabe, *Chem. Lett.*, 1979, 1427.
104. *F. &F.*, **10**, 307; R. E. Murray and G. Zweifel, *Synthesis*, 1980, 150.
105. *F. &F.*, **7**, 20; M. Makosza, J. Czyewski, and M. Jawdosiuk, *Org. Synth.*, 1976, **55**, 99.
106. *F. &F.*, **7**, 303; G. W. Gokel, S. A. DiBiase, and B. A. Lipisko, *Tetrahedron Lett.*, 1976, 3495.
107. M. Ochai, K. Sumi, K. Takaoka, M. Kurishima, Y. N. M. Shiro, and E. Fujita, *Tetrahedron*, 1988, **44**, 4095.
108. *F. &F.*, **5**, 274; D. N. Brattesam and C. H. Heathcock, *Tetrahedron Lett.*, 1974, 2279.
109. *F. &F.*, **15**, 325; Y. Z. Huang, Y. Shen, and C. Chen, *Synth. Commun.*, 1989, **19**, 83.
110. *F. &F.*, **12**, 391; T. Ono, T. Tamaoka, Y. Yuasa, T. Matsuda, J. Nokami, and S. Wakabayashi, *J. Am. Chem. Soc.*, 1984, **105**, 7890.
111. *F. &F.*, **11**, 418; T. Nakami, T. Mandai, Y. Imakura, N. Nishiuchi, M. Kawada, and S. Wakabayashi, *Tetrahedron Lett.*, 1981, **22**, 4489; T. Mandai, S. Hashhio, J. Goto, and M. Kawada, *Tetrahedron Lett.*, 1981, **22**, 2187.
112. *F. & F.*, **11**, 571; I. Matsada, S. Murata, and Y. Ishii, *J. Chem. Soc., Chem. Commun.*, 1979, 16.
113. R. Haruta, M. Ishiguro, K. Furnta, A. Mori, N. Ikeda, and Y. Yamamoto, *Chem. Lett.*, 1982, 1093.
114. *F. &F.*, **11**, 157; Y. Sakakihara, N. Yadani, I. Ibuki, M. Sakai, and N. Uchino, *Chem. Lett.*, 1982, 1565.
115. *F. &F.*, **11**, 291; P. T. Lansburg and J.-P. Vacca, *Tetrahedron Lett.*, 1982, **23**, 2623.
116. *F. &F.*, **11**, 416; S. Tomoda, Y. Takeuchi, and Y. Nomura, *Chem. Lett.*, 1981, 1069.
117. *F. &F.*, **11**, 433; D. F. Taber and S. A. Saleh, *J. Org. Chem.*, 1981, **46**, 4817.
118. *F. &F.*, **11**, 503; W. R. Jackson and C. G. Lovel, *J. Chem. Soc., Chem. Commun.*, 1982, 1231.
119. *F. &F.*, **11**, 591; Y. Sato and Y. Niinomi, *J. Chem. Soc., Chem. Commun.*, 1982, 56.
120. *F. &F.*, **12**, 503; H. Okada, I. Matusda, and Y. Izumi, *Chem. Lett.*, 1983, 97.
121. *F. &F.*, **1**, 1133; R. A. Carboni, *Org. Synth., Coll. Vol. 4*, 1963, 877.
122. W. J. Middleton, R. E. Heckert, E. L. Little, and C. G. Kespan, *J. Am. Chem. Soc.*, 1958, **80**, 2783; C. A. Stewart, Jr, *J. Org. Chem.*, 1963, **28**, 3320.
123. B. C. McKusick, R. E. Heckart, T. L. Cairns, D. D. Coffman, and H. F. Mower, *J. Am. Chem. Soc.*, 1958, **80**, 2806.
124. A. Jonczyc, M. Fedorynski, and M. Makoszaa, *Tetrahedron Lett.*, 1972, 1637.
125. J. M. Cox and R. Ghosh, *Tetrahedron Lett.*, 1969, 3351.
126. *F. &F.*, **11**, 536; D. Kahne and D. B. Colium, *Tetrahedron Lett.*, 1981, **22**, 5011.
127. M. Ohno, N. Naruse, S. Torimitsu, and I. Terasawa, *J. Am. Chem. Soc.*, 1966, **88**, 3168; M. Ohno, N. Naruse, and I. Terasawa, *Org. Synth.*, 1969, **49**, 27; *F. &F.*, **3**, 214.
128. *F. &F.*, **12**, 182; E. J. Corey, D. H. Hua, and S. P. Setz, *Tetrahedron Lett.*, 1984, **25**, 3.
129. Review: L. I. Krimen and D. J. Cota, *Org. React.*, 1969, **17**, 213.
130. R. R. Ritter and P. P. Minieri, *J. Am. Chem. Soc.*, 1948, **70**, 4045.
131. I. Ikeda, Y. Machii, and M. Okahara, *Synthesis*, 1978, 301.
132. J. Anatol and A. Medete, *Bull. Soc. Chim. Fr.*, 1972, 189.

133. J. Anatol and J. Berecoechea, *Bull. Soc. Chim. Fr.*, 1975, 395; *Synthesis*, 1975, 111.
134. W. Haaf, *Chem. Ber.*, 1963, **96**, 3359.
135. V. I. Sokolov and O. A. Reutov, *Bull. Acad. Sci. USSR, Div. Chem. Sci.*, 1968, 225; V. J. Berger and D. Vogel, *J. Prakt. Chem.*, 1969, **311**, 737; H. C. Brown and J. T. Kurek, *J. Am. Chem. Soc.*, 1969, **91**, 5647; D. Chow, J. H. Robson, and G. F. Wright, *Can. J. Chem.*, 1965, **43**, 312.
136. *F. &F.*, **10**, 13; S. Top and G. Jaouen, *J. Org. Chem.*, 1981, **46**, 78.
137. *F. &F.*, **15**, 339; A. G. Martinez, R. M. Alvarez, E. T. Vilar, A. F. Fraile, M. Hanack, and L. R. Subramanian, *Tetrahedron Lett.*, 1989, **30**, 581.
138. *F. &F.*, **5**, 694; J. L. Fry, *J. Chem. Soc., Chem. Commun.*, 1974, 45.
139. M. L. Scheinbaum and M. B. Dines, *Tetrahedron Lett.*, 1971, 2205; *F. &F.*, **4**, 360.
140. *F. &F.*, **13**, 300; F. Scavo and P. Helquist, *Tetrahedron Lett.*, 1985, **26**, 2603.
141. Preparation: E. L. Manzer, *Inorg. Chem.*, 1977, **16**, 525.
142. *F. &F.*, **14**, 214; E. J. Roskamp and S. F. Pedersen, *J. Am. Chem. Soc.*, 1987, **109**, 3152.
143. P. E. Spoerri and A. S. Du Bois, *Org. React.*, 1945, **5**, 387.
144. C. R. Hauser and D. S. Hoffenberg, *J. Org. Chem.*, 1955, **20**, 1482; *F. &F.*, **1**, 69.
145. *F. &F.*, 7, 304; J. Hall and M. Gisler, *J. Org. Chem.*, 1976, **41**, 3769.
146. Br. Radziszewski, *Ber.*, 1885, **18**, 355; K. B. Wiberg, *J. Am. Chem. Soc.*, 1953, **75**, 3961; J. E. McIsaac Jr, R. E. Ball, and E. J. Behrman, *J. Org. Chem.*, 1971, **36**, 3048.
147. *F. &F.*, **1**, 470; G. B. Payne, *Tetrahedron*, 1962, **18**, 763.
148. *F. &F.*, **8**, 387; *F. &F.*, **13**, 135; S. Rozen and M. Brand, *Angew. Chem., Int. Ed. Engl.*, 1986, **25**, 554.
149. *F. &F.*, **15**, 300; J. T. Eaton, W. D. Rounds, J. H. Urbanowicz, and G. W. Gribble, *Tetrahedron Lett.*, 1988, **29**, 6557.
150. *F. &F.*, **10**, 305; C. R. Noller, *Org. Synth., Coll. Vol. 2*, 1943, 587, note 5.
151. C. Cacci, D. Misiti, and F. La Torre, *Synthesis*, 1980, 243.
152. *F. &F.*, **15**, 295; J. Jammot, R. Pascal, and A. Commeyras, *Tetrahedron Lett.*, 1989, **30**, 563.
153. *F. &F.*, **11**, 142; M. Ravindranathan, N. Kalyanam, and S. Sivaram, *J. Org. Chem.*, 1982, **47**, 4812.
154. *F. &F.*, **10**, 323; S. A. Dibiase, R. P. Wolak, Jr, D. M. Dishong, and G. W. Gokel, *J. Org. Chem.*, 1980, **45**, 3630.
155. *F. &F.*, **14**, 70; A. Armstrong, I. Brackenridge, R. W. F. Jackson, and J. M. Kirk, *Tetrahedron Lett.*, 1988, **29**, 2483.
156. *F. &F.*, **11**, 109; J. W. Cornforth and R. H. Cornforth, *J. Chem. Soc.*, 1947, 96; A. McKillop, A. Henderson, P. S. Ray, C. Avendano, and E. G. Molinero, *Tetrahedron Lett.*, 1982, **23**, 3357.
157. *F. &F.*, **15**, 286; M. Shimizu and H. Yoshioka, *Heterocycles*, 1988, **27**, 2527.
158. *F. &F.*, **4**, 502; E. C. Taylor and J. A. Zoltewicz, *J. Am. Chem. Soc.*, 1960, *82*, 2656.
159. *F. &F.*, **9**, 438; B. Singh and G. Y. Lesher, *Synthesis*, 1978, 829.
160. *F. &F.*, **1**, 1075; T. Kauffmann, S. Spaude, and D. Wolf, *Ber.*, 1964, **97**, 3436.
161. *F. &F.*, **11**, 1; A. Rahm, R. Guilhemat, and M. Pereyre, *Synth. Commun.*, 1982, **12**, 485.
162. R. F. Nystrom, *J. Am. Chem. Soc.*, 1955, **77**, 2544; W. G. Brown, *Org. React.*, 1951, **6**, 469.
163. *F. &F.*, **1**, 980; M. Freifelder, *J. Am. Chem. Soc.*, 1960, **82**, 2386.

164. *F. &F.*, **13**, 278; J. Mauger and A. Robert, *J. Chem. Soc., Chem. Commun.*, 1986, 395.
165. H. C. Brown and C. P. Garg, *J. Am. Chem. Soc.*, 1964, **86**, 1085.
166. H. Stephen, *J. Chem. Soc.*, 1925, **127**, 1874.
167. E. N. Zil'bermann and P. S. Pyryalova, *J. Gen. Chem., USSR*, 1964, **33**, 3348.
168. *F. &F.*, **4**, 245; T. Cuvigny, M. Larchevéque, and H. Normant, *Compt. Rendu.*, 1972, **274**, 797.
169. *F. &F.*, **5**, 543; T. Cuvigny, M. Larchevéque, and H. Normant, *Bull. Soc. Chim. Fr.*, 1978, 1174.
170. W. Buchner and R. Dufaux, *Helv. Chim. Acta*, 1966, **49**, 1145.
171. P. G. Arapakos, *J. Am. Chem. Soc.*, 1967, **89**, 6794; P. G. Arapakos, M. K. Scott, and F. E. Huber, Jr. *J. Am. Chem. Soc.*, 1969, **91**, 2059.
172. S. Yamada, K. Tomioka, and K. Koga, *Tetrahedron Lett.*, 1976, 61.
173. E. E. van Tamelen, H. Rudler, and C. Bjorklund, *J. Am. Chem. Soc.*, 1971, **93**, 7113.
174. P. L. Pickard and T. L. Tolbert, *J. Org. Chem.*, 1961, **26**, 4886.
175. *F. &F.*, **14**, 99.
176. F. J. Weiberth and S. S. Hall, *J. Org. Chem.*, 1987, **52**, 3901.
177. *F. &F.*, **12**, 235; L. R. Krepski, S. M. Heilmann, and J. K. Rasmussen, *Tetrahedron Lett.*, 1983, **24**, 4075; *F. &F.*, **13**, 138; M. Gill, M. J. Keifel, and D. A. Lally, *Tetrahedron Lett.*, 1986, **27**, 1933.
178. *F. &F.*, **1**, 1197; H. Reinheckel and D. Jahnke, *Ber.*, 1964, **97**, 2661.
179. E. E. Blaise, *Compt. Rend.*, 1901, **132**, 478.
180. *F. &F.*, **12**, 568; S. M. Hannick and Y. Kishi, *J. Org. Chem.*, 1983, **49**, 3833.
181. J. F. Thorpe, *J. Chem. Soc.*, 1909, **95**, 1901.
182. K. Ziegler and H. Ohlinger, *Annalen*, 1932, **495**, 84.
183. K. Ziegler, H. Eberle, and H. Ohlinger, *Annalen*, 1933, **504**, 94.
184. K. Ziegler, H. Eberle, and H. Ohlinger, *Annalen*, 1933, **504**, 94; K. Ziegler and R. Aumhammer, *Annalen*, 1934, **513**, 43.
185. *F. &F.*, **1**, 1095; E. M. Fry and L. F. Fieser, *J. Am. Chem. Soc.*, 1940, **62**, 3489; A. C. Cope and R. J. Cotter, *J. Org. Chem.*, 1964, **29**, 3467.
186. J.-A. MacPhee and J.-E. Dubois, *Tetrahedron*, 1980, **36**, 775.
187. J. P. Abarella, *J. Org. Chem.*, 1977, **44**, 2009; A. I. Meyers and P. D. Pansegrau, *Tetrahedron Lett.*, 1984, 2941.
188. *F. &F.*, **14**, 189; B. M. Trost, J. Florez, and D. J. Jebaratnam, *J. Am. Chem. Soc.*, 1987, **109**, 613.
189. *F. &F.*, **15**, 16; E. P. Kündig, V. Desobry, D. P. Simmons, and E. Wenger, *J. Am. Chem. Soc.*, 1989, **111**, 1804.
190. *F. &F.*, **13**, 315; B. A. Pearlman, R. S. Putt, and J. A. Fleming, *J. Org. Chem.*, 1983, **50**, 3625.
191. *F. &F.*, **12**, 193; L. E. Overman and R. M. Buck, *Tetrahedron Lett.*, 1984, **25**, 5737.
192. *F. &F.*, **11**, 186; H. L. Goerning and C. C. Tseng, *J. Org. Chem.*, 1981, **46**, 5250.
193. *F. &F.*, **11**, 299; D. Enders and H. Lotter, *Tetrahedron Lett.*, 1982, **23**, 639.
194. *F. &F.*, **8**, 405; D. Savoia, C. Trombini, and A. Umani-Ronchi, *Tetrahedron Lett.*, 1977, 653.
195. E. J. Corey and J. Das, *Tetrahedron Lett.*, 1982, **23**, 4217.
196. *F. &F.*, **8**, 415; E. Marchand, G. Morel, and A. Foucaud, *Synthesis*, 1978, 360.
197. *F. &F.*, **5**, 523; S. J. Selikson and D. S. Watt, *Tetrahedron Lett.*, 1974, 3029.
198. *F. &F.*, **9**, 159; C. F. Allen and J. A. van Allen, *Org. Synth. Coll. Vol. 3*, 1955, 275.

199. G. Stork, A. Aozorio, and A. Y. W. Leong, *Tetrahedron Lett.*, 1978, 5175.
200. *F. &F.*, **11**, 23; F. Texier-Boullet and A. Foucaud, *Tetrahedron Lett.*, 1982, **23**, 4927.
201. M. Schmeisser, P. Sarttoni, and B. Lippsmeier, *Ber.*, 1970, **103**, 868.
202. *F. &F.*, **8**, 515; H. Emde and G. Simchen, *Synthesis*, 1977, 636.
203. G. Stork and L. Maldonado, *J. Am. Chem. Soc.*, 1971, **93**, 5286; 1974, **96**, 5272; *F. &F.*, **4**, 300.
204. G. Stork, J. C. Depezay, and J. D'Angelo, *Tetrahedron Lett.*, 1975, 389.
205. S. Takakashi, M. Uno, and K. Seto, *Japan Kokai Tokyo Koho*, 1975, **JP 60**, 204, 753; *Chem. Abstr.*, 1985, **104**, 6862f.
206. A. Jonczyc, D. Lipiak, and T. Zdrojewski, *Tetrahedron*, 1990, **46**, 1025.
207. *F. &F.*, **13**, 150; H. Suginome, C. F. Lin, M. Tokuda, and A. Furusaki, *J. Chem. Soc., Perkin Trans. I*, 1985, 327.
208 *F. &F.*, **12**, 525; E. R. F. Gesing, U. Groth, and K. C. P. Vollhardt, *Synthesis*, 1984, 351.
209. *F. &F.*, **12**, 180; R. E. Greiger, M. Lalonde, H. Stoller, and K. Schleich, *Helv. Chim. Acta*, 1984, **67**, 1274.
210. *F. &F.*, **15**, 79; K. M. Doxee and J. B. Farahi, *J. Am. Chem. Soc.*, 1988, **110**, 7239.
211. E.-J. Tillmans and J. J. Ritter, *J. Org. Chem.*, 1957, **22**, 839.
212. B. E. Tate and P. D. Bartlett, *J. Am. Chem. Soc.*, 1956, **78**, 5575.
213. K. Haase and H. M. R. Hoffmann, *Angew. Chem., Int. Ed. Engl.*, 1982, **21**, 83.
214. J. F. Normant and C. Piechucki, *Bull. Soc. Chim. Fr.*, 1972, 2402; *F. &F.*, **5**, 166.
215. *F. &F.*, **8**, 476; E. C. Taylor, J. C. Andrade, K. C. John, and A. McKillop, *J. Org. Chem.*, 1978, **43**, 2280.
216. *F. &F.*, **13**, 54; S.-I. Murahashi, T. Naota, and N. Nakajima, *Tetrahedron Lett.*, 1985, **26**, 925.
217. *F. &F.*, **11**, 26; Z. M. Ismail and H. M. R. Hoffmann, *Angew. Chem., Int. Ed. Engl.*, 1982, **21**, 859.
218. *F. &F.*, **10**, 1; G. A. Kraus and M. Shimagaki, *Tetrahedron Lett.*, 1981, **22**, 1171.
219. *F. &F.*, **9**, 127, S. Hünig and G. Whener, *Synthesis*, 1979, 522; J. K. Rasmussen and S. M. Heilmann, *Synthesis*, 1979, 523.
220. P. Cazeau, F. Moulines, O. Lamporte, and F. Duboudinn, *J. Organomet. Chem.*, 1980, **201**, C9; F. Duboudin, P. Cazeau, F. Moulines, and O. Lamporte, *Synthesis*, 1982, 213; M. T. Reetz and I. Chatziiosifidis, *Synthesis*, 1982, 330.
221. K. Herrman and G. Simchen, *Synthesis*, 1979, 204.
222. *F. &F.*, **15**, 15; K. Homma and T. Mukiyama, *Chem. Lett.*, 1989, 259.
223. P. G. Gassman and T. L. Guggenheim, *J. Am. Chem. Soc.*, 1982, **104**, 5849.
224. J. C. Mullis and W. P. Weber, *J. Org. Chem.*, 1982, **47**, 2873.
225. *F. &F.*, **4**, 542.
226. U. Hertenstein, S. Hünig, and M. Öller, *Ber.*, 1980, **113**, 3783.
227. K. Utimoto, Y. Wakabayashi, Y. Shishiyama, M. Inoue, and H. Nozaki, *Tetrahedron Lett.*, 1981, **22**, 4279.
228. J. L. Belletire, H. Howard, and K. Donahue, *Synth. Commun.*, 1982, **12**, 763.
229. *F. &F.*, **12**, 148.
230. S. Kiyooka, R. Fujiyama, and K. Kawaguchi, *Chem. Lett.*, 1984, 1979.
231. M. T. Reetz and H. Müller-Starke, *Tetrahedron Lett.*, 1984, **25**, 3301.
232. *F. &F.*, **1**, 1113; **5**, 647; D. N. Dhar, *Chem. Rev.*, 1967, **67**, 611.
233. *F. &F.*, **4**, 155; R. W. Begland, A. Cairncross, D. S. Donald, D. R. Hartter, W. A. Sheppard, and O. W. Webster, *J. Am. Chem. Soc.*, 1971, **93**, 4953.
234. O. W. Webster, D. R. Hartter, R. W. Begland, W. A. Sheppard, and A. Cairncross,

J. Org. Chem., 1972, **37**, 4133; R. W. Begland and D. R. Hartter, *J. Org. Chem.*, 1972, **37**, 4136.
235. *F. & F.*, **11**, 44; T. Iverson and D. R. Bundle, *J. Chem. Soc., Chem. Commun.*, 1981, 1240.
236. F. Cramer, K. Pawelik, and H. J. Baldauf, *Ber.*, 1958, **91**, 1049.

5

N-OXIDES AND NITRONES

The *N*-oxide functional group (**1**) occurs in three important structures, viz. in *N*-oxides of tertiary amines (**2**), in *N*-oxides of imines (nitrones) (**3**), and in *N*-oxides of nitriles (**4**).

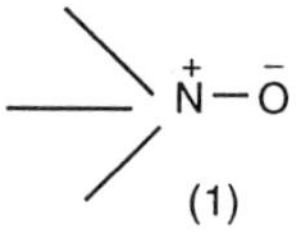

(1)

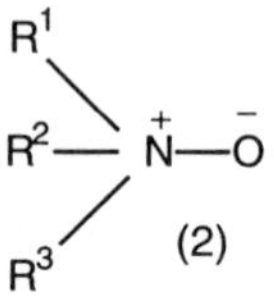

(2) 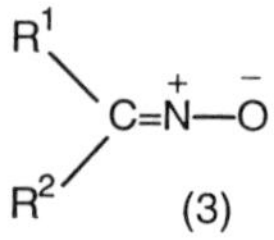 (3) R-C≡N⁺—O⁻ (4)

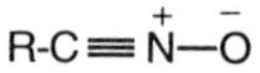

N-OXIDES OF TERTIARY AMINES (2)

5.1 Preparation

5.1.1 *From amines by oxidation*

Tertiary amines are readily oxidized in good yields to *N*-oxides (**2**) by hydrogen peroxide, for example, *N*-methylmorpholine *N*-oxide (**5**) is prepared by the oxidation of *N*-methylmorpholine with 50 per cent aqueous hydrogen peroxide at 50–75 °C. They can also be prepared in anyhydrous conditions by treating a tertiary amine with *t*-butyl hydroperoxide, tertiary butanol, and vanadium oxyacetylacetonate (Scheme 5.1).[1]

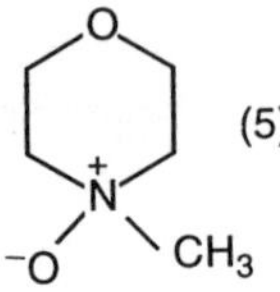

(5)

$$R_3N \xrightarrow{H_2O_2 / CH_3CO_2H} R_3\overset{+}{N}\text{—}\overset{-}{O}$$

$$R_3N \xrightarrow[VO(CH_2COCH_2CO)_2]{Bu^tO_2H / Bu^tOH} R_3\overset{+}{N}\text{—}\overset{-}{O} \qquad (5.1)$$

5.1.2 *By distillation with dimethylformamide*

Another way of obtaining anhydrous *N*-oxides is by the removal of water from the commercially available hydrates by distillation with dimethylformamide.[2]

5.2 Reactivity[3]

5.2.1 *Deoxygenation to tertiary amines*

With acetic-formic anhydride

Acetic-formic anhydride has been applied to the deoxygenation of amine oxides,[4] for example, Scheme 5.2.

$$C_6H_{11}\text{—}\overset{+}{N}Me_2(O^-) + CH_3CO_2CHO \xrightarrow[25^\circ C]{CH_2Cl_2} C_6H_{11}\text{—}NMe_2\ (100\ \%) + CH_3CO_2H + CO_2$$

Mechanism

$$R_3\overset{+}{N}\text{—}\overset{-}{O} + HC(=O)\text{—}O\text{—}C(=O)\text{—}CH_3 \longrightarrow R_3\overset{+}{N}\text{—}O\text{—}C(=O)H + {}^-O\text{—}C(=O)CH_3 \longrightarrow R_3N + CH_3CO_2H + CO_2 \qquad (5.2)$$

Catalytic reduction

N-Oxides can be reduced under mild conditions by hydrogen transfer using Pd/C, hydrogen, and NaH_2PO_2 to yield tertiary amines,[5] for example, Scheme 5.3.

$$\text{quinoline } N\text{-oxide} \xrightarrow[\text{Pd/C}]{NaH_2PO_2} \text{quinoline (83\%)}$$

$$N\text{-methylpiperidine } N\text{-oxide} \xrightarrow[\text{Pd/C}]{NaH_2PO_2} N\text{-methylpiperidine (90\%)} \quad (5.3)$$

With iron pentacarbonyl

The conversion of *N*-oxides to amines is possible by the use of iron pentacarbonyl as a deoxygenator (Scheme 5.4).[6]

$$R^1R^2R^3\overset{+}{N}{-}O^- \xrightarrow[\text{reflux}]{Fe(CO)_5\,/\,Bu^n_2O} R^1R^2R^3N \quad 45\text{-}50\% \quad (5.4)$$

Using samarium diiodide

Another method for the deoxygenation of *N*-oxides is by the use of samarium diiodide for efficient electron transfer,[7] for example, Scheme 5.5.

$$\text{quinoline } N\text{-oxide} \xrightarrow[20^\circ \text{ to } 65^\circ C]{SmI_2\,/\,THF\,/\,HMPT} \text{quinoline (96\%)} \quad (5.5)$$

With silyl triflates

N-Oxides of tertiary amines react with a silyl trifluoromethanesulphonate (a silyl triflate) (**6**) to yield intermediates which can be converted into secondary[8] or tertiary amines[9] (Scheme 5.6).

$$R^1R^2\overset{+}{N}(O^-)CH_2\text{-}R^3 + Bu^tMe_2Si\text{-}O\text{-}SO_2CF_3\ (6) \xrightarrow{CH_2Cl_2} R^1R^2\overset{+}{N}(O\text{—}SiMe_2Bu^t)CH_2\text{—}R^3\ TfO^-\ (100\%) \xrightarrow{CH_3Li\ in\ THF} R^1R^2N\text{—}CH(R^3)\text{-}O\text{—}SiMe_2Bu^t$$

$$R^1R^2N\text{—}CH(R^3)\text{-}O\text{—}SiMe_2Bu^t \xrightarrow{R^4\text{- Hal, }110^\circ C} \left[R^1R^2R^4\overset{+}{N}\text{—}CH(R^3)\text{-}O\text{—}SiMe_2Bu^t\ Hal^-\right] \xrightarrow{Bu^n{}_4N^+F^-} R^1R^2N\text{—}R^4\ (35\text{-}60\%)$$

$$R^1R^2N\text{—}CH(R^3)\text{-}O\text{—}SiMe_2Bu^t \xrightarrow{HOAc} R^1R^2N\text{—}H\ (c.\ 70\%) \quad (5.6)$$

5.2.2 *As oxidizing agents*

N-Oxides act as oxidizing agents, being reduced to tertiary amines.

Conversion of alkyl halides to aldehydes

N-Oxides, acting as oxidizing agents, convert alkyl halides of suitable structure into aldehydes in moderate yields,[10] for example, Scheme 5.7.

$$n\text{-}C_7H_{15}\text{-}CH_2\text{—}I + {}^-O\text{—}\overset{+}{N}Me_3 \xrightarrow{CHCl_3} \left[n\text{-}C_7H_{15}\text{-}CH(H)\text{—}O\text{—}\overset{+}{N}Me_3\ I^-\right] \longrightarrow n\text{-}C_7H_{15}CHO + (CH_3)_3\overset{+}{N}H + I^-\ (41.5\text{ - }43\%) \quad (5.7)$$

Preparation of alcohols

Two methods are available for the preparation of alcohols, one by the action of trimethylamine *N*-oxide on a trifluorosilylalkane[11] and another by its reaction with an alkyl aluminium compound[12] (Scheme 5.8).

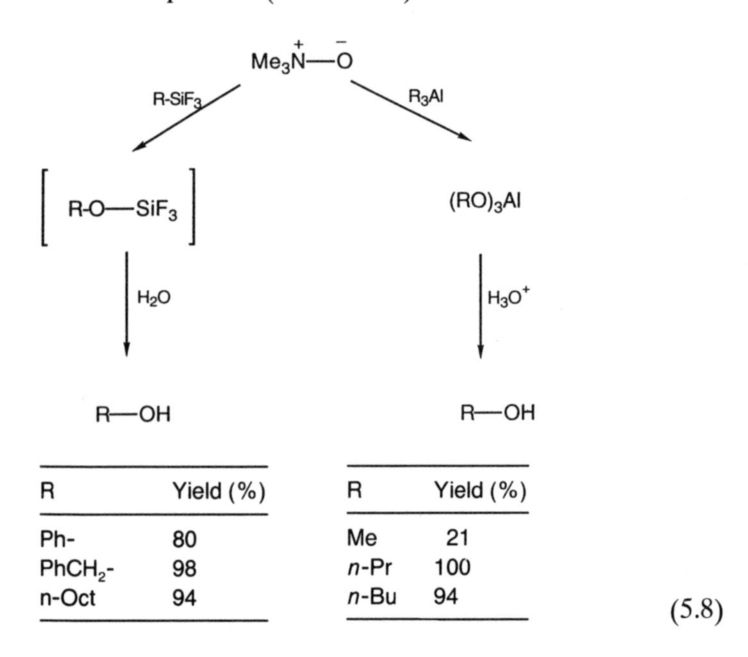

R	Yield (%)
Ph-	80
PhCH$_2$-	98
n-Oct	94

R	Yield (%)
Me	21
n-Pr	100
n-Bu	94

(5.8)

Use to prepare diols from alkenes

N-Oxides are stable solids which have found use in specific oxidations, for example, the *N*-oxide of *N*-methyl-morpholine (**5**) is used in conjunction with a catalytic amount of osmium tetroxide, thus saving expense, to oxidize an alkene to a 1,2-diol (Scheme 5.9).[13]

OsO_4 (catalytic amount)
N-methylmorpholine *N*-oxide
(5)
HO OH

(5.9)

5.2.3 *As 1,3-dipoles*

1,3-Dipolar addition of alkenes

A strong base (LDA in excess) can convert suitable *N*-oxides of tertiary amines into ylids, which in the presence of alkenes yield heterocyclic compounds by

$(CH_3)_3\overset{+}{N}-\overset{-}{O}$ —LDA→ [CH_3, $CH_2=N^+$, CH_2^-] + Ph(H)C=C(H)Ph → CH_3–N, H_2C, CH_2, Ph–C–H, H–C–Ph

72%

$(CH_3)_3\overset{+}{N}-\overset{-}{O}$ —i, LDA / THF, -78°C; ii, cyclohexene→ N—CH_3

90%

(5.10)

1,3-dipolar additions [3 + 2] (Scheme 5.10).[14] The same ylid has been used for the preparation of β-amino ketones by reaction with suitable ketones (Scheme 5.11). *N*-Methylpiperidine *N*-oxide similarly gives *trans*-octahydroindolizines on reaction of its ylide with alkenes. (Scheme 5.12),[15] as does trimethylamine *N*-oxide (Scheme 5.13).[16]

CH_3 H

$CH_3-N^+-CH_2$

O^-Li

—LDA, THF→ $CH_2-\overset{+}{N}(CH_3)=CH_2$, H —LDA→ $\overset{-}{C}H_2-\overset{+}{N}(CH_3)=CH_2$

F_3C-CO_2^-

O

O—H, $\overset{-}{C}H_2$, $\overset{+}{N}$—Me, CH_2 → O, CH_2-NMe_2

(5.11)

$\overset{+}{N}$, O^-, CH_3 → $\overset{+}{N}$, CH_2 → N

(5.12)

$$(CH_3)_3\overset{+}{N}-\overset{-}{O} \xrightarrow{LiNEt_2} CH_3\text{-}\overset{+}{N}(=CH_2)-\overset{-}{C}H_2$$

75-85 % (5.13)

NITRONES (**3**)

5.3 Preparation[17]

5.3.1 *By oxidation of imines*

Several methods have been developed for the oxidation of imines to nitrones. For example, *N*-methylhydroxylamine-*O*-sulphonic acid[18] will oxidize imines to nitrones (Scheme 5.14).[19] Nitrones can also be prepared by the oxidation of imines with potassium permanganate in a phase-transfer system, with yields of between 15 and 90 per cent of the nitrone present in the product mixture,[20] for example, Scheme 5.15.

$$CH_3NHOH + Cl\text{-}SO_3H \longrightarrow CH_3NH\text{-}O\text{-}SO_3H \quad 85\%$$

$$R^1\text{-}CH{=}N\text{-}R^2 \xrightarrow[\text{MeOH, } 0\text{-}10^\circ C]{CH_3NH\text{-}O\text{-}SO_3H} R^1\text{-}CH{=}\overset{+}{N}(O^-)\text{-}R^2 \qquad (5.14)$$

$$Ph{-}CH{=}N{-}Bu^t \xrightarrow[Bu^n_4N^+Cl^- \text{ or } (n\text{-}C_8H_{17})_4N^+Br^-]{KMnO_4 \text{ / aqueous } CH_2Cl_2} Ph{-}CH{=}\overset{+}{N}(O^-){-}Bu^t \;\; 89\% \; + \; Ph{-}C(=O){-}NHBu^t \;\; 11\% \qquad (5.15)$$

5.3.2 *From silyl oximes*

Trimethylsilyl oximes react with Meerwein's reagent ($Me_3O^+BF_4^-$) or with methyl trifluoromethanesulphonate (MeO_3SCF_3) and the products yield nitrones after the reaction with potassium fluoride or with $Bu_4^nN^+F^-$ (Scheme 5.16).[21]

$$\mathrm{RHC{=}N{-}OSiMe_3} \xrightarrow{Me_3O^+\,BF_4^-} \left[\mathrm{RHC{=}\overset{+}{N}(Me){-}O{-}SiMe_3}\right] \xrightarrow[KF]{F^-} \mathrm{RHC{=}\overset{+}{N}(Me){-}O^-} + \mathrm{Me_3SiF} \quad (5.16)$$

5.3.3 *In situ preparation from secondary amines*

Nitrones may be prepared *in situ* for use in 1,3-dipolar additions by the oxidation of secondary amines using 30 per cent hydrogen peroxide and a catalytic amount of selenium dioxide (Scheme 5.17).[22]

$$\mathrm{Et_2NH} \xrightarrow[SeO_2\,/\,CH_3OH]{30\,\%\ H_2O_2} \mathrm{Et{-}\overset{+}{N}(O^-){=}CHCH_3}\quad 74\,\% \quad (5.17)$$

5.3.4 *From aldehydes or ketones*

Nitrones can be prepared in good to high yields from aldehydes or ketones as shown in Scheme 5.18.[23]

5.3.5 *By oxidation of secondary amines*

Oxidation of secondary amines with hydrogen peroxide catalysed by sodium tungstate, $Na_2WO_4.2H_2O$, gives good yields of nitrones (Scheme 5.19).[24] In the presence of an activated double carbon–carbon bond a 1,3-dipolar addition occurs (Scheme 5.20). Electrophiles may then react at the α-position (Scheme 5.21).

$CH_3\overset{+}{N}H_2OH\ Cl^- + 2\ Me_3SiCl \xrightarrow[\text{ether, }25^\circ C]{\text{excess }Et_3N} CH_3N(SiMe_3)(OSiMe_3)$

$\xrightarrow{R^1R^2C{=}O\text{ in benzene}} [R^1R^2C(OSiMe_3)N(OSiMe_3)CH_3] \xrightarrow{50^\circ C} R^1R^2C{=}\overset{+}{N}(O^-)CH_3 + O(SiMe_3)_2$

R^1	R^2	Yield (%)
H	C_6H_5–	97
H	Me_2N–C_6H_4–	93

(5.18)

$Bu^n\text{-}NH\text{-}Bu^n \xrightarrow[Na_2WO_4]{H_2O_2/MeOH} Bu^n\text{-}\overset{+}{N}(O^-){=}CH(CH_2)_2CH_3$ 89 %

2-methylpiperidine $\xrightarrow[Na_2WO_4]{H_2O_2/MeOH}$ 2-methyl-3,4,5,6-tetrahydropyridine 1-oxide 76 %

(5.19)

CH_3NHCH_2– $+ CH_2{=}CH{-}OBu^n \xrightarrow[Na_2WO_4.2H_2O]{H_2O_2}$ 2-methyl-5-butoxyisoxazolidine (N–O ring, OBu^n) (5.20)

Another oxidizing agent which can be used to prepare nitroxides is dimethyldioxirane ($Me_2C(O)_2$) in acetone, though its action is restricted to cyclic secondary amines. Yields are about 95 per cent,[25] for example, Scheme 5.22.

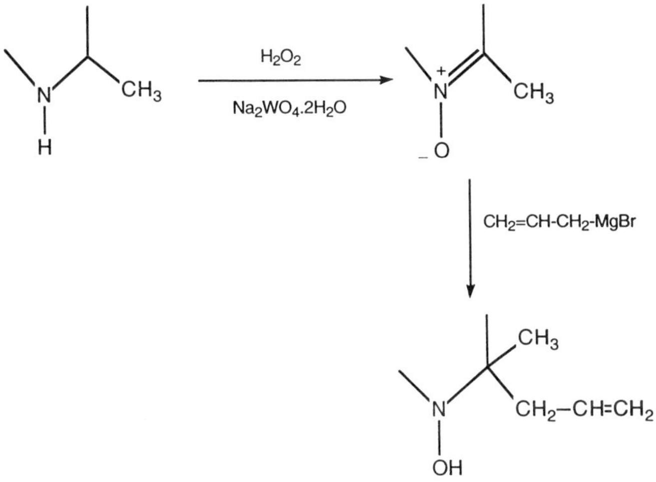

H_2O_2 / $Na_2WO_4.2H_2O$; $CH_2{=}CH{-}CH_2{-}MgBr$

(5.21)

$(CH_3)_2CO_2$ / $(CH_3)_2C{=}O$ / $0^{\circ}C$; 98%

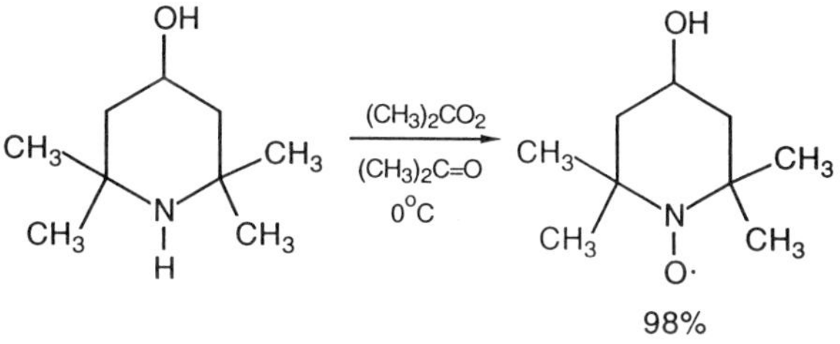

(5.22)

5.4 Reactivity

5.4.1 *As 1,3-dipoles*

The main use of nitrones is for [3 + 2] 1,3-dipolar additions with alkenes and other dipolarophiles yielding various heterocycles,[26] for example, Scheme 5.23. The reaction is regiospecific and leads to the production of 2-substituted adducts.[27]

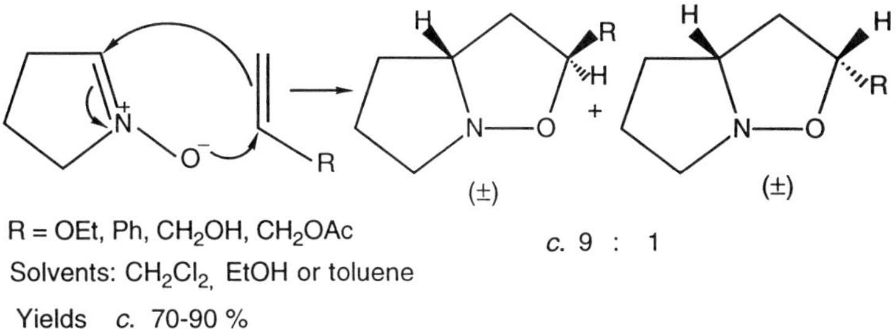

R = OEt, Ph, CH_2OH, CH_2OAc

Solvents: CH_2Cl_2, EtOH or toluene

Yields *c.* 70-90 %

(±) (±) *c.* 9 : 1

(5.23)

By generating nitrones *in situ* by one of the methods described above[23] the heterocycles shown in Scheme 5.24 have been synthesized by 1,3-dipolar addition.

CH_2-CHO / CH_2 / CH_2–CH=CH_2 —[$Si(CH_3)_3$–CH_3N-O-$Si(CH_3)_3$, 50° C]→ [CH_2-CH=$N^+(CH_3)$–O^- ... CH=CH_2] → bicyclic isoxazolidine (H, H, N–CH_3, O) 78 %

cyclohexanone —[$Si(CH_3)_3$–CH_3N-O-$Si(CH_3)_3$, 50° C]→ [cyclohexylidene=$N^+(CH_3)$–O^- + Ph-N=C=O] → spiro compound (CH_3–N–O, N–Ph, C=O) 76 % (5.24)

Nitrones can be utilized in a 1,3-dipolar addition, followed by acidic decomposition of the resulting isoxazolidine (**7**), to provide a synthesis of *E*-α,β-unsaturated aldehydes (**8**) (Scheme 5.25).[28]

N-Oxides of imines react with alkenes in a catalysed reaction to form heterocycles,[29] for example, Scheme 5.26.

R–CH=O → (MeNHOH, $-H_2O$) → R(H)C=N^+(Me)–O^- → (CH_2=CH–$SiMe_3$) → (7)

Me–N, O, R, $SiMe_3$ (7)

HF / CH_3CN

H(R)C=C(H)–C(H)=O (7) + CH_3NH_2 + Me_3SiF

R	Yield %
phenyl	81
2-furyl	43

(5.25)

Me(H)C=C(H)Me + MeCH(Cl)–CH=N^+(O^-)–cyclohexyl → (i, $AgBF_4$, SO_2; ii, KCN, H_2O) → Me, Me, Me, CN, O–N–cyclohexyl

88 %

(5.26)

5.4.2 *As a nucleophile*

With acyl chlorides

N-Tertiary-butyl hydroxylamine[30] can be caused to react with aldehydes to yield nitrones which, by reaction with an acyl chloride and a base, are converted, after hydrolysis, into α-acyloxyaldehydes (Scheme 5.27).[31]

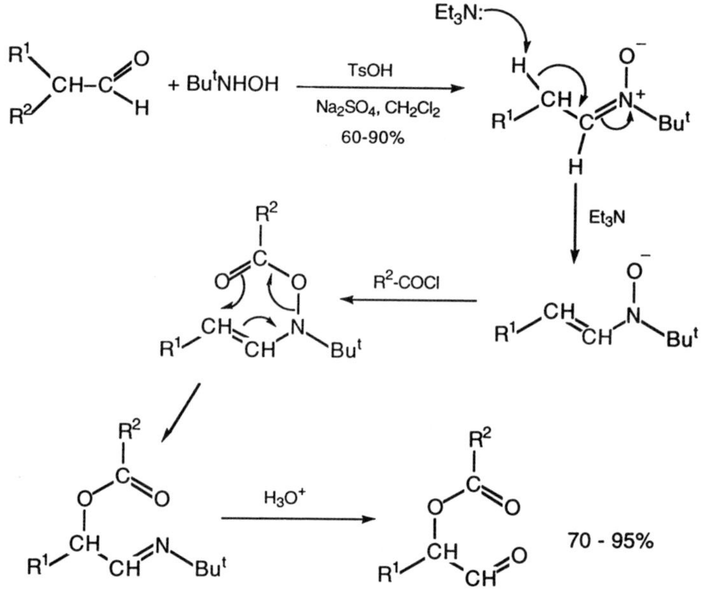

TsOH; Na_2SO_4, CH_2Cl_2; 60-90%; Et_3N; R^2-COCl; H_3O^+; 70 - 95%

(5.27)

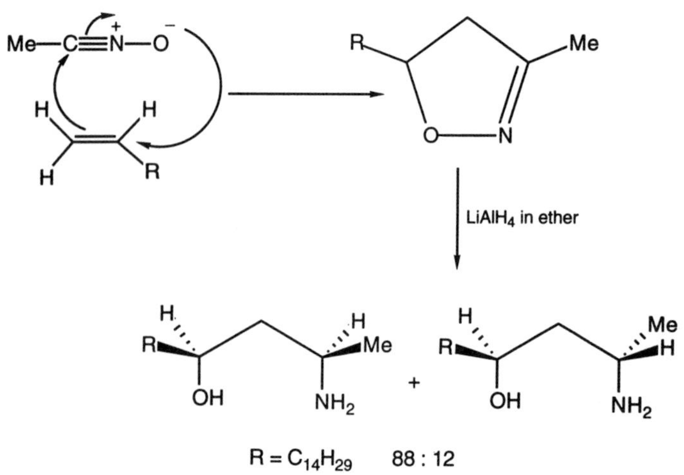

LiAlH$_4$ in ether

R = $C_{14}H_{29}$ 88 : 12

(5.28)

NITRILE *N*-OXIDES

5.5 Generation and reaction with alkenes

Nitrile *N*-oxides react with alkenes by 1,3-dipolar addition to yield isoxazolines which can be reduced to yield β-amino alcohols (Scheme 5.28).[32] If the reduction is done with Raney nickel and hydrogen instead of lithium aluminium hydride, the product is a β-hydroxy ketone,[33] for example, Scheme 5.29.

CH_3-C≡$\overset{+}{N}$–$\overset{-}{O}$ + (H)(CH$_3$)C=C(H)(CH$_3$) ⟶ isoxazoline (CH$_3$, CH$_3$, CH$_3$)

H$_2$ / Ni / aq. MeOH
Methyl borate buffer

⟶ CH_3-CO-CH(CH$_3$)-CH(OH)-CH$_3$ 88 % (±) (5.29)

$Ph{-}SO_2{-}CH_2{-}NO_2$ —CH_2N_2→ $Ph{-}SO_2{-}CH{=}\overset{+}{N}(O^-){-}OCH_3$ —Na_2SiO_3, CH_2Cl_2→ $Ph{-}SO_2{-}C{\equiv}\overset{+}{N}{-}O^-$

$Ph{-}SO_2{-}C(Br){=}N{-}OH$ —$AgNO_3$→ $Ph{-}SO_2{-}C{\equiv}\overset{+}{N}{-}O^-$

$Cl_2C{=}NOH$ —$AgNO_3$→ $Cl{-}C{\equiv}\overset{+}{N}{-}O^-$ (9) (5.30)

Wade and his co-workers have used the methods shown in Scheme 5.30 for the preparation of nitrile oxides *in situ* for reactions with alkenes.[34] The chloronitrile *N*-oxide (**9**) reacts with alkenes in a 1,3-dipolar addition to provide heterocyclic compounds (isoxazolines),[35] for example, Scheme 5.31.

A nitrile oxide derived from phenyl isocyanate has been trapped in a 1,3-dipolar addition by alkenes with the formation of isoxazolines (Scheme 5.32).[36]

Another way to generate the nitrile oxide is by the action of chloramine-T on an aldoxime. Again the oxide has been trapped in a 1,3-dipolar addition by an alkene (Scheme 5.33).[37]

40%

66 : 44

(5.31)

i, KOH; ii, NH_2OH

$Ph\text{-}S\text{-}CH_2\text{-}NO_2$ 92%

Et_3N / PhH

p-$ClC_6H_4CO_3H$

97 %

80 %

(5.32)

R-CH=NOH —(p-$CH_3C_6H_4SO_2NHCl$, (Chloramine-T))→ R-C≡$\overset{+}{N}$-O^- + >C=C< → isoxazoline (5.33)

Primary nitroalkanes can be dehydrated by p-toluene sulphonic acid to give nitrile oxides which as 1,3-dipoles have been trapped by multiple carbon–carbon bonds to yield isoxazolines,[38] for example, Scheme 5.34.

C_6H_5-CO-CH_2-NO_2 → [C_6H_5-CO-C≡$\overset{+}{N}$-O^-] —(Ph-C≡CH)→ 3-PhCO-5-Ph-isoxazole 46% (5.34)

References

1. M. N. Sheng and J. G. Zajacek, *J. Org. Chem.*, 1968, **33**, 588.
2. C. L. Anderson, *Tetrahedron Lett.*, 1986, **27**, 3961.
3. A. R. Katritzky and J. N. Lagowski, *Chemistry of heterocyclic* N*-oxides*, Academic Press, New York, 1971, pp. 279 and 362.
4. *F. &F.*, **13**, 1; 2; P. Strazzolini, A. G. Giumanini, and S. Cauci, *Tetrahedron*, 1990, **46**, 1097; N. Tokitch and R. Okazaki, *Chem. Lett.*, 1985, 1517.
5. *F. &F.*, **13**, 230; S. K. Boyer, J. Bach, J. McKenna, and E. Jagdmann, Jr, *J. Org. Chem.*, 1985, **50**, 3408.
6. *F. &F.*, **4**, 270; H. Alper and J. T. Edwards, *Can. J. Chem.*, 1970, **48**, 1543.
7. *F. &F.*, **15**, 284; Y. Handa, J. Inanaga, and M. Yamaguchi, *J. Chem. Soc., Chem. Commun.*, 1989, 298.
8. *F. &F.*, **12**, 86; R. O. Kazaki and N. Tokitoh, *J. Chem. Soc., Chem. Commun.*, 1984, 192.
9. R. O. Kazaki and N. Tokitoh, *Chem. Lett.*, 1984, 1937.
10. V. Franzen, *Org. Synth.*, 1967, **47**, 96; V. Franzen, *Org. Synth. Coll. Vol. V*, 1973, 872.
11. K. Sato, M. Kira, and H. Sakurai, *Tetrahedron Lett.*, 1989, 4375.
12. G. W. Kabalka and R. J. Newton, Jr, *J. Organomet. Chem.*, 1978, **156**, 65.

13. *F. &F.*, **7**, 256; V. van Rheenen, R. C. Kelly, and D. Y. Cha, *Tetrahedron Lett.*, 1976, 1973; V. van Rheenen, R. C. Kelly, and D. Y. Cha, *Org. Synth.*, 1977, 2042.
14. *F. &F.*, **13**, 326; R. Beugelmans, G. Negron, and G. Roussi, *J. Chem. Soc., Chem. Commun.*, 1983, 31; R. Beugelmans, L. Benaadjila-Iguertsira, J. Chastanet, G. Negron, and R. Roussi, *Can. J. Chem.*, 1985, **63**, 725.
15. J. Chastanet and G. Roussi, *J. Org. Chem.*, 1985, **50**, 2910.
16. B. De, J. F. DeBernadis, and R. Prasad, *Synth. Commun.*, 1988, **18**, 481; *F. &F.*, **14**, 329.
17. J. Hamer and A. Macaluso, *Chem. Rev.*, 1964, **64**, 473; W. Rundel, in Houben-Weyl, *Methoden der organischen Chemie*, 4th edn, Georg Thieme Verlag, Stuttgart, 1968, Vol. X/4, p. 309.
18. Preparation: E. Schmitz, R. Ohme, and D. Murawski, *Ber.*, 1965, **98**, 2516.
19. *F. &F.*, **8**, 241; M. Abou-Gharbia and M. M. Joullié, *Synthesis*, 1977, 318.
20. *F. &F.*, **15**, 273; D. Christensen and K. A. Jargensen, *J. Org. Chem.*, 1989, **54**, 126.
21. *F. &F.*, **13**, 327; N. A. Le Bel and N. Balasubramanian, *Tetrahedron Lett.*, 1985, **26**, 4331.
22. *F. &F.*, **14**, 177; S. I. Murahashi and T. Shiota, *Tetrahedron Lett.*, 1987, **28**, 2383.
23. *F. &F.*, **13**, 187; J. A. Robl and J. R. Hwu, *J. Org. Chem.*, 1985, **50**, 5913.
24. *F. &F.*, **15**, 295; H. Mitsui, S. Zenki, T. Shiota, and S.-I. Murahashi, *J. Chem. Soc., Chem. Commun.*, 1984, 874.
25. *F. &F.*, **15**, 144; R. W. Murray and M. Singh, *Tetrahedron Lett.*, 1988, **29**, 4677.
26. J. J. Tufariello, in *1,3-Dipolar cycloaddition chemistry* (ed. A. Padwa), Wiley-Interscience, New York, 1984, Vol. 2, Chapter 9; S. A. Ali, J. H. Kahn, and M. I. M. Wazeer, *Tetrahedron*, 1988, **44**, 5911.
27. R. H. Huisgen, *J. Org. Chem.*, 1976, **41**, 403.
28. *F. &F.*, **12**, 566; P. DeShong and J. M. Leginus, *J. Org. Chem.*, 1984, **49**, 3421.
29. M. Petrzilka, D. Felix, and A. Eschenmoser, *Helv. Chim. Acta*, 1973, **56**, 2950; *F. & F.*, **5**, 110.
30. Preparation: A. Calder, A. R. Forrester, and S. P. Hepburn, *Org. Synth.*, 1972, **52**, 77.
31. *F. &F.*, **12**, 95; C. H. Cummins and R. M. Coates, *J. Org. Chem.*, 1983, **48**, 2070.
32. *F. &F.*, **11**, 289; V. Jäger, W. Schwab, and V. Buss, *Angew. Chem., Int. Ed. Engl.*, 1981, **20**, 601; W. Schwab and V. Jäger, *Angew. Chem., Int. Ed. Engl.*, 1981, **20**, 603.
33. *F. &F.*, **11**, 457; D. P. Curran, *J. Am. Chem. Soc.*, 1982, **104**, 4024; see also A. P. Kazikowski and M. Adamczyk, *Tetrahedron Lett.*, 1982, **23**, 3123; R. Wollenberg and J. E. Goldstein, *Synthesis*, 1980, 757.
34. *F. &F.*, **11**, 40; P. A. Wade and H. R. Hinney, *J. Am. Chem. Soc.*, 1979, **101**, 1319; P. A. Wade and M. K. Pillay, *J. Org. Chem.*, 1981, **46**, 5425.
35. *F. &F.*, **11**, 146; P. A. Wade, M. K. Pillay, and S. M. Singh, *Tetrahedron Lett.*, 1982, **23**, 4563.
36. F. G. Bordwell and J. E. Bartmess, *J. Org. Chem.*, 1978, **43**, 3101; *F. &F.*, **15**, 264; D. P. Curran and J. F. C. Chao, *J. Org. Chem.*, 1988, **53**, 5369.
37. *F. &F.*, **15**, 78; A. Hassner and K. M. L. Rai, *Synthesis*, 1989, 57.
38. *F. &F.*, **12**, 507; T. Shimizu, Y. Hayashi, and K. Teramura, *Bull. Chem. Soc. Japan*, 1984, **57**, 2531.

6

ALKYL ISONITRILES (ISOCYANIDES)[1]

X-ray measurements show that the isonitrile nitrogen–carbon bond has a length of 1.15 Å and that the atoms C—N—C are linear which indicates that the structure can be represented as a hybrid of (**1**) and (**2**) (Scheme 6.1) with (**1**) being the much more important contributor (dipole moment *c.* 3.6 D).

$$-\overset{|}{\underset{|}{C}}\overset{1.43\,Å}{—}N\overset{1.15\,Å}{\equiv}C \quad \text{Linear}$$

$$\underset{(1)}{R—\overset{+}{N}\equiv\overset{-}{C}} \longleftrightarrow \underset{(2)}{R—\ddot{N}=C:} \qquad (6.1)$$

The infra-red band of the nitrogen–carbon multiple bond (N═C) is at *c.* 2150 cm^{-1} in isonitriles.

6.1 Preparation

6.1.1 *From alkyl halides*

Silver and cuprous cyanides, AgCN and CuCN, react with alkyl iodides (**3**) to give isonitriles (compare KCN, Chapter 4, p. 175), after decomposition of the silver iodide–nitrile complex (**4**) with aqueous potassium cyanide[2] (Scheme 6.2). The yields are never greater than 50 per cent.

$$\underset{(3)}{R\text{-}I + AgCN} \xrightarrow{\text{DMSO}} \underset{(4)}{R\text{-}NC.AgI} \xrightarrow{\text{aqueous KCN}} R—\overset{+}{N}\equiv\overset{-}{C} \qquad (6.2)$$

An improvement can be made by the use of tetramethylammonium dicyanoargentate (**5**), $Me_4N^+Ag(CN)_2^-$ in acetonitrile.[3] Good yields can be obtained from alkyl halides (**6**) (Scheme 6.3), whose reactivities in this context increase as follows: primary < secondary < tertiary; chloride < bromide < iodide.

$$\underset{(6)}{\text{R-Hal}} + \underset{(5)}{\text{Ag(CN)}_2^-} \xrightarrow{\text{CH}_3\text{CN}} \text{R}-\overset{+}{\text{N}}\equiv\overset{-}{\text{C}} + \text{AgHal} + \text{CN}^-$$

R-Hal	Temp. °C	Yield (%)
CH_3I	80	80
Ph_2CHBr	20	92
Ph_3CCl	20	97

(6.3)

Also noteworthy is the fact that good yields of acyl isonitriles (**7**) can be obtained by the reaction of silver cyanide[4] with an acyl halide (**8**) (Scheme 6.4).

$$\underset{(8)}{\text{Bn}-\overset{\overset{\text{O}}{\|}}{\text{C}}-\text{I}} + \text{AgCN} \xrightarrow{\text{CH}_2\text{Cl}_2,\ 20^\circ\text{C}} \underset{(7)\quad 73\,\%}{\text{Bn}-\overset{\overset{\text{O}}{\|}}{\text{C}}-\text{NC}} + \text{AgI} \tag{6.4}$$

6.1.2 *From amines*

The original method of A. W. Hofmann,[5] the carbylamine reaction, which involves synthesis from a primary amine (**9**), chloroform (**10**), and potassium hydroxide involves dichlorocarbene (**11**) as an intermediate (Scheme 6.5).[6]

$$\underset{(9)}{\text{R-NH}_2} + \underset{(10)}{\text{CHCl}_3} \xrightarrow[\text{Bu}^t\text{O}^-\ \text{in pentane}]{\text{EtO}^-/\text{EtOH or}} \text{R}-\overset{+}{\text{N}}\equiv\text{C}^-$$

$$\underset{(9)}{\text{R-}\ddot{\text{N}}\text{H}_2} + \underset{(11)}{:\text{CCl}_2} \longrightarrow \text{R-}\overset{\overset{\text{H}}{|}}{\underset{\underset{\text{H}}{|}}{\overset{+}{\text{N}}}}\text{-}\underset{\underset{\text{Cl}}{|}}{\overset{-}{\text{C}}}\text{-Cl} \xrightarrow{-2\text{HCl}} \text{R}-\overset{+}{\text{N}}\equiv\text{C}^- \tag{6.5}$$

The yields of isonitriles are low (20–40 per cent) but can be improved by the use of a phase transfer method[7,8] in which the method of Makoska and Wawrzyniensicz[9] is used to generate dichlorocarbene, $:CCl_2$, from chloroform and 50 per cent aqueous sodium hydroxide plus a catalytic amount of benzyltriethylammonium chloride ($Et_3NCH_2Ph^+Cl^-$). Yields of 40–60 per cent (R = Bu^n, Bn, cyclohexyl, *n*-dodecyl) were achieved and tertiary butyl isonitrile can be prepared in 50 per cent yield in this way.[10]

6.1.3 *Dehydration of formamides*

The dehydration of formamides (**12**) is the most important method for preparation of isonitriles and a variety of reagents have been used usually in conjunction with a base. The mechanism of these dehydrations often occurs by the reagent esterifying the enolic form of the amide (**13**) and thus making the hydroxyl group into a better leaving group as in Scheme 6.6.

R-NH-CHO (12) ⇌ R-N=CHOH (13) —$POCl_3$→ R—N̈=C(H)—O—P(Cl)(Cl)=O ↓ $R—\overset{+}{N}≡C—H$ (Pr^i_2NH) → $R—\overset{+}{N}≡C^-$

Reagent	Base and Conditions	R	Yield (%)	Ref
SO_2Cl in DMF	Na_2CO_3,−50°C	n-C_6H_{13},c-C_6H_{11}	55–96	11
($Me_2N{=}CHCl^+.Cl^-$)*		Bu^t,Bn,$CH_2{=}CH$,+		12,13
		$Me_2CHCH_2CMe_2CH_2$		
Ph_3PBr_2	El_3N, 80 °C	Me,Et,c-C_8H_{11}	56–73	14
$COCl_2$	Et_3N in CH_2Cl_2	c-C_6H_{11}	70	12
$POCl_3$	Pr^i_2NH	CH_2CO_2Et, o-Me-C_6H_4	60–84	13
	(2,4,6^{i}Pr)-C_6H_2			
PPh_3,CCl_4,	Et_3N in $CHCl_3$	Ph, Bn, Bu^n	89–91	16
	or CH_2Cl_2			
PPh_3	Non-basic‡	Ph, Bn, (2,6-Me)-C_6H_3	25–60	15
EtOOC-N=N-COOOEt,		Cyclohexyl, (2,4,6-Mc)-C_6H_2		
2-chloro-3-ethylbenzoxazolium BF_4^- (structure: Et on N, Cl at C-2)	Et_3N in CH_2Cl_2, 20°C	Bn, Ph, m-NO_2-C_6H_4	70–85	17

* See Chapter 8, Amides, p. 377

+ Some optically active formamides were used

‡ Useful for the preparation of base-sensitive isonitriles

** See p. 280

(6.6)

** Preparation[17]

NH$_2$ / OH (o-aminophenol) —[i, EtO—C(S$^-$)=S K$^+$; ii, HCl]→ 2-chlorobenzoxazole —[$EtO_3^+BF_4^-$]→ N-ethyl-2-chlorobenzoxazolium BF_4^-

(6.6 cont.)

Formamides can be converted into isonitriles by dehydration in yields of 75 to 90 per cent by the use of the compound (**14**), (prepared as shown in Scheme 6.7) without racemization of chiral substrates.[18]

O=C(imidazolyl)$_2$ + 2 CH_3SO_3H ⟶ O=C(imidazolium-H$^+$)$_2$ (14) + 2 $CH_3SO_3^-$

(6.7)

6.1.4 *From tertiary alkenes*

Tertiary alkyl isonitriles (**15**) can be prepared by the reaction of tertiary alkenes (**16**) with hydrogen cyanide in the presence of a cuprous halide.[19] A variety of tertiary alkenes have been used to prepare tertiary alkyl isonitriles in yields which are virtually quantitative based upon the amount of cuprous ion (Cu^+) used, for example, Scheme 6.8.

$$(CH_3)_2C{=}CH_2 \ (16) + HCN \xrightarrow[\text{15 hours}]{\text{CuBr, 70°C}} \left[\begin{array}{c}(CH_3)_3C\text{-}NC.CuBr \\ + \\ (CH_3)_3C\text{-}NC.CuCN\end{array}\right] \xrightarrow{\text{aqueous KCN, 20°C}} (CH_3)_3C{-}\overset{+}{N}{\equiv}\overset{-}{C} \ (15) \quad (6.8)$$

6.1.5 *Reduction of isocyanates or urethanes*

The reduction of isocyanates (**17**) or urethanes (**18**) with trichlorosilane and triethylamine in dichloromethane gives isonitriles in yields of 40 to 70 per cent (Scheme 6.9).[20]

$$R^1\text{-}N{=}C{=}O \ (17) \text{ or } R^1\text{-}NH\text{-}CO_2R^2 \ (18) \xrightarrow[CH_2Cl_2]{Cl_3SiH,\ Et_3N} R^1{-}\overset{+}{N}{\equiv}\overset{-}{C} \quad (6.9)$$

Isocyanates (**19**) are also reduced to isonitriles by 2-phenyl-3-methyl-1,3,2-oxazaphospholine (**20**), the preparation of which is shown in Scheme 6.10.[21,22]

$$Ph\text{-}PCl_2 + CH_3\text{-}NH\text{-}CH_2\text{-}CH_2\text{-}OH \xrightarrow[\text{heat}]{2\ Et_3N} \text{Ph-P(N(CH}_3\text{)-CH}_2\text{-CH}_2\text{-O) } (20) \ 41\%$$

$$R\text{-}N{=}C{=}O \ (19) \xrightarrow[20\text{-}30°C]{(20)} R{-}\overset{+}{N}{\equiv}C^{-} \quad 50\text{-}90\% \quad (6.10)$$

6.2 Reactivity

Because the most important resonance hybrid in the sturcture of isonitriles is zwitterionic, they are very reactive compounds towards electrophiles and nucleophiles. This reactivity has been widely applied in recent years for organic synthesis. Since the terminal carbon atom in the zwitterionic form (**1**) carries a negative charge, it is expected to react readily with electrophiles. Conversely, isonitriles are reluctant to react with nucleophiles and hence they are usually stable under alkaline conditions. However, powerful nucleophiles can attack the carbon atom, presumably by virtue of the contribution of (**2**) to the structure.

6.2.1 *Reaction on carbon with electrophiles*

Hydrolysis

Protonation of the carbon occurs readily and opens the way for weak nucleophiles to attack the same carbon atom, for example, in the hydrolysis of isonitriles with aqueous acid. The reaction proceeds via a formamide (**21**), which is itself hydrolysed by aqueous acid (Scheme 6.11).

$$\text{R-}\overset{+}{\text{N}}\equiv\overset{-}{\text{C}} + H_3O^+ \rightleftharpoons \text{R-}\overset{+}{\text{N}}\equiv\text{CH} + H_2\ddot{O} \longrightarrow \text{R-}\overset{+}{\text{N}}\text{H=CH-O-H}$$

$$\text{R-NH}_2 + \text{H.CO}_2\text{H} + \text{H}^+ \longleftarrow \text{R-}\overset{+}{\text{N}}\text{H}_2\text{-C(=O)-H} \; (+\, H_2\ddot{O}) \xleftarrow{H_3O^+} \underset{(21)}{\text{R-NH-CHO}} \qquad (6.11)$$

Reaction with carboxylic acids and amines

A carboxylic acid (**22**) can provide a proton and attack of the resulting weakly nucleophilic carboxylate anion (**23**) on the protonated isonitrile (**24**) enables the formamide (**25**) to be isolated[23] (Scheme 6.12).

$$\text{R}^1\text{-}\overset{+}{\text{N}}\equiv\overset{-}{\text{C}} + \underset{(22)}{\text{R}^2\text{-CO}_2\text{H}} \longrightarrow \underset{(24)}{\text{R}^1\text{-}\overset{+}{\text{N}}\equiv\text{CH}} + \underset{(23)}{{}^-\text{O}_2\text{C-R}^2}$$

$$\left[\text{R}^1\text{-N=CH-O-C(=O)-R}^2\right] \xrightarrow{\text{R}^2\text{-CO}_2\text{H}} \left[\text{R}^1\text{-}\overset{+}{\text{N}}\text{H=CH-O-C(=O)-R}^2 \;\; \text{R}^2\text{-C(=O)-O}^-\right]$$

$$\longrightarrow \underset{(25)}{\text{R}^1\text{-NH-CHO}} + (\text{R}^2\text{-CO})_2\text{O} \qquad (6.12)$$

Isonitriles have been used to facilitate the reaction between a carboxylic acid (**26**) and an amine (**27**) to yield an amide (Scheme 6.13).[24] The mechanism is the same as in Scheme 6.12.

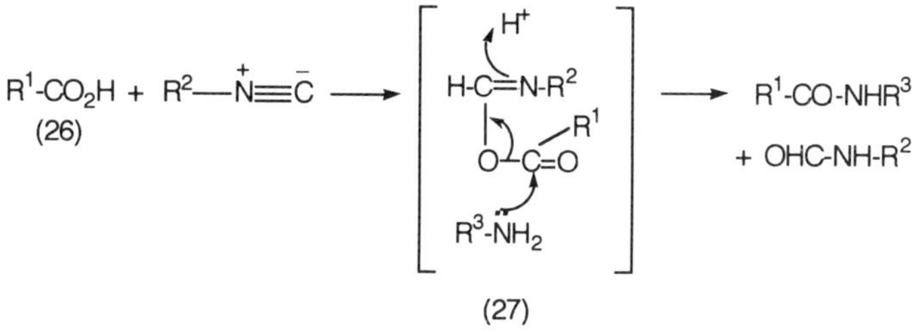

(6.13)

Tertiary butyl isocyanide (**28**)[27] is used to facilitate the reaction of a carboxylic acid (**29**) and an amine (**30**) to form an amide and the method has been applied to the synthesis of peptides[24] (Scheme 6.14).

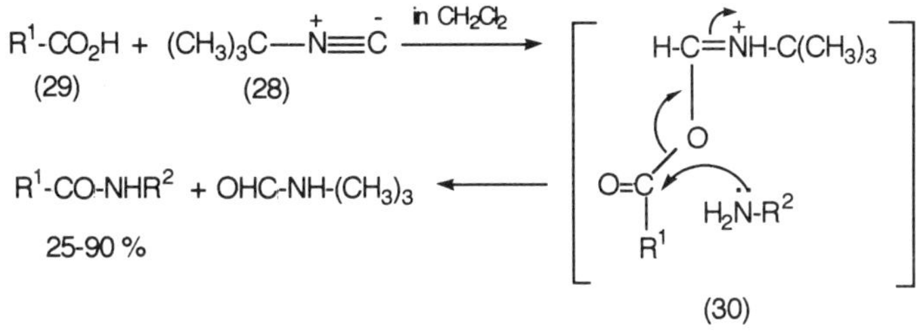

(6.14)

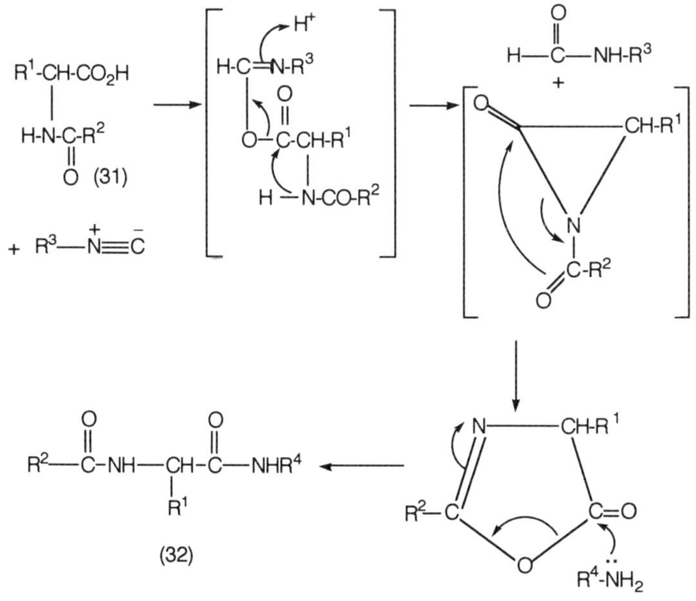

(6.15)

Synthesis of peptides

N-Acylamino acids (**31**) can be used in this way for the synthesis of peptides (**32**) and a number of small peptides were obtained in yields of 18–90 per cent (Scheme 6.15).[28]

Tertiary butyl isonitrile (**28**) can be used for various synthetic purposes. It undergoes cycloaddition to α,β-enones (**33**) to give iminolactones (**34**) which can be converted into γ-butyrolactones (**35**)[25] (Scheme 6.16).

$$R^1R^2C{=}C(R^3)COR^4 \; (33) + (CH_3)_3C{-}\overset{+}{N}{\equiv}\overset{-}{C} \; (28) \xrightarrow[65\%]{(C_2H_5)_2AlCl} (34) \xrightarrow{1)\,H_2O,\,Pd/C;\; 2)\,H_3O^+} (35) \qquad (6.16)$$

In the presence of $TiCl_4$, the isocyanide (**28**) causes conjugate hydrocyanation to α,β-enones with elimination of isobutene (**36**) and functions as a masked hydrogen cyanide[26] (Scheme 6.17).

$$R^1R^2C{=}C(R^3)COR^4 + (CH_3)_3C{-}\overset{+}{N}{\equiv}\overset{-}{C} \; (28) \xrightarrow[CH_2Cl_2,\, 0^\circ C]{TiCl_4} R^2{-}C(R^1)(CN){-}CH(R^3)COR^4 + H_2C{=}C(CH_3)_2 \; (36) \qquad (6.17)$$

Reaction with an imine

The reaction between an isonitrile (**37**), an amine (**38**), an aldehyde (**39**) or ketone, and a carboxylic acid (**40**) has been applied to the synthesis of derivatives of α-amino acids (Scheme 6.18).[29] This method has been applied to the synthesis of an antibiotic (+)-furanomycin from glucose, tertiary butyl isocyanide being the compound used.[30]

R^1-CHO + R^2-NH_2 → R^1-CH=$\overset{+}{N}$-R^2 → $R^3\overset{+}{N}$≡C—C(H)(R^1)—NR^2

(39) (38)

${}^{-}C≡\overset{+}{N}$-R^3

(37)

O=C(—O—H)—R^4

(40)

R^3-N=C—C(H)(-NHR^2)(R^1), O, R^4—C=O

Intramolecular acylation

R^4-C(=O)—N(R^3)-C(=O)-C(H)(R^1)-NHR^2

(6.18)

With acetals and enols of aldehydes

Trimethoxysilanes of enols of aldehydes (**41**) react with acetals (**42**) and tertiary butyl isonitrile (**43**) in the presence of titanium tetrachloride to yield (20–80 per cent) γ-alkoxy-α-hydroxyamides (**44**)[31] (Scheme 6.19).

R^1—C(H)(OR^2)(OR^2)

(42)

(R^3)(R^3)C=C(R^4)—O—$SiMe_3$

(41)

→ R^1—C(H)(OR^2)—C(R^3)(R^3)—C(=O)—R^4 (Cl—$TiCl_3$)

${}^{-}C≡\overset{+}{N}$—Bu^t

(43)

↓

R^1—C(H)(OR^2)—C(R^3)(R^3)—C(O—$TiCl_3$)(R^4)—$\overset{+}{C}$=N—Bu^t

$\xleftarrow{H_2O}$

R^1—C(H)(R^2O)—C(R^3)(R^3)—C(OH)(R^4)—C(=O)—$NHBu^t$

(44)

R^1	R^2	R^3	R^4	Yield (%)
H	Me	Me	H	54
Me	H	Me	H	61

(6.19)

With thallium(III) nitrate

Thallium(III) nitrate, $Tl(NO_3)_3$ in methanol, acting as an electrophile, converts isonitriles (**45**) into urethanes (**46**)[32] in a rapid reaction, liberating thallous nitrate (**47**) and nitric acid (**48**) (Scheme 6.20).

$$\underset{(45)}{R-\overset{+}{N}\equiv\overset{-}{C}} \xrightarrow[\text{ii) } H_2O]{\text{i) } Tl(NO_3)_3/MeOH} \underset{(46)}{R\text{-}NH\text{-}CO_2CH_3} + \underset{(47)}{TlNO_3} + \underset{(48)}{2\,HNO_3}$$

R	Yield (%)
$EtCOCH_2$	84
cyclohexyl	90
Bu^t	35

Mechanism:

$$\underset{(45)}{R\text{-}\overset{+}{N}\equiv\overset{-}{C}} \quad Tl(NO_3)_2\text{-}NO_3 \longrightarrow R\text{-}\overset{+}{N}\equiv C\text{-}Tl(NO_3)_2\ NO_3^- \ (+\ MeOH)$$

$$\longrightarrow R\text{-}N{=}C(OMe)\text{-}Tl(NO_3)\text{-}NO_3 \ + HNO_3\ (48)$$

$$\longrightarrow R\text{-}\overset{+}{N}\equiv C\text{-}OMe\ NO_3^- \ (+\ H_2O) \ + TlNO_3\ (47)$$

$$\longrightarrow R\text{-}N{=}C(OMe)\text{-}O\text{-}H \ + \underset{(48)}{HNO_3} \xrightarrow{\text{tautomerism}} \underset{(46)}{RNH\text{-}C(=O)\text{-}O\text{-}Me} \qquad (6.20)$$

6.2.2 *Cyclic addition reactions*

With ketones

Isonitriles undergo a large number of addition reactions, for example, their reaction with ketones (**49**) in the presence of boron trifluoride gives good (40–90 per cent) yields of four-membered cyclic compounds (**50**), the initial attack being that of the electrophilic carbon atom of the carbonyl compound (enhanced by boron trifluoride) on the electron-rich carbon atom of the isonitrile, for example, with acetone[33] (Scheme 6.21).

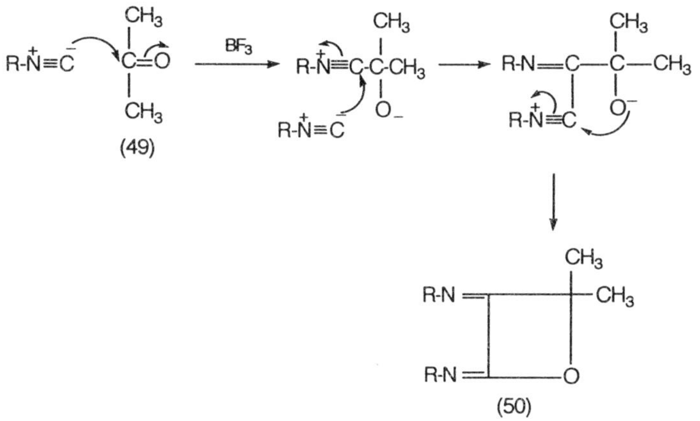

(49)

(50)

(6.21)

With α,β-unsaturated ketones

Isonitriles react with certain α,β-unsaturated ketones (**51**), in the presence of diethylaluminium chloride or ethylaluminium dichloride in a 1,4-cycloaddition, the products of which can be reduced to give cyclic lactones (**52**)[34] for example, Scheme 6.22.

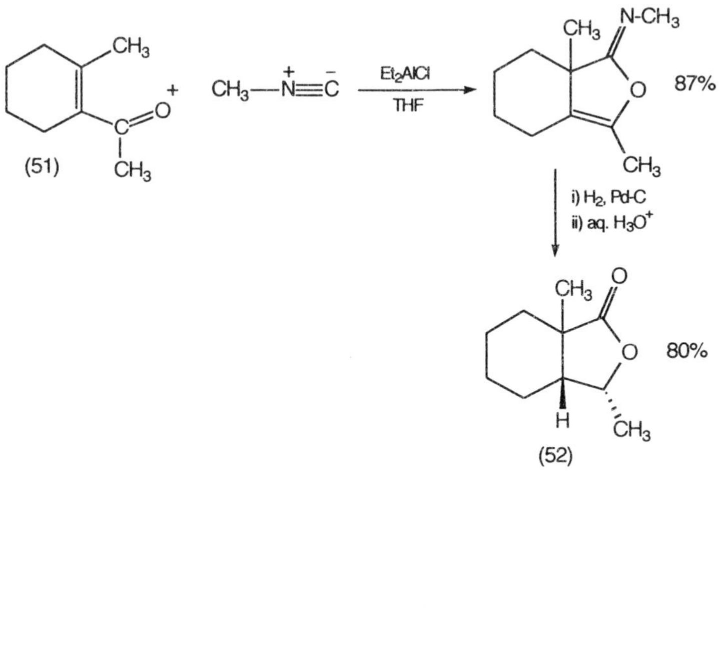

(51)

87%

i) H_2, Pd-C
ii) aq. H_3O^+

80%

(52)

(6.22)

6.2.3 *Reaction on carbon with powerful nucleophiles*

With Grignard reagents and lithium alkyls

Grignard reagents and lithium alkyls react with isonitriles and the intermediate organometallic compounds react with a variety of reagents to furnish valuable syntheses.[35] However, the isonitriles used for this purpose must not contain an α-hydrogen atom since, as will be seen later, organometallic compounds such as lithium alkyls can remove an α-proton and this would lead to complicating side reactions in this context.

Preparation of aldehydes and α-ketoacids 1,1,3,3-Tetramethylbutyl isonitrile (**53**) has been used[36,37] to prepare aldehydes (**54**) from Grignard reagents (**55**). The intermediate imino-Grignard reagent (**56**) may alternatively react with carbon dioxide to yield, after hydrolysis, an α-keto acid (**57**) (Scheme 6.23).

$$\underset{(55)}{R{-}MgBr} + \underset{(53)}{\bar{C}{\equiv}\overset{+}{N}{-}CMe_2{-}CH_2{-}CMe_3} \xrightarrow{\text{in THF}} \underset{(56)}{R{-}\underset{\underset{MgBr}{|}}{C}{=}N{-}CMe_2{-}CH_2{-}CMe_3}$$

$$(56) \xrightarrow[100^\circ C]{\text{aqueous oxalic acid}} \underset{(54)}{R{-}CHO}$$

$$(56) \xrightarrow{CO_2} R{-}\underset{\underset{CO_2^-}{|}}{C}{=}N{-}CMe_2{-}CH_2{-}CMe_3 \xrightarrow{\text{aqueous } H_3O^+} \underset{(57)}{R{-}\underset{\underset{O}{||}}{C}{-}CO_2H}$$

R	Yield α-keto acid (%)	Yield aldehyde (%)
Bu^n	34	—
Bu^s	47	67
n-hexyl	26	62

(6.23)

Synthetic uses of lithium aldimines Lithium aldimines (**58**) (prepared by reaction with lithium alkyls, as above) are extremely useful synthetically, since they react with a large number of compounds to produce a variety of useful compounds[38] (Scheme 6.24). The yields of the carbonyl containing compounds were 40–92 per cent for a variety of alkyl groups (R^1Li, R^1 = $EtBu^n, Bu^s$) and isonitriles (R = 1,1,3,3-tetramethylbutyl and tertiary butyl).[38] An example[39] of one of the above reactions in which an alkyllithium is used is shown in Scheme 6.25.

(6.24)

(6.25)

The intermediate lithium aldimines (**58**) have been used in two ways[40] for the synthesis of chiral α-amino acids (in maximum optical yields of 63 per cent); for example, the lithium aldimine (**59**) obtained from the racemic isonitrile (**60**) is carbonated and then reduced asymmetrically with (+)-di-isopinocamphyl borane (**61**) prepared from (−)-α-pinene (**62**) (Scheme 6.26). The use of (+)-α-pinene gave the enantiomer of valine (**63**) with 35 per cent optical purity.

Alternatively the use of the optically active *R*-(+)-isonitrile (**64**) in this procedure with reduction by non-chiral complex metal hydrides (e.g. B_2H_6 or $LiBH_4$) or by hydrogenation ($H_2/Pd(OH)_2$) gave 20–80 per cent yields of α-amino acids with a maximum optical purity of 63 per cent.

(64)

(62) + B_2H_6 ⟶ (61) [BH_2, H]

(±)-Ph—C(CH_3)(CH_2CH_3)—$\overset{+}{N}\equiv\overset{-}{C}$ (60) $\xrightarrow[\text{ether}]{Pr^iLi,\ 0^\circ C}$ [Ph-C(CH_3)(CH_2CH_3)-N=C(Li)-Pr^i] (59) $\xrightarrow{CO_2}$ Ph-C(CH_3)(CH_2CH_3)-N=C(CO_2^-)-Pr^i

i, (61) in THF
ii, aqueous HCl

Ph-C(CH_3)(CH_2CH_3)-$\overset{+}{N}H_2$-CH(CO_2H)-Pr^i $\xrightarrow{Pd/H_2}$ Ph-CH(CH_3)-CH_2CH_3 + $(CH_3)_2$CH-CH(NH_2)-CO_2H (63)

42 % yield; optical purity 40 % (6.26)

Synthesis of hindered nitriles and ketones Triphenylmethyl isonitrile (**65**) is the isonitrile of choice for the synthesis of hindered nitriles (**66**) and ketones (**67**)[41] (Scheme 6.27). For the electronic and steric mechanism see Ref. 41.

Ph_3C—$\overset{+}{N}\equiv\overset{-}{C}$ (65) + R^1-Li ⟶ [Ph_3C-N=C(Li)-R^1] $\xrightarrow{\text{spontaneous}}$ Ph_3C-Li + R^1-C≡N (66) $\xrightarrow{R^2\text{-Li}}$ R^1-C(=O)-R^2 (67)

R^1	Yield (nitrile) (%)
Bu^t	88
MeCH(Et)	24

R^1	R^2	Yield (ketone) (%)
Bu^t	Bu^t	75
MeCH(Et)	MeCH(Et)	63

(6.27)

Conversion of isonitriles to nitriles Grignard reagents (**68**), instead of lithium alkyls, can also be used to convert isonitriles to nitriles (**69**)[41] (Scheme 6.28).

$$R^1R^2HC\text{-}MgBr \; (68) + Ph_3C\text{—}\overset{+}{N}{\equiv}\overset{-}{C} \rightleftharpoons R^1R^2HC\text{—}C(MgBr){=}N\text{-}CPh_3$$

$$\downarrow$$

$$Ph_3C\text{-}MgBr + R^1R^2HC\text{—}C{\equiv}N \; (69)$$

R^1	R^2	Yield (%)
Ph	H	78
Cyclohexyl		100
Me	Et	79

(6.28)

With oxidizing agents

Oxidizing agents, usually being nucleophilic, react on the carbon atom of isonitriles to yield isocyanates (see p. 302) or urethanes (carbamates) (see p. 395). See the paper by Kienzle[32] for references to ozone, tertiary butyl hypochlorite, dimethylsulphoxide, and 'nitrile oxide', none of which are generally applicable.

6.2.4 *Reduction*

With tributyl tin hydride

The reaction of isonitriles with tributyl tin hydride has been used for the conversion of esters of α-amino acids (**70**) into the corresponding carboxylic ester (**71**) with loss of the amino group[42] (Scheme 6.29). The method has been shown to be general and a variety of aliphatic isonitriles (steroidal, long chain, sugars, and peptides) have been reduced with tri-*n*-butyl tin hydride (**72**) in refluxing benzene. The mechanism is that proposed in Scheme 6.30. The reaction requires the presence of AIBN as a radical initiator and procedes with yields of 55–90 per cent.

$$R^1\text{—}CH(CO_2R^2)\text{—}NH_2 \ (70) \xrightarrow[\text{ether or THF}]{CH_3CO_2CHO} R^1\text{-}CH(CO_2R^2)\text{-}NH\text{-}CHO \ (76\text{-}86\ \%)$$

$$\xrightarrow[\text{in } CH_2Cl_2,\ 20^\circ C]{POCl_3\,/\,Et_3N} \left[R^1\text{—}CH(CO_2R^2)\text{—}\overset{+}{N}\equiv\overset{-}{C}\right] \ (79\text{-}85\ \%) \xrightarrow[80^\circ C]{Bu^n_3SnH,\ AIBN/PhH} R^1\text{-}CH_2\text{-}CO_2R^2 \ (70\text{-}81\ \%) \quad (6.29)$$

$$R_3SnH \ (72) \xrightarrow{AIBN} R_3Sn\cdot$$

$$R^1\text{—}\overset{+}{N}\equiv\overset{-}{C} + R_3Sn\cdot \longrightarrow R^1N\!=\!\dot{C}\text{—}SnR_3 \longrightarrow R^1\cdot + R_3SnCN$$

$$R^1\cdot \xrightarrow{R_3SnH} R_3Sn\cdot + R^1H \qquad R = Bu^n \quad (6.30)$$

VINYL ISOCYANIDES

6.3 Preparation of vinyl isocyanides

6.3.1 *Via isocyanomethane-phosphoric acid diethyl ester*

Vinyl isocyanides (**73**)[43] are produced via the isocyanomethane-phosphoric acid diethyl ester (**74**). This reagent is prepared as shown in Scheme 6.31. Reaction of the lithium salt of (**74**) with carbonyl compounds (**75**) gives vinyl isocyanides (**73**) in 70–85 per cent yield (Scheme 6.32).

$$P(OEt)_3 + [(Me)_3\overset{+}{N}CH_2NHCHO]Br^- \longrightarrow (EtO)_2P(=O)\text{—}CH_2NHCHO \xrightarrow{POCl_3,\ NEt_3} (EtO)_2P(=O)\text{—}CH_2\text{—}\overset{+}{N}\equiv\overset{-}{C} \ (74) \quad (6.31)$$

$$(74) \xrightarrow[\text{THF, -70°C}]{\text{Bu}^n\text{Li}} (EtO)_2P(=O)\text{—}\overset{-}{C}H\text{-}\overset{+}{N}{\equiv}\overset{-}{C} \;\; R^1R^2C{=}O \;(75) \longrightarrow \left[(EtO)_2P(=O)\text{—}CH(\text{-}\overset{+}{N}{\equiv}\overset{-}{C})\text{—}C(R^1)(R^2)\text{—}\overset{-}{O}\right] \longrightarrow R^1R^2C{=}CH\text{—}\overset{+}{N}{\equiv}\overset{-}{C}\;(73) \quad (6.32)$$

6.4 Reactivity of α-hydrogen atoms of vinyl isocyanides

The isonitrile group activates an α-hydrogen atom and enables it to be removed as a proton thus leaving a nucleophilic carbon atom at the α-position. In this way derivatives of methyl isonitrile (**76**) have been applied to the synthesis of a variety of organic compounds. By the action of *n*-butyl lithium on methyl isonitrile a proton can be removed to yield isocyanomethyl lithium (**77**) (Scheme 6.33).[44] This lithium compound can be used for the synthesis of primary amines (**78**), for example, Scheme 6.34.

$$CH_3\text{—}\overset{+}{N}{\equiv}\overset{-}{C}\;(76) \xrightarrow[\text{under } N_2,\ -60°C]{\text{Bu}^n\text{Li/THF}} \text{Li-}CH_2\text{—}\overset{+}{N}{\equiv}\overset{-}{C}\;(77) \quad (6.33)$$

$$\text{cyclohexanone} \xrightarrow{\text{Li-}CH_2\text{—}\overset{+}{N}{\equiv}\overset{-}{C}} \text{1-(HO)-1-(}CH_2\text{—}\overset{+}{N}{\equiv}\overset{-}{C}\text{)cyclohexane} \xrightarrow[\text{MeOH}]{\text{HCl}} \text{1-(HO)-1-(}CH_2\text{-}\overset{+}{N}H_3\ Cl^-\text{)cyclohexane}\;(78)\ 63\% \quad (6.34)$$

6.4.1 *Tosyl methyl isonitrile*

From isocyanomethyl lithium (**77**), tosyl methyl isonitrile (**79**) can be obtained (Scheme 6.35), and this derivative of methyl isonitrile is useful for the conversion of an aldehyde or a ketone (**80**) into a carboxylic acid (**81**) or a nitrile (**82**)[45,46] (Scheme 6.36).

$$\text{Li-CH}_2\text{-}\overset{+}{\text{N}}\equiv\overset{-}{\text{C}}\ (77) + p\text{-CH}_3\text{C}_6\text{H}_4\text{SO}_2\text{F} \xrightarrow{66\%} p\text{-CH}_3\text{C}_6\text{H}_4\text{SO}_2\text{CH}_2\text{-}\overset{+}{\text{N}}\equiv\overset{-}{\text{C}}\ (79) \quad (6.35)$$

$$\text{Ts-CH}_2\text{-}\overset{+}{\text{N}}\equiv\overset{-}{\text{C}}\ (79) \xrightarrow[\text{THF}]{\text{KOBu}^t} \text{Ts-}\overset{-}{\text{C}}\text{H-}\overset{+}{\text{N}}\equiv\overset{-}{\text{C}} + \text{R}^1\text{R}^2\text{C=O}\ (80) \longrightarrow \left[\text{Ts-CH(}\overset{+}{\text{N}}\equiv\overset{-}{\text{C}}\text{)-C(R}^1\text{)(R}^2\text{)-O}^-\right] \xrightarrow{\text{aqueous H}_3\text{O}^+} \text{R}^1\text{R}^2\text{C=C(NH-CHO)(Ts)}$$

$$\text{R}^1\text{R}^2\text{C=C(NH-CHO)(Ts)} \xrightarrow[\text{ii, H}_2\text{O}]{\text{i, MeONa / MeOH}} \text{R}^1\text{R}^2\text{CH-C}\equiv\text{N}\ (82) + \text{H.CO}_2\text{Me} + \text{TsNa}$$

$$\text{R}^1\text{R}^2\text{C=C(NH-CHO)(Ts)} \xrightarrow{\text{aqueous HO}^- \text{ or H}_3\text{O}^+} \text{R}^1\text{R}^2\text{CH-CO}_2\text{H}\ (81)$$

R^1	R^2	Yield (nitrile) (%) (**82**)	Yield (acid) (%) (**81**)
Ph	H	81	72
Bu^t	H	72	63
Ph	Me	69	42
Me	Me	93	83

(6.36)

Tosyl methyl isonitrile (**79**)[47] can be alkylated and reduced by lithium in liquid ammonia to hydrocarbons (**83**) in good yield (45–95 per cent)[48] (Scheme 6.37).

$$\text{Ts-CH}_2\text{-}\overset{+}{\text{N}}\equiv\overset{-}{\text{C}}\ (79) \xrightarrow[\text{R}^1\text{-Hal}]{\text{HO}^-\text{/ PTC}} \text{Ts-CHR}^1\text{-}\overset{+}{\text{N}}\equiv\overset{-}{\text{C}}$$

$$\text{Ts-CHR}^1\text{-}\overset{+}{\text{N}}\equiv\overset{-}{\text{C}} \xrightarrow{\text{Li / liquid NH}_3} \text{R}^1\text{-CH}_3\ (83)$$

$$\text{Ts-CHR}^1\text{-}\overset{+}{\text{N}}\equiv\overset{-}{\text{C}} \xrightarrow{\text{NaH / R}^2\text{-Hal}} \text{Ts-CR}^1\text{R}^2\text{-}\overset{+}{\text{N}}\equiv\overset{-}{\text{C}} \xrightarrow{\text{Li / liquid NH}_3} \text{R}^1\text{-CH}_2\text{-R}^2\ (83)$$

PTC = Phase transfer catalyst

(6.37)

Preparation of α-hydroxaldehydes

Tosylmethyl isonitrile has been used for the preparation of monomers of α-hydroxyaldehydes (**84**)[49] (Scheme 6.38).

(6.38)

Synthesis of oxazoles, imidazoles, and thiazoles

Tosyl methyl isonitrile (**79**) has also been applied to the synthesis of a variety of heterocyclic compounds, a few examples of which follow.

Tosyl methyl isonitrile is used in the synthesis of oxazoles (**85**)[50] (Scheme 6.39), while *p*-tolylthiomethyl isonitrile, $p\text{-}CH_3C_6H_4{-}S{-}CH_2{-}N^+{\equiv}C^-$, can be used similarly for the synthesis of oxazoles with a sulphur substituent.[51]

(6.39)

Imidazoles (**86**) can also be synthesized[52] by reaction with imidoyl halides (**87**)[53] (Scheme 6.40).

A simpler method for the synthesis of imidazoles from imines of aldehydes

Ts-CH⁻—N⁺≡C⁻ + R¹—CH=N—R² (87) (Cl leaving), Na⁺; in diglyme or THF, added to NaH in DMSO, 20° C under N_2 → [Ts—CH—N⁺≡C⁻ / R¹—C=N-R² ⇌ Ts—C—N⁺≡C⁻ / R¹—C—N(R²)—H] → (86)

R^1	R^2	Yield (%)
Ph	Ph	60
p-NO_2-C_6H_4	Ph	85
Ph	p-NO_2-C_6H_4	88
Ph	c-C_6H_{11}	80

(6.40)

has been developed by the same authors using base-induced cycloaddition[54] (Scheme 6.41).

The use of phase transfer catalysis furnished a good synthesis of thiazoles (**88**) from tosylmethyl isonitrile and carbon disulphide (**89**)[55] (Scheme 6.42).

Ts-CH_2—N⁺≡C⁻ + R¹–CH=N—R²; A: K_2CO_3 / MeOH, DME; B: NaH / DME; C: K_2CO_3, MeOH

A		
R^1	R^2	Yield (%)
Ph	p-NO_2-C_6H_4	34
p-Cl-C_6H_4	p-Cl-C_6H_4	43
Ph	CH_3	10
p-NO_2-C_6H_4	Ph	82

B+C		
R^1	R^2	Yield (%)
Ph	p-NO_2-C_6H_4	65
p-Cl-C_6H_4	p-Cl-C_6H_4	40
Ph	CH_3	3(B only)

(6.41)

Ts-CH2—N≡C + CS2 (89) —[Bu$^n_4$N$^+$Br$^-$/CHCl3, 10% NaOH, Phase transfer]→ 95%; R-Hal → (88) 90-95%; R-COCl → (88) 85-90% (6.42)

Synthesis of α-hydroxyamides

α-Hydroxyamides (**90**) can be obtained from the reaction between isonitriles (**91**) and acetals (**92**) in the presence of titanium tetrachloride[56] (Scheme 6.43). By the same reaction methyl isonitrile (**93**) has been applied to the preparation of α-hydroxyamides (**94**) in yields of 17–95 per cent directly from aldehydes and ketones[57] (Scheme 6.44).

R1R2C(OR3)2 (92) + R4—N≡C (91) —[TiCl4, CH2Cl2]→ R2—C(R1)(OR3)—C(=O)—NHR4 (90) 80-90% (6.43)

CH3—N≡C (93) —[TiCl4, CH2Cl2]→ [CH3—N=C(TiCl3)—Cl] —[R1R2C=O, CH2Cl2]→ [R1R2C(OTiCl3)—C(Cl)=N-CH3] —[aq. H3O$^+$]→ R1R2C(OH)—C(=O)-NH-CH3 (94) (6.44)

References

1. P. T. Hoffmann, G. Gokel, D. Marquarding, and I. Ugi, Academic Press, New York, 1971.
2. A. Gautier, *Annalen*, 1867, **142**, 289.
3. L. B. Engemyr, A. Martinsen, and J. Sonstad, *Acta Chem. Scand.*, 1974, **A28**, 255.
4. G. Höfle and B. Lange, *Angew. Chem., Int. Ed. Engl.*, 1977, **16**, 262; *F. &F.*, **8**, 442.
5. A. W. Hofmann, *Annalen*, 1867, **144**, 114.
6. I. U. Nef, *Annalen*, 1892, **270**, 267.
7. W. P. Weber and G. W. Gokel, *Tetrahedron Lett.*, 1972, 1637; *F. &F.*, **1**, 28.
8. W. D. Ollis and D. H. R. Barton *Comprehensive organic chemistry*, Pergamon Press, Oxford, 1979, Vol. 2, ref. 618, p. 590.
9. M. Makoska and M. Wawrzyniensicz, *Tetrahedron Lett.*, 1969, 4659.
10. P. Weber, G. Gokel, and I. Ugi, *Angew. Chem., Int. Ed. Engl.*, 1972, **11**, 530.
11. *F. &F.*, **4**, 186; H. M. Walborsky and G. E. Niznik, *J. Org. Chem.*, 1972, **37**, 187.
12. I. Ugi, W. Betz, U. Fetzer, and K. Offermann, *Ber.*, 1961, **94**, 2814; I. Ugi, U. Fetzer, U. Eholzer, H. Knupfer, and K. Offermann, *Angew. Chem. Int. Ed. Engl.*, 1965, **4**, 472 (this reference contains a preparation of a formamide).
13. *F. &F.*, **13**, 249; R. Obrecht, R. H. Herrmann, and I. Ugi, *Synthesis*, 1985, 400.
14. *F. &F.*, **3**, 321; H. J. Bestmann, J. Lienert, and L. Mott, *Annalen*, 1968, **718**, 24.
15. *F. &F.*, **4**, 554; B. Beijer, E. von Hinrichs, and I. Ugi, *Angew. Chem.*, 1972, **11**, 929.
16. R. Appel, R. Kleinstück, and K.-D. Ziehn, *Angew. Chem., Int. Ed. Engl.*, 1971, **10**, 132; *F. &F.*, **4**, 549.
17. *F. &F.*, **8**, 91; Y. Echigo, Y. Watanabe, and T. Mukaiyama, *Chem. Lett.*, 1977, 697.
18. *F. &F.*, **11**, 112; G. Giesman, E. von Hinrichs, and I. Ugi, *J. Chem. Res.(S)*, 1982, 79.
19. S. Otsuki, K. Mori, and K. Yamagami, *J. Org. Chem.*, 1966, **31**, 4170.
20. *F. &F.*, **11**, 553; J. E. Baldwin, J. C. Battaro, P. D. Riordan, and A. E. Derome, *J. Chem. Soc., Chem. Commun.*, 1982, 942.
21. H. Gilman, D. J. Peterson, and D. Wittenberg, *Chem. &Ind.*, 1958, 1479.
22. T. Mukaiyama and Y. Yokota, *Bull. Chem. Soc. Japan*, 1965, **38**, 858; *F. &F.*, **2**, 323.
23. W. D. Ollis & D. H. R. Barton *Comprehensive organic chemistry*, Pergamon Press, Oxford, 1979, Vol. 2, p. 571.
24. D. Rehn, and I. Ugi, *J. Chem. Res. (S)*, 1977, 119.
25. *F. &F.*, **11**, 99; Y. Ito, H. Kato, and T. Saegusa, *J. Org. Chem.*, 1982, **47**, 741.
26. *F. &F.*, **11**, 99; Y. Ito, H. Kato, H. Imai, and T. Saegusa, *J. Am. Chem. Soc.*, 1982, **104**, 6449.
27. *F. &F.*, **2**, 51; I. Ugi, R. Meyr, M. Lipinski, F. Bodesheim, and F. Rosendahl, *Org. Synth.*, 1961, **41**, 13.
28. A. Aigner and D. Marquarding, *Tetrahedron Lett.*, 1978, 3325.
29. I. Ugi, *Angew. Chem., Int. Ed. Engl.*, 1975, **14**, 61.
30. *F. &F.*, **10**, 67; H. R. Divanford, Z. Lysenko, P.-C. Wang, and M. M. Joullié, *Synth. Commun.*, 1978, **8**, 269; M. M. Joullié, P.-C. Wang, and J. E. Semple, *J. Am. Chem. Soc.*, 1980, **102**, 887.
31. H. Pelliser and G. Gil, *Tetrahedron*, 1989, **45**, 3415.
32. *F. &F.*, **4**, 495; F. Kienzle, *Tetrahedron Lett.*, 1972, 1771.
33. W. D. Ollis and D. H. R. Barton *Comprehensive organic chemistry*, Pergamon Press, Oxford, 1979, Vol. 2, p. 575.

34. *F. &F.*, **11**, 11; Y. Ito, H. Kato, and T. Saegusa, *J. Org. Chem.*, 1982, **47**, 741.
35. W. D. Ollis & D. H. R. Barton *Comprehensive organic chemistry*, Pergamon Press, Oxford, 1979, Vol. 2, p. 573.
36. *F. &F.*, **4**, 503 (this reference contains the preparation of 1,1,3,3-tetramethylbutyl isonitrile).
37. *F. &F.*, **4**, 480; H. M. Walborsky, W. H. Morrison III, and G. E. Niznik, *J. Am. Chem. Soc.*, 1970, **92**, 6675.
38. *F. &F.*, **5**, 650; G. E. Niznik, W. H. Morrison III, and H. M. Walborsky, *J. Org. Chem.*, 1974, **39**, 600.
39. G. E. Niznik, W. H. Morrison III, and H. M. Walborsky, *Org. Synth.*, 1971, **51**, 31.
40. N. Hirowatari and H. M. Walborsky, *J. Org. Chem.*, 1974, **39**, 604.
41. M. P. Periasamy and H. M. Walborsky, *J. Org. Chem.*, 1974, **39**, 611.
42. *F. &F.*, **10**, 413; D. H. R. Barton, G. Bringmann, G. Lamotte, R. H. S. Motherwell, and W. B. Motherwell, *Tetrahedron Lett.*, 1979, 2291; D. H. R. Barton, G. Bringmann, and W. B. Motherwell, *Synthesis*, 1980, 68.
43. *F. &F.*, **4**, 271; U. Schröder and R. Schröder, *Tetrahedron Lett.*, 1973, 633.
44. U. Schöllkopf and P. Böhme, *Angew. Chem., Int. Ed. Engl.*, 1971, **10**, 491.
45. U. S. Schröder and R. Schröder, *Angew. Chem., Int. Ed. Engl.*, 1972, **11**, 311; A. M. van Leusen, G. J. M. Boerman, R. B. Helmholt, H. Siderrius, and J. Strating, *Tetrahedron Lett.*, 1972, 2369; B. E. Hoogenboom, O. H. Oldenziel, and A. M. van Leusen, *Org. Synth.*, 1977, **57**, 102; *F. &F.*, **5**, 684; **6**, 600.
46. *F. &F.*, **5**, 685; U. S. Schröder and R. Schröder, *Angew. Chem., Int. Ed. Engl.*, 1973, **12**, 407; 1972, **11**, 311; O. H. Oldenziel and A. M. van Leusen, *Tetrahedron Lett.*, 1973, 1357.
47. J. S. Yadav, P. S. Reddy, and A. B. Sahasrabuhde, *Synth. Commun.*, 1983, **13**, 379.
48. J. S. Yadav, P. S. Reddy, and B. V. Jashi, *Tetrahedron*, 1988, **44**, 7243.
49. *F. &F.*, **5**, 685; O. H. Oldenziel and A. M. van Leusen, *Tetrahedron Lett.*, 1974, 163; 167.
50. A. M. van Leusen, B. E. Hoogenboom, and H. Siderius, *Tetrahedron Lett.*, 1972, 2369.
51. A. M. van Leusen and H. E. Gennep, *Tetrahedron Lett.*, 1973, 627; U. Schöllkopf and E. Blume, *Tetrahedron Lett.*, 1973, 627.
52. A. M. van Leusen and O. H. Oldenziel, *Tetrahedron Lett.*, 1972, 2373.
53. F. Cramer and U. Baer, *Ber.*, 1960, **93**, 1231.
54. *F. &F.*, **8**, 493; A. M. van Leusen, J. Wildeman, and O. H. Oldenziel, *J. Org. Chem.*, 1977, **42**, 1153; see also D. van Leusen and A. W. van Leusen, *Tetrahedron Lett.*, 1977, 4233.
55. *F. &F.*, **8**, 494; A. M. van Leusen and J. Wildeman, *Synthesis*, 1977, 501.
56. *F. &F.*, **12**, 498; T. Mukaiyama, K. Watanabe, and M. Shiono, *Chem. Lett.*, 1974, 1457. (For the structure of the actual reagent involved see B. Crociani, M. Nicolini, and R. L. Richards, *J. Organomet. Chem.*, 1975, **101**, C1.)
57. *F. &F.*, **12**, 354; M. Schiess and D. Seebach, *Helv. Chim. Acta*, 1983, **66**, 1618.

7

CYANATES, ISOCYANATES, THIOCYANATES, AND ISOTHIOCYANATES

The structures of these related compounds are as follows: cyanate (**1**), R—O—C≡N; isocyanate (**2**), R—N=C=O; thiocyanate (**3**), R—S—C≡N; isothiocyanate (**4**), R—N=C=S.

ALKYL CYANATES (**1**)

$$\text{>C—O—C}\overset{1.13\ \text{Å}}{\equiv}\text{N}$$

7.1 Preparation

7.1.1 *From alkoxides*

Certain alkyl cyanates can be prepared by the reaction of an alkoxide (**5**) with a cyanogen halide (**6**) (Scheme 7.1).[1]

$$\underset{(5)}{\text{RO}^-\text{K}^+} + \underset{(6)}{\text{Hal—CN}} \longrightarrow \text{R—O—C}\equiv\text{N} \qquad (7.1)$$

7.1.2 *From alkyl halides*

When an alkali metal cyanate reacts with an alkyl halide (**7**), the major and sometimes only product is an alkyl isocyanate (**8**) (Scheme 7.2). With secondary alkyl iodides a mixture of the two often results.[2]

$$\left[\text{N}\equiv\text{C—O}^- \longleftrightarrow {}^-\text{N=C=O}\right] \xrightarrow[(7)]{\text{R-Hal}} \text{N}\equiv\text{C—OR} + \underset{\substack{(8)\\ \text{major product}}}{\text{R—N=C=O}} \qquad (7.2)$$

7.2 Reactivity

Alkyl cyanates are unstable compounds, tending to isomerize rapidly to alkyl isocyanates. As a result they are not as yet widely used in synthesis and therefore limited research has been carried out into their reactivity.

ALKYL ISOCYANATES[3] (**2**)

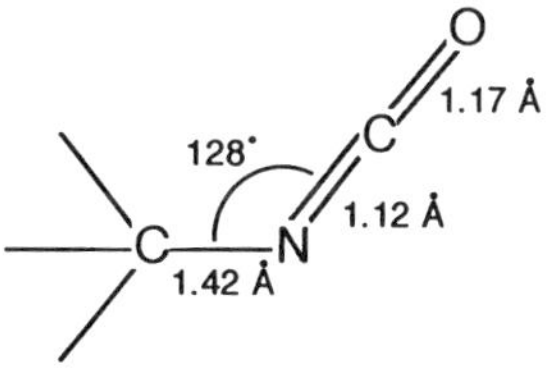

$$\text{C(R}_3\text{)}\overset{1.42\ \text{Å}}{—}\text{N}\overset{1.12\ \text{Å}}{=}\text{C}\overset{1.17\ \text{Å}}{=}\text{O}\qquad \angle\text{C–N=C} = 128^\circ$$

Alkyl isocyanates (**2**) can be prepared by a variety of methods. They are much more stable than alkyl cyanates and hence they prove to be useful compounds for synthetic purposes.

7.3 Preparation[4]

7.3.1 *From metal cyanates*

Alkyl isocyanates are the major products obtained from the reaction of a metal cyanate (**9**) with an alkyl halide (**10**), a dialkyl sulphate (**11**) or *p*-toluene-sulphonic ester (Scheme 7.3).[5]

$$\underset{(9)}{O{=}C{=}N{—}Ag} + \underset{(10)}{R{—}Hal} \longrightarrow O{=}C{=}N{—}R + AgHal$$

$$\underset{(9)}{O{=}C{=}N^- K^+} + \underset{(11)}{(CH_3)_2SO_4} \xrightarrow{Na_2CO_3} O{=}C{=}N{—}CH_3 + CH_3SO_4K \qquad (7.3)$$

7.3.2 *From primary amines*

Another general method is from the reaction between primary amines (**12**) and phosgene (**13**) (Scheme 7.4).[6]

Alkyl isocyanates can also be obtained from the reaction between primary

$$Cl{-}C({=}O){-}Cl \;(13) + H_2N{-}R \;(12) \longrightarrow \left[Cl{-}C({=}O){-}N(H){-}R \right] + HCl \xrightarrow{-HCl} O{=}C{=}N{-}R \qquad (7.4)$$

amines and carbon monoxide when catalysed by palladium chloride,[7] for example, Scheme 7.5.

$$Bu^n{-}NH_2 + CO \xrightarrow[\text{1 atm., 2 days}]{PdCl_2} Bu^n{-}N{=}C{=}O \;\; 49\% \qquad (7.5)$$

7.3.3 *By oxidation of isonitriles*

Isonitriles (**14**) have been oxidized by a variety of reagents (mercury oxide, ozone, bromine/dimethylsulphoxide) to yield isocyanates,[8] e.g. Scheme 7.6. Good yields are also obtained by the use of dimethylsulphoxide in the presence of a Lewis acid (Scheme 7.7).[9]

$$R{-}N{=}C{:} \;(14) \xrightarrow[Me_2SO,\ \text{Reflux 1 day}]{Br_2 \text{ in } CHCl_3} \left[R{-}N{=}CBr_2\right] \longrightarrow R{-}N{=}C{=}O + Me_2S + Br_2 \qquad (7.6)$$

$$R{-}N{=}C{:} \;(14) + Me_2SO \xrightarrow[\text{or } Ph_3C^+ClO_4^-,\ 50^\circ - 80^\circ C]{TsOH \text{ or } HCl} R{-}N{=}C{=}O + Me_2S \qquad (7.7)$$

R	Yield (%)
Bu^n	85
Cyclohexyl	92

Nitrile *N*-oxides (**15**) have been used for the oxidation of isonitriles (**16**). The isocyanates (**17**) were converted into ureas (**18**) after hydrolysis (yield as ureas after being boiled with water = 28–62 per cent) (Scheme 7.8).[10]

$$Ph{-}C{\equiv}\overset{+}{N}{-}\overset{-}{O} + R{-}N{=}C: \xrightarrow[\text{reflux, 5 hours}]{\text{ether}} Ph{-}C{\equiv}N + R{-}N{=}C{=}O$$

(15) (16) (17)

$$R{-}N{=}C{=}O \xrightarrow{H_2O} [R{-}NHCO_2H] \longrightarrow R{-}NH_2 + CO_2 \xrightarrow{R{-}N{=}C{=}O\ (17)} (R{-}NH)_2C{=}O\ (18)$$

R = Bu^t or cyclohexyl (7.8)

7.3.4 *By oxidation of an isonitrile dichloride–ferric chloride complex*

Isocyanates have been prepared by a reaction in which an isonitrile dichloride–ferric chloride complex (**19**), prepared from cyanogen chloride (**20**), an alkyl chloride (**21**), and ferric chloride (**22**), is oxidized by zinc oxide e.g. Scheme 7.9.[11] When the reaction is performed without isolation of the complex, the yield of isocyanate is 85 per cent. With isolation the yield falls to 38 per cent. A similar reaction is shown in Scheme 7.10 with subsequent formation of a urea.

$$3\,ClCN\ (20) + 3\,Me_2CHCl\ (21) + 2\,FeCl_3\ (22) \longrightarrow (Me_2CH{-}N{=}CCl_2)_3(FeCl_3)_2\ (19) \xrightarrow{ZnO} 3\,Me_2CH{-}N{=}C{=}O \quad (7.9)$$

$$(Bu^t{-}N{=}CCl_2)(FeCl_3) \xrightarrow[\text{or DMSO}]{F_3C{-}CO_2H\,/\,Bu^n_3N} Bu^t{-}N{=}C{=}O \xrightarrow{Pr^i{-}NH_2} Bu^t{-}NH{-}CO{-}NH{-}Pr^i$$

Conditions	Yield (%)
$F_3C{-}CO_2H/Bu^t_3N$	30
DMSO	18

(7.10)

7.3.5 *From carbamates*

Isocyanates have been prepared from carbamates (**23**) (urethanes) and from carbamyl chlorides (**24**) by the use of trimethylsilyl compounds as shown in Scheme 7.11.[12]

$$R^1\text{—NH—C(=O)—OR}^2\ (23) \xrightarrow[Et_3N]{Me_3SiCl} R^1\text{—N(SiMe}_3\text{)—C(=O)—OR}^2 \xrightarrow[2\ hours]{PhCH_3,\ reflux} R^1\text{—N=C=O} + R^2O\text{-}SiMe_3$$

$$R^1\text{—NH—C(=O)—Cl}\ (24) \xrightarrow{CH_3CONHSiMe_3} R^1\text{—N(SiMe}_3\text{)—C(=O)—Cl} \xrightarrow{-20^\circ\ to\ 0^\circ C} R\text{-N=C=O} + Me_3SiCl \qquad (7.11)$$

7.3.6 *From methyl thiocarbamates*

By the use of a derivative of benzoxazole (**25**), methyl thiocarbamates (**26**) can be converted into isocyanates (Scheme 7.12).[13]

$$\text{2-aminophenol} \xrightarrow[ii,\ HCl]{i,\ EtOCO_2^-\ K^+} \text{2-chlorobenzoxazole} \xrightarrow{Et_3O^+\ BF_4} \text{3-ethyl-2-chlorobenzoxazolium}\ BF_4^-\ (25)$$

$$(25) + R\text{-NH-C(=S)-OMe}\ (26) + Et_3N \xrightarrow{CH_2Cl_2} \text{3-ethylbenzoxazole-2-thione} + R\text{-N=C=O} + CH_3Cl + Et_3NH^+\ BF_4^- \qquad (7.12)$$

65-85 %

7.3.7 *From pyridinium or pyrrilium salts*

Katritzky and his co-workers have obtained good yields of isocyanates by the pyrolysis of pyridinium salts (**27**) derived from acyl chlorides (**28**) and *N*-aminopyridinium salts (**29**) or from acyl hydrazides (**30**) and 2,4,6-pyrrilium salts (**31**) (Scheme 7.13).[14]

Ph Ph Ph
220°C
Ph N⁺ Ph Ph N⁺ Ph Ph N Ph
(29) NH$_2$ (27) $^-$N C R +
+ O R—N=C=O
R—C(=O)—Cl (28)

Ph
R—C(=O)—NHNH$_2$ + (30) Ph O⁺ Ph
(31) (7.13)

7.3.8 *From nitrile N-oxides*

Isocyanates can be obtained from nitrile *N*-oxides (**32**) by the action of heat.[15] Heat often causes the nitrile oxides to dimerize, thus reducing the yield of isocyanate. A modification, involving carrying out the reaction in boiling benzene in the presence of sulphur dioxide avoids this difficulty and enables good yields to be attained (Scheme 7.14).[16]

$R{-}C{\equiv}\overset{+}{N}{-}\overset{-}{O}$ + SO_2 ⟶ (cyclic adduct, R, N—O, S=O, O) $\xrightarrow[\text{heat}]{\text{PhH}}$ R—N=C=O + SO_2
(32) 76 - 85 % (7.14)

7.3.9 *Via the Curtius rearrangement*

Isocyanates are intermediates in several related rearrangements to electron deficient nitrogen atoms, viz. the Hofmann, the Curtius, and the Lossen rearrangements. However, the only one from which they are easily isolated is the Curtius rearrangement, in which an acyl azide (**33**) is heated in an inert solvent such as ether, benzene, or toluene[17] (Scheme 7.15).[18]

$$n\text{-}C_{11}H_{23}\text{—C(=O)—Cl} + NaN_3 \longrightarrow n\text{-}C_{11}H_{23}\text{—C(=O)—N=}\overset{+}{N}\text{=}\overset{-}{N}\ \mathbf{(33)} \xrightarrow[\text{inert solvent}]{\text{heat}} n\text{-}C_{11}H_{23}\text{—C(=O)—}\ddot{N}\text{:} + N_2 \longrightarrow \underset{85\%}{O\text{=C=N—}n\text{-}C_{11}H_{23}} \qquad (7.15)$$

7.3.10 *Modifications of the Curtius reaction*

Near quantitative yields of acyl azides (**34**) for the Curtius preparation of isocyanates can be obtained by treating an acyl chloride (**35**) with a mixture of pyridine (**36**) and hydrazoic acid (**37**) (Scheme 7.16).[19]

$$\left[\underset{(36)}{\text{pyridine}} + \underset{(37)}{HN_3} \rightleftharpoons \text{pyridinium (N}^+\text{—H)}\ N_3^-\right] \xrightarrow[0^\circ\,C]{\text{R-COCl, (35)}} \underset{(34)}{\text{R-C(=O)-N=}\overset{+}{N}\text{=}\overset{-}{N}} \xrightarrow[\text{heat}]{\text{Curtius}} \text{R—N=C=O} \qquad (7.16)$$

An alternative preparation of acyl azides (**38**) for the Curtius reaction utilizes a quaternary azide (**39**) as a source of azide ion (Scheme 7.17).[20]

In another modification of the Curtius reaction an acyl chloride (**40**) or an acid anhydride (**41**) is treated with trimethylsilyl azide (**42**) in carbon tetrachloride or in mesitylene at 30–65 °C (Scheme 7.18).[21]

Tri-*n*-butyltin azide (**43**), has been used to produce the acyl azide in yet another adaptation of the Curtius reaction (Scheme 7.19).[22]

$$Bu^n_4N\text{-}OH + NaN_3 \xrightarrow{\text{aq. NaOH}} Bu^n_4N^+ N_3^- \ \ (39)$$

100 %, extracted into CH_2Cl_2

$$Bu^n_4N^+N_3^- \xrightarrow[\text{PhH or PhCH}_3]{\text{RCOCl}} R\text{—}C(=O)\text{—}N_3 \ (38) \xrightarrow{50^\circ\text{-}90^\circ C} R\text{—}N{=}C{=}O \ \ (60 - 90\ \%) \tag{7.17}$$

$$Me_3SiCl + NaN_3 \xrightarrow[20^\circ C,\ 1\ hour]{\text{HMPT}} Me_3SiN_3$$

$$\underset{(40)}{R\text{—}C(=O)\text{—}Cl} \text{ or } \underset{(41)}{(R\text{-}CO)_2O} + \underset{(42)}{Me_3SiN_3} \longrightarrow \left[R\text{—}C(=O)\text{—}N_3\right] \xrightarrow{30^\circ - 65^\circ C} R\text{—}N{=}C{=}O \ \ (70 - 80\ \%) \tag{7.18}$$

$$Bu^n_3SnCl + \underset{\text{(conc. aqueous)}}{NaN_3} \xrightarrow{0^\circ C} \underset{(43)\ 91\ \%}{Bu^n_3SnN_3} + NaCl$$

$$\underset{(43)}{Bu^n_3SnN_3} + R\text{—}C(=O)\text{—}Cl \longrightarrow \underset{40 - 60\ \%}{R\text{—}N{=}C{=}O} + Bu^n_3SnCl \tag{7.19}$$

7.3.11 *From amides*

An alternative to the Curtius method involves the oxidation of a primary amide (**44**) with lead tetraacetate at 50–60 °C in dimethylformamide or benzene followed by rearrangement (Scheme 7.20).[23]

$$\underset{(44)}{R\text{—}C(=O)\text{—}NH_2} \xrightarrow[\substack{\text{DMF or PhH} \\ 50^\circ - 60^\circ C}]{Pb(OAc)_4} \left[R\text{—}C(=O)\text{—}\ddot{N}{:}\right] \longrightarrow O{=}C{=}N\text{—}R \tag{7.20}$$

7.3.12 *Preparation of substituted alkyl isocyanates*

α-Alkoxy isocyanates

α-Alkoxy isocyanates (**45**) result from the addition of isocyanic acid (prepared by the depolymerization of cyanuric acid) across the carbon–carbon double bond of enol ethers (**46**) in a reaction which may be catalysed by *p*-toluenesulphonic acid (Scheme 7.21).[24] This addition will also occur across certain terminal alkenes (**47**) which contain an electron releasing group,[25] for example, Scheme 7.22.

$$\underset{(46)}{R^1O{-}C(R^2)H{=}CHR^3} \xrightarrow[\text{Ether, 40 °C, TsOH}]{\text{H-N=C=O}} \underset{(45)}{R^1O{-}C(R^2)(N{=}C{=}O){-}CH_2R^3} \tag{7.21}$$

$$\underset{(47)}{Ph{-}C(CH_3){=}CH_2} + H{-}N{=}C{=}O \xrightarrow{\text{Lewis acid}} Ph{-}C(CH_3)_2{-}N{=}C{=}O \quad 41\ \% \tag{7.22}$$

α-Hydroxy isocyanates

Isocyanic acid (**48**) adds across aldehydes and ketones (**49**) to produce α-hydroxy isocyanates (**50**) (Scheme 7.23).[26]

$$\underset{(49)}{R^1R^2C{=}O} + \underset{(48)}{H{-}N{=}C{=}O} \xrightarrow[\text{then 0°C}]{-78\text{°C}} \underset{(50)}{R^1R^2C(OH){-}N{=}C{=}O} \tag{7.23}$$

α-Iodo isocyanates

In a regiospecific reaction which has been much studied, iodine isocyanate (**51**) which is usually prepared *in situ* from silver cyanate and iodine, adds across alkenes (**52**) (Scheme 7.24).[27] The α-iodo isocyanate is an intermediate in a useful synthesis of aziridines (see Chapter 21, p. 709).

The addition of iodine isocyanate to allenes and to alkynes has also been studied.[28]

$$>C{=}C< \;(52) + I{-}N{=}C{=}O\;(51) \longrightarrow \text{[iodonium ion]} + {}^{-}N{=}C{=}O \longrightarrow \text{I–C–C–N=C=O} \qquad (7.24)$$

7.4 Reactivity

The chemistry of isocyanates is dominated by the attack of nucleophiles on the carbon of the carbonyl group.

7.4.1 *Hydrolysis*

Hydrolysis is promoted by aqueous acid or alkali or simply by boiling with water (Scheme 7.25). In the absence of base the primary amine produced by hydrolysis reacts with unchanged isocyanate to give a substituted urea (**53**) (Scheme 7.26).[29]

$$R{-}N{=}C{=}O + H_2O \xrightarrow[\text{or aqueous } H_3O^+]{\text{aqueous } HO^-} R{-}NH_2 + CO_2 \qquad (7.25)$$

$$R^1{-}\ddot{N}H_2 + R^2{-}N{=}C{=}O \longrightarrow \left[R^2{-}N{=}C(O^-){-}\overset{+}{N}H_2{-}R^1\right] \longrightarrow R^2HN{-}C(=O){-}NHR^1\;(53) \qquad (7.26)$$

7.4.2 *With alcohols*

In a general reaction which usually gives good yields, alcohols (**54**) react with isocyanates to yield carbamates (urethanes) (**55**). This reaction can be used to characterize alcohols (Scheme 7.27). The reaction is catalysed by organometallic compounds.[30] It has been found that boron fluoride etherate or aluminium chloride are good catalysts for reaction with secondary alcohols (**56**) (Scheme 7.28).[31]

$$R^1{-}N{=}C{=}O + \underset{(54)}{R^2OH} \longrightarrow \underset{(55)}{R^1{-}N(H){-}C(=O){-}OR^2} \qquad (7.27)$$

$$R^1{-}N{=}C{=}O + \underset{(56)}{R^2R^3CHOH} \xrightarrow[\text{PhH or } Et_2O,\ 25^\circ C \text{ for } 0.5-2 \text{ hours}]{BF_3{\cdot}Et_2O \text{ or } AlCl_3} R^1{-}N(H){-}C(=O){-}OCHR^2R^3 \qquad (7.28)$$

Optically active isocyanates, for example, (**57**) have been used for the resolution of racemic alcohols.[32]

$$1\text{-}C_{10}H_7{-}CH(CH_3){-}N{=}C{=}O \quad (57)$$

Similarly the compound $Cl_3C{-}C(O){-}N{=}C{=}O$ has been used for the classification of alcohols.[33]

7.4.3 *With amines*

As indicated above, isocyanates react with amines to yield ureas and this reaction has been used for the resolution of racemic amines, the isocyanate used being one of the following.[34]

$$C_6H_5{-}CH(CH_3){-}N{=}C{=}O \qquad CH_3CH_2{-}CH(CH_3){-}N{=}C{=}O$$

In a further extension of these resolution techniques, isopropyl isocyanate, $Me_2CH{-}N{=}C{=}O$, has been used to form ureas (**58**), carbamates (urethanes)

(**59**), and amides (**60**) from (±)-amines (**61**), alcohols (**62**), and carboxylic acids (**63**) which can then be resolved by gas chromatography on chiral polymers (Scheme 7.29).[35]

$$R^1\text{—}CH(R^2)\text{-}NH_2 \;(61) \xrightarrow[20^\circ C,\ CH_2Cl_2]{(CH_3)_2CH\text{—}N{=}C{=}O} R^1\text{—}CH(R^2)\text{-}NH\text{—}C(=O)\text{—}NH\text{—}CH(CH_3)_2 \;(58)$$

$$R^1\text{—}CH(R^2)\text{-}OH \;(62) \xrightarrow[100^\circ C]{(CH_3)_2CH\text{—}N{=}C{=}O} R^1\text{—}CH(R^2)\text{-}O\text{—}C(=O)\text{—}NH\text{—}CH(CH_3)_2 \;(59)$$

$$CH_3\text{—}NH\text{—}CH(R^1)\text{-}CO_2H \;(63) \xrightarrow[100^\circ C]{(CH_3)_2CH\text{—}N{=}C{=}O} (CH_3)_2CH\text{—}NH\text{—}C(=O)\text{—}N(CH_3)\text{—}CH(R^1)\text{-}C(=O)\text{—}NH\text{—}CH(CH_3)_2 \;(60) \quad (7.29)$$

7.4.4 *With ketones*

Isocyanates (**64**) react with ketones (**65**) in a reaction catalysed by magnesium chloride and sodium iodide to give low yields of β-keto amides (**66**) (Scheme 7.30).[36]

$$R^1\text{—}C(=O)\text{—}CH_2\text{—}R^2 \;(65) + R^3\text{—}N{=}C{=}O \;(64) \xrightarrow[Et_3N\ /\ CH_3CN]{MgCl_2\ /\ NaI} R^1\text{—}C(=O)\text{—}CH(R^2)\text{-}C(=O)\text{—}NHR^3 \;(66)$$

R^1=Ph, R^2=Me, R^3=Bu^n, Yield = 26% (7.30)

7.4.5 *With organometallics*

Isocyanates react with Grignard reagents or with alkyl lithiums (**67**) to give good yields of amides (**68**) (Scheme 7.31).[37]

$$R^1\text{—N=C=O} + R^2Li \xrightarrow{\text{ether}} R^1\text{—NH—C(=O)—}R^2$$

(67) (68) 60 - 95 % (7.31)

Triethylaluminium reacts with isocyanates (**69**) to yield amides of propanoic acid (**70**) (Scheme 7.32).[38]

$$R\text{—N=C=O} \xrightarrow{Et_3Al} CH_3CH_2\text{—C(=O)—NHR}$$

(69) (70) 90 -95 % (7.32)

The reaction of an isocyanate with an alkyl copper reagent prepared from an alkyne (**71**) yields an α,β-unsaturated amide (**72**) (Scheme 7.33).[39]

$$Bu^n\text{—C≡CH} \xrightarrow[\text{ether - HMPT}]{EtCu\ MgBr_2} (Bu^n)(Et)C\text{=}C(H)(CuMgBr_2)\ (71)$$

$$\xrightarrow[\text{HMPT}]{R^3\text{-N=C=O},\ P(OEt)_3\ (\text{catalyst})} (Bu^n)(Et)C\text{=}C(H)\text{—C(=O)—}NHR^3\ (72)$$

(7.33)

7.4.6 *Oxidation*

The oxidation of isocyanates (**73**) to nitroalkanes (**74**) can be efficiently achieved by the use of oxone (2 $KHSO_5$. $KHSO_4$. K_2SO_4)[40] in acetone containing water. The rate of reaction can be increased by the addition of Triton B[41] (Scheme 7.34).

$$R\text{—N=C=O} \xrightarrow[H_2O,\ \text{Triton B}]{\text{Oxone in acetone}} R\text{—}NO_2$$

(73) (74) (7.34)

7.4.7 *Reduction*

Reduction of isocyanates with lithium aluminium hydride yields secondary methylamines (**75**) (Scheme 7.35).[42]

$$R\text{—}N\text{=}C\text{=}O \xrightarrow{LiAlH_4} R\text{—}NH\text{—}CH_3 \quad (75) \qquad (7.35)$$

7.4.8 *Deoxygenation*

Deoxygenation of isocyanates to give isonitriles (**76**) can be achieved by the use of a heterocyclic phosphorus compound, 2-phenyl-3-methyl-1,3,2-oxazaphospholine (**77**), prepared as shown in Scheme 7.36.[43]

Ph—P(NMe$_2$)$_2$ + H—N(CH$_3$)—CH$_2$—CH$_2$—OH $\xrightarrow{Et_3N}$ (77) 83%

Cyclohexyl—N=C=O + (77) ⟶ Cyclohexyl—N=C (76) 82% + Ph—P(=O) oxazaphospholidine (7.36)

Another method of deoxygenation utilizes trimethylsilyl chloride and triethylamine (Scheme 7.37).[44]

$$R\text{—}N\text{=}C\text{=}O \xrightarrow{Me_3SiCl\ /\ Et_3N} R\text{—}N\text{=}C \quad 40\text{ - }70\ \% \qquad (7.37)$$

7.4.9 *With phosphorus ylides (Wittig reaction)*

Isocyanates react with phosphorus ylides (**78**) in a specific version of the general Wittig reaction, by nucleophilic attack on the carbonyl carbon atom (Scheme 7.38).[45]

$$R^1{-}N{=}C{=}O + Ph_3\overset{+}{P}{-}\overset{-}{C}(R^3){-}R^2 \ \textbf{(78)} \longrightarrow R^1{-}N{=}C(O^-){-}C(R^2)(R^3){-}\overset{+}{P}Ph_3 \longrightarrow R^1{-}N{=}C{=}C(R^3){-}R^2 + Ph_3P{=}O \qquad (7.38)$$

7.4.10 *With bisulphite ions*

Isocyanates form bisulphite addition compounds (**79**) which can be used for purification of the isocyanate (Scheme 7.39).[46]

$$R{-}N{=}C{=}O + NaHSO_3 \longrightarrow R{-}N{=}C(OH){-}SO_3^- \ Na^+ \ \textbf{(79)} \qquad (7.39)$$

7.4.11 *Carbodiimide formation*

The compound 3-methyl-1-ethylphospholene-1-oxide (**80**) catalyses the formation of carbodiimides (**81**) from isocyanates (Scheme 7.40).[47] For the mechanism of the reaction see Ref. 48.

(80)

$$R{-}N{=}C{=}O \xrightarrow{(80)} R{-}N{=}C{=}N{-}R \ \textbf{(81)} \qquad (7.40)$$

Yields *c.* 95 %

7.4.12 *Polymerization*

Under the catalytic influence of phosphines, isocyanates give cyclic dimers (**82**) and trimers (**83**) (Scheme 7.41).[49]

$R^1—N=C=O \xrightarrow{R^2_3P}$ (82) + (83) (7.41)

7.4.13 *Formation of lactams*

Suitably substituted arylethyl isocyanates (**84**) ring close under the influence of boron trifluoride etherate to give lactams (**85**) (Scheme 7.42).[50]

(84) $\xrightarrow[20^\circ C]{BF_3 \cdot Et_2O}$ (85) *c.* 88 % (7.42)

7.4.14 *With alkynes*

Alkynes with an isocyanate group in the 5-position (**86**) undergo a [2+2+2] cycloaddition to other alkynes catalysed by dicarbonylcyclopentadienyl cobalt to yield pyridones (**87**) (Scheme 7.43).[51]

(86) + $(CH_2)_2CH_3$–C≡C–$SiMe_3$ $\xrightarrow{CpCo(CO)_2}$ (87) 60% (7.43)

ACYL ISOCYANATES

$$R-C(=O)-N=C=O$$

7.5 Preparation[52]

7.5.1 *From acid chlorides*

Acyl isocyanates (**88**) can be prepared from acid chlorides (**89**) and sodium cyanate (**90**) in a reaction catalysed by stannic chloride (Scheme 7.44).[53]

$$\underset{(89)}{R-C(=O)-Cl} + \underset{(90)}{NaOCN} \xrightarrow[\text{in 1,2-dichlorobenzene, 8 hours at } 80^\circ C]{SnCl_4} \underset{(88)}{R-C(=O)-N=C=O} \qquad (7.44)$$

R	Yield (%)
Me	36
Et	52
Cl_3C*	92

* No catalyst, 130–170°C; 1,2,4-trichlorobenzene as solvent

In a similar reaction between isocyanic acid (**91**) and an acyl chloride (**92**) in the presence of pyridine, good yields of acyl isocyanates (**93**) are obtained (Scheme 7.45).[54]

$$\underset{(92)}{R-C(=O)-Cl} + \underset{(91)}{H-N=C=O} \xrightarrow[-10^\circ\text{-}0^\circ C,\ \text{pyridine}]{\text{Ether or THF}} \underset{(93)}{R-C(=O)-N=C=O} \qquad (7.45)$$

7.5.2 *From amides*

Amides (**94**) react with oxalyl chloride (**95**) to yield isocyanates (Scheme 7.46).[55]

$$\mathrm{Cl{-}C({=}O){-}C({=}O){-}Cl} \;(95) + \mathrm{H_2N{-}C({=}O){-}R} \;(94) \longrightarrow \left[\mathrm{Cl{-}C({=}O){-}C({=}O){-}N(H){-}C({=}O){-}R}\right] \longrightarrow \mathrm{O{=}C{=}N{-}C({=}O){-}R} + \mathrm{CO} + \mathrm{HCl} \qquad (7.46)$$

CHLOROSULPHONYL ISOCYANATE[56] (**96**)

$$\mathrm{Cl{-}S({=}O)_2{-}N{=}C{=}O} \quad (96)$$

7.6 Preparation

Chlorosulphonyl isocyanate (**96**) is prepared from cyanogen chloride (**97**) and sulphur trioxide (Scheme 7.47).[57]

$$\underset{(97)}{\mathrm{Cl{-}CN}} + \mathrm{SO_3} \longrightarrow \underset{(96)}{\mathrm{O{=}C{=}N{=}SO_2Cl}} \qquad (7.47)$$

7.7 Reactivity

Chlorosulphonyl isocyanate is a reactive compound which has been used in a variety of synthetic procedures.

7.7.1 *With nucleophilic aromatic compounds*

Chlorosulphonyl isocyanate has been used for the preparation of nitriles (**98**) from highly nucleophilic aromatic compounds such as indole,[58] 1,3-dimethoxybenzene, or anthracene (**99**) (Scheme 7.48).[59]

(99)

$O{=}C{=}N{-}SO_2Cl$

$N{-}SO_2Cl$

DMF

(98)

CN

$+ SO_3 + HCl$

90%

$O{=}C{-}N(H){-}SO_2Cl$

(7.48)

7.7.2 *With carboxylic acids*

In a synthetically useful reaction chlorosulphonyl isocyanate has been used to convert carboxylic acids (**100**) into nitriles (**101**) (Scheme 7.49).[60]

$$\underset{(100)}{R\text{-}CO_2H} + ClSO_2N{=}C{=}O \longrightarrow R{-}C({=}O){-}N(H){-}SO_2Cl + CO_2 \xrightarrow{DMF} \underset{(101)}{R{-}C{\equiv}N} + SO_3 + HCl \quad (7.49)$$

7.7.3 *With ketones*

A similar reaction with an excess of a ketone (**102**) yields a β-cyano ketone (**103**) (Scheme 7.50).[61]

7.7.4 *With alkenes*

Chlorosulphonyl isocyanate reacts with unsaturated carbon–carbon bonds; with alkenes (**104**)[62] and with enol esters[63] β-lactams (**105**) are produced (Scheme 7.51). An alternative reductive work up involves the use of aqueous

$$R^1{-}\underset{\underset{O}{\parallel}}{C}{-}\underset{R^3}{\overset{R^2}{C}}{-}H + ClSO_2N{=}C{=}O \longrightarrow R^1{-}\underset{\underset{O}{\parallel}}{C}{-}\underset{R^3}{\overset{R^2}{C}}{-}\underset{\underset{O}{\parallel}}{C}{-}\overset{H}{N}{-}SO_2Cl \longrightarrow R^1{-}\underset{\underset{O}{\parallel}}{C}{-}\underset{R^3}{\overset{R^2}{C}}{-}C{\equiv}N$$

(102) → 60-90% (103) (7.50)

$$Me_2CH{=}CH_2 + ClSO_2N{=}C{=}O \longrightarrow CH_2{=}\overset{CH_3}{C}{-}CH_2{-}\overset{O}{\overset{\parallel}{C}}{-}\underset{H}{N}{-}SO_2Cl$$

(104) 30 % + 4,4-dimethyl-1-chlorosulphonylazetidin-2-one (Me, Me, $ClSO_2$) 70 % —NaOH→ 52 % overall (105) (7.51)

sodium sulphite and ether (Scheme 7.52).[64] Allenes[65] and dienes[66] react similarly.

$$R^1{-}\underset{\underset{O}{\parallel}}{C}{-}O{-}CH{=}C(R^2)(R^3) \xrightarrow[\text{ii, } Na_2SO_3/NaHCO_3]{\text{i, } ClSO_2N{=}C{=}O} R^1{-}\underset{\underset{O}{\parallel}}{C}O\text{-azetidinone (H, } R^2, R^3\text{, N–H)}$$

35 -65 % (7.52)

7.7.5 *With alkynes*

Acetylenes (**106**) react with chlorosulphonyl isocyanate eventually to yield ketones (**107**) (Scheme 7.53).[67]

$$\text{Et—C≡C—Et (106)} + \text{ClSO}_2\text{N=C=O} \longrightarrow [\text{cyclic adduct}] \xrightarrow{H_2O} \text{Et—C(=O)—CH}_2\text{CH}_2\text{CH}_3 \ (107) \quad (7.53)$$

7.7.6 *For dehydration of amides*

For the use of chlorosulphonyl isocyanate for the dehydration of amides to yield nitriles without racemization of an α-carbon atom carrying a hydrogen atom, see Chapter 4, Table 4.1.[68]

THIOCYANATES

The molecular dimensions of the thiocyanate group are shown below.

C—S 1.81 Å; S—C 1.61 Å; C≡N 1.17 Å; C—S—C angle 142°

7.8 Preparation

Most synthetic routes to thiocyanates involve nucleophilic attack of the thiocyanate anion at a carbon bonded to an electron-withdrawing group, in a straightforward nucleophilic substitution reaction.

7.8.1 *From alkyl halides*

Thiocyanates (**108**) result from the reaction between an alkyl halide (**109**) and an alkali metal thiocyanate (**110**) (Scheme 7.54).[69] In this reaction tertiary alkyl halides and aryl methyl halides give either a mixture of thiocyanate and isothiocyanate (R—N=C=S) or exclusively the iso compound.

$$\underset{(109)}{\text{R—Hal}} + \underset{(110)}{{}^{-}\text{S—C≡N}} \xrightarrow[\text{acetone}]{\text{EtOH or}} \underset{(108)\ 70\text{ - }90\ \%\ \text{yields}}{\text{R—S—C≡N}} + \text{Hal}^{-} \quad (7.54)$$

In a specific version of this reaction, thiocyanates have been formed from the reaction of tertiary alkyl bromides with potassium thiocyanate (Scheme 7.55).[70]

$$Bu^t{-}Br + KSCN \xrightarrow{EtOH} Bu^t{-}S{-}C{\equiv}N \qquad (7.55)$$

7.8.2 *From sulphonates*

Alkyl sulphonates (**111**)[71] and *p*-toluenesulphonates (**112**)[72] react similarly by substitution with thiocyanate ions (Scheme 7.56).

$$\underset{(111)}{(CH_3)_2SO_4} + {}^{-}S{=}C{=}N \longrightarrow CH_3{-}S{-}C{\equiv}N + CH_3SO_4^{-}$$

$$\underset{(112)}{Ts{-}O{-}R} + {}^{-}S{=}C{=}N \longrightarrow R{-}S{-}C{\equiv}N + TsO^{-} \qquad (7.56)$$

7.8.3 *From sulphenyl chlorides*

A good synthesis of thiocyanates results from the reaction of cyanotrimethylsilane (see Chapter 4, p. 245) with a sulphenyl chloride (**113**). Sulphenyl chlorides are prepared from the reaction of sulphuryl chloride with a thiol (**114**) or disulphide (**115**) (Scheme 7.57).[73]

$$\underset{(114)}{R{-}SH} \text{ or } \underset{(115)}{R{-}S{-}S{-}R} \xrightarrow[CCl_4,\ NEt_3,\ 30min]{SO_2Cl_2,\ 0^{\circ}C} \underset{(113)}{R{-}SCl} \xrightarrow[1\ hour]{SiMe_3CN,\ acetonitrile} R{-}S{-}C{\equiv}N + Me_3SiCl$$

R	Yield of thiocyanate from sulphenylchloride (%)
$-CH_2-CH_2-$	85
Bu^n	76
Bn	86

(7.57)

7.8.4 *From amines*

Thiocyanates can be prepared from amines (**116**) using the thiocyanating reagent (**117**) shown in Scheme 7.58.[74]

R—NH$_2$ (116) + 2,4,6-triphenylpyrylium SCN^- (117) → 1-R-2,4,6-triphenylpyridinium SCN^- →(Benzene or ethanol, heat) R—S—C≡N + 2,4,6-triphenylpyridine (7.58)

7.8.5 *From β-propiolactones*

Sodium or potassium thiocyanates react with β-propiolactones (**118**) to form thiocyanates,[75] for example, Scheme 7.59.

$$\text{(118)} + NaSCN_{(aq)} \xrightarrow{H^+} NCS{-}CH_2{-}CH_2{-}CO_2H \quad (7.59)$$

7.8.6 *From epoxides*

Hydrogen thiocyanide (**119**) reacts with epoxides (**120**), opening the epoxide to form thiocyanates (Scheme 7.60).[76]

$$\underset{(120)}{H_2C{-}CH_2\ (\text{-O-})} + \underset{(119)}{HSCN} \longrightarrow HOCH_2{-}CH_2{-}SCN \quad (7.60)$$

7.8.7 *From halogeno-orthocarbonic esters*

A mixture of thiocyanates (**121**) and isothiocyanates (**122**) can be prepared from halogeno-orthoformic esters (**123**) by reaction with potassium or ammonium thiocyanate (**124**) (Scheme 7.61). The thiocyanate isomerizes to the isothiocyanate on treatment with Zn catalysts at 160 °C.

$$RO{-}C({=}O){-}O{-}CH(Cl){-}CH_3\ (123) + NH_4^+SCN^-\ (124) \xrightarrow[Bu^n_4P^+Br^-]{Me_2CO\ reflux} RO{-}C({=}O){-}CH(S{-}C{\equiv}N){-}CH_3 + RO{-}C({=}O){-}CH(N{=}C{=}S){-}CH_3\ (122)$$

Decalin / $ZnCl_2$ 160°C

Isomerization

(7.61)

7.9 Reactivity

The nitrile group in a thiocyanate undergoes many of the reactions exhibited by ordinary nitriles.

7.9.1 *Hydration*

The similarity of reactions between thiocyanates and nitriles (see page 309) is demonstrated by the hydration of thiocyanates to amidines (**125**),[77] for example, Scheme 7.62.

$$Ph{-}S{-}C{\equiv}N \xrightarrow[ii,\ H_2O]{i,\ conc.\ H_2SO_4} Ph{-}S{-}C({=}O){-}NH_2$$

87%

(125)

(7.62)

7.9.2 *With alcohols*

A similar reaction can occur with alkyl thiocyanates (**126**) on addition of an alcohol (**127**) (Scheme 7.63).[78]

$$R^1{-}S{-}C{\equiv}N \;(126) + R^2OH \;(127) \xrightarrow[\text{ii, } H_2O]{\text{i, } H_2SO_4} R^1{-}S{-}C(=O){-}NHR^2$$

R^1	R^2	Yield (%)
Et	Pr^i	25
Me	Bu^t	25
Me	Cyclohexyl	45
Et	Cyclohexyl	60
Pr^n	Cyclohexyl	67
Bu^n	Cyclohexyl	66

(7.63)

Thiocyanates have been used in a reaction with alcohols and a base to produce compounds which have found use in synthesis (Scheme 7.64).[79]

i, NaH
ii, R-N=C=S
iii, CH_3I
100%
NaH
O^- R—N=C=S
S^- CH_3—I
O—C=NR
SMe
O—C=NR

(7.64)

7.9.3 *Reduction*

Thiocyanates are reduced to thiols (**128**) by the action of lithium aluminium hydride (Scheme 7.65).[80]

$$R{-}S{-}C{\equiv}N \xrightarrow[\text{ether}]{LiAlH_4} R{-}SH \;(128)$$

(7.65)

7.9.4 *Isomerism*

Due to their relative instability, thiocyanates (**129**) isomerize to isothiocyanates (**130**) on heating (Scheme 7.66).[81]

$$R{-}S{-}C{\equiv}N \xrightarrow[\text{1 hour}]{40^{\circ}C} S{=}C{=}N{-}R \qquad (7.66)$$

(129) (130)

Certain allylic thiocyanates (**131**) have been shown to perform the isomerization to isothiocyanates (**132**) via a six-membered cyclic transition state,[82] for example, Scheme 7.67.

(131) → (132) (7.67)

ISOTHIOCYANATES

The molecular dimensions of the isothiocyanate group are shown below.

C—N 1.28 Å, N=C 1.33 Å, C=S 1.56 Å

7.10 Preparation

7.10.1 *From amines*

Primary aliphatic amines (**133**) react with carbon disulphide and dicyclohexylcarbodiimide (**134**) to give isothiocyanates (**135**) (Scheme 7.68).[83]

Isothiocyanates can be formed from primary amines with the reagent (**136**) formed from thiophosgene (**137**) (Scheme 7.69).[84]

$$R{-}NH_2 \;(133) + CS_2 + C_6H_{11}{-}N{=}C{=}N{-}C_6H_{11} \;(134) \xrightarrow{\text{ether, } 0^\circ C} R{-}N{=}C{=}S \;(135) + C_6H_{11}{-}NH{-}C(=S){-}NH{-}C_6H_{11}$$

R	Yield (%)
$(EtO)_2CH_2CH_2$	87
Ph_2CH	81
Bn	90

(7.68)

2 (2-hydroxypyridine) + $S{=}CCl_2$ (137) → di(2-pyridyl) thionocarbonate, 85 % (136)

$$R{-}NH_2 \xrightarrow[CH_2Cl_2]{(136)} R{-}N{=}C{=}S + 2\ \text{(2-hydroxypyridine)}$$

R	Yield (%)
Bu^i	85
Bu^n	94
Cyclohexyl	85
$CH_2{=}CHCH_2$	85
Bn	90

(7.69)

7.10.2 *From alkenes*

Isothiocyanates are prepared by addition of hydrogen thiocyanate across a double bond,[85] for example, Scheme 7.70. However, if substituents attached to the double bond are electron withdrawing then the thiocyanate will be formed, for example, Scheme 7.71.

$$\mathrm{Bu^t(H)C{=}C(CH_3)_2} \xrightarrow[45^\circ C]{H_2SO_4\ /\ NaSCN} \mathrm{Bu^t{-}CH_2{-}C(CH_3)_2{-}NCS} \qquad (7.70)$$

$$\mathrm{CH_2{=}CHCO_2Et} \xrightarrow{H_2SO_4,\ NaSCN} \mathrm{N{\equiv}C{-}S{-}CH_2{\cdot}CH_2{-}CO_2Et} \qquad (7.71)$$

7.10.3 *From glucosides*

Isocyanates occur naturally as glucosides in mustard oil, e.g. (**138**). These naturally occurring glucosides are enzymically cleaved to form the isothiocyanates,[86] for example, Scheme 7.72.

$$\underset{(138)}{\text{Glucosyl}{-}\mathrm{S{-}C({=}N{-}OSO_3^-K^+){-}CH_2{-}CH{=}CH_2}} \xrightarrow{\text{enzyme, } H_2O} \mathrm{CH_2{=}CH{-}CH_2{-}N{=}C{=}S} + \mathrm{KHSO_4} + \text{Glucose} \qquad (7.72)$$

7.11 Reactivity

7.11.1 *Reduction*

Isothiocyanates are reduced to secondary amines (**139**) with lithium aluminium hydride (Scheme 7.73).[87]

$$\mathrm{R{-}N{=}C{=}S} \xrightarrow{LiAlH_4} \underset{(139)}{\mathrm{R{-}NH{-}CH_3}} \qquad (7.73)$$

7.11.2 *With alcohols*

Isothiocyanates have been used in a reaction with alcohols and a base to produce quantitative yields of compounds (**140**) which are of use in the synthesis of antibodies,[88] for example, Scheme 7.74.

OH
i, NaH
ii, CH_3-N=C=S
iii, CH_3I
SMe
O—C=NMe
(140)

(7.74)

References

1. J. C. Kauer and W. W. Henderson, *J. Am. Chem. Soc.*, 1964, **86**, 4732.
2. A. Holm and C. Wentrup, *Acta Chem. Scand.*, 1966, **20**, 2123.
3. S. Ozaki, *Chem. Rev.*, 1972, **72**, 457.
4. H. J. Twitchett, *Chem. Soc. Rev.*, 1974, **3**, 209.
5. K. H. Slotta and L. Lorenz, *Ber.*, 1925, **58**, 1320.
6. W. Siefken, *Annalen*, 1949, **562**, 75.
7. E. W. Stern and M. L. Spector, *J. Org. Chem.*, 1966, **31**, 596.
8. *F. &F.*, **1**, 308; H. W. Johnson, Jr and P. H. Daughhetee, Jr, *J. Org. Chem.*, 1964, **29**, 246; H. W. Johnson, Jr and H. Krutzsch, *J. Org. Chem.*, 1967, **32**, 1939.
9. *F. &F.*, **3**, 121; D. Martin and A. Weise, *Angew. Chem., Int. Ed. Engl.*, 1967, **6**, 168.
10. P. V. Finzi and M. Arbasino, *Tetrahedron Lett.*, 1965, 4645.
11. *F. &F.*, **4**, 110; R. Fuks and M. Hartemink, *Tetrahedron*, 1973, **29**, 297.
12. *F. &F.*, **2**, 438; G. Greber and H. R. Kricheldorf, *Angew. Chem., Int. Ed. Engl.*, 1968, **7**, 941.
13. *F. &F.*, **8**, 91; T. Mukaijama and Y. Edigo, *Chem. Lett.*, 1977, 383.
14. *F. &F.*, **8**, 520; J. B. Bapat, R. J. Blade, A. J. Boulton, J. Epsztajn, A. R. Katritzky, J. Lewis, P. Molina-Buendia, P. L. Nie, and C. A. Ramsden, *Tetrahedron Lett.*, 1976, 2691.
15. C. Grundmann, P. Kochs, and J. R. Boal, *Annalen*, 1972, **762**, 162.
16. *F. &F.*, **8**, 464; G. Trickes and H. Meier, *Angew. Chem., Int. Ed. Engl.*, 1977, **16**, 555.
17. Reviews: P. A. S. Smith, *Org. React.*, 1946, **3**, 337; D. V. Banthorpe, in *The chemistry of the azido group* (ed. S. Patai) Interscience, New York, 1971, p. 397.
18. C. F. H. Allen and A. Bell, *Org. Synth. Coll. Vol. 3*, 1955, 846; *F. &F.*, **1**, 1041.
19. *F. &F.*, **5**, 330; J. W. van Reijendam and F. Baardmann, *Synthesis*, 1973, 413.
20. *F. &F.*, **6**, 564; A. Brändström, B. Lamm, and I. Palmertz, *Acta Chem. Scand.*, 1974, **28B**, 699.
21. *F. &F.*, **5**, 719; S. S. Washburne and W. R. Petersen, Jr, *J. Organomet. Chem.*, 1971, **33**, 153; S. S. Washburne and W. R. Petersen, Jr, *Synth. Commun.*, 1972, **2**, 227; H. R. Kricheldorf, *Synthesis*, 1972, 551; J. H. MacMillan and S. S. Washburne, *J. Org. Chem.*, 1973, **38**, 2982.
22. *F. &F.*, **7**, 377; H. R. Kricheldorf and E. Leppert, *Synthesis*, 1976, 329.
23. J. B. Aylward, *Quart. Rev. Chem. Soc.*, 1971, **25**, 407; H. E. Baumgarten, H. L. Smith, and A. Staklis, *J. Org. Chem.*, 1975, **40**, 3554.
24. F. W. Hoover and H. S. Rothrock, *J. Org. Chem.*, 1963, **23**, 2082; J. L. McClanahan and J. I. Harper, *Chem. &Ind.*, 1963, 1280; *F. &F.*, **1**, 172.

25. F. W. Hoover and H. S. Rothrock, *J. Org. Chem.*, 1964, **29**, 143.
26. *F. &F.*, **1**, 171; F. W. Hoover, H. B. Stevenson, and H. S. Rothrock, *J. Org. Chem.*, 1963, **28**, 1825.
27. *F. &F.*, **1**, 501; **3**, 161; A. Hassner and C. Heathcock, *Tetrahedron*, 1964, **20**, 1037; A. Hassner and C. Heathcock, *J. Org. Chem.*, 1965, **30**, 1748; A. Hassner, M. E. Lorber and C. Heathcock, *J. Org. Chem.*, 1967, **32**, 540; A. Hassner, R. P. Hoblitt, C. Heathcock, J. E. Kropp, and M. E. Lorber, *J. Am. Chem. Soc.*, 1970, **92**, 1326; C. G. Gebelein, *Chem. &Ind.*, 1970, 57; R. M. Carlson and S. Y. Lee, *Tetrahedron Lett.*, 1969, 4001; L. A. Paquette, D. E. Kuhla, J. H. Barrett, and R. J. Haluska, *J. Org. Chem.*, 1969, **34**, 2866.
28. C. G. Grebelein and D. Swern, *J. Org. Chem.*, 1968, **33**, 2758; *F. &F.*, **2**, 223; T. Giebrokk, *Acta Chem. Scand.*, 1973, **27**, 3368.
29. R. G. Arnold, J. A. Nelson, and J. J. Verbarc, *Chem. Rev.*, 1957, **57**, 47.
30. A. G. Davies and R. J. Puddephatt, *J. Chem. Soc. (C)*, 1967, 2663; 1968, 1479.
31. *F. &F.*, **13**, 46; T. Ibaka, G.-N. Chu, T. Aoyagi, K. Kitada, T. Tsukida, and F. Yoneda, *Chem. Pharm. Bull.*, 1985, **33**, 451.
32. W. H. Pirkle and R. W. Anderson, *J. Org. Chem.*, 1974, **39**, 3901; W. H. Pirkle and M. S. Hockstra, *J. Org. Chem.*, 1974, **39**, 3904; *F. &F.*, **6**, 416.
33. Preparation: A. J. Speziale and L. R. Smith, *J. Org. Chem.*, 1962, **27**, 3742. Use: I. R. Trehan, C. Mondo, and A. K. Bose, *Tetrahedron Lett.*, 1968, 67.
34. *F. &F.*, **15**, 256; B. Schönenberger and A. Brossi, *Helv. Chim. Acta*, 1986, **69**, 1486; L. A. Chrisey and A. Brossi, *Org. Synth.*, 1989; *F. &F.*, **6**, 416.
35. *F. &F.*, **11**, 277; I. Benecke and W. A. König, *Angew. Chem., Int. Ed. Engl.*, 1982, **21**, 609.
36. *F. &F.*, **15**, 196; L. C. Lasley and B. B. Wright, *Synth. Commun.*, 1989, **19**, 59.
37. N. A. LeBel, R. M. Cherluck, and E. A. Curtis, *Synthesis*, 1973, 678.
38. *F. &F.*, **1**, 1197; H. Reinheckel and D. Jahnke, *Ber.*, 1964, **97**, 2661.
39. J.-F. Normant, G. Cahiez, C. Chuit, and J. Villieras, *J. Organomet. Chem.*, 1973, **54**, C53; *Tetrahedron Lett.*, 1973, 2407.
40. R. J. Kennedy and A. M. Stock, *J. Org. Chem.*, 1960, **25**, 1901.
41. *F. &F.*, **15**, 144; P. E. Eaton and G. E. Wicks, *J. Org. Chem.*, 1988, **53**, 5353.
42. A. E. Finholt, C. D. Anderson, and C. L. Agre, *J. Org. Chem.*, 1953, **18**, 1338; H. C. Brown, P. M. Weissman, and N. M. Yoon, *J. Am. Chem. Soc.*, 1966, **88**, 1458.
43. *F. &F.*, **2**, 323; T. Mukaiyama and Y. Yokota, *Bull. Chem. Soc. Japan*, 1965, **38**, 858.
44. *F. &F.*, **11**, 553; J. E. Baldwin, J. C. Bottaro, and A. E. Derome, *J. Chem. Soc., Chem. Commun.*, 1982, 942.
45. P. Frøyen, *Acta Chem. Scand.*, 1974, **B28**, 586.
46. *Org. Synth., Collected Vol.*, **III**, 438; **IV**, 903; **V**, 437.
47. *F. &F.*, **1**, 679; T. W. Campbell, J. J. Monagle, and V. S. Foldi, *J. Am. Chem. Soc.*, 1962, **84**, 3673.
48. J. J. Monagle, T. W. Campbell, and H. F. McShane, *J. Am. Chem. Soc.*, 1962, 4288; J. J. Monagle and J. V. Mengenhauser, *J. Org. Chem.*, 1966, **31**, 2321.
49. H. Ulrich, *Cycloaddition reactions of heterocumulenes*, Academic Press, New York, 1967, pp. 122–6; H. Ulrich, *Acc. Chem. Res.*, 1969, **2**, 186.
50. *F. &F.*, **6**, 69; B. S. Ohta and S. Kimoto, *Tetrahedron Lett.*, 1975, 2279.
51. *F. &F.*, **12**, 162; R. A. Earl and K. P. C. Volhardt, *J. Am. Chem. Soc.*, 1983, **105**, 6991.
52. A. J. Speziale and L. R. Smith, *J. Org. Chem.*, 1962, **27**, 4361; 1963, **28**, 1805. Houben–Weyl, *Methoden der organischen Chemie*, Georg Thieme Verlag, Stuttgart,

Band 4E, p. 803; *Chem. Abstr.*, 1970, **72**, 120628u; M. O. Lozinskii and P. S. Pel'kis, *Russ. Chem. Rev.*, 1968, **37**, 363; K. A. Nuridzhanyan, *Russ. Chem. Rev.*, 1970, **39**, 130.

53. *F. &F.*, **15**, 313; M.-Z. Deng, P. Caubere, J. P. Senet, and S. Lecolier, *Tetrahedron*, 1988, **44**, 6079.
54. *F. &F.*, **1**, 172; P. R. Steyermark, *J. Org. Chem.*, 1963, **28**, 586.
55. A. J. Speziale and L. R. Smith, *J. Org. Chem.*, 1962, **27**, 3742; A. J. Speziale, L. R. Smith, and J. E. Fedder, *J. Org. Chem.*, 1965, **30**, 4306; J. Goerdeler and H. Schenk, *Angew. Chem., Int. Ed. Engl.*, 1963, **2**, 552; *Ber.*, 1965, **98**, 2954.
56. Reviews: R. Graf, *Angew. Chem., Int. Ed. Engl.*, 1968, **7**, 172; J. K. Rasmussen and A. Hassner, *Chem. Rev.*, 1976, **76**, 389; D. N. Dhar and K. S. K. Murthy, *Synthesis*, 1986, 437.
57. *F. &F.*, **1**, 117; R. Graf, *Ber.*, 1956, **89**, 1071; *Org. Synth.*, 1966, **46**, 23.
58. *F. &F.*, **8**, 106; G. Mehta, D. J. Dhar, and S. C. Sur, *Synthesis*, 1978, 574.
59. G. Lohaus, *Ber.*, 1967, **100**, 2719; *Org. Synth.*, 1970, **50**, 52; *F. &F.*, **2**, 70.
60. G. Lohaus, *Org. Synth.*, 1970, **50**, 18.
61. *F. &F.*, **5**, 136; J. K. Rasmussen and A. Hassner, *Synthesis*, 1973, 682.
62. R. Graf, *Org. Synth.*, 1966, **46**, 51.
63. *F. &F.*, **6**, 122; K. Claus, D. Grimm, and G. Prossel, *Annalen*, 1974, 539.
64. T. Durst and M. J. O'Sullivan, *J. Org. Chem.*, 1970, **35**, 2043.
65. J. D. B. Rao, R. Y. Chandrasekaran, E. Haley, P. de Meester and S. C. Chu, *Tetrahedron Lett.*, 1985, **26**, 5001.
66. *F. &F.*, **10**, 94; F. W. Hauser and R. P. Rhee, *J. Org. Chem.*, 1981, **46**, 227.
67. E. J. Moriconi and J. F. Kelly, *J. Am. Chem. Soc.*, 1966, **88**, 3657; *F. &F.*, **12**, 122; J. D. Buynak, H. Pajonhesh, D. Z. Liveley, and Y. R. Ramalakshmi, *J. Chem. Soc., Chem. Commun.*, 1984, 948; E. J. Moriconi, J. C. White, R. W. Franck, J. Jansing, J. F. Kelly, R. A. Salamone, and Y. Shimakawa, *Tetrahedron Lett.*, 1970, 27.
68. *F. &F.*, **11**, 125; G. Botteghi, G. Chelucci, and M. Marchetti, *Synth. Commun.*, 1982, **12**, 125.
69. R. L. Shriner, *Org. Synth. Coll. Vol. 2*, 1943, 366.
70. H. L. Wheler and T. B. Johnson, *J. Am. Chem. Soc.*, 1902, **24**, 680.
71. P. Walden, *Ber.*, 1907, **40**, 3214.
72. W. F. H. Jackman and J. Kenyon, *J. Am. Chem. Soc.*, 1937, **59**, 2473.
73. D. N. Harpp, B. T. Friedlander, and R. A. Smith, *Synthesis*, 1979, 181.
74. A. R. Katrintilay, U. Gruntz, N. Mongelli, and M. C. Rezeda, *J. Chem. Soc., Chem. Commun.*, 1978, 133.
75. M. R. Frederick, F. T. Friedorek, B. A. Bamkert, J. T. Gregory, and W. L. Beears, *J. Am. Chem. Soc.*, 1952, **74**, 1323.
76. T. Wagner-Jauregg, *Annalen*, 1949, **561**, 87.
77. R. Riernschmeider, W. Wojan, and G. Orlich, *J. Am. Chem. Soc.*, 1951, **73**, 5903.
78. R. Reinschneider, *J. Am. Chem. Soc.*, 1956, **78**, 844; A. Knorr, *Ber.*, 1916, **49**, 1735.
79. *F. &F.*, **12**, 78; S. Knapp and D. V. Patel, *J. Am. Chem. Soc.*, 1983, **105**, 6985.
80. G. W. Glazebrook and R. W. Saville, *J. Chem. Soc., Chem. Commun.*, 1954, 2094.
81. R. G. R. Brown, *Organic sulphur compounds*, (ed. M. S. Kharasch), Pergamon Press, Oxford, 1961, Vol. 1, p. 312.
82. O. Mumm and H. Richter, *Ber.*, 1940, **73**, 843.
83. *F. &F.*, **3**, 92; J. C. Jochims and A. Seelinger, *Angew, Chem., Int. Ed. Engl.*, 1967, **6**, 1974.
84. S. Kim and K. Y. Yi, *Tetrahedron Lett.*, 1985, **26**, 1661.

85. L. S. Lushin, G. E. Gaitot, and W. E. Craig, *J. Am. Chem. Soc.*, 1956, **78**, 4965; M. S. Kharasch, E. M. May, and F. R. Mayo, *J. Am. Chem. Soc.*, 1937, **59**, 1580.
86. M. G. Elthinger and A. J. Lundeen, *J. Am. Chem. Soc.*, 1956, **78**, 4172; 1957, **79**, 1764.
87. R. G. R. Brown, *J. Am. Chem. Soc.*, 1957, **79**, 1764.
88. *F. &F.*, **12**, 78; S. Knapp and D. V. Patel, *J. Am. Chem. Soc.*, 1983, **105**, 6985.

8

AMIDES, THIOAMIDES, AND CARBAMATES (URETHANES)

AMIDES[1]

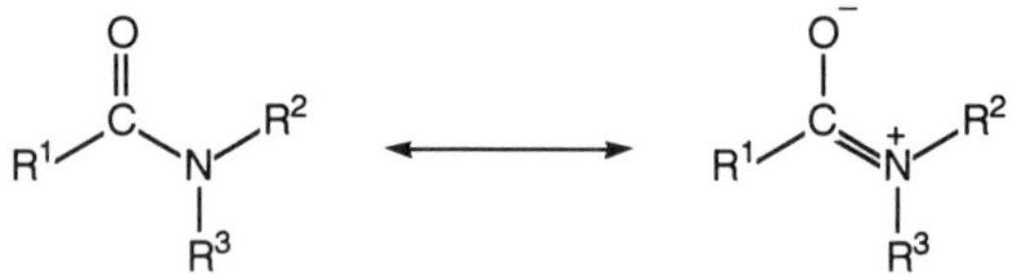

The bond lengths and bond angles of amides (**1**) are as follows.

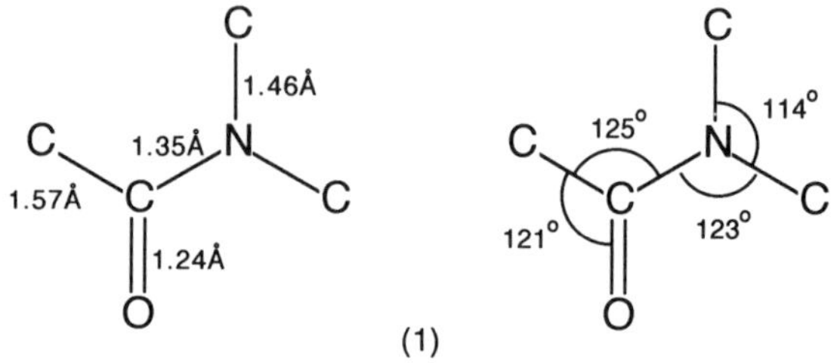

(1)

These values should be compared with those for the C=N bond of an imine (**2**) and those for a keto carbonyl group (**3**).

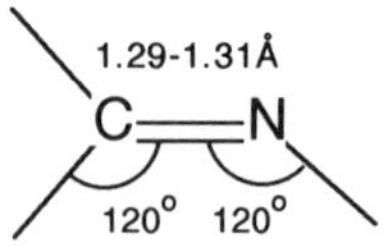

(2) 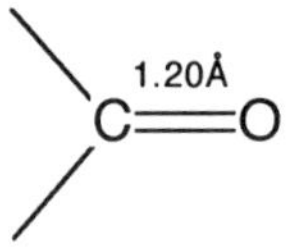 (3)

The infra-red absorptions of the carbonyl groups (>C=O) and of the amino groups (>N—H) in primary and secondary amines vary with structure and the extent of hydrogen bonding (for a discussion of the details see Ref. 2). The carbonyl group absorptions in solution are as follows:

primary amides: *c.* 1690 cm^{-1};
secondary amides: 1670–1700 cm^{-1};
tertiary amides: 1630–1670 cm^{-1}.

In small cyclic structures these values change to higher frequencies, e.g. 1730–1760 cm^{-1} for β-lactams (four-membered rings, see Chapter 22) in solution.[2] The >N—H stretching frequencies in primary and secondary amides occur between 3100 and 3400 cm^{-1} depending upon the extent of hydrogen bonding. For example, acetamide, CH_3CONH_2, (KCl disc) shows two absorptions, one at 3190 cm^{-1} and another at 3330 cm^{-1}.

There have been many studies of the NMR spectra of amides (**4**).[3] These studies have shown that for a secondary/tertiary amide (**4**) two rotational isomers (**5a**) and (**5b**) exist in solution with a rotational energy barrier of 7–18 kcal/mole for most R substituents (Scheme 8.1). Assignments of configurations to the isomers (**5a**) and (**5b**) were made from long-range couplings and from solvent effects.

$$CH_3-C(=O)-N(CH_3)R$$

(4)

$$\underset{(5a)}{CH_3(^-O)C{=}\overset{+}{N}(CH_3)R} \rightleftharpoons \underset{(5b)}{CH_3(^-O)C{=}\overset{+}{N}(R)CH_3} \quad (8.1)$$

8.1 Preparation

8.1.1 *From carboxylic acid salts*

Primary amides (**6**) can be obtained by the dehydration of ammonium salts of carboxylic acids (**7**) (Scheme 8.2).

Carboxylic salts of amines (**8**) also give substituted amides (**9**) on dehydration (Scheme 8.3).

$$\underset{(7)}{R\text{-}CO_2^-\ NH_4^+} \longrightarrow \underset{(6)}{R-C(=O)-NH_2} + H_2O \quad (8.2)$$

$$\underset{(8)}{R^1\text{-}CO_2^-\ R^2\text{-}NH_3^+} \longrightarrow \underset{(9)}{R^1-C(=O)-NHR^2} + H_2O \quad (8.3)$$

8.1.2 *From carboxylic acids and amines*

Many methods have been developed for the preparation of amides (**10**) from amines (**11**) and carboxylic acids (**12**) without preforming an ammonium salt (Scheme 8.4) and several examples are collected in Table 8.1.

$$\underset{(12)}{R^1\text{-}CO_2H} + \underset{(11)}{HNR^2R^3} \xrightarrow{\text{Reagent}} \underset{(10)}{R^1\text{—}C(=O)\text{—}NR^2R^3} + H_2O \qquad (8.4)$$

Table 8.1 Preparation of amides from carboxylic acids and amines

Reagent	Conditions	Nature of Acid	Nature of Amine	Peptide synthesis	Yield (%)	Ref.
$SiCl_4$	Pyridine, 25 °C or reflux	Al., Ar	1^y, 2^y Al., Ar	Y	25–90	4[f], 5[g]
$BF_3.Et_2O$	NEt_3 or DBU in C_6H_6 or $CH_3C_6H_5$	Al., Ar	1^y, 2^y	N	28–100	6[h]
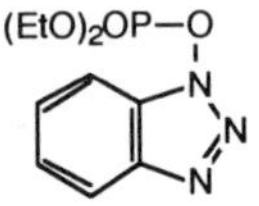 $(EtO)_2OP\text{—}O$-benzotriazole	NEt_3 in DMF	Peptide synthesis only		Y	91–96	7[i]
$PhPO(OAr)_2$[a] $Ar = p\text{-}NO_2C_6H_4$	$Bu^nN^+HSO_4^-$ in aq. KOH and CH_2Cl_2	Acid or ester Al., Ar	1^y, 2^y Al., Ar	Y	82–100	8[j], 9[k]
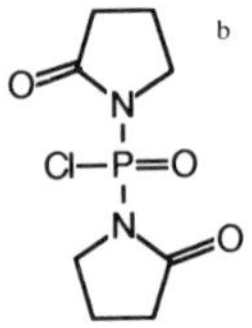 Cl—P(=O)(2-oxopyrrolidin-1-yl)$_2$ [b]	NEt_3 in CH_2Cl_2	Al., Ar	1^y	—	85–99	10[l]
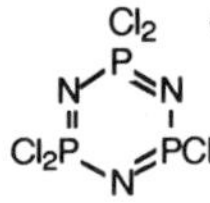 $(NPCl_2)_3$ [c]	RCO_2Na in THF, C_6H_6 or c-C_6H_{10}	Al., Ar	NH_3, 1^y, 2^y Al., Ar	—	63–83	11, 12[m]
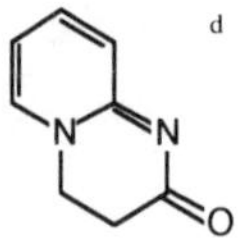 bicyclic pyrido-pyrimidinone with N and C=O groups [d]	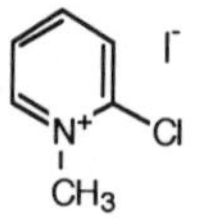 2-chloro-1-methylpyridinium iodide (CH_3, N^+, Cl, I^-) + Bu_3^nN	Al., Ar	1^y, 2^y Al., Ar	—	70–100	13, 14[n]
$Ph_3P(OCH_2CF_3)_2$[e]	−60 °C, 1 h	Al., Ar, Vinyl	1^y Al., Ar	—	75–85	15[o]

Reagent	Conditions	Nature of Acid	Nature of Amine	Peptide synthesis	Yield (%)	Ref.
$N_3PO(OPh)_2$	Mild	Peptide synthesis only		Y	69–90	16[p]
$Ph_3P{=}O$ (2 mole) + $(F_3C{-}SO_2)_2O$	25 °C	Al., Ar	1[y], 2[y] Al., Ar	—	71–95	17[q]
$TiCl_4$		Al., Ar	1[y], 2[y] Al., Ar	—	66–86	18[r]
Ph-O—P(=O)(Cl)—NH-Ph	Et_3N in CH_2Cl_2 25 °C	Ar	1[y] Al., Ar	—	20–98	19[s]
$BH_3.NMe_3$	Xylene, reflux	Ar	1[y], 2[y] Al., Ar	—	60–99	20[t]

Al. = aliphatic, Ar = aromatic, 1[y] = primary, 2[y] = secondary.

[a] peptide synthesis but with complete racemization.

[b] $H_2NCH_2CH_2CO_2H$ (β-alanine) ⟶ cyclic lactam; i) PCl_5, CH_3NO_2 ii) H_2O

[c] $PCl_5 + NH_4Cl \xrightarrow[\text{as solvent}]{ClCH_2CH_2Cl}$ $(NPCl_2)_3$

[d]

[e] $Ph_3PBr_2 + 2NaO\text{-}CH_2CF_3 \longrightarrow Ph_3P(O\text{-}CH_2CF_3)_2$

[f] e.g. $CH_3COOH + PhNH_2 \longrightarrow CH_3CONHPh$ 60%

Table 8.1 (cont.)

g	e.g. Z-Gly + Gly-OEt	$\xrightarrow[\text{1.25 hr}]{\text{60°C, pyridine}}$	Z-Gly-Gly-OEt 70%
h	e.g. Bu^nCOOH + $PhCH_2NH_2$	→	$Bu^nCONHCH_2Ph$ 100%
i	e.g. Z-Phe + Ser-OEt	→	Z-Phe-Ser-OEt 96%
j	e.g. Bu^tCOOH + $PhCH(Me)NH_2$	→	$Bu^tCONHCH(Me)Ph$ 90%
k	e.g. Z-Gly-L-Phe-OBzl + Gly-OEt	→	Z-Gly-L-Phe-Gly-OEt 73%
l	e.g. PhCOOH + 2-naphthylamine (NH_2)	→	2-naphthyl NH-COPh 90%
m	e.g. PhCOOH + $PhNH_2$	→	PhCONHPh 83%
n	e.g. PhCOOH + $Bu^n{}_2NH$	→	$PhCONBu^n{}_2$ 90%
o	e.g. CH_3COOH + $PhNH_2$	→	$CH_3CONHPh$ 81%
p	e.g. Boc-Trp + Gly-OEt	→	Boc-Trp-Gly-OEt 89.5%
q	e.g. PhCOOH + o-phenylenediamine (NH_2, NH_2)	→	2-phenylbenzimidazole (N, C—Ph, N-H) 85%
r	e.g. PhCOOH + morpholine (HN, O)	→	Ph—C(=O)—N(morpholine, O) 85%
s	e.g. Bu^tCOOH + $PhNH_2$	→	$Bu^tCONHPh$ 91%
t	e.g. CH_3CH_2COOH + Ph(Me)NH	→	$CH_3CH_2CON(Me)Ph$ 91%

A two step preparation from acids (**13**) and amines (**14**) gives good yields of amides, using compound (**15**) which may be synthesized *in situ* from the oxazolone (**16**) and the phosphorus compound (**17**) (Scheme 8.5).[21] This method has been applied to the synthesis of peptides.

Amides (**18**) have been prepared from amines (**19**) and acids or acid chlorides by the action of 1,3-thiazolidine-2-thione (**20**) (Scheme 8.6).[22]

The formation of cyclic amides (**21**) (γ- and δ-lactams) from ω-amino acids (**22**) is facilitated by titanium(IV) isopropoxide (Scheme 8.7).[23]

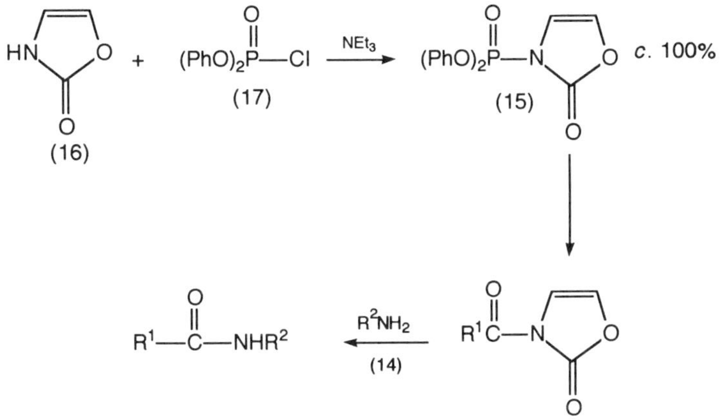

(8.5)

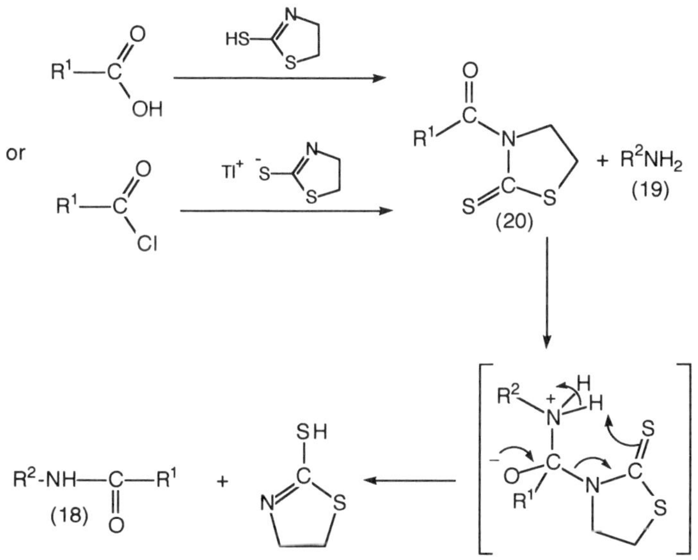

(8.6)

8.1.3 *From carboxylic esters and amines*

Esters (**23**) react with ammonia or amines (**24**) to yield amides (**25**) (Scheme 8.8).

Various modifications to the direct reaction have been made to enhance the reactivity of the amine towards the ester. The Grignard reagent of the amine (**26**) (prepared *in situ* from the amine and methyl magnesium iodide) can be used (Scheme 8.9).[24] To obtain better yields, the aluminium derivative of the amine (**27**) has been used (Scheme 8.10). This compound is also prepared *in*

$$\underset{(22)}{RNH(CH_2)_nCHR^1COOH} \xrightarrow[\text{in } ClCH_2CH_2Cl]{25\%\ Ti(OPr^i)_4} \underset{(21)}{\text{cyclic lactam: } (CH_2)_n\text{–CH}(R^1)\text{–C(=O)–N(R)}}$$

R	R^1	n	Yield (%)
H	H	2	93
H	CH_3	2	81
CH_3	H	2	85
H	H	3	75
CH_3	H	3	86
H	H	4	35

(8.7)

$$\underset{(23)}{R^1\text{—}\overset{O}{\overset{\|}{C}}\text{—OEt}} + \underset{(24)}{R\text{-}\ddot{N}H_2} \longrightarrow \underset{(25)}{R^1\text{—}\overset{O}{\overset{\|}{C}}\text{—NHR}^2} + EtOH \qquad (8.8)$$

$$R^2\text{-}NH_2 \xrightarrow{CH_3MgI} CH_4 + \underset{(26)}{R^2\text{-}NHMgI} \xrightarrow{R^1\text{-}CO_2Et} R^1\text{—}\overset{O}{\overset{\|}{C}}\text{—NHR}^2 \qquad (8.9)$$

situ, as a product of the reaction between aluminium chloride and an ammonium hydrochloride.[25]

Unreactive esters (**28**) can be readily converted to the amide by condensation with the alkali-metal derivative of an arylamine (**29**) (Scheme 8.11).[26]

Sodium hydride in dimethylsulphoxide has been used to prepare amides (**30**) from primary amines (**31**) and esters (**32**) (Scheme 8.12).[27]

Esters (**33**) react with primary or secondary amines (**34**) in the presence of boron tribromide to produce amides (**35**) (Scheme 8.13).[28]

$$(CH_3)_3Al + R^2R^3\overset{+}{N}H_2\ Cl^- \xrightarrow{\text{In toluene}} CH_3\text{—}Al(Cl)\text{—}NR^2R^3\ (27) \xrightarrow{R^1\text{-}CO_2Et} R^1\text{—}C(=O)\text{—}NR^2R^3$$

Ester	R^2	R^3	Yield (%)
$C_6H_5CO_2Me$	H	H	83
	H	CH_3	91
	CH_3	CH_3	100
$C_6H_{11}CO_2Me$ (cyclohexyl)	H	H	93
	H	CH_3	81
	CH_3	CH_3	76
$Cl(CH_2)_4CO_2Et$	H	H	61
	H	CH_3	75
	CH_3	CH_3	64

(8.10)

$$H_2C(NEt_2)\text{—}CO_2Et\ (28) + ArNH.Li\ (29) \longrightarrow H_2C(NEt_2)\text{—}CONHAr$$

Ar	Yield (%)
$C_6H_5NH_2$	
C_6H_5NHEt	65–90
2,6-$(CH_3)_2C_6H_3NH_2$	

(8.11)

$$\underset{(31)}{\text{R-NH}_2} + \underset{(32)}{\text{R}^1\text{CO}_2\text{R}^2} \xrightarrow[\text{12 hours, 25°C}]{\text{NaH / DMSO}} \underset{(30)}{\text{R}^1\text{—C(=O)—NHR}^2} \quad (8.12)$$

R	Ester	Yield (%)
Ph	CH_3CO_2Et	85
Ph	*p*-$CH_3OC_6H_4COOEt$	92
c-C_6H_{11}	*p*-$CH_3OC_6H_4COOEt$	82

$$\underset{(33)}{\text{RCOOR}^1} + \underset{(34)}{\text{HNR}^2\text{R}^3} \xrightarrow{\text{BBr}_3} \underset{(35)}{\text{RCONR}^2\text{R}^3} \quad (8.13)$$

R	R^1	R^2	R^3	Yield (%)
Ph	Me	Ph	H	81
Me	Bu	Ph	Me	35
$PhCH_2$	$PhCH_2$	Ph	H	74
$PhCH_2$	$PhCH_2$	CH_2COOH	H	62

8.1.4 *From acyl chlorides and amines*

Acyl halides (**36**) readily react with amines (**37**) to yield amides (**38**) (Scheme 8.14). Often the free carboxylic acid (**39**) is boiled under reflux with thionyl chloride ($SOCl_2$), the excess of which is removed by distillation before addition of the amine (Scheme 8.15).

$$\underset{(36)}{\text{R}^1\text{COCl}} + \underset{(37)}{2\text{HNR}^2\text{R}^3} \longrightarrow \underset{(38)}{\text{R}^1\text{—C(=O)—NR}^2\text{R}^3} + \text{R}^2\text{R}^3\overset{+}{\text{N}}\text{H}_2\,\text{Cl}^- \quad (8.14)$$

$$\underset{(39)}{\text{R}^1\text{CO}_2\text{H}} \xrightarrow{\text{SOCl}_2} \text{R}^1\text{COCl} \xrightarrow{\text{R}^2\text{NH}_2} \text{R}^1\text{—C(=O)—NHR}^2 \quad (8.15)$$

By adding an amine to thionyl chloride and the carboxylic acid in hexamethylphosphine triamide (HMPT) (**40**) as solvent at −10 °C very high yields of amides have been obtained (Scheme 8.16)[29] (cf. reaction with HMPT at higher temperature).

$$Me_2N-P(=O)(NMe_2)-NMe_2 \quad (40)$$

Acid	Amine	Yield (%)
Ph	Bu^nNH_2	78
Ph	Et_2NH	78
$CHCl_2$	Ph_2NH	70

(8.16)

Another method of preparing amides (**41**) from acid chlorides (**42**) utilizes a derivative (**43**) of trimethylsilane (**44**) prepared as shown in Scheme 8.17.[30]

$$2\,Me_3SiCl + NH_3 \longrightarrow (Me_3Si)_2NH$$
(44) (43)

$$R\text{-}CO_2H \longrightarrow R\text{-}COCl\ (42) \xrightarrow[\text{ii) } H_2SO_4/MeOH]{\text{i) } HN(SiMe_3)_2} R-C(=O)-NH_2\ (41)$$

R	Yield (%)
$ClCH_2$	87
Ph	87
$PhCH=CHCH_2$	92
(bicyclic)CH_2CH_2	60

(8.17)

The thionyl chloride–amine method for preparing amides has been improved by the use of dicyclohexylamine salts of carboxylic acids (**45**) (Scheme 8.18).[31] The method has been used successfully for the synthesis of dipeptides with less racemization than the old method.

Amides (**46**) can be prepared from acyl chlorides (**47**) using triethylphosphite (**48**) (Scheme 8.19).[32]

R^1-CO_2H + $[C_6H_{11}]_2NH$ → $[C_6H_{11}]_2\overset{+}{N}H_2$ R^1-CO_2^- (45)

i) $SOCl_2$ / pyridine in CH_2Cl_2 under Air, 1 minute, 20° C
ii) R^2R^3NH + base in CH_2Cl_2

→ R^1—C(=O)—N(R^2)(R^3)

R^1	R^2	R^3	Base	Yield (%)
$Ph(CH_2)_2$	H	Bn	DMAP	93
$Ph(CH_2)_2$	H	Bu^t	DBU	86
$Ph(CH_2)_2$	Bu^n	Bu^n	DBU	85
Cyclohexyl	H	Ph	DMAP	90
Cyclohexyl	H	Bn	DMAP	92
Ph	H	Bu^s	DBU	86
Cyclohexyl	$-(CH_2)_4-$		DMAP	95
Ph	$-(CH_2)_4-$		DBU	91

(8.18)

PhCOCl (47) + $P(OEt)_3$ (48) → Ph—C(=O)—P(=O)$(OEt)_2$

R^2R^3NH → R^2R^3NCOPh (46) + H—P(=O)$(OEt)_2$

R^1	R^2	R^3	Yield (%)
Ph	Et	Et	47
Ph	c-C_6H_{11}	H	58
Ph	$PhCH_2$	H	82
Ph	$-(CH_2)_2-O(CH_2)_2-$		74
Ph	Ph	H	no reaction

(8.19)

8.1.5 *From nitriles*

The controlled hydration of nitriles (**49**) can be used to prepare amides (**50**). Many different methods for effecting this transformation have been investigated.

Complexes of transition metals and triphenylphosphine have been used to convert nitriles into amides, for example, the rhodium complex *trans*-$Rh(OH)(CO)(PPh_3)_2$ (Scheme 8.20).[33]

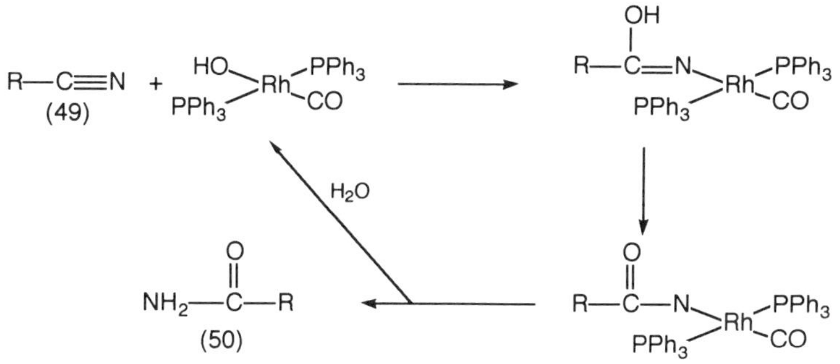

$$\underset{(49)}{R-C\equiv N} + Rh(OH)(CO)(PPh_3)_2 \longrightarrow R-C(OH)=N-Rh(CO)(PPh_3)_2 \longrightarrow R-C(=O)-N-Rh(CO)(PPh_3)_2 \xrightarrow{H_2O} \underset{(50)}{NH_2-C(=O)-R} \quad (8.20)$$

Both the rhodium complex shown in Scheme 8.20 and a platinum complex ($PtC_6H_8(PPh_3)_2$) have been used. The platinum catalyst produced 58 moles acetamide per mole of catalyst from acetonitrile while the rhodium catalyst produced 150 moles acetamide per mole of catalyst from acetonitrile and 200 moles propamide per mole of catalyst from propionitrile.

Hydrogen peroxide with formic acid has been used for the hydrolysis of nitriles.[34] Formic acid can be used without hydrogen peroxide,[35] and formic acid in a tantalum or silver vessel has been found to give good yields of amides (**50**) from alkyl nitriles (**49**) (Scheme 8.21).[36]

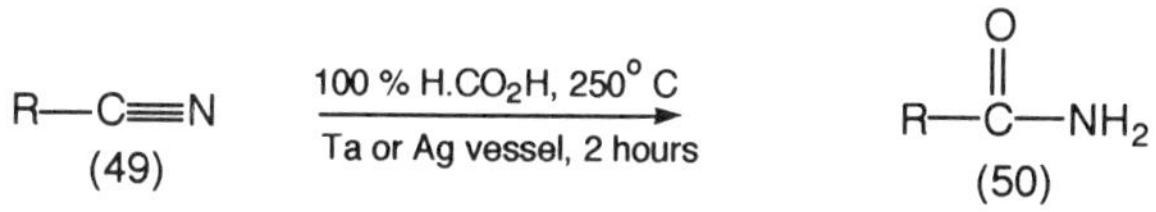

$$\underset{(49)}{R-C\equiv N} \xrightarrow[\text{Ta or Ag vessel, 2 hours}]{100\ \%\ H.CO_2H,\ 250^\circ C} \underset{(50)}{R-C(=O)-NH_2}$$

R	Yield (%)
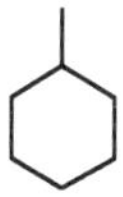	99
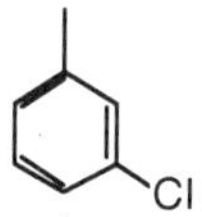	90
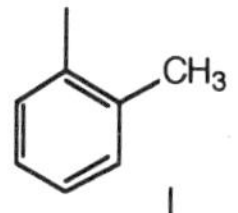	87
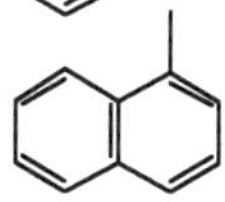	89

(8.21)

Formic acid and hydrogen chloride or hydrogen bromide have been used for the preparation of amides (**50**) from nitriles (**49**) in yields of 85–99 per cent (Scheme 8.22).[37]

$$\underset{(49)}{R{-}C{\equiv}N} \xrightarrow[30^\circ \text{ to } 50^\circ C]{H.CO_2H / HCl} \underset{(50)}{R{-}C({=}O){-}NH_2}$$

R	Yield (%)
cyclohexyl	92
phenyl	84
4-$CH_3C_6H_4$	85
$PhCH_2$–	99
$CH_3CH_2C(H)(CH_3)$–OH	89

(8.22)

$$R{-}C{\equiv}N \xrightarrow[\text{20\% aq. } HO^- / \text{excess } H_2O_2 \text{ Phase transfer}]{Bu^n_4N^+\ HSO_4^- / CH_2Cl_2} R{-}C({=}O){-}NH_2$$

R	Yield (%)
Ph	92
$Ph(CH_2)_3$	95
cyclohexyl	79
cyclopentyl	85
$PhCH{=}CHCH_2$	80

(8.23)

Hydrogen peroxide with alkali in aqueous ethanol can be used for this hydrolysis,[38] but better yields have since been obtained by the use of phase transfer methods (Scheme 8.23).[39]

Powdered sodium or potassium hydroxide in tertiary butanol also convert nitriles (**49**) into amides.[40] The reaction has been extended to the synthesis of secondary amides[41] (Scheme 8.24).

$$R^1\text{—C}\equiv\text{N} \; (49) \xrightarrow[\text{ii) } R^2\text{NH}_2\text{, 4 hours}]{\text{i) KOH / Bu}^t\text{OH / reflux, 20 minutes}} R^1\text{—C(=O)—NHR}^2 \; (50)$$

R^1	R^2	Yield (%)
Bn	Me	52
Bn	Pr^n	70
Pr^n	Me	56
Pr^n	Pr^n	78

(8.24)

Another method for the controlled hydrolysis of a nitrile (**49**) involves the use of manganese dioxide on silica gel, which is prepared by treating silica gel with manganese sulphate and aqueous potassium permanganate (Scheme 8.25).[42]

$$\text{R—C}\equiv\text{N} \; (49) \xrightarrow[\text{heat in } n\text{-octane}]{MnO_2 - SiO_2} \text{R—C(=O)—NH}_2$$

R	Yield (%)
Bu^t	83
Bn	35

(8.25)

A route to secondary and tertiary amides (**51**) is provided by the reaction between nitriles (**52**) and primary or secondary amines (**53**), catalysed by a rhodium triphenylphosphine complex (Scheme 8.26).[43]

8.1.6 *From thioamides*

Many useful methods exist for the conversion of thioamides into amides. For instance alkylthioamides (**54**) are converted into amides (**55**) by reaction with nitrous acid (Scheme 8.27).[44]

Another method for the conversion of thioamides (**54**) into amides (**55**) involves the use of *m*-chloroperbenzoic acid (Scheme 8.28).[45]

$$R^1-C\equiv N + R^2R^3NH \xrightarrow[RhH_2(PPh_3)_4 \text{ (catalyst)}]{H_2O / 160^\circ C} R^1-C(=O)-NR^2R^3$$

(52) (53) (51)

Nitrile	Amine	Yield (%)
CH_3CN	Bu^tNH_2	93
CH_3CN	piperidine	97

(8.26)

$$R^1R^2N(R^3)C=S \xrightarrow{HNO_2} R^1R^2N(R^3)C=O$$

(54) → $[R^1R^2N(R^3)C^+-S-NO]$ → (55)

R^1	R^2	R^3	Yield (%)
Ph	$-(CH_2)_2-O-(CH_2)_2-$		88
CH_3	$-(CH_2)_4-$		92
H	Ph	CH_3	97
Ph	H	Ph	65
2-pyridyl	H	2,6-dimethylphenyl (CH_3, CH_3)	60

(8.27)

Potassium superoxide, KO_2, reacts with *o*-nitrobenzenesulphonyl chloride (**56**) producing a peroxysulphur reagent that converts thioureas, thioamides (**54**), and thiocarbamates into the corresponding carbonyl compounds,[46] for example, Scheme 8.29.

$$R^1R^2N\text{-}C(=S)\text{-}R^3 \ (54) \xrightarrow[CH_2Cl_2,\ 25^\circ C]{m\text{-}ClC_6H_4CO_3H} R^1R^2N\text{-}C(=O)\text{-}R^3 \ (55)$$

Thioamide	Yield of amide (%)
HCSNHMe	76
N-methylpyrrolidine-2-thione	82

(8.28)

$$\text{(56) } o\text{-}NO_2C_6H_4SO_2Cl + KO_2 + R\text{—}C(=S)\text{—}NHR^1 \ (54) \xrightarrow[-35^\circ C]{CH_3CN} R\text{—}C(=O)\text{—}NHR^1$$

R	R^1	Yield (%)
Ph	CH_3	86
CH_3	p-$NO_2C_6H_4$	95
Ph	Ph	91

(8.29)

8.1.7 *From α,β-unsaturated aldehydes*

Oxidation of an α,β-unsaturated aldehyde (**57**) by manganese dioxide in the presence of sodium cyanide and an amine or ammonia gives high yields of α,β-unsaturated amides (**58**). The reaction proceeds via the α,β-unsaturated carbinolamine (**59**) (Scheme 8.30).[47]

$$\text{Ph-CH=CH-CHO (57)} \xrightarrow{NaCN,\ HCN} \text{Ph-CH=CH-CH(OH)-CN (59)} \xrightarrow{MnO_2} \text{Ph-CH=CH-C(=O)-CN} \xrightarrow[\text{in } Pr^iOH]{NH_3} \text{Ph-CH=CH-C(=O)-NH}_2 \text{ (58) 100\%}$$

(8.30)

In a reaction which occurs through an arsenic ylide (**60**), good yields of the *E*-isomers of α,β-unsaturated amides (**61**) have been obtained,[48] for example, Scheme 8.31.

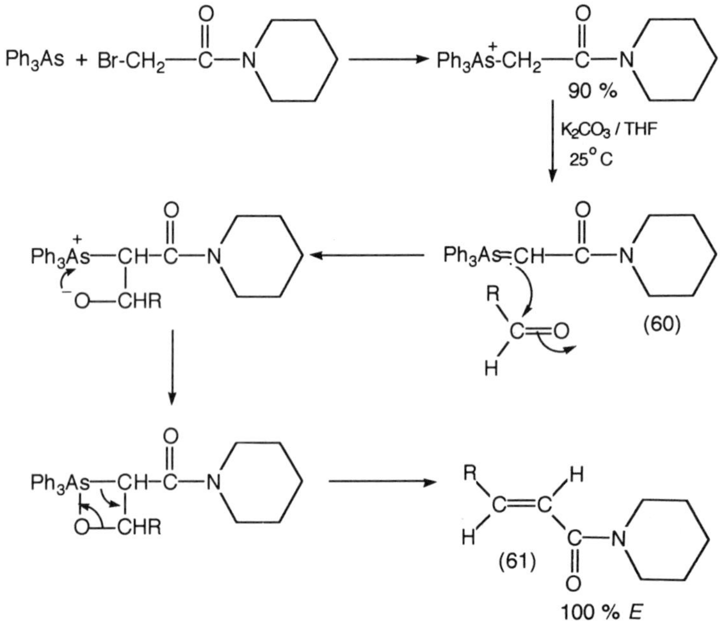

R	Yield (%)
Ph	98
$n\text{-}C_6H_{13}$	99
$CH_3CH{=}CH$	88
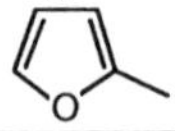	98

(8.31)

8.1.8 *From alkenes*

Nitronium tetrafluoroborate (**62**) reacts with alkenes (**63**) in acetonitrile at −15 °C to give, after hydrolysis, modest yields of nitroacetamides (**64**) (Scheme 8.32).[49] This is an extension of the Ritter reaction in which nitriles (**65**) add to alkenes (**66**) via the formation of the carbocation (**67**). Only alkenes, or alcohols, that form a stable carbocation react (Scheme 8.33).[50]

R	R^1	R^2	Yield (%)
CH_3	H	H	50
CH_3	CH_3	H	23
CH_3	H	CH_3	13

(8.32)

R	R^1	R^2	R^3	R^4	R^5	Yield (%)
H_2N	Me	Me	H	n-C_3H_7	Me_2C-C_4H_9	12
Me	Ph	H	H	H	PhCHMe	43
$PhCH_2$	Me	Me	H	Et	$Me_2CC_3H_7$	69

(8.33)

8.1.9 *From alcohols and nitriles*

Derivatives of antimony pentachloride (**68**), prepared *in situ*, have been used by two schools for the conversion of mixtures of aliphatic alcohols (**69**) and nitriles (**70**) into amides (**71**) (Scheme 8.34).[51] The reaction of the alcohol (**69**) and the nitrile (**70**) with the salt (**68**) is extremely fast. D. H. R. Barton and his co-workers[52] have extended the reaction and have suggested the mechanism shown in Scheme 8.35.

$$Ph_2CCl_2 + SbCl_5 \xrightarrow{\text{in } FCl_2C\text{-}CClF_2} Ph_2CCl^+ \; SbCl_6^-$$

(crystaline salt)
(68)

i) R^1OH(69) in R^2CN(70)
ii) aq. work-up

$$R^1\text{—NH—C(=O)—}R_2$$

c. 60 %
(71)

(8.34)

$$R^1\text{-OH} + Ph_2\overset{+}{C}\text{-Cl}\; SbCl_6^- \longrightarrow \left[R^1\text{—}\overset{+}{O}(H)\text{—CPh}_2(Cl)\; SbCl_6^-\right]$$

(69) (68)

$$\longrightarrow \left[R^1\text{—}\overset{+}{O}\text{=CPh}_2\, SbCl_6^- \quad R^2\text{—C≡N:}\right]$$

(70)

$$\longrightarrow R^2\text{—C≡}\overset{+}{N}\text{—}R^1\; SbCl_6^- + Ph_2C\text{=O} \xrightarrow{H_2O} R^2\text{—C(=O)—NHR}^1$$

(71)

(8.35)

8.1.10 *From isocyanates*

In a recent procedure amides are obtained by reaction of an isocyanate (**72**) with an alkyl- or aryl-lithium (**73**) (Scheme 8.36).[53]

Diphenylphosphorylazide (**74**) is prepared as shown in Scheme 8.37 and it can be used to prepare urethanes (**75**) by its reaction under reflux with a carboxylic acid (**76**), triethylamine, and an alcohol (**77**).[54]

Reaction of trimethylsilyl azide (**78**) with acid chlorides (**79**) yields the azide which rearranges to the isocyanate (**80**) in 69–94 per cent yield. Reaction of the isocyanate (**80**) with an alcohol yields a urethane. (Scheme 8.38).[55]

$$R{-}N{=}C{=}O \;(72) + R^1{-}Li \;(73) \longrightarrow R{-}N{=}C(R^1){-}O^- \xrightarrow{H_3O^+} R{-}NH{-}C(R^1){=}O$$

R	R¹	Yield (%)
c-C_6H_{11}	CH_3	63
Ph	CH_3	73
c-C_6H_{11}	Ph	89
n-C_5H_{11}	CH_3	57

(8.36)

$$(PhO)_2P(Cl){=}O + NaN_3 \xrightarrow[20^\circ C]{\text{acetone}} (PhO)_2P(N_3){=}O \;(74) \;>90\%$$

$$R^1\text{-}CO_2H\;(76) + R^2\text{-}OH\;(77) \xrightarrow[Et_3N,\ 5\text{-}25\text{ hours}]{(PhO)_2PON_3} R^1\text{-}NH\text{-}CO_2R^2\;(75) \qquad (8.37)$$

$$R^1{-}C(=O){-}Cl\;(79) \xrightarrow[(78)]{Me_3SiN_3} R^1{-}N{=}C{=}O\;(80) \xrightarrow{R^3\text{-}OH} R^1{-}NH{-}C(=O){-}OR^3 \qquad (8.38)$$

8.1.11 *From ketones*

The Willgerodt reaction[56]

The reaction discovered by Willgerodt[57] over a hundred years ago involves the heating of an alkyl aryl ketone and ammonium polysulphide (**81**) in aqueous solution at a high temperature (typically 210–230 °C) for several hours in a sealed tube. The products are an amide and a smaller amount of the ammonium salt of the corresponding acid, for example, acetophenone (**82**) yields the amide (**83**) and a smaller amount of the ammonium salt of phenylacetic acid (**84**) (Scheme 8.39). Longer chain alkyl aryl ketones (**85**) react similarly, for

$$Ph{-}C(=O){-}CH_3\;(82) \xrightarrow{(81)} Ph{-}CH_2{-}C(=O){-}NH_2\;(83)\;50\% + Ph{-}CH_2{-}CO_2^-\,\overset{+}{N}H_4\;(84)\;13.5\% \qquad (8.39)$$

example, Scheme 8.40. The yields when the original inconvenient experimental procedure is used are seldom greater than 50 per cent, and the total products are often hydrolysed without isolation and the carboxylic acid is isolated.

$$Ph{-}C(=O){-}CH_2{-}CH_2{-}CH_3\ (85) \longrightarrow Ph{-}CH_2{-}CH_2{-}CH_2{-}C(=O){-}NH_2\ \ 37\ \% \tag{8.40}$$

Modifications of the Willgerodt reaction

Modifications of the original procedure have enabled greater yields to be obained. For example, diluting the reaction mixture with dioxan or using aqueous ammonia, pyridine, and sulphur increase the yield of the phenylacetic acid from acetophenone to 70 and 86 per cent respectively. It has been found that certain purely aliphatic ketones also undergo the reaction, for example, with pinacolone (**86**) (Scheme 8.41). Phenylacetylene (**87**) reacts to give a good yield of phenylacetamide (**88**) (Scheme 8.42).[58]

$$(CH_3)_3C{-}C(=O){-}CH_3\ (86) \xrightarrow[\text{dioxan}]{NH_4S_n\ +\ S} (CH_3)_3C{-}CH_2{-}C(=O){-}CH_3\ \ 58\ \% \tag{8.41}$$

$$Ph{-}C{\equiv}C{-}H\ (87) \xrightarrow[\text{pyridine}\ +\ S]{\text{aqueous}\ NH_3} Ph{-}CH_2{-}C(=O){-}NH_2\ (88)\ \ 80\ \% \tag{8.42}$$

The Kindler method

Kindler[59] introduced a modification of the Willgerodt reaction in which thioamides (**89**) are produced from a carbonyl compound (**90**), an amine (**91**), and sulphur, for example, Scheme 8.43.[60] The use of dimethylformamide as a solvent for the Kindler reaction enables good yields to be obtained at lower temperatures (50–60 °C) (Scheme 8.44).[61]

$$Ph{-}C(=O){-}CH_3\ (90) + S + \text{HN(morpholine)}\ (91) \xrightarrow[\text{reflux}]{\text{pyridine, }60^{\circ}C} Ph{-}CH_2{-}C(=S){-}N(\text{morpholino})\ (89)\ \ 94\% \tag{8.43}$$

morpholine (91)

$$R^1\text{—}\underset{\underset{O}{\|}}{C}\text{—}CH_3 + S + R^2R^3NH \xrightarrow[50^\circ\text{-}60^\circ C]{DMF} R^1\text{-}CH_2\text{—}\underset{\underset{S}{\|}}{C}\text{—}NR^2R^3 \qquad (8.44)$$

The addition of *p*-toluenesulphonic acid as a catalyst (to aid the formation of enamines which are thought to be intermediates in the reaction) allows the reaction to take place at room temperature (Scheme 8.45).[62]

$$R\text{—}\underset{\underset{O}{\|}}{C}\text{—}CH_3 + Me_2NH \xrightarrow[20^\circ C]{TsOH\,/\,DMF} H_2O + R\text{—}\underset{\underset{NMe_2}{|}}{C}{=}CH_2 \xrightarrow{S} R\text{-}CH_2\text{-}\underset{\underset{S}{\|}}{C}\text{—}NMe_2$$

for example

$$Ph\text{—}C(=CH_2)\text{—}N(\text{morpholino}) \longrightarrow Ph\text{—}CH_2\text{—}C(=S)\text{—}N(\text{morpholino}) \quad 75\% \qquad (8.45)$$

The mechanisms of the Willgerodt reaction and of the Kindler modification are not known with certainty though many possibilities have been advanced. The reviews should be consulted for more details.

8.1.12 *Acylation of secondary amides*

Secondary amides (**92**) are acylated by acid chlorides (**93**) in the presence of a dicyclopentadienyl tin compound (**94**) (Scheme 8.46).[63]

$$\underset{(92)}{R^1\text{-NH—}\overset{\overset{O}{\|}}{C}\text{—}R^2} + \underset{(93)}{R^3\,COCl} \xrightarrow[\text{THF, HMPT, }60^\circ C]{[C_5H_5^-]_2\,Sn^{2+}\ (94)} R^1\text{—}N(\text{—}\overset{\overset{O}{\|}}{C}\text{—}R^2)(\text{—}\underset{\underset{O}{\|}}{C}\text{—}R^3) \qquad (8.46)$$

8.1.13 *Alkylation of primary amides for preparation of secondary and tertiary amides*

Phase transfer catalysis has been used in a method for the *N*-alkylation of amides (**95**) (Scheme 8.47).[64]

Primary amides (**96**) can be alkylated by alkenes (**97**) in the presence of mercuric nitrate (**98**), followed by reductive removal of mercury by sodium

$$R{-}C({=}O){-}NH_2 \ (95) \xrightarrow[\text{Phase Transfer Catalysis}]{R^1X} R{-}C({=}O){-}NHR^1$$

$$R{-}C({=}O){-}NH_2 \ (95) \xrightarrow[\text{Phase Transfer Catalysis}]{\text{excess } R^1X} R{-}C({=}O){-}NR^1_2$$

R	R^1	primary amide Yield (%)	secondary amide Yield (%)
Ph	Et	73	51
Ph	$PhCH_2$	80	—
Ph	i-C_4H_9	32	—
Ph	Et	84	—
Et	n-C_4H_9	93	94
Et	$PhCH_2$	92	—

(8.47)

borohydride, to yield secondary amides (**99**) in good yield (Scheme 8.48).[65] Urethanes (**100**) and ureas (**101**) react similarly (Scheme 8.49).

Formaldehyde (**102**) and an aluminium trialkyl convert a primary amide (**103**) into a secondary amide (**104**) (Scheme 8.50).[66]

$$R^1R^2C{=}CHR^3 \ (97) + Hg(NO_3)_2 \ (98) \longrightarrow \left[R^1R^2\overset{+}{C}{-}CH(R^3)(HgNO_3)\right] NO_3^- \xrightarrow{R^4{-}C(=O){-}\ddot{N}H_2 \ (96)} R^2R^1C(NHC(=O)R^4){-}CH(R^3){-}HgNO_3 + H^+$$

$$\xrightarrow{NaBH_4} R^2R^1C(NHC(=O)R^4){-}CH_2R^3 \quad (99) \quad 45 - 99\%$$

(8.48)

$$R^1R^2C{=}CHR^3 + EtO_2C\text{-}NHR^4\ (100) \xrightarrow[\text{ii) } NaBH_4]{\text{i) } Hg(NO_3)_2} EtO_2C\text{—}NR^4\text{—}C(R^1)(R^2)\text{—}CH_2R^3$$

25 - 100 %

$$R^1CH{=}CHR^2 + H_2N\text{—}C(=O)\text{—}NH_2\ (101) \xrightarrow[\text{ii) } NaBH_4]{\text{i) } Hg(NO_3)_2} R^1\text{—}CH(CH_2R^2)\text{-}NH\text{—}C(=O)\text{—}NH_2$$

$$\xrightarrow[\text{ii) } NaBH_4]{\text{i) } R^3CH{=}CHR^4\ /\ Hg(OAc)_2} R^1\text{—}CH(CH_2R^2)\text{-}NH\text{—}C(=O)\text{—}NH\text{—}CH(CH_2R^4)\text{—}R_3 \quad (8.49)$$

$$R^1\text{—}C(=O)\text{—}NH_2\ (103) + H_2C{=}O\ (102) \longrightarrow R^1\text{—}C(=O)\text{—}NH\text{—}CH_2\text{-}OH \xrightarrow{AlR^2_3\ /\ PhH,\ \text{heat}} R^1\text{—}C(=O)\text{—}NH\text{—}CH_2R^2\ (104)$$

R^1	R^2	Yield (%)
Ph	CH_3	81
n-C_5H_{11}	CH_3	70
Ph	CH_3CH_2	71
n-C_5H_{11}	CH_3CH_2	86
1-(hydroxymethyl)pyrrolidin-2-one (N–CH₂OH)	CH_3	25
1-(hydroxymethyl)pyrrolidin-2-one (N–CH₂OH)	CH_3CH_2	29

(8.50)

8.1.14 *From oximes by the Beckmann rearrangement*

The well known Beckmann rearrangement of oximes (**105**)[67] yields amides (**106**) (Scheme 8.51) and many convenient methods of effecting the rearrangement have been developed, some of which are illustrated by what follows.

(8.51)

p-Toluenesulphonates of oximes (**107**) on silica gel also exhibit the rearrangement (Scheme 8.52).[68] By comparison other authors[69] have used alumina (Al_2O_3) in place of silica gel to produce amides in 40–90 per cent yield.

R^1	R^2	Yield (%)
-$(CH_2)_4$-		45
CH_3	Ph	70
Ph	Ph	86
CH_3	Bu^t	20

(8.52)

A similar rearrangement of *N*-methyl nitrones (**108**) in the presence of *p*-toluenesulphonyl chloride (tosyl chloride) also yields *N*-methylamides (**109**) (Scheme 8.53).[70]

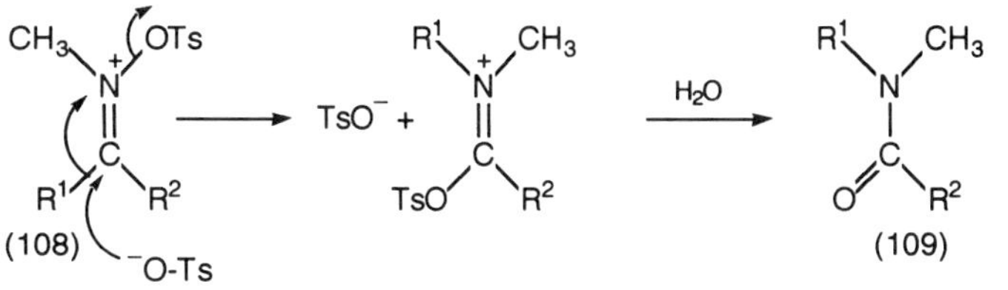

(8.53)

8.1.15 *From N-protected primary amines*

Protected primary amines (**110**) are oxidized by ruthenium tetroxide, prepared *in situ*, to yield protected amides (**111**) which can be deprotected with TFA in dichloromethane (Scheme 8.54).[71]

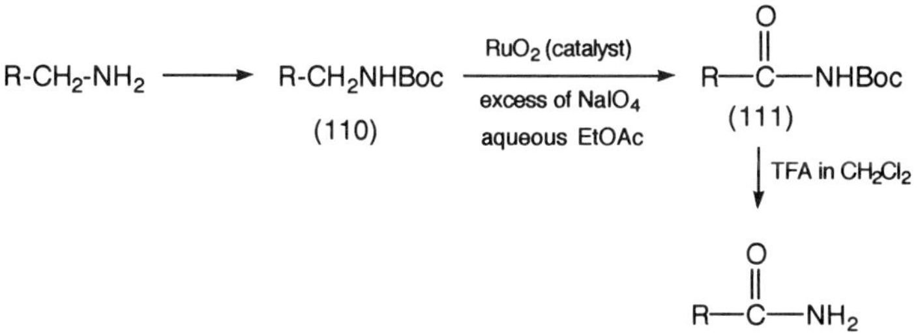

Amine	Yield-NHBoc (%)	Yield-NH_2 (%)
$EtNH_2$	90	95
$H_2N(CH_2)_3NH_2$	90	98

(8.54)

8.1.16 *From tertiary amines*

By the oxidation of suitable tertiary amines (**112**) with the quaternary permanganate (**113**),[72] amides (**114**) result in good yields (Scheme 8.55).[73]

8.1.17 *From azides and carboxylic acids*

The reaction of an azide (**115**) with a carboxylic acid (**116**) in the presence of a derivative of triphenylphosphine yields amides (**117**). The method is applicable to the synthesis of peptides[74] (Scheme 8.56).

8.1.18 *Preparation of β-ketoamides from β-diketones*

Tosyl azide (**118**) reacts with 1,3-diketones (**119**) to give 2-diazo-1,3-diketones (**120**) in high yield. These products undergo Wolff rearrange-

$$\text{Ph-CH}_2\text{-}\overset{+}{\text{N}}\text{Et}_3\ \text{Cl}^- \xrightarrow{\text{KMnO}_4} \text{Ph-CH}_2\text{-}\overset{+}{\text{N}}\text{Et}_3\ \text{MnO}_4^- \quad (113)$$

$$\underset{(112)}{\text{R}^1\text{-CH}_2\text{-NR}^2\text{R}^3} \xrightarrow{\text{Ph-CH}_2\text{-}\overset{+}{\text{N}}\text{Et}_3\ \text{MnO}_4^-} \underset{(114)}{\text{R}^1\text{—C(=O)—NR}^2\text{R}^3}$$

R^1	R^2	R^3	Yield (%)
Pr^i	Bu^n	Bu^n	93
Ph	Me	Me	70
H	Me	Ph	78
Me	Et	Ph	63

(8.55)

$$\underset{(115)}{\text{R}^1\text{-N}_3} + \underset{(116)}{\text{R}^2\text{-CO}_2\text{H}} \xrightarrow[\text{EtOAc}]{\text{Ph}_2\text{POEt in}} \left[\text{R}^1\text{-N=P(Ph)}_2\text{OEt} \longleftrightarrow \text{R}^1\text{-}\bar{\text{N}}\text{-}\overset{+}{\text{P}}\text{(Ph)}_2\text{OEt}\right] \xrightarrow{\text{R}^2\text{-CO}_2\text{H}} \text{R}_1\text{—N(H)—}\overset{+}{\text{P}}\text{(Ph)}_2\text{OEt} + \text{R}^2\text{—C(=O)—O}^-$$

$$\longrightarrow \text{R}^1\text{—}\overset{+}{\text{N}}\text{(H)—}\overset{+}{\text{P}}\text{(Ph)}_2\text{OEt}\ /\ \text{R}^2\text{—C(O}^-\text{)(O}^-\text{)} \longrightarrow \text{R}^1\text{—}\overset{+}{\text{N}}\text{(H)—P(Ph)}_2\text{OEt}\ /\ \text{R}^2\text{—C(O}^-\text{)—O (cyclic)}$$

$$\longrightarrow \text{R}^1(\text{H})\overset{+}{\text{N}}\text{=C(R}^2\text{)O}^- + \text{O=P(Ph)}_2\text{OEt} \longrightarrow \underset{(117)}{\text{R}^1\text{-NH—C(=O)—R}^2}$$

(8.56)

ment on irradiation to form α-keto ketenes (**121**), which can be trapped by a primary or secondary amine to provide α-alkyl-β-ketoamides (**122**) (Scheme 8.57).[75]

$$R^1-\overset{O}{\overset{\|}{C}}-CH_2-\overset{O}{\overset{\|}{C}}-R^2 \ (119) \xrightarrow[(118)]{TsN_3} R^1-\overset{O}{\overset{\|}{C}}-\underset{N^+=N^-}{\underset{\|}{C}}-\overset{O}{\overset{\|}{C}}-R^2 \ (120)$$

$$(120) \xrightarrow[\text{Wolff rearrangement}]{h\nu\ (254\ nm)} \left[R^1-\overset{O}{\overset{\|}{C}}-\underset{R^2}{\underset{|}{C}}=C=O \right] (121) \xrightarrow{HNR^3R^4} R^1-\overset{O}{\overset{\|}{C}}-\underset{R^2}{\underset{|}{CH}}-\overset{O}{\overset{\|}{C}}-NR^3R^4 \ (122)$$

R^1	R^2	R^3	R^4	Yield (%)
CH_3	CH_3	CH_2=CH	CH_2=CH	65
-CH_2-C(CH_3)$_2$-CH_2-		-(CH_2)$_4$-		74
-CH_2-C(CH_3)$_2$-CH_2-		H	H	62
-CH_2-C(CH_3)$_2$-CH_2-		-CO-(CH_2)$_3$-		67

(8.57)

8.1.19 *Preparation of N-methoxyamides*

N-Methoxyamides (**123**) which act as useful synthetic intermediates, can be prepared by the reaction between acid chlorides (**124**) and *N*,*O*-dimethylhydroxylamine (**125**) (Scheme 8.58).[76]

$$\underset{(124)}{R\text{-}COCl} + \underset{(125)}{CH_3O\text{-}NHCH_3} \longrightarrow \underset{(123)}{R\text{-}CO-\underset{OCH_3}{\underset{|}{N}}-CH_3} \qquad (8.58)$$

8.2 Reactivity

The delocalization in amides furnishes an electrophilic carbon atom in the carbonyl group which can be attacked by a variety of nucleophiles.

$$Nu^{-} \quad R^1—C(=O)—NR^2R^3$$

The nitrogen atom is somewhat nucleophilic and can be attacked by strong electrophiles. The hydrogen atoms on the nitrogen atom in primary and secondary amides are another source of reactivity since they can be removed as protons by bases. Furthermore, hydrogen atoms α to the carbonyl group can be removed by strong bases and reactions can occur through the 'enol' form of a primary or a secondary amide.

8.2.1 *Reaction with nucleophiles*

Hydrolysis with acids and alkali

Water in acidic solution, and the hydroxide ion in alkaline solution function as nucleophiles and lead to hydrolysis of the amide (**126**) to the corresponding carboxylic acid (**127**) (Scheme 8.59).

With acid:

$$\underset{(126)}{R—C(=O)—NH_2} + H_3O^+ \rightleftharpoons R—\overset{+}{C}(OH)—NH_2 \overset{H_2O}{\rightleftharpoons} R—C(\overset{+}{O}H_2)(O—H)—NH_2$$

$$\rightleftharpoons R—C(OH)(O—H)—\overset{+}{N}H_3 \rightleftharpoons \underset{(127)}{R\text{-}CO_2H} + NH_3 \rightleftharpoons R\text{-}CO_2^- + \overset{+}{N}H_4$$

With alkali:

$$\underset{(126)}{R—C(=O)—NH_2} + HO^- \longrightarrow R—C(OH)(O^-)—NH_2 \longrightarrow \underset{(127)}{R\text{-}CO_2H} + \overset{-}{N}H_2 \longrightarrow R\text{-}CO_2^- + NH_3 \qquad (8.59)$$

Sodium peroxide which provides a highly nucleophilic anion is a very efficient hydrolysing agent for water-soluble amides (**128**) (Scheme 8.60).[77] The reaction is carried out in aqueous solution and yields of *c.* 85 per cent result in a short time (*c.* 1 h) at 50–100 °C.

$$\underset{(128)}{R{-}C({=}O){-}NH_2} + HO{-}O^- \longrightarrow NH_3 + R{-}C({=}O){-}O{-}O^- \xrightarrow{H_2O} R{-}CO_2^- + H_2O_2 \qquad (8.60)$$

R^1	R^2	Yield (%)
n-C_5H_{11}	H	87
c-C_6H_{11}	H	90
Bu^t	H	68
Ph	H	83
Ph	Ph	89

(8.61)

In an efficient method for the hydrolysis of amides which gives good yields (*c.*90 per cent) of carboxylic acids, the amide is boiled under reflux in water containing Amberlite resins (15 × excess of Amberlyst 15 or Amberlite-IR-120 resins).[78] Acyl hydrazides are also hydrolysed in this way.

Amides (**129**) can be converted into carboxylic acids (**130**) by the action of tetrachloro- or tetrafluoro-phthalic anhydride (**131**) at 135–170 °C (Scheme 8.61).[79]

With Meerwein's reagent

Amides (**132**) react with triethyloxonium fluoroborate and sodium ethoxide to yield ketals.[80] These may react with nitroalkanes to give nitroenamines (**135**)[81] (Scheme 8.62).

$$R{-}C({=}O){-}NMe_2\ (132) \xrightarrow{Et_3O^+\ BF_4^-} R{-}C({=}\overset{+}{O}Et){-}NMe_2\ BF_4^-\ (133) \xrightarrow{NaOEt,\ 0^\circ C} R{-}C(OEt)_2{-}NMe_2\ (134) \xrightarrow{CH_3NO_2} (Me_2N)(R)C{=}C(H)(NO_2)\ (135)$$

R	Yield of (**133**) (%)	Yield of (**134**) (%)	Yield of (**135**) (%)
EtO_2CH	100	76	—
Bu^t	97	100	60
$NCCH_2$	100	51	51
EtO_2C	98	72	—
p-$NO_2C_6H_4$	100	53	—
Me_2NCO	95	60	96

(8.62)

Reduction

To amines One of the particularly valuable uses of lithium aluminium hydride is the reduction of an amide (**136**) to an amine (**137**) in which the hydride ion can be regarded as the nucleophile. Better yields are obtained, however, by the use of diborane in tetrahydrofuran (Scheme 8.63).[82]

Primary amides can be reduced to amines by lithium borohydride in a mixture of diglyme and methanol. Carboxylate anions or secondary amides in the same molecule are unaffected by the reducing agent.[83]

$$R{-}C(=O){-}NH_2 \;(136) + H{-}Al^{-} \xrightarrow[B_2H_6\,(THF)]{LiAlH_4\,(ether)\ or} R{-}CH(O{-}Al){-}NH_2 \rightleftharpoons R{-}CH{=}NH + HO^{-};\quad R{-}CH{=}NH + H{-}Al^{-} \rightleftharpoons R{-}CH_2{-}\bar{N}H \xrightarrow{aq.\ H_3O^+} R{-}CH_2{-}NH_2\;(137) \quad (8.63)$$

To aldehydes Aliphatic amides (**138**) in which the alkyl group contains six or fewer carbon atoms can be converted into aldehydes (**139**) in good yields by the Sonn–Müller reaction (cf. the Stephen reduction of nitriles in Chapter 4, p. 225) (Scheme 8.64).

$$R{-}CONHPh\;(138) \xrightarrow[in\ benzene]{PCl_5} R{-}C(Cl){=}N{-}Ph \xrightarrow[in\ ether]{SnCl_2/HCl} R{-}CH{=}N{-}Ph \xrightarrow{hot\ H_2O} R{-}CHO\;(139) \quad (8.64)$$

The reduction of tertiary amides (**140**) to aldehydes (**141**) has been achieved by the use of lithium aluminium hydride,[84] though modified lithium aluminium hydrides, e.g. ethoxy derivatives, are better and sodium aluminium hydride is also superior to lithium alumunium hydride for this purpose[85] (Scheme 8.65).

$$R{-}C(=O){-}N(CH_3)_2\;(140) \xrightarrow{LiAlH_2(OEt)_2} R{-}CH(O^-){-}N(CH_3)_2 \xrightarrow{aq.\ H_3O^+} R{-}CHO\;(141);\qquad LiAlH_4 + 2\,EtOH \xrightarrow{in\ ether} LiAlH_2(OEt)_2 \quad (8.65)$$

Tertiary amides (**142**) are reduced by one equivalent of lithium *n*-butyldiisobutylaluminium hydride to aldehydes (**143**) in high yield (Scheme 8.66).[86]

To alcohols The reduction of tertiary amides (**144**) (but not primary or secondary) with 'superhydride' ($LiEt_3BH$), yields 50–100 per cent alcohols (**145**) (Scheme 8.67).[87]

$$\mathrm{R{-}C({=}O){-}NR^1R^2}\ (142) \xrightarrow{\mathrm{LiBu^nBu^i_2AlH}} \mathrm{R{-}CH(O^-){-}NR^1R^2} \xrightarrow{\mathrm{aq.\ H_3O^+}} \mathrm{R{-}CHO}\ (143) \qquad (8.66)$$

$$\mathrm{R{-}C({=}O){-}NR^1R^2}\ (144) + \mathrm{H{-}BEt_3\,Li} \xrightarrow[0^\circ \text{ to } 25^\circ\mathrm{C}]{\mathrm{LiEt_3BH\,/\,THF}} \mathrm{R{-}CHO} + {}^{-}\mathrm{NR^1R^2} \longrightarrow \mathrm{R{-}CH_2OH}\ (145) \qquad (8.67)$$

Tertiary amides (**146**) are reduced either to primary alcohols (**147**) or to tertiary amines (**148**), depending upon their structure, by sodium borohydride substituted by amines (Scheme 8.68).[88]

$$\mathrm{R{-}CO{-}NMe_2}\ (146) \xrightarrow[\text{or}]{\mathrm{Na\ Me_2NBH_3}} \mathrm{R{-}CH_2{-}OH}\ (147)$$

$$\text{but}\quad \mathrm{R{-}CO{-}NPr^i_2}\ (146) \xrightarrow{\mathrm{Na\ Bu^tNBH_3}} \mathrm{R{-}CH_2{-}NPr^i_2}\ (148) \qquad (8.68)$$

Tertiary amides (**149**) can be reduced to alcohols (**150**) in yields of about 70 per cent by sodium or potassium and *n*-butyl alcohol in an electron transfer reaction followed by proton capture (Scheme 8.69). However, complex mixtures of compounds result from primary and secondary amides, presumably through the ready ionizability of hydrogen atoms on nitrogen which causes side reactions.[89]

$$\mathrm{R{-}C({=}O){-}NMe_2}\ (149) \xrightarrow[\text{HMPT and ether}]{\mathrm{Na\ or\ K\ and\ Bu^nOH\ in}} \mathrm{R{-}CH_2OH}\ (150)$$

Mechanism:

$$\mathrm{R{-}C({=}O){-}NMe_2} \xrightarrow{2e} \mathrm{R{-}\dot{C}(O^-){-}NMe_2} \xrightarrow{2\,\mathrm{Bu^nOH}} \mathrm{R{-}CH(OH){-}NMe_2} \rightleftharpoons \mathrm{R{-}CHO + HNMe_2} \xrightarrow{2e} \mathrm{R{-}\bar{C}H{-}O^-} \xrightarrow{2\,\mathrm{Bu^nOH}} \mathrm{R{-}CH_2OH} \qquad (8.69)$$

Formation of acetylenic ketones

Acetylenic ketones (**151**) can be prepared as shown in Scheme 8.70 from acetylenes (**152**) and tertiary amides (**153**).[90] The reaction mixture is worked up with aqueous ammonium chloride for 30 minutes and the yields are in the range 80–100 per cent.

$$R^1{-}C{\equiv}C{-}H \ (152) \longrightarrow R^1{-}C{\equiv}C^- \xrightarrow{R^2{-}C(=O){-}NR^3_2\ (153)} R^1{-}C{\equiv}C{-}\bar{B}F_3 \longrightarrow BF_3 + \bar{N}R^3_2 + R^1{-}C{\equiv}C{-}C(=O){-}R^2\ (151) \quad (8.70)$$

8.2.2 *Reaction with electrophiles*

Hofmann degradation

Electrophiles are able to attack amides on the nitrogen atom. The best known example of this is the Hofmann degradation and its ramifications. Bromine and alkali cause a rearrangement and degrade a primary amide (**154**) to a primary amine (**155**) (see Chapter 1, p. 25). The electrophile is the bromonium ion (**156**) derived from bromine and alkali[91] (Scheme 8.71). Isocyanates (**157**) can be isolated from the reaction by use of phase transfer catalysis.[92]

$$R{-}C(=O){-}NH_2\ (154) + \overset{+}{Br}\ (156) \underset{\text{or } Bu^tOBr}{\overset{NaOBr}{\rightleftharpoons}} R{-}C(=O){-}\overset{+}{N}H_2{-}Br \longrightarrow H^+ + R{-}C(=O){-}NH{-}Br \xrightarrow{^-OH} R{-}C(=O){-}\bar{N}{-}Br \longrightarrow R{-}N{=}C{=}O\ (157) \xrightarrow{\text{aq. } HO^-} R{-}NH_2 + CO_2\ (155) \quad (8.71)$$

With lead tetra-acetate

Related to the Hofmann degradation is the oxidation of a primary amide (**158**) with lead tetra-acetate, a reaction which involves a similar migration of an alkyl group from carbon to an electron deficient nitrogen atom. When the reaction is carried out in tertiary butanol in the presence of a catalyst (Et_3N or $SnCl_4$), it gives 65–85 per cent yields of urethanes (**159**) (carbamates), which can, if required, be hydrolysed and decarboxylated in the usual way to yield amines[93] (Scheme 8.72).

$$R{-}C(=O){-}NH_2 \ (158) \xrightarrow{Pb(OAc)_4} R{-}C(=O){-}\ddot{N}: \longrightarrow R{-}N{=}C{=}O \xrightarrow[Et_3N \text{ or } SnCl_4]{Bu^tOH} R{-}NH{-}CO_2Bu^t \ (159) \quad (8.72)$$

8.2.3 *Reaction through the 'enol'*

Dehydration to nitriles (see Chapter 4, p. 183)

The dehydration of primary amides (**160**) to nitriles (**161**) can be achieved by heating the amide with phosphorus pentoxide. An improved method involves dehydration with a 3:1 ratio of an amide to the reagent (**162**). The reaction takes place through the 'enol' (**163**) (Scheme 8.73).[94]

$$R{-}C(=O){-}NH_2 \ (160) \rightleftharpoons R{-}C(OH){=}NH \ (163) \xrightarrow{(162)\ (NPCl_2)_3} R{-}C(=NH){-}O{-}P(Cl)(NPCl_2)_2N \longrightarrow R{-}C{\equiv}N \ (161)\ 75\text{–}100\% + H^+ + {}^-O{-}P(Cl)(NPCl_2)_2N \quad (8.73)$$

O-Methylation

O-Methylation of amides and thioamides can readily be carried out using diazomethane in the presence of silica gel (Scheme 8.74).[95]

Substrate	Product	Yield % X = O or S
2-pyrrolidinone	2-methoxy-1-pyrroline (OMe)	100 -
2-piperidinone	2-methoxy-tetrahydropyridine (OMe)	100 -
azepan-2-one (X)	XMe imino ether	100, 78
$PhCXNH_2$	Ph–C(=NH)–XMe	83, 78

Reagents: CH_2N_2, silica gel (8.74)

8.2.4 *Acidity of hydrogen attached to nitrogen*

With strong bases and carbon disulphide

Hydrogen atoms on nitrogen in primary (**164**) and secondary amides are acidic and can be removed by strong bases, for example, by sodium hydride (Scheme 8.75).[96] The reactions are carried out in 1:1 DMA (dimethylacetamide) and

$$R^1{-}C(=O){-}NH_2\ (\mathbf{164}) \xrightarrow{NaH} R^1{-}C(=O){-}\bar{N}H + S{=}C{=}S \longrightarrow R^1{-}C(=O)(NH{-}C(=S){-}S^-) \longrightarrow R^1{-}C(O^-)(NH)(S{-}C{=}S) \longrightarrow R^1{-}C(=O){-}S^- + H{-}N{=}C{=}S \longrightarrow R^1{-}COSH + {}^-N{=}C{=}S\ \ (65\text{-}80\ \%)$$

$$R^1{-}CO{-}NHR^2 \xrightarrow{NaH} R^1{-}CO{-}\bar{N}R^2 \xrightarrow{CS_2} R^1{-}COSH + R^2{-}N{=}C{=}S\ \ (60\text{-}95\ \%) \qquad (8.75)$$

benzene. Tertiary amides ($RCONR_2$) are inert because they do not contain ionizable hydrogen atoms.

Michael addition to α,β-unsaturated amides

Michael addition to α,β-unsaturated amides is catalysed by caesium fluoride plus tetraalkoxysilanes (**165**).[97] Many types of compound can be added, for example, ketones, nitroalkanes, ethyl cyanoacetates (**166**), and diethyl malonate (Scheme 8.76).

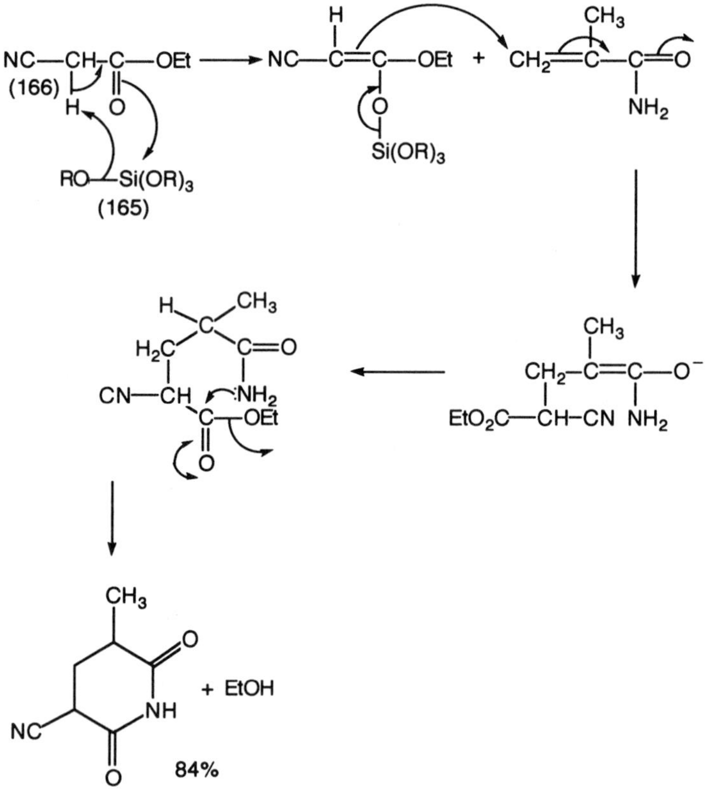

(8.76)

8.2.5 *Acidity of α-hydrogen atoms*

α-Nitration of amides

Because the polarization of the carbonyl group in an amide is opposed by back co-ordination from the amino group, the hydrogens on the α-carbon atom are only weakly acidic. However, a strong base like lithium diisopropylamide (LDA) can remove a proton and thus allow base-catalysed condensations, for example, the α-nitration of amides is possible in this way, to form α-nitroamides (**167**) (Scheme 8.77).

$$R^1\text{—}CH_2\text{—}C(=O)\text{—}NR^2R^3 \xrightarrow{LiNPr^i_2} R^1\text{—}CH{=}C(O^-)\text{—}NR^2R^3 + Pr^nO\text{—}NO_2$$

$$\xrightarrow{\text{i) } Pr^nONO_2 \text{; ii) aq. HOAc}} R^1\text{—}CH(NO_2)\text{—}C(=O)\text{—}NR^2R^3 \quad (167)$$

(8.77)

α-Hydroxylation of amides

α-Hydroxyamides (**168**) have been prepared by reaction of the anion of an amide with bis(trimethylsilyl) peroxide (Scheme 8.78).[98]

$$R^1R^2CH\text{—}C(=O)\text{—}N(Me)Ph \xrightarrow{\text{i) LDA; ii) } (Me_3SiO)_2\text{; iii) } H_3O^+} R^1R^2C(OH)\text{—}C(=O)\text{—}N(Me)Ph \quad (168)$$

Me_3SiCl H_2O_2-(DABCO)

R^1	R^2	Yield (%)
CH_3	H	51
n-C_3H_7	H	56
n-C_4H_9	H	58
n-$C_{14}H_{29}$	H	52

(8.78)

Deprotection of N-benzyl protected amides

N-Benzyl groups have been used for protection of amides. Hydrogenation to remove benzyl groups fails with amides hence another method using either molecular oxygen or oxodiperoxymolybdenum (hexamethylphosphorictri-amide)(pyridine) (MoOPH) has to be used (Scheme 8.79).[99]

$$\mathrm{Ph{-}C(=O){-}N(CH_3){-}CH_2Ph} \xrightarrow[\text{ii) } O_2 \text{ or MoOPH}]{\text{i) } Bu^tLi} \left[\mathrm{Ph{-}C(=O){-}N(CH_3){-}CH(OH)Ph}\right] \longrightarrow \mathrm{Ph{-}C(=O){-}N(CH_3){-}H} + \mathrm{Ph{-}CHO} \quad (79\,\%) \qquad (8.79)$$

8.2.6 *Hydroformylation and amidocarbonylation*

Applied to N-heterocycle synthesis

Rhodium-catalysed hydroformylation and amidocarbonylation of unsaturated amides (**169**) has been used in the production of nitrogen heterocycles (**170**)[100] (Scheme 8.80).

$$\underset{(169)}{\mathrm{CH_2{=}CH{-}CH_2{-}CH_2{-}C(=O){-}NH_2}} \xrightarrow[\mathrm{RhCl(PPh_3)_3}]{H_2/CO} \mathrm{OHC{-}CH(CH_3){-}CH_2{-}CH_2{-}C(=O)NH_2} \longrightarrow \text{6-hydroxy-5-methylpiperidin-2-one} \longrightarrow \text{5-methyl-3,4-dihydro-pyridin-2-one (imine)} \rightleftharpoons \underset{(170)}{\text{5-methyl-3,4-dihydro-1}H\text{-pyridin-2-one}} \qquad (8.80)$$

Applied to chiral α-amino acid synthesis

The method has been developed for asymmetric synthesis of, for example, a chiral α-amino acid (**171**) (Scheme 8.81).[101]

$CH_2{=}C(CO_2CH_3)(NHCOCH_3)$ —[CO / H_2, 100 atm, 80°C, 70 hours; $HRh(CO)(PPh_3)_3$ / chelating diphosphine]→ $H–C(CO_2CH_3)(NHCOCH_3)(CH_3)$ + $H–C(CO_2CH_3)(CH_2CHO)(NHCOCH_3)$ + $OHC–C(CO_2CH_3)(NHCOCH_3)(CH_3)$ (*R*)

—[i) $NaBH_4$ / THF, 25°C, 4 hours; ii) HCl, reflux, 6 hours]→ $HOH_2C–C(COO^-)(\overset{+}{N}H_3)(CH_3)$ (*R*) (171) (8.81)

DIMETHYLFORMAMIDE (DMF)

8.3 As a solvent and as a catalyst

Dimethylformamide (**172**)[102] (b.p. 153 °C) has become one of the most useful amides in organic synthesis both as an aprotic, water-soluble solvent and as a reactant. As a solvent it has been used to advantage to facilitate many reactions,[103] for example, for the generation of anions from β-dicarbonyl compounds such as malonate derivatives.[104] It has been used as a solvent for the demethylation of aryl methyl ethers (**173**) with sodium thioethoxide (**174**) (Scheme 8.82).[105] Yields of about 90 per cent are obtained.

$(CH_3)_2N–C(=O)–H$ (172)

Dimethylformamide is also a good solvent for the demethylation of quaternary heterocyclic iodides (**175**),[106] for example, Scheme 8.83. The rate of this reaction is increased by the addition of triphenylphosphine to the reaction mixture.[107]

Dimethylformamide can be used as a catalyst for the decarboxylation of acids.[108]

OCH3 NaSEt DMF reflux OCH3 aq. H_3O^+ OCH3

O—CH3 (173) S—Et (174) O^- + CH_3-S-Et OH

(8.82)

$\overset{+}{N}$ CH_3 I^- (175) Reflux in DMF for several hours N + CH_3I

(8.83)

8.4 Reactivity

8.4.1 *With organometallic reagents*

The reaction of dimethylformamide (**176**) with organometallic compounds (**177**) has been utilized for the synthesis of aldehydes (**178**) through α-hydroxyamino compounds (**179**) as unstable intermediates, for example, from lithium alkyls[109] yields are *c.* 60 per cent (Scheme 8.84).

R—Li + H—C(=O)—NMe_2 ⟶ [R—CH(O^-)-NMe_2] (177) (176) (179) $\xrightarrow{H_2O}$ R-CHO + $HNMe_2$ (178)

(8.84)

Enamines (**180**) can be prepared in 60–80 per cent yields by the reaction of dimethylformamide (**181**) with a Grignard reagent (**182**) in tetrahydrofuran and ether (Scheme 8.85).[110] Acetylenic Grignard reagents (**183**) have been used in this way to prepare acetylenic aldehydes (**184**).[111] Yields are of the order of 50 per cent (Scheme 8.86).

Aldehydes (**185**) result from the formylation of tertiary Grignard reagents (**186**) with dimethylformamide (**187**) in the Bouveault reaction (Scheme 8.87).[112]

$$\text{R-CH}_2\text{—MgBr} + \text{H—C(=O)—NMe}_2 \xrightarrow[\text{Et}_2\text{O}]{\text{THF}} \left[\text{RCH(H)—CH(O}^-\text{)—NMe}_2\right] \xrightarrow{-\text{HO}^-} \text{R-CH=CH-NMe}_2$$

(182) (181) (180) (8.85)

$$\text{R-C}\equiv\text{C—MgBr} + \text{H—C(=O)—NMe}_2 \longrightarrow \left[\text{R-C}\equiv\text{C—CH(O}^-\text{)—NMe}_2\right] \xrightarrow{\text{aq. H}_2\text{SO}_4} \text{R-C}\equiv\text{C—CHO} + \text{HNMe}_2$$

(183) (184) (8.86)

$$\text{Me}_3\text{C-MgCl} + \text{OHC-NMe}_2 \xrightarrow[\text{acid work-up}]{\text{ether}} \text{Me}_3\text{C-CHO} \quad 60\ \%$$

(186) (187) (185) (8.87)

8.4.2 *With dimethylsulphate*

Meyers and Jagdmann have used dimethylformamide (**188**) and dimethylsulphate (**189**) in a sequence of reactions to prepare in good yields amines (**190**), aldehydes (**191**), and ketones (**192**) via a common intermediate (Scheme 8.88).[113]

8.4.3 *With acid chlorides*

Dimethylformamide (**193**) reacts with acid chlorides (**194**) at 0 °C to give iminium ions as intermediates (**195**) and (**196**), which can be isolated and react with compounds containing a variety of nucleophilic functional groups for examples see Scheme 8.89.[114] The yields are high (80–95 per cent).

8.4.4 *Preparation of epoxides*

Dimethylformamide and chlorine has been used to prepare epoxides (**197**) from phenanthrene, acenaphthene, indene, and stilbenes (**198**),[115] for example, Scheme 8.90.

$(CH_3)_2N\text{-}CHO$ (188) + $(CH_3)_2SO_4$ (189) $\xrightarrow{\text{heat}}$ $(CH_3)_2\overset{+}{N}{=}C(OCH_3)H$ + $CH_3SO_4^-$

$\xrightarrow{H_2N\text{-}C(CH_3)_3,\ CH_2Cl_2}$ $(CH_3)_2N\text{-}CH{=}N\text{-}C(CH_3)_3$

$\xrightarrow{Bu^sLi,\ Me_3SiCl}$ $Me_3Si\text{-}CH_2(CH_3)N\text{-}CH{=}N\text{-}C(CH_3)_3$

$\xrightarrow{\text{i) } Bu^nLi;\ \text{ii) } R^1COR^2}$ $R^1R^2C{=}CH\text{-}N(CH_3)\text{-}CH{=}N\text{-}C(CH_3)_3$

For amines:

$R^1R^2C{=}CH\text{-}N(CH_3)\text{-}CH{=}N\text{-}C(CH_3)_3$ $\xrightarrow{NaBH_4}$ $R^1R^2CH\text{-}CH_2\text{-}N(CH_3)\text{-}CH_2\text{-}NH\text{-}C(CH_3)_3$ $\xrightarrow{\text{aq. } H_3O^+}$ $R^1R^2CH\text{-}CH_2\text{-}NH\text{-}CH_3$

(190)

50 - 70%

For aldehydes:

$R^1R^2C{=}CH\text{-}N(CH_3)\text{-}CH{=}N\text{-}C(CH_3)_3$ $\xrightarrow{H_2N\text{-}N(CH_3)_3}$ $[R^1R^2CH\text{-}CH{=}N\text{-}N(CH_3)_2]$ $\xrightarrow{Cu(OAc)_2,\ THF / H_2O}$ $R^1R^2CH\text{-}CHO$

(191)

55-85 %

cont......

For ketones:

$$R^1R^2C{=}CH{-}N(CH_3){-}CH{=}N{-}C(CH_3)_3 \xrightarrow[\text{ii) } R^3\text{-I}]{\text{i) Bu}^t\text{Li}} \left[R^1R^2C{=}C(R_3){-}N(CH_3){-}CH{=}NC(CH_3)_3\right] \longrightarrow R^1R^2CH_2{-}C(=O){-}R^3$$

50-75 % (192) (8.88)

$$PhO{-}C(=O){-}Cl\ (194) + H{-}C(=O){-}\ddot{N}Me_2\ (193) \xrightarrow{0^\circ C} PhO{-}C(=O){-}O{-}CH{=}\overset{+}{N}Me_2\ Cl^-\ (195)$$

ROH: $\left[PhO{-}C(=O){-}OR\right] \longrightarrow R\text{-}O\text{-}Ph + CO_2$

$R\text{-}\ddot{N}H_2$: $RNH\text{-}CO_2Ph$ A urethane (carbamate)

$R\text{-}CO_2H$: $\left[PhO{-}C(=O){-}O{-}C(=O){-}R\right] \longrightarrow R\text{-}CO_2Ph + CO_2$

$$PhO{-}C(=O){-}Br + H{-}C(=O){-}\ddot{N}Me_2 \xrightarrow{0^\circ C} PhO{-}C(=O){-}O{-}CH{=}\overset{+}{N}Me_2\ Br^-\ (196) \xrightarrow{ROH} R\text{-}Br$$

(8.89)

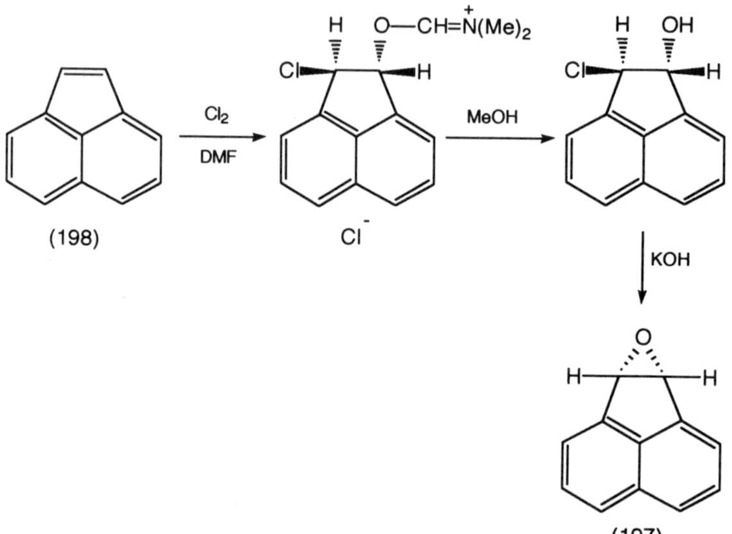

Hydrocarbon	Yield of iminium salt (%)	Yield of chlorohydrin	Yield of epoxide (%)
Phenanthrene	68	68–65`	60
Acenaphthene	58	"	78
Indene	77	"	45
Stilbene *trans*	82	"	80
cis	65	"	—

(8.90)

8.4.5 *With alcohols and benzoyl chloride*

In a one pot reaction, formamide reacts with an alcohol (**199**) and benzoyl chloride (**200**) to give trialkyl orthoformates (**201**) (45–60 per cent for lower alcohols) (Scheme 8.91).[116]

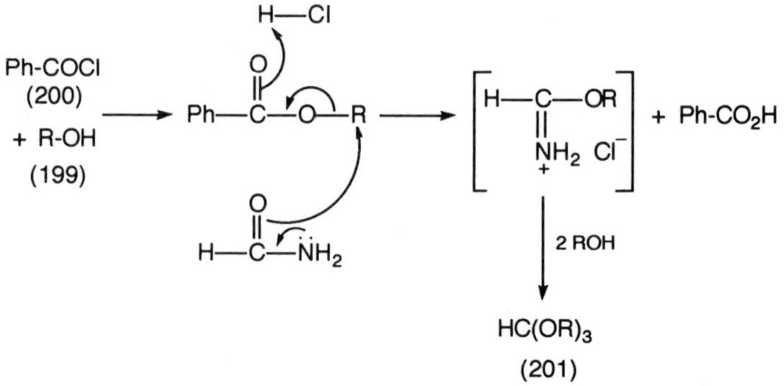

(8.91)

8.4.6 *Synthesis of N,N-dimethylamides*

N,*N*-Dimethylamides (**202**) can be prepared by the reaction of dimethylformamide with a carboxylic acid (**203**) in the presence of half a mole of phosphorus pentoxide (Scheme 8.92).[117]

$$\underset{(203)}{\text{R-CO}_2\text{H}} \xrightarrow[\text{0.5 mole}]{\text{P}_2\text{O}_5} \text{R—C(=O)—O—C(=O)—R} \xrightarrow{\text{H—C(=O)—}\ddot{\text{N}}\text{Me}_2} \text{R—C(=O)—O—C(=O)—H} + \underset{(202)}{\text{R—C(=O)—NMe}_2} \quad (8.92)$$

Acid chlorides (**204**) react similarly by refluxing in dimethylformamide (**205**) (Scheme 8.93).[118]

$$\underset{(204)}{\text{R—C(=O)—Cl}} + \underset{(205)}{\text{Me}_2\ddot{\text{N}}\text{—C(=O)—H}} \xrightarrow{\text{reflux}} \left[\text{R—C(=O)—}\overset{+}{\text{N}}\text{Me}_2\text{(—C(=O)—H)}\right]\text{Cl}^- \longrightarrow \text{R—C(=O)—NMe}_2 + \text{HCl} + \text{CO} \quad (8.93)$$

8.4.7 *The Vilsmeier reaction*

Dimethylformamide (**206**) (and other disubstituted formamides) reacts with phosphoryl chloride (**207**), phosgene (**208**), thionyl chloride (**209**), oxalyl chloride (**210**), or phosphorus pentachloride (**211**) to yield an iminium ion (**212**) (Scheme 8.94). This can be employed in the varied ramifications of the Vilsmeier Reaction,[119] and acts like a Mannich reagent, as an electrophile capable of attacking electron-rich centres in activated aromatic compounds, enols, etc.

The iminium ion (**212**) reacts with a carboxylic acid (**213**) to form an intermediate (**214**) which is not isolated but can be reduced with tri-*tert*-butoxy lithium aluminium hydride (**215**) to an aldehyde (**216**) (Scheme 8.95).[120]

With phosphoryl chloride:

$$Me_2N{-}CH{=}O\ (206) + POCl_3\ (207) \longrightarrow Me_2\overset{+}{N}{=}CH{-}O{-}P(O)Cl_2\ \ Cl^- \longrightarrow Me_2\overset{+}{N}{=}CH{-}Cl\ \ Cl^-\ (212)$$

With phosgene:

$$Me_2N{-}CH{=}O\ (206) + COCl_2\ (208) \longrightarrow Me_2\overset{+}{N}{=}CH{-}O{-}C(O){-}Cl\ \ Cl^- \longrightarrow (212)$$

With thionyl chloride:

$$Me_2N{-}CH{=}O\ (206) + SOCl_2\ (209) \longrightarrow Me_2\overset{+}{N}{=}CH{-}O{-}S(O){-}Cl\ \ Cl^- \longrightarrow (212)$$

With oxalyl chloride:

$$Me_2N{-}CH{=}O\ (206) + ClOC{-}COCl\ (210) \longrightarrow Me_2\overset{+}{N}{=}CH{-}O{-}C(O){-}COCl\ \ Cl^- \longrightarrow (212)$$

With phosphorus pentachloride:

$$Me_2N{-}CH{=}O\ (206) + PCl_5\ (211) \longrightarrow Me_2\overset{+}{N}{=}CH{-}O{-}PCl_4\ \ Cl^- \longrightarrow (212) \quad (8.94)$$

$$\underset{(212)}{\mathrm{Cl(H)C{=}\overset{+}{N}Me_2\ Cl^-}} \xrightarrow[(213)]{\mathrm{R{\cdot}CO_2H}} \left[\underset{(214)}{\mathrm{(COR)O(H)C{=}\overset{+}{N}Me_2\ Cl^-}}\right] \xrightarrow[(215),\mathrm{CuI},\ -78^{\circ}\mathrm{C}]{\mathrm{Li\ AlH(OBu^t)_3}} \underset{(216)}{\mathrm{RCHO}} \qquad (8.95)$$

For references to NMR studies of the electrophilic intermediates see Ref. 121. See Ref. 122 for the oxalyl chloride method.

Some further examples of the Vilsmeier reaction are shown below.

A new variation of the Vilsmeier formylation reaction involves the reaction of triphenylphosphine dibromide (**217**) with dimethylformamide in the presence of a suitably substituted substrate (Scheme 8.96).[123]

Smith[124] investigated the reaction with indole (**218**) and used phosphoryl chloride to generate the electrophile (**219**) (Scheme 8.97). For other examples of reactions with electron-rich aromatic compounds see Ref. 125.

It has been possible to generate the iminium ion intermediate (**220**) from dimethylformamide and phosgene and use it to react with an enol ether (**221**) (Scheme 8.98).[126]

Formylation of alkenes can be accomplished with *N*-disubstituted formamides and phosphorus oxychloride, for example, Scheme 8.99.[127]

Vilsmeier formylation may also be performed on the α-position of acetals and ketals (**222**), so that hydrolysis of the products gives keto aldehydes or dialdehydes (**223**),[128] for example, Scheme 8.100.

A formyl group and a halogen can be added to triple bond containing compounds (**224**) by treatment with *N*,*N*-disubstituted formamides and phosphorus oxychloride (Scheme 8.101).[129]

Hepburn and Hudson generated the iminium salt (88 per cent) by adding phosphorus pentachloride to an excess of dimethylformamide, and used it to obtain 75–85 per cent yields of alkyl halides (**225**) from alcohols (**226**) (Scheme 8.102).[130] These authors converted the chloro-chloride (**227**) into the bromo-bromide (**228**) by the action of hydrogen bromide (**229**) and used the bromo-bromide to obtain alkyl bromides in the same way (Scheme 8.103).

Dimethylformamide (**230**) reacts with dimethyl sulphate (**231**) to yield another resonance stabilized iminium salt (**232**) (Scheme 8.104) which has been used for the formation of aminal esters (**233**), acetals (**234**), and methyl esters (**235**).[131] This iminium salt (**232**) reacts with an amine (**236**) and a sodium alkoxide (RO^-Na^+) to yield an aminal ester (**233**) (Scheme 8.105).[132]

With an aldehyde (**237**) and methanol, the iminium ion (**232**) yields an acetal (**234**). (Scheme 8.106).[133] Yields are in the range 55–80 per cent.

$(Ph_3\overset{+}{P}\text{-}Br)\ Br^-$ (217) + $OHC\text{-}N(CH_3)_2$ → $[Ph_3P(Br)\text{—}O\text{—}\overset{+}{C}H\text{—}N(CH_3)_2]$

H_2O

$+\ (CH_3)_2NH\ +\ HBr$

Alkene	Yield %
N–H	40
S	45
N–H	78

(8.96)

Reaction of the iminium ion (**232**) with carboxylic acids (**238**) gives high yields of esters (**235**), and reaction with sodium methoxide or ethoxide (**239**) gives good yields of the corresponding acetals (**234**) (Scheme 8.107).[134]

$CH{=}\overset{+}{N}Me_2$

H

N

H

$\overset{+}{C}H{=}NMe_2$

Cl

(218) (219)

$CH{=}\overset{+}{N}Me_2$

H

$\overset{+}{N}$

H

Cl^-

CHO

N

H

H_2O

$CH{=}\overset{+}{N}Me_2$

N

H

Cl^-

95 %

$LiAlH_4$ / ether

$CH_2{-}NMe_2$

N

H

Gramine (8.97)

MeO

(221)

$CH{=}\overset{+}{N}Me_2$

Cl (220)

$Me\overset{+}{O}$

H

$CH{=}\overset{+}{N}Me_2$

O

CHO

aq. H_3O^+

MeO

$CH{=}\overset{+}{N}Me_2$ (8.98)

DMF, $POCl_3$; $PO_2Cl_2^-$; Cl ; CH–CH–$\ddot{N}Me_2$; H

H_2O, Na_2CO_3 ; CH—CHO ; 81% ; Cl^- ; CH–CH=$\overset{+}{N}Me_2$; 69%

(8.99)

CH_3 ; H ; O ; (222) ; DMF, $POCl_3$; Cl ; O ; C ; H ; CH_3 ; O ; O—CHO ; 84% ; (223)

(8.100)

$$\underset{(224)}{Ar—C{\equiv}CH} \xrightarrow{Ph\text{-}N(CH_3)CHO,\ POCl_3} Ar—\underset{Cl}{C}H{=}CH–\overset{O}{\overset{\|}{C}}—H$$

R	Yield (%)
Ph	39
p-Br-C_6H_4	41
p-CH_3-C_6H_4	41
p-CH_3O-C_6H_4	67

(8.101)

$$\underset{(226)}{R\text{-}OH} \xrightarrow[75^\circ \text{ to } 100^\circ C]{Me_2\overset{+}{N}{=}CHCl\ Cl^-} \underset{(225)}{R\text{-}Cl}$$

(8.102)

$$Me_2\overset{+}{N}{=}CHCl\ \ Cl^- \xrightarrow{HBr} Me_2\overset{+}{N}{=}CHBr\ \ Br^-$$
(227) (228)

$$R\text{-}OH \xrightarrow[\text{Dioxan, } 100^\circ C]{Me_2\overset{+}{N}{=}CHBr\ \ Br^-} R\text{-}Br$$
(229)

R	Yield (%) (from chloro compound)	Yield (%) (from bromo compound)
$CH_3(CH_2)_3CH_2$	77	89
$(CH_3)_3CCH_2$	73	—
$(CH_3CH_2)_2CH$	78	82
$c\text{-}C_6H_{11}CH_2$	85	82

(8.103)

$$Me_2\ddot{N}\text{-}CH{=}O + (MeO)_2SO_2 \xrightarrow{60^\circ - 80^\circ C} Me_2\overset{+}{N}{=}CH\text{-}O\text{-}S(OMe)(=O)_2\ \ MeO^-$$
(230) (231)

$$\left[Me_2\ddot{N}\text{-}CH{=}\overset{+}{O}Me \longleftrightarrow Me_2\overset{+}{N}{=}CH\text{-}\ddot{O}Me \right]\ MeOSO_3^-$$
(232)

100 %

(8.104)

$$Me_2\overset{+}{N}{=}CH\text{-}OMe\ (232) + H\text{-}\ddot{N}Me_2\ (236) \longrightarrow MeOH + Me_2\overset{+}{N}{=}CH\text{-}NMe_2 \xrightarrow{^-OR} Me_2N\text{-}CH(OR)\text{-}NMe_2$$
(233)

(8.105)

$$\text{RCHO (237)} + \text{CH}_3\text{-O-CH=}\overset{+}{\text{N}}\text{Me}_2\ \text{MeOSO}_3^-\ (232) \longrightarrow \text{R-}\overset{+}{\text{C}}\text{H-OCH}_3\ \text{MeOSO}_3^- + \text{Me}_2\text{N-CHO} \xrightarrow{\text{CH}_3\text{OH}} \text{R-CH(OCH}_3)\text{-OCH}_3\ (234) \qquad (8.106)$$

$$[\text{H-C(OR)=}\overset{+}{\text{N}}\text{Me}_2]\ \text{ROSO}_3^-\ (232) + \text{R}^1\text{CO}_2\text{H}\ (238) \longrightarrow \text{R}^1\text{CO}_2\text{R}\ (235) + \text{ROSO}_3\text{H} + \text{H-C(=O)NMe}_2$$

$$[\text{H-C(OR)=}\overset{+}{\text{N}}\text{Me}_2]\ \text{ROSO}_3^-\ (232) + \text{NaOR}\ (239) \longrightarrow \text{H-C(OR)}_2\text{-NMe}_2\ (234) + \text{ROSO}_3\text{Na}$$

R^1	Yield (%) ($R = CH_3$)	Yield (%) ($R = CH_3CH_2$)
H	97	98
CH_3	97	98
Ph	93	89
Ph-CH=CH	97	94

(8.107)

ACETALS OF DIMETHYLFORMAMIDE

8.5 Preparation

The so-called acetals of dimethylformamide have been utilized for a variety of synthetic purposes. The dimethyl and diethyl acetals (**240**) (R = Me or Et) can be prepared from dimethylformamide as shown in Scheme 8.108.[135]

$$Me_2N{-}\overset{H}{\overset{|}{C}}{=}O \quad R_2\overset{+}{O}(R)\ BF_4^- \longrightarrow Me_2\overset{+}{N}{=}CH{-}OR + BF_4^- + R_2O$$

$$\updownarrow$$

$$\underset{(240)}{Me_2N{-}CH(OR)_2} \xleftarrow{RONa} Me_2N{-}CH{=}\overset{+}{O}R \quad {}^-OR \qquad (8.108)$$

8.6 Reactivity

8.6.1 *With active methylene groups*

They react with active methylene groups in many compounds, for example, ethyl acetoacetate, acetophenone, and nitromethane (**241**) (Scheme 8.109).[135]

$$Me_2N{-}CH(OEt)_2 + \underset{(241)}{CH_3{-}NO_2} \longrightarrow Me_2N{-}C(H){=}C(H)NO_2 \qquad (8.109)$$

8.6.2 *With o-hydroxyacetophenone*

The reaction of the dimethyl acetal with *ortho*-hydroxyacetophenone (**242**) is of interest because the immediate product (**243**) can easily be converted into a heterocyclic compound called chromone (**244**) (Scheme 8.110).[136]

$$\underset{(242)}{\text{2-}HOC_6H_4COCH_3} \xrightarrow[120^\circ C]{(MeO)_2CH{-}NMe_2} \underset{(243)}{\text{2-}HOC_6H_4COCH{=}CHNMe_2}\ (80\ \%) \xrightarrow{aq.\ H_2SO_4} \underset{(244)}{\text{chromone}}\ (71\ \%) + Me_2NH \qquad (8.110)$$

8.6.3 *Preparation of α,β-unsaturated ketones*

The condensation of a ketone with DMF dimethyl acetal can be used to prepare α,β-unsaturated ketones (**245**). The condensation product may be treated with an alkyllithium compound for example, Scheme 8.111.[137]

$(MeO)_2CH\text{-}NMe_2$, 110°C → $=CH\text{-}NMe_2$ 58 %

Bu^nLi → $=CH\text{-}Bu^n$ 75 %

(245)

(8.111)

8.6.4 *As alkylating reagents*

The acetals (**246**) act as good alkylating agents, for example, for the production of esters (**247**) from acids (**248**) (Scheme 8.112) or of ethers (**249**) from phenols (**250**) (Scheme 8.113).[138]

$(CH_3CH_2O)_2CH\text{-}NMe_2$ (246) + $H\text{-}O\text{-}C(=O)\text{-}Ph$ (248) → (CH_3CN, 20°C, 36 hours) → CH_3CH_2OH + $CH_3CH_2\text{-}O\text{-}CH{=}\overset{+}{N}Me_2$ (251) + $Ph\text{-}CO_2^-$

→ $Ph\text{-}CO_2CH_2CH_3$ 90 % (247) + $Me_2N\text{-}CHO$

(8.112)

It should be noted that the iminium ion intermediate (**251**) postulated here is the same as the one produced from dimethylformamide and dimethyl sulphate (see p. 373) which can also be used to alkylate carboxylic acids in good yield. See earlier.

$$\text{PhOH (250)} + (EtO)_2CH\text{-}NMe_2 \xrightarrow[\text{Reflux, 46 hours}]{ClCH_2\text{-}CH_2Cl} \text{PhOEt (249)} + EtOH + Me_2N\text{-}CHO \quad (8.113)$$

Alkylation similarly occurs with heterocyclic thiols (**252**),[139] for example, Scheme 8.114.

$$\text{(252)} + (MeO)_2CH\text{-}NMe_2 \xrightarrow[\text{1 hour}]{\text{PhH, reflux}} \text{2-SMe-pyrimidine, 86 \%} \quad (8.114)$$

8.6.5 *Trans-acetalization*

Trans-acetalization occurs easily under mild conditions and has been used for the regiospecific preparation of steroidal ketals (**253**),[140] for example, Scheme 8.115. A modification of this reaction involves methylation of the acetal (**254**) and thermal decomposition of the resulting quaternary salt (**255**) to yield an alkene (**256**) from a *cis*-diol,[141] for example, Scheme 8.116.

$$Me_2N\text{—}CH(OEt)_2 + HO\text{-}CH_2\text{-}CH_2\text{-}OH \longrightarrow Me_2N\text{-}(1,3\text{-dioxolan-2-yl})$$

$$\text{steroidal ketone} \xrightarrow[\text{CH}_3\text{CO}_2\text{H catalytic amount}]{Me_2N\text{-dioxolane}} \text{(253) 83\%} \quad (8.115)$$

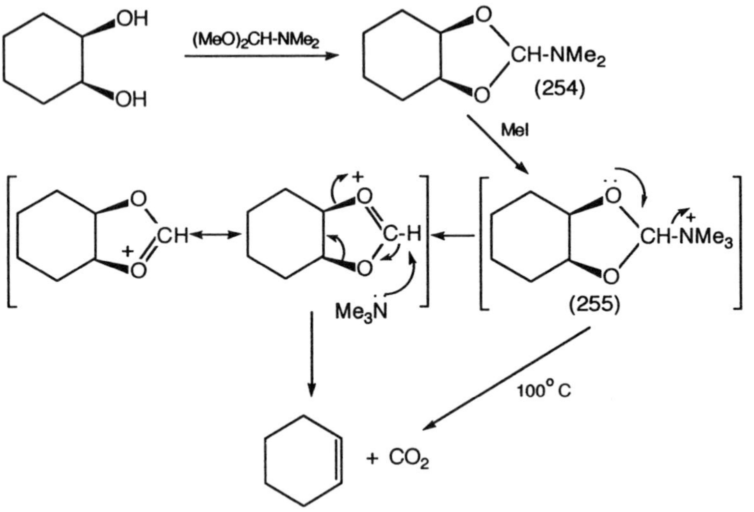

(8.116)

8.6.6 *With 1,2-diols*

The dimethyl acetal reacts with 1,2-diols to yield either an epoxide or the dimethylformamide acetal of the diol, for example, *cis*-cyclohexan-1,2-diol (**257**) forms an acetal (**258**) (Scheme 8.117),[142] whereas the *trans*-diol (**259**) gives a high yield (88 per cent) of the epoxide (**260**) (Scheme 8.118).

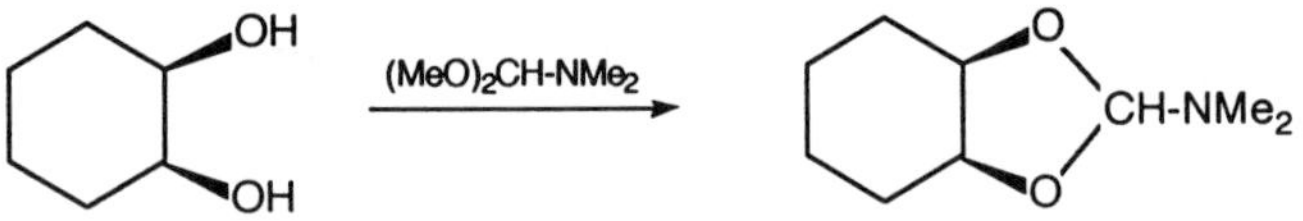

(8.117)

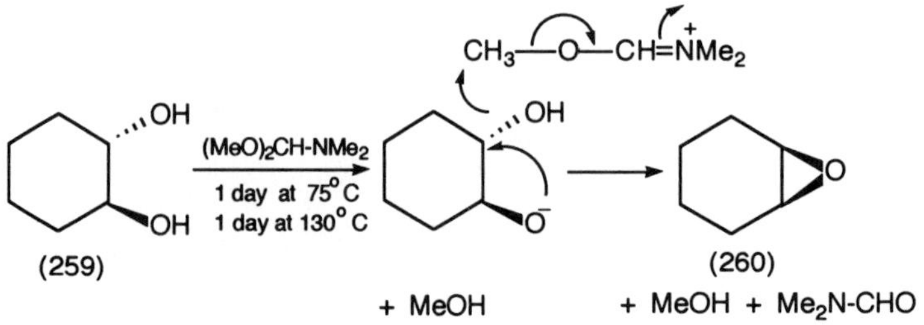

(8.118)

meso-Hydrobenzoin (**261**) gives 73 per cent of the epoxide (**262**) (Scheme 8.119). However, it has been found that both isomers of hydrobenzoin could be converted into alkenes (**263**) via the acetals (**264**) by heating the acetals with

Ph—CH-OH | Ph—CH-OH (261) + $(MeO)_2CH\text{-}NMe_2$ —heat→ (262)

(8.119)

acetic anhydride (Scheme 8.120).[143] The mechanism of this reaction is probably similar to that previously described above for cyclohexan-*cis*-diol involving thermal *cis*-elimination of carbon dioxide, i.e. Scheme 8.121.

$(MeO)_2CH\text{-}NMe_2$

racemic hydrobenzoin → (264) —Ac_2O, 165° to 180° C→ *trans*-stilbene 80 % (263)

meso hydrobenzoin → (264) —Ac_2O, 90° C→ *cis*-stilbene 75 % (263)

(8.120)

Ac_2O; CH-NMe$_2$; CH_3—C—O—C—CH_3; $+ CH_3CO_2^-$; $COCH_3$; $+ CO_2$

(8.121)

A similar alternative method of converting a *cis*-diol into an alkene by *cis*-elimination involves the use of the anhydride of trifluoromethylsulphonic acid (**265**) (Scheme 8.122).[144]

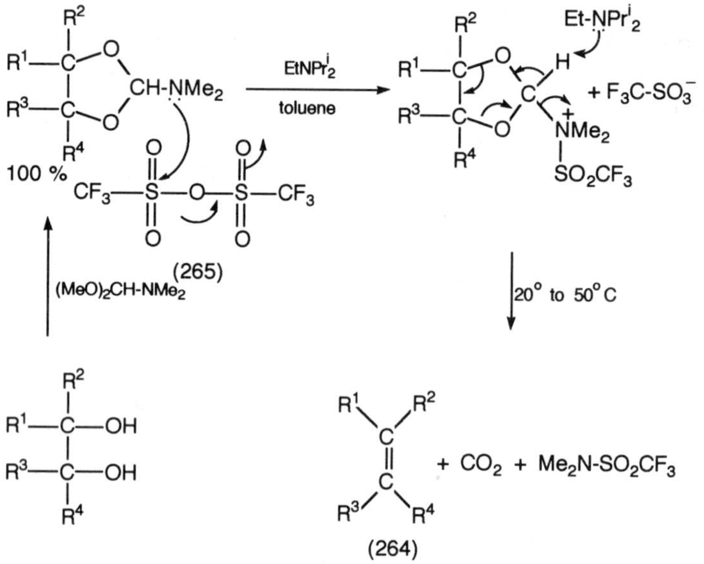

(8.122)

8.6.7 *Regiospecific formation of Mannich bases*

Dimethylformamide dimethyl acetal (**266**) can be used for the regiospecific formation of Mannich bases (**267**) in alkaline media (Scheme 8.123).[145] Reagents other than dimethylformamide dimethyl acetal, but less readily available, are described in Ref. 145 and they give better yields of products than does dimethylformamide dimethyl acetal.

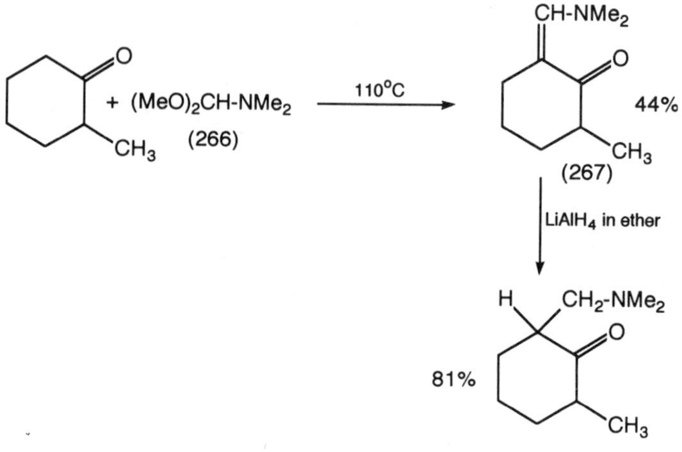

(8.123)

8.6.8 *With β-hydroxycarboxylic acids*

The di-neopentyl acetal of dimethylformamide has been used to obtain the *Z* isomer of a butadiene (**268**) from a *β*-hydroxycarboxylic acid (**269**) which by dehydration through a *β*-lactone (**270**) yields the *E* isomer of the butadiene (**271**) (Scheme 8.124).[146]

Amides react with azides (**272**) in a cyclic aza-Wittig reaction to yield heterocyclic compounds (**273**). An example is illustrated in Scheme 8.125.[147]

PhSO$_2$Cl, pyridine, 0° C
(269) → (270)
Me$_2$N-CH(O-CH$_2$-CMe$_3$)$_2$, CH$_2$Cl$_2$, 0° C
(268) Z-isomer 81 %
DMF, 100° C
(271) E-isomer 93 %

(8.124)

(272)
SOCl$_2$, reflux
R-CONHMe, Et$_3$N / PhH, reflux
Ph$_3$P, 20° C in xylene
(273)
R = CH$_3$, cyclopropyl, CH=CH-Ph
Yields 45 - 85 %

(8.125)

THIOAMIDES

8.7 Preparation

8.7.1 *From nitriles*

Thioamides, and from them amines, have been prepared from nitriles through their reaction with diphenylphosphino-dithioic acid (**274**), which can be prepared in a yield of about 50 per cent from the reaction between phosphorus pentasulphide (**275**) and benzene in the presence of an excess of aluminium chloride (Scheme 8.126).[148] This compound acts as a nucleophile and reacts at the electrophilic carbon atom of the nitrile (**276**) to yield the thioamide (**277**) (Scheme 8.127).[149] The thioamide can be converted to an amine (**278**) by the method of Borch.[150]

$$PhH + P_4S_{10}\ (275) \xrightarrow{AlCl_3} Ph_2PSSH\ (274) \qquad (8.126)$$

R—C≡N (276) + HS—P(=S)Ph₂ (274) —(40° C, 12 hours)→ [R—C(=NH)—S—P(=S)Ph₂ + HS—P(=S)Ph₂ (274)] → Ph₂P(=S)—S—P(=S)Ph₂ + R—C(=S)—NH₂ (277)

(277) —(i) $Et_3O^+BF_4^-$; ii) $NaBH_4$)→ $R\text{-}CH_2\text{-}NH_2$ (278)

R	Yield (%)
Et	83
$PhCH_2$	91
Ph	76

(8.127)

8.7.2 *From amides*

The preparation of thioamides (**279**) from amides (**280**) with phosphorus pentasulphide, P_4S_{10}, has been improved by use of ultrasound (Scheme 8.128).[151]

$$\underset{(280)}{R-\overset{\overset{O}{\|}}{C}-NR^1R^2} \xrightarrow[\text{ultrasound}]{P_4S_{10}} \underset{(279)}{R-\overset{\overset{S}{\|}}{C}-NR^1R^2}$$

R	R^1	R^2	Yield (%)
Ph	Me	Me	78
$PhCH_2$	Me	Me	84
Me	Me	Ph	97

(8.128)

CARBAMATES (URETHANES)

8.8 Preparation

8.8.1 *From primary amines*

Urethanes (**281**) result from the reaction of a primary amine (**282**) with carbon monoxide, oxygen, and an alcohol (**283**). The catalyst is either 5 per cent rhodium on carbon or palladium black and an inorganic halide. e.g. Kl or Csl. The intermediate dialkylureas (**284**) can be isolated when catalysts of lower activity are used (Scheme 8.129).[152]

$$\underset{(282)}{R^1\text{-}NH_2} + CO + \underset{(283)}{R^2\text{-}OH} + O_2 \xrightarrow[\text{CsI}]{\text{Rh-C}} \underset{(281)}{R^1\text{-}NH\text{-}CO_2R^2} + \underset{(284)}{R^1\text{-}NH-\underset{\underset{O}{\|}}{C}-NH\text{-}R^1}$$

(8.129)

8.8.2 *From secondary amines*

Vinyl carbamates (**285**) can be prepared from secondary amines (**286**), acetylene (**287**), and carbon dioxide in a reaction catalysed by ruthenium chloride (Scheme 8.130).[153] Small amounts of butadiene esters (**288**) are sometimes produced.

$$R^1R^2N{-}H + CO_2 + HC{\equiv}CH \xrightarrow[CH_3CN,\ 90\,^\circ C]{RuCl_3} R^1R^2N{-}C(=O){-}O{-}CH{=}CH_2$$

(286) (287) (285)

$$R^1R^2N{-}C(=O){-}O{-}C(=CH_2){-}CH{=}CH_2$$

(288)

Secondary amine	pyrrolidine (N–H)
Yield %	46

(8.130)

8.8.3 *From primary or secondary amines*

Methyl carbamates (**289**) can be prepared from primary or secondary amines (**290**), alkyl halides, and carbon dioxide in a reaction promoted by copper(I) *t*-butoxide (**291**) (Scheme 8.131).[154]

$$CuOC(CH_3)_3 + R^1R^2NH + CO_2 \xrightarrow{Bu^tOH} R^1R^2N{-}COOCu$$

(291) (290)

$$R^1R^2N{-}COOCu \xrightarrow{Bu^t\text{-}NC} R^1R^2N{-}COOCu[CNCBu^t]$$

$$R^1R^2N{-}COOCu[CNCBu^t] \xrightarrow{CH_3I} R^1R^2N{-}COOCH_3$$

(289)

R^1	R^2	Yield (%)
H	H	56
Bu^t	H	99
Et	Et	94
$-(CH_2)_5-$		83
Ph	H	76

(8.131)

8.8.4 *From secondary alcohols and isocyanates*

For the preparation of urethanes (**292**) from secondary alcohols (**293**) and isocyanates (**294**), Lewis acids, viz. boron trifluoride etherate or aluminium chloride, are superior to bases as catalysts (Scheme 8.132).[155]

Good yields (45–96 per cent) of urethanes can also be obtained by the reaction of isocyanates with alcohols in the presence of one equivalent of cuprous chloride in DMF at 25 °C.[156]

$$R_1-CH(OH)-R^2 \; (293) + R^3\text{-N=C=O} \; (294) \xrightarrow[\text{30 minutes to 2 hours in ether or benzene}]{BF_3 / Et_2O} R_1-CH(R^2)-O-C(=O)-NHR^3 \; (292)$$

Alcohol	Yield (%)
3-hydroxy-cyclohex-1-ene-1-CO_2Et	97
cyclopentanol (OH)	86
3-((CO_2Et)(H)methylidene)cyclohexanol (OH)	95

(8.132)

8.8.5 *From carboxylic acids*

Urethanes (**295**) can be prepared from carboxylic acids (**296**) and alcohols (**297**) in the presence of triethylamine and diphenylphosphoryl azide (**298**) (Scheme 8.133).[157]

8.8.6 *Preparation of allylic urethanes*

Allylic urethanes (**299**) can be prepared as shown in Scheme 8.134.[158] The products can be hydrolysed to give allylic amines (**300**) or reduced with lithium aluminium hydride yielding allylic methylamines (**301**).

$$R^1\text{-}CO_2H + N_3PO(OPh)_2 + R^2\text{-}OH \xrightarrow[\text{several hours}]{Et_3N,\ \text{reflux}} R^1\text{-}NH\text{-}CO_2R^2$$

(296) (298) (297) (295)

$N_3PO(OPh)_2$ (298) ← NaN_3 in acetone ← $Cl\text{—}P(=O)(OPh)_2$ ← 180° C ← $POCl_3 + 2PhOH$ (8.133)

$CH_3CH{=}C(CH_3)_2$ + $CH_3O_2C\text{-}N{=}S{=}N\text{-}CO_2CH_3$ → $(CH_3)(H)C{=}C(CH_3)(CH_2\text{·}N(CO_2CH_3)\text{—}SNHCO_2CH_3)$

100 %

$H_2N\text{-}CO_2CH_3 \xrightarrow{Cl_2} Cl_2N\text{-}CO_2CH_3$ (73 %) $\xrightarrow{SCl_2\ /\ \text{pyridine}}$ $CH_3O_2C\text{-}N{=}S{=}N\text{-}CO_2CH_3$

KOH / MeOH / H_2O → $(CH_3)(H)C{=}C(CH_3)(CH_2\text{·}NHCO_2CH_3)$ (299), 73 %

(299) → more vigorous hydrolysis → $(CH_3)(H)C{=}C(CH_3)(CH_2\text{·}NH_2)$ (300)

(299) → $LiAlH_4$ in ether → $(CH_3)(H)C{=}C(CH_3)(CH_2\text{·}NHCH_3)$ (301) (8.134)

8.9 Reactivity

8.9.1 *Degradation to amines*

Urethanes (**302**) are degraded by halogenoboron catechols (**303**) prepared as shown in Scheme 8.135 by the action of a boron trihalide on the diphenol (**304**). The reaction is mild and selective with the bromide being more reactive.[159]

1-Chlorocarbonylbenzotriazole (**305**), prepared from the reaction of benzotriazole and phosgene, reacts with an alcohol (**306**) to form a stable 1-alkoxy-carbonyltriazole (**307**), which reacts with amines (**308**) to form carbamates (**309**) (Scheme 8.136).[160]

Catechol (304) + $BHal_3$ → (303)

Hal = Cl or Br

$$R^1\text{-NH-}CO_2R^2 \xrightarrow{2 \times (303)} R^1\text{-}NH_2 + CO_2 + R^2\text{-OH}$$

(302)

for example

N-BOC-4-($OSiBu^tMe_2$)piperidine → (25°C, 0.5 hours, CH_2Cl_2) → 4-($OSiBu^tMe_2$)piperidine (N–H)

(8.135)

(305) COCl + ROH (306) → (307) COOR + HNR^1R^2 (308) → RO–C(=O)–N(R_1)(R^2) (309) + benzotriazole (N–H)

R	R^1	R^2	Yield (%)
EtEt	Et	49	
Et	H	$c\text{-}C_6H_{11}$	67
Ph	H	$c\text{-}C_6H_{11}$	95
$PhCH_2$	H	$c\text{-}C_6H_{11}$	74

(8.136)

8.9.2 *Generation of a nitrene*

Carboethoxynitrene (**310**) is generated from *N*-*p*-nitrobenzenesulphoxyurethane (**311**) in the singlet state and decays to the triplet state in competition with reaction with an alkene by insertion into a carbon–carbon double bond or a carbon–hydrogen bond (Scheme 8.137).[161]

EtO–C(=O)–NHOH + Cl-SO2-C6H4-NO2-p —(Et3N)→ EtO–C(=O)–NH–O–SO2-C6H4-NO2-p (311) —(Et3N)→ [EtO–C(=O)–N: ⟷ EtO–C(O⁻)=N⁺] singlet state (310)

(310) + cyclohexene → EtO–C(=O)–N(aziridine) 12 % + EtO–C(=O)–N(H)–cyclohexenyl 2%

(310) → EtO–C(=O)–N· triplet state

(8.137)

8.9.3 *Cyclization*

Suitable unsaturated benzylurethanes (**312**) can be caused to cyclize onto the double bond by the action of iodine,[162] for example, Scheme 8.138.

(312) Ph-CH2-O–C(=O)–NMe + I2 —(CH2Cl2, 25°C)→ [iodonium intermediate, Ph-CH2-O–C(=O)–NMe] + I⁻ → cyclized oxonium, CH2Ph, I⁻ → PhCH2I + cyclic carbamate 92 %

(8.138)

DITHIOCARBAMATES

Aryllithiums react with tetraisopropylthiuram disulphide (**313**) to form crystalline dithiocarbamates (**314**), which are cleaved to aryl thiols (**315**) on alkaline hydrolysis (Scheme 8.139).[163]

$$\mathrm{Pr^i_2NH} \xrightarrow[\text{aq. HO}^-]{CS_2} \mathrm{Pr^i_2N{-}C({=}S){-}S^-\ Na^+} \xrightarrow{K_3Fe(CN)_6} \underset{(313)}{\mathrm{Pr^i_2N{-}C({=}S){-}S{-}S{-}C({=}S){-}NPr^i_2}} \xrightarrow{\text{Ar-Li}} \underset{(314)}{\mathrm{Ar{-}S{-}C({=}S){-}NPr^i_2}}\ 82\ \% \xrightarrow{\text{aq. KOH}} \underset{(315)}{\mathrm{ArSH}} \qquad (8.139)$$

SULPHONAMIDES

Sulphonyl groups can be removed from sulphonamides (**316**) to yield amines (**317**) by reduction. A sodium–potassium alloy in isopropyl alcohol has been used,[164] but this mixture is liable to explode. A safer method has been developed in which potassium and dicylcohexyl-18-crown-6 ether are used (Scheme 8.140).[165]

$$\underset{(316)}{\mathrm{R^1R^2N{-}SO_2R^3}} \xrightarrow[\text{dicyclohexyl-18-crown-6}]{K} \left[\mathrm{R^1R^2N^-\ \ K^+}\right] \xrightarrow{\text{HCl}} \underset{(317)}{\mathrm{R^1R^2NH}}$$

$$\left[\mathrm{R^1R^2N^-\ \ K^+}\right] \xrightarrow{Ac_2O\ /\ \text{pyridine}} \mathrm{R^1R^2N{-}COCH_3} \qquad (8.140)$$

References

1. J. Zabicky (ed.), *The chemistry of the amides* (series ed. S. Patai), Wiley, London, 1970.
2. L. J. Bellamy, *The infrared spectra of complex molecules*, 2nd edn, Chapman and Hall, London, 1980.
3. W. D. Phillips, *J. Chem. Phys.*, 1955, **23**, 1363; H. S. Gutowsky, *Discussions Faraday Soc.*, 1955, **19**, 247; H. S. Gutowsky and C. H. Holm, *J. Chem. Phys.*, 1956, **25**, 1228; R. A. Ogg, J. D. Ray, and L. Piette, *J. Mol. Spectrosc.*, 1958, **2**, 66; C. Franconi, *Z. Elektrochem.*, 1961, **65**, 645; D. G. de Kowalewski, *Arkiv Kemi*, 1960, **16**, 373; V. J. Kowalewski and D. G. de Kowalewski, *J. Chem. Phys.*, 1960, **32**, 1272; E. W. Randall and J. D. Baldeschewieler, *J. Mol. Spectrosc.*, 1962, **8**, 365; J. V. Hatton and R. E. Richards, *Mol. Phys.*, 1960, **3**, 253; 1962, **5**, 139; M. T. Rogers and J. C. Woodberry, *J. Phys. Chem.*, 1962, **66**, 540; L. A. LaPlanche and M. T. Rogers, *J. Am. Chem. Soc.*, 1963, **85**, 3728; 1964, **86**, 337; R. M. Moriarty, *J. Org. Chem.*, 1963, **28**, 1296.
4. *F. &F.*, **4**, 242; T. H. Chan and L. T. Wong, *J. Org. Chem.*, 1969, **34**, 2766.
5. T. H. Chan and L. T. Wong, *J. Org. Chem.*, 1971, **36**, 850.
6. *F. &F.*, **6**, 67; J. Tani, O. Oine, and I. Inoue, *Synthesis*, 1975, 714.
7. S. Kim, H. Chang, and Y. K. Ko, *Tetrahedron Lett.*, 1985, **26**, 1341.
8. *F. &F.*, **9**, 50; T. Mukaiyama, N. Morito, and Y. Watanabe, *Chem. Lett.*, 1979, 1305.
9. *F. &F.*, **10**, 41; Y. Watanabe and T. Mukaiyama, *Chem. Lett.*, 1981, 285.
10. *F. &F.*, **10**, 41; J. Diago-Meseguer, A. L. Palomoo-Coll, J. R. Fernandez-Lizarbe, and A. Zugaza-Bibao, *Synthesis*, 1980, 547.
11. M. L. Nielsen and G. Crawford, *Inorg. Synth.*, 1960, **6**, 94.
12. L. Caglioti, M. Poloni, and G. Rosini, *J. Org. Chem.*, 1968, **33**, 2979.
13. R. Adams and I. J. Pachter, *J. Am. Chem. Soc.*, 1952, **74**, 5491.
14. *F. &F.*, **7**, 110; T. Mukaiyama, Y. Aikawa, and S. Kobayashi, *Chem. Lett.*, 1976, 57.
15. T. Kubota, S. Miyashita, T. Kitazume, and N. Ishikawa, *J. Org. Chem.*, 1980, **45**, 5052.
16. T. Shioiri, N. Ninomiya, and S. Yamada, *J. Am. Chem. Soc.*, 1972, **94**, 6203.
17. J. B. Hendrickson and M. S. Hussoin, *J. Org. Chem.*, 1987, **52**, 4137.
18. *F. &F.*, **15**, 319; A. Nordahl and R. Carlson, *Acta Chem. Scand.*, 1988, **B42**, 28.
19. *F. &F.*, **11**, 416; R. Mestrres and C. Palomo, *Synthesis*, 1981, 218; 1982, 288.
20. *F. &F.*, **12**, 65; G. Trapani, A. Reho, and A. Latrofa, *Synthesis*, 1983, 1013.
21. *F. &F.*, **11**, 220; T. Kunieda, Y. Abe, T. Higuchi, and M. Hirobe, *Tetrahedron Lett.*, 1981, **22**, 1257.
22. *F. &F.*, **11**, 518; Y. Nagao, K. Seno, K. Kawabata, T. Miyasaka, S. Takao, and E. Fujita, *Tetrahedron Lett.*, 1980, **21**, 841.
23. *F. &F.*, **14**, 312; M. Mader and P. Helquist, *Tetrahedron Lett.*, 1988, **29**, 3049.
24. H. L. Bassett and C. R. Thomas, *J. Chem. Soc.*, 1954, 1188.
25. *F. &F.*, **11**, 121; J. I. Levin, E. Turos, and S. M. Weinreb, *Synth. Commun.*, 1982, 989.
26. E. S. Stern, *Chem. &Ind.*, 1956, 277.
27. B. Singh, *Tetrahedron Lett.*, 1971, 321.
28. H. Yazawa, K. Tanaka, and K. Kariyama, *Tetrahedron Lett.*, 1974, 3995.

29. *F. &F.*, **5**, 664; J. F. Normant and H. Deshayes, *Bull. Soc. Chim. Fr.*, 1972, 2854.
30. *F. &F.*, **13**, 141; R. Pellegata, A. Italia, M. Villa, G. Palmisano, and G. Lesma, *Synthesis*, 1985, 517; R. O. Sauer, *J. Am. Chem. Soc.*, 1944, **66**, 1707.
31. *F. &F.*, **13**, 297; F. Matsuda, S. Itoh, N. Hattori, M. Yanagiya, and T. Matsumoto, *Tetrahedron*, 1985, **41**, 3625.
32. *F. &F.*, **11**, 178; M. Sekine, M. Satoh, H. Yamagata, and T. Hata, *J. Org. Chem.*, 1980, **45**, 4162.
33. *F. &F.*, **5**, 172; preparation: G. Gregorio, G. Pregaglia, and R. Ugo, *Inorg. Chim. Acta*, 1969, **3**, 89; use: L. Vaska and J. Peone, Jr, *J. Chem. Soc., Chem. Commun.*, 1971, 418; M. A. Bennett and T. Yoshida, *J. Am. Chem. Soc.*, 1973, **95**, 3030.
34. *F. &F.*, **5**, 172; 319; **10**, 305.
35. *F. &F.*, **7**, 304.
36. *F. &F.*, **3**, 147; F. Becke and J. Gnad, *Annalen*, 1968, **713**, 212.
37. *F. &F.*, **5**, 319; F. Becke, H. Fleig, and P. Pässler, *Annalen*, 1971, **749**, 198.
38. *F. &F.*, **1**, 469; C. R. Noller, *Org. Synth. Coll. Vol. 2*, 1943, 586; J. S. Buck, *Org. Synth. Coll. Vol. 2*, 1943, 44.
39. *F. &F.*, **10**, 305; C. Cacchi, D. Misiti, and F. La Torre, *Synthesis*, 1980, 243.
40. *F. &F.*, **7**, 304; J. H. Hall and M. Gisler, *J. Org. Chem.*, 1976, **41**, 3769.
41. S. Linke, *Synthesis*, 1978, 303.
42. *F. &F.*, **15**, 197; K.-T. Lin, M.-H. Shih, H.-W. Huang, and C.-J. Hu, *Synthesis*, 1988, 715.
43. S.-I. Murahashi, T. Naota, and E. Saito, *J. Am. Chem. Soc.*, 1986, **108**, 7846; *F. & F.*, **14**, 136.
44. *F. &F.*, **11**, 491; K. A. Jørgensen, A.-B. A. G. Ghattas, and S.-O. Lawesson, *Tetrahedron*, 1982, **38**, 1163.
45. *F. &F.*, **12**, 120; K. S. Kochhar, D. A. Cottrell, and H. W. Pinnick, *Tetrahedron Lett.*, 1983, **24**, 1323.
46. *F. &F.*, **13**, 260; Y. H. Kim, B. C. Chung, and H. S. Chang, *Tetrahedron Lett.*, 1985, **26**, 1075.
47. *F. &F.*, **4**, 317; N. W. Gilman, *J. Chem. Soc., Chem. Commun.*, 1971, 733.
48. *F. &F.*, **15**, 350; Y. Z. Huang, L. Shi, J. Yang, and J. Zhang, *Tetrahedron Lett.*, 1987, **28**, 2159.
49. *F. &F.*, **4**, 358; M. L. Scheinbaum and M. Dines, *J. Org. Chem.*, 1971, **36**, 3641.
50. J. J. Ritter and P. P. Minieri, *J. Am. Chem. Soc.*, 1948, **70**, 4045.
51. *F. &F.*, **5**, 115; G. B. Olah and J. J. Svoboda, *Synthesis*, 1972, 307.
52. D. H. R. Barton, P. D. Magnus, and R. N. Young, *J. Chem. Soc., Chem. Commun.*, 1973, 331; *F. &F.*, **6**, 108; D. H. R. Barton, P. D. Magnus, J. A. Garbarino, and R. N. Young, *J. Chem. Soc., Perkin Trans. I*, 1974, 2101.
53. *F. &F.*, **5**, 454; N. A. Le Bel, R. M. Cherluck, and E. A. Curtis, *Synthesis*, 1973, 678.
54. *F. &F.*, **4**, 210; T. Shioiri, K. Ninomiya, and S. Yamada, *J. Am. Chem. Soc.*, 1972, **94**, 6203.
55. H. R. Kricheldorf, *Synthesis*, 1972, 551.
56. Review with practical details: M. Carmack and M. A. Spielman, *Org. React.*, 1947, **3**, 83. A more modern review: E. V. Brown, *Synthesis*, 1975, 358.
57. C. Willgerodt, *Ber.*, 1887, **20**, 2467.
58. D. F. De Tar and M. Cormack, *J. Am. Chem. Soc.*, 1946, **68**, 2025; 2029.
59. K. Kindler, *Annalen*, 1923, **431**, 193; 222; K. Kindler and T. Li, *Ber.*, 1941; **74**, 321; for a review see F. Azinger, W. S. Schafer, K. Halcour, A. Saus, and H. Treim, *Angew. Chem., Int. Ed. Engl.*, 1964, **3**, 19.
60. E. Schwenk and E. Bloch, *J. Am. Chem. Soc.*, 1942, **64**, 3051.

61. *F. &F.*, **1**, 279; A. Carayon-Gentil, M. Minot, and P. Charbier, *Bull. Soc. Chim. Fr.*, 1964, 1420.
62. R. Mayer and J. Wehl, *Angew. Chem., Int. Ed. Engl.*, 1964, **3**, 705.
63. *F. &F.*, **12**, 202; T. Mukaiyama, J. Ichikawa, and M. Asami, *Chem. Lett.*, 1983, 293; 683.
64. *F. &F.*, **11**, 405; T. Gajda and A. Zwierzak, *Synthesis*, 1981, 1005.
65. *F. &F.*, **11**, 317; J. Barluenga, C. Jiménez, C. Nájera, and M. Yus, *J. Chem. Soc., Chem. Commun.*, 1981, 670; *J. Chem. Soc., Perkin Trans. I*, 1983, 591.
66. H. E. Zaugg and W. B. Martin, *Org. React.*, 1965, **14**, 52; *F. &F.*, **8**, 506; A. Basha and S. M. Weinreb, *Tetrahedron Lett.*, 1977, 1465.
67. J. Zabicky (ed.), *The chemistry of the carbon–nitrogen double bond* (series ed. S. Patai), Wiley, London, 1970.
68. A. Costa, R. Mestres, and J. M. Riego, *Synth. Commun.*, 1982, 1003.
69. J. Cymerman Craig and A. R. Naik, *J. Am. Chem. Soc.*, 1962, **81**, 3410.
70. *F. &F.*, **6**, 598; **4**, 510; D. H. R. Barton, M. J. Day, R. H. Hesse, and M. M. Pechet, *J. Chem. Soc., Perkin Trans. 1*, 1975, 1764.
71. *F. &F.*, **15**, 281; K. Tanaka, S. Yoshifugi, and Y. Nitta, *Chem. Pharm. Bull.*, 1988, **36**, 3125.
72. *F. &F.*, **9**, 43.
73. *F. &F.*, **10**, 28; H.-J. Schmidt and H. J. Schäfer, *Angew. Chem., Int. Ed. Engl.*, 1981, **20**, 109.
74. *F. &F.*, **13**, 132; J. Garcia, F. Urpi, and J. Vilcrasa, *Tetrahedron Lett.*, 1985, **25**, 4841 (amides); J. Zaloon, M. Calandra, and D. C. Roberts, *J. Org. Chem.*, 1985, **50**, 2601 (peptides).
75. *F. &F.*, **15**, 323; R. Moriarty, B. R. Baily III, O. Prakash, and I. Prakash, *J. Am. Chem. Soc.*, 1985, **107**, 1375; J. Crossy, D. Belotti, A. Thellend, and J. B. Pete, *Synthesis*, 1988, 720.
76. *F. &F.*, **11**, 201; **12**, 179; F. Z. Basha and J. F. DeBernadis, *Tetrahedron Lett.*, 1984, **25**, 5271.
77. *F. &F.*, **6**, 548; H. L. Vaughan and M. D. Robbins, *J. Org. Chem.*, 1975, **40**, 1187.
78. *F. &F.*, **11**, 276; W. J. Greenlee and E. D. Thorsett, *J. Org. Chem.*, 1981, **46**, 5351.
79. *F. &F.*, **15**, 300; J. T. Eaton, W. D. Rounds, J. H. Urbanowicz, and G. W. Gribble, *Tetrahedron Lett.*, 1988, **29**, 6653.
80. *F. &F.*, **4**, 527.
81. H. Bredereck, W. Kantlehner, and D. Schweizer, *Ber.*, 1971, **104**, 3475.
82. *F. &F.*, **1**, 202; H. C. Brown and P. Heim, *J. Am. Chem. Soc.*, 1964, **86**, 3566; Z. B. Papanastassiou and R. J. Brumi, *J. Org. Chem.*, 1964, **29**, 2870.
83. *F. &F.*, **12**, 2176; K. Soai, A. Ookawa, and H. Hayashi, *J. Chem. Soc., Chem. Commun.*, 1983, 668.
84. H. C. Brown and A. Tsukamoto, *J. Am. Chem. Soc.*, 1964, **86**, 1089.
85. L. I. Zakharkin, D. N. Maslin, and V. V. Gavrilenko, *Tetrahedron*, 1969, **25**, 5555; *F. &F.*, **3**, 259.
86. *F. &F.*, **12**, 276; S. Kim and K. H. Ahn, *J. Org. Chem.*, 1984, **49**, 1717.
87. *F. &F.*, **8**, 310; H. C. Brown and S. C. Kim, *Synthesis*, 1977, 635.
88. *F. &F.*, **12**, 446; R. O. Hutchins, K. Learn, F. El-Telbany, and Y. P. Stercho, *J. Org. Chem.*, 1984, **49**, 2438.
89. *F. &F.*, **5**, 324; M. M. Larchevêque and T. Cuvigny, *Compt. Rend.*, 1973, **276(C)**, 209.
90. M. Yamaguchi, T. Waseda, and I. Hirao, *Chem. Lett.*, 1983, 35.
91. Review: E. S. Wallis and J. F. Lane, *Org. React.*, 1946, **3**, 267.

92. A. O. Sy and J. W. Raksis, *Tetrahedron Lett.*, 1980, **21**, 2223.
93. *F. &F.*, **6**, 316; H. E. Baumgarten, H. L. Smith, and A. Staklis, *J. Org. Chem.*, 1975, **40**, 3554.
94. *F. &F.*, **4**, 386; 508; *Can. J. Chem.*, 1972, **50**, 3857.
95. H. Nishiyama, H. Nagase, and K. Ohno, *Tetrahedron Lett.*, 1979, 4671.
96. I. Shahak and Y. Sasson, *J. Am. Chem. Soc.*, 1973, **95**, 3440.
97. *F. &F.*, **13**, 69; C. Chuit, R. J. P. Corriu, R. Perz, and C. Reye, *Tetrahedron*, 1986, **42**, 2293.
98. *F. &F.*, **15**, 41; M. Pohmakotr and C. Winotai, *Synth. Commun.*, 1988, **18**, 2141.
99. *F. &F.*, **15**, 64; R. M. Williams and E. Kwarst, *Tetrahedron Lett.*, 1989, **30**, 451.
100. Review: I. Ojima, *Chem. Rev.*, 1988, **88**, 1011; I. Ojima and A. Korda, *Tetrahedron Lett.*, 1989, 6283.
101. S. Gladiali and L. Pinna, *Tetrahedron Asymmetry*, 1990, **1**, 693.
102. *F. &F.*, **13**, 341–342.
103. Review: *F. &F.*, **1**, 278.
104. H. E. Zaugg, D. A. Dunnigan, R. J. Michaels, L. R. Swett, T. S. Wang, A. H. Summers, and R. W. De Net, *J. Org. Chem.*, 1961, **26**, 2522.
105. *F. &F.*, **3**, 115; G. I. Feutrill and R. N. Mirrington, *Tetrahedron Lett.*, 1970, 1327.
106. *F. &F.*, **4**, 184; D. Aumann and L. W. Deady, *J. Chem. Soc., Chem. Commun.*, 1973, 32.
107. U. Berg, R. Gallo, and J. Metzger, *J. Org. Chem.*, 1976, **41**, 2621; T.-L. Ho, *Synth. Commun.*, 1973, **3**, 99.
108. *F. &F.*, **12**, 203; R. Richter and B. Tucker, *J. Org. Chem.*, 1983, **43**, 2625.
109. E. A. Evans, *J. Chem. Soc.*, 1956, 4691.
110. C. Hansson and B. Wickberg, *J. Org. Chem.*, 1973, **38**, 3074.
111. E. R. H. Jones, L. Skattebøl, and M. C. Whiting, *J. Chem. Soc.*, 1958, 1054.
112. *F. &F.*, **14**, 148.
113. *F. &F.*, **11**, 347; A. I. Meyers and G. E. Jagdmann, Jr, *J. Am. Chem. Soc.*, 1982, **104**, 877.
114. R. R. Koganty, M. B. Shambu, and G. A. Digenis, *Tetrahedron Lett.*, 1973, 4511.
115. *F. &F.*, **5**, 249; M.-C. Lasne, S. Masson and A. Thuillier, *Bull. Soc. Chim. Fr.*, 1973, 1751.
116. *F. &F.*, **3**, 147; R. Ohme and E. Schmitz, *Annalen*, 1968, **716**, 207.
117. *F. &F.*, **2**, 163; H. Schindlbauer, *Monatshefte*, 1968, **99**, 1799.
118. G. M. Coppinger, *J. Am. Chem. Soc.*, 1954, **76**, 1372.
119. A. Vilsmeier and A. Haack, *Ber.*, 1927, **60**, 119; for a review see D. Burn, *Chem. & Ind.*, 1973, **3**, 870.
120. *F. &F.*, **12**, 201; T. Fujisawa, T. Mori, S. Tsuge, and T. Sato, *Tetrahedron Lett.*, 1983, **24**, 1543.
121. S. Patai, *The chemistry of the nitrogen–carbon double bond*, Wiley, London, 1970, p. 651.
122. C. Reichardt and K. Schagerer, *Angew. Chem., Int. Ed. Engl.*, 1973, **12**, 323.
123. *F. &F.*, **3**, 321; H. J. Bestmann, J. Lienert, and L. Mott, *Annalen*, 1968, **718**, 24.
124. G. F. Smith, *J. Chem. Soc.*, 1954, 3842.
125. H. W. Moore and H. R. Snyder, *J. Org. Chem.*, 1964, **29**, 97.
126. D. Burn, G. Cooley, M. T. Davies, J. W. Ducker, P. Feather, B. Ellis, A. K. Hiscock, D. Kirk, A. P. Leftwick, V. Petrow, and D. M. Williamson, *Tetrahedron*, 1964, **20**, 597.
127. D. Burn, *Chem. & Ind.*, 1973, 870; C. Junz and W. Muller, *Ber.*, 1967, **100**, 1536.

128. R. D. Youssefyeh, *Tetrahedron Lett.*, 1964, 2161.
129. V. Q. Yen, *Annales de Chimie [13]*, 1962, **7**, 785.
130. *F. &F.*, **6**, 220; D. R. Hepburn and H. R. Hudson, *Chem. &Ind.*, 1974, 664.
131. *F. &F.*, **5**, 250; H. Bredereck, F. Effenberger, and G. Simchen, *Ber.*, 1963, **96**, 1350.
132. H. Bredereck, G. Simchen, S. Rebsdat, W. Kåntlehner, P. Horn, R. Whal, H. Hoffmann, and P. Grieshaber, *Ber.*, 1968, **101**, 41.
133. W. Kantlehner, H.-D. Gutbrod, and P. Gross, *Annalen*, 1974, 690; 1979, 522; 1362.
134. W. Kantlehner and B. Funke, *Ber.*, 1971, **104**, 3711.
135. *F. &F.*, **1**, 282; H. Meerwein, W. Florian, G. Stopp, and N. Schon, *Annalen*, 1961, **641**, 12.
136. *F. &F.*, **4**, 185; B. Fröhlisch, *Ber.*, 1971, **104**, 348.
137. *F. &F.*, **9**, 183; K. H. Fuhr, *J. Org. Chem.*, 1978, **43**, 4248.
138. H. Vorbrüggen, *Angew. Chem., Int. Ed. Engl.*, 1963, **2**, 211; H. Brechbühler, H. Büchi, E. Hatz, J. Schreiber, and A. Eschenmoser, *Angew. Chem., Int. Ed. Engl.*, 1963, **2**, 212; *Helv. Chim. Acta*, 1965, **48**, 1746.
139. A. Holy, *Tetrahedron Lett.*, 1972, 585.
140. H. Vorbrüggen, *Steroids*, 1963, **1**, 45. For the preparation of the acetal with ethylene glycol see H. Meerwein, W. Florian, G. Stopp, and N. Schon, *Annalen*, 1961, **641**, 2.
141. *F. &F.*, **8**, 192; S. Hanessian, A. Bargiotti, and M. LaRue, *Tetrahedron Lett.*, 1978, 737.
142. *F. &F.*, **3**, 115; H. Neumann, *Chimia*, 1969, **23**, 267.
143. F. W. Eastwood, K. I. Harrington, J. S. Josan, and J. L. Pura, *Tetrahedron Lett.*, 1970, 5223.
144. *F. &F.*, **14**, 324; J. L. King, B. A. Posner, K. T. Mak, and N. C. Yang, *Tetrahedron Lett.*, 1987, **28**, 3919.
145. *F. &F.*, **13**, 120; P. F. Schuda, C. B. Ebner, and T. M. Morgan, *Tetrahedron Lett.*, 1986, **27**, 1567.
146. *F. &F.*, **12**, 204; J. I. Luengo and M. Koreeda, *Tetrahedron Lett.*, 1984, **25**, 4881; M. Koreeda and J. I. Luengo, *J. Org. Chem.*, 1984, **49**, 2079.
147. H. Takeuchi, S. Hagiwara, and S. Eguchi, *Tetrahedron*, 1989, **45**, 6375.
148. *F. &F.*, **10**, 172; W. Higgins, P. W. Vogel, and W. G. Craig, *J. Am. Chem. Soc.*, 1955, **77**, 1864.
149. S. A. Benner, *Tetrahedron Lett.*, 1981, **22**, 1851.
150. *F. &F.*, **2**, 430; S. Raucher and P. Klein, *Tetrahedron Lett.*, 1980, **21**, 4061.
151. *F. &F.*, **11**, 428; S. Raucher and P. Klein, *J. Org. Chem.*, 1981, **46**, 3558.
152. *F. &F.*, **12**, 426; S. Fukuoka, M. Chono, and M. Kohno, *J. Org. Chem.*, 1984, **49**, 1458.
153. *F. &F.*, **14**, 272; Y. Sasaki and P. H. Dixneuf, *J. Org. Chem.*, 1987, **52**, 314.
154. *F. &F.*, **9**, 122; T. Tsuda, H. Washita, K. Watanabe, M. Miwa, and T. Saegusa, *J. Chem. Soc., Chem. Commun.*, 1978, 815.
155. *F. &F.*, **13**, 46; T. Ibuka, G.-N. Chu, T. Aoyagi, K. Kitada, T. Tsukida, and F. Yoneda, *Chem. Pharm. Bull.*, 1985, **33**, 451.
156. *F. &F.*, **15**, 101; M. E. Duggan and J. S. Imagire, *Synthesis*, 1989, 131.
157. *F. &F.*, **4**, 210; T. Schioiri, K. Ninomiya, and S. Yamada, *J. Am. Chem. Soc.*, 1972, **94**, 6203.
158. *F. &F.*, **12**, 57; G. Kresze and H. Münsterer, *J. Org. Chem.*, 1983, **48**, 3561.
159. *F. &F.*, **13**, 47; W. Gerrard, M. F. Lappert, and B. A. Mountfield, *J. Chem. Soc.*,

1959, 1529; R. K. Boekman, Jr and J. C. Potenza, *Tetrahedron Lett.*, 1985, **26**, 1411.
160. *F. & F.*, **8**, 87; I. Butula, L. Curkovic, M. V. Prostenik, V. Vela, and F. Zorko, *Synthesis*, 1977, 704.
161. *F. &F.*, **1**, 736; *F. &F.*, **2**, 296; W. Lwowski and T. J. Maricich, *J. Am. Chem. Soc.*, 1965, **87**, 3630; J. S. McConaghy, Jr and W. Lwowski, *J. Am. Chem. Soc.*, 1967, **89**, 2357; 4450.
162. *F. &F.*, **11**, 266; S. Takano and S. Hatakeyama, *Heterocycles*, 1982, **19**, 1243.
163. *F. &F.*, **12**, 466; R. Rothstein and K. Binovic, *Recl. Trav.*, 1954, **73**, 561; K.-Y. Jen and M. P. Cava, *Tetrahedron Lett.*, 1982, 2001.
164. *F. &F.*, **1**, 1102.
165. *F. &F.*, **11**, 431; T. Ohsawa, T. Takagaki, F. Ikehara, Y. Takahashi, and T. Oishi, *Chem. Pharm. Bull. Japan*, 1982, **30**, 3178.

9

UREAS, THIOUREAS, AND CARBODIIMIDES

UREAS AND THIOUREAS

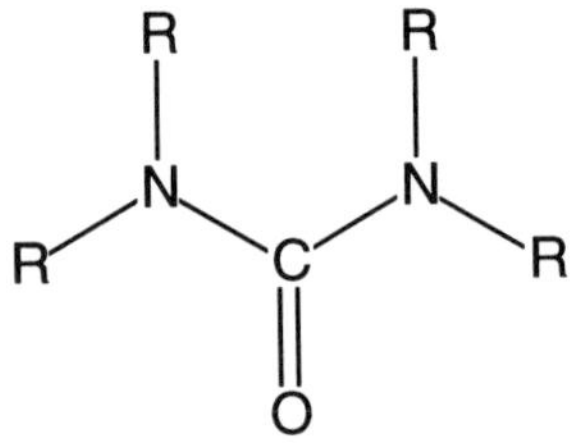

R = hydrogen or alkyl

Urea (**1**) is a typical amide in that, as in other amides, there is delocalization of electrons between the nitrogen atoms and the carbon and oxygen atoms (Scheme 9.1). It differs from ordinary amides since the presence of the second nitrogen atom leads to high resonance stability and hence urea is a weak monacidic base, forming stable salts containing a symmetrical cation (**2**) with, for example, nitric acid (Scheme 9.1).

(1)

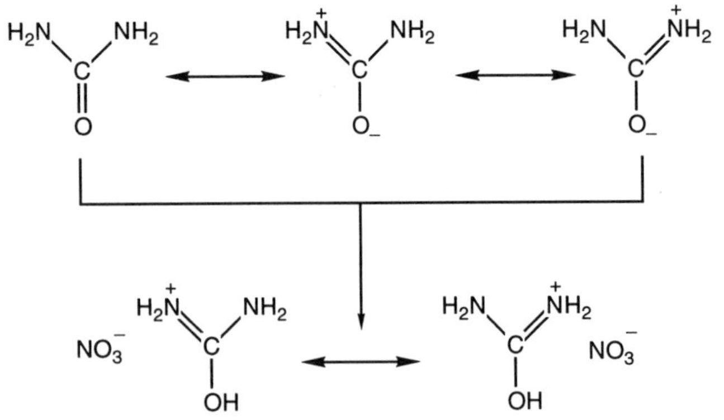

(2)

(9.1)

This structure for urea is confirmed by the shortening of the carbon–nitrogen bond to 1.35 Å which indicates some double bond character. In addition, the two nitrogen atoms, the carbon atom, and the oxygen atom are coplanar (**3**).

$$\text{(CH}_3)_2\text{N–C(=O)–N(CH}_3)_2 \quad \text{C=O } 1.26\ \text{Å},\ \text{C–N } 1.35\ \text{Å} \qquad (3)$$

Urea (m.p. 133 °C) crystallizes from solvents containing long chain hydrocarbons to form inclusion compounds (clathrate compounds), which contain elongated channels into which straight chain hydrocarbons fit but not branched chain hydrocarbons. Thus a separation can be achieved.[1]

9.1 Preparation

9.1.1 *From cyanic acid or isocyanates*

The classical synthesis of urea by Whöler in 1828 was the first synthesis of a naturally occurring organic compound from inorganic materials (Scheme 9.2).[2]

$$NH_4^+ + CNO^- \xrightarrow[\text{heat}]{\text{EtOH}} H_2N\text{—}\overset{\overset{\displaystyle O}{\|}}{C}\text{—}NH_2 \qquad (9.2)$$

The synthesis can be applied to unsymmetrical ureas by using salts of amines (Scheme 9.3), and to thioureas from ammonium thiocyanates (Scheme 9.4).

$$R\overset{+}{N}H_3 + CNO^- \xrightarrow[\text{heat}]{\text{EtOH}} RNH\text{—}\overset{\overset{\displaystyle O}{\|}}{C}\text{—}NH_2 \qquad (9.3)$$

$$R\overset{+}{N}H_3 + CNS^- \xrightarrow{140^\circ C} RNH\text{—}\overset{\overset{\displaystyle S}{\|}}{C}\text{—}NH_2 \qquad (9.4)$$

9.1.2 *By the Curtius, Lossen, and Schmidt reactions*

The Curtius, Schmidt, and Lossen reactions all yield organic isocyanates which react with amines to give ureas (Scheme 9.5). This reaction is usually carried out in the presence of an amine without isolation of the intermediate isocyanate.

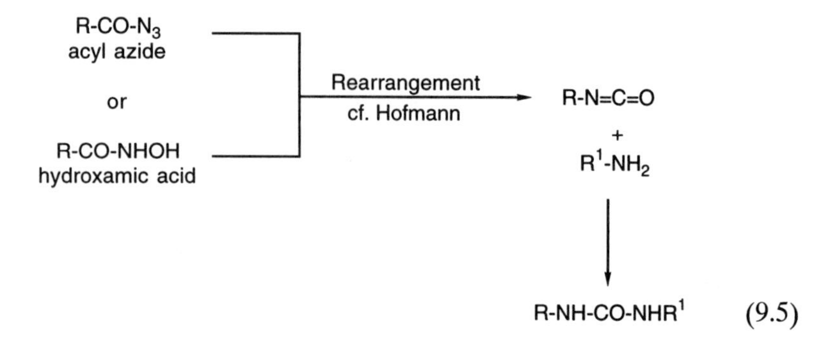

(9.5)

9.1.3 *By oxidation of a primary amide*

Oxidation of a primary amide with lead tetraacetate gives a nitrene (**4**) which rearranges to an isocyanate (**5**). When the reaction is carried out in dimethylformamide in the presence of an amine, a urea is formed (Scheme 9.6). The rearrangement is catalysed by triethylamine or by stannic chloride ($SnCl_4$).[3]

$$\text{R-CO-NH}_2 \xrightarrow[\text{Bu}^t\text{OH}]{\text{Pb(OAc)}_4} \underset{(4)}{\text{R-CO-}\ddot{\text{N}}\text{:}} \longrightarrow \underset{(5)}{\text{R-N=C=O}} \xrightarrow[\text{in DMF}]{\text{Bu}^t\text{NH}_2} \text{RNH-CO-NHBu}^t$$

R	Bu^t	c-C_6H_{11}	$CH_3C(CH_3)(Ph)CH_2$
Yield (%)	96	88	97

(9.6)

9.1.4 *Preparation of substituted ureas*

The action of heat on urea (**6**) yields ammonia and cyanic acid (**7**) which will react *in situ* with an added amine to form a substituted urea (Scheme 9.7).[4]

$$\underset{(6)}{\text{H}_2\text{N-CO-NH}_2} \xrightarrow{\text{heat}} \underset{(7)}{\text{HN=C=O}} + \text{NH}_3 \xrightarrow{\text{R-NH}_2} \text{H}_2\text{N-CO-NHR} \xrightarrow{\text{heat}} \text{R-N=C=O} + \text{NH}_3 \xrightarrow{\text{R-NH}_2} \text{RNH-CO-NHR} \qquad (9.7)$$

9.1.5 *From amines*

Amines react with Grignard reagents to give 'nitrogen' Grignards (**8**) which with aliphatic nitro compounds (nitroalkanes) in the presence of iron pentacarbonyl give, on work-up with dilute sulphuric acid, ureas in good yield (Scheme 9.8).[5]

$$R^1\text{-}NH_2 \xrightarrow{CH_3MgBr} \underset{(8)}{R^1\text{-}NHMgBr} \xrightarrow[\text{ii) dil. } H_2SO_4]{\text{i) } R^2\text{-}NO_2 / Fe(CO)_5} R^1\text{-}NH\text{-}CO\text{-}NHR^2$$

R^1	$Me(CH_2)_{11}$	$c\text{-}C_6H_{11}$	$c\text{-}C_6H_{11}$
R^2	Ph	Ph	$c\text{-}C_6H_{11}$
Yield (%):	42	72	55

(9.8)

Aliphatic amines react with carbon monoxide and oxygen in the presence of selenium as catalyst to give good yields of ureas (Scheme 9.9).[6] Tellurium can also be used as a catalyst in the reaction (Scheme 9.10).[7]

$$2\,R\text{-}NH_2 + CO + Se \xrightarrow{THF} R\text{—}NH\text{-}C(H)\text{—}O\text{—}Se^- + R\overset{+}{N}H_3 \longrightarrow R\text{-}NH_2 + H_2O_2$$

$$R\text{—}NH\text{-}C(H)\text{—}O\text{—}Se^- \xrightarrow{O_2} R\text{-}N{=}C{=}O + Se + HO_2^-$$

$$R\text{-}N{=}C{=}O \xrightarrow{R\text{-}NH_2} RNH\text{-}CO\text{-}NHR$$

R	*n*-butyl	*n*-hexyl	*n*-octyl	cyclohexyl
Yield (%)		95–100		

(9.9)

$$2\ R\text{-}NH_2\ +\ CO \xrightarrow{Te} (R\text{-}NH)_2CO\ +\ H_2 \qquad (9.10)$$

Amines react with carbon dioxide, with elimination of water in the presence of dicyclohexylcarbodiimide (**9**) (which absorbs the water), and a tertiary amine at −75 °C to yield ureas (**10**), (Scheme 9.11).[8] The mechanism is as shown in Scheme 9.12.

$$R\text{-}NH_2 + CO_2 \xrightarrow{-H_2O} R\text{-}C{=}N{=}O$$

NEt3 · RNH2 · RNH2 · (9) · (10)

$R\text{-}NH\text{-}COO^- + H\overset{+}{N}Et_3$

$R\text{-}NH\text{-}C(=O)\text{-}O\text{-}C(NH\text{-}c\text{-}C_6H_{11})(=N\text{-}c\text{-}C_6H_{11})$

$R\text{-}NH\text{-}C(=O)\text{-}NH\text{-}R$ (10)

R	A	B	C	D
		Yield (%) Amine		
Ph_2CH	91	81	85	91
c-C_6H_{11}	98	75	89	91
Ph	—	31	—	—
i-C_3H_7	48	—	—	—
$PhCH_2$	—	—	80	—

Amine: A=NEt_3, B=*N*-methylmorpholine, C=2,6-lutidine, D=pyridine

(9.11)

$O{=}C{=}O$ + $R\ddot{N}H_2$ → c-C_6H_{11}—N=C=N—c-C_6H_{11} / $O{=}C(\text{—}O^-)\text{—}\overset{+}{N}H_2\text{—}R$

→ $O{=}C(\text{—}NH\text{—}R)\text{—}O\text{—}C(\text{—}NH\text{—}c\text{-}C_6H_{11}){=}N\text{—}c\text{-}C_6H_{11}$ + $R\ddot{N}H_2$

→ $HRN\text{—}C(=O)\text{—}NHR$ + c-C_6H_{11}—NH—C(=O)—NH—c-C_6H_{11}

(9.12)

9.1.6 *From thioureas and thioamides*

Ureas (**11**) and amides can be obtained from thioureas (**12**) and thioamides by desulphurization with potassium superoxide in dimethylformamide,[9] for example, Scheme 9.13.

$$\underset{(12)}{\text{Ph—NH—}\overset{\text{S}}{\overset{\|}{\text{C}}}\text{—NH-Ph}} \xrightarrow[\text{18-crown-6-ether}]{\text{KO}_2\text{, DMF, 25}^\circ\text{C}} \underset{(11)}{\text{Ph—NH—}\overset{\text{O}}{\overset{\|}{\text{C}}}\text{—NH-Ph}} \quad 71\% \qquad (9.13)$$

The reaction of KO_2 with *o*-nitrobenzenesulphonyl chloride (**13**) results in a peroxysulphur reagent that converts thioureas, thioamides, and thiocarbamates into the corresponding carbonyl compounds (Scheme 9.14).[10]

$$\underset{(13)}{o\text{-O}_2\text{N-C}_6\text{H}_4\text{-SO}_2\text{Cl}} + \text{KO}_2 + \text{RNR}^1\text{—}\overset{\text{S}}{\overset{\|}{\text{C}}}\text{—NHR}^2 \xrightarrow[-35^\circ\text{C}]{\text{CH}_3\text{CN}} \text{RNR}^1\text{—}\overset{\text{O}}{\overset{\|}{\text{C}}}\text{—NHR}^2$$

R:	Ph	Ph
R^1:	H	Et
R^2:	Et	Ph
Yield (%):	91	88

(9.14)

9.1.7 *Preparation of thioureas*

Tetrasubstituted thioureas (**14**) can be prepared from ureas as shown in Scheme 9.15.[11]

$$\text{R}^1\text{—}\underset{\text{R}^2}{\text{N}}\text{—}\overset{\text{O}}{\overset{\|}{\text{C}}}\text{—}\underset{\text{R}^3}{\text{N}}\text{—R}^4 \xrightarrow[\text{reflux for 2 to 18 hours}]{\text{P}_4\text{S}_{10}\text{, benzene or pyridine}} \underset{(14)}{\text{R}^1\text{—}\underset{\text{R}^2}{\text{N}}\text{—}\overset{\text{S}}{\overset{\|}{\text{C}}}\text{—}\underset{\text{R}^3}{\text{N}}\text{—R}^4}$$

R^1	R^2	R^3	R^4	Yield (%)
CH_3	CH_3	CH_3	CH_3	78
CH_3	CH_3	CH_3	$CH(CH_3)_2$	66
CH_3	CH_3	CH_3	$c\text{-}C_6H_{11}$	56
CH_3	CH_3	CH_3	Ph	46
CH_3	CH_3	Ph	Ph	43
CH_3	Ph	Ph	Ph	6

(9.15)

9.2 Reactivity

9.2.1 *Protonation and alkylation*

Protonation of a urea occurs on oxygen and so does alkylation, for example, with dimethyl sulphate (**15**) (Scheme 9.16). The resulting compound (**16**) is a moderately strong base because with a proton it can form a symmetrical cation (**17**).

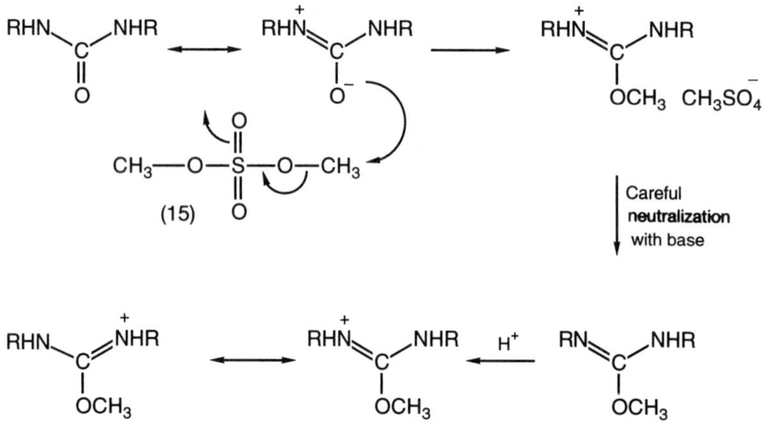

(9.16)

Thiourea reacts similarly with alkylating agents on the sulphur atom and treatment of the resulting thiouronium ion (**18**) with aqueous alkali yields thiols (mercaptans) (**19**).[12] Thiouronium salts are also used for the characterization of carboxylic acids as crystalline salts (**20**) (Scheme 9.17).

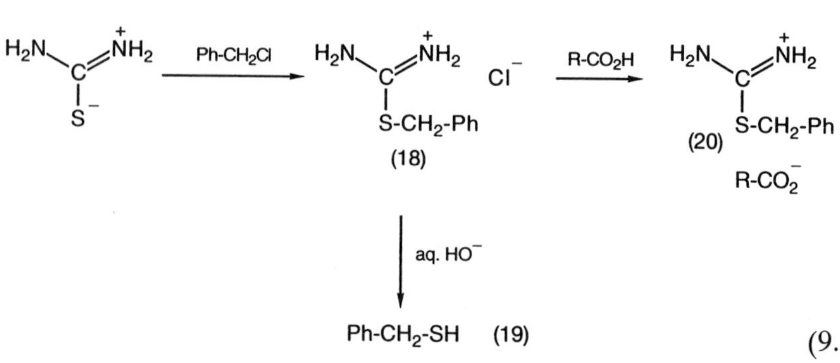

(9.17)

9.2.2 *Normal amide reactions*

Ureas show the usual amide reactions, for example, they are hydrolysed by acids and bases at 100 °C, but rather reluctantly because they are resonance stabilized.

The reaction with nitrous acid is that of a normal amide (Scheme 9.18), but it should be noted that monoalkylated ureas (**21**) yield nitrosoureas (**22**) (Scheme 9.19).

$$H_2N\text{-}CO\text{-}NH_2 \xrightarrow{HNO_2} CO_2 + N_2 + H_2O \quad (9.18)$$

$$\underset{(21)}{RNH\text{-}CO\text{-}NH_2} \xrightarrow{HNO_2} \underset{(22)}{R\text{-}N(N{=}O)\text{-}CO\text{-}NH_2} \quad (9.19)$$

Substituted hydrazines (**23**) are obtained from monoalkylated ureas by Hofmann degradation with bromine and alkali (Scheme 9.20).

$$R\text{-}NH\text{-}CO\text{-}NH_2 \xrightarrow{Br_2\,/\,aq.\,HO^-} \underset{(23)}{R\text{-}NH\text{-}NH_2} \quad (9.20)$$

9.2.3 *Reactions through decomposition to isocyanates*

By the action of heat, ureas decompose and the intermediate isocyanates (**24**) can be trapped with a variety of compounds (Scheme 9.21).

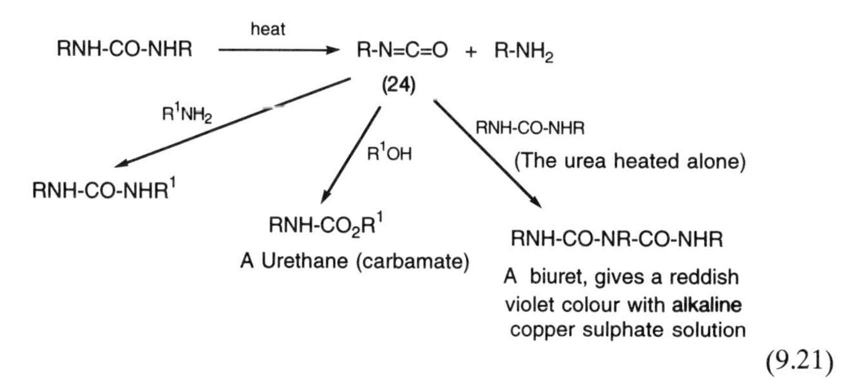

(9.21)

9.2.4 *With β-dicarbonyl compounds to give pyrimidines*

An example of this reaction is the preparation of barbituric acid (**25**, R = H) and substituted barbituric acids (**16**, R = alkyl) from malonic ester and its derivatives (Scheme 9.22).

$$O{=}C(NH_2)_2 + (EtO_2C)_2CH{-}R \xrightarrow{EtO^- / EtOH} \text{(25)} \qquad (9.22)$$

The mechanism of acid-catalysed reaction between urea and pentan-2,4-dione has been investigated by Butler and Leitch,[13] who concluded from kinetic and other data that the rate-determining step is the reaction of the free urea with the protonated enol of pentan-2,4-dione followed by rapid ring closure yielding the pyrimidone (Scheme 9.23).

$$H_2\overset{+}{O}(Me)C{=}CH{-}C(O){-}Me + H_2N{-}C(O){-}NH_2 \xrightarrow{r.d.s.} Me(HN{-}C(O){-}NH_2)C{=}CH{-}C(O){-}Me + H_2O \xrightarrow{fast} \text{pyrimidone} + H_3O^+ \qquad (9.23)$$

9.2.5 *With (−)-ephedrine hydrochloride*

Urea reacts with (−)-ephedrine hydrochloride (**26**) to yield a chiral imidazolidinone (**27**) (Scheme 9.24),[14] which has been used to synthesize several naturally occurring chiral compounds.[15]

$$\underset{}{H_2N-\overset{O}{\overset{\|}{C}}-NH_2} + \underset{(26)}{H_3C-CH(\overset{+}{N}H_2CH_3)-CH(OH)-Ph\ \ Cl^-} \longrightarrow \underset{(27)}{\text{1-methyl-4-methyl-5-phenyl-imidazolidin-2-one},\ \alpha_D -44.5^\circ} \quad (9.24)$$

CARBODIIMIDES

$$R^1\text{—N=C=N—}R^2 \qquad (19)$$

Carbodiimides (**28**) can be regarded as dehydration products of ureas. They find use as dehydrating agents since they easily accept water and are transformed into ureas. In the infra-red the N=C=N bond shows a very strong absorption at 2020–2120 cm^{-1} for both aliphatic and aromatic carbodiimides. The bond angles vary depending on the substitutents.[16]

$$Ar\text{—}N{=}C{=}N\text{—}Ar$$

α = 166 - 170°, β = 123 - 180°, γ = *c.* 90°

9.3 Preparation

9.3.1 *From 1,3-disubstituted ureas and thioureas*

1,3-Disubstituted ureas and thioureas are converted into carbodiimides by triphenylphosphine, carbon tetrachloride, and triethylamine (Scheme 9.25).[17]

$$\begin{matrix} R^1NH\text{-}CO\text{-}NHR^2 \\ \text{or} \\ R^1NH\text{-}CS\text{-}NHR^2 \end{matrix} \xrightarrow[\text{at } 40^\circ C]{Ph_3P\,/\,CCl_4\,/\,CH_2Cl_2} R^1\text{-N=C=N-}R^2$$

R^1	Ph	Ph	Ph
R^2	$C(CH_3)_3$	Ph	$c\text{-}C_6H_{11}$
Yield (%):		85–92	

(9.25)

1,3-Disubstituted ureas can also be converted to carbodiimides in somewhat lower yields by the action of triphenylphosphine dibromide and triethylamine (Scheme 9.26).[18]

R^1-NH-CO-NHR^2 $\xrightarrow[Et_3N]{Ph_3PBr_2}$ [R^1—N(H)—C$^+$(—O—PPh$_3$—Br)—N(H)—R_2 Br^-] ⟶ R^1-N=C=N-R^2

R^1:	Ph	c-C_6H_{11}	Ph	Ph
R^2:	Ph	c-C_6H_{11}	c-C_6H_{11}	*n*-C_4H_9
Yield (%):	75	70	72	66

(9.26)

9.3.2 *From isocyanates*

Carbodiimides are formed from isocyanates by treatment with 3-methyl-1-phenyl-3-phospholene-1-oxide (**29**) (Scheme 9.27).[19]

Primary amines (**30**) react with isonitriles (**31**) catalysed by palladium chloride to yield carbodiimides. The reaction procedes via a carbene-metal complex (**32**) which is converted to the carbodiimide by treatment with silver oxide (Scheme 9.28).[20]

9.3.3 *From thioureas*

Carbodiimides can be prepared from thioureas by treatment with 2 mol of butyllithium in the presence of carbon disulphide. Decomposition of the intermediate yields a carbodiimide (Scheme 9.29).[21]

9.4 Reactivity

Dicyclohexylcarbodiimide (DCC) is the most used carbodiimide.[22] Its reactivity as a dehydrating agent and as a reactant is typical of the general class of carbodiimides, examples of which are to be found in other chapters of this book. For example, dicyclohexylcarbodiimide (**33**) in the presence of copper(I) chloride has been used to invert the stereochemistry of secondary alcohols (Scheme 9.30).[23]

$$2\ R\text{-}N{=}C{=}O \xrightarrow{(29)} R\text{-}N{=}C{=}N\text{-}R + CO_2$$

Mechanism

R:	Ph	o-$CH_3C_6H_4$	m-CH_3-C_6H_4	p-CH_3-C_6H_4	o-NO_2-C_6H_4	n-C_4H_9
Yield (%):	94	87	98	88	95	60

(9.27)

$$R^1\text{-}NH_2\ (30) + 2\ R^2\text{—}\overset{+}{N}{\equiv}\overset{-}{C}\ (31) \xrightarrow{PdCl_2} PdCl_2R^2\text{—}\overset{+}{N}{\equiv}\overset{-}{C}\left[:C(NHR^1)NHR^2\right]\ (32) \xrightarrow{Ag_2O} R^1\text{-}N{=}C{=}N\text{-}R^2$$

R^1:	n-C_4H_9	c-C_6H_{11}	Ph	$CH_2{=}CHCH_2$	$CH_3CH(CO_2C_2H_5)$	c-C_6H_{11}
R^2:	t-C_4H_9	t-C_4H_9	t-C_4H_9	t-C_4H_9	c-C_6H_{11}	
Yield (%):	93	94	72	77	80	77

(9.28)

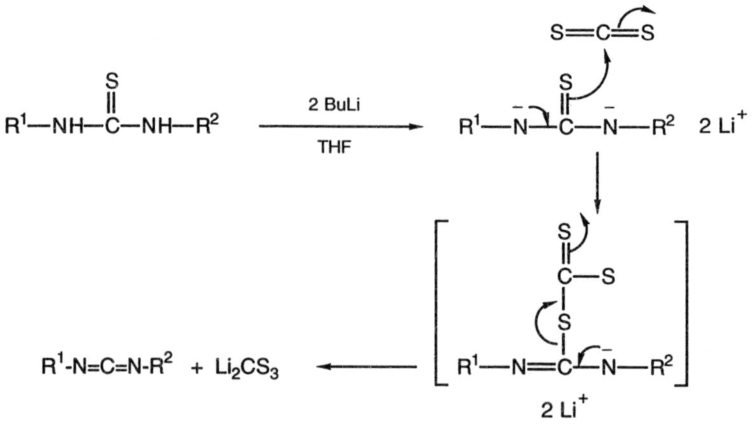

$$R^1{-}NH{-}C({=}S){-}NH{-}R^2 \xrightarrow[\text{THF}]{\text{2 BuLi}} R^1{-}\bar{N}{-}C({=}S){-}\bar{N}{-}R^2 \ \ 2\,Li^+ \xrightarrow{S=C=S} \left[R^1{-}N{=}C(S{-}C({=}S){-}S)(\bar{N}{-}R^2)\right] 2\,Li^+ \longrightarrow R^1\text{-}N{=}C{=}N\text{-}R^2 + Li_2CS_3$$

R^1	R^2
2,6-$Me_2C_6H_3$	2,6-$Me_2C_6H_3$
2-Me-C_6H_4	2-Me-C_6H_4
Me_3C	Me_3C
c-C_6H_{11}	c-C_6H_{11}
4-Me-C_6H_4	4-Me-C_6H_4
2,6-$Me_2C_6H_3$	Ph
2-Me-C_6H_4	Ph

Yields 65–85%

(9.29)

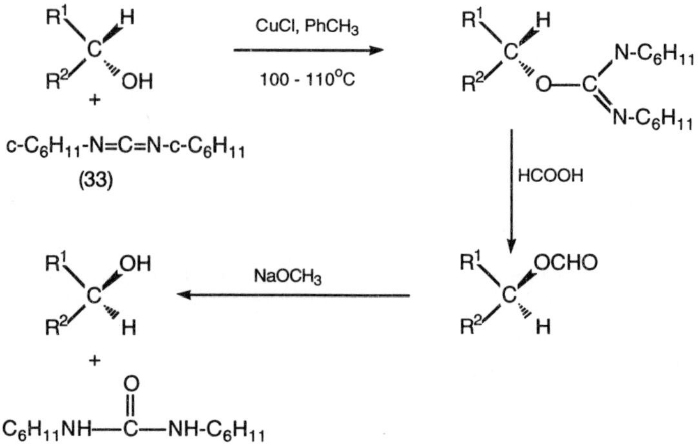

$$R^1R^2CH{-}OH + \text{c-}C_6H_{11}\text{-}N{=}C{=}N\text{-c-}C_6H_{11}\ (33) \xrightarrow[100-110^{\circ}C]{CuCl,\ PhCH_3} R^1R^2CH{-}O{-}C({=}N\text{-}C_6H_{11})(N\text{-}C_6H_{11}) \xrightarrow{HCOOH} R^1R^2CH{-}OCHO \xrightarrow{NaOCH_3} R^1R^2CH{-}OH + C_6H_{11}NH{-}C({=}O){-}NH\text{-}C_6H_{11}$$

Alcohol	Yield%
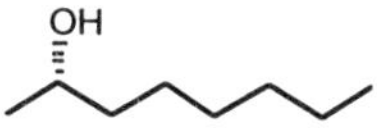	90
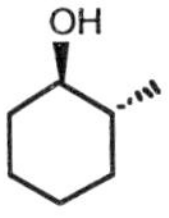	91
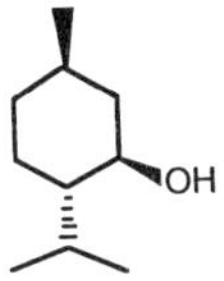	90

(9.30)

References

1. *F. &F.*, **1**, 1262; W. Schlenk, Jr, *Annalen*, 1949, **565**, 204; W. Schlenk, Jr, *Annalen*, 1951, **573**, 142; W. Schlenk, Jr, *Fortschr. Chem. Forsch.*, 1951, **2**, 92; R. W. Schiessler and D. Flitter, *J. Am. Chem. Soc.*, 1952, **74**, 1720; A. E. Smith, *J. Chem. Phys.*, 1950, **18**, 150.
2. Review: J. Shorter, *Quarterly Rev.*, 1978, 1.
3. *F. &F.*, **6**, 316; H. E. Baumgarten, H. L. Smith, and A. Stacklis, *J. Org. Chem.*, 1975, **40**, 3554.
4. A. Sonn, *Ber.*, 1914, **47**, 2437.
5. *F. & F.*, **7**, 183; M. Yamashita, K. Mizushima, Y. Watanabe, T. Mitsudo, and Y. Takegami, *J. Chem. Soc., Chem. Commun.*, 1976, 670.
6. *F. &F.*, **4**, 422; N. Sonada, T. Yasuhara, K. Kondo, T. Ikeda, and T. Tsutsumi, *J. Am. Chem. Soc.*, 1971, **93**, 6344.
7. *F. &F.*, **11**, 498; N. Kambe, K. Kondo, H. Ishii, and N. Sonoda, *Bull. Chem. Soc. Japan*, 1981, **54**, 1460.
8. *F. &F.*, **8**, 163; H. Ogura, K. Takeda, R. Tokue, and T. Kobayashi, *Synthesis*, 1978, 394.
9. *F. &F.*, **11**, 443; E. Katori, T. Nagaro, T. Kunieda, and M. Hirobe, *Chem. Pharm. Bull. Japan*, 1981, **29**, 3075.
10. *F. &F.*, **13**, 260; Y. H. Kim, B. C. Chung, and H. S. Chang, *Tetrahedron Lett.*, 1985, **26**, 1075.
11. *F. &F.*, **5**, 534; Z. Voss, *Annalen*, 1971, **746**, 92.
12. *F. &F.*, **1**, 1165; G. G. Urquhart, J. W. Gates, Jr, and R. Connor, *Org. Synth., Coll. Vol. 3*, 1955, 363; A. J. Speziale, *Org. Synth., Coll. Vol. 4*, 1963, 401; 1963, 491; P. R. Shildneck and W. Wimdus, *Org. Synth., Coll. Vol. 2*, 1943, 411.
13. A. R. Butler and E. Leitch, *J. Chem. Soc., Perkin Trans. II*, 1976, 832.
14. *F. &F.*, **15**, 154; H. Roder, G. Helmchen, E. M. Peters, and H. G. von Schnering, *Angew. Chem., Int. Ed. Engl.*, 1984, **23**, 898.
15. E. Stephan, G. Pourcelot, and P. Cresson, *Chem. &Ind.*, 1988, 562 [(−)-citronellic acid]; C. Fehr and J. Galindo, *J. Am. Chem. Soc.*, 1988, **110**, 6909 [(*R*)- and (*S*)-damascone].

16. F. Kurzer and K. Douraghi-Zadel, *Chem. Rev.*, 1967, **67**, 107; A. Williams and I. T. Ibrahim, *Chem. Rev.*, 1981, **81**, 599.
17. *F. &F.*, **4**, 552; R. Appel, R. Kleinstück, and K. D. Ziehen, *Ber.*, 1971, **104**, 1335.
18. *F. &F.*, **3**, 321; H. J. Bestmann, J. Lienert, and L. Mott, *Annalen*, 1968, **718**, 24.
19. *F. &F.*, **1**, 337; T. W. Campbell and J. Monagle, *Org. Synth.*, 1963, **43**, 31; T. W. Campbell, J. Monagle, and V. S. Foldi, *J. Am. Chem. Soc.*, 1962, **84**, 3673; 4288.
20. *F. &F.*, **6**, 450; Y. Ito, T. Hirao, and T. Saegusa, *J. Org. Chem.*, 1975, **40**, 2981.
21. *F. &F.*, **7**, 45; S. Sakai, T. Fujinama, N. Otani, and T. Aizawa, *Chem. Lett.*, 1976, 811.
22. *F. &F.*, **1**, 231; **2**, 126; **3**, 91; **4**, 141; **5**, 206; **6**, 174; **7**, 100; **8**, 162; **9**, 156; **10**, 142; **11**, 173; **13**, 333; **14**, 131; **15**, 146.
23. *F. &F.*, **14**, 131; J. Kaulen, *Angew. Chem., Int. Ed. Engl.*, 1987, **26**, 773.

10

AMIDINES

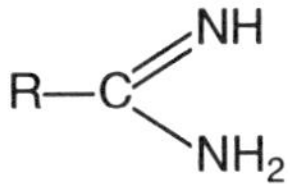

$$R{-}C{\overset{\displaystyle\diagup\!\!\!\diagup NH}{\underset{\displaystyle\diagdown NH_2}{}}}\quad (1)$$

Amidines (**1**)[1] are molecules which contain two nitrogen atoms joined to one carbon atom. Their stability arises from the double bond which gives rise to resonance possibilities (**2**). They can be regarded as analogous to the amides with the C=NH group replacing the carbonyl group, C=O. Amidines are strong monacidic bases ($pK_a = 8.3$–12.4) because by accepting a proton they form a symmetrical cation which is stabilized by delocalization (**3**).

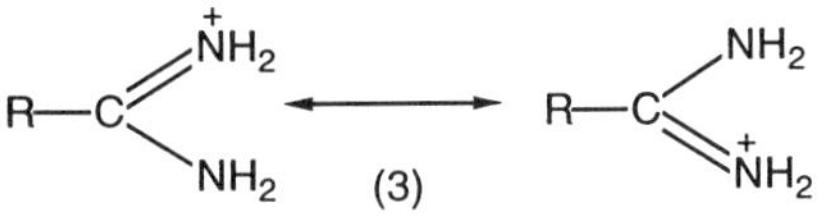

$$R{-}C(={\overset{+}{N}H_2})(NH_2) \longleftrightarrow R{-}C(NH_2)(={\overset{+}{N}H_2})\quad (3)$$

$$R{-}C(=NH)(NH_2) \longleftrightarrow R{-}C(\overset{-}{N}H)(={\overset{+}{N}H_2})\quad (2)$$

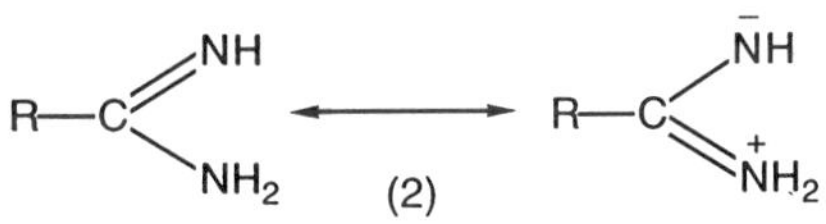

No direct structural determinations have been performed on unsubstituted amidine compounds. However, investigation of formamidoxime (**4**)[2] in which the oxygen substituent on nitrogen does not greatly affect the π-system, has shown the molecule to be planar with the molecular dimensions shown. The C=N double bond is only slightly longer than an unconjugated double bond (1.29 Å cf. 1.27 Å), while the C—N single bond is appreciably shorter than a pure single bond (1.33 Å cf. 1.47 Å).

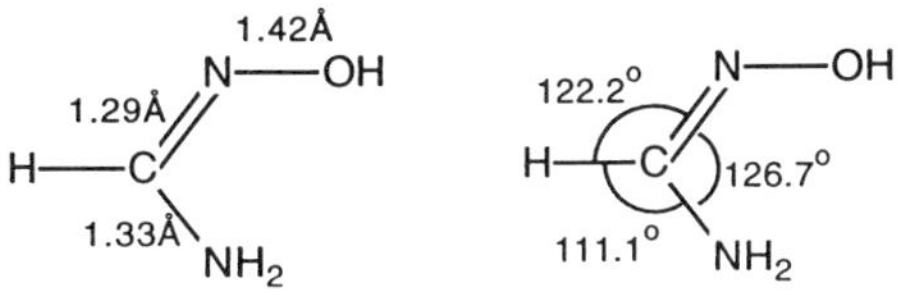

H—C(=N—OH)(NH₂): C=N 1.29Å, N—OH 1.42Å, C—NH₂ 1.33Å; angles 122.2°, 126.7°, 111.1°

(4)

The infra-red spectrum shows an NH vibration at 3249 cm^{-1}, a C=N vibration at 1650 cm^{-1}, and asymmetric and symmetric NH_2 vibrations in the range 3226 cm^{-1} to 3330 cm^{-1}.

Amidines can show geometrical isomerism about the double bond and *E* and *Z* isomers (**5**) have been observed. The *Z* isomer is thermodynamically less stable.

$$\underset{Z}{\mathrm{H(NR_2)C{=}N{-}R}} \qquad \underset{E}{\mathrm{H(NR_2)C{=}N{-}R}}$$

(5)

10.1 Preparation

10.1.1 *The Pinner synthesis*

The classical synthesis of amidines, by the reaction of a nitrile with ethanol and dry hydrogen chloride to form the hydrochloride of an imino ether, was discovered by Pinner.[3] The intermediate imino ether hydrochloride (**6**) crystallizes and is then treated with ammonia (Scheme 10.1).

$$\mathrm{R{-}C{\equiv}N} \underset{}{\overset{HCl}{\rightleftharpoons}} \mathrm{R{-}C{\equiv}\overset{+}{N}{-}H} \xrightarrow[HCl]{EtOH} \underset{(6)}{\mathrm{R{-}C(OEt){=}\overset{+}{N}H_2\,Cl^-}} \xrightarrow{\text{Excess } NH_3} \mathrm{R{-}C(NH_2){=}NH} \quad (10.1)$$

10.1.2 *Direct reaction of ammonia with nitriles*

Amidines can be prepared by the direct addition of ammonia across the carbon–nitrogen triple bond of a nitrile. This can be achieved by heating the nitrile with an ammonium salt (Scheme 10.2).[4]

$$HCl + NH_3 \rightleftharpoons$$

$$R{-}C{\equiv}N + NH_4Cl \rightleftharpoons R{-}C{\equiv}\overset{+}{N}{-}H \;(:NH_3) \xrightarrow{\text{heat}} R{-}C(=\overset{+}{N}H_2){-}NH_2\ Cl^- \xrightarrow{\text{aq. } HO^-} R{-}C(=NH){-}NH_2 \qquad (10.2)$$

10.1.3 *From a nitrile, an alkyl halide, and an amine*

A nitrile reacts with an alkyl halide and an amine in the presence of ferric chloride in a multi-step reaction to yield an amidine after treatment of the reaction mixture with alkali. The ferric ion acts as a Lewis acid and enables an amine to attack the electrophilic carbon atom of the nitrile generated in this way (Scheme 10.3).[5]

$$Me{-}\overset{+}{C}{=}\overset{-}{N} + FeCl_3 \longrightarrow Me{-}\overset{+}{C}{=}N{-}\overset{-}{F}eCl_3 \xrightarrow{+\ R^3{-}Cl} Me{-}C{\equiv}\overset{+}{N}R^3.\overset{-}{F}eCl_4 \xrightarrow{H\ddot{N}R^1R^2} Me{-}\overset{+}{C}(NR^3H)(NR^1R^2) + FeCl_4^- \xrightarrow{\text{aq. NaOH}} Me{-}C(=NR^3){-}NR^1R^2$$

R^1	R^2	R^3	Yield (%)
H	H	$CHMe_2$	30
H	H	CMe_3	30
Et	Et	CMe_3	34
$-(CH_2)_5-$		$CHMe_2$	75
$-(CH_2)_4-$		CMe_3	30
$-(CH_2)_5-$		CMe_3	40
$-(CH_2)_2O(CH_2)_2-$		CMe_3	38

(10.3)

10.1.4 *From an isonitrile and an amine*

The reaction of an isonitrile with an amine in the presence of silver chloride at low temperature yields the *Z* isomer of an amidine which may be converted to the more thermodynamically stable *E* isomer at higher temperatures (Scheme 10.4; the authors recorded no yields).[6]

$$R^1—\overset{+}{N}≡\overset{-}{C} + HNR^2R^3 \xrightarrow[0^\circ C]{AgCl} (Z)\text{-}R^1N=CH(NR^2R^3) \xrightarrow{25^\circ C} (E)\text{-}R^1N=CH(NR^2R^3)$$

Mechanism:

$$R^1—\overset{+}{N}≡\overset{-}{C} \xrightarrow{AgCl} R^1—\overset{+}{N}≡C—\overset{-}{Ag}—Cl \xrightarrow{HNR^2R^3} R^1—N=C(—\overset{-}{Ag}—Cl)(—\overset{+}{N}HR^2R^3) \xrightarrow{-H^+} R^1—N=C(—\overset{-}{Ag}—Cl)(—NR^2R^3) \longrightarrow AgCl + (Z)\text{-}R^1N=CH(NR^2R^3)$$

R^1 = *m*-ClC_6H_4, *p*-$NO_2C_6H_4$, C_6H_5

R^2 = R^3 = $(CH_2)_2O$

and

R^1 = *o*-ClC_6H_4, *m*-ClC_6H_4, *p*-ClC_6H_4, *o*-$NO_2C_6H_4$, *p*-$NO_2C_6H_4$, C_6H_5

R^2 = Me Et; R^3 = Me Et; R^2R^3 = CH_2CH_2, $(CH_2)_4$, $(CH_2)_5$ (10.4)

10.1.5 *The action of HMPT on secondary amides*

The useful reagent hexamethyl phosphoric triamide (HMPT) (**7**) can be used to prepare *N,N*-dimethylamidines from secondary amides (Scheme 10.5).[7]

10.1.6 *Preparation of dichloroamidines*

N,N-Diethyl-1,2,2-trichlorovinylamine (**8**), prepared as shown in Scheme 10.6, converts amines into dichloroamidines in good yields.[8]

$$R^1-C(=O)-NHR^2 \rightleftharpoons R^1-C(-OH)=NR^2 \xrightarrow{(7)\ O=P(NMe_2)_3} R^1-C(-O-P(=O)(NMe_2)_2)=NR^2 \xrightarrow{HNMe_2} R^1-C(-NMe_2)=NR^2 + HO-P(=O)(NMe_2)_2 \quad (10.5)$$

$$Cl_3C-C(=O)-NEt_2 + (EtO)_3P \longrightarrow \underset{(8)}{Cl_2C=C(Cl)-NEt_2} + (EtO)_3P=O$$

$$R\text{-}NH_2 + \underset{(8)}{Cl_2C=C(Cl)-NEt_2} \longrightarrow Cl_2CH-C(=N-R)-NEt_2 + HCl \quad (10.6)$$

10.1.7 *Amidine salts from DMF and thionyl chloride and their uses*

Dimethylformamide reacts with thionyl chloride to give an amidine salt (**9**) (Scheme 10.7).[9] This salt reacts readily with nucleophilic anions to yield interesting compounds. The amidinium salt (**9**) reacts with hydrogen cyanide as shown in Scheme 10.8.[10]

With lithium dimethylamide the reaction is as expected (Scheme 10.9).[11]

$$2\ H-C(=O)-NMe_2 \xrightarrow[\substack{110^\circ C \\ 2\ hours}]{SOCl_2} \underset{(9)}{H-C(NMe_2)=\overset{+}{N}Me_2\ Cl^-} \quad (10.7)$$

$$\underset{(9)}{H-C(NMe_2)=\overset{+}{N}Me_2\ Cl^-} \xrightarrow{HCN} \left[H-C(NMe_2)(CN)-NMe_2 + HCl \quad {}^-CN \right] \xrightarrow{HCN} (NC)_2CH-NMe_2 + H_2\overset{+}{N}Me_2\ Cl^- \quad (10.8)$$

$$\text{H}-\overset{\overset{\text{NMe}_2}{|}}{\text{C}}=\overset{+}{\text{N}}\text{Me}_2 \;(9) + \text{Li}^+ \, {}^-\text{NMe}_2 \xrightarrow{-20^\circ\text{C}} \text{H}-\text{C}(\text{NMe}_2)_3 \;\; 67\% \qquad (10.9)$$

10.1.8 *From acids and amines*

Amidines can be prepared from acids and amines by the use of phosphorus pentoxide with hexamethyldisiloxane, $(CH_3)_3SiOSi(CH_3)_3$, in methylene dichloride at 160 °C (Scheme 10.10).[12]

$$\text{R}^1\text{-CO}_2\text{H} + 2\ \text{R}^2\text{—NH}_2 \xrightarrow[\text{in CH}_2\text{Cl}_2\text{ at }160^\circ\text{C}]{\text{P}_2\text{O}_5\text{ / hexamethyldisiloxane}} \text{R}^1\text{—}\overset{\overset{\text{NR}^2}{\|}}{\text{C}}\text{—NHR}^2 \;\; 80\text{-}90\% \qquad (10.10)$$

10.1.9 *From cyanuric chloride, DMF, and amines*

Cyanuric chloride (**10**) reacts with dimethylformamide to yield a resonance stabilized compound (**11**),[13] which reacts with amines to yield amidines (**12**) and with amides giving acylamidines (**13**)[14] (Scheme 10.11).

10.2 Reactivity

10.2.1 *Synthesis of pyrimidines and imidazoles*

The nucleophilic reactivity of nitrogen in amidines provides a means of synthesizing pyrimidines and imidazoles. Pyrimidines (**14**) result from the reaction of amidines with β-dicarbonyl compounds,[15] for example, with acetoacetic ester (**15**) (Scheme 10.12). A similar reaction with α-halogenoketones yields imidazoles (**17**) (Scheme 10.13).[16]

Cl, N, N, Cl, N, Cl (10) + 6 $Me_2N{-}C({=}O)H$ DMF $\xrightarrow[35^{\circ}C]{\text{dioxan}}$ $Me_2\ddot{N}{-}CH{=}N{-}CH{=}\overset{+}{N}Me_2$ Cl^-

⟷ $Me_2\overset{+}{N}{=}CH{-}N{=}CH{-}\ddot{N}Me_2$ (11)

(11) $\xrightarrow{\text{R-NH}_2,\ \text{MeO}^-/\text{MeOH, under reflux}}$ $R{-}N{=}CH{-}NMe_2$ (12)

(11) $\xrightarrow{\text{R-CO-NH}_2,\ \text{NaOPr / PrOH},\ 90^{\circ}C}$ $R{-}CO{-}N{=}CH{-}NMe_2$ (13) (10.11)

$CH_3{-}C({=}O){-}CH_2{-}C({=}O){-}OEt$ + $:NH_2{-}C(R){=}NH:$ (15) → CH_3, N, R, O, N, H (14) (10.12)

$R^1{-}C({=}O){-}CH_2{-}Hal$ + $H_2\ddot{N}{-}C(R^2){=}NH:$ (16) $\xrightarrow{-H_2O,\ -HHal}$ R^1, H, N, R^2, N (17) (10.13)

10.2.2 *Reaction with hydrazine and synthesis of 1,2,4-triazines*

The presence of a carbon–nitrogen double bond in the amidines makes them susceptible to attack by nucleophiles on the carbon atom. For example, hydrazine will react in this way in a fast reaction in methanol (Scheme 10.14).[17] The yields are nearly quantitative when $R = H, CH_3, C_2H_5, C_3H_7$. The products of this reaction react with α-diketones to furnish a general synthesis of 1,2,4-triazines (**18**) (Scheme 10.15).

$$R{-}C(=NH){-}NH_2 + H_2\ddot{N}{-}NH_2 \xrightarrow[0\text{-}5^{\circ}C]{MeOH} R{-}C(\bar{N}H)(NH_2){-}\overset{+}{N}H(H){-}NH_2 \longrightarrow R{-}C(NH_2){=}\overset{+}{N}H{-}NH_2 + NH_2^- \longrightarrow R{-}C(NH_2){=}N{-}NH_2 + NH_3 \qquad (10.14)$$

$$R^1{-}C(NH_2){=}N{-}NH_2 + O{=}C(R^2){-}C(R^3){=}O \longrightarrow \text{(18)}\ 80\text{-}90\ \% \qquad (10.15)$$

10.2.3 *DBU and DBN*

Preparation and uses

As a result of amidines being strong monacidic bases, two bicyclic amidines, 1,5-diazabicyclo[4.3.0]nonene-5 (DBN) (**19**)[18] and 1,5-diazabicyclo[5.4.0]-undecene-5 (DBU) (**20**),[19] find application as strong bases but weak (hindered) nucleophiles for elimination reactions of alkyl halides, often being superior for this purpose to other bases (e.g. $KOBu^t$). (For a review of their applications see Ref. 20.)

DBN (**19**) can be prepared by a Michael reaction between pyrrolidone and acrylonitrile, and reduction of the resulting nitrile to a primary amine which loses water and cyclizes spontaneously to yield the cyclic amidine (Scheme 10.16).[21] DBU (**20**) is prepared by a similar method (Scheme 10.17).[22]

A typical example of their use is in the preparation of hexatrienes for an elimination in which the use of potassium *tert*-butoxide gives only polymers (Scheme 10.18).[23]

It has been found that DBU is ofter superior to DBN for the elimination of hydrogen halides from alkyl halides. For example, in the reaction shown in Scheme 10.19, DBN gives a yield of 60 per cent of the alkene whereas DBU gives a yield of 91 per cent.[20]

$$\text{2-pyrrolidinone (N–H)} + CH_2{=}CH{-}C{\equiv}N \longrightarrow \text{N}{-}CH_2CH_2CN \xrightarrow[100^\circ C\ 200\ atm.]{H_2/\text{Raney Co}} [\text{N}{-}CH_2CH_2CH_2NH_2] \xrightarrow{-H_2O} \text{DBN (19)} \quad (10.16)$$

$$\text{caprolactam} \xrightarrow{CH_2{=}CHCN} \text{N}{-}CH_2CH_2CN \xrightarrow[\text{catalyst}]{H_2} \text{N}{-}CH_2CH_2CH_2NH_2 \xrightarrow{-H_2O} \text{DBU (20)} \quad (10.17)$$

$$R^1\text{-CH=CH-CH(Br)-CH}_2\text{-C(R}^3\text{)=CHR}^2 \xrightarrow[\text{DMSO}]{\text{DBN}} R^1\text{-CH=CH-CH=CH-C(R}^3\text{)=CHR}^2$$

Yields are 30–40% when R^1, R^2, R^3 = H, CH_3 (10.18)

$$CH_3\text{-}CH_2\text{-}CH_2\text{-}CH(Br)\text{-}CH_2\text{-}CH_2\text{-}CH_3 \longrightarrow CH_3\text{-}CH_2\text{-}CH_2\text{-}CH{=}CH\text{-}CH_2\text{-}CH_3 \quad (10.19)$$

10.2.4 *Guanidines*

Hindered guanidines as catalysts

As proton acceptors Guanidines have in recent times been used as catalysts and reagents for a variety of reactions. Barton and his co-workers have developed hindered guanidines as strong bases (proton acceptors).[24] A hindered guanidinium salt of *m*-iodylbenzoic acid (**21**) has been used by Barton and his co-workers for a variety of useful oxidations, for example, for the cleavage of 1,2-diols to yield aldehydes and ketones, and for the oxidation of secondary nitroalkanes to ketones (the Nef reaction, see Chapter 11, p. 443).[25]

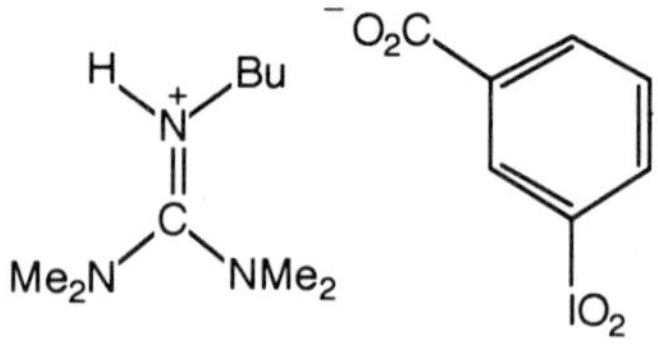

$(Me_2N)_2C{=}N^+(H)Bu \;\; ^-O_2C\text{-}C_6H_4\text{-}IO_2$ (*m*)

(21)

Methylation of phenols Pentaalkylguanidines (**22**) act as catalysts for the *o*-methylation of phenols,[26] for example, Scheme 10.20.

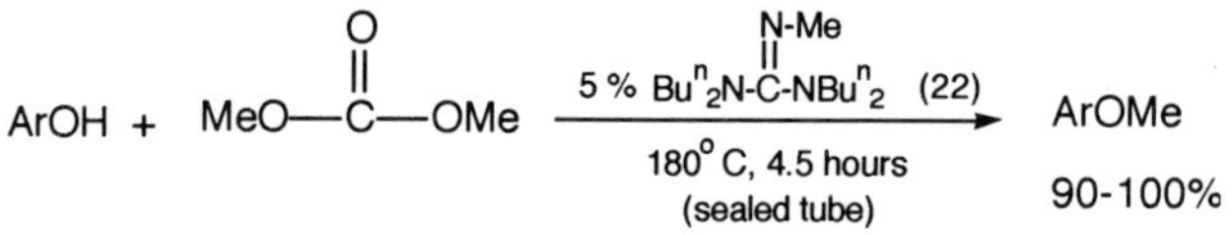

$$ArOH + MeO{-}\overset{O}{\overset{\|}{C}}{-}OMe \xrightarrow[\text{180°C, 4.5 hours (sealed tube)}]{5\%\ Bu^n_2N\text{-}C(=N\text{-}Me)\text{-}NBu^n_2\ (22)} \underset{90\text{-}100\%}{ArOMe} \qquad (10.20)$$

Tetramethylguanidine Tetramethylguanidine (**23**) has found use as a catalyst for Michael reactions.[27]

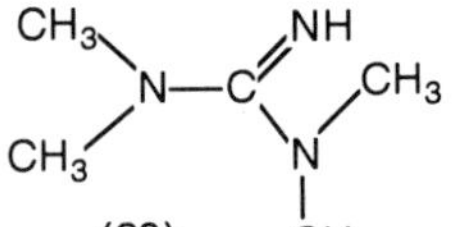

$(CH_3)_2N{-}C({=}NH){-}N(CH_3)_2$

(23)

It has been used to promote several other reactions: the preparation of i-steroids,[28] in solid-phase peptide synthesis,[29] for conversion of hydrazones into iodides,[30] and as a catalyst for the preparation of silyl ethers from alcohols.[31]

References

1. S. Patai (ed.), *The chemistry of amidines and imidates*, Wiley, London, 1975.
2. D. Hall and F. J. Llewellyn, *Acta Crystallogr.*, 1956, **9**, 108; D. Hall, *Acta Crystallogr.*, 1965, **18**, 955.
3. A. Pinner and Fr. Klein, *Ber.*, 1877, **10**, 1889; C. Liebermann and A. Hagen, *Ber.*, 1883, **16**, 1647; A. Pinner, *Ber.*, 1884, **17**, 179.
4. H. Loewe, J. Urbanietz, and H. Mieth, *Int. Congr. Chemother. Proc. 5th*, 1967, **2(2)**, 645; *Chem. Abstr.*, 1969, **70**, 47042a.
5. *F. &F.*, **5**, 307; R. Fuks, *Tetrahedron*, 1973, **29**, 2147.
6. *F. &F.*, **10**, 347; A. F. Hagerty and A. Chandler, *Tetrahedron Lett.*, 1980, **21**, 885.
7. *F. &F.*, **4**, 244; E. B. Pedersen, N. O. Vesterager, and S.-O. Lawesson, *Synthesis*, 1972, 547.
8. *F. &F.*, **1**, 253; A. J. Speziale and R. C. Freeman, *J. Am. Chem. Soc.*, 1960, **82**, 903; 909.
9. *F. &F.*, **5**, 652. Preparation: W. Kantlehner and P. Speh, *Ber.*, 1971, **104**, 3714. *F. & F.*, **12**, 477.
10. H. Gold and O. Bayer, *Ber.*, 1961, **94**, 2594.
11. H. Brendereck, F. Effenberger, and T. Brendle, *Angew. Chem., Int. Ed. Engl.*, 1966, **5**, 132; H. Brendreck, F. Effenberger, T. Brendle, and H. Muffler, *Ber.*, 1968, **101**, 1885.
12. *F. &F.*, **15**, 269; S. Ogata, A. Mochizuki, M. Kakimoto, and Y. Imai, *Bull. Chem. Soc. Japan*, 1986, **59**, 2171.
13. *F. &F.*, **11**, 194; H. Gold, *Angew. Chem.*, 1960, **72**, 956.
14. J. T. Gupton, C. Colon, C. R. Harrison, M. J. Lizzi, and D. Polk, *J. Org. Chem.*, 1980, **45**, 4522.
15. H. R. Snyder and H. M. Foster, *J. Am. Chem. Soc.*, 1954, **76**, 118; F. E. King, T. J. King, and I. H. Muir, *J. Chem. Soc.*, 1946, 5.
16. F. Kunckell, *Ber.*, 1901, **34**, 637.
17. *F. &F.*, **5**, 327; H. Neunhoffer and F. Weischedel, *Annalen*, 1971, **749**, 16.
18. *F. &F.*, **1**, 189; *F. &F.*, **2**, 98.
19. *F. &F.*, **2**, 101.
20. *F. &F.*, **4**, 116; 244; H. Oedinger, F. Müller, and K. Eiter, *Synthesis*, 1972, 591.
21. W. Reppe *et al.*, *Annalen*, 1955, **596**, 210.
22. H. Oediger and F. Möller, *Angew. Chem., Int. Ed. Engl.*, 1967, **6**, 76.
23. C. W. Spangler, R. Eichen, K. Silver, and B. Butzlaff, *J. Org. Chem.*, 1971, **36**, 1695.
24. D. H. R. Barton, J. D. Elliott, and S. D. Géro, *J. Chem. Soc., Chem. Commun.*, 1981, 1136; *J. Chem. Soc., Perkin Trans. I*, 1982, 2085; *F. &F.*, **11**, 105; 249.
25. *F. &F.*, **12**, 102; D. H. R. Barton, C. R. A. Godfrey, J. W. Morzycki, W. B. Motherwell, and A. Stobie, *Tetrahedron Lett.*, 1982, **23**, 957; D. H. R. Barton, W. B. Motherwell, and S. Z. Zard, *Tetrahedron Lett.*, 1983, 5227.
26. G. Barcelo, D. Grenouillat, J.-P. Senet, and G. Sennyey, *Tetrahedron*, 1990, **46**, 1839.
27. *F. &F.*, **1**, 1145; L. N. Nysted and R. R. Burtner, *J. Org. Chem.*, 1962, **27**, 3175; *F. & F.*, **4**, 490; G. P. Pollini, A. Barco, and G. DeGuilli, *Synthesis*, 1972, 44; F. S. Alvarez and D. Wren, *Tetrahedron Lett.*, 1973, 569; *F. &F.*, **12**, 477; A. T. Hewson and D. T. MacPherson, *Tetrahedron Lett.*, 1983, **24**, 647.

28. M. Anastasia, P. Allevi, P. Cuiffreda, and A. Fiecchi, *Synthesis*, 1983, 123.
29. D. B. Whitney, J. P. Tam, and R. B. Merrifield, *Tetrahedron*, 1984, **40**, 4237.
30. D. H. R. Barton, G. Barshiardes, and J.-L. Fourrey, *Tetrahedron Lett.*, 1983, **24**, 1605.
31. S. Kim and H. Chang, *Synth. Commun.*, 1984, **14**, 899.

11

NITROALKANES AND NITROALKENES

NITROALKANES[1]

The nitroalkanes (**1**) have a nitro group connected to an alkyl group. They are polar substances (dipole moment *c.* 3.5 D) and are usually liquids of high boiling point (CH_3NO_2, b.p. 101 °C) which are stable to distillation. The infra-red frequencies of the nitro group are in the following ranges:

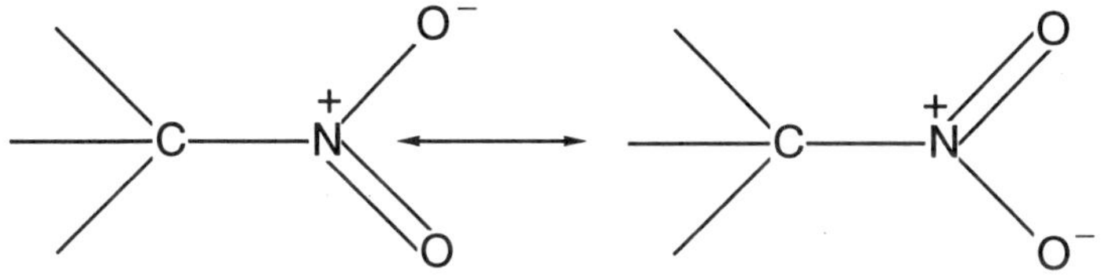

(1)

symmetrial NO_2 stretch, 1360–1380 cm^{-1};
antisymmetrical NO_2 stretch, 1550–1570 cm^{-1};
C—NO stretch, 830– 930 cm^{-1}.

The molecular dimensions of nitromethane are as shown in the following diagram, its calculated dipole moment (μ) being 4.54 Debye units (cf. above).

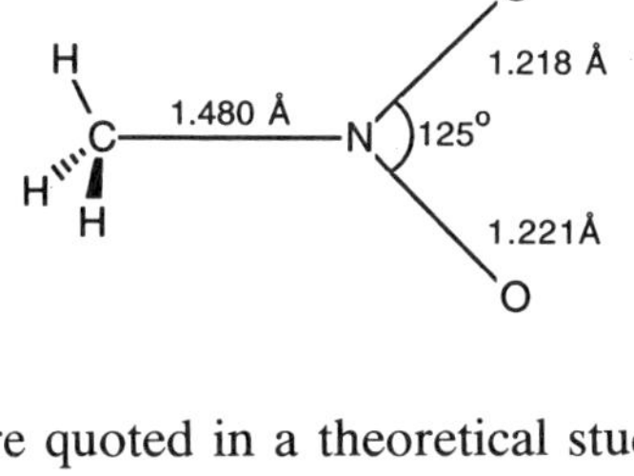

The values given (Å) are quoted in a theoretical study[2] and are taken from Ref. 3.

11.1 Preparation

11.1.1 *From alkyl halides and metal nitrites*

Alkyl halides react with metal nitrites to yield nitroalkanes.[4] Often mixtures of nitroalkanes and alkyl nitrites are obtained, though experimental procedures are available for the preparation of nitroalkanes in good yield, for example, by the method shown in Scheme 11.1 from primary alkyl halides.[4,5,6]

$$\text{R-Hal} + NaNO_2 \text{ or } AgNO_2 \longrightarrow \text{R-}NO_2$$

R	Yield (Hal=Br)	Yield (Hal=I) (%)
Bu^n	73	74
$Hexyl^n$	76	78
$Heptyl^n$	79	82
$Octyl^n$	80	83
Bu^i	17	59
$Amyl^i$	72	78

$$CH_3(CH_2)_7\text{-Br} + AgNO_2 \xrightarrow[\text{66 hours}]{\text{Ether}} CH_3(CH_2)_7\text{-}NO_2 \quad \text{75-80\%} \tag{11.1}$$

The use of crown ethers produces similar yields with primary halides but can increase the yield of nitroalkane produced from secondary halides in such reactions (Scheme 11.2).[7] The yields are about 65 per cent and some alkyl nitrite is produced which is removed by distillation. The crown ether is used in catalytic quantities.

$$CH_3(CH_2)_7\text{-Br} + AgNO_2 \xrightarrow[\text{in ether, 3 days}]{\text{18-Crown-6}} CH_3(CH_2)_7\text{-}NO_2 \quad \text{65-70\%} \tag{11.2}$$

18-Crown-6

11.1.2 *By addition of dinitrogen tetroxide to alkenes*[8]

A series of investigations[9] into the addition of dinitrogen tetroxide (N_2O_4), either as a liquid or dissolved in ether, sometimes with added oxygen, to alkenes in the liquid phase has shown that two modes of addition occur, yielding dinitroalkanes (**2**) and 2-nitroalkyl nitrates (**3**) which on work up give 2-nitro-alcohols (**4**) (Scheme 11.3).

$$R^1R^2C{=}CHR^3 \xrightarrow[\text{0°C}]{\text{liq. } N_2O_4 \text{ or } N_2O_4 \text{ in ether}} R^1C(NO_2)(R^2){-}CH(R^3){-}NO_2 \; (2) + R^1C(O{-}N{=}O)(R^2){-}CH(R^3){-}NO_2$$

$$R^1C(O{-}N{=}O)(R^2){-}CH(R^3){-}NO_2 \xrightarrow{O_2 \text{ or } N_2O_4} R^1C(O{-}NO_2)(R^2){-}CH(R^3){-}NO_2 \; (3) \longrightarrow R^1C(OH)(R^2){-}CH(R^3){-}NO_2 \; (4)$$

R^1	R^2	R^3	Yield (%) of (**2**)	Yield (%) of (**3**)	Yield (%) of (**4**)
H	H	H	39	28	20
H	Me	H	21	21	33
Me	Me	H	43	2	28
H	Et	H	39	—	33
H	Me	Me	30	—	35
Me	Me	Bu^t	48	—	32
-Cyclohexene-			42	18	25

(11.3)

The use of dinitrogen tetroxide, oxygen, and water followed by base-catalysed elimination gives nitroalkenes in good yields (Scheme 11.4).[10] This reaction has also been used in the modification of steroids.[11]

$$RCH{=}CH_2 \xrightarrow[\text{ii) } H_2O]{\text{i) } N_2O_4} R{-}CH(NO_2){-}CH_2NO_2 + R{-}CH(OH){-}CH_2NO_2 + \text{byproducts}$$

$$\xrightarrow{NH_3 \text{ (from dinitro); phthalic anhydride or base (from nitro alcohol)}} RCH{=}C(H)NO_2 \qquad (11.4)$$

11.1.3 *By oxidation of primary amines*

Several oxidizing agents have been applied to the oxidation of primary amines to nitroalkanes.

With potassium permanganate

Primary amines having the amino group on a tertiary carbon atom (**5**) can be oxidized to trialkyl nitroalkanes by potassium permanganate and magnesium sulphate in aqueous acetone at 25–30 °C (Scheme 11.5).[12]

$$R^1R^2R^3C{-}NH_2 \ (5) \xrightarrow[25-30^{\circ}C,\ 2\text{ days}]{KMnO_4 \text{ in acetone / aq. } MgSO_4} R^1R^2R^3C{-}NO_2$$

R^1	R^2	R^3	Yield (%)
Me	Me	Et	71
Me	Me	Me	83
Me	Me	Pr	82
Me	$(CH_2)_4$	$(CH_2)_4$	72
Me	$(CH_2)_5$	$(CH_2)_5$	73

(11.5)

By ozone

Ozonization on dry silica gel at −78 °C and then at 20 °C gives good yields of nitroalkanes from primary amines (Scheme 11.6).[13]

With dimethyldioxirane

For details of the preparation of dimethyldioxirane (**6**) from acetone and similar derivatives from other aliphatic ketones, see Ref. 14. Good yields of nitroalkanes are obtained by the oxidation of primary amines with these compounds (Scheme 11.7).[15]

$$\text{R-NH}_2 \xrightarrow[\text{-78}^\circ\text{C then 20}^\circ\text{C}]{\text{O}_3,\ \text{SiO}_2} \text{R-NO}_2$$

R	Yield (%)
Bu^s	70
n-C_6H_{13}	69
Bn	66
Bu^t	70

(11.6)

$$\text{R-NH}_2 + \underset{(6)}{(\text{H}_3\text{C})_2\text{C}(\text{O-O})} \longrightarrow \left[\text{R-NHOH} \longrightarrow \text{R-N(OH)}_2 \xrightarrow{\text{-H}_2\text{O}} \text{R-N=O}\right] \longrightarrow \text{R-NO}_2 \quad 85\text{-}97\%$$

R	Yield (%)
Bu^n	84
Bu^s	87
Bu^t	90
cyclohexyl-CH_2	96
adamantyl-CH_2	95
benzyl	97

(11.7)

11.1.4 *By oxidation of oximes*

Several oxidizing agents have been used for the preparation of nitroalkanes from oximes. For instance, peracids in buffered solvents oxidize oximes (**7**) to nitroalkanes (Scheme 11.8).[16]

Cyclic nitroalkanes have been prepared from oximes of cyclic ketones (**8**) by chlorination with hypochlorites followed by reductive removal of chlorine (Scheme 11.9).[17]

$$R^1\text{—}C(R^2)\text{=N-OH}\ (7) \xrightarrow[\text{Na}_2\text{HPO}_4\text{ (buffer) in CH}_3\text{CN}]{\text{F}_3\text{CCO}_3\text{H}} R^1\text{—CH}(R^2)\text{—NO}_2$$

R^1	R^2	Yield (%)
H	Ph	77
COOEt	COOEt	55

Oxidative mechanism:

$$R_2C\text{=N-OH} + CF_3CO_3H \longrightarrow R_2C(OH^+)\text{—N-OH} \longrightarrow R_2C\text{=}N^+(OH)\text{—OH} \longrightarrow R_2C\text{=}N^+(OH)\text{—}O^- \longrightarrow R_2CHNO_2 \quad (11.8)$$

$$\text{cyclohexanone oxime (8)} \xrightarrow[\text{ii) Bu}_4{}^n\text{N}^+\text{OCl}^-]{\text{i) HOCl, PhH - H}_2\text{O, pH 5.5}} \text{1-chloro-1-nitrocyclohexane (93\%)} \xrightarrow{\text{Mg, THF}} \text{nitrocyclohexane (95\%)} \quad (11.9)$$

11.1.5 *From carboxylic acids*

Substitution of the halogen in an α-halogenocarboxylic acid (**9**) by a nitro group, using sodium nitrite, yields an α-nitrocarboxylic acid (**10**) which is easily decarboxylated at 100 °C to give a nitroalkane (Scheme 11.10).[18] Yields are low, being 35–38 per cent for nitromethane (**11**).

$$NaNO_2 + ClCH_2CO_2^- \xrightarrow{\text{aqueous}} O_2N^+CH_2CO_2^- \; (10) \xrightarrow{\text{heat in aq. soln.}} CH_2{=}N^+(O^-)_2 + CO_2 \xrightarrow{\text{aq. } H_3O^+} CH_3NO_2 \; (11) \qquad (11.10)$$

(9) (10) (11)

A somewhat better method for the preparation of nitroalkanes from carboxylic acids depends upon the introduction of a nitro group into a carboxylic acid by removal of a proton from the α-position and reaction of the resulting anion with n-propyl nitrate (Scheme 11.11).[19]

$$R\text{-}CH_2CO_2H \xrightarrow[\text{HMPT - THF}]{2\ LiNPr^i_2} R\text{-}\bar{C}H\text{-}CO_2^- + 2Li^+ \xrightarrow[Pr^n\text{-}O\text{-}NO_2]{LiNPr^i_2} R\text{-}CH(N^+O_2^-)\text{-}CO_2^- \longrightarrow RCH{=}N^+(O^-)_2 + CO_2 \xrightarrow{\text{aq. } H_3O^+} R\text{-}CH_2NO_2$$

R	Yield (%)
$CH_3(CH_2)_7$	68
$CH_3(CH_2)_{10}$	53
$CH_3(CH_2)_8CH{=}CH(CH_2)_6$	45

(11.11)

11.1.6 *From acyl nitrates*

Acids and their anhydrides can be converted by the action of nitric acid into acyl nitrates (**12**) which on being heated decarboxylate and yield nitroalkanes. Some of the corresponding alcohol and some ester are also produced (Scheme 11.12).[20] The mechanism of nitroalkane and by-product formation appears to involve free-radicals (Scheme 11.13).[20]

$$(R\text{-}CO)_2O + HNO_3\ (90\%) \xrightarrow{20^\circ C} \underset{(12)}{R\text{-}CO_2NO_2} + R\text{-}CO_2H$$

10:1

or

$$\underset{(12)}{R\text{-}CO_2H} + 2(CH_3CO)_2O + 2\ HNO_3 \longrightarrow \underset{(12)}{R\text{-}CO_2NO_2} + 3CH_3CO_2H + CH_3CO_2NO_2$$

$$R\text{-}CO_2NO_2 \xrightarrow[CH_3CN \text{ or } RNO_2]{270\text{-}300^\circ C} R\text{-}NO_2 + CO_2$$

R	Yield of nitroalkane (%)
Bu^n	57
Me	54
$CH_3(CH_2)_6$	60
$(CH_2)_6$	48
Bu^t	30

(11.12)

11.1.7 *From isocyanates*

Oxone ($2KHSO_5.KHSO_4.K_2SO_4$)[21] in aqueous acetone decomposes to form an intermediate, probably the dioxirane (**14**), which reacts with alkyl isocyanates (**13**) to give high yields of primary, secondary and tertiary nitroalkanes (Scheme 11.14). The most likely reaction path is via the carbamate and the amine. Triton B increases the rate of the reaction.[22]

11.1.8 *From azides*

Primary and secondary nitroalkanes can be prepared by the conversion of azides into phosphine imines (**15**) followed by ozonolysis (Scheme 11.15).[23]

$$RCO_2NO_2 \longrightarrow RCO_2^{\cdot} + NO_2^{\cdot}$$

$$RCO_2^{\cdot} \longrightarrow R^{\cdot} + CO_2$$

$$R + 2RCO_2NO_2 \longrightarrow RNO_2 + RONO + RCO_2^{\cdot} + CO_2$$

$$R + RCO_2NO_2 \longrightarrow RCO_2R + NO_2^{\cdot}$$

$$R + NO_2^{\cdot} \longrightarrow RNO_2$$

$$R + RCO_2^{\cdot} \longrightarrow RCO_2R$$

$$RONO \longrightarrow RO^{\cdot} + NO^{\cdot} \longrightarrow \text{alcohol, aldehyde, acid etc} \qquad (11.13)$$

$$\underset{(13)}{R\text{-}N{=}C{=}O} \xrightarrow[\text{aq. acetone, Triton B}]{(CH_3)_2CO_2\ (14)} R\text{-}NO_2 \qquad (11.14)$$

$$R^1N_3 + R^2_3P \longrightarrow \underset{(15)}{R^1\text{-}N{=}PR^2_3} \xrightarrow[-78^{\circ}C]{O_3 / CH_2Cl_2} R^1NO_2 + R^2_3\text{-}P{=}O$$

$$\underset{(15)}{R^1\text{-}N{=}PR^2_3} \xrightarrow{\text{through ozonide}} R^1\text{-}N{=}O \longrightarrow R^1NO_2 \qquad (11.15)$$

11.1.9 *From nitroalkenes*

Nitroalkanes can be prepared by the selective reduction of nitroalkenes (**16**) with lithium tri-*sec*-butylborohydride or with lithium triethylborohydride (Scheme 11.16).[24]

$$R^1\text{-}CH{=}C(NO_2)R^2 \ (16) \xrightarrow{LiR_3BH} \left[R^1\text{-}CH_2(R^2)C{=}\overset{+}{N}(O^-)_2 \right] \xrightarrow[\text{or } SiO_2]{\text{aq. } H_3O^+} R^1\text{-}CH_2\text{—}CH(R^2)\text{-}NO_2 \quad 65\text{-}80\%$$

R^2 = H or Me

R = H or Me

(11.16)

11.1.10 *α-Nitro esters*

α-Nitro esters can be readily prepared by reaction with the anions formed from diethyl malonate or from ethyl acetoacetate (Scheme 11.17).[25]

$$(CH_3)_2C{=}O + KCN + NaNO_3 \longrightarrow O_2N\text{—}O\text{—}C(CH_3)_2CN$$

$$O_2N\text{—}O\text{—}C(CH_3)_2CN + {}^-CR(CO_2Et)R^1 \longrightarrow O_2N\text{—}C(CO_2Et)(R)\text{—}R^1 + (CH_3)_2C{=}O + CN^-$$

R^1	Yield (%) R=-CO_2Et	Yield (%) R=-COMe
H	42	52
Me	56	—
Et	51	46
Pr^n	48	—
Bu^n	54	46
Bu^i	47	40
n-C_5H_{11}	46	—
Ph	57	70
Bn	67	—

(11.17)

11.1.11 *α-Nitrocyclohexanones*

Certain *α*-nitrocyclohexanones (**17**) can be prepared by the reaction of a cyclohexanone enol acetate (**18**) with trifluoroacetyl nitrate (**19**), which is prepared *in situ* (Scheme 11.18).[26]

$$NH_4NO_3 + (F_3C\text{-}CO_2)_2O \longrightarrow F_3C\text{-}CO_2NO_2 \ \ (19)$$

$$(18) + F_3C\text{-}CO_2NO_2\ (19) \longrightarrow (17) \quad c.\ 100\% \qquad (11.18)$$

11.1.12 *Nitrofluorocompounds*

Nitrofluorination of alkenes by nitronium tetrafluoroborate $NO_2^+BF_4^-$ in 70 per cent HF–pyridine gives moderate yields (40–70 per cent) of nitrofluoro compounds (**20**) (Scheme 11.19).[27]

$$RHC{=}CHR \xrightarrow[70\%\ HF\ \text{-}\ pyridine]{NO_2BF_4} R\text{—}CH(NO_2)\text{-}CH(F)\text{—}R \ \ (20) \qquad (11.19)$$

11.2 Reactivity

11.2.1 *Acidity of α-hydrogen atoms*

Tautomerism

A hydrogen atom on the α-position of a nitroalkane is activated by the nitro group and its removal as a proton yields an anion (**22**) which is stabilized by resonance. This results in the pK_a of nitromethane in DMSO being 17.2.[28] Due to this acidity of the α-hydrogen atom tautomerism can arise between a nitroalkane (**21**) and its nitronic acid (**23**) which is often stable at 0 °C but changes to the true nitroalkane at higher temperatures (Scheme 11.20).

The Nef reaction

The well known Nef reaction (hydrolysis)[29] depends upon the equilibrium between a nitroalkane and its nitronic acid. The nitronic acid can be hydrolysed in acid solution since it contains a carbon–nitrogen double bond (an imino bond), and the product is an aldehyde or a ketone (Scheme 11.21).

$$R\text{-}CH_2\text{—}\overset{+}{N}(=O)O^- \xrightarrow{\text{base}} R\text{-}\overset{-}{C}H\text{—}\overset{+}{N}(=O)O^- \longleftrightarrow R\text{-}CH{=}\overset{+}{N}(O^-)O^-$$

(21) (22)

$$R\text{-}CH{=}\overset{+}{N}(O^-)O^- \xrightarrow{\text{aq. } H_3O^+,\ 0^\circ C} R\text{-}CH{=}\overset{+}{N}(OH)O^-$$

$$R\text{-}CH{=}\overset{+}{N}(OH)O^- \xrightarrow[\text{slow}]{>0^\circ C} R\text{-}CH_2NO_2$$

(23) nitronic acid (21)

(11.20)

$$R^1\text{—}CH(R^2)\text{-}NO_2 \xrightarrow[\text{ii) dil. } H_2SO_4]{\text{i) aq. NaOH}} R^1R^2C{=}O + [HN(OH)_2]$$

$$R^1\text{—}CH(R^2)\text{-}NO_2 \xrightarrow{HO^-,\ H_2O} R^1R^2C{=}\overset{+}{N}(O^-)O^- \xrightarrow{HO^-} R^1\text{—}C(OH)(R^2)\text{—}N(O^-)O^- \xrightarrow{-H^+} R^1R^2C{=}O$$

$\overset{+}{H}B$

(11.21)

$$\text{Ph—CH(OCH}_3\text{)—CH(CH}_3\text{)—NO}_2 \xrightarrow[\text{DBU, } CH_2Cl_2]{(CH_3)_3SiCl,} \left[\text{Ph—CH(OCH}_3\text{)—C(CH}_3\text{)}{=}\overset{+}{N}(OSi(CH_3)_3)\text{—}O^-\right]$$

(24) (25)

$$\xrightarrow{ClC_6H_4CO_3H} \text{Ph—CH(OCH}_3\text{)—C(=O)CH}_3 \quad 95\%$$

(11.22)

Modifications of the Nef reaction

A new Nef reaction has been reported in which secondary nitro-compounds (**24**) are converted to trialkylsilyl nitronates (**25**) in >90 per cent yield by reaction with trialkylsilyl chloride. The silylnitronates are then oxidized by MCPBA at 25 °C to ketones (Scheme 11.22). Primary nitro-compounds cannot be converted into aldehydes by this method.[30]

Synthesis of benzaldehyde (**26**) and its derivatives from the relevant halo-compound (**27**) can be carried out with the deprotonated nitronic acid of 2-nitropropane (**28**) (Scheme 11.23).[31]

A modification of the Nef reaction which initially produces an acetal (**29**), gives better yields of aldehydes than the direct Nef reaction (Scheme 11.24).[32]

$$C_6H_5CH_2Cl\ (27) + (CH_3)_2C{=}N^+(O^-)O^-\ Na^+\ (28) \xrightarrow{EtOH} (CH_3)_2C{=}N{-}OH + C_6H_5CHO\ (26) \qquad (11.23)$$

$$R\text{-}CH_2\text{-}NO_2 \xrightarrow[\text{ii, } H_2SO_4 / MeOH,\ -35^{\circ}C]{\text{i, } MeO^- / MeOH} R{-}CH(OMe){-}OMe\ (29) \xrightarrow{aq.\ H_3O^+} R\text{-}CHO$$

Mechanism:

$$R\text{-}CH_2\text{-}NO_2 \longrightarrow RCH{=}N^+(O^-)O^-\ (+\ MeO^-) \xrightarrow{2H^+} RCH(:OMe){-}N(OH)_2 \xrightarrow{H^+} R(H)C{=}O^+Me\ (+\ MeOH) + [HN(OH)_2] \longrightarrow RCH(OMe)_2 \xrightarrow{H_2O} R\text{-}CHO$$

R	Yield (%) acetal or aldehyde
$Ph(CH_2)_2$	78
$CH_3(CH_2)_6$	99

(11.24)

Basic silica gel has been found to give high yields of aldehydes or ketones from nitroalkanes in a mild version of the Nef reaction (Scheme 11.25).[33]

$$R\text{-}CH_2\text{-}NO_2 \xrightarrow[\text{MeOH, 25}^\circ\text{C}]{SiO_2\,/\,NaOH} R\text{-}CHO$$

Nitroalkane	Yield (%) of aldehyde or ketone
Nitrocyclohexane (NO_2 on cyclohexane)	99
2-Nitropropane (NO_2)	97
$CH_3(CH_2)_6NO_2$	87
Keto nitroalkane (O, NO_2)	81
Dioxolane-protected nitroalkane (O, O, NO_2)	81

(11.25)

A reductive method using titanium trichloride, instead of the hydrolytic method of the original Nef reaction, has been used to convert nitroalkanes to aldehydes and ketones (Scheme 11.26).[34] Further examples have been reported (Scheme 11.27).[35] The reaction involves reduction and hydrolysis (Scheme 11.28).

A Nef reaction can also be carried out using vanadyl acetylacetonate[36] or molybdenum hexacarbonyl as a catalyst in a reaction in which the potassium nitronate is oxidized by *tert*-butyl hydroperoxide to the carbonyl compound (Scheme 11.29).[37]

$$R^1\text{—}CH(R^2)\text{-}NO_2 \xrightarrow{NaOMe\,/\,MeOH} \left[R^1R^2C{=}\overset{+}{N}(O^-)_2 \right] \xrightarrow[\text{pH 5 - 6}]{\text{aq. }TiCl_3\,/\,NH_4OAc} R^1R^2C{=}O$$

Nitroalkane	Yield (%) of aldehyde or ketone
$CH_3COCH_2CH_2CH(NO_2)CH_2CH_3$	85
$(MeO)_2CHCH_2CH(NO_2)CH_3$ (MeO, MeO, NO_2)	90
1-methyl-2-nitrocyclohexane (NO_2)	60
$PhCH_2CH(NO_2)CH_2Ph$ (NO_2, Ph, Ph)	70
2-(1-nitroethyl)cyclohexanone ethylene ketal (O, O, NO_2)	90

(11.26)

$$\text{cyclopentanone or cyclohexanone} \xrightarrow[\text{piperidine}]{CH_3NO_2} \text{1-}(H_2C\text{-}NO_2)\text{cyclopentene or cyclohexene} \xrightarrow{TiCl_3} \text{H-C=O (aldehyde) cyclopentene or cyclohexene} \quad (11.27)$$

$$R_2CH\text{-}NO_2 \longrightarrow R_2C{=}\overset{+}{N}(O^-)(O^-)\ Na^+ \xrightarrow{TiCl_3} R_2C{=}\overset{+}{N}(O^-)OTiCl_2 \longrightarrow R_2C{=}NOH \longrightarrow R_2C{=}NH \xrightarrow{H_2O} R_2C{=}O \quad (11.28)$$

R^1—CH(R^2)–NO_2 $\xrightarrow{Bu^tO^-K^+}$ [$R^1R^2C{=}\overset{+}{N}(O^-)O^-$ K^+] $\xrightarrow[VO(acac)_2 \text{ or } Mo(CO)_6]{Me_3C\text{-}OOH}$ $R^1R^2C{=}O$

Nitroalkane	Yield of aldehyde or ketone (%)
NO_2	82(V)
NO_2	86(V)
NO_2	45(V)
O, NO_2	60(Mo)

(11.29)

Oxidative methods for the conversion of nitroalkanes to aldehydes and ketones have also been used and a number of examples follow. Ozone[38] and singlet oxygen, which is useful in the presence of carbon–carbon double bonds,[39] have been used as oxidants (Scheme 11.30).

R^1—CH(R^2)–NO_2 ⟶ $R^1R^2C{=}\overset{+}{N}(O^-)O^-$ $\xrightarrow[\text{or } ^1O_2 \text{ (singlet oxygen)}]{\text{i) } O_3, \text{ ii) } Me_2S}$ $R^1R^2C{=}O$

R^1	R^2	Yield (%) for O_3 oxidation	Yield (%) for O oxidation
$CH_3CO(CH_2)_2$	Et	83	60
Ph	H	49	68
$CH_3(CH_2)_6$	H	67	65

(11.30)

Cerium(IV) ammonium nitrate (CAN) ($Ce(NH_4)_2(NO_3)_6$) has also been used as the oxidant (Scheme 11.31).[40]

Secondary nitro-compounds (**30**) are oxidized by *n*-propyl nitrite (**31**)[41] and sodium nitrite to ketones. Primary nitro-compounds are oxidized under the same conditions to acids (Scheme 11.32).[42] The mechanism for the conversion

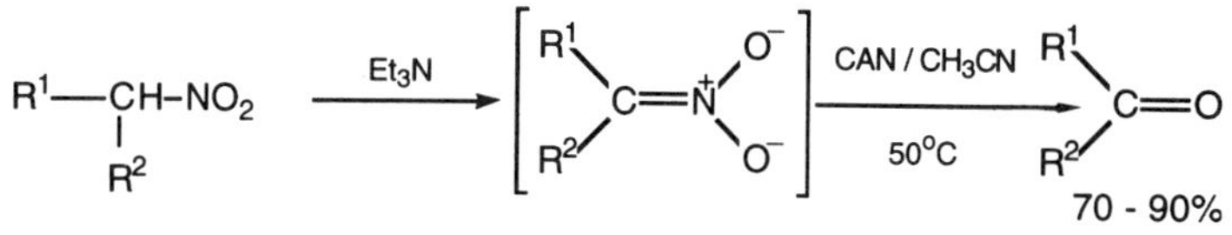

$$R^1\text{-}CH(R^2)\text{-}NO_2 \xrightarrow{Et_3N} \left[R^1R^2C{=}\overset{+}{N}(O^-)_2 \right] \xrightarrow[50^\circ C]{CAN / CH_3CN} R^1R^2C{=}O \quad 70 - 90\% \qquad (11.31)$$

$$Pr^n\text{-}OH \xrightarrow{HNO_2} \underset{(31)}{Pr^n\text{-}O\text{-}N{=}O}$$

$$\underset{(30)}{R^1\text{-}CH(R^2)\text{-}NO_2} + \underset{(31)}{CH_3CH_2CH_2ONO} + NaNO_2 \xrightarrow{DMSO} R^1R^2C{=}O + CH_3CH_2CH_2OH + NaNO_2 + N_2O$$

$$R\text{-}CH_2\text{-}NO_2 \xrightarrow[NaNO_2]{RO\text{-}N{=}O} R\text{-}CO_2H$$

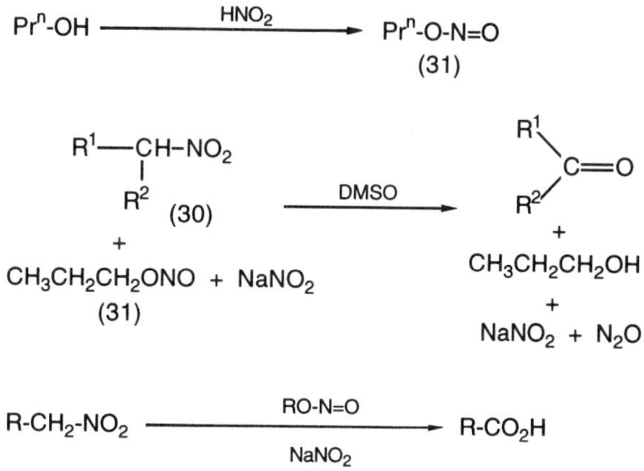

Nitroalkane	Yield (%) of aldehyde or ketone
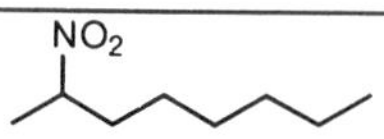 NO_2	83
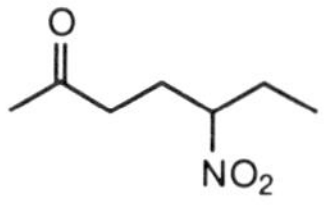 O, NO_2	76
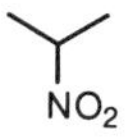 NO_2	70
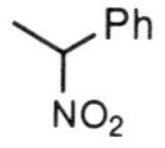 Ph, NO_2	79
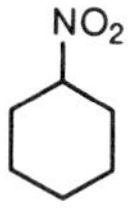 NO_2	67

(11.32)

for both the primary and secondary nitro-compounds is initially the same, but the second step is then dependent on the nature of R^1 and R^2 (Scheme 11.33).

The alkali metal salts of straight chain primary and secondary nitroalkanes are readily selectively oxidized by potassium permanganate to the corresponding aldehyde or ketone, leaving many other functional groups intact (Scheme 11.34).[43] The oxidation of primary nitroalkanes by this method is carried out

$$R^1\text{—}CH(R^2)\text{—}NO_2 + NO_2^- \longrightarrow R^1\text{—}\bar{C}(R^2)\text{—}NO_2 + HNO_2$$

$$R^1\text{—}\bar{C}(R^2)\text{—}NO_2 \xrightarrow{CH_3CH_2CH_2\text{-}ON{=}O} R^1\text{—}C(NO)(R^2)\text{—}NO_2 + CH_3CH_2CH_2O^-$$

$$CH_3CH_2CH_2O^- \xrightarrow{HONO} CH_3CH_2CH_2\text{-}ON{=}O$$

If R^1 = alkyl and R^2 = H

$$R^1\text{—}CH(NO_2)\text{—}N{=}O \rightleftharpoons R^1\text{—}C(NO_2){=}N\text{—}OH \quad \text{a nitrolic acid}$$

$$R^1\text{—}C(NO_2){=}N\text{—}OH \xrightarrow{\text{hydrolysis}} R^1\text{—}C({=}O)OH$$

If $R^1 = R^2$ = alkyl

$$R^2\text{—}C(R^1)(NO_2)\text{—}N{=}O + NO_2^- \longrightarrow R^1R^2C{=}N\text{—}O^- + N_2O_4$$

$$R^1R^2C{=}N\text{—}O^- \xrightarrow{O{=}NONO_2} \left[R^1R^2C{=}\overset{+}{N}(O^-)\text{—}N{=}O \longleftrightarrow R^1R^2\overset{+}{C}\text{—}N(O^-)\text{—}N{=}O \right] + NO_3^-$$

$$\left[\ \cdots\ \right] \longrightarrow R^1\text{—}C(R^2)(\text{—}O\text{—})N\text{—}N{=}O \longrightarrow R^1R^2C{=}O + N_2O \qquad (11.33)$$

with 70–90 per cent of the theoretical amount of potassium permanganate in the presence of magnesium sulphate to prevent the oxidation of the aldehyde produced to the carboxylic acid.

In another alternative to the Nef reaction, a nitroalkane is treated with a base and then oxidized with a molybdenum oxide, pyridine, HMPT complex to yield

$$R^1{-}CH(R^2){-}NO_2 \xrightarrow{NaH,\ Bu^tOH} \left[R^1R^2C{=}\overset{+}{N}(O^-)_2 \right] \xrightarrow{KMnO_4} R^1R^2C{=}O$$

Nitroalkane	Yield of aldehyde (%)
$CH_3(CH)_2)_2$-NO_2	83–97
$CH_3(CH_2)_6$-NO_2	85
$CH_2{=}CH(CH_2)_7CH_2CH_2NO_2$	59
$PhCH_2CH_2CH_2NO_2$	82

Nitroalkane	Yield of ketone (%)
$(CH_3)_2CHNO_2$	96
nitrocyclobutane	94
$CH_3CH(NO_2)CH_2Ph$	90
nitrocyclododecane	91

(11.34)

$$R{-}CH_2{-}NO_2 \xrightarrow{\text{i) base; ii) } MoO_5\text{, pyridine, HMPT}} R{-}CO_2H$$

$$R^1{-}CH(R^2){-}NO_2 \longrightarrow R^1R^2C{=}O$$

Nitroalkane	Yield (%) of acid or ketone
$CH_3(CH_2)_2$-NO_2	73
$CH_3CH_2CH(NO_2)CH_2CO_2Et$	73
$PhCH_2CH_2NO_2$	74
nitrocyclohexane	86

(11.35)

carboxylic acids from primary nitroalkanes and ketones from secondary nitroalkanes (Scheme 11.35).[44]

11.2.2 *Reaction with electrophiles*

Alkylation

Phenylsulphonylnitromethane (**32**) can be alkylated with primary alkyl and benzyl halides to yield α-nitrosulphones (**33**), which can be converted into various other useful compounds (Scheme 11.36).[45]

$PhSO_2CH_2NO_2$ (32) $\xrightarrow{MeONa}$ $PhSO_2\text{-}\bar{C}HNO_2\ Na^+$ $\xrightarrow[\text{Hal = Br or I}]{\text{R-Hal}}$ $R\text{—}CH(SO_2Ph)\text{—}NO_2$ (33) 50 - 75%

aq. $TiCl_3$ → R—C≡N 75%

excess of $KMnO_4$ in aq. HO^- → $R\text{-}CO_2H$ 80 - 90%

1-Bn-3-carbamoylpyridinium ($CONH_2$, $\overset{+}{N}$, Bn), $h\nu$ → $R\text{-}CH_2NO_2$ 60 - 70%

R	Yield RCN (%)
$PhCH_2CH_2$	74
$PhCH{=}CHCH_2CH_2$	74
	Yield RCO_2H (%)
$CH_3(CH_2)_4CH_2$	80
$CH_3(CH_2)_8CH_2$	82
	Yield of RCH_2NO_2 (%)
$PhCH{=}CHCH_2$	69
$PhCH_2CH_2$	62
2-$CH_3C_6H_4CH_2$	61

(11.36)

Formation of silyl derivatives

A silyl derivative of the nitronic acid of nitroethane (**34**)[46] (Scheme 11.37) has found use for various synthetic purposes.[47,48]

$$CH_3CH_2NO_2 \xrightarrow{LDA} \left[CH_3\text{-}CH{=}\overset{+}{N}(O^-)_2\ Li^+ \right] \xrightarrow{Cl\text{-}SiMe_2Bu^t} \underset{(34)}{(CH_3)(H)C{=}\overset{+}{N}(O^-)(O\text{-}SiMe_2Bu^t)} \quad (11.37)$$

Henry reaction

Since an α-hydrogen can be removed as a proton, nitroalkanes can undergo base-catalysed condensation reactions with electrophiles. In the Henry reaction[49] a nitroalkane reacts with aldehydes and ketones to form β-nitro alcohols (**35**) (Scheme 11.38).[50] Dicarbonyl compounds (**36**) may be used to form cyclic compounds (**37**) by double condensation (Scheme 11.39).[51]

$$CH_3CH_2CH_2{-}NO_2 \xrightarrow{BuLi} \left[CH_3CH_2CH{=}N^+(O^-)(OLi) \right] \xrightarrow{PhCHO,\ Cl_3TiO\text{-}Pr^i} \underset{(35)\quad 60\%}{HO(Ph)CH{-}CH(CH_2CH_3)NO_2} \quad (11.38)$$

$$\underset{\substack{(36)\\ \text{glutaraldehyde}}}{H_2C(CH_2CHO)_2} \xrightarrow[\text{aq. } Na_2CO_3]{CH_3NO_2} \underset{(37)}{\text{HO, OH, } NO_2 \text{ cyclohexane}}\ 51\% \quad + \text{ other stereoisomers} \quad (11.39)$$

Lower nitroalkanes have been converted to higher members of the series by an extension of the initial base-catalysed condensation of nitromethane with an aldehyde.[52] More recently improvements to the method have been made with resulting better yield of products (Scheme 11.40).[53] The mild conditions required are compatible with various functional groups often present in natural product synthesis.

$$\text{R-CHO} + CH_3NO_2 \xrightarrow[\text{18-Crown-6}]{\text{Base, KF - Pr}^i\text{OH}} \text{R—}\underset{\underset{\displaystyle O^-}{|}}{\text{CH}}CH_2NO_2$$

$$\xrightarrow{25^\circ C,\ \text{4-NMe}_2\text{-pyridine} / Ac_2O} \text{R—}\underset{\underset{\displaystyle \text{OAc}}{|}}{\text{CH}}CH_2NO_2 \xrightarrow{NaBH_4 / \text{EtOH}} \text{R-}CH_2\text{-}CH_2\text{-}NO_2 \quad \text{65 - 90\% overall} \qquad (11.40)$$

Steric hindrance in the Henry Reaction has been overcome by the use of special conditions as illustrated in Scheme 11.41 for methylcyclohexanones (**38**).[54]

$$\underset{(38)}{R^3\text{-cyclohexanone}} + R^1\text{—}\overset{H}{\underset{R^2}{C}}\text{—}NO_2 \xrightarrow[F^-\ \text{(catalyst) } 30^\circ C]{\text{pressure of 9 kbar}} R^3\text{-cyclohexan-1-ol-1-}C(NO_2)(R^1)(R^2) \quad 50\%$$

R^1	R^2	R^3	Yield (%)
H	H	2-CH_3	50
		3-CH_3	61
CH_3	H	2-CH_3	41
		3-CH_3	87
		4-CH_3	60
CH_3	CH_3	H	74

(11.41)

The Henry Reaction followed by a Tiffeneau-Demyanov ring expansion has been used to homologize ketones,[55] for example, the conversion of cyclohexanone (**39**) to cycloheptanone (**40**) (Scheme 11.42).

The Henry Reaction can also be used to convert a nitroalkane into an α-nitro ketone (**41**) by condensing it with an aldehyde in a reaction catalysed by

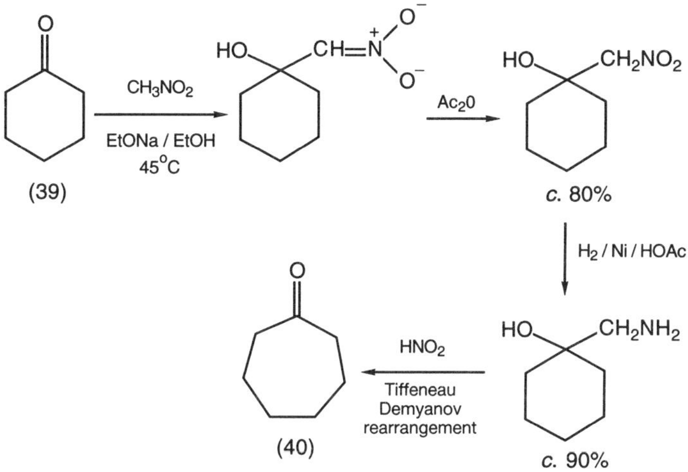

(39) *c.* 80% *c.* 90% (40) (*c.* 40% overall) (11.42)

alumina, followed by oxidation of the secondary alcohol so formed with acidified potassium dichromate under phase transfer catalysed conditions (Scheme 11.43).[56]

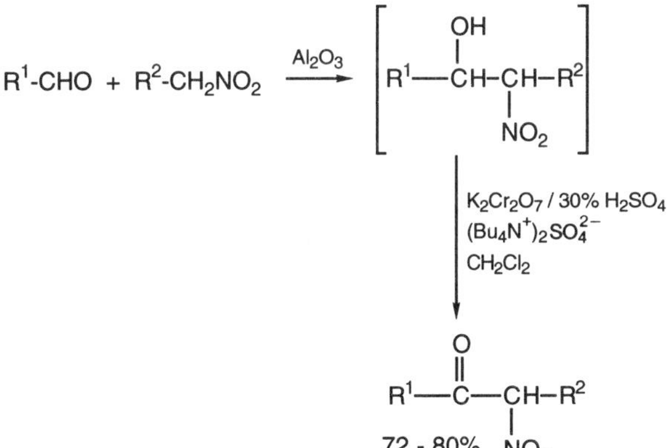

(41) (11.43)

Secondary or tertiary α-nitro ketones, formed by a Henry reaction of an aldehyde with a nitroalkane can be converted as their tosylhydrazones (**42**), into the corresponding ketones by reduction with lithium aluminium hydride (Scheme 11.44).[57]

The Henry reaction followed by Nef hydrolysis has been used to ascend the sugar series (e.g. aldopentose to aldohexose) by reaction of a monosaccharide with nitromethane or nitroethane e.g. (Scheme 11.45).[58]

A dianion can be generated from nitroethane by treatment with *n*-butyl lithium (2 equiv.) in an excess of HMPT. The dianion can then react with an

$$R^1CHO + H{-}C(NO_2)(R^2){-}R^3 \xrightarrow[\text{Henry}]{NaOH} R^1{-}CH(OH){-}C(NO_2)(R^2){-}R^3 \xrightarrow{Na_2Cr_2O_7} R^1{-}C(=O){-}C(NO_2)(R^2){-}R^3 \xrightarrow[MeOH]{Ts\text{-}NH\text{-}NH_2} R^1{-}C(=N{-}NH{-}Ts){-}C(NO_2)(R^2){-}R^3$$

(42)

80 - 95%

$$\xrightarrow[THF]{LiAlH_4} R^1{-}C(=O){-}CH(R^2){-}R^3$$

80 - 95% (11.44)

L-Arabinose $\xrightarrow[\text{MeONa / MeOH}]{CH_3NO_2}$ CH_2NO_2–HC–OH–HC–OH–HO–CH–HO–CH–CH_2OH + CH_2NO_2–HO–CH–HC–OH–HO–CH–HO–CH–CH_2OH

i) aq. NaOH; ii) aq. H_2SO_4 → L-Mannose

i) aq. NaOH; ii) aq. H_2SO_4 → L-Glucose

(11.45)

aldehyde, for example, with benzaldehyde. The specificity of the reaction can be reversed by using HMPT[59] (Scheme 11.46). In the absence of HMPT (**43**) is the only product (57 per cent).

$$CH_3\text{-}CH_2\text{-}NO_2 \xrightarrow{Bu^nLi:\ HMPT\ (2:1)} \left[(CH_3)(Li)C{=}\overset{+}{N}(OLi)(O^-) \right] \xrightarrow{PhCHO} Ph\text{—}CH(OH)\text{-}CH(CH_3)\text{-}NO_2 \quad (43)$$

$$CH_3\text{-}CH_2\text{-}NO_2 \xrightarrow[\text{excess of HMPT}]{Bu^nLi\ (2mol)} \bar{C}H_2\text{-}\bar{C}H\text{-}NO_2 \xrightarrow{PhCHO} Ph\text{—}CH(OH)\text{-}CH_2CH_2NO_2$$

(inverse addition) (11.46)

Nitroalkanes can be converted into α-methylene ketones (**44**) by acylation,[60] reaction with formaldehyde, and acetylation followed by denitration and elimination (Scheme 11.47).[61]

$$R^1\text{-}CH_2\text{-}NO_2 \xrightarrow{\text{acylation}} R^1\text{—}CH(COR^2)\text{-}NO_2 \ (45) \xrightarrow[\text{ii) } Ac_2O\text{ /pyridine}]{\text{i) } CH_2O\ /\ PPh_3,\ Pr^iOH} R^2\text{-CO—}C(R_1)(CH_2\text{-}OAc)\text{—}NO_2$$

$$\xrightarrow{Bu_3SnH,\ AIBN,\ C_6H_6} R^2\text{-CO—}CH(R^1)\text{-}CH_2OAc \xrightarrow{DBU\ /\ PhH} (R^1)(R^2\text{-CO})C{=}CH_2 \ (44) + HOAc$$

(11.47)

By a similar method in which sodium hydroxide replaces triphenylphosphine, α-nitro esters (**46**) give α-methylene esters (**47**) (Scheme 11.48).

$$R^1O\text{-}\overset{O}{\overset{\|}{C}}\text{—}CH(NO_2)\text{-}CH_2CH_3 \ (46) \xrightarrow[\text{instead of } PPh_3 \text{ in the hydroxymethylation step}]{\text{as above except use NaOH}} R^1O\text{-}\overset{O}{\overset{\|}{C}}\text{—}\overset{CH_2}{\overset{\|}{C}}\text{—}CH_2CH_3 \ (47)$$

(11.48)

Michael addition

Nitroalkanes (**48**) and α,β-unsaturated ketones (**49**) react in the presence of alumina (Michael addition), and the intermediates may be oxidized *in situ* to γ-diketones (**50**) by hydrogen peroxide in alkaline solution (Scheme 11.49).[62]

$$R^2\text{—}CH(R^1)CH_2\text{–}NO_2 \ (48) + HR^3C{=}CR^5\text{–}C(=O)\text{—}R^4 \ (49) \xrightarrow{Al_2O_3} R^2\text{—}CH(R^1)\text{–}C(=N^+(O^-)_2)\text{—}CH(R^3)\text{–}CH(R^5)\text{—}C(=O)\text{—}R^4$$

$$\xrightarrow{H_2O_2/aq.K_2CO_3} R^2\text{—}CH(R^1)\text{–}C(=O)\text{—}CH(R^3)\text{–}CH(R^5)\text{—}C(=O)\text{—}R^4 \ (50)$$

R^1	R^2	R^3	R^4	R^5	Yield (%)
CH_3	H	H	CH_3	H	73
CH_2CH_3	H	H	CH_3	H	52
CH_3	H	H	H	CH_3	64
CH_3	H	-$(CH_2)_3$-		H	70
4-CH_3-C_6H_4	H	H	CH_3	H	58
2-(1,3-dioxolan-2-yl)CH_2 [H, CH_2]	H	H	CH_3	H	80

(11.49)

$$CH_3NO_2 \ (51) + CH_3CH_2\text{—}C(=O)\text{—}CH{=}CH_2 \ (52) \longrightarrow CH_3CH_2\text{—}C(=O)\text{—}CH_2CH_2CH_2NO_2$$

$$\xrightarrow{+\ CH_3CH_2\text{—}C(=O)\text{–}CH{=}CH_2} CH_3CH_2\text{—}C(=O)\text{–}CH_2{\cdot}CH_2\text{–}C(=N^+(O^-)_2)\text{–}CH_2{\cdot}CH_2\text{–}C(=O)\text{—}CH_2CH_3$$

$$\xrightarrow{H_2O_2\ /\ aq.\ K_2CO_3} CH_3CH_2\text{—}C(=O)\text{–}CH_2{\cdot}CH_2\text{–}C(=O)\text{–}CH_2{\cdot}CH_2\text{–}C(=O)\text{—}CH_2CH_3 \ (53)$$

(11.50)

Nitromethane (**51**) when used in this way reacts with 2 moles of an α,β-unsaturated ketone (**52**) to yield a triketone (**53**) (Scheme 11.50).

The anion of nitromethane, acting as a nucleophile, adds to an α,β-unsaturated dinitrile (**54**) and the intermediate spontaneously undergoes a ring closure with elimination of the nitrite ion to yield a derivative of cyclopropane (**55**) (Scheme 11.51). Yields are low in simple systems, but good in complex systems.[63]

$R^1R^2C{=}C(CN)C{\equiv}N$ (54) + $^{-}CH_2NO_2$ $\xrightarrow[CH_3NO_2]{NaOMe / MeOH}$ $[R^1R^2C(CH_2NO_2)-C(CN){=}C{=}N^{-}]$ ⟶ (55) + NO_2^-

(54)

(55)

α,β-Unsaturated dinitrile	Yield (%)
NC CN (cyclohexylidene)	33
CN CN	10
CN CN	22
NC CN MeO	77

(11.51)

Other similar examples of this reaction have been reported. Potassium salts of nitroalkanes (**56**) react with alkenes particularly those substituted with electron-withdrawing groups to give cyclopropanes for example (Scheme 11.52).[64]

$$(CH_3)_2CH\text{-}CH_2(H)C{=}C(CN)(CO_2Me) + (CH_3)_2\bar{C}\text{-}NO_2\ K^+ \ (\mathbf{56}) \longrightarrow \text{cyclopropane: } (CH_3)_2CH\text{-}CH_2(H)C\text{—}C(CN)(CO_2Me),\ C(CH_3)_2$$

Yield 61%, 100% *E* (11.52)

A Michael addition followed by a Nef reaction enables 1,4-diketones (**57**) to be prepared from α,β-unsaturated ketones and nitroalkanes (Scheme 11.53).[65]

$$R^1\text{—}C(=O)\text{—}C(R^2){=}CHR^3 + R^4CH_2NO_2 \xrightarrow[\text{18-Crown-6}]{\text{KF}} R^1\text{—}C(=O)\text{—}CH(R^2)\text{-}CH(R^3)\text{-}CH(R^4)\text{-}NO_2 \quad c.\ 80\%$$

$$\xrightarrow{\text{KMnO}_4\text{ (stoichiometric amount) on SiO}_2\text{ (silica gel)}} R^1\text{—}C(=O)\text{—}CH(R^2)\text{-}CH(R^3)\text{-}C(=O)R^4 \ (\mathbf{57})$$

overall yield *c.* 70% (11.53)

The Mannich reaction[66]

Since nitroalkanes contain an activated (acidic) hydrogen atom at the α-position, they undergo a variety of Mannich reactions. For details see under the Mannich reaction in Chapter 1, p. 50.

Arylation

The reaction of a nitroalkane with an aryl lead triacetate in dimethylsulphoxide gives good yields of α-arylated nitroalkanes (**58**) (Scheme 11.54).[67]

$$\text{1-H-1-NO}_2\text{-cyclopentane} \xrightarrow{PhPb(OAc)_3} \text{1-Ph-1-NO}_2\text{-cyclopentane}$$

$$CH_3\text{-}CH_2\text{-}NO_2 + PhPb(OAc)_3 \longrightarrow CH_3\text{-}C(Ph)(H)\text{-}NO_2\ (58A) + CH_3\text{-}C(Ph)_2\text{-}NO_2\ (58B)$$

equivalents of $PhPb(OAc)_3$	Yield of (**58A**) (%)	Yield of (**58B**) (%)
1.1	58–65	0–5
2.2	0–5	66–74

(11.54)

With acetylenes

A lead compound of an acetylene reacts with the anion of a nitroalkane (**60**), (**61**) yielding a nitroacetylene (**59**) (Scheme 11.55).[68]

With aromatic aldehydes

Nitroethane is used in an efficient preparation of aryl nitriles (**62**) from aromatic aldehydes in a reaction in which an aldoxime is an intermediate. Other nitroalkanes can be used but reaction times are longer than with nitroethane[69] (Scheme 11.56). Yields are *c.* 80 per cent.

11.2.3 *Reactions of the nitro group with nucleophiles*

Reductions to amines

Reduction of the nitro group in a nitroalkane can be achieved by the action of nucleophilic reducing agents, for example by lithium aluminium hydride (Scheme 11.57). High yields of amines are obtained.[70]

This reduction can be stopped at the stage of the oxime[71] or carried out with aqueous chromium chloride to produce the oxime (**63**) (Scheme 11.58).[72]

It should be noted that aromatic nitro compounds are reduced to azo compounds, not to amines, with lithium aluminium hydride.

Tertiary and aryl nitroalkanes (**64**) can be reduced to amines in good yield by

(60) + $R{-}C{\equiv}C{-}Pb(OAc)_3$ → (59)

$R{-}C{\equiv}C{-}SnMe_3 + Pb(OAc)_4$ → $R{-}C{\equiv}C{-}Pb(OAc)_3$

(61) → (59)

R	Yield from (**60**) (%)	Yield from (**61**) (%)
Ph	72	62
$CH_3(CH_2)_5$	47	53
H	48	—
$(CH_3)_3Si$	60	—

(11.55)

$$\text{Ar-CHO} \xrightarrow[\text{pyridinium chloride, reflux}]{CH_3CH_2NO_2} \text{Ar-CN} \quad (62)$$

(11.56)

$$CH_3{-}CH(NO_2){-}CH_2{-}CH_3 \xrightarrow{LiAlH_4} CH_3{-}CH(NH_2){-}CH_2{-}CH_3 \quad 85\%$$

(11.57)

sodium borohydride in the presence of cupric acetate, $Cu(OAc)_2$ (Scheme 11.59).[73]

Sodium borohydride in methanol has been used to reduce aliphatic nitroalkanes to amines when catalysed by nickel boride (Ni_2B) (Scheme 11.60).[74]

CH_3 NO_2 —aq. $CrCl_2$→ CH_3 N–OH

(63)

70%

(11.58)

R–C(CH_3)(CH_3)–NO_2 (64) —$NaBH_4$, $Cu(OAc)_2$→ R–C(CH_3)(CH_3)–NH_2

R	Yield (%)
OAc	75 - 80
NO_2 (both nitro groups reduced)	75 - 80

(11.59)

			Yield (%)
NO_2	$NaBH_4$, Ni_2B MeOH →	NH_2	68
NO_2			66
$C_8H_{17}NO_2$	→	$C_8H_{17}NH_2$	61
O_2N	→	H_2N	64
O_2N CO_2CH_3	→	N O	50

(11.60)

Another combination which has been used is an excess of sodium borohydride with tellurium in ethanol (in which the reducing species is NaHTe).[75] Yields are in the range 80–100 per cent.

A mixture of Raney nickel and aqueous sodium hypophosphite has been used to reduce nitroalkanes to amines.[76]

Formation of α-amino nitriles via reduction to imines

α-Amino nitriles (**65**) can be prepared by the reduction of secondary nitroalkanes (**66**) to imines (**67**) with tri-*n*-butylphosphine and diphenyldisulphide and reaction of the imine, without isolation, with hydrogen cyanide (Scheme 11.61).[77] Yields are in the region of 70 per cent.

$$R^1R^2CH{-}NO_2\ (66) \xrightarrow[\text{Ph-S-S-Ph}]{Bu^n_3P} \left[R^1R^2C{=}NH\right]\ (67) \xrightarrow{\text{NaCN / HOAc}} R^2{-}C(R^1)(CN){-}NH_2\ (65) \quad (11.61)$$

Reduction to thiols

High yields of thiols (**68**) result from the reduction of tertiary nitroalkanes with sodium sulphide and aluminium amalgam (Scheme 11.62). The mechanism is complicated but proceeds via electron-transfer substitution and involves a polysulphide species.[78]

$$R{-}NO_2 \xrightarrow[\text{ii) Al / Hg}]{\text{i) }Na_2S.5H_2O\text{ - DMSO}} R{-}SH\ (68)$$

R	Yield (%)
$PhC(Me)_2-$	92
1-methylcyclohexyl	85
$PhC(Me)_2C(Me)_2-$	78
1-adamantyl	38

(11.62)

Reduction to alkanes[79]

Tributyl tin hydride with a radical initiator (AIBN) can be used to reduce tertiary nitroalkanes and secondary nitroalkanes provided they contain an electron-attracting group (Scheme 11.63). Nitro groups are reduced but not keto, ester, chloro, or sulphur groups.

$$Ph\text{-}CH_2\text{—}CH(NO_2)\text{-}C(=O)\text{—}CH_2CH_3 \xrightarrow[\text{AIBN - PhH}]{Bu^n{}_3SnH} Ph\text{-}CH_2CH_2\text{—}C(=O)\text{—}CH_2CH_3 \quad 78\%$$

$$Ph\text{—}C(CH_3)_2\text{—}NO_2 \longrightarrow Ph\text{—}C(CH_3)_2\text{—}H \quad 92\% \qquad (11.63)$$

Denitration of tertiary nitroalkanes in this way yields a radical which may attack activated carbon–carbon double bonds (the Giese reaction) to form a new carbon–carbon bond in good yield (Scheme 11.64).[80]

$$(CH_3)_2C(NO_2)\text{—}(CH_2)_2\text{-}CO_2Et + CH_2{=}CH\text{-}CO_2Me \xrightarrow[\text{AIBN, PhH}]{Bu^n{}_3SnH} (CH_3)_2C(CH_2\text{-}CH_2\text{-}CO_2Me)\text{—}(CH_2)_2\text{-}CO_2Et \quad 60\% \qquad (11.64)$$

Reduction to alkenes

1,2-Dinitroalkanes (**69**) are converted into alkenes by treatment with tri-*n*-butyltinhydride and AIBN (to accelerate the reaction) (Scheme 11.65). Yields are in the range 70–95 per cent of *c.* 1:1 mixtures of *E* and *Z* alkenes.[81] This type of reduction to an alkene has also been performed with sodium thiomethoxide.[82]

Reduction to oximes or hydroxylamines[83]

Nitroalkanes are reduced by $Et_3NH^+.Sn(SPh)_3^-$. Primary and secondary nitroalkanes yield oximes (Scheme 11.66). Tertiary nitroalkanes give hydroxylamines (Scheme 11.67). The reaction is thought to proceed through nitroso compounds which are reduced faster than nitro compounds (Scheme 11.68).

Primary and secondary nitroalkanes react with iodotrimethylsilane (two moles) to yield nitriles and oximes, respectively (Scheme 11.69).[84]

$$Bu^n{}_3Sn^{\cdot} + R^2R^1C(NO_2)\text{—}CR^3R^4X\ (69) \longrightarrow Bu^n{}_3Sn^+ + R^2R^1C(\dot{N}O_2^{\cdot-})\text{—}CR^3R^4X \longrightarrow NO_2^- + R^2R^1\dot{C}\text{—}CR^3R^4X \longrightarrow R^1R^2C{=}CR^3R^4 + X^{\cdot}$$

$$X^{\cdot} \longrightarrow Bu^n{}_3Sn^{\cdot} + XH$$

X	R^1	R^2	R^3	R^4	Yield (%)	ratio of *E*/*Z*
NO_2	$-(CH_2)_4-$		$-(CH_2)_4-$		72	—
NO_2	C_3H_7	Me	Me	C_3H_7	78	*E*/*Z* mixture
NO_2	H	Ph	Ph	H	98	*E* only
Ts	Me	Me	Et	CO_2Et	81	—

(11.65)

$$CH_3(CH_2)_3NO_2 \xrightarrow[CH_3CN,\ 1\ hour,\ 20^\circ C]{SnCl_2\,/\,3PhSH\,/\,3Et_3N} CH_3(CH_2)_2CH{=}NOH \quad E{:}Z\ \text{ratio } 1{:}1 \quad 82\%$$

or

$$CH_3\text{—}CH(NO_2)\text{-}CO_2Et \xrightarrow[\text{in benzene}]{SnCl_2/3PhSH/3Et_3N,\ 10\ min.\ at\ 20^\circ C} CH_3\text{—}C({=}N\text{-}OH)\text{—}CO_2Et \quad 100\%\ Z \qquad (11.66)$$

$$Ph\text{—}C(CH_3)_2\text{—}NO_2 \xrightarrow[\text{in benzene}]{SnCl_2/3PhSH/3Et_3N} Ph\text{—}C(CH_3)_2\text{—}NHOH \quad 91\% \qquad (11.67)$$

A nitroalkane can be converted into a ketoxime (**70**) as shown in Scheme 11.70 by the action of the Vilsmeier reagent and a Grignard reagent on the anion.[85]

$$R\text{-}NO_2 \xrightarrow{Sn(SPh)_3^-} R\text{-}N(O^-)\text{-}O\text{-}Sn(SPh)_3 \longrightarrow R\text{-}N{=}O + (PhS)_3SnO$$

$$R\text{-}N{=}O \xrightarrow{Sn(SPh)_3^-} R\text{-}\overset{-}{N}\text{-}O\text{-}Sn(SPh)_3 \xrightarrow{Et_3\overset{+}{N}H} R\text{-}NHO\text{-}Sn(SPh)_3 \longrightarrow R\text{-}NHOH$$

$$R_2CH\text{-}NHOH \longrightarrow R_2C{=}NOH \qquad (11.68)$$

$$R^1\text{-}CH(R^2)\text{-}NO_2 \xrightarrow{2\ ISiMe_3} R^1R^2C{=}N\text{-}OH \text{ or } R^2\text{-}CN$$

R^1	R^2	Yield of oxime (%)	Yield of nitrile (%)
Me	Me	89	—
Et	Me	68	—
$-(CH_2)_5-$		84	—
H	Ph	—	96
H	PhCH=CH	—	71

(11.69)

$$R^1\text{-}CH_2\text{-}NO_2 \xrightarrow{Bu^nLi} R^1\text{—}CH{=}\overset{+}{N}(O^-)_2\ Li^+$$

$$\xrightarrow[\text{Vilsmeier reagent}]{Me_2\overset{+}{N}{=}CH\text{-}Cl\ Cl^-} R^1\text{—}CH{=}\overset{+}{N}(O^-)\text{—}O\text{—}CH_2\overset{+}{N}Me_2 \xrightarrow{R^2\text{-}MgBr / CuI} R^1R^2C{=}N\text{—}OH\ \ (70)$$

70 - 95% (11.70)

Autoxidation

Autoxidation of a nitroalkane with oxygen and cuprous chloride gives a peroxide (**71**) which reacts with tertiary amines yielding tertiary amine *N*-oxides (**72**). These rearrange on addition of nitrous acid yielding *N*-nitrosamines (**73**) and aldehydes (Scheme 11.71).[86]

$$Me_2CH\text{-}NO_2 + O_2 \xrightarrow{Cu_2Cl_2} Me_2C(NO_2)\text{—O-OH} \quad (71)$$

$$R^1_2N\text{-}CH_2\text{-}R^2 + Me_2C(NO_2)\text{—O-OH} \longrightarrow R^1_2\overset{+}{N}(O^-)\text{—}CH_2\text{-}R^2 \quad (72)$$

$$(72) \xrightarrow{\text{rearrangement, } HNO_2} \left[R^1_2N\text{—}CH(R^2)\text{-OH}\right] \longrightarrow R^1_2N\text{-}N{=}O \; (73) + R^2\text{-CHO} \quad 15-65\%$$

(11.71)

Conversion to nitriles

By the action of phosphorus trichloride and pyridine on a primary nitroalkane, moderate yields of nitriles can be obtained (Scheme 11.72).[87] This single-step method may be used to replace the normal procedure which would be by conversion to an aldehyde by the Nef reaction followed by formation and dehydration of an oxime.

$$R\text{-}CH_2\text{-}NO_2 \rightleftharpoons R\text{-}CH{=}\overset{+}{N}(OH)O^- \xrightarrow[\text{pyridine}]{PCl_3} H(R)C{=}\overset{+}{N}(O\text{-}PCl_2)O^- + HCl \longrightarrow R\text{—}C{\equiv}\overset{+}{N}\text{—}O^- \xrightarrow{PCl_3} R\text{—}C{\equiv}N + POCl_3$$

R	Yield (%)
Bn	60
CH_2 (cyclohexenyl-1-methyl)	43
AcO, CH_2 (1-acetoxycyclohexyl-1-methyl)	52
$(CH_2)_3\text{-}CO_2Me$	43

(11.72)

Conversion of tertiary nitroalkanes to primary nitroalkanes, then to aldehydes

By condensation of a primary nitroalkane with a tertiary nitroalkane, a larger primary nitroalkane with a quaternary centre results. Oxidation of this yields an aldehyde (Scheme 11.73).[88]

$$R{-}C(Me)_2{-}NO_2 + Na^+ \, ^-CH_2{-}NO_2 \xrightarrow[25^\circ C]{NaH\text{ - }DMSO} R{-}C(Me)_2{-}CH_2{-}NO_2$$

$$\xrightarrow{NaOBu^t / Bu^tOH} R{-}C(Me)_2{-}CH{=}\overset{+}{N}(O^-)O^- \; Na^+ \xrightarrow[0^\circ C]{KMnO_4} R{-}C(Me)_2{-}CHO \quad (80\text{ - }95\%) \tag{11.73}$$

Conversion to iodides

Tertiary nitroalkanes are converted into tertiary alkyl iodides (**74**) in high yield by the action of iodotrimethylsilane (**75**) (Scheme 11.74).[89]

$$R^2{-}C(R^1)(R^3){-}NO_2 + \underset{(75)}{Me_3SiI} \longrightarrow \left[R^1R^2R^3C{-}\overset{+}{N}(=O){-}O{-}SiMe_3 \right] \xrightarrow{I^-} Me_3Si{-}O{-}N{=}O + \underset{(74)}{R^2{-}C(R^1)(R^3){-}I}$$

$$Me_3Si{-}O{-}N{=}O \xrightarrow{Me_3SiI} Me_3Si{-}O{-}SiMe_3 + NO + 0.5\,I_2$$

Nitroalkane	Yield (%)
$(CH_3)_3C{-}NO_2$	94
1-nitroadamantane (NO_2)	98

(11.74)

NITROALKENES AND NITROALKYNES

(76)

α,β-Nitroalkenes (**76**) have a nitro group attached to a carbon–carbon double bond, and because the two groups are in conjugation these nitroalkenes are stable compounds and they display characteristic infra-red frequencies (1353 and 1424 cm^{-1} for monoalkyl and 1346 and 1515 cm^{-1} for di- and tri-alkyl).

11.3 Preparation

11.3.1 *Henry reaction (from carbonyl compounds)*

Nitroalkenes can be prepared by the condensation of nitroalkanes with carbonyl compounds in the presence of a base (the Henry Reaction, see p. 00) (Scheme 11.75).[90] An efficient way of synthesizing nitroalkenes by this method involves a base-catalysed condensation and dehydration of the hydroxynitro compound (**77**) so formed (Scheme 11.76).[91] Yields are in the range 65–95 per cent.

$$\text{R-CHO} + \text{CH}_3\text{-NO}_2 \xrightarrow{\text{base}} \text{R-CH=CH-NO}_2 + \text{H}_2\text{O} \qquad (11.75)$$

$$\text{R}^1\text{-CHO} + \text{R}^2\text{CH}_2\text{NO}_2 \xrightarrow{\text{base}} \underset{(77)}{\text{R}^1\text{—CH(OH)-CH(NO}_2\text{)-R}^2} \xrightarrow[\text{DCC, 25}^\circ\text{C}]{\text{Cu}_2\text{Cl}_2} (\text{R}_1)(\text{H})\text{C=C}(\text{NO}_2)(\text{R}^2) \qquad (11.76)$$

Nitroethylene (**78**) can be readily prepared from 2-nitroethanol (**79**)[92] treated with phthalic anhydride (Scheme 11.77).[93]

Similarly nitroethane and formaldehyde react to form 2-nitropropene (**80**) (Scheme 11.78).[94]

$$HO\text{-}CH_2\text{-}CH_2NO_2 \ (79) \xrightarrow[\text{175 - 180°C, 80 mm Hg}]{\text{phthalic anhydride}} CH_2{=}CH\text{-}NO_2 \ (78)$$

$$CH_2O + CH_3NO_2 \xrightarrow{\text{i) KOH; ii) } H_2SO_4} (79) \qquad (11.77)$$

$$CH_3\text{—}CH(NO_2)\text{-}CH_2OH \ 84\% \xrightarrow[\text{175 - 180°C, 80 mm Hg}]{\text{phthalic anhydride}} CH_3\text{—}C(NO_2){=}CH_2 \ 72\% \ (80)$$

$$CH_2O + CH_3CH_2NO_2 \xrightarrow{\text{i) KOH; ii) } H_2SO_4} CH_3\text{—}CH(NO_2)\text{-}CH_2OH \qquad (11.78)$$

11.3.2 *From aldehydes and ketones*

Ethylenediamine has been found to be a good catalyst for the preparation of nitroalkenes from aldehydes and ketones and nitromethane, and though the method is not generally useful the reaction has been used in steroid synthesis (Scheme 11.79).[95]

11.3.3 *From alkenes*

The reaction of mercuric nitrite, $Hg(NO_2)_2$, with alkenes has been shown to be a general method for the preparation of nitroalkenes[96] (Scheme 11.80). The nitroalkene produced may subsequently be used in the preparation of α,β-unsaturated nitriles and ketones. The unsaturated cyanide is produced by Michael addition of hydrogen cyanide followed by elimination of nitrous acid and the unsaturated ketone results from a Nef reaction.

Analogous to the above nitromercuration, nitroalkenes can be prepared by nitroselenation in somewhat lower yield, but under nonaqueous conditions (Scheme 11.81).[97]

$H_2NCH_2CH_2NH_2$

CH_3NO_2, heat

95%

(11.79)

$HgCl_2 + NaNO_2$

$[Hg(NO_2)_2]$

aq. HO^-

HCN

>98%

KOBu

DBU

dil. H_2SO_4 (Nef)

85%

70%

(11.80)

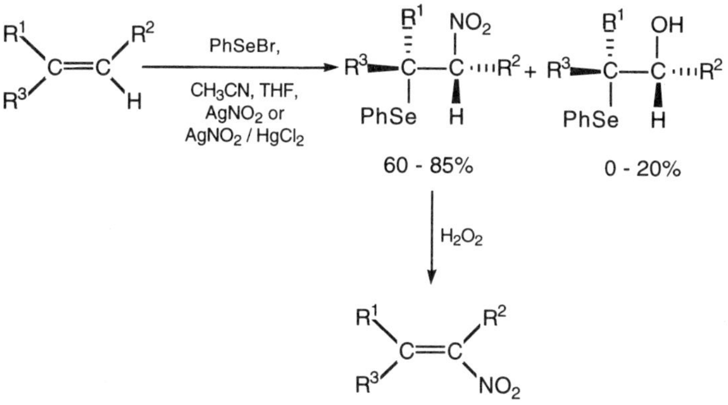

$R^1R^3C{=}CHR^2$ —(PhSeBr, CH_3CN, THF, $AgNO_2$ or $AgNO_2$ / $HgCl_2$)→ $R^3R^1C(SePh)$—$CH(NO_2)R^2$ (60 - 85%) + $R^3R^1C(SePh)$—$CH(OH)R^2$ (0 - 20%)

—(H_2O_2)→ $R^1R^3C{=}C(R^2)NO_2$

Alkene	Yield (%)
	75
	77
	70
	81
	76

(11.81)

The action of dinitrogen tetroxide and iodine on an alkene (Scheme 11.82) or an alkyne (Scheme 11.83) constitutes a preparation of nitroalkenes and nitroalkynes and was the first preparation of a nitroalkyne (nitroacetylene (**81**)).[98]

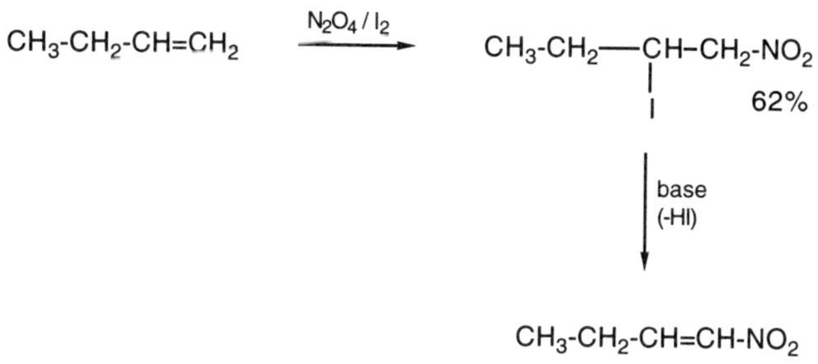

$CH_3\text{-}CH_2\text{-}CH{=}CH_2$ —(N_2O_4 / I_2)→ $CH_3\text{-}CH_2\text{—}CH(I)\text{-}CH_2\text{-}NO_2$ 62%

—(base (-HI))→ $CH_3\text{-}CH_2\text{-}CH{=}CH\text{-}NO_2$ (11.82)

Nitroalkynes are dienophiles, polarophiles, and electrophiles and as such they have been used in the preparation of nitroalkenes e.g. Scheme 11.83.[99]

$$(CH_3)_3C\text{-}C{\equiv}CH \xrightarrow{N_2O_4 / I_2} (CH_3)_3C\text{-}C(I){=}CH\text{-}NO_2 \xrightarrow[\text{reduced pressure}]{KOH,\ 100^{\circ}C} (CH_3)_3C\text{-}C{\equiv}C\text{-}NO_2 \quad \textbf{(81)} \qquad (11.83)$$

A nitroalkene has been prepared from a silylated alkene (**82**) through an alkene(phenyl)iodonium tetrafluoroborate (**83**) which in the presence of copper salts reacts with nucleophiles to cause substitution. The preparation of only one nitroalkene has been reported (Scheme 11.84).[100]

SiMe$_3$ / CMe$_3$ (82) $\xrightarrow[CH_2Cl_2,\ 0^{\circ}C,\ 15\ min.]{PhIO,\ BF_3,\ Et_2O}$ IPhBF$_4$ / CMe$_3$ (83) 80% $\xrightarrow[DMF - H_2O,\ 25^{\circ}C,\ 28\ hours]{NaNO_2,\ CuSO_4}$ NO$_2$ / CMe$_3$ 55% (11.84)

11.3.4 *From nitroalkanes*

A nitroalkene can be prepared in good yield from an appropriate nitroalkane by the action of phenylselenium bromide on the lithium anion of the nitroalkane followed by removal of the selenium by oxidation with hydrogen peroxide (Scheme 11.85).[101]

$$R\text{-}CH_2\text{-}CH_2\text{-}NO_2 \xrightarrow{Bu^nLi} R\text{-}CH_2\text{—}CH{=}\overset{+}{N}(O^-)O^-\ \ Li^+ \quad (Br\text{—}Se\text{-}Ph)$$

$$\xrightarrow{PhSeBr} R\text{-}CH_2\text{—}CH(NO_2)\text{-}Se\text{-}Ph \xrightarrow{aq.H_2O_2} RHC{=}CHNO_2$$

Nitroalkene	Yield (%)
C_6H_{13}–CH=CH–NO_2	80
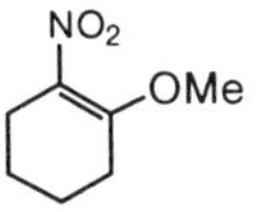 2-methoxy-1-nitrocyclohexene (ring with NO_2 and OMe substituents)	61

(11.85)

11.3.5 *From cyclic nitro ketones*

Cyclic 3-substituted 1-nitroalkenes (**85**) can be prepared from cyclic α-nitro ketones (**84**) (Scheme 11.86).[102] The use of chiral amines to form the imine (where R^1 is asymmetric) leads to asymmetric alkylation.

11.3.6 *By pyrolysis of nitro-substituted Mannich bases*

Nitroalkenes as their crystalline aromatic amine adducts (**86**) have been prepared from nitro-substituted Mannich bases by pyrolysis (Scheme 11.87).[103]

The use of a boron trifluoride complex of the Mannich base is more useful since it allows lower temperatures (*c.* 100 °C) to be used.[104] Yields are 80–90 per cent.

11.3.7 *Preparation of 1-nitro-1,3-dienes*

1-Nitro-1,3-dienes (**87**) can be prepared by the addition of trifluoracetyl nitrate ($CF_3CO—ONO_2$) to a 1,3-diene followed by the elimination of trifluoroacetic acid. The nitrate is prepared *in situ* from ammonium nitrate and trifluoroacetic acid (Scheme 11.88).[105]

NO_2 (84) $\xrightarrow{R^1\text{-}NH_2}$

or $R^1\text{-}NHNH_2$ ⟶ hydrazone

2 x BuLi, R^2Hal
THF / HMPT
-78°C

2 x $NaBH_4$
2 x $CeCl_3$
MeOH, 20°C

(85)

R^1	R^2	Yield (%)
NMe_2	CH_3	44
C_6H_{13}	CH_3	34
NME_2		43
C_6H_{13}		37
NMe_2		35
C_6H_{13}		19
NMe_2		42
C_6H_{13}		35

(11.86)

$$R^1\text{—}\underset{\displaystyle NO_2}{\underset{|}{C}}H\text{-}CH_2\text{-}\overset{+}{N}HR^2_2\ Cl^- \xrightarrow{\text{heat}} R^1\text{—}\underset{\displaystyle {}^-O\text{—}\overset{+}{N}\text{=}O}{\underset{|}{C}}\text{=}CH_2 \quad Ar\ddot{N}H_2$$

$$R^1\text{—}\underset{\displaystyle NO_2}{\underset{|}{C}}H_2 + CH_2O + R^2_2\overset{+}{N}H_2Cl^- \longrightarrow R^1\text{—}\underset{\displaystyle NO_2}{\underset{|}{C}}H\text{-}CH_2\text{-}\overset{+}{N}HR^2_2\ Cl^-$$

Michael reaction

Ar = (4-methylphenyl, CH_3)

$$R^1\text{—}\underset{\displaystyle NO_2}{\underset{|}{C}}H\text{-}CH_2\text{-}NHAr$$

(crystalline derivative)

(86) (11.87)

$$CH_2\text{=}CH\text{-}CH\text{=}CH_2 + [F_3C\text{-}CO\text{-}ONO_2] \longrightarrow \left[CH_2\text{=}CH\text{—}\underset{}{\overset{O\text{—}CO\text{-}CF_3}{\overset{|}{C}}}HCH_2\text{—}NO_2 \; + \; \underset{CF_3\text{-}C\text{=}O}{\underset{|}{O}}\text{—}CH_2\text{-}CH\text{=}CH\text{-}CH_2\text{—}NO_2 \right]$$

$$NH_4NO_3 + F_3C\text{-}CO_2H \xrightarrow{CH_2Cl_2} [F_3C\text{-}CO\text{-}ONO_2]$$

$$\xrightarrow{KOAc\ /\ Et_2O} CH_2\text{=}CH\text{-}CH\text{=}CH\text{—}NO_2$$

(87) 89%

Similarly:

$$\text{cyclohexene} \xrightarrow[\text{ii) } KOAc/Et_2O]{\text{i) } F_3CO\text{—}ONO_2,\ CH_2Cl_2} \text{1-nitrocyclohexene}$$

70% (11.88)

11.4 Reactivity

11.4.1 *Reduction to alkenes*

Reduction to alkenes by hydrogenolysis of the nitro group can be effected by the action of sodium sulphide and phenyl mercaptan in dimethylformamide (Scheme 11.89).[106] At least one aryl group is needed to achieve a good yield in this reaction.

$$R^1R^2C{=}C(R^3)NO_2 \xrightarrow[\text{DMF}]{Na_2S.9H_2O\text{ - }PhSH} R^1R^2C{=}C(R^3)H$$

R^1	R^2	R^3	Yield (%)
Ph	Ph	H	94
Ph	H	Ph	93
4-Cl-C_6H_4	H	Et	60
Ph	H	COOEt	95

(11.89)

11.4.2 *Conversion to aldehydes and ketones*

Nitroalkenes are converted into aldehydes or ketones when they are treated with Raney nickel and an aqueous solution of sodium hypophosphite (Scheme 11.90).[107]

$$R^1\text{-}CH{=}C(R^2)NO_2 \xrightarrow[\text{aq. }NaH_2PO_2]{\text{Raney Ni}} R^1\text{-}CH_2\text{—}C({=}O)\text{—}R^2 \qquad (11.90)$$

Nitroalkenes react with trialkyl aluminium compounds (**88**) to yield either ketones (**89**) or nitroalkanes (**90**) depending upon the work-up procedure (Scheme 11.91).[108]

$$R^1(O_2N)C{=}CHR^2 + R^3\text{-}AlR^3_2\ (88) \xrightarrow{\text{hexane}} \left[R^1C(CHR^2R^3){=}N^+(O^-)(O\text{-}AlR^2_2)\right]$$

$$\xrightarrow{\text{conditions1,2 or 3}} R^1\text{—}C({=}O)\text{—}CHR^2R^3\ (89) \quad\text{or}\quad R^1\text{—}CH(NO_2)\text{-}CHR^2R^3\ (90)$$

Conditions: 1 Et_2O, 0.1 M HCl
2 Et_2O, M HCl
3 0.2 M NaOH, $KMnO_4$

R^1	R^2	R^3	conditions	Yield of (**89**) (%)	Yield (**90**) (%)
-$(CH_2)_4$-		Et	1	21	80
-$(CH_2)_4$-		Et	2	96	4
-$(CH_2)_4$-		Bu^i	1	26	74
-$(CH_2)_4$-		Bu^i	2	94	6
-$(CH_2)_4$-		Ph	1	4	96
-$(CH_2)_4$-		Ph	2	95	5
$CH_3(CH_2)_5$-	H	Bu^i	1	0	100
$CH_3(CH_2)_5$-	H	Bu^i	3	93	7
Ph	H	Et	1	6	94
Ph	H	Et	2	81	19
Ph	H	Bu^i	1	7	93
Ph	H	Bu^i	3	95	5

(11.91)

11.4.3 *Michael addition*

Michael additions to nitroalkenes are catalysed by nickel acetonylacetonate, resulting in better yields than in the uncatalysed reaction (Scheme 11.92).[109]

$$CH_2(CO_2Me)_2 + Ph\text{-}CH{=}CH\text{-}NO_2 \xrightarrow{Ni(acac)_2} (MeO_2C)_2CH\text{—}CH(Ph)\text{-}CH_2\text{-}NO_2 \quad 62\% \qquad (11.92)$$

11.4.4 *Diels–Alder reactions*

The Diels–Alder reaction of nitroalkenes has been used by Corey and Moinet[110] to produce a lactone which was used for the synthesis of a prostoglandin (Scheme 11.93).

Nitroalkenes undergo Diels–Alder reactions with 1,3-dienylsilyl ethers (**91**) to give nitrocyclohexanones (**92**) which can be reductively denitrated by trialkyltin hydrides (**93**) (Scheme 11.94).[111] The addition is regio-specific, but results in a mixture of stereoisomers.

2-Nitropropene (**94**) and the silyl ether of cyclohexanone (**95**) undergo a Diels–Alder reaction in the presence of stannic chloride to yield on work up, a γ-diketone (**96**) which can be cyclized by alkali to yield an α,β-unsaturated ketone (**97**) (Scheme 11.95).[112]

NO2 + ether, -20°C → 100%; NaOMe, TiCl3 → 56%; i) aq. HO⁻, ii) dil. HCl → 30%

(11.93)

Me3SiO, CH2, H, OMe (91) + H, Et, Me, NO2 — i) PhH, heat; ii) aq. H3O+ → MeO, Me, NO2, Et (92) 71%; Bu3nSnH (93) → MeO, Me, Et 82%

(11.94)

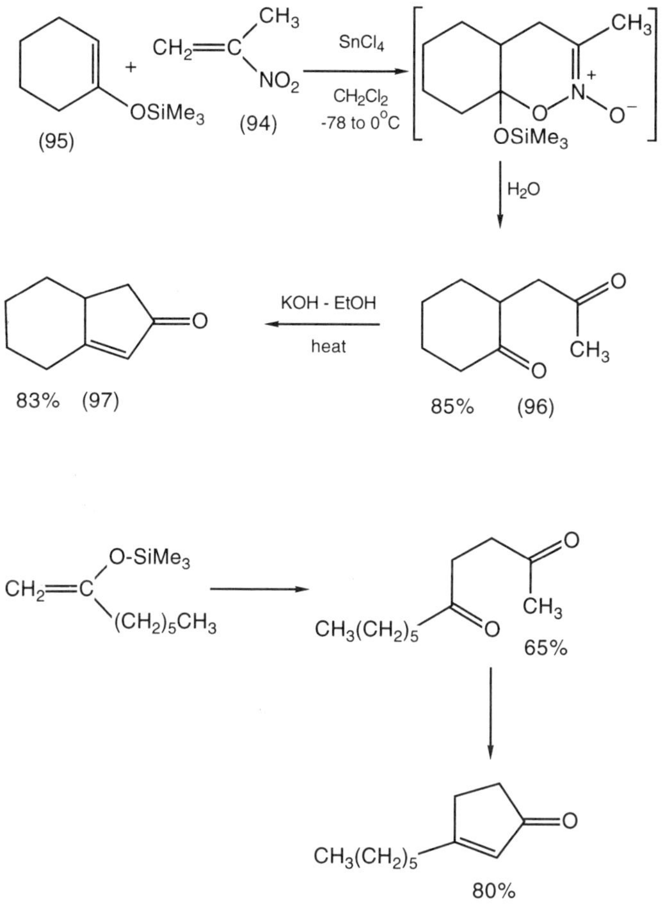

(11.95)

11.4.5 *Conversion into allylic alcohols*

Nitroalkenes are converted into allylic alcohols (**99**) by addition of phenyl mercaptan as its lithium salt and formaldehyde across the carbon–carbon double bond followed by radical elimination with tri-*n*-butyl tin hydride and AIBN (Scheme 11.96).[113]

$$R\text{—}CH{=}CH\text{-}NO_2 \xrightarrow[\text{ii) aq. } CH_2O]{\text{i) PhSLi}} R\text{—}\underset{SPh}{CH}\text{-}\overset{NO_2}{CH}\text{-}CH_2OH \; (98) \xrightarrow[110^{\circ}C]{Bu^n_3SnH,\ AIBN} (E)\text{-}RCH{=}CHCH_2OH \; (99)$$

R	Yield of (**98**) (%)	Yield of (**99**) (%)	*E*:*Z*
Me	85	75	99:1
Et	91	76	95:5
Pr^n	66	85	99:1
Pr^i	57	78	99:1
Bu^n	63	58	99:1
Ph	78	72	95:5
PhCH$_2$CH$_2$	79	80	99:1

(11.96)

11.4.6 *Use in preparation of aromatic ketones*

Diprotonation of nitroalkenes using trifluoromethanesulphonic acid converts them into electrophiles which react with aromatic compounds eventually to yield ketones (Scheme 11.97).[114]

$$R^2(H)C{=}C(R^1)\text{-}N^+(O^-){=}O \xrightarrow{CF_3SO_3H,\ H^+} \left[R^2(H)C^+\text{-}C(R^1){=}N^+(OH)\text{-}OH\right] \xrightarrow{C_6H_6} \left[R^2\text{—}CH(C_6H_6^+)\text{-}C(R^1){=}N^+(O^-)OH\right] \xrightarrow{H_2O} C_6H_5\text{-}CH(R^2)\text{-}C(=O)R^1$$

R^1	R^2	Yield (%)
CH_3	H	85
CH_2CH_3	H	87
CH_3	Ph	65
Ph	H	37
$-(CH_2)_4-$		72

(11.97)

11.4.7 *Conversion into dithioacetals*

Nitroalkenes react with aluminium chloride in the presence of ethane thiol to give dithioacetals (**100**) (Scheme 11.98).[115]

$R^1CH{=}C(NO_2)R^2 \xrightarrow[EtSH]{AlCl_3} R^1{-}CH(:SEt){-}C(NO_2)(R^2){-}H \longrightarrow R^1CH{=}\overset{+}{S}Et + R^2CH{=}\overset{+}{N}(O^-)_2$

$R^1CH{=}\overset{+}{S}Et \xrightarrow{EtSH} R^1\text{-}CH(SEt)_2$ (100)

$R^2CH{=}\overset{+}{N}(O^-)_2 \xrightarrow{H^+} R^2CH_2NO_2$ (11.98)

$R^1(H)C{=}C(R^2)\overset{+}{N}(O^-){=}O \underset{}{\overset{EtXH}{\rightleftharpoons}} R^1{-}CH(\overset{+}{X}HEt){-}C(R^2){=}\overset{+}{N}(O^-){-}O^- \xrightarrow{SnCl_2.2H_2O} R^1{-}CH(XEt){-}C(R^2){=}N{-}OH$

R^1	R^2	X	Yield (%)
CH_3	CH_3	O	89
CH_3CH_2	CH_3	O	90
CH_3CH_2	CH_3	S	90
CH_3	H	O	83
CH_3CH_2	H	O	86
CH_3CH_2	H	S	71

(11.99)

11.4.8 *Reduction to alkoxyoximes*

Nitroalkenes are reduced to alkoxyoximes by stannous chloride in alcoholic solution (Scheme 11.99).[116]

ALLYLIC NITRO COMPOUNDS

11.5 Reactivity

11.5.1 *Reduction to alkenes*

Tetrakis(triphenylphosphine)palladium, $Pd(PPh_3)_4$,[117] complexes with allylic nitro compounds (**101**) and the resulting complexes can be reduced by a variety of reducing agents whose nature determines the structure of the resulting alkene, for example, Scheme 11.100.[118,119]

$$PhCH_2C(NO_2)(CH_3){-}CH{=}CH_2 \ \textbf{(101)} \xrightarrow[\text{reducing agent}]{Pd(PPh_3)_4\,/} PhCH_2CH(CH_3){-}CH{=}CH_2 \ \textbf{(102)} + (PhCH_2)(CH_3)C{=}C(H)(CH_3) \ \textbf{(103)}$$

reducing agent	Yield (%)	ratio (**102**):(**103**)
HCO_2NH_4	55	92: 8
$NaBH_3CN$	98	32: 68

(11.100)

11.5.2 *Conversion to allylic phenyl sulphones*

Two similar procedures have been described for the conversion of allylic nitroalkanes (**104**) into allylic phenyl sulphones (**105**). In one[120] sodium benzene sulphinate ($PhSO_2Na.2H_2O$) is used in presence of a palladium–triphenylphosphine catalyst, $(PPh_3)_4Pd$ (Scheme 11.101).

In the second variant by the same authors, a base (Et_3N) is used to isomerize an α,β-unsaturated nitroalkene to an allylic compound and a second phosphorus compound ($Ph_2PCH_2CH_2PPh_2$) is added to the second stage along with the triphenylphosphine–palladium complex and the sodium benzene sulphinate (Scheme 11.102).[121]

$$CH_3CH{=}C(CH_3)CH_2{-}NO_2 \ (104) \xrightarrow[\text{DMF, }70^\circ C]{PhSO_2Na,\ Pd(PPh_3)_4} CH_3CH{=}C(CH_3)CH_2{-}SO_2Ph \ (105) \qquad (11.101)$$

$$\text{1-nitrocyclohexene} \xrightarrow[\text{DMF}]{Et_3N} [\text{3-nitrocyclohexene}] \xrightarrow[Pd(PPh_3)_4,\ Ph_2P{-}CH_2{-}CH_2{-}PPh_2]{PhSO_2Na.2H_2O} \text{3-(phenylsulphonyl)cyclohexene } 70\% \qquad (11.102)$$

11.5.3 *Allylic alkylation*

Allylic nitro compounds (**106**), prepared via the Michael addition of nitroalkanes to phenyl vinyl sulphoxide (**107**), readily undergo palladium catalysed

$$R^2{-}C(R^1)(H){-}NO_2 + CH_2{=}CH{-}SOPh \ (107) \xrightarrow[25^\circ C,\ 24hr]{\text{1,5-Diazabicyclo[5.4.0]unde-5-ene}} R^2{-}C(R^1)(NO_2){-}CH{=}CH_2 \ (106)$$

$$(106) \xrightarrow[\text{THF, }60^\circ C]{NaCH(CO_2Me)_2,\ Pd(PPh_3)_4} R^1R^2C{=}CH{-}CH(CO_2Me)_2 \ (109) + R^2{-}C(R^1)(CH(CO_2Me)_2){-}CH{=}CH_2 \ (108)$$

R^1	R^2	Yield of (**106**) (%)	Yield of (**108**) and (**109**) (%)	ratio (**108**):(**109**)	*E*:*Z* of (**109**)
CH_3	CH_3	73	60	73:27	—
$-(CH_2)_5-$		67	73	88:12	—
CH_3	$(CH_3)_2CHCH_2$	70	48	100:0	88:12
CH_3	$CH_3COCH_2CH_2$	74	51	100:0	62:38
CH_3	$PhCH_2$	60	77	100:0	82:18

(11.103)

allylic alkylations, in a similar reaction to the above reduction (Scheme 11.103).[122]

11.5.4 *Conversion to allylic thioethers*

In a similar way allylic thioethers can be prepared from allylic nitroalkanes (Scheme 11.104).[123]

$R-C(CH_3)(NO_2)-CH=CH_2$ —(conditions 1 or 2, $Pd(PPh_3)_4$, HMPT, rearrangement)→ $R-C(CH_3)(SPh)-CH=CH_2$ (110) + $(CH_3)(R)C=C(H)CH_2\text{-}SO_2Ph$ (111)

Conditions: 1 PhSNa, HMPA, 50°C, 15 hours
2 $PhSO_2Na.2H_2O$, DMF, 25°C, 10 hours

R	Conditions	Yield of (**110**)	Yield of (**111**)	*E*:*Z* of (**111**)
n-C_6H_{13}	1	0	77	77:23
Ph	1	0	75	96:4
$(CH_2)_2CO_2Me$	1	0	65	70:30
n-C_6H_{13}	2	97	—	—
$(CH_2)_2CO_2Me$	2	76	—	—

(11.104)

NITROSOALKANES AND ALKENES

11.6 Preparation

11.6.1 *From alkylhydroxylamines*

Nitrosoalkanes (**112**) can be prepared in 55–95 per cent yields by the oxidation of alkylhydroxylamines with silver carbonate on Celite in halogenated solvents (Scheme 11.105).[124]

$$\text{R-NHOH} \xrightarrow[\text{in } CH_2Cl_2 \text{ or } CFCl_3,\ <20^{\circ}C]{Ag_2CO_3\,/\,\text{Celite}} \underset{(112)}{\text{R-N=O}}$$

R	Yield (%)
Ph	87
cyclohexyl	95
Pr^i	90
2-adamantyl	89
1-cyclobutylethyl	93

(11.105)

11.6.2 *From β-chloro-oximes*

A general preparation of nitrosoalkenes (**113**), which are unstable, but can be used *in situ* for [4 + 2] cycloadditions, is by the dehydrohalogenation of β-chloro-oximes (**114**) (Scheme 11.106).[125]

$$\underset{(114)}{R^2R^1C(Cl)\text{—}C(=NOH)R^3} \xrightarrow[\text{MeOH or EtOH}]{NaHCO_3} \underset{(113)}{R^1R^2C\text{=}C(N\text{=}O)R^3} \qquad (11.106)$$

11.6.3 *From α-chloroketones*

Another useful method for the preparation of these nitrosoalkanes is by reaction of α-chloroketones (**115**) with compound (**116**)[126] followed by reaction with fluoride ions (Scheme 11.107).[127]

$$\underset{(115)}{R^1C(=O)\text{—}CH(Cl)R^2} + \underset{(116)}{H_2N\text{-O-SiMe}_2Bu^t} \longrightarrow R^1C(=N\text{—O-SiMe}_2Bu^t)\text{—}CH(Cl)R^2 \xrightarrow{F^-} R^1HC\text{=}C(N\text{=}O)R^2 \qquad (11.107)$$

When nitrosoalkenes have electron-withdrawing groups they can then act as dienophiles in Diels–Alder reactions.

11.7 Reactivity

11.7.1 *With trialkylphosphites*

Some alkyl nitroso compounds (e.g. **117**) react with trialkylphosphites to yield imines (**118**)[128] through a mechanism in which a rearrangement occurs,[129] for example, Scheme 11.108.

$$Me_2C(Ph){-}N{=}O \ (117) \xrightarrow[\text{[fast]}]{P(OEt)_3} Me_2C(Ph){-}\overset{-}{N}{-}O{-}\overset{+}{P}(OEt)_3 \xrightarrow{\text{[slow]}} Me_2C{=}N{-}Ph \ (118) + O{=}P(OEt_3)$$

$$Me_2C{=}N{-}Ph \xrightarrow{H_2O} Me_2C{=}O \ (79\ \%) + PhNH_2 \qquad (11.108)$$

The migration is also known to occur with 2,4,4-trimethyl-2-nitrosopentane ($Bu^tCH_2CMe_2NO$) (**119**), but both alkyl groups can migrate as shown in Scheme 11.109.

$$Bu^t{-}CH_2{-}C(CH_3)_2{-}N{=}O \ (119) \xrightarrow{PO(OEt)_3} Bu^t{-}CH_2{-}C(CH_3){=}N{-}CH_3 + CH_3{-}C(CH_3){=}N{-}CH_2{-}Bu^t \qquad (11.109)$$

References

1. Review: N. Levy and J. D. Rose, *Chem. Soc. Quarterly Rev.*, 1947, **1**, 358; reference to symposium, *Nitroalkanes and nitroalkenes in synthesis* in *Tetrahedron*, 1990, **46**, 7313. (ed. A. G. M. Barrett).
2. J. P. Ritchie, *Tetrahedron*, 1988, **44**, 7467.
3. J. R. Murdoch, A. Streitweiser, Jr, and S. Gabriel, *J. Am. Chem. Soc.*, 1978, **100**, 6338; A. Pross, D. J. DeFrees, B. A. Levi, S. K. Pollack, L. Radon, and W. Hehre, *J. Org. Chem.*, 1981, **46**, 1693.

4. *F. &F.*, **1**, 1011; N. Kornblum *Org. React.*, 1962, **12**, 101.
5. N. Kornblum, B. Taub, and H. E. Ungnade, *J. Am. Chem. Soc.*, 1954, **76**, 3209.
6. N. Kornblum and H. E. Ungnade, *Org. Synth., Coll. Vol. 4*, 1963, 724.
7. *F. &F.*, **6**, 135; J. W. Zubrick, B. I. Dunbar, and H. D. Durst, *Tetrahedron Lett.*, 1975, 71.
8. *F. &F.*, **1**, 324.
9. N. Levy and C. W. Scaife, *J. Chem. Soc.*, 1946, 1093; 1100; N. Levy, C. W. Scaife, and A. E. Wilder-Smith, *J. Chem. Soc.*, 1946, 1096; 1948, 52; H. Baldock, N. Levy, and C. W. Scaife, *J. Chem. Soc.*, 1949, 2627.
10. W. K. Seifert, *J. Org. Chem.*, 1963, **28**, 125.
11. C. E. Anagnostopoulos and L. F. Fieser, *J. Am. Chem. Soc.*, 1954, **76**, 532.
12. N. Kornblum and W. J. Jones, *Org. Synth.*, 1963, **43**, 87.
13. *F. &F.*, **8**, 375; E. Keinan and Y. Mazur, *J. Org. Chem.*, 1977, **42**, 844.
14. *F. &F.*, **11**, 442; R. Curci, M. Fiorentino, L. Triosi, J. O. Edwards, and R. H. Pater, *J. Org. Chem.*, 1980, **45**, 4758; G. Cicala, R. Curci, M. Fiorentino, and O. Laricchiuta, *J. Org. Chem.*, 1982, **47**, 2670.
15. *F. &F.*, **13**, 120; R. W. Murray, R. Jeyeraman, and L. Mohan, *Tetrahedron Lett.*, 1986, **27**, 2335.
16. *F. &F.*, **1**, 822; W. D. Emmons and A. S. Pagano, *J. Am. Chem. Soc.*, 1955, **77**, 4557.
17. *F. &F.*, **10**, 209; E. J. Corey and H. Estreicher, *Tetrahedron Lett.*, 1980, **21**, 1117.
18. F. C. Whitmore and M. G. Whitmore, *Org. Synth., Coll. Vol. I*, 1941, 401.
19. *F. &F.*, **3**, 185; P. E. Pfeffer and L. S. Silbert, *Tetrahedron Lett.*, 1970, 699.
20. *F. &F.*, **4**, 356; G. B. Bachman and T. F. Bierman, *J. Org. Chem.*, 1970, **35**, 4229.
21. *F. &F.*, **11**, 442.
22. *F. &F.*, **15**, 144; P. E. Eaton and G. E. Wicks, *J. Org. Chem.*, 1988, **53**, 5353.
23. *F. &F.*, **12**, 365; E. J. Corey, B. Samuelsson, and F. A. Luzzio, *J. Am. Chem. Soc.*, 1984, **106**, 3682.
24. *F. &F.*, **12**, 288; R. S. Varma and G. W. Kabalka, *Synth. Commun.*, 1984, **14**, 1093.
25. W. D. Emmons and J. P. Freeman, *J. Am. Chem. Soc.*, 1955, **77**, 4391.
26. *F. &F.*, **12**, 531; P. Dampawan and W. W. Zajac, Jr, *Synthesis*, 1983, 545.
27. *F. &F.*, **5**, 539; G. A. Olah and M. Nojima, *Synthesis*, 1973, 785.
28. F. G. Bordwell, J. E. Bartmess, and J. A. Hautala, *J. Org. Chem.*, 1978, **43**, 3095.
29. J. U. Nef, *Annalen*, 1894, **280**, 263; *F. &F.*, **8**, 461; W. E. Noland, *Chem. Rev.*, 1955, **55**, 137.
30. *F. &F.*, **14**, 86; J. M. Aizpurua, M. Oiarbide, and C. Palomo, *Tetrahedron Lett.*, 1987, **28**, 5361.
31. H. B. Hass and M. L. Bender, *Org. Synth., Coll. Vol. 4*, 1963, 932; H. B. Hass and M. L. Bender, *J. Am. Chem. Soc.*, 1949, **71**, 1767; B. H. Klandermann, *J. Org. Chem.*, 1966, **31**, 2618.
32. *F. &F.*, **6**, 545; R. M. Jacobson, *Tetrahedron Lett.*, 1974, 3215.
33. *F. &F.*, **8**, 461; E. Keinan and Y. Mazur, *J. Am. Chem. Soc.*, 1977, **99**, 3861.
34. *F. &F.*, **4**, 506; J. E. McMurry and J. Melton, *J. Am. Chem. Soc.*, 1971, **93**, 5309 (also imines); *F. &F.*, **5**, 670; definitive paper; J. E. McMurry and J. Melton, *J. Org. Chem.*, 1973, **38**, 4367.
35. *F. &F.*, **5**, 671; T.-L. Ho and L. M. Wong, *Synthesis*, 1974, 196.
36. *F. &F.*, **5**, 75.
37. *F. &F.*, **8**, 63; P. A. Bartlett, F. R. Green III, and T. R. Webb, *Tetrahedron Lett.*, 1977, 331.

38. *F. &F.*, **5**, 495; J. E. McMurry, J. Melton, and H. Padgett, *J. Org. Chem.*, 1974, **39**, 259; J. E. McMurry and J. Melton, *Org. Synth.*, 1977, **56**, 36.
39. *F. &F.*, **8**, 370; J. R. Williams, L. R. Unger, and R. H. Moore, *J. Org. Chem.*, 1978, **43**, 1271.
40. *F. &F.*, **10**, 81; G. A. Olah and B. G. B. Gupta, *Synthesis*, 1980, 44.
41. Preparation: W. A. Noyes, *Org. Synth. Coll. Vol. 2*, 1943, 108.
42. *F. &F.*, **5**, 565; N. Kornblum and P. A. Wade, *J. Org. Chem.*, **38**, 1418, 1973; N. Kornblum, R. K. Blackwood, and D. D. Mooberry, *J. Am. Chem. Soc.*, 1956; **78**, 1501.
43. *F. &F.*, **11**, 440; N. Kornblum, A. S. Erikson, W. J. Kelly, and B. Henggeler, *J. Org. Chem.*, 1982, **47**, 4534; *F. &F.*, **1**, 950; H. Schechter and F. T. Williams, Jr, *J. Org. Chem.*, 1962, **27**, 3699.
44. *F. &F.*, **11**, 218; M. R. Gabobardes and H. W. Pinnick, *Tetrahedron Lett.*, 1981, **22**, 5235.
45. *F. &F.*, **11**, 419; P. A. Wade, H. R. Hinney, N. V. Amin, P. D. Vail, S. D. Morrow, S. A. Hardinger, and M. S. Saft, *J. Org. Chem.*, 1981, **46**, 765.
46. Preparation: E. W. Calvin, A. K. Beck, B. Bastani, D. Seebach, Y. Kai, and J. D. Dunitz, *Helv. Chim. Acta*, 1980, **63**, 697.
47. *F. &F.*, **12**, 84; E. Vedejs, T. H. Eberlein, and D. L. Varie, *J. Am. Chem. Soc.*, 1982, **104**, 1445.
48. E. Vedejs and D. A. Perry, *J. Org. Chem.*, 1984, **49**, 573.
49. Review: H. B. Hass and E. F. Riley, *Chem. Rev.*, 1943, **32**, 406; *F. &F.*, **15**, 335.
50. A. G. M. Barrett, C. Robyr, and C. D. Spilling, *J. Org. Chem.*, 1989, **54**, 1233.
51. *F. &F.*, **1**, 740; review: F. W. Lichtenthaler, *Angew. Chem., Int. Ed. Engl.*, **3**, 211.
52. *F. &F.*, **9**, 323; G. B. Bachman and R. J. Maleski, *J. Org. Chem.*, 1972, **37**, 2810.
53. R. H. Wollenberg and S. J. Miller, *Tetrahedron Lett.*, 1978, 3219.
54. *F. &F.*, **12**, 462; K. Matsumoto, *Angew. Chem., Int. Ed. Engl.*, 1984, **23**, 617.
55. *F. &F.*, **1**, 1072; H. J. Dauben, Jr, H. J. Ringold, R. H. Wade, D. L. Pearson, and A. G. Anderson, Jr, *Org. Synth., Coll. Vol. 4*, 1963, 221.
56. *F. &F.*, **12**, 405; G. Rosini, R. Ballini, P. Sorrenti, and M. Petrini, *Synthesis*, 1984, 607.
57. *F. &F.*, **12**, 274; G. Rosini, R. Ballini, and V. Zanotti, *Synthesis*, 1983, 137.
58. J. C. Sowden and H. O. L. Fischer, *J. Am. Chem. Soc.*, 1945, **67**, 1713; 1947, **69**, 1963; J. C. Sowden, *J. Am. Chem. Soc.*, 1950, **72**, 3325; review: J. C. Sowden, *Adv. Carbohydr. Chem.*, 1951, **6**, 291.
59. *F. &F.*, **15**, 61; D. Seebach, A. K. Beck, T. M. Mukhopadbyay, and E. Thomas, *Helv. Chim. Acta*, 1982, **65**, 1101; K. Yamada, S. Tanaka, S. Kohmoto, and Y. Yamamoto, *J. Chem. Soc., Chem. Commun.*, 1989, 110.
60. R. L. Crombie, J. S. Nimitz, and H. S. Mosher, *J. Org. Chem.*, 1982, **47**, 4094.
61. *F. &F.*, **12**, 233; N. Ono, M. Miyake, M. Fujii, and A. Kaji, *Tetrahedron Lett.*, 1983, **24**, 3477.
62. *F. &F.*, **13**, 15; E. Marotta, R. Ballini, and M. Petrini, *Synthesis*, 1986, 237; *F. &F.*, **14**, 20; R. Ballini, M. Petrini, E. Marcantoni, and G. Rosini, *Synthesis*, 1988, 231.
63. K. Annen, H. Hofmeister, H. Laurent, A. Seeger, and R. Wiechert, *Ber.*, 1978, **111**, 3094.
64. N. Ono, T. Yanai, I. Hamamoto, A. Kamimura, and A. Kaji, *J. Org. Chem.*, 1985, **50**, 2806; *F. &F.*, **13**, 200.
65. *F. &F.*, **11**, 440; J. H. Clark and D. G. Cork, *J. Chem. Soc., Chem. Commun.*, 1982, 1215; *F. &F.*, **9**, 446–7 (18-crown-6).
66. W. D. Ollis and D. H. R. Barton, *Comprehensive organic chemistry*, Pergamon

Press, Oxford, 1979, Vol. 2, p. 351; B. Reichert, *Die Mannich-Reaktion*, Springer Verlag, Berlin, 1959, p. 68.
67. *F. &F.*, **11**, 32; J. T. Pinhey and B. A. Rowe, *Tetrahedron Lett.*, 1981, **22**, 783; R. P. Khozyrod and J. P. Pinhey, *Aust. J. Chem.*, 1985, **38**, 713; R. P. Khozyrod and J. P. Pinhey, *Tetrahedron Lett.*, 1982, **23**, 5365.
68. *F. &F.*, **14**, 188; M. G. Maloney, J. T. Pinhey, and E. G. Roche, *Tetrahedron Lett.*, 1986, **27**, 5025.
69. *F. &F.*, **11**, 359; D. Dauzonne, P. Demereseman, and R. Royer, *Synthesis*, 1981, 739.
70. R. F. Nystrom and W. G. Brown, *J. Am. Chem. Soc.*, 1948, **70**, 3738; A. Dornow and M. Gellrich, *Annalen*, 1955, **594**, 177; for reviews see W. G. Brown, *Org. React.*, 1951, **6**, 469; J. S. Pizey, *Synthetic reagents: Dimethylformamide, Lithium Aluminium Hydride, mercuric oxide, thionyl chloride*, Ellis Horwood, Chichester, 1974, p. 105.
71. W. D. Ollis and D. H. R. Barton, *Comprehensive organic chemistry*, Pergamon Press, Oxford, Vol. 2, p. 343.
72. J. R. Hanson, *Synthesis*, 1974, 1.
73. *F. &F.*, **13**, 279; J. A. Cowan, *Tetrahedron Lett.*, 1986, **27**, 1205.
74. *F. &F.*, **13**, 280; J. O. Osby and B. Ganem, *Tetrahedron Lett.*, 1985, **26**, 6413.
75. *F. &F.*, **13**, 282; H. Suzuki, K. Takaoka, and A. Osuka, *Bull. Chem. Soc. Japan*, 1985, **58**, 1067.
76. *F. &F.*, **12**, 422; M. Gaudemar, *Tetrahedron Lett.*, 1983, **24**, 2749.
77. *F. &F.*, **12**, 514; D. H. R. Barton, W. B. Motherwell, and S. Z. Zard, *Tetrahedron Lett.*, 1984, **25**, 3707.
78. *F. &F.*, **9**, 434; N. Kornblum and J. Widmer, *J. Am. Chem. Soc.*, 1978, **100**, 7086.
79. *F. &F.*, **10**, 413; **11**, 547; **12**, 517; N. Ono, H. Mujake, R. Tamura, and A. Kaji, *Tetrahedron Lett.*, 1981, **22**, 1705; D. D. Tanner, E. V. Blackburn, and G. E. Diaz, *J. Am. Chem. Soc.*, 1981, **103**, 1557; N. Ono, H. Mikaye, and A. Kaji, *J. Org. Chem.*, 1984, **49**, 4997.
80. *F. &F.*, **13**, 318; N. Ono, H. Miyake, and A. Kaji, *Chem. Lett.*, 1985, 635.
81. N. Ono, H. Miyake, R. Tamura, I. Hamamoto, and A. Kaji, *Chem. Lett.*, 1981, 1139.
82. *F. &F.*, **11**, 547; N. Kornblum, S. C. Carlson, and R. G. Smith, *J. Am. Chem. Soc.*, 1979, 647.
83. M. Bartra, P. Romea, F. Urpi, and J. Vilarrasa, *Tetrahedron*, 1990, **46**, 587.
84. *F. &F.*, **12**, 263; G. A. Olah, S. G. Narang, L. D. Field, and A. P. Fung, *J. Org. Chem.*, 1983, **48**, 2766.
85. *F. &F.*, **12**, 564; T. Fujisawa, Y. Kurita, and T. Sato, *Chem. Lett.*, 1983, 1537.
86. *F. &F.*, **3**, 213; B. Frank, J. Conrad, and P. Misbach, *Angew. Chem., Int. Ed. Engl.*, 1970, **9**, 892.
87. *F. &F.*, **8**, 400; P. A. Wehrli and B. Schaer, *J. Org. Chem.*, 1977, **42**, 3956.
88. *F. &F.*, **10**, 330; N. Kornblum and A. S. Erickson, *J. Org. Chem.*, 1981, **46**, 1037.
89. *F. &F.*, **12**, 263; G. A. Olah, S. C. Narang, L. D. Field, and A. P. Fung, *J. Org. Chem.*, 1983, 2766.
90. W. Lehnert, *Tetrahedron*, 1972, **28**, 663.
91. *F. &F.*, **11**, 173; P. Knochel and D. Seebach, *Synthesis*, 1982, 1017.
92. W. E. Noland, *Org. Synth.*, 1961, **41**, 67.
93. *F. &F.*, **5**, 476; S. Ranganathan, D. Ranganathan, and A. K. Mehratra, *J. Am. Chem. Soc.*, 1974, **96**, 5261.
94. *F. &F.*, **7**, 253; G. D. Buckley and C. W. Scaife, *J. Chem. Soc.*, 1947, 1471.

95. *F. &F.*, **12**, 337; D. H. R. Barton, W. B. Motherwell, and S. Z. Zard, *Bull. Soc. Chim. Fr.*, 1983, II-61.
96. *F. &F.*, **9**, 292; E. J. Corey and H. Estreicher, *J. Am. Chem. Soc.*, 1978, **100**, 6294.
97. *F. &F.*, **11**, 467; T. Hayama, S. Tamoda, Y. Takeuchi, and Y. Nomura, *Tetrahedron Lett.*, 1982, **23**, 4733.
98. T. E. Stevens and W. D. Emmons, *J. Am. Chem. Soc.*, 1958, **80**, 338.
99. V. Jager and H. G. Viehe, *Angew. Chem., Int. Ed. Engl.*, 1969, **8**, 273.
100. *F. &F.*, **15**, 176; M. Ochai, K. Sumi, Y. Takaoka, M. Kunishima, Y. Nagao, M. Shiro, and E. Fujita, *Tetrahedron*, 1988, **44**, 4095.
101. *F. &F.*, **10**, 17; T. Sakakibara, I. Takai, E. Ohara, and R. Sudoh, *J. Chem. Soc., Chem. Commun.*, 1981, 261.
102. *F. &F.*, **15**, 215; S. E. Denmark, J. A. Sternberg, and R. Lueoend, *J. Org. Chem.*, 1988, **53**, 1251.
103. A. T. Blomquist and T. H. Shelley, *J. Am. Chem. Soc.*, 1948, **70**, 147.
104. W. D. Emmons, W. N. Cannon, J. W. Dawson, and R. W. Ross, *J. Am. Chem. Soc.*, 1953, **75**, 1993.
105. A. J. Bloom and J. M. Mellor, *Tetrahedron Lett.*, 1986, **27**, 873.
106. *F. &F.*, **9**, 434; N. Ono, S. Kawai, K. Tanaka, and A. Kaji, *Tetrahedron Lett.*, 1979, 1733.
107. *F. &F.*, **12**, 422; M. Gaudemar, *Tetrahedron Lett.*, 1983, **24**, 2749.
108. *F. &F.*, **14**, 217; A. Pecunioso and R. Menicagli, *J. Org. Chem.*, 1988, **53**, 45.
109. J. H. Nelson, P. N. Howells, G. C. DeLullo, and G. L. Landen, *J. Org. Chem.*, 1980, **45**, 1246; *F. &F.*, **10**, 42.
110. E. J. Corey and G. Moinet, *J. Am. Chem. Soc.*, 1973, **95**, 6831; 6832.
111. *F. &F.*, **11**, 547–548; N. Ono, H. Miyake, and A. Kaji, *J. Chem. Soc., Chem. Commun.*, 1982, 33.
112. M. Miyashita, T. Yananu, and A. Yoshikoshi, *J. Am. Chem. Soc.*, 1976, **98**, 4679.
113. *F. &F.*, **15**, 328; A. Kamimura and N. Ono, *J. Chem. Soc., Chem. Commun.*, 1988, 1278.
114. *F. &F.*, **15**, 339; K. Okabe, T. Ohwada, T. Ohta, and K. Shudo, *J. Org. Chem.*, 1989, **54**, 733.
115. *F. &F.*, **11**, 28; K. Fuji, K. Kawabata, M. Node, and E. Fujita, *Tetrahedron Lett.*, 1981, **22**, 875.
116. *F. &F.*, **13**, 299; R. S. Varma and G. W. Kabalka, *Chem. Lett.*, 1985, 243.
117. Preparation: N. Ono, I. Hamamoto, T. Yanai, and A. Kaji, *J. Chem. Soc., Chem. Commun.*, 1985, 523.
118. *F. &F.*, **14**, 298.
119. N. Omo, I. Hamamoto, A. K. Kamimura, and A. Kaji, *J. Org. Chem.*, 1986, **51**, 3734.
120. *F. &F.*, **13**, 290; R. Tamura, K. Hayashi, M. Kakihana, M. Tsuji, and D. Oda, *Tetrahedron Lett.*, 1985, **26**, 851.
121. R. Tamura, K. Hayashi, M. Kakihara, M. Tsuji, and D. Oda, *Chem. Lett.*, 1985, 229.
122. N. Ono, I. Hamamoto, and A. Kaji, *J. Chem. Soc., Chem. Commun.*, 1982, 821.
123. N. Ono, I. Hamamoto, T. Yanai, and A. Kaji, *J. Chem. Soc., Chem. Commun.*, 1985, 523.
124. *F. &F.*, **4**, 427; J. A. Massen and Th. J. de Beer, *Recl. Trav.*, 1971, **90**, 373.
125. *F. &F.*, **11**, 476; E. Francotte, R. Merényi, B. Vandenbrucke-coyette, and H.-G. Viehe, *Helv. Chim. Acta*, 1981, **64**, 1208.

126. Preparation: R. West and P. Boudjouk, *J. Am. Chem. Soc.*, 1973, **95**, 3983; F. Douboudin, E. Fraimaet, G. Vinon, and F. Dabescat, *J. Organomet. Chem.*, 1974, **82**, 41.
127. S. E. Denmark, M. S. Dappen, and J. A. Sternberg, *J. Org. Chem.*, 1984, **49**, 4741.
128. B. Sklarz and M. K. Sultan, *Tetrahedron Lett.*, 1972, 1319.
129. R. A. Abramovitch, J. Crust, and E. P. Kyba, *Tetrahedron Lett.*, 1972, 4059.

12

N-NITROSO COMPOUNDS: NITROSAMINES, NITROSAMIDES, NITROSOUREAS, AND NITROSOURETHANES

NITROSAMINES

The substitution of a nitroso group (N=O) for a hydrogen atom of an amine leads to the class of compounds known as nitrosamines. The nitrosamines derived from primary amines (**1**) are tautomeric and are only stable at low temperatures (Scheme 12.1). Nitrosamines derived from secondary amines (**2**) are stable at ordinary temperatures and their chemistry has become very important because of their established carcinogenic properties.[1]

$$R\text{-}NH_2 \xrightarrow{HNO_2} R\text{—}N(H)\text{—}N{=}O \ (1) \rightleftharpoons R\text{—}N{=}N\text{—}OH \xrightarrow{>0^\circ C} R\text{-}OH \text{ etc.} + N_2$$

$$R^1R^2N\text{—}H \xrightarrow{HNO_2} R^1R^2N\text{—}N{=}O \ (2) \qquad (12.1)$$

12.1 Nitrosamines derived from primary amines

Methylnitrosamine (**3**) may be prepared indirectly from methylamine (Scheme 12.2), and has been identified from the correspondence between its ultraviolet spectrum and that of dimethylnitrosamine (**4**).[2]

Apart from the decomposition of methylnitrosamine (**3**), nothing is known of its chemistry, though that of potassium diazotate has been studied in some detail (see Chapter 16, p. 636).

$$CH_3NH_2 + NOCl \xrightarrow[-80^\circ C]{KOEt - Et_2O} CH_3{-}N{=}N{-}O^- K^+ \quad 15\%$$

50 % precipitated

$\downarrow H_3O^+$

$$CH_2N_2 \xleftarrow{H_2O} (CH_3)(H)N{-}N{=}O \ (3)$$

$$(CH_3)_2N{-}N{=}O \ (4) \qquad (12.2)$$

12.2 Nitrosamines derived from secondary amines

Bond distances and bond angles have been determined by electron diffraction.[3]

C–N: 1.42 - 1.47 Å; N–N: 1.30–1.34 Å; N–O: 1.22 - 1.26 Å

$\alpha = 120^\circ$, $\beta = 117^\circ$, $\gamma = 123^\circ$, $\delta = 114^\circ$,

X-ray diffraction shows that nitrosamines from secondary amines are better represented by the zwitterionic structure (**5**) rather than by the uncharged structure (**6**) (Scheme 12.3).[4]

$$(CH_3)(NH_2COCH_2)\overset{+}{N}{=}N{-}O^- \ (5) \longleftrightarrow (CH_3)(NH_2COCH_2)N{-}N{=}O \ (6)$$

$$(5) \rightleftharpoons (CH_3)(NH_2COCH_2)\overset{+}{N}{=}N{-}O^- \ (7) \longleftrightarrow (6) \qquad (12.3)$$

The zwitterionic structure is further confirmed by the observation of non-equivalent NMR signals for α-hydrogen atoms which are interpretable in terms of E and Z geometrical isomers (**5** and **7**).[5] This is demonstrated by the NMR spectrum of dimethylnitrosamine which has hydrogen signals at δ_H 3.77 and 3.02 ppm. The energy required to rotate about the nitrogen–nitrogen bond is 23.4 kcal/mole. E and Z isomers have been isolated and separated,[6] and it is sometimes possible to detect them on thin layer chromatographic plates. Both forms occur in some crystals,[4] for example (**8**).

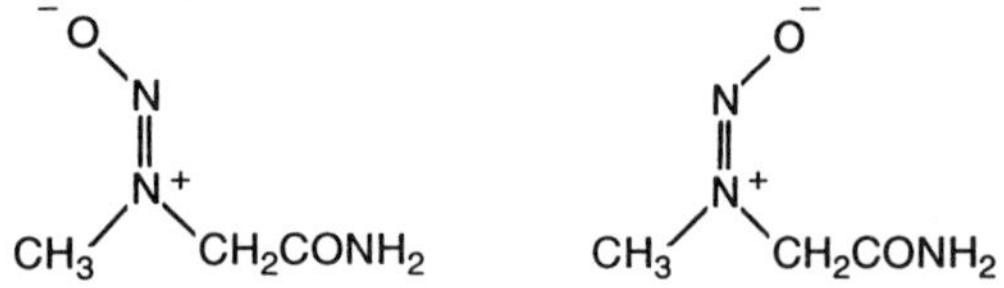

E (8) *Z*

The ultraviolet spectra of nitrosamines in cyclohexane show a maximum at *c.* 220 mμ (ε = *c.*7000, π–π^* transition) and at *c.*360 mμ (ε = *c.*100, n–π^* transition).[7] By adding trichloroacetic acid to solutions of nitrosamines in cyclohexane these authors were able to determine the basic dissociation constants for the formation of the cations (**9**) and (**10**) from the changes produced in the ultraviolet spectra.

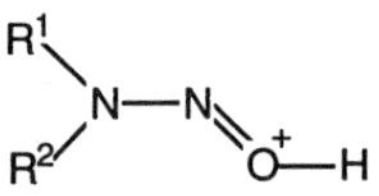

(8) 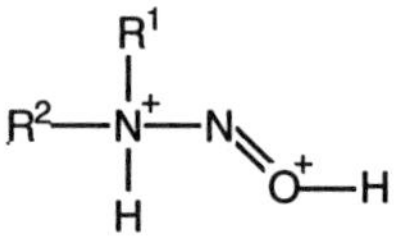 (9)

The nitroso groups of nitrosamines have a strong characteristic absorption in the infra-red at 1450–1480 cm^{-1}.[8] Nitrosamines also show characteristic behaviour in a mass spectrometer, the strongest peak usually being at M-30, which results from the loss of the nitroso group (NO = 30) from the parent nitrosamine.[9]

12.3 Preparation

12.3.1 *From secondary amines*

Use of nitrous acid

The classical and most widely used preparation of nitrosamines is by the action of a nitrosating agent upon a secondary amine. The reaction between nitrous acid, generated from sodium nitrite and an acid (aqueous HCl, aqueous H_2SO_4, or aqueous acetic acid), and secondary amines is the most important preparation of nitrosamines.[10] The kinetics of the reaction,[11] show that the rate is dependent upon the first power of the concentration of secondary amine and the square of the nitrous acid concentration, i.e.

$$\text{Rate} = k\,[HNO_2]^2[R_2NH]$$

a result first explained by Hammett[12] in terms of a mechanism involving dinitrogen trioxide as the nitrosating agent (Scheme 12.4).

$$2\,HNO_2 \rightleftharpoons H_2O + N_2O_3$$

$$R_2NH + N_2O_3 \xrightarrow{\text{r.d.s.}} R_2HN^+{-}N{=}O + NO_2^-$$

$$R_2HN^+{-}N{=}O \longrightarrow R_2N{-}N{=}O + H^+ \qquad (12.4)$$

Since the free amine is involved in the rate-determining step, the optimum pH for the nitrosation depends upon the dissociation constant of the amine. Various compounds[13] have been found to catalyse this reaction some of which are listed below:

(1) Halide ions, in particular iodide and bromide;
(2) thiocyanate (CNS^-, a pseudohalide).

The ions catalyse the reaction by forming an active compound, such as nitrosyl bromide, which may then either react with the amine to yield the nitrosamine, or react with nitrous acid to give dinitrogen trioxide, which then itself reacts with the amines (Scheme 12.5). Note that the reaction of nitrous acid with itself to form the chemically reactive dinitrogen trioxide is a relatively slow reaction and is rate limiting under many circumstances. For the reaction of very reactive amines, when this occurs $[R_2NH]$ disappears from the rate equation.

$$HNO_2 + X^- + H^+ \rightleftharpoons NOX + H_2O$$

$$NOX + RNH_2 \xrightarrow{\text{rate limiting}} R\overset{+}{N}H_2\text{-}N{=}O \xrightarrow{\text{fast}} RHN\text{-}N{=}O$$

where X^- = nitrite or other ion, such as Cl^-, Br^-, I^- or CNS^- (12.5)

An example of electrophilic catalysis of the nitrosation reaction has been reported in which certain carbonyl compounds render amine–nitrite interactions possible according to a somewhat different mechanism of charge compensation. The hydrogen ion appears not to be involved and measurable nitrosation was observed under neutral and alkaline conditions.

Use of formaldehyde

Unhindered secondary amines are readily nitrosated in the presence of formaldehyde (Scheme 12.6).[14]

$$R_2NH + H_2C{=}O \rightleftharpoons \left[R_2N{-}\overset{+}{C}H_2 \longleftrightarrow R_2\overset{+}{N}{=}CH_2\right] + HO^-$$

$$\left[R_2N{=}CH_2\right]^+ \xrightleftharpoons{ONO^-} \left[R_2N{-}CH_2{-}O{-}N{=}O\right] \longrightarrow R_2N{-}N{=}O + H_2C{=}O \qquad (12.6)$$

Production in vivo

As a result of the carcinogenic effects of nitrosamines, much attention has been directed towards the study of reactions which lead to their production *in vivo*. Secondary amine nitroso compounds (**10**) have significance *in vivo* due to their mutagenic properties.[15] They are thought to be carcinogenic because of the enzymatic formation of an alkylating agent (Scheme 12.7). This hypothesis has led to the development of methods for the synthesis of α-acylnitrosamines (**11**) (Scheme 12.8).[16]

$$\underset{(10)}{R^1N(NO){-}CR^2R^3{-}H} \xrightarrow[\text{hydroxylation}]{\text{Enzymatic}} R^1N(NO){-}CR^2R^3{-}OH \longrightarrow R^1N(H){-}NO + R^2R^3C{=}O$$

$$R^1N(H){-}NO \rightleftharpoons R^1{-}N{=}N{-}OH \longrightarrow R^1N_2^+ \quad \text{(Formation of an alkylating agent)} \qquad (12.7)$$

$$\text{R-CH=N-R}^1 \xrightarrow[\text{ii) AcOH, Et}_3\text{N or Ag}^+\text{OAc}^-]{\text{i) NOCl, -30}^\circ\text{C}} \text{R—CH(OAc)—N(NO)—R}^1 \quad \mathbf{(11)} \qquad (12.8)$$

Use of nitrogen oxides

Various gaseous nitrogen oxides can be used under alkali conditions for the preparation of secondary amine *N*-nitrosamines (**12**),[17] for example, Scheme 12.9.

$$\text{piperazine} + N_2O_3 \xrightarrow{\text{NaOH (aq)}} \mathbf{(12)}\ \ 68\% \qquad (12.9)$$

Use of nitrosyl chloride

Nitrosyl chloride (**13**) in aprotic solvents can be used for the preparation of *N*-nitrosamines from secondary amines (Scheme 12.10).[18]

$$\text{piperidine (NH)} + 2\text{NOCl}\ \mathbf{(13)} + \text{pyridine} \xrightarrow{\text{THF}} \text{piperidine N—NO}\ \ 100\% \qquad (12.10)$$

Use of nitronium tetrafluoroborate

Nitrosonium tetrafluoroborate (**14**) can be used as a nitrosating agent to form *N*-nitrosamines from secondary amines (Scheme 12.11).[19]

By trans-nitrosation

Trans-nitrosation, as shown in Scheme 12.12 has been used for the synthesis of *N*-nitrosamines.[20]

N-Nitrosodiphenylamine (**15**) has also been used in *trans*-nitrosation reactions which are believed to proceed via a free-radical chain mechanism,[21] for example, Scheme 12.13.

Fremy's salt (**16**) has been used as a *trans*-nitrosating agent for the preparation of *N*-nitrosamines from secondary amines,[22] for example, Scheme 12.14.

$$R^1R^2NH + NO^+BF_4^- \ (14) \longrightarrow R^1R^2N{-}NO + R^1R^2\overset{+}{N}H_2.BF_4^- \quad (12.11)$$

R^1	R^2	Yield (%)
Et	Et	88
Me	$PhCH_2$	86
Piperidine		88

$$\text{N-nitrosocarbazole} + R^1R^2NH \xrightarrow{\text{pH 7}} \text{carbazole (N–H)} + R^1R^2N{-}NO \quad (12.12)$$

$$Ph_2N{-}N{=}O \ (15) + R^1R^2NH \xrightarrow[H^+]{\text{dioxan, 3 days, 50}^\circ\text{C}} Ph_2NH + R^1R^2N{-}N{=}O \quad (12.13)$$

$$Et_2NH + {}^{\cdot}O{-}N(SO_3^-)_2(K^+)_2 \ (16) \xrightarrow{2\%\ NaNO_3\ (aq)} Et_2N{-}NO \ \ 40\% \quad (12.14)$$

From dialkyl-N-chloroamines

Dialkyl-*N*-nitrosamines (**17**) can be prepared from compounds containing a nitrogen–chlorine bond, as shown in Scheme 12.15.[23]

Using sodium nitroprusside

Sodium nitroprusside, $Na_2Fe(CN)_5NO$ (**18**), has been used to nitrosate amines under aqueous alkali conditions (Scheme 12.16).[24]

$$NO_2^- + (PhCH_2)_2NCl \longrightarrow NO_2Cl + (PhCH_2)_2N^-$$
$$NO_2Cl + NO_2^- \longrightarrow N_2O_4 + Cl^-$$
$$(PhCH_2)_2N^- + N_2O_4 \longrightarrow \underset{\substack{90\ \% \\ (17)}}{(PhCH_2)_2N\text{-}NO} + NO_3^- \qquad (12.15)$$

$$(CH_3CH_2)_2NH + \underset{(18)}{Fe(CN)_5NO^{2-}} \xrightarrow{\text{aq. } HO^-} \underset{44\ \%}{(CH_3CH_2)_2N\text{-}N{=}O} \qquad (12.16)$$

12.3.2 *From tertiary amines*

Use of nitrosating agents

N-Nitrosamines can be obtained from the cleavage of tertiary amines (**19**) with a nitrosating agent such as sodium nitrite in aqueous acetic acid, or tetranitromethane, but this method has not found much practical application (Scheme 12.17).[25]

$$\underset{(19)}{R^1{-}CH_2{-}NR^2R^3} \xrightarrow{NO^+} R^1{-}CH_2{-}\overset{+}{N}R^2R^3(NO) \longrightarrow R^1{-}CH{=}\overset{+}{N}R^2R^3 \longrightarrow R^1{-}CH(O{-}N{=}O){-}NR^2R^3 \longrightarrow R^1CHO + O{=}N{-}NR^2R^3 \qquad (12.17)$$

By oxidation with copper chloride

N-Nitrosamines are formed by the copper chloride catalysed oxidation of tertiary amines (**19**) shown in Scheme 12.18.[26]

Fremy's salt

Fremy's salt, shown above (**16**), can also be used to form *N*-nitrosamines from tertiary amines (**20**) by the mechanism shown in Scheme 12.19.[27]

$$R^1R^2N{-}CH_2{-}R^3 \; (19) \xrightarrow[O_2,\ CuCl,\ \text{in pyridine}]{Me_2CHNO_2} R^1R^2N{-}NO + R^3{-}CHO + Me_2CO$$

$$R^1R^2N{-}CH_2{-}R^3 \xrightarrow{Me_2C(OOH)NO_2} R^1R^2N^+(O^-){-}CH_2{-}R^3 + Me_2C(OH)NO_2 \longrightarrow R^1R^2N{-}CH(OH){-}R^3 + HNO_2$$

R^1	R^2	R^3	R^4	Yield (%)
Et	Et	Me	Me	59
Me	Pr^n	H	H	59
Me	Me	Et	Et	17
Me	Bu^t	H	H	60
$-(CH_2)_5-$		H	H	58

(12.18)

$$Me_2N{-}CH_2Bu^n \; (20) \xrightarrow[(16)]{(K^+)_2NO(SO_3)_2^{2-}} Me_2N^{+\bullet}{-}CH_2Bu^n \xrightarrow{-H^+} Me_2N^+{=}CHBu^n \xrightarrow{H_2O,\ -Bu^nCHO} Me_2N{-}H \xrightarrow{NOSO_3^{2-}} Me_2N{-}NO \; (23\ \%)$$

(12.19)

12.3.3 *From primary amines*

By the action of nitrous acid on a primary amine it has been reported that up to 10 per cent of a secondary nitrosamine is produced (Scheme 12.20).[28]

$$RNH_2 + HNO_2 \longrightarrow R^+ + N_2 + H_2O$$

$$RNH_2 + R^+ \longrightarrow R_2\overset{+}{N}H_2 \xrightarrow{N_2O_3} R_2\overset{+}{N}H{-}N{=}O \longrightarrow R_2N{-}N{=}O \quad (12.20)$$

12.4 Reactivity

Consideration of the zwitterionic (**21**) and uncharged (**22**) structures of a nitrosamine leads to the expectations that

(i) electrophiles (E^+) would attack the molecule at the amino nitrogen atom;
(ii) nucleophiles (Nu) would attack the nitroso nitrogen atom; and
(iii) the positive charge on the amino nitrogen atom would enable a base to remove a proton from an α-carbon atom (Scheme 12.21).

R^1—CH(H)—$\overset{+}{N}(R^2)$=N—O^- (21) ⟷ R^1—CH(H)—N(R^2)—N=O (22)

i, R^1—CH_2—N(R^2)—N=O + E^+

ii, R^1—CH(H)—$\overset{+}{N}(R^2)$=N—O^- + :Nu

iii, R^1—CH(H)—$\overset{+}{N}(R^2)$=N—O^- + :B

(12.21)

Furthermore, the polarity of the zwitterionic structure (1,3-dipole) might result in cycloaddition reactions with suitable acceptors.

12.4.1 *Decomposition by acids*

Nitrosamines are easily decomposed by hot concentrated acids (e.g. conc. HCl in acetic acid; 3 per cent HBr in HOAc[29]) often in the presence of nitrous acid traps, such as urea or sulphamic acid, to remove nitrous acid as it is formed. This

R^1R^2N—N=O $\overset{H_3O^+}{\rightleftharpoons}$ $R^1R^2\overset{+}{N}$(H)—N=O ⟶ R^1R^2NH (23) + $H_2\overset{+}{O}$—N=O ⟶ HO-N=O (12.22)

is the usual way in which they are converted to the parent amines (**23**) (Scheme 12.22).[30]

Gas-phase protonolysis reactions of *N*-nitrosamines have also been studied, thus allowing comparison with solution chemistry occurring in acidic media.[31]

12.4.2 *With transition metal carbonyl complexes*

Secondary amines are formed in the reaction of various transition metal carbonyl complexes with aryl nitrosamines (Scheme 12.23).[32]

$$\mathrm{Ph{-}N(Bn){-}NO} \xrightarrow{M(CO)_n} \mathrm{Ph{-}N(Bn){-}H}$$

$M(CO)_n$	Yield (%)
$Mo(CO)_6$	61
$Fe(CO)_5$	85

(12.23)

12.4.3 *Reaction with nucleophilic organometallic reagents*

Grignard reagents and alkyllithium compounds attack the nitroso nitrogen, as expected in the manner shown in Scheme 12.24.[33]

$$R^1R^2N{-}N{=}O + CH_3{-}Li \xrightarrow{H^+} R^1R^2N{-}N(OH){-}CH_2{-}H \longrightarrow R^1R^2N{-}N{=}CH_2$$

for example:

$$Et_2\text{-}NNO + MeLi \xrightarrow{-40^\circ C} Et_2NN{=}CH_2\ \ 80\% \qquad (12.24)$$

Secondary nitrosamines react with phenyl- or *tert*-butyl-lithium to give *N′*-alkylation lithium salts, which undergo elimination to form azomethine imines (**24**) when treated with water or ethanol. The azomethine imines form *syn*-hexahydrotetrazines (**25**) on standing or can be trapped with *N*-phenylmaleimide (**26**) for example, Scheme 12.25.

Reaction of aliphatic and alicyclic nitrosamines with excess phenyl-, cyclohexyl-, or *tert*-butyl-magnesium halide gave trisubstituted hydrazines (**27**)

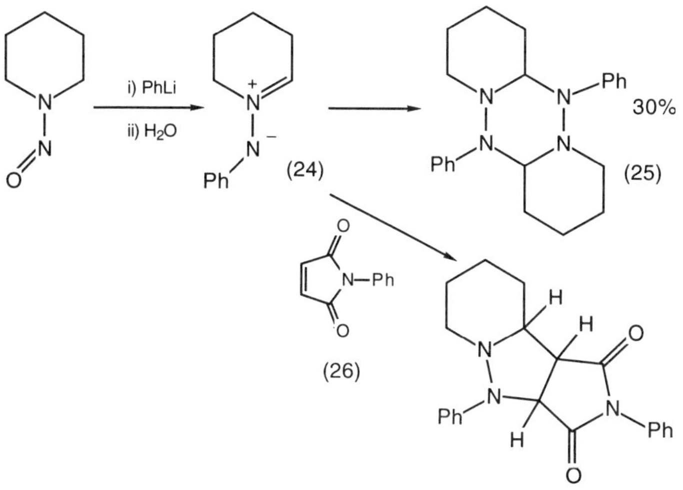

(12.25)

resulting from α-carbon and nitroso nitrogen alkylation for example, Scheme 12.26.

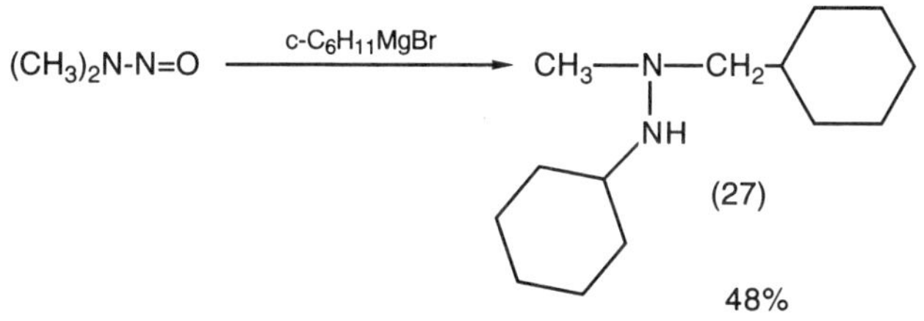

48% (12.26)

Denitrosation can also be achieved using other organometallic reagents, for example, copper, lithium, or magnesium reagents but yields are less than 25 per cent, with the nitrosamine being the main product recovered[36] (Scheme 12.27).

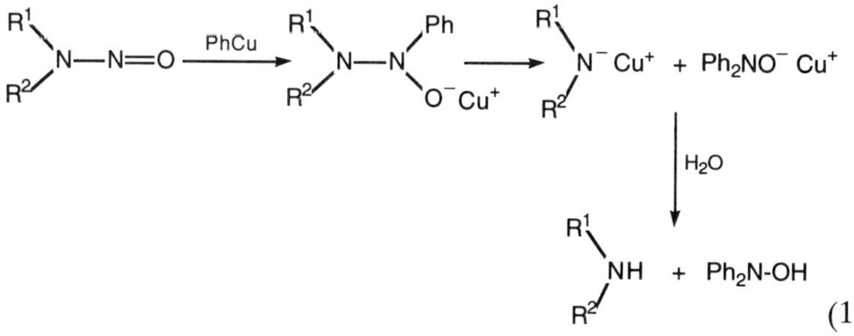

(12.27)

12.4.4 *Reaction with bases*

Removal of the α-carbon atom proton may be followed by reaction with an electrophile (Scheme 12.28).

i, $LiPr^n NH$
ii, E^+

H_3O^+

R^1—CH(E)—NHR^1 + HONO (12.28)

Isomerization

Consideration of the zwitterionic contribution to the structure of a nitrosoamine indicates that α-hydrogen atoms should show acidic properties. This has been verified experimentally and used for synthetic purposes. The *trans*-nitrosopyrrolidine (**28**) has been shown to isomerize in basic solution to the *cis* compound (**29**) (Scheme 12.29).[37] Likewise, deuterium exchange has been observed (Scheme 12.30) and alkylation of the intermediate anions has been achieved.[38]

(28) $\xrightleftharpoons{OH^- \text{ aq.}}$ (29)

(12.29)

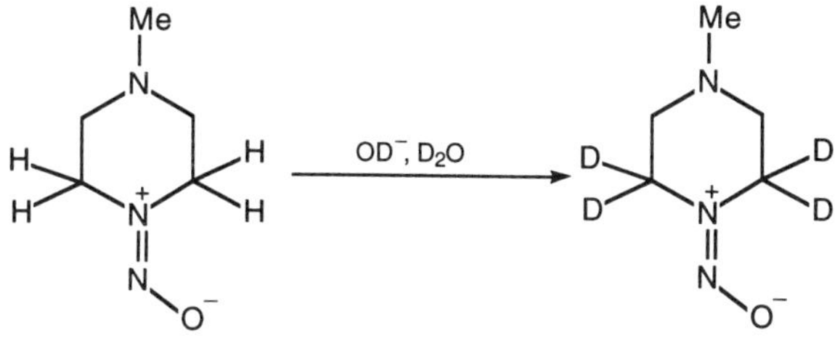

(12.30)

Umpolung

The advantages of this acidity of the α-hydrogen have been used by Seebach and co-workers to develop a synthetic procedure, which depends on the polarity of the parent amine (**30**) being reversed in the nitrosamine (**31**) (Scheme 12.31), an example of the phenomenon which Seebach called 'Umpolung'.[39] This is demonstrated in Scheme 12.32.

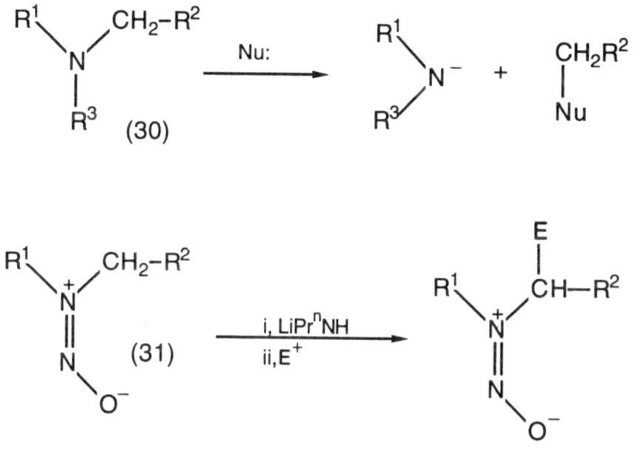

(12.31)

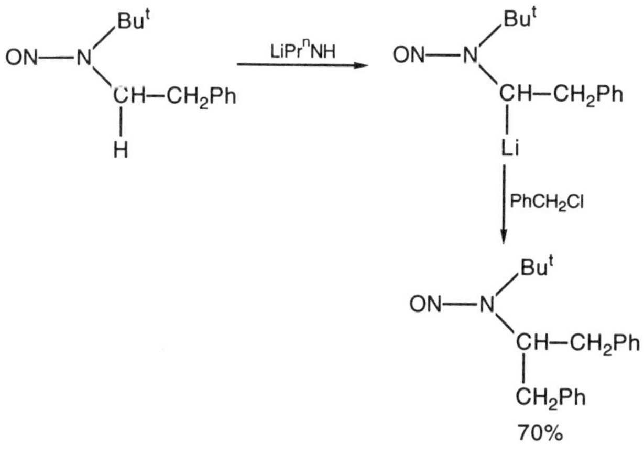

(12.32)

Anomalous reactions of the anions

Depending upon the structure of the nitrosamine, anomalous reactions of the anions have been reported by Seebach.[40] For example, elimination can occur (Scheme 12.33).

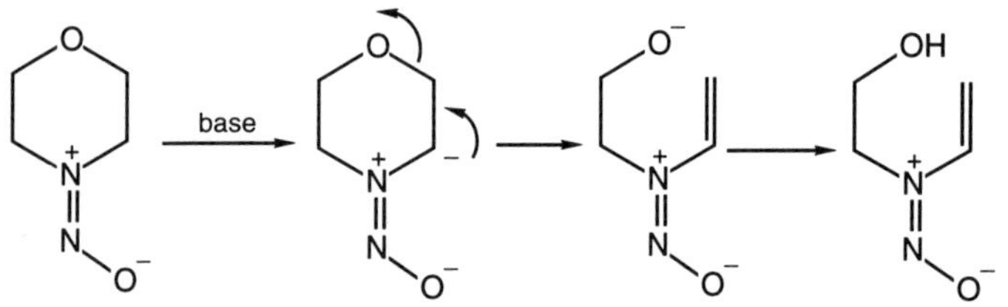

(12.33)

Formation of imines

Imines are sometimes produced by an alternative elimination (Scheme 12.34).[41]

Secondary amine *N*-nitroso compounds (**32**) rearrange and eliminate to form imines (**33**) in basic conditions by the mechanism shown in Scheme 12.35.[42]

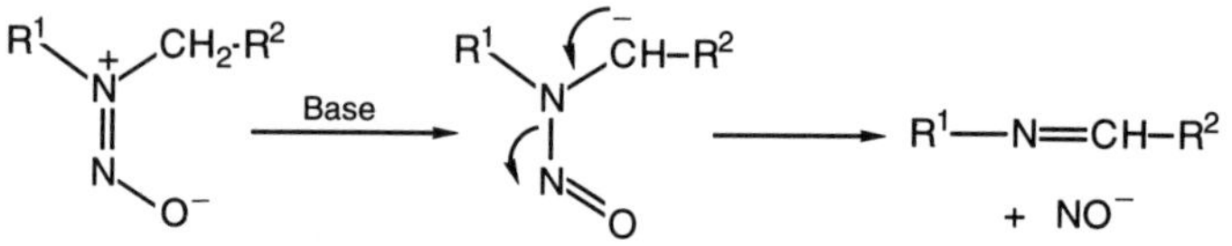

(12.34)

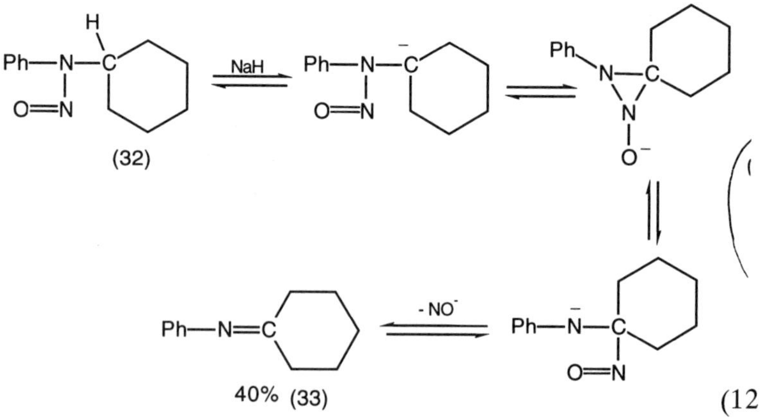

(12.35)

Formation of oximes

Under basic conditions the rearrangement of α-nitrile *N*-nitrosamines (**34**) proceeds as shown in Scheme 12.36,[43] for example.

$$Bn—N(N{=}O)—CH_2—CN \xrightarrow[CH_3OH]{base} Bn—NH—C(=N\text{-}OH)—CN \; (45\ \%) \xrightarrow{H_3O^+} Bn—NH—C(=N\text{-}OH)—CONH_2 \; (75\ \%)$$

(34)

(12.36)

12.4.5 *With reducing agents*

Nitrosamines react with reducing agents to form hydrazines,[44] for example, Scheme 12.37.

$$(Bu^i)_2N—N{=}O \xrightarrow{Ti(II)\ catalyst,\ CH_2Cl_2,\ Et_2O} (Bu^i)_2N—NH_2 \quad 95\%$$

(12.37)

Nitrosamines (**35**) can be reduced to secondary amines (**36**) using a slight excess of titanium tetrachloride and sodium borohydride in the molar ratio of 1:2 (Scheme 12.38).[45]

$$R^1R^2N—N{=}O \; (35) \xrightarrow{TiCl_4\ /\ NaBH_4} R^1R^2NH \; (36)$$

R^1	R^2	Yield (%)
Ph	Me	92
$PhCH_2$	Me	75
Ph	$PhCH_2$	93
N-nitroso-1,2,3,4-tetrahydroquinoline (structure)		87

(12.38)

12.4.6 *Reaction with oxidizing agents*

The oxidation of nitrosamines with a peroxide (e.g. hydrogen peroxide or methyl ethyl ketone peroxide) is referred to in Chapter 19, p. 683.

Trifluoroacetic peracid has been used as an oxidizing agent, forming nitro-compounds (**37**) from nitroso-compounds (**38**),[46] for example, Scheme 12.39.

$$Et_2N\text{-}N{=}O \xrightarrow{CF_3CO_3H} Et_2N\text{-}NO_2 \quad 76\ \%$$

(38) (37) (12.39)

12.4.7 *Reaction with aryl halides*

Dimethylnitrosamine (**39**) has been found to react with aryl halides to form dimethylarylamines (**40**),[47] for example, Scheme 12.40.

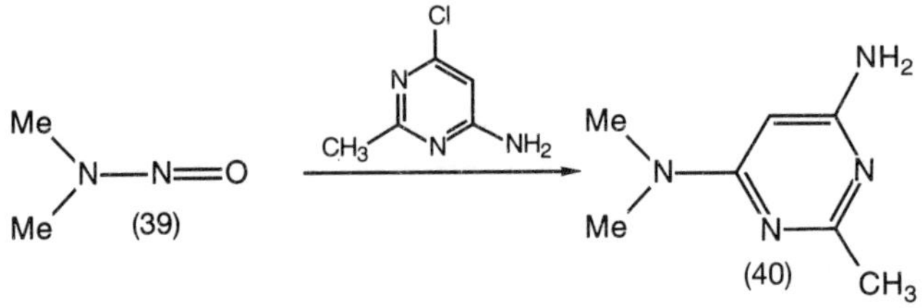

Me₂N—N=O (39) → (with 4-chloro-2-methyl-6-aminopyrimidine) → (40) (12.40)

12.4.8 *Deoxygenation*

Since the zwitterionic structure (**41**) of the nitrosamine can be regarded as an *N*-oxide, it is not surprising that several reagents are able to cause deoxygenation to yield reactive intermediates. For example, ethyl diphenylphosphinite ($EtOPPh_2$) may be used to convert 1-nitrosobenzotriazol (**41**) to the nitrene (**42**) with subsequent decomposition to benzyne (**43**) (Scheme 12.41).[48]

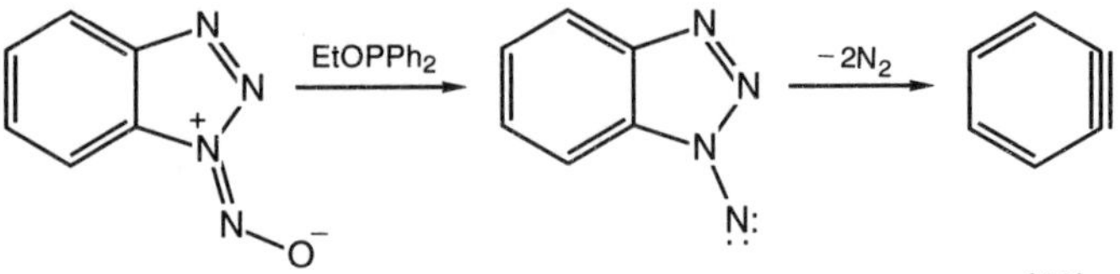

(41) $\xrightarrow{EtOPPh_2}$ (42) $\xrightarrow{-2N_2}$ (43) (12.41)

Iron pentacarbonyl, $Fe(CO)_5$, has been found to deoxygenate nitrosamines (Scheme 12.42).[49]

$$R^1R^2N{-}NO \xrightarrow[\text{reflux, 16 h}]{Fe(CO)_5,\ \text{Butyl ether}} R^1R^2NH$$

R^1	R^2	Yield (%)
Ph	Ph	91
Bn	Ph	85
Me	Ph	90

(12.42)

The reaction of *N*-nitrosodibenzylamine (**44**) with phenacyl bromides (**45**) in the presence of hexafluoroantimonate proceeds by deoxygenation, via the intermediate *N*-dibenzylaminonitrene (**46**), and in benzene gives diphenylmethane (**47**) as the major product (Scheme 12.43).[50]

$$\underset{(45)}{ArCOCH_2Br} + AgSbF_6 \longrightarrow [ArCOCH_2^+\ SbF_6^-] + AgBr$$

$$+\ \underset{(44)}{(PhCH_2)_2\overset{+}{N}{=}N{-}O^-} \longrightarrow (PhCH_2)_2\overset{+}{N}{=}N{-}O{-}CH(H){-}COAr \quad (\text{Et}_3\text{N:})$$

$$\longrightarrow [ArCOCHO] + \underset{(46)}{\left[(PhCH_2)_2\overset{+}{N}{=}\overset{-}{N}{:}\right]} \longrightarrow \underset{\substack{31\% \\ (47)}}{PhCH_2Ph} + \underset{9.7\%}{PhCH_2CH_2Ph}$$

(12.43)

12.4.9 *Photochemical reactions*

Intramolecular photochemical reactions occur in acidic solution for example (Scheme 12.44).[51]

$$R^1{-}CH_2{-}N(N{=}O){-}R^2 \xrightarrow{h\nu,\ H_3O^+} R^1{-}C({=}N{-}OH){-}NH{-}R^2$$

(12.44)

12.4.10 *Reactions with radicals*

Nitrosamines react with thiols via a radical mechanism to yield radical intermediates (Scheme 12.45).[52]

$$2\ R^1R^2N{-}N{=}O + R^3SH \longrightarrow R^1R^2N{-}N(O^\cdot)(SR^3) + R^1R^2NH + NO^\cdot \qquad (12.45)$$

12.4.11 *Overberger–Lombardino reaction*

Benzyl substituted nitrosamines (**48**) are reduced by sodium dithionate (NaS_2O_4) to form a hydrocarbon. This is known as the Overberger–Lombardino reaction[53] (Scheme 12.46).

$$\underset{\textbf{(48)}}{(Ph{-}CH_2)_2N{-}N{=}O} \xrightarrow{NaS_2O_4,\ aq.\ OH^-} Ph{-}CH_2{-}CH_2{-}Ph \qquad (12.46)$$

NITROSAMIDES

$$R{-}C({=}O){-}N(N{=}O){-}R$$

The molecular dimensions of nitrosamides are as follows:

C, a, α, β, d, O, N, c, C, γ, b, δ, ε, H, e, ϕ, N, C, f, η, N, σ, g, O

Bond lengths: a, 1.46 Å; b, 0.95 Å; c, 1.32 Å; d, 1.22 Å; e, 1.46 Å; f, 1.34 Å; g, 1.22 Å.
Bond angles: α, 122°; β, 125°; γ, 116°; δ, 116°; ε, 122°, ϕ, 117°; η, 121°; σ, 113°.

From these dimensions it can be seen that the contributors to the state of the molecule are much the same as those described above for nitrosamines (Scheme 12.47).

$$R(H)N{-}C(=O){-}N(R){-}N{=}O \longleftrightarrow R(H)N{-}C(=O){-}\overset{+}{N}(R){=}N{-}O^{-} \quad (12.47)$$

12.5 Preparation

Nitrosamides (**49**) can be prepared as intermediates in the White procedure for the deamination of amines.[54] Dinitrogen tetroxide reacts with a secondary amide to yield a nitrosamide (**49**), which decomposes on heating to yield an ester and nitrogen, as shown in Scheme 12.48.

$$PhCH_2CH_2NH_2 \xrightarrow[\text{pyridine}]{PhCOCl} PhCH_2CH_2NH{-}C(=O)Ph$$

$$PhCH_2CH_2NH{-}C(=O)Ph \xrightarrow{N_2O_4,\ CH_3CO_2Na/CH_3CO_2H} \underset{(49)}{PhCH_2CH_2{-}N(N{=}O){-}C(=O)Ph}$$

$$\underset{(49)}{PhCH_2CH_2{-}N(N{=}O){-}C(=O)Ph} \xrightarrow{\text{heat}} PhCH_2CH_2OCOPh + N_2 \quad (12.48)$$

12.6 Reactivity

Nitrosamides react with primary amines in an alkyl group exchange, as shown in Scheme 12.49.[55]

Nitrosamides can be used for the preparation of diazoalkanes, see Chapter 13.

$$R^1{-}C(=O){-}N(NO){-}R^2 \xrightarrow{R^3NH_2,\ CH_2Cl_2,\ 40^\circ C} R^1{-}C(=O){-}N(H){-}R^3 \quad (12.49)$$

NITROSOUREAS AND NITROSOURETHANES

Nitroso-compounds have been extensively studied in carcinogen and mutagen research.[55]

2-Haloethylnitrosoureas (**50**) (Scheme 12.50) have been shown to be antitumour agents.[56]

$$\text{Hal–CH}_2\cdot\text{CH}_2\text{–N(NO)–C(=O)–NHR}$$

(50)

Hal	R
Cl	cyclohexyl
Cl	CH_2CH_2Cl
F	CH_2CH_2F
Cl	3-methylcyclohexyl
Cl	H

(12.50)

Nitrosourethanes have been found to alkylate nucleophilic groups in DNA. For example, *N*-methyl-*N*-nitro-*N*-nitrosoguanidine methylates cysteine and rapidly methylates DNA *in vitro* and in cultured mammalian cells.[57]

References

1. P. N. Magee and J. M. Barnes, *Adv. Cancer Res.*, 1967, **10**, 163; H. Druckrey, R. Preussmann, S. Ivanovic, and D. Schmähl, *Krebsforsch.*, 1967, **69**, 103.
2. E. H. Muller and W. Rundel, *Ber.*, 1960, **93**, 1541.
3. R. Rademadnere, R. Stølevic, and W. Lüttle, *Angew. Chem., Int. Ed. Engl.*, 1968, **7**, 806; R. Rademadnere and R. Stølevic, *Acta Chem. Cryst. Scand.*, 1969, **23**, 660.
4. B. Krebs and J. Manett, *Ber.*, 1975, **108**, 1130; L. K. Templeton, D. H. Templeton, and A. Zalkin, *Acta Cryst. Allogr.*, 1973, **B29**, 50.
5. C. E. Looney, W. D. Phillips, and E. L. Rilley, *J. Am. Chem. Soc.*, 1957, **79**, 435; G. J. Karabatos and R. A. Taller, *J. Am. Chem. Soc.*, 1964, **86**, 4376.
6. A. Manscheck, H. Muncsch, and A. Matthaus, *Angew. Chem. Int. Ed. Engl.*, 1966, **5**, 728.
7. W. S. Layne, H. H. Jaffé, and H. Zimmer, *J. Am. Chem. Soc.*, 1963, **84**, 435.
8. W. S. Layne, H. H. Jaffé, and H. Zimmer, *J. Am. Chem. Soc.*, 1963, **85**, 435.
9. H. Budzikiewicz, C. Djerassi, and D. Williams, *Mass spectroscopy of organic compounds*, Holden-Day, San Fransisco, 1967, p. 329; J. Collin, *Bull. Soc. Roy. Sci. Liège*, 1954, **23**, 201.

10. D. L. H. Williams, *Adv. Phys. Org. Chem.*, 1983, **19**, 381.
11. T. W. J. Taylor, *J. Chem. Soc.*, 1928, 1099; 1987; T. W. J. Taylor and L. S. Price, *J. Chem. Soc.*, 1929, 2052.
12. L. P. Hammett, *Physical organic chemistry*, McGraw-Hill, New York, 1940.
13. P. Bogovski, R. Preussman, and E. A. Walker (ed.) N-*Nitroso compounds: analysis and formation*, International Agency for Research on Cancer (WHO), Lyond, 1972.
14. L. K. Keefer and P. P. Roller, *Science*, 1973, **181**, 1245.
15. D. Seebach and D. Enders, *Angew. Chem., Int. Ed. Engl.*, 1975, **14**, 15.
16. M. A. Barton and B. R. Brown, *J. Chem. Soc., Chem. Commun.*, 1980, 59; S. R. Tannenbaum, P. Kraft, J. E. Baldwin, and S. Branz, *Cancer Lett.*, 1977, **2**, 305; L. Butula and D. Kolbah, *Acta Pharm. Jug.*, 1966, **16**, 173; M. Weissler, *Angew. Chem., Int. Ed. Engl.*, 1974, **13**, 743; M. Weissler, *Tetrahedron Lett.*, 1975, 2575; M. Weissler, in N-*Nitrosamines*, ACS Symposium Series 101 (ed. J.-P. Anselme), American Chemical Society, Washington DC, 1979, p. 64; J. E. Baldwin, A. Scott, S. E. Branz, S. R. Tannenbaum, and L. Green, *J. Org. Chem.*, 1978, **43**, 2427; M. Mochizuki, T. Anjo, and M. Okada, *Chem. Pharm. Bull.*, 1978, **26**, 3905; P. P. Roller, D. R. Shimp, and L. K. Keffer, *Tetrahedron Lett.*, 1975, 2065; J. E. Baldwin, S. E. Branz, R. F. Gomez, P. L. Kraft, A. J. Sinskey, and S. R. Tannenbaum, *Tetrahedron Lett.*, 1976, 333; S. S. Hecht and C. B. Chen, *J. Org. Chem.*, 1979, **44**, 1563; J. E. Saavedra, *Tetrahedron Lett.*, 1978, 1923.
17. B. C. Challis and J. A. Challis, *The chemistry of functional groups, Supplement F, Part II* (ed. S. Patai), Wiley, NY, 1982, pp. 1151–223; B. C. Challis and D. E. G. Shuker, *J. Chem. Soc., Perkin Trans. II*, 1979, 1020; B. C. Challis and S. A. Kyrtopoulos, *J. Chem. Soc., Perkin Trans I*, 1979, 299.
18. R. E. Lyle, J. E. Saavedra, and G. C. Lyle, *Synthesis*, 1976, 462.
19. L. Noszko, S. Kuhn, and M. Selke, *Ber.*, 1956, **89**, 2374.
20. J.-P. Anselme (ed.), N-*Nitrosamines*, ACS Symposium Series 101, American Chemical Society, Washington DC, 1979; N. L. Drake, H. J. S. Winkler, C. M. Kraebel, and T. D. Smith, *J. Org. Chem.*, 1962, **27**, 1026.
21. B. C. Challis and M. R. Osborne, *J. Chem. Soc., Perkin Trans. II*, 1973, 1526.
22. L. Castedo, R. Riguera, and M. P. Vázquez, *J. Chem. Soc., Chem. Commun.*, 1983, 301.
23. M. Nakajima, J. C. Warner, and J.-P. Anselme, *J. Chem. Soc., Chem. Commun.*, 1984, 451.
24. H. Maltz, M. A. Grant, and M. C. Navaroli, *J. Org. Chem.*, 1971, **36**, 363.
25. P. A. S. Smith and R. N. Loeppky, *J. Am. Chem. Soc.*, 1967, **89**, 1147; E. Schmidt and H. Fischer, *Ber.*, 1920, **53**, 1537; E. Schmidt and R. Schumacher, *Ber.*, 1921, **54**, 1414; G. E. Hein, *J. Chem. Educ.*, 1963, **40**, 181.
26. B. J. Franck, J. Conrad, and P. Misbach, *Angew. Chem., Int. Ed. Engl.*, 1970, **9**, 892.
27. L. Castedo, R. Riguera, and M. P. Vásquez, *J. Chem. Soc., Chem. Commun.*, 1983, 301.
28. A. L. Fridman, F. M. Muthametshin, and S. S. Navikov, *Russ. Chem. Rev.*, 1971, **40**, 34; E. Linneman, *Annalen*, 1872, **161**, 45.
29. D. L. H. Williams, *Nitrosation*, Cambridge University Press, 1988.
30. W. Macmillan and T. H. Reade, *J. Chem. Soc.*, 1929, 585.
31. R. H. Fish, R. L. Holmstead, and W. Gaffield, *Tetrahedron*, 1976, **32**, 2689.
32. *F. &F.*, **3**, 207; H. Alper, *Organometal. Chem. Synth.*, 1970, **1**, 69.
33. C. J. Michejda and R. W. Schulmenz, *J. Org. Chem.*, 1973, **38**, 2412.
34. P. R. Ferina and H. Frickelmann, *J. Org. Chem.*, 1973, **38**, 4259.
35. P. R. Ferina and H. Frickelmann, *J. Org. Chem.*, 1975, **40**, 1070.

36. M. T. Rahmn, I. Ara, and A. F. M. Salahaddin, *Tetrahedron Lett.*, 1976, 1235.
37. C. G. Overberger, M. Valentine, and J.-P. Anselme, *J. Am. Chem. Soc.*, 1969, **91**, 687.
38. L. K. Keffer and C. H. Fodor, *J. Am. Chem. Soc.*, 1970, **92**, 5747.
39. D. Seebach and D. Enders, *Angew. Chem., Int. Ed. Engl.*, 1975, **14**, 15.
40. D. Seebach, D. Enders, B. Renger, and W. Brugel, *Angew. Chem., Int. Ed. Engl.*, 1973, **12**, 495.
41. D. Seebach, D. Enders, B. Renger, and W. Brugel, *Angew. Chem., Int. Ed. Engl.*, 1973, **12**, 495.
42. J. E. Baldwin, D. H. R. Barton, N. J. A. Gutteridge, and R. J. Martin, *J. Chem. Soc. C*, 1971, 2184.
43. H. U. Daeniker, *Helv. Chim. Acta*, 1964, **47**, 33.
44. I. D. Entwistle, R. A. W. Johnstone, and A. H. Wilby, *Tetrahedron*, 1982, **38**, 419.
45. *F. &F.*, **10**, 404; S. Kano, Y. Tanaka, E. Sugino, and H. Hibino, *Synthesis*, 1980, 741.
46. W. D. Emmons, *J. Am. Chem. Soc.*, 1954, **76**, 3468.
47. *F. &F.*, **5**, 479; F. Yoneda, K. Senga, and S. Nishigaki, *Chem. Pharm. Bull. Soc. Japan*, 1973, **21**, 260.
48. J. I. G. Cadogan and J. B. Thomas, *J. Chem. Soc., Chem. Commun.*, 1969, 770.
49. H. Alper and J. T. Edward, *Can. J. Chem.*, 1970, **48**, 1543; A. Tanaka and J.-P. Anselme, *Tetrahedron Lett.*, 1971, 3567.
50. K. Nishiyama and J.-P. Anselme, *J. Org. Chem.*, 1978, **43**, 2045.
51. E. M. Burgess and J. M. Lavanish, *Tetrahedron Lett.*, 1964, 1227; Y. C. Chow, *Tetrahedron Lett.*, 1964, 2333.
52. W. A. Waters, *J. Chem. Soc., Chem. Commun.*, 1978, 741.
53. C. G. Overberger, J.-P. Anselme, and J. G. Lombardino, *Organic compounds with nitrogen–nitrogen double bonds*, The Ronald Press Co., New York, 1966, Chapter 7.
54. E. H. White, *Org. Synth.*, 1967, **47**, 44.
55. U. Schulz and D. R. McCalla, *Can. J. Chem.*, 1969, **47**, 2021; P. D. Lawley, *Nature*, 1968, **218**, 580; D. R. McCalla, A. Reuvers, and R. Kitai, *Can. J. Biochem.*, 1968, **46**, 807; G. P. Wheeler and B. J. Bowdon, *Biochem. Pharmacol.*, 1972, **21**, 265; P. D. Lawley, *Topics in chemical carcinogenics*, Jap, 1972, p. 237.
56. G. A. Digenis and C. H. Issidorides, *Bioorg. Chem.*, 1979, **8**, 97.
57. P. D. Lawley and C. J. Thatcher, *Biochem. J.*, 1970, **116**, 693.

13

DIAZOALKANES[1]

For diazoalkanes R—CH=N=N the measured bond lengths are in the following ranges: C—N, 1.28–1.33 Å; N—N, 1.12–1.13 Å. Diazomethane, CH_2N_2, is planar with the molecular dimensions shown, and the infra-red stretching band of the N—N bond is at 2088 cm^{-1} in carbon tetrachloride.

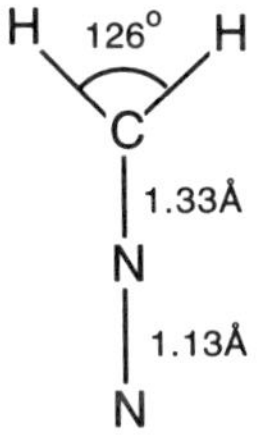

H 126° H
C
1.33Å
N
1.13Å
N

The two main contributors to the structure are (**1**) and (**2**) with a minor contribution from (**3**).

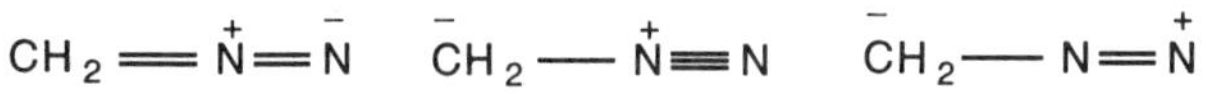

$CH_2{=}\overset{+}{N}{=}\overset{-}{N}$ (1) $\overset{-}{C}H_2{-}\overset{+}{N}{\equiv}N$ (2) $\overset{-}{C}H_2{-}N{=}\overset{+}{N}$ (3)

13.1 Preparation[2]

13.1.1 *By the action of alkali*

Any compound containing the structure (**4**) on treatment with strong concentrated bases (aq. HO^-) gives the diazoalkane, R—CH=N^+=N^-. Thus

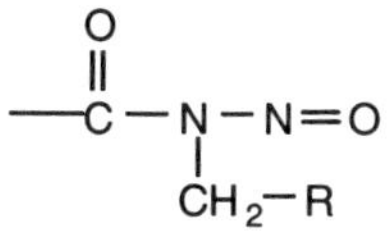

—C(=O)—N(CH$_2$—R)—N=O (4)

diazomethane itself can be prepared from *N*-methyl-*N*-nitrosourea (**5**) in this way or from *N*-methyl-*N*-nitroso-*p*-toluenesulphonamide (**6**). *N*-Nitrosourethanes (**7**) can also be used (Scheme 13.1). The mechanism of this reaction is in doubt. Two mechanisms have been proposed (Scheme 13.2).

$H_2N\text{-}CO\text{-}N(CH_3)\text{-}N{=}O$

(5)

$p\text{-}CH_3C_6H_4\text{-}SO_2\text{-}N(CH_3)\text{-}N{=}O$

(6)

$$R-CH_2-N(CO_2Et)-N{=}O \;(7) \xrightarrow[\text{in EtOH}]{EtO^-} R-CH{=}\overset{+}{N}{=}\overset{-}{N} \qquad (13.1)$$

1. $R-C(=O)-\overset{+}{N}(CH_3)=N\text{-}O^- + HO^- \rightleftharpoons R-C(OH)(O^-)-\overset{+}{N}(CH_3)=N\text{-}O^- \rightarrow R\text{-}CO_2H + CH_2-N{=}N-O^-$

$R\text{-}CO_2^- + CH_2-\ddot{N}{=}N-OH \underset{H^+}{\overset{-H^+}{\rightleftharpoons}}$ $R\text{-}CO_2H + CH_2-N{=}N-O^-$; both, with OH^-, give $CH_2{=}\overset{+}{N}{=}\overset{-}{N}$

2. $R\text{-}C(=O)\text{-}N(CH_3)\text{-}N{=}O \xrightarrow[\text{thermally}]{\text{Rearrangement as}} CH_2-\ddot{N}{=}N-O-C(=O)\text{-}R \xrightarrow{OH^-}$

$$CH_2{=}\overset{+}{N}{=}\overset{-}{N} + H_2O + R\text{-}CO_2^- \qquad (13.2)$$

A related preparation of diazoalkanes (**8**), which enables almost any primary amine (**9**) to be used as starting material, utilizes mesityl oxide (**10**) as reactant to which amines add in a Michael reaction, and the resulting secondary amine (**11**) can be nitrosated and finally decomposed by the action of alkoxides (**12**) (Scheme 13.3).[3]

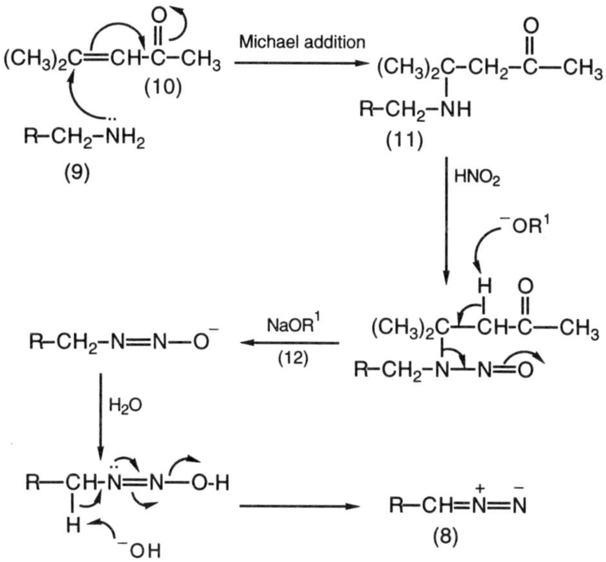

$(CH_3)_2C{=}CH{-}C(=O){-}CH_3$ (10) + $R{-}CH_2{-}NH_2$ (9) —Michael addition→ $(CH_3)_2C(NH{-}CH_2{-}R){-}CH_2{-}C(=O){-}CH_3$ (11)

(11) —HNO_2→ $(CH_3)_2C(N(N{=}O){-}CH_2{-}R){-}CH(H){-}C(=O){-}CH_3$ + $^{-}OR^1$ —$NaOR^1$ (12)→ $R{-}CH_2{-}N{=}N{-}O^-$

$R{-}CH_2{-}N{=}N{-}O^-$ —H_2O→ $R{-}CH(H){-}N{=}N{-}O{-}H$ + ^{-}OH ⟶ $R{-}CH{=}\overset{+}{N}{=}\overset{-}{N}$ (8)

(13.3)

13.1.2 *Preparation of diazomethane from chloroform and hydrazine*

The reaction of chloroform with hydrazine (**13**) and a base, dichlorocarbene (**14**) being an intermediate, gives a low yield of diazomethane (**15**) (Scheme 13.4).[4] By the use of a crown ether as a phase transfer catalyst the yield in this reaction can be increased to 87 per cent.[5]

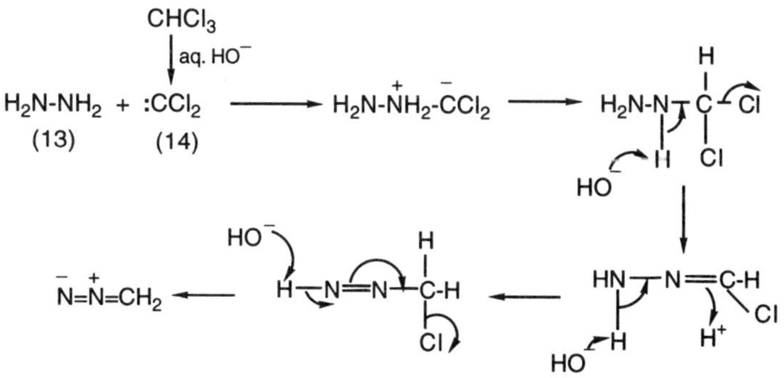

$CHCl_3$ —aq. HO^-→ :CCl_2 (14)

$H_2N{-}NH_2$ (13) + :CCl_2 (14) ⟶ $H_2N{-}\overset{+}{N}H_2{-}\overset{-}{C}Cl_2$ ⟶ $H_2N{-}NH{-}CHCl{-}Cl$ + HO^- ⟶ $HN{-}N{=}C(H){-}Cl$ (HO^-, H^+) ⟶ $H{-}N{=}N{-}CH_2{-}Cl$ + HO^- ⟶ $\overset{-}{N}{=}\overset{+}{N}{=}CH_2$

(13.4)

13.1.3 *From primary amines with an electron-attracting group on the α-carbon atom*

Diazoalkanes (**16**) can be obtained directly from primary amines which contain an electron-attracting group on the α-carbon atom, for example, from glycine ester (**17**) by the action of nitrous acid (Scheme 13.5).

$$2\,HNO_2 \longrightarrow N_2O_3$$

$$\underset{(17)}{EtO_2C\text{·}CH_2\text{—}NH_2} + NO^+ \xrightarrow{H^+} EtO_2C\text{·}CH_2\text{—}\overset{H}{N}\text{—}N\text{=}O \longrightarrow EtO_2C\text{·}CH_2\text{·}N\text{=}N\text{—}OH$$

$$\underset{c.\ 80\%\quad (16)}{EtO_2C\text{-}CH\text{=}\overset{+}{N}\text{=}\overset{-}{N}} \longleftarrow EtO_2C\text{-}\underset{H}{CH}\text{-}\overset{+}{N}\equiv N \quad {}^{-}OH \qquad (13.5)$$

Easier practical manipulation with good yields (40–80 per cent) can be attained by the use of isoamyl nitrite and a small amount of acetic acid in refluxing chloroform or benzene.[6] Diazoketones (**18**) can be prepared in a similar way (Scheme 13.6).

$$R\text{-CO-}CH_2\text{-}NH_2 \xrightarrow{HNO_2} \underset{(18)}{R\text{-CO-CH=}\overset{+}{N}\text{=}\overset{-}{N}} \qquad (13.6)$$

13.1.4 *From hydrazones*

The preparation of diazoalkanes (**19**) from hydrazones (**20**) by oxidation enables high yields of products to be obtained in high purity,[7] for example, Scheme 13.7.

$$(EtO_2C)_2C\text{=}O \xrightarrow[HOAc]{H_2N\text{-}NH_2} \underset{(20)}{(EtO_2C)_2C\text{=}N\text{-}NH_2} \xrightarrow[THF]{Ag_2O} \underset{(19)\quad 89\%}{(EtO_2C)_2C\text{=}\overset{+}{N}\text{=}\overset{-}{N}}$$

$$\underset{(20)}{Ph\text{—}\overset{O}{\overset{\|}{C}}\text{—}\underset{Ph}{C}\text{=}N\text{—}NH_2} \xrightarrow[THF]{Ag_2O} \underset{(19)\quad 87\%}{Ph\text{—}\overset{O}{\overset{\|}{C}}\text{—}\underset{Ph}{C}\text{=}\overset{+}{N}\text{=}\overset{-}{N}} \qquad (13.7)$$

By the oxidation of the hydrazone of acetone (**21**) with mercuric oxide in the presence of catalytic quantities of potassium hydroxide a yield of *c.* 80 per cent of diazopropane (**22**) can be obtained (Scheme 13.8).[8]

$$(CH_3)_2C{=}N{-}NH_2 \ (21) \xrightarrow[KOH]{HgO} (CH_3)_2C{=}\overset{+}{N}{=}\overset{-}{N} \ (22) \ 80\% \qquad (13.8)$$

Another method for the preparation from hydrazones utilizes tosylhydrazine (**23**) to form tosyl hydrazones (**24**) from aldehydes (**25**) and ketones. These hydrazones react with *n*-butyllithium which acts as a base to remove a proton from a nitrogen atom. Pyrolysis of the lithium salt (**26**) under reduced pressure gives good yields of diazoalkanes (**27**) (Scheme 13.9).[9]

$$\underset{(25)}{R{-}CHO} + \underset{(23)}{H_2N{-}NHTs} \longrightarrow \underset{(24)}{R{-}CH{=}N{-}NHTs} \xrightarrow{Bu^nLi} \underset{(26)}{R{-}CH{=}\ddot{N}{-}\overset{-}{N}{-}Ts} \ Li^+$$

$$\xrightarrow[\text{under reduced pressure}]{70^\circ - 120^\circ C, \text{ 1 hour}} R{-}CH{=}\overset{+}{N}{=}\overset{-}{N} \quad 85\% \ (27) \qquad (13.9)$$

13.1.5 *By diazo transfer*

The diazo transfer reaction[2] from azides, often tosyl azide (**28**), reacting with carbanions gives good yields of diazo compounds (**29**) from malonic ester and similar easily deprotonated compounds, for example, Scheme 13.10.

$$\underset{(30)}{(EtO_2C)_2CH_2} \xrightarrow{EtO^- / EtOH} (EtO_2C)_2\overset{-}{C}H + \underset{(28)}{N{\equiv}\overset{+}{N}{-}\overset{-}{N}Ts}$$

$$\longrightarrow (EtO_2C)_2C(H){-}N{=}N{-}\overset{-}{N}{-}Ts \longrightarrow (EtO_2C)_2C{=}\ddot{N}{-}\overset{-}{N}{-}\overset{-}{N}{-}Ts \ (+H^+)$$

$$\longrightarrow \underset{(29)}{(EtO_2C)_2C{=}\overset{+}{N}{=}\overset{-}{N}} + H\overset{-}{N}Ts \qquad (13.10)$$

The diazo transfer reaction has been extended by the discovery that a β-ketoaldehyde (**31**), produced by Claisen condensation of methyl formate (**32**) with a ketone (**33**), reacts with tosyl azide (**34**) to give good yields (50–70 per cent) of diazoaldehydes and diazoketones (**35**) (Scheme 13.11). A similar reaction starting from acetic esters yields diazoacetic esters.[10,11]

$$R^1\text{-CO-}CH_2\text{-}R^2 \ (33) \xrightarrow[MeO^-]{HCO_2Me\ (32)} R^1\text{-CO-CH}(CHO)\text{-}R^2 \ (31) \xrightarrow{TsN_3\ (34)} \ldots \longrightarrow R^1\text{-CO-}CR^2{=}\overset{+}{N}{=}\overset{-}{N} \ (35) + Ts\text{-}NHCHO \qquad (13.11)$$

13.1.6 *The Forster synthesis*

Diazoketones (**36**) can be prepared by the Forster synthesis[12] from the reaction of the monoxime of an α-diketone (**37**) with chloramine (**38**) (Scheme 13.12).[13]

$$R^1\text{-CO-}C(R^2){=}N\text{-OH} \ (37) + H_2N\text{-Cl} \ (38) \longrightarrow \ldots \xrightarrow{-H_2O,\ -H^+} R^1\text{-CO-}C(R^2){=}\overset{+}{N}{=}\overset{-}{N} \ (36) \qquad (13.12)$$

$$NaOCl + NH_3 \xrightarrow{aq.\ NaOH} H_2N\text{-Cl}\ (38)$$

13.1.7 *Preparation of diazoketones from diazomethane and acyl chlorides*

The preparation of diazoketones (**39**) from diazomethane (**40**) and an acid chloride (**41**) is the first stage in the Arndt–Eistert synthesis (see p. 538) in which an excess of diazomethane is used to react with the hydrogen chloride produced and prevent its reaction with the required diazoketone (Scheme 13.13).

$$\underset{(41)}{R\text{-}CO\text{-}Cl} + \underset{(40)}{\overset{-}{C}H_2\text{-}\overset{+}{N}\equiv N} \xrightarrow{\text{ether}} R\text{-}CO\text{-}CH(H)\text{-}\overset{+}{N}\equiv N\ Cl^- \longrightarrow HCl + \underset{(39)}{R\text{-}CO\text{-}CH{=}\overset{+}{N}{=}\overset{-}{N}}; \quad HCl \xrightarrow{CH_2N_2} CH_3Cl + N_2 \qquad (13.13)$$

13.1.8 *Preparation of substituted diazoketones*

Diazoketones (**42**) and diazoesters can be alkylated by allylic and benzylic halides after treatment with silver oxide (Scheme 13.14).[14]

$$\underset{(42)}{R^1\text{-}CO\text{-}CH{=}\overset{+}{N}{=}\overset{-}{N}} \xrightarrow{Ag_2O} R^1\text{-}CO\text{-}C(Ag){=}\overset{+}{N}{=}\overset{-}{N} \xrightarrow{R^2\text{-}Hal} R^1\text{-}CO\text{-}C(R^2){=}\overset{+}{N}{=}\overset{-}{N}$$

R^1	R^2	Hal	Yield (%)
EtO	$CH_2CH{=}CH_2$	I	66
EtO	$PhCH_2$	Br	59
Ph	$CH_2CH{=}CH_2$	I	55

(13.14)

Cleavage of diazoacetic esters/amides

Diazoacetoacetic esters (**43**) or amides may be cleaved by base in the same solvent used for their preparation by the diazo transfer route (Scheme 13.15).[15]

$$\underset{(43)}{CH_3\text{-}CO\text{-}CH_2\text{-}CO\text{-}X} + TosN_3 \xrightarrow[Et_3N]{MeCN} CH_3\text{-}CO\text{-}C(=N_2)\text{-}CO\text{-}X \xrightarrow{KOH} HC(=N_2)\text{-}CO\text{-}X + CH_3CO_2K \qquad (13.15)$$

X= Oalkyl, NEt_2, N-piperidinyl

13.2 Reactivity

From the structures shown above [(**1**), (**2**), (**3**)] (p. 519) one can expect a diazoalkane to show a variety of reactions[16] it can act:

(1) as a nucleophile or as a base;
(2) as an electrophile or as an acid;
(3) as a source of carbenes; and
(4) as a 1,3-dipole.

The two canonical forms most important in reactions are:

$$\underset{(1)}{CH_2{=}\overset{+}{N}{=}\overset{-}{N}} \quad \text{and} \quad \underset{(2)}{\overset{-}{C}H_2\text{-}\overset{+}{N}{\equiv}N}$$

13.2.1 *As a nucleophile or as a base*

The canonical form involved in these reactions is (**2**).

Alkylation

Of alcohols For the alkylation of hydroxyl groups in alcohols, a dry Lewis acid catalyst, for example boron trifluoride, is required in order to cause alkylation; in aqueous acid water is the nucleophile captured by the cation and an alcohol results (**44**); in a dry alcohol the product is an ether (**45**) (Scheme 13.16). For the alkylation of alcohols, rhodium(II) acetate [$Rh_2(OAc)_4$] is a good catalyst.[17]

$$R\text{-}\overset{-}{C}H\text{-}\overset{+}{N}{\equiv}N + H^+ \rightleftharpoons R\text{-}CH_2\text{-}\overset{+}{N}{\equiv}N \longrightarrow R\text{-}CH_2^+ + N_2$$
$$R\text{-}CH_2^+ \xrightarrow{H_2O} \underset{(44)}{R\text{-}CH_2OH} \qquad R\text{-}CH_2^+ \xrightarrow{EtOH} \underset{(45)}{R\text{-}CH_2OEt} \qquad (13.16)$$

Of enols When the hydroxy group is acidic, for example, in enols, phenols, and carboxylic acids, it is not necessary to use an acidic catalyst in order to alkylate it with a diazoalkane. The acidic proton can attack the diazoalkane to produce a diazonium ion which then decomposes and reacts with a nucleophile.

The enolic forms of β-dicarbonyl compounds are alkylated to yield an enol ether (**46**), as shown in Scheme 13.17 for ethyl acetoacetate.[18] From acetylacetone (**47**), by using this reaction, a synthesis of furans (**48**) has been developed (Scheme 13.18).[19]

$$CH_3-\underset{\underset{O}{\|}}{C}-CH_2-CO_2Et \rightleftharpoons \underset{\text{Enol}}{CH_3-\underset{\underset{OH}{|}}{C}=CH-CO_2Et} \xrightarrow{CH_2N_2} CH_3-\underset{\underset{O^-}{|}}{C}=CH-CO_2Et + CH_3-\overset{+}{N}\equiv N \longrightarrow \underset{(46)}{CH_3-\underset{\underset{OCH_3}{|}}{C}=CH-CO_2Et} \quad (13.17)$$

$$\underset{(47)}{CH_3-CO-CH_2-CO-CH_3} \rightleftharpoons CH_3-C(OH)=CH-CO-CH_3 \xrightarrow[Rh_2(OAc)_4,\ 25^\circ C]{N_2CH-CO_2Et} [CH_3-C(O-CH_2-CO_2Et)=CH-CO-CH_3] \xrightarrow{-H_2O} \underset{(48)\ 68\ \%}{\text{ethyl 5-methyl-3-methylfuran-2-carboxylate}} \quad (13.18)$$

Of carboxylic acids Carboxylic acids (**49**) are esterified by diazoalkanes (Scheme 13.19).

$$\overset{-}{C}H_2-\overset{+}{N}\equiv N + \underset{(49)}{R-CO_2H} \longrightarrow CH_3-\overset{+}{N}\equiv N + R-CO_2^- \longrightarrow N_2 + R-CO_2CH_3 \quad (13.19)$$

With hydrogen halides

The reaction with dry hydrogen halides (**50**) results in the formation of an alkyl halide (**51**) (Scheme 13.20).

$$\overset{-}{C}H_2-\overset{+}{N}\equiv N + \underset{(50)}{H-Hal} \longrightarrow CH_3-\overset{+}{N}\equiv N + X^- \longrightarrow N_2 + \underset{(51)}{CH_3-Hal} \quad (13.20)$$

With aldehydes, ketones, acetals, and acyl chlorides

Reaction with aldehydes and ketones (**52**) gives epoxides (**53**) and homologues of the aldehydes and ketones (**54**) (Scheme 13.21). (For a review of the

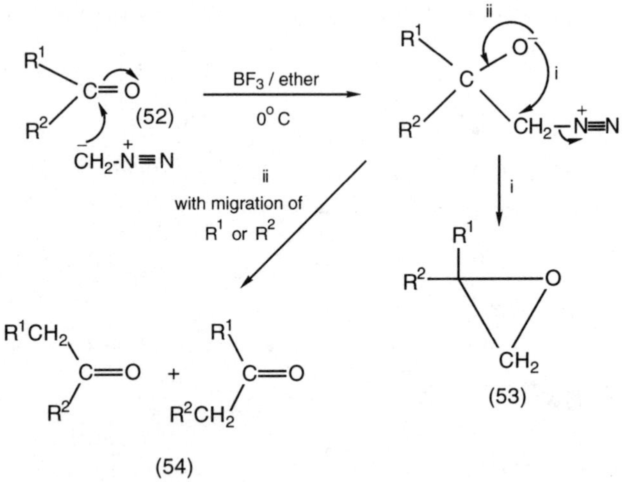

(13.21)

reactions of diazomethane with carbonyl compounds see Ref. 20.) The migration is not regiospecific. However, the reaction can be used for the ring expansion of cyclic ketones (**55**), and increased yields are obtained, especially if the catalyst is triethyloxonium tetrafluoroborate instead of boron trifluoride, for example, Scheme 13.22. Another example when specificity is not a problem is shown in Scheme 13.23.[21] Also intramolecular cyclizations go well, for example, Scheme 13.24.

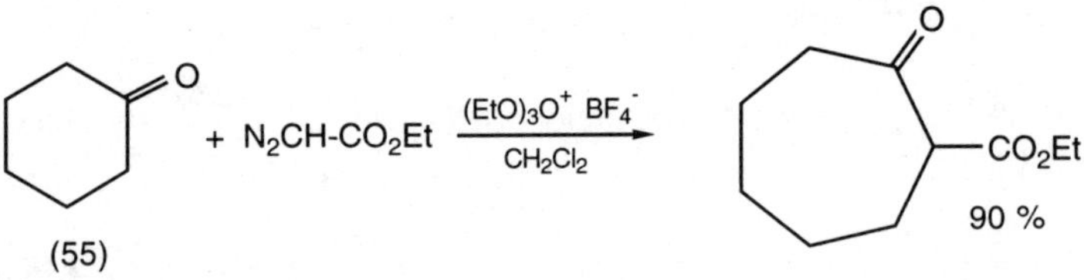

(13.22)

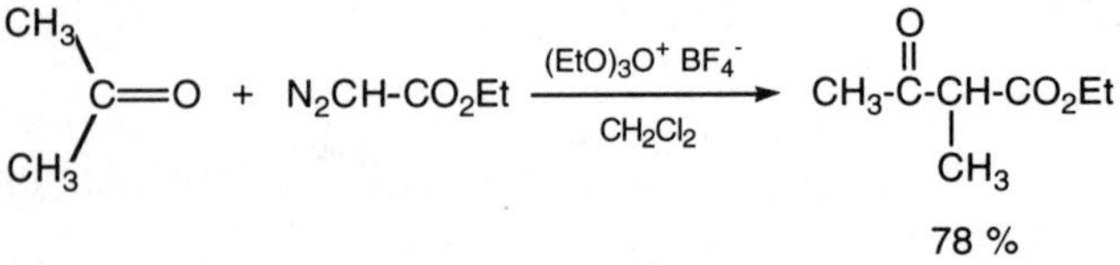

(13.23)

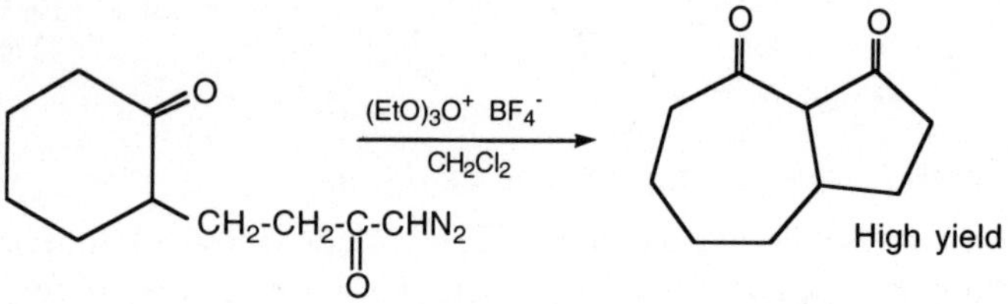

(13.24)

The best diazoalkane for the conversion of ketones to the next higher homologue is ethyl diazoacetate (**56**), catalysed by boron trifluoride etherate, BF_3–Et_2O, or by triethyloxonium tetrafluoroborate, $Et_3O^+BF_4^-$, but again the reaction is not regiospecific. A modification in which an α-halogeno ketone (**57**) is used enables regiospecificity to be achieved (Scheme 13.25). Steric and electronic factors lead to the migration of the unsubstituted methylene (CH_2) group.[22]

$$\underset{(57)}{Ph{-}\overset{O}{\overset{\|}{C}}{-}CH_2Br} + \underset{(56)}{EtO{-}\overset{O}{\overset{\|}{C}}{-}CH{=}\overset{+}{N}{=}\overset{-}{N}} \xrightarrow[\text{ii. 230°C or } CaCl_2.2H_2O \text{ in DMSO, 150°C}]{\text{i. Zn}} \underset{98\%}{Ph{-}CH_2{-}\overset{O}{\overset{\|}{C}}{-}CH_3} \qquad (13.25)$$

Ethyl diazoacetate reacts with dimethyl acetals of α,β-unsaturated aldehydes (**58**) to give acetals of β,γ-unsaturated aldehydes (**59**) (Scheme 13.26).[23]

$$\underset{(58)}{CH_2{=}CH{-}CH(OMe)_2} \xrightarrow[BF_3.Et_2O]{N_2CH{-}CO_2Et} \underset{(59)\ 68\%}{CH_2{=}CH{-}\underset{EtO_2C}{\underset{|}{CH}}{-}CH(OMe)_2} \qquad (13.26)$$

The reaction with acid chlorides (**60**) has already been dealt with in describing preparations of diazoketones (Scheme 13.27).

$$\underset{(60)}{R{-}\overset{O}{\overset{\|}{C}}{-}Cl} \xrightarrow{CH_2Cl_2} R{-}\overset{O}{\overset{\|}{C}}{-}CH{=}\overset{+}{N}{=}\overset{-}{N} + HCl \qquad (13.27)$$

The Simmons–Smith reaction

Diazoalkanes (**61**) react with electrophilic halides, for example, with iodine (Scheme 13.28). This reactivity is used in the Simmons–Smith reaction.[24] Diazomethane reacts with zinc iodide in ether generating iodomethylzinc iodide (**62**) which reacts with an added alkene to yield a cyclopropane derivative (**63**) (Scheme 13.29). The reaction is stereospecific with the stereochemistry of the alkene being retained, as in the epoxidation of alkenes by peracids.

$$\underset{(61)}{N{\equiv}\overset{+}{N}{-}\overset{-}{C}H_2} + I{-}I \longrightarrow N{\equiv}\overset{+}{N}{-}CH_2I + I^- \longrightarrow N_2 + CH_2I_2 \qquad (13.28)$$

$N\equiv\overset{+}{N}-\overset{-}{C}H_2$ + ZnI_2 ⟶ $N\equiv\overset{+}{N}-CH_2-Zn-I$ + I^- ⟶ $I-CH_2-Zn-I + N_2$ (62)

(62) + $R^1-CH{=}CH-R^2$ (*cis*-or *trans*-) ⟶ transition state [R^1, H–C=C–H, R^2; CH_2, ZnI, I] ⟶ (63) (cyclopropane: H–C(R^1), H–C(R^2), CH_2)

(63) (13.29)

With benzene for the synthesis of tropylium salts

The reaction of diazomethane with an excess of benzene in the presence of cuprous bromide as catalyst yields tropylidene (**64**). Addition of phosphorus pentachloride and then perchloric acid gives tropylium perchlorate (**65**) in 85 per cent yield (Scheme 13.30).[25]

$N\equiv\overset{+}{N}-\overset{-}{C}H_2$ + Cu-Br ⟶ $N\equiv\overset{+}{N}-CH_2-Cu-Br$ —($-N_2$)⟶ $\overset{+}{C}H_2-Cu-Br$

+ benzene ⟶ Wheland intermediate (H, $CH_2-Cu-Br$, H, +)

Wheland intermediate —(- CuBr, rearrangement)⟶ (64)

(64) —(PCl_5, CCl_4)⟶ tropylium Cl^- —($HClO_4$)⟶ tropylium ClO_4^- (65)

(13.30)

As a nitrogen nucleophile

Preparation of diazoacetaldehyde Occasionally a diazoalkane reacts as a nitrogen nucleophile, for example, diazoacetaldehyde (**66**), which may be prepared using excess *p*-toluenesulphonyl azide as shown in Scheme 13.31,[26] or more conveniently from diazomethane (**67**) and formic-acetic anhydride (**68**) (Scheme 13.32).[27]

$$\text{Ph-N(CH}_3\text{)-CH=CH-CHO} \xrightarrow[70^{\circ}-80^{\circ}\text{C in vacuo}]{\text{Excess of } p\text{-CH}_3\text{C}_6\text{H}_4\text{SO}_2\text{N}_3} \overset{-}{\text{N}}\text{=}\overset{+}{\text{N}}\text{=CH-CHO (distils out)} \; \textbf{(66)} \qquad (13.31)$$

$$\underset{(67)}{\text{CH}_2\text{=}\overset{+}{\text{N}}\text{=}\overset{-}{\text{N}}} + \underset{(68)}{\text{CH}_3\text{C(=O)-O-C(=O)-H}} \xrightarrow{\text{Et}_2\text{O}} \underset{(66)\ 46\ \%}{\overset{-}{\text{N}}\text{=}\overset{+}{\text{N}}\text{=CHCHO}} + \text{CH}_3\text{CO}_2\text{CH}_3 + \text{N}_2 \qquad (13.32)$$

Reaction with organoboranes Diazoacetaldehyde reacts with unhindered trialkylorganoboranes (**69**) in the presence of water to provide a good synthesis of aldehydes (**70**) with yields of 77–98 per cent (Scheme 13.33).[28] The proposed mechanism and the role played by water is shown in Scheme 13.34.

$$\underset{(69)}{\text{R}_3\text{B}} + \text{N}_2\text{CH-CHO} \xrightarrow[\text{H}_2\text{O}]{\text{THF}} \underset{(70)}{\text{R-CH}_2\text{-CHO}} \qquad (13.33)$$

$$\text{R}_3\text{B}\ \ \overset{-}{\text{N}}\text{=}\overset{+}{\text{N}}\text{=CH-CHO} \longrightarrow \text{R}_2\overset{-}{\text{B}}(\text{R})\text{-N=}\overset{+}{\text{N}}\text{=CH-CHO} \longrightarrow \text{R}_2\text{B-N=N-CH(R)-CHO}\ (\text{H-OH},\ :\text{OH}_2) \longrightarrow \text{R}_2\text{BOH} + \text{N}_2 + \text{R-CH}_2\text{CHO} + \text{H}_2\text{O} \qquad (13.34)$$

13.2.2 *As an electrophile*

With powerful nucleophiles

Grignard reagents Diazoalkanes (**71**) are attacked, on nitrogen, by good nucleophiles, for example, by Grignard reagents (**72**) to give hydrazones (**73**) in yields of 30–70 per cent (Scheme 13.35).[29]

$$R^1_2C\text{-}\overset{+}{N}{\equiv}\overset{-}{N} \;(71) + R^2\text{-}MgX \;(72) \longrightarrow R^1_2\overset{-}{C}\text{-}N{=}N\text{-}R^2 + MgX^+ \xrightarrow{\text{aq. } H_3O^+} R^1_2C{=}N\text{-}NH\text{-}R^2 \;(73) \quad (13.35)$$

Lithium aluminium hydride Lithium aluminium hydride gives amines (**74**) by hydrogenolysis,[30] for example, Scheme 13.36. Other functional groups are reduced accordingly, for example, Scheme 13.37.[31]

$$CH_2N_2 \xrightarrow{LiAlH_4} CH_3NH_2 \;(74) \quad (13.36)$$

$$EtO_2C\text{-}CHN_2 \xrightarrow[\text{ether}]{LiAlH_4} HO\text{-}CH_2CH_2NH_2 \quad 72\,\%$$

$$Ph\text{-}C(=O)\text{-}CHN_2 \xrightarrow[\text{ether}]{LiAlH_4} \underset{93\,\%}{Ph\text{-}CH(OH)\text{-}CH_2NH_2} + \underset{3\,\%}{Ph\text{-}C(=O)\text{-}CH_3} \quad (13.37)$$

Hydrogenation Hydrogenation using platinum catalyst causes loss of nitrogen (Scheme 13.38).[32]

$$N_2CH\text{-}CO_2Et \xrightarrow{H_2/Pt} CH_3CO_2Et + N_2 \quad \text{Quantitative yield} \quad (13.38)$$

13.2.3 *As a source of carbenes*

By copper catalysis

Many of the reactions of diazoalkanes occur through carbenes. Usually, for synthetic purposes, carbene generation is promoted by metal catalysts,[33] for example, Scheme 13.39.

Copper-catalysed addition of diazoalkanes to alkenes occurs with no insertion; instead addition is *cis*, and with cyclic alkenes (**75**) usually yields *exo* products.[34] Thus diazomethane adds to alkenes under the influence of cuprous chloride to give cyclopropanes (**76**) e.g. Schemes 13.40[35] and 13.41.[36]

$N_2C(CO_2Me)_2$; cupric acetylacetonate * (catalyst) or $h\nu$; CO_2Me, CO_2Me

* $[CH_3\text{-}C{=}CH\text{-}C(=O)\text{-}CH_3\text{, O}]_2\text{-}Cu^{II}(OH)_2$

(13.39)

(75) $\xrightarrow[Cu_2Cl_2,\ 0\,^{\circ}C]{CH_2N_2}$ + (76)

(13.40)

OH $\xrightarrow[Cu_2Cl_2]{CH_2N_2}$ OH (76) 40 %

(13.41)

The copper-catalysed reaction of diazoketones with alkenes has been shown to occur both *inter*- and *intra*-molecularly, for example (Scheme 13.42).[37]

Intra-molecularly:

$CH_2{=}CH(CH_2)_3C(=O)\text{-}CHN_2 \xrightarrow{Cu}$ O; 75 %

(13.42)

Synthesis of cycloheptatrienes

Alkyl diazotates (**77**) can be used in a similar way except using a rhodium(II) catalyst, to synthesize cycloheptatrienes (**78**) from benzene (**79**) (Scheme 13.43).[38]

Synthesis of cyclopropanes

The formation of cyclopropanes (**80**) by the insertion of carboethoxycarbenes formed alkyldiazoesters (**81**) into olefinic double bonds (**82**) is efficiently catalysed by rhodium(II) *n*-butanate,[39] for example, Scheme 13.44.

(79) + $N_2CH\text{-}CO_2Me$ (77) $\xrightarrow[20^\circ C]{Rh(OCOCF_3)_2}$ (78) *c.* 100 % (13.43)

(82) + $N_2CH\text{-}CO_2Bu^n$ (81) $\xrightarrow{Rh(O_2CBu^n)_2}$ (80) 90 % (13.44)

13.2.4 *Cycloadditions of diazoalkanes acting as 1,3-dipoles*

There have been many studies of the reactions of diazoalkanes as 1,3-dipoles with alkynes (acetylenes) and with alkenes (olefins).[40] The reactions of simple diazoalkanes have been shown to be HO(1,3-dipole)LU(dipolarophile) controlled.[41] (For a theoretical study of the reaction between diazomethane and ethylene by the LCAO-SCF-MO method see Ref. 42). With alkynes (**83**) good syntheses of pyrazoles (**84**) have been achieved, for example, Schemes 13.45[43] and 13.46.[44]

(83) $R^1\text{-}C\equiv C\text{-}R^2$ + $(MeO)_2CH\text{-}\bar{C}H\text{-}\overset{+}{N}\equiv N$ $\xrightarrow{\text{ether}}$ $\xrightarrow{\text{aq. } H_3O^+}$ (84) (13.45)

(83) $HC\equiv C\text{-}OEt$ + CH_2N_2 $\xrightarrow[\text{Dark, 2 weeks}]{Et_2O,\ 20^\circ C}$ (84) 4-Ethoxypyrazole 72 % (13.46)

When alkenes instead of alkynes are used, the regiospecificity of the reaction changes, for example, (Scheme 13.47).[45] This reversal of regiospecificity has been treated theoretically by Sustmann and his fellow workers[46] who provide a perturbation treatment as an explanation. This amounts to frontier orbital theory in which distortions (closed shell repulsions) in the transition state are taken into account. With alkenes (**85**) reduced pyrazoles (**86**) result which

EtO-C-H, CH$_2$, N, N$^+$, CH$_2$ → EtO, H, N, N (13.47)

R, H, C=C(CO$_2$CH$_3$)$_2$ (85), (MeO)$_2$CH-$\bar{C}$H-$\overset{+}{N}$≡N — ether, -20° C → R, H, CO$_2$CH$_3$, CO$_2$CH$_3$, (MeO)$_2$CH, N, N (86)

$h\nu$, -N$_2$

R, H, CO$_2$CH$_3$, CO$_2$CH$_3$, CH(OMe)$_2$ (87) — aq. H.CO$_2$H → R, H, CO$_2$CH$_3$, CO$_2$CH$_3$, OHC, H

Yields *c.* 80 %
when R = Ph, Et or CO$_2$Me (13.48)

Me, Me, C, N+, N$_-$ (88); R^1, C, C, SO$_2$R^2 (89) — ether, -78° C, 30 min then at 25° C for 12 hours → Me, Me, N, N, R^1, SO$_2$R^2

benzene, $h\nu$

R^1 and R^2 = CH$_3$ or Ph

Me, Me, R^{2}SO$_2$, R^1 + N$_2$ (13.49)

under photolysis eliminate nitrogen to give good yields of cyclopropanes (**87**), for example, Scheme 13.48.[47]

The 1,3-dipolar addition of 2-diazopropane (**88**) to sulphonyl substituted alkynes (**89**) has also been studied,[48] for example, Scheme 13.49.

These authors have also shown that a reversal of regiospecificity in this reaction occurs when trimethylsilyl derivatives of sulphonyl substituted alkynes (**90**) are used, and they attribute this to control of the reaction by steric rather than stereoelectronic factors,[49] for example, Scheme 13.50.

A similar reaction of 2-diazopropane with *p*-tolyl(2-trimethylsilyl)ethynylsulphone (**91**) yielded, on desilylation without methylation, the compound (**92**) with the same regiospecificity.

Me Me SO_2Ar C N+ N_ C C $SiMe_3$ (90)

Me Me N N SO_2Ar $SiMe_3$

KI, 18-crown-6, 3 hours at 25° C

Ar = CH_3 *p* - tolyl

Me Me N N SO_2Ar 93 %

H_2C N+ N_ SO_2Ar C C $SiMe_3$ (91)

THF- ether 25° C, 12 hours

[H H N N SO_2Ar $SiMe_3$]

MeI / K_2CO_3 / DMF 25° C, 6 hours

CH_3—N N SO_2Ar 95 %

$Bu^n_4N^+F^-$/THF 25° C

CH_3—N N SO_2Ar $SiMe_3$ 100 %

(13.50)

SO_2Ar

CH_3CH_2— N

(92)

Diazoacetaldehyde dimethylacetal (**93**)[50] is a powerful 1,3-dipole and reacts with alkenes (**94**) and alkynes (**95**) (Scheme 13.51).[51]

$H_2N\text{-}CH_2\text{-}CH(OMe)_2 \xrightarrow[\text{ii, } H_2O]{\text{i, KCN}} H_2N\text{-}C(=O)\text{-}NH\text{-}CH_2CH(OMe)_2$

N_2O_4

$H_2N\text{-}C(=O)\text{-}N(N{=}O)\text{-}CH_2CH(OMe)_2 \xrightarrow{\text{NaOH}} \bar{N}{=}\overset{+}{N}{=}CH\text{-}CH(OMe)_2$ (93)

Overall yield = 45 %

$R(H)C{=}C(CO_2CH_3)_2$ (94)

$(MeO)_2CHCH{=}\overset{+}{N}{=}\bar{N}$ (93)

ether, -20° C

R, CO_2CH_3, H, CO_2CH_3, N, N, $CH(OMe)_2$ $\xrightarrow{h\nu}$ R, CO_2CH_3, H, CO_2CH_3, $CH(OMe)_2$ + N_2

$H.CO_2H$

R, CO_2CH_3, H, CO_2CH_3, H, CHO

Yields >80 %

$$R^1\text{-}C{\equiv}C\text{-}R^2\ (95) + (MeO)_2CHCH{=}\overset{+}{N}{=}\overset{-}{N}\ (93) \xrightarrow{\text{ether}} \left[\text{pyrazoline: } R^1, R^2, \text{N=N}, CH(OMe)_2\right] \xrightarrow{\text{aq. } H_3O^+} \text{pyrazole: } R^1, R^2, CHO, \text{N, NH}$$

(13.51)

DIAZOALKANES CONTAINING OTHER FUNCTIONAL GROUPS

Because of the high and varied reactivity of diazoalkanes and their subsequent wide use in synthesis, attention has been directed towards the preparation of diazoalkanes which contain other functional groups. Thus diazoesters and diazoketones are used widely for synthetic purposes.

13.3 Reactivity

13.3.1 *Wolff rearrangement of diazoketones*[52]

Under the influence of silver oxide or platinum, diazoketones (**96**) rearrange with loss of nitrogen to yield ketenes (**97**) and this is a step in the Arndt–Eistert homologation of carboxylic acids (**98**).[53] The mechanism is related to that of the Hofmann rearrangement of amides, the Curtius rearrangement of acyl azides, and the Lossen rearrangement of hydroxamic acids, though it involves migration to an electron-deficient CH group instead of to an electron-deficient nitrogen atom as in those rearrangements (Scheme 13.52).

13.3.2 *Rhodium catalysed decomposition of diazoesters and diazoketones*

The decomposition of diazoesters (**99**) or diazoketones catalysed by rhodium(II) acetate, $Rh_2(OAc)_4$, in the presence of a trialkylsilane (**100**) gives silyl esters (**101**) or silyl ketones,[54] for example, Scheme 13.53.

R^1–C(=O)–$\overset{-}{C}H$–$\overset{+}{N}\equiv N$ (96) $\xrightarrow[\text{Pt or } Ag_2O \text{ as catalysts}]{\text{heat / light}}$ R^1-CH=C=O + N_2 (97)

R^1-CO_2H (98) $\xrightarrow{SOCl_2}$ R^1-COCl $\xrightarrow{CH_2N_2}$ (96)

(97) $\xrightarrow{H_2O}$ $R^1-CH_2-CO_2H$ (98)

(97) $\xrightarrow{R^2OH}$ $R^1-CH_2CO_2R^2$

(97) $\xrightarrow{R^2NH_2}$ R^1-CH_2-C(=O)-NHR^2

(13.52)

N_2CH-CO_2Et (99) + Et_3SiH (100) $\xrightarrow{Rh_2(OAc)_4}$ $Et_3Si-CH_2-CO_2Et$ (101) (85-90%) (13.53)

The rhodium acetate catalysed decomposition of α-diazoesters (**102**) yields α,β-unsaturated esters (**103**),[55] for example, Scheme 13.54.

Me_2CH-CH_2-C($=\overset{+}{N}=\overset{-}{N}$)-$CO_2Me$ (102) $\xrightarrow{Rh_2(OAc)_4}$ [H_A, CO_2Me, $CHMe_2$, H, $Rh(OAc)_n$]

migration of H_A furnishes the *cis*- isomer

Me_2CH(H)C=C(H)CO_2Me + N_2

99%

(103) (13.54)

13.3.3 *Synthesis of ketones*

A synthesis of ketones (**104**) has been developed in which a diazoketone (**105**) is treated with chlorothiobenzene in ether, then with benzene and stannic chloride and finally reduced with zinc and acetic acid,[56] for example, Scheme 13.55. Nine examples are reported.

(13.55)

13.3.4 *Synthesis of α-halo-α,β-unsaturated ketones*

Diazoketones (**106**) react with phenylselenium chloride or bromide (PhSeHal) in methylene dichloride at room temperature to give an adduct (**107**) which with hydrogen peroxide can be converted in good yield into an α-chloro- or bromo-α,β-unsaturated ketone (**108**),[57] for example, Scheme 13.56.

(13.56)

13.3.5 *Synthesis of β-diketones*

Diazoketones (**109**) can lose a proton to lithium diisopropylamide and the lithium anion (**110**) so formed can act as a nucleophile and attack an aldehyde, the resulting diazo compound (**111**) being transformed into a β-diketone (**112**) by the action of rhodium tetraacetate (Scheme 13.57).[58]

13.3.6 *Formation of cyclic systems*

Protonated diazoketones (**113**) can cyclize onto suitably placed unsaturated systems,[59] a reaction which has found application for the synthesis of natural products[60] (Scheme 13.58).

$$CH_3-\underset{\underset{O}{\|}}{C}-CH=\overset{+}{N}=\overset{-}{N} \xrightarrow[\text{THF, -78}^\circ\text{C}]{\text{LDA}} CH_3-\underset{\underset{O}{\|}}{C}-\underset{\underset{Li}{|}}{C}=\overset{+}{N}=\overset{-}{N}$$

(109) (110)

R-CHO

$$R-\underset{\underset{O}{\|}}{C}-CH_2-\underset{\underset{O}{\|}}{C}-CH_3 \xleftarrow{Rh(OAc)_4} R-\overset{OH}{\underset{H}{C}}-\overset{\overset{+}{N}=\overset{-}{N}}{C}-\underset{\underset{O}{\|}}{C}-CH_3$$

(112) (111) (13.57)

(113) $\xrightarrow{F_3C\text{-}CO_2H}$ 80% + $F_3C\text{-}CO_2Me$

H^+; $F_3C\text{-}CO_2^-$; + N_2; +OMe; :OMe

(13.58)

13.3.7 *Diazoacetyl chloride for the formation of ester or amide derivatives*

Diazoacetyl chloride (**114**), prepared as shown in Scheme 13.59,[61] reacts with alcohols, phenols, thiols, and amines in the presence of triethylamine to give diazoacetic ester or amide derivatives (**115**) in high yield (Scheme 13.60).[62]

$$CH_2N_2 + COCl_2 \longrightarrow \underset{(114)}{N_2CH\text{-}COCl} \qquad (13.59)$$

$$\underset{(114)}{N_2CH\text{-}COCl} \xrightarrow[Et_3N]{R\text{-}XH} \underset{(115)}{N_2CH\text{-}C(=O)\text{-}X\text{-}R}$$

$$X = O, S, NR^1, \quad R^1 = \text{alkyl or aryl} \qquad (13.60)$$

13.3.8 *Trimethylsilyldiazomethane*

Preparation[63]

Trimethylsilyldiazomethane (**116**) can be obtained by the treatment of *N*-nitroso-*N*-(trimethylsilylmethyl)urea, $(CH_3)_3SiCH_2N(NO)CONH_2$, with potassium hydroxide (Scheme 13.61).[64]

$$(CH_3)_3Si\text{-}CH_2\text{-}N(NO)\text{-}C(=O)\text{-}NH_2 \xrightarrow{KOH} \underset{(116)}{(CH_3)_3Si\text{-}CH{=}\overset{+}{N}{=}\overset{-}{N}} \qquad (13.61)$$

Trimethylsilyldiazomethane (**116**) can also be prepared by a diazo transfer reaction between trimethylsilylmagnesium chloride (**117**) and diphenyl phosphoroazidate (**118**) (Scheme 13.62).[65] Another preparation by reaction of diazomethane with trimethylsilyl trifluoromethanesulphonate has been reported (Scheme 13.63).

$$\underset{(117)}{Me_3Si\text{-}CH_2\text{-}MgCl} + \underset{(118)}{N{\equiv}\overset{+}{N}\text{-}\overset{-}{N}\text{-}P(=O)Ph_2} \longrightarrow Me_3Si\text{-}CH(H)\text{-}N{=}N\text{-}\overset{-}{N}\text{-}P(=O)Ph_2$$

$$\longrightarrow Me_3Si\text{-}CH{=}N\text{-}\overset{-}{N}\text{-}\overset{-}{N}\text{-}P(=O)Ph_2 \;(H^+) \longrightarrow \underset{(116)}{Me_3Si\text{—}CH{=}\overset{+}{N}{=}\overset{-}{N}} + H\overset{-}{N}\text{—}P(=O)Ph_2 \qquad (13.62)$$

$$Me_3Si\text{-}SO_2CF_3 + CH_2N_2 \xrightarrow[Pr^i_2NEt]{ether} Me_3Si\text{-}CH{=}\overset{+}{N}{=}\overset{-}{N} \quad 74\% \qquad (13.63)$$

Synthetic applications

Replacement of diazomethane Trimethylsilyldiazomethane is supplied by Aldrich as a stable solution in a mixture of hexanes. It can replace diazomethane for several purposes, for example, the conversion of phenols and enols into methyl ethers;[67] for the formation of methyl esters (**119**) from carboxylic acids (**120**) (Scheme 13.64);[68] and for the ring expansion of cyclic ketones (**121**) (Scheme 13.65).[69]

$$\text{(120) } o\text{-}NO_2C_6H_4CO_2H \xrightarrow[\text{MeOH / PhH / 20}^\circ\text{C}]{Me_3Si\text{-}CH{=}\overset{+}{N}{=}\overset{-}{N}} \text{(119) } o\text{-}NO_2C_6H_4CO_2CH_3 \quad 95\ \% \qquad (13.64)$$

$$\text{(121) 4-}Bu^t\text{-cyclohexanone} \xrightarrow[BF_3\text{ - }Et_2O / CH_2Cl_2,\ 0^\circ C]{Me_3Si\text{-}CH{=}\overset{+}{N}{=}\overset{-}{N}} \text{4-}Bu^t\text{-cycloheptanone} \quad 70\ \% \qquad (13.65)$$

Conversion of aldehydes into methyl ketones It has also been applied to the conversion of an aldehyde (**122**) into a methyl ketone (**123**) (Scheme 13.66).[70]

$$\underset{(122)}{Bu^t\text{-}CHO} \xrightarrow[MgBr_2 / 0^\circ C]{Me_3Si\text{-}CH{=}\overset{+}{N}{=}\overset{-}{N}} \underset{(123)}{Bu^t\text{-}\overset{O}{\overset{\|}{C}}\text{-}CH_3} \quad 89\ \% \qquad (13.66)$$

Synthesis of alkynes Trimethylsilyldiazomethane has also found use in the synthesis of alkynes (**124**) (Scheme 13.67).[71]

$$Me_3Si\text{-}CH{=}\overset{+}{N}{=}\overset{-}{N} \xrightarrow[0^\circ C]{Bu^nLi} \left[Me_3Si\text{-}\overset{-}{C}{=}\overset{+}{N}{=}\overset{-}{N} \ Li^+ \right] \xrightarrow{Ph_2C{=}O} \underset{(124)\ 80\ \%}{Ph\text{-}C{\equiv}C\text{-}Ph} \qquad (13.67)$$

1,3-Dipolar additions Trimethylsilyldiazomethane undergoes 1,3-dipolar additions to alkenes (**125**) with subsequent loss of nitrogen in the presence of $PdCl_2$ (Scheme 13.68).[72] However, in the absence of a catalyst 1,3-dipolar addition affords a pyrazole (**126**) (Scheme 13.69).[73]

Preparation of cyclobutanones Diazopropane (**127**)[74] reacts in an interesting way with tropone (**128**) and with tropolone (**129**) to yield cyclobutanones (**130**) (Scheme 13.70).[75]

$C_6H_5CH{=}CHCO_2Et$ (125) + $Me_3Si{-}CH{=}\overset{+}{N}{=}\overset{-}{N}$ $\xrightarrow[C_6H_6,\ 60^{\circ}C]{PdCl_2}$

C_6H_5 / CO_2Et / Me_3Si / H — 34%

+

C_6H_5 / CO_2Et / H / $SiMe_3$ — 45%

(13.68)

$C_6H_5CH{=}CHCO_2CH_3$ + $Me_3Si{-}CH{=}\overset{+}{N}{=}\overset{-}{N}$ $\longrightarrow$ C_6H_5 / CO_2CH_3 / N / N / H (126) 38%

(13.69)

Tropone (128) $\xrightarrow[(127)]{(CH_3)_2CN_2}$ [N, N, CH_3, CH_3, O $\longrightarrow$ CH_3, CH_3, O] $\downarrow$ CH_3, CH_3, O (130) 40 %

Tropolone (129) (O, OH) $\xrightarrow[(127)]{(CH_3)_2CN_2}$ CH_3, CH_3, O, $OCH(CH_3)_2$ (130) 54 %

(13.70)

13.3.9 *Dimethyl diazomalonate*

With vinyl ethers

Dimethyl diazomalonate (**131**) reacts by addition–elimination with vinyl ethers (**132**) in the presence of varied copper catalysts. The mixture of products is very dependent upon the composition of the catalyst[76] (Scheme 13.71).

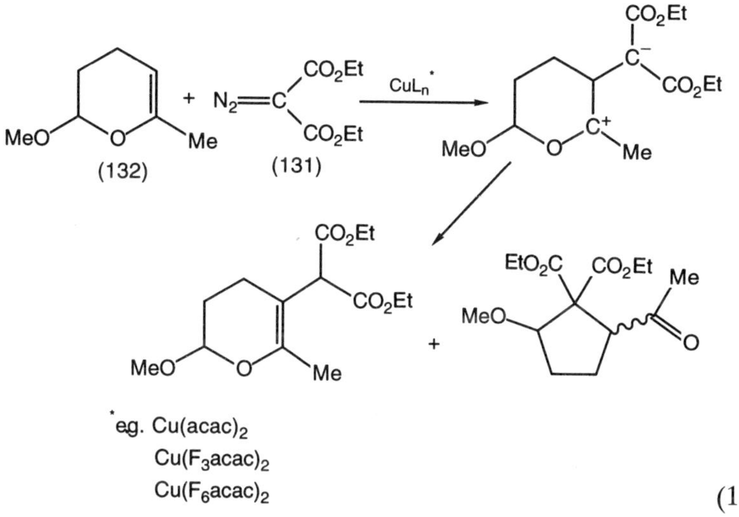

(13.71)

With epoxides

Dimethyl diazomalonate (**131**) in the presence of rhodium(II) acetate converts epoxides (**133**) into the corresponding alkenes (**134**) in good (>80 per cent) yields (Scheme 13.72).[77]

(13.72)

13.3.10 *Darzen's reaction on α-halogenodiazoketones*

The Darzens reaction carried out on an α-halogenodiazoketone (**135**) and a non-enolizable (no α-hydrogen atom) aldehyde proceeds normally to yield an α,β-epoxy diazomethyl ketone (**136**), which gives a keto acetal (**137**) on reaction with methanol or undergoes Wolff rearrangement under photolysis (Scheme 13.73).[78]

$$Cl{-}CH_2{-}C(=O){-}Cl + CH_2N_2 \longrightarrow Cl{-}CH_2{-}C(=O){-}CH{=}\overset{+}{N}{=}\overset{-}{N} \quad (135)$$

Ph-CHO
NaOH/MeOH

(136) (Ph, H, H on epoxide ring) $C(=O){-}CH{=}\overset{+}{N}{=}\overset{-}{N}$

hν
MeOH
Wolff Rearrangement

$CH_3O{-}C(=O){-}CH{=}CH{-}CH(OH){-}Ph$

62%

reflux
CH_3OH

$Ph{-}CH{=}CH{-}C(=O){-}CH(OCH_3)_2$ (137) (13.73)

13.3.11 *α,β-Epoxydiazomethyl ketones*

Preparation

Other authors[79] prepared a wider range of α,β-epoxydiazomethyl ketones (**138**) as shown in Scheme 13.74.

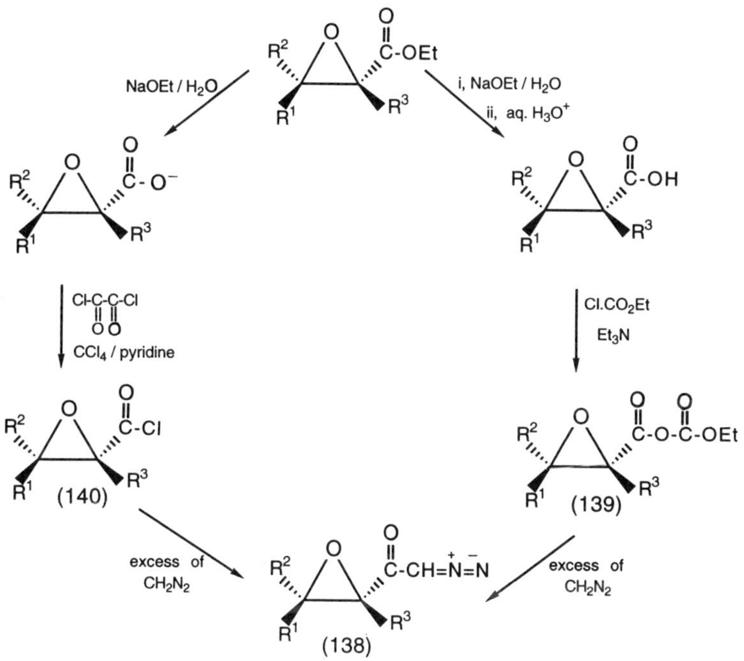

R^1	R^2	R^3	anhydride (**139**)	Yield (%) via acyl chloride (**140**)
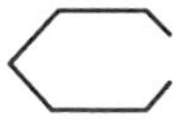		H	60	60
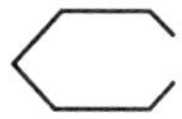		CH_3	17	—
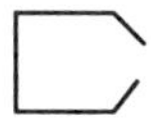		H	23	—
CH_3	CH_3	CH_3	—	64
CH_3	CH_3	Ph	68	—
H	H	Ph-CH_2	51	—

(13.74)

Synthesis of alkene ketoacetals

These authors by the action of activated copper powder or copper sulphate[80] converted the epoxyalkanes (**141**) into alkene ketoacetals (**142**). The mechanism of the reaction involves an oxygen transfer and a carbene intermediate (Scheme 13.75).

Another reaction studied[81] was that with alkenes catalysed by palladium acetate, for example, Scheme 13.76. These reactions were followed by interesting epoxide rearrangements induced by Lewis acids.

Cu or $CuSO_4$ / R^4-OH
Reflux 1 - 6 hours

(141) (142)

R^4-OH

- H^+

R^1	R^2	R^3	R^4	Yield (%) Cu	CuSO$_4$
H	H	Bn	Me	76	—
H	H	Bn	Et	78	—
-$(CH_2)_5$-	H	Me		32	75
-$(CH_2)_4$-	H	Me		—	46
Me	Me	Me	Et	21	61

(13.75)

$Pd(OAc)_2$
(catalytic amount)
30 min to 5.5 hours

R^1	R^2	R^3	Yield (%)
$-(CH_2)_4-$		Me	48.5
$-(CH_2)_4-$		H	62.5
Ph	Me	Me	38.5

(13.76)

References

1. M. Regitz (ed.), *Diazoalkanes*, Thieme, Stuttgart, 1977; Review of diazomethane: T. H. Black, *Aldrichim. Acta*, 1983, **16**, 36; *F. &F.*, **12**, 156.
2. Review: M. Regitz, *Synthesis*, 1972, 351.
3. E. C. S. Jones and J. Kenner, *J. Chem. Soc.*, 1933, 363; D. W. Anderson and J. Kenner, *J. Chem. Soc.*, 1935, 286.
4. H. Staudinger and O. Kupfer, *Ber.*, 1912, **45**, 401.
5. H. Ledon, *Synthesis*, 1974, 347.
6. *F. &F.*, **4**, 270; N. Takamura, T. Mizoguchi, K. Koga, and S. Yamada, *Tetrahedron Lett.*, 1971, 4495.
7. *F. &F.*, **8**, 187; **5**, 244; E. Giganek, *J. Org. Chem.*, 1965, **30**, 4366.
8. S. D. Andrews, A. C. Day, P. Raymond, and M. C. Whiting, *Org. Synth.*, 1970, **50**, 27.
9. T. Sasaki, S. Eguchi, I. H. Ryu, and Y. Hirako, *Tetrahedron Lett.*, 1974, 2011.
10. M. Regitz and J. Rüter, *Ber.*, 1968, **101**, 1263; *F. &F.*, **3**, 291.
11. M. Regitz and F. Menz, *Ber.*, 1968, **101**, 2622.
12. M. O. Forster, *J. Chem. Soc.*, 1915, **107**, 260.
13. *F. &F.*, **1**, 122; M. P. Cava, R. L. Little, and D. R. Napier, *J. Am. Chem. Soc.*, 1958, **80**, 2257; M. P. Cava and P. M. Wentraub, *Steroids*, 1964, **4**, 41; J. Meinwald and P. O. Gassman, *J. Am. Chem. Soc.*, 1960, **82**, 2857.
14. *F. &F.*, **3**, 253; U. Schöllkopf and N. Rieber, *Ber.*, 1969, **102**, 488.
15. M. Regitz, J. Hockner, and A. Liedhegener, *Org. Prep. Proc. Int.*, 1969, **1**, 99.
16. G. W. Cowell and A. Ledwith, *Quart. Rev.*, 1970, **24**, 119.
17. *F. &F.*, **5**, 571; R. Paulissen, H. Reimlinger, E. Hayez, A. J. Hubert, and P. Teyssié, *Tetrahedron Lett.*, 1973, 2233; R. Paulissen, E. Hayez, A. J. Hubert, and P. Teyssié, *Tetrahedron Lett.*, 1974, 607.
18. R. Paulissen, H. Reimlinger, E. Hayez, A. J. Hubert, and P. Teyssié, *Tetrahedron Lett.*, 1973, 2233.
19. R. Paulissen, E. Hayez, A. J. Hubert, and P. Teyssié, *Tetrahedron Lett.*, 1974, 607.
20. C. D. Gutsche, *Org. React.*, 1954, **8**, 364; *F. &F.*, **1**, 195; **1**, 369.
21. *F. &F.*, **3**, 139; W. L. Mock and M. E. Hartman, *J. Am. Chem. Soc.*, 1970, **92**, 5767.
22. *F. &F.*, **12**, 223; V. Dave and E. W. Warnoff, *J. Org. Chem.*, 1983, **48**, 2590.
23. *F. &F.*, **12**, 224; M. P. Doyle, M. L. Trudell, and J. W. Tempstra, *J. Org. Chem.*, 1983, **48**, 5146.
24. *F. &F.*, **1**, 194; G. Wittig and K. Schwarzenbach, *Annalen*, 1961, **650**, 1; H. Hoberg, *Annalen*, 1962, **656**, 1; 15.
25. *F. &F.*, **1**, 165; E. Müller and H. Fricke, *Annalen*, 1963, **661**, 38.
26. *F. &F.*, **2**, 101–2; Z. Arnold, *J. Chem. Soc., Chem. Commun.*, 1967, **6**, 299.
27. *F. &F.*, **4**, 5; J. Hooz and G. F. Morrison, *Org. Prep. Proc. Int.*, 1971, **3**, 227.

28. *F. &F.*, **3**, 73; J. Hooz and G. F. Morrison, *Can. J. Chem.*, 1970, **48**, 868.
29. M. S. Kharasch and O. Reinmuth, *Grignard reactions of nonmetallic substances*, Constable and Co, London, 1954, p. 1223.
30. K. Clusius and F. Endtinger, *Helv. Chim. Acta*, 1958, **41**, 182.
31. W. Ried and F. Müller, *Chem. Ber.*, 1952, **85**, 470.
32. W. Gruber and H. Renner, *Monatshefte.*, 1950, **81**, 751.
33. *F. &F.*, **5**, 244; for a review see B. W. Pearce and D. S. Wulfman, *Synthesis*, 1973, 137; *F. &F.*, **3**, 6.
34. W. von E. Doering and T. Mole, *Tetrahedron*, 1960, **10**, 65; P. S. Shell and R. M. Etter, *Proc. Chem. Soc.*, 1961, 443.
35. W. von E. Doering and W. R. Roth, *Tetrahedron*, 1963, **19**, 715.
36. R. E. Pincock and J. I. Wells, *J. Org. Chem.*, 1964, **29**, 965; *F. &F.*, **3**, 67; J. P. Chesick, R. E. Pinock, and J. I. Wells, *J. Am. Chem. Soc.*, 1962, **84**, 3250; P. G. Gassman, A. Topp, and J. W. Keller, *Tetrahedron Lett.*, 1969, 1093.
37. *F. &F.*, **2**, 82; G. Stork and J. Ficini, *J. Am. Chem. Soc.*, 1961, **83**, 4678.
38. *F. &F.*, **10**, 340; A. J. Anciaux, A. Demonceau, A. F. Noels, A. J. Hubert, R. Warin, and P. Teyssié, *J. Org. Chem.*, 1981, 873; Buchner reaction: E. W. Warnhoff, *Org. React.*, 1970, **17**, 239; *F. &F.*, **1**, 368.
39. *F. &F.*, **7**, 313; A. J. Hubert, A. F. Noels, A. J. Anciaux, and P. Teyssié, *Synthesis*, 1976, 600.
40. R. Huisgen, in *1,3-Dipolar cycloaddition chemistry* (ed. A. Padwa), Wiley, New York, 1984, Vol. 1, p. 1; M. Regitz and H. Heydt, *ibid.*, p. 393.
41. R. Sustmann, *Tetrahedron Lett.*, 1971, 2717; K. N. Houk, J. Sims, C. R. Watts, and L. J. Luskus, *J. Am. Chem. Soc.*, 1973, **95**, 7301.
42. G. Leyroy and M. Sana, *Tetrahedron*, 1975, **31**, 2091.
43. H. Abdallah and R. Grée, *Tetrahedron Lett.*, 1980, **21**, 2239.
44. S. H. Groen and J. F. Arens, *Recl. Trav. Chim. Pays-Bas*, 1961, **80**, 879.
45. A. Firestone, *J. Org. Chem.*, 1976, **41**, 2212.
46. R. Sustmann, W. Sicking, and M. Felderhoff, *Tetrahedron*, 1990, **46**, 783.
47. H. Abdullah and R. Grée, *Tetrahedron Lett.*, 1980, **21**, 2239.
48. A. Padwa, M. W. Wannamaker, and A. D. Dyszlewski, *J. Org. Chem.*, 1987, **52**, 4760.
49. A. Padwa and M. W. Wannamaker, *Tetrahedron*, 1990, **46**, 1145.
50. Preparation: W. Kirmse and M. Burchhoff, *Ber.*, 1967, **100**, 1491.
51. H. Abdallah and R. Grée, *Tetrahedron Lett.*, 1980, **21**, 2239.
52. L. Wolff, *Annalen*, 1912, **394**, 25; *F. &F.*, **1**, 1004; M. S. Newman and P. F. Beal III, *J. Am. Chem. Soc.*, 1950, **72**, 5163; J. Klein and E. D. Bergman, *J. Org. Chem.*, 1957, **22**, 1019.
53. F. Arndt and B. Eistert, *Ber.*, 1935, **68**, 200; review: W. E. Bachmann and W. S. Struve, *Org. React.*, 1942, **1**, 38.
54. *F. &F.*, **15**, 280; V. Bagheri, M. P. Doyle, J. Taunton, and E. E. Claxton, *J. Org. Chem.*, 1988, **53**, 6158.
55. *F. &F.*, **11**, 459; N. Ikota, N. Takamura, S. D. Young, and B. Ganem, *Tetrahedron Lett.*, 1981, **22**, 4163.
56. M. A. McKervey and P. Ratananukul, *Tetrahedron Lett.*, 1983, **24**, 117; *F. &F.*, **12**, 43; 44.
57. *F. &F.*, **13**, 27; D. J. Buckley and M. A. McKervey, *J. Chem. Soc., Perkin Trans. I*, 1985, 2193.
58. *F. &F.*, **11**, 156; R. Pellicciari, R. Fringuelli, E. Sisani, and M. Curini, *J. Chem. Soc., Perkin Trans. I*, 1981, 2566.

59. *F. &F.*, **11**, 557; I. A. Blair, A. Ellis, D. W. Johnson, and L. N. Mander, *Aust. J. Chem.*, 1978, **31**, 405.
60. L. Lombardo, L. N. Mander, and J. V. Turner, *J. Am. Chem. Soc.*, 1980, **102**, 6628; K. C. Nicolaou and R. E. Zipkin, *Angew. Chem., Int. Ed. Engl.*, 1981, **20**, 785.
61. *F. &F.*, **9**, 133.
62. H. J. Bestmann and F. M. Saliman, *Angew. Chem., Int. Ed. Engl.*, 1979, **18**, 947.
63. *F. &F.*, **10**, 431.
64. D. Seyforth, M. Menzel, A. W. Dow, and T. Flood, *J. Organomet. Chem.*, 1972, **44**, 279.
65. *F. &F.*, **11**, 573; S. Mori, T. Aoyama, and T. Shioiri, *Chem. Pharm. Bull.*, **30**, 3380.
66. *F. &F.*, **12**, 538; M. Martin, *Synth. Commun.*, 1983, **13**, 809.
67. S. Terasawa, K. Sudo, and T. Shioiri, *Chem. Pharm. Bull. Japan*, 1984, 3759.
68. *Chem. Abstr.*, 1982, **97**, 181296x (Japanese Patent).
69. N. Hashimoto, T. Aoyama, and T. Shioiri, *Tetrahedron Lett.*, 1980, **21**, 4619.
70. T. Aoyama and T. Shioiri, *Synthesis*, 1988, 228.
71. E. W. Colvin and B. J. Hamill, *J. Chem. Soc., Perkin Trans. I*, 1977, **8**, 869.
72. *F. &F.*, **15**, 344; T. Aoyama, Y. Iwamoto, S. Nishigaki, and T. Shioiri, *Chem. Pharm. Bull. Japan*, 1989, **37**, 253; T. Aoyamaa and T. Shioiri, *Tetrahedron Lett.*, 1988, **29**, 6295.
73. T. Aoyama and T. Shioiri, *Tetrahedron Lett.*, 1988, **29**, 6295.
74. *F. &F.*, **3**, 74.
75. M. Franck-Neumann, *Tetrahedron Lett.*, 1968, 3451.
76. M. E. Alonso and R. Fernández, *Tetrahedron*, 1989, **45**, 3313.
77. *F. &F.*, **12**, 203; M. G. Martin and B. Ganem, *Tetrahedron Lett.*, 1984, **25**, 251.
78. *F. &F.*, **5**, 114; N. F. Woolsey and M. H. Khalil, *J. Org. Chem.*, 1973, **38**, 4216; 1975, **40**, 3521.
79. L. Thijs, F. L. M. Smeats, P. J. M. Cillisen, J. Harmsen, and B. Zwanenburg, *Tetrahedron*, 1980, **36**, 2141.
80. L. Thijs and B. Zwanenburg, *Tetrahedron*, 1980, **36**, 2145.
81. F. L. M. Smeets, L. Thijs, and B. Zwanenburg, *Tetrahedron*, 1980, **36**, 3269.

14

HYDROXYLAMINES

The compounds discussed in this chapter are derivatives of hydroxylamine (**1**). Aldoximes (**2**) result from condensation with aldehydes and ketoximes (**3**) from ketones. Substitution on the oxygen atom of an oxime yields *O*-alkyloximes (**4**).

Substitution of hydroxylamine (**1**) can occur on the nitrogen atom, yielding *N*-alkylhydroxylamines (**5**) or on the oxygen atom, yielding *O*-alkylhydroxylamines (**6**). Substitution of both the nitrogen atom and the oxygen atom gives *N*-alkyl-*O*-alkylhydroxylamines (**7**). The *N*-acylhydroxylamines or hydroxamic acids (**8**) are derivatives of hydroxylamine which have some importance.

NH_2OH (1)

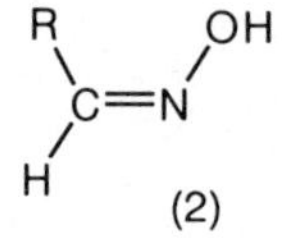

$(R)(H)C{=}N{-}OH$ (2) $(R^1)(R^2)C{=}N{-}OH$ (3) $(R^1)(R^2)C{=}N{-}OR^3$ 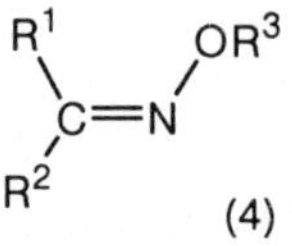 (4)

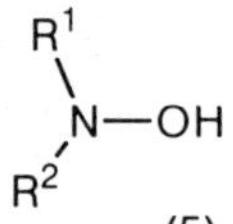

$(R^1)(R^2)N{-}OH$ (5) $(H)(H)N{-}OR$ (6) $(R^1)(R^2)N{-}OR^3$ (7)

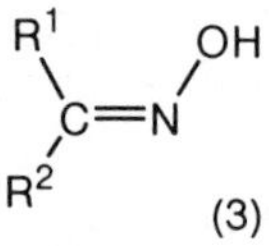

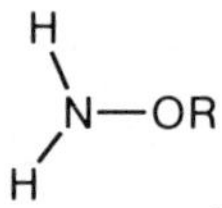

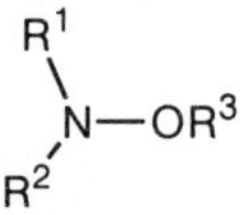

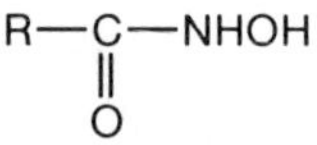

$R{-}C({=}O){-}NHOH$ (8)

OXIMES AND *O*-ALKYLOXIMES

Oximes are the most important derivatives of hydroxylamine. As a result of the rigidity of the carbon–nitrogen double bond (an imino bond), they can occur in two stereoisomeric forms, (**9**, *E*) and (**10**, *Z*). For aldoximes the two isomers can be distinguished by the $^{15}N{-}C{-}H$ coupling constants in the NMR spectra ($J_E = 3$ Hz and $J_Z = 16$ Hz). The infra-red bands of oximes are in the following regions: —OH at c. 3600 cm^{-1} in dilute solution and c. 3250 cm^{-1} in the solid state and C=N at 1620–1670 cm^{-1}.

$$\begin{array}{c} R \\ H \end{array}\!\!>C=N\backslash OH \quad (9,\ E) \qquad \begin{array}{c} R \\ H \end{array}\!\!>C=N/OH \quad (10,\ Z)$$

Oximes are used to characterize aldehydes and ketones and as intermediates in synthesis.

14.1 Preparation

14.1.1 *From hydroxylamine and aldehydes or ketones*

The classical general preparation is from aldehydes and ketones by condensation with hydroxylamine hydrochloride (Scheme 14.1).

$$(CH_3)_3C-\underset{\underset{O}{\|}}{C}-CH_3 + \overset{+}{N}H_3OH\ Cl^- \xrightarrow[\text{aq. EtOH, } 100^\circ C]{NaOAc} (CH_3)_3C-\underset{\underset{NOH}{\|}}{C}-CH_3 \tag{14.1}$$

14.1.2 *By the action of 'nitrous acid' on reactive methylene groups*

Ethyl acetoacetate (**11**) and diethyl malonate both react with sodium nitrite and acetic acid to form oximes (isonitroso derivatives) by attack at the acidic methylene group, for example, Scheme 14.2.[1]

$$\underset{(11)}{CH_3-\underset{\underset{O}{\|}}{C}-CH_2\text{-}CO_2Et} \xrightarrow[\text{aq. HOAc}]{NaNO_2} CH_3-\underset{\underset{O}{\|}}{C}-\underset{\underset{NOH}{\|}}{C}-CO_2Et \tag{14.2}$$

Amyl nitrite and sodium ethoxide, or hydrochloric acid, will react at an α-methylene group of some ketones to give a monoxime of a ketaldehyde or of a diketone, for example, Scheme 14.3.[2]

$$CH_3-\underset{\underset{O}{\|}}{C}-CH_3 \xrightarrow[HCl]{C_5H_{11}O\text{-}N=O} CH_3-\underset{\underset{O}{\|}}{C}-\underset{\underset{NOH}{\|}}{CH} \tag{14.3}$$

14.1.3 *By reduction of allylic nitro compounds with carbon disulphide*

Oximes may be prepared by reducing allylic nitro compounds (**12**) with carbon disulphide and triethylamine in acetonitrile or methylene dichloride (Scheme 14.4).[3]

(12) → CS_2 / NEt_3; CH_3CN or CH_2Cl_2; 0.5-84 hours at 20° C

R^1	R^2	R^3	Solvent	Yield (%)
H	H	Me	MeCN	64
Me	H	H	MeCN	72
Me	H	Me	MeCN	77
H	Bu^t	H	MeCN	85

(14.4)

14.1.4 *From styrenes by reaction with ethyl nitrite*

Oximes have been prepared from styrenes (**13**) in very high yield in a reaction with ethyl nitrite catalysed by a cobalt complex (Scheme 14.5).[4]

$$\text{Ar-CH=CH}_2 + \text{Et-O-N=O} \xrightarrow[\text{Et}_4\text{N}^+\text{BH}_4^- \text{ in C}_6\text{H}_{10}]{\text{Co(DGH)}_2\text{Cl Pyridine}} \text{CH}_3\text{—C(Ar)=NOH}$$

(13)

DGH = the monoanion of dimethylglyoxime: $CH_3—C{=}NO^-$ / $CH_3—C{=}NOH$ (14.5)

14.2 Reactivity

14.2.1 *Reduction*

To amines

Amines are obtained by catalytic reduction of oximes. Several reducing agents are effective, including lithium aluminium hydride, sodium borohydride, or a rhodium–alumina catalyst and hydrogen (Scheme 14.6).[5]

$$\text{cyclohexanone oxime} \xrightarrow[0.7\text{-}1.0\ \text{atm. at}\ 60^\circ\text{C}]{Rh / Al_2O_3 / H_2} \text{cyclohexylamine (80\%)} \qquad (14.6)$$

The reaction of aliphatic ketoximes with diisobutylaluminium hydride (DIBAH) yields secondary amines by combined Beckmann rearrangement and reduction (Scheme 14.7).[6]

$$R^1R^2C{=}N{-}OH \xrightarrow{Bu^i_2AlH} R^1{-}NH{-}CH_2{-}R^2$$

Starting Material	Product	Yield (%)
Ph(Me)C=N–OH	Ph–NH–CH$_2$–Me	92
1-tetralone oxime	2,3,4,5-tetrahydro-1H-1-benzazepine (NH)	92

(14.7)

To N-substituted hydroxylamines

The reduction of aldoximes or ketoximes with diborane in tetrahydrofuran at 25 °C constitutes a good synthesis of *N*-mono-substituted hydroxylamines (**14**)[7] (Scheme 14.8). At higher temperatures in diglyme and tetrahydrofuran the reduction proceeds further to give a primary amine (**15**) in yields of about 70 per cent for a range of oximes. In tetrahydrofuran, *O*-alkyl or *O*-acyl oximes (**16**) are reduced directly to the amine, with yields of 50–90 per cent (Scheme 14.8).

To imines

The reduction of oximes with aqueous titanium trichloride, in aqueous solvents (dioxan, acetic acid, dimethylformamide, or acetone), yields imines (**17**), which can be isolated when they are stable, for example, the imines from cyclohexanone

$$R^1R^2C{=}NOH \xrightarrow[\text{ii, basic hydrolysis}]{\text{i, } B_2H_6 \text{ in diglyme and THF, } 105-110^\circ C \text{ for 20 hours}} R^1R^2CH{-}NH_2 \quad (15)$$

$$R^1R^2C{=}NOH \xrightarrow[\text{ii, aq. 10\% NaOH at } 0^\circ C \text{ then reflux}]{\text{i, } B_2H_6 \text{ in THF at } 25^\circ C} R^1R^2CH{-}NHOH \quad (14)\ 50-70\%$$

$$R^1R^2C{=}NOR^3\ (16) \xrightarrow[50-90\%]{\text{i, } B_2H_6 \text{ in THF, reflux; ii, basic hydrolysis}} R^1R^2CH{-}NH_2\ (15) \qquad (14.8)$$

oxime, 2,6-dimethylcyclohexanone oxime, diisopropyl ketoxime, and the oxime of mesityl oxide were isolated in yields of about 80 per cent. Hydrolysis of the imine *in situ* or subsequently yields aldehydes or ketones (Scheme 14.9).[9]

$$R^1R^2C{=}NOH \xrightarrow[20^\circ C]{\text{aq. } TiCl_3} R^1R^2C{=}NH\ (17) \xrightarrow{H_2O} R^1R^2C{=}O \qquad (14.9)$$

Oximes can also be reduced to imines with *n*-tributylphosphine and diphenyldisulphide. The imine can then be converted by various reagents into other compounds, including the ketone (**18**), the amine (**19**), and the amino nitrile (**20**) (Scheme 14.10).[10]

14.2.2 *Reactivity towards nucleophiles on carbon*

Like imines, oximes contain a polarized carbon–nitrogen double bond and this shows reactivity towards nucleophiles at the carbon atom.

$$R^1R^2C^{\delta+}{=}N^{\delta-}{-}OH \quad \leftarrow Nu^-$$

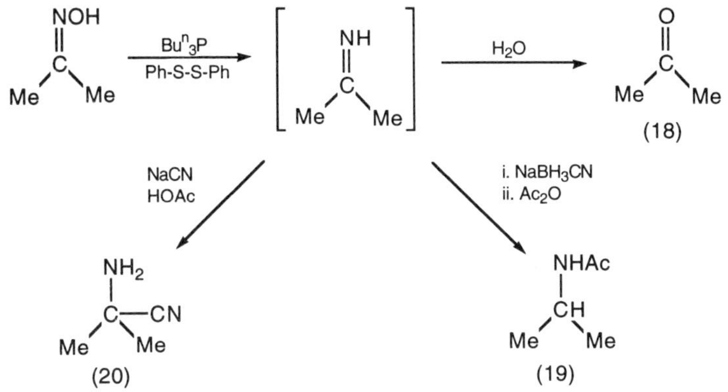

(14.10)

14.2.3 *Conversion to aldehydes and ketones*

Aqueous acid can be used for the hydrolysis of oximes to aldehydes and ketones, but the reaction is slow and the yields are variable.[11] Much better results are obtained by the use of thallium(III) nitrate in methanol at room temperature for a few minutes (Scheme 14.11). This gives ketones in yields of 70 to 95 per cent from a range of aliphatic oximes.[12] There is some e.s.r. evidence for a 1-electron reaction mechanism as well as the one shown in Scheme 14.11.

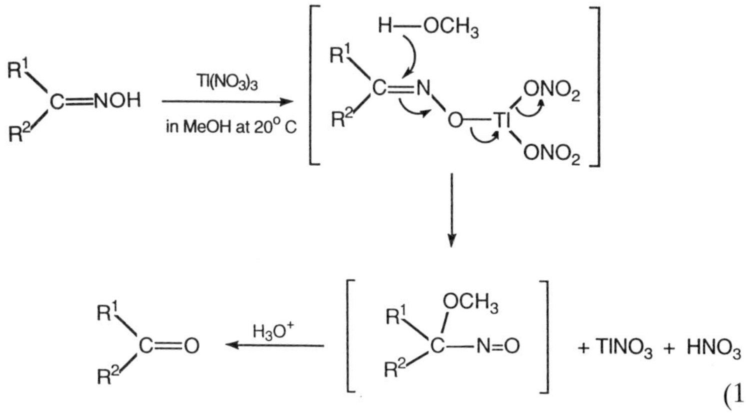

(14.11)

The treatment of oximes with aluminium isopropoxide and then with 2 M hydrochloric acid gives good yields of ketones from ketoximes but less good yields from aldoximes (Scheme 14.12).[13]

The use of a peroxy complex of palladium and triphenyl phosphine (**21**)[14] is also a good method for the generation of some ketones (pinacolone,

$$R^1R^2C{=}NOH \xrightarrow[\text{ii, aq. HCl}]{\text{i, Al(OPr}^i)_3} R^1R^2C{=}O \qquad (14.12)$$

D-camphor, and cyclohexanone) from ketoximes (Scheme 14.13).[15] The mechanism is probably a 1,3-dipolar addition, as shown in Scheme 14.14.

$$(Ph_3P)_4Pd \xrightarrow[\text{in PhH at } 20^\circ\text{ C}]{O_2} (Ph_3P)_2Pd(O_2) \ (21)$$

$$R^1R^2C{=}NOH \xrightarrow[\text{in PhH at } 25^\circ\text{C for a few minutes}]{(21)} R^1R^2C{=}O + 2Ph_3P{=}O + Pd[H] \quad 50-98\% \qquad (14.13)$$

$$R^1R^2C^{\delta+}{=}N^{\delta-}OH \xrightarrow{(21)} [R^2R^1C(-N(OH)-Pd(PPh_3)_2-O-O-)] \longrightarrow R^1R^2C{=}O \qquad (14.14)$$

Another good method for the conversion of ketoximes into ketones, in which chromous diacetate is used, has been developed.[16] The yields are between 74 and 95 per cent for a range of ketoxime acetates, including camphor, cyclohexanone, and progesterone (Scheme 14.15).

$$R^1R^2C{=}NOH \xrightarrow[20^\circ\text{ C}]{Ac_2O} R^1R^2C{=}NOAc \xrightarrow[\text{ii, hydrolysis at pH 5}]{\text{i, >2 equiv } Cr(OAc)_2\text{, THF - } H_2O\text{, 9:1 at 25-65}^\circ\text{ C}} R^1R^2C{=}O \qquad (14.15)$$

Cyclohexanone oxime (**22**) has been converted into cyclohexanone (**23**) in 72 per cent yield, by exchange with acetone at 80 °C in a sealed tube, but there are no reports of other ketones being used (Scheme 14.16).[17]

Good yields of aldehydes and ketones can also be obtained by the oxidation of oximes (and of semicarbazones) with ceric ammonium nitrate (CAN) in methanol, ethanol, or acetonitrile at −20 to 0 °C (Scheme 14.17).[18]

$$\text{cyclohexanone oxime (22)} \xrightarrow[\text{sealed tube at 80}^\circ\text{ C for 100 hours}]{\text{acetone}} \text{cyclohexanone (23)} \quad 72\ \% \qquad (14.16)$$

$$R^1R^2C{=}NOH \xrightarrow[\text{in MeOH, EtOH or } CH_3CN \text{ at } -20^\circ \text{ to } 0^\circ C]{\text{excess of CAN}} R^1R^2C{=}O$$

Oxime	Yield (%)
Cyclohexanone	88
Cyclopentanone	83
Camphor	27
Carvone	71
Cycloheptanone	82
Heptanal	72

(14.17)

There are many other methods for effecting the conversion of oximes to ketones. For instance, ion exchange resins are very efficient in facilitating the recovery of aldehydes and ketones from oximes, hydrazones, and semicarbazones. Exchange with acetone in the presence of Amberlyst 15 gives yields in the range 85–95 per cent.[19] Bispyridinesilver permanganate[20] has been used to convert oximes into aldehydes or ketones in yields of 50–90 per cent.[21]

Ketoximes are converted into ketones by reaction with $Fe_2(CO)_9$ in methanol at 60 °C. Better yields are obtained with acetates of the oximes.[22] Ketones can also be recovered from ketoximes by the action of pyridinium dichromate.[23] Conversions in the range of 80–100 per cent were achieved.

Hydrogenation catalysed by Raney nickel in aqueous tetrahydrofuran and methanol or in aqueous boric acid and methanol gives 70–100 per cent yields of ketones from ketoximes following hydrolysis of an intermediate imine (Scheme 14.18).[24]

$$R^1R^2C{=}NOH \xrightarrow{H_2} \left[R^1R^2C{=}NH\right] \xrightarrow{H_2O} R^1R^2C{=}O$$

70 - 100 % (14.18)

14.2.4 *The Beckmann rearrangement*

Lewis acids (usually PCl_5) can accept electrons from the oxygen atom of the hydroxyl group of a ketoxime and transform it into a very good leaving group thus initiating the Beckmann rearrangement to isomeric amides (Scheme 14.19).[25]

Polyphosphoric acid (PPA)[26] is a good reagent for effecting the Beckmann rearrangement, the yields being in the range 75–90 per cent, for example, Scheme 14.20.

$$R^1R^2C{=}N{-}OH \xrightarrow{PCl_5} \left[R^1(Cl)C{=}NR^2 \right] \longrightarrow R^1C({=}O){-}NHR^2 \qquad (14.19)$$

$$CH_3(CH_3CH_2)C{=}N{-}OH \xrightarrow[\text{for 10 minutes}]{\text{PPA at 120°C}} CH_3{-}C({=}O){-}NHCH_2CH_3 \quad 99\,\%$$

$$Ph_2C{=}NOH \xrightarrow[\text{for 10 minutes}]{\text{PPA at 130° C}} PhNH{-}C({=}O){-}Ph \quad 99\,\% \qquad (14.20)$$

Stereochemistry

The rearrangement is stereochemically *trans*, as shown in Schemes 14.19 and 14.20, and the migrating group usually retains its configuration. However, it has been found, in certain circumstances, that the configuration of the migrating group can be reversed.[27] An alternative mechanism is possible, if the catalyst is a strong acid e.g. (Scheme 14.21).

14.2.5 *Variants of the Beckmann rearrangement*

Sulphonates of oximes (**24**) can be made to undergo a Beckmann type rearrangement by treatment with a trialkylaluminium. The resulting imine (**25**) may be reduced to an amine (**26**) *in situ* with diisobutylaluminium hydride (DIBAH) (Scheme 14.22).[28]

Another variation of this procedure has used mesylates of ketoximes (**27**) with diethylaluminium halides to effect rearrangement with ring expansion. The α-halogeno imine produced may then be arylated and reduced[29] for example, Scheme 14.23.

A further example of this procedure is reported (Scheme 14.24).[30]

Beckmann rearrangements can be induced by Grignard reagents. Sulphonate esters of oximes (**28**) in dry toluene at −78 °C, give imines with simple Grignard reagents, which can be converted into several other compounds either by reduction or with Grignard reagents. (Scheme 14.25).[31]

The Beckmann rearrangement can be effected by stirring the *p*-toluenesulphonate of a ketoxime (**29**) with silica at room temperature (Scheme 14.26).[32]

The Beckmann rearrangement of ketoximes to amides also occurs when the ketoximes are treated with triphenylphosphine and carbon tetrachloride (Scheme 14.27).[33]

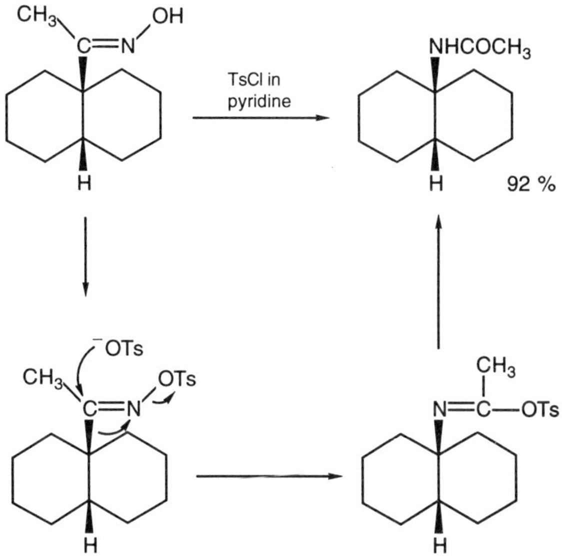

TsCl in pyridine

92 %

In the presence of strong acid:

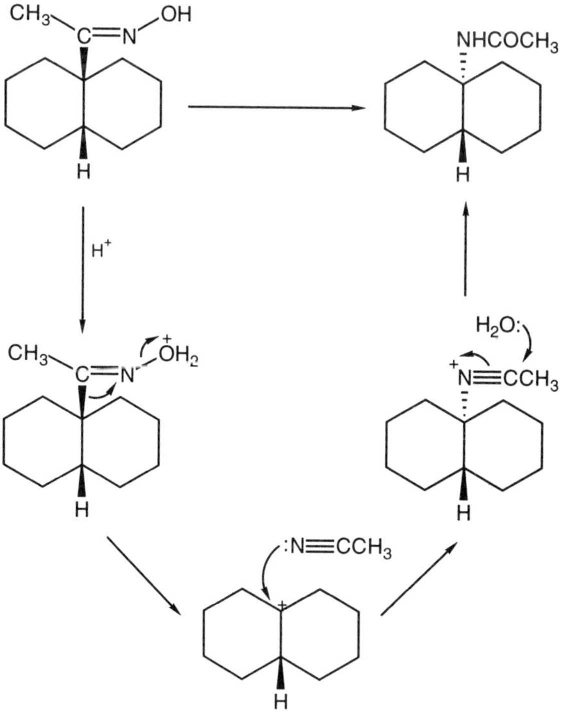

(14.21)

$R^1R^2C{=}N{-}OSO_2R^3$ (24) $\xrightarrow[CH_2Cl_2]{(R^4)_3Al}$ $[R^1{-}\overset{+}{N}{\equiv}CR^2 \longrightarrow R^1{-}N{=}C(R^2)R^4$ (25)$]$ $\xrightarrow{DIBAH}$ $R^1{-}NH{-}CH(R^2)R^4$ (26) 45-85%

Starting material	R^4	Product	Yield (%)
N-OMs	Me	R^4, NH	70
	Pr^n		64
Me, C, N, OMs	$-C{\equiv}C{-}Bu$	R^4, CH, CH_2, Me	67
	Me		83
Me, H, H, N—OTs		Me, H, N, R^4, H, H	57
	Pr^n		60

(14.22)

(27) N-OMs $\xrightarrow[CH_2Cl_2]{Et_2AlI}$ [N=, I] $\xrightarrow[\text{ii, DIBAH}]{\text{i, PHMgBr}}$ N-H, Ph 81% (14.23)

14.2.6 *O-Alkyloximes in radical reactions*

An oxime ether can serve as a radical trap.[34] The intermolecular addition of radicals to *O*-benzylformaldoxime (**30**) has been studied,[35] for example,

Et_2AlCl — silyl enol ether — 95%

H_2, Pd-C

90%

15 : 1

(14.24)

(28)

R^3MgHal — R^4MgX — DIBAH Bu^iAlH

Starting material	RMgX	R^3MgX	Product	Yield(%)
cyclohexanone O-mesyl oxime (N–OMs)	BuMgBr	–	azepane, 2-R (NH)	63
	MeMgI	$CH_2{=}CHCH_2MgBr$	azepane, 2-R, 2-R^3 (NH)	72
	MeMgI	$HC{\equiv}CCH_2MgBr$		66
acetophenone O-mesyl oxime (Ph–C(Me)=N–OMs)	MeMgI	–	Ph–NH–CH(Me)–R	52
	BuMgBr	–		55

(14.25)

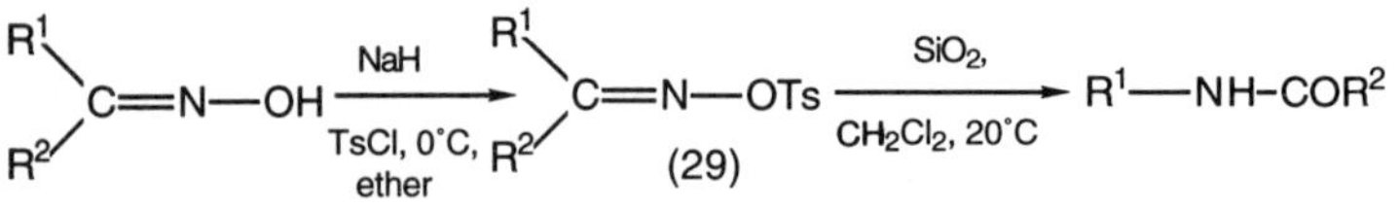

$$R^1R^2C{=}N{-}OH \xrightarrow[\text{TsCl, 0°C, ether}]{\text{NaH}} R^1R^2C{=}N{-}OTs\ (29) \xrightarrow[\text{CH}_2\text{Cl}_2,\ 20°\text{C}]{\text{SiO}_2,} R^1{-}NH{-}COR^2$$

R^1	R^2	Yield (%)
$-(CH_2)_4-$		45
$-(CH_2)_5-$		70
Me	Ph	70
Ph	Ph	86
Me	Pr^i	60

(14.26)

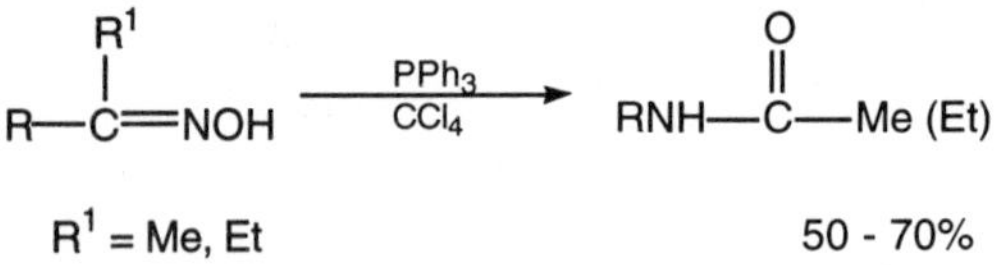

$$R{-}C(R^1){=}NOH \xrightarrow[\text{CCl}_4]{\text{PPh}_3} RNH{-}C(=O){-}Me\ (Et)$$

R^1 = Me, Et 50 - 70% (14.27)

Scheme 14.28. The use of *O*-alkylethers of oximes as reagents in radical cyclization has also been investigated.[36] This is a useful method for conversion of

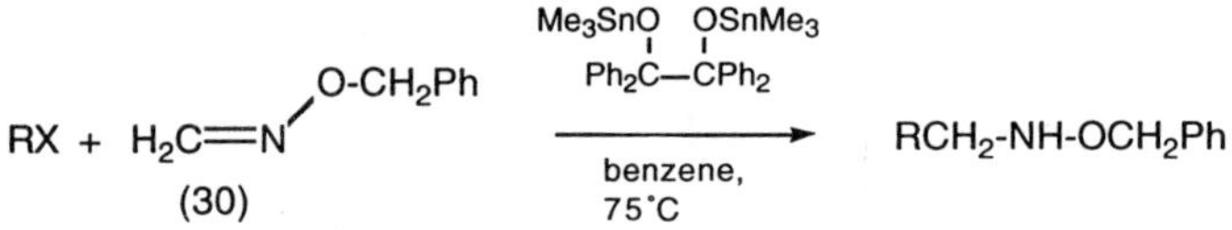

$$RX + H_2C{=}N{-}O{-}CH_2Ph\ (30) \xrightarrow[\text{benzene, 75°C}]{Ph_2C(OSnMe_3){-}C(OSnMe_3)Ph_2} RCH_2{-}NH{-}OCH_2Ph$$

R	X	Yield (%)
n-Oct	I	77
Bu^t	Br	84
Ph	I	67
$c-C_6H_{11}$	I	76
$c-C_6H_{11}$	Br	56
$c-C_6H_{11}$	Ph-Se	78

(14.28)

carbohydrate precursors to carbocycles, for example, Scheme 14.29. When R = CH_2Ph, only two out of the four possible cyclization products are formed, in the yields shown in Scheme 14.29.

14.2.7 *Cleavage*

Ketoximes (**31**) may be cleaved using aqueous bromine with a sodium hydrogen carbonate buffer. Sodium hypobromate may be the actual reagent (Scheme 14.30).[37]

$PhCH_2O\text{-}NH_3^+Cl^-$

pyr, CH_2Cl_2
21°C, 4-8 hrs

PhOC(=S)Cl,
pyr, 21°C,
2-4 hrs

AIBN, $(Bu^n)_3SnH$,
benzene, reflux, 10-14 hrs

93%

62 : 38 (14.29)

$NaHCO_3$, H_2O, CH_2Cl_2, Br_2

$R^1R^2C{=}NOH$ (31) → $R^1R^2C{=}O$

Oxime	Yield of ketone (%)
Ph–C(=NOH)–Me	72
n-C_5H_{11}–C(=NOH)–Me	86
cyclohexanone oxime (=NOH)	80
cycloheptanone oxime (=NOH)	84

(14.30)

Another reagent for the cleavage of oximes is sodium dithionate,[38] for example, Scheme 14.31.

OH
N
C
H
$Na_2S_2O_4$ (aq)
pH7, 20°C
O
C
H
93%

OH
N
O
95%

(14.31)

14.2.8 *Aldoximes under Beckmann conditions*

Unlike ketoximes, aldoximes undergo dehydration to nitriles or to isonitriles on treatment with Lewis acids.[39] Methylketene diethyl acetal $CH_3CH{=}C(OEt)_2$[40] forms adducts (**32**) (60–95 per cent yields) with *syn*-aromatic aldoximes. These adducts decompose in the presence of catalytic amounts boron trifluoride and mercuric oxide, the oxime *E* isomer producing an isonitrile (**33**) and the *Z* isomer giving the nitrile (**34**) (Scheme 14.32). The only aliphatic oxime studied, heptanaldoxime, formed an adduct in 74 per cent yield which decomposed to yield 67 per cent of the nitrile.[41]

14.2.9 *Reduction of O-alkylbenzaldoximes*

The reduction of *O*-alkylbenzaldoximes (**35**) to *O*-alkylhydroxylamines (**36**) can be effected with sodium cyanoborohydride (Scheme 14.33).[42]

14.2.10 *Conversion of oximes into enamides and enimides*

Ketoximes are converted to (**37**) on heating with acetic anhydride–acetyl chloride (Scheme 14.34).[43]

Ketoximes, when boiled under reflux with acetic anhydride in pyridine for 10 h, give enimides (**38**) in yields of 50–93 per cent for a range of ketoximes, including butan-2-one, cyclohexanones, and steroidal ketones. If the reaction mixture is worked up on a column of alumina, enamides (**39**) are obtained.

(32) *E* adduct

$BF_3.HgO$

$:C{=}NR + CH_3CH_2CO_2Et$

isonitrile (33)

(32) *Z* adduct

$BF_3.HgO$

$RC{\equiv}N$ nitrile (34)

R	Isomer	Yield adduct (%)	Yield isonitrile (%)	Yield nitrile (%)
benzene	*E*	83	50	45
p-toluene	*E*	95	73	23
p-nitrobenzene	*E*	52	70	—
benzene	*Z*	61	—	60
p-toluene	*Z*	61	—	80
heptan-	*Z*	74	—	67

(14.32)

(35) $\xrightarrow[\text{CH}_3\text{OH, pH3}]{\text{NaBH}_3\text{CN}}$ (36) 35-65%

R^1	R^2	Configuration	Yield (%)
H	Me	*E*	58
H	Me	*Z*	56
H	Bu^t	*E*	67
H	Bn	*E*	51
Pr^i	Me	*E*	64
Cl	Me	*E*	32

(14.33)

$\xrightarrow[\text{Py, 100°C}]{\text{Ac}_2\text{O, AcCl}}$ (37) 55%

(14.34)

Alternatively, enamides can be prepared directly from the oximes by the action of acetic anhydride in dimethylformamide followed by reduction with titanium triacetate (Scheme 14.35).[44] A free radical reductive mechanism for the action of titanium triacetate has been suggested (Scheme 14.36).

R^1—CH_2(R²)C=N—OH

Ac_2O, Py reflux, 10hrs → R^1–CH=C(R^2)–N($COCH_3$)—$COCH_3$ enimide (38)

enimide (38) ⇄ enamide (39): Ac_2O, Py; NaOMe Al_2O_3

i, Ac_2O, DMF, 0°C ii, aq. $TiCl_3$ + NaOAc -> $Ti(OAc)_3$, 20°C → R^1–CH=C(R^2)–$NHCOCH_3$

enamide (39)

80% with cyclohexanone oxime (14.35)

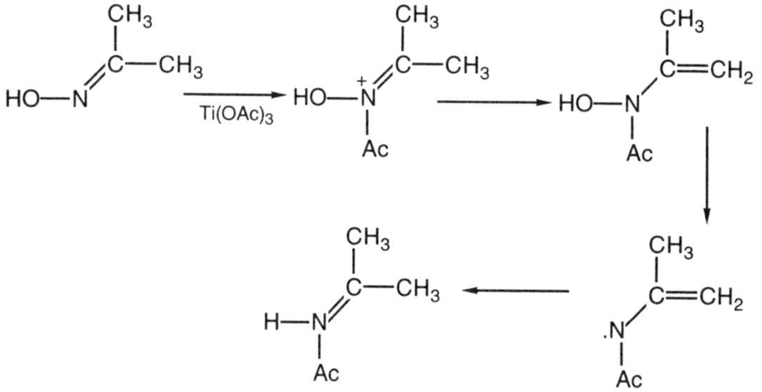

(14.36)

14.2.11 *Synthesis of benzofurans*

Acetone oxime, functioning as a nucleophile, reacts with aryl halides complexed by chromium tricarbonyl [an arene/tricarbonyl chromium complex] (**40**) under phase transfer conditions to yield *O*-aryloximes (**41**) after liberation from the chromium by iodine. *O*-Aryloximes may be converted into benzofurans (**42**) on being heated with ethanolic sulphuric acid (Scheme 14.37).[45]

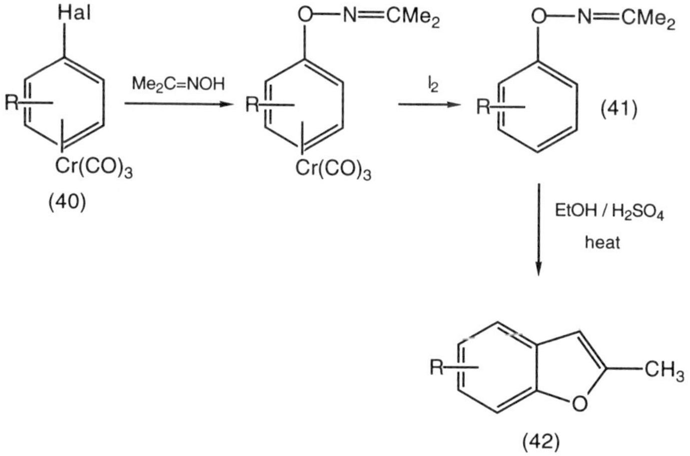

R	Hal	Yield (%)
H	F	78
o-Me	F	56
m-Me	F	55
p-Me	F	81
m-Me	Cl	74

(14.37)

14.2.12 *Synthesis of pyrroles*

The reaction of ketoximes with acetylene in potassium hydroxide/dimethyl sulphoxide 'superbase' forms pyrroles (**43**) and *N*-vinyl pyrroles (**44**) with yields of 70–80 per cent,[46] for example, Scheme 14.38.

$H_2C{=}CH{-}CH_2{-}CH_2{-}C(Me){=}N{\sim}OH$ + $HC{\equiv}CH$ → (KOH/DMSO, 100°C, 50-100 psig) → (43) 30%

$H_2C{=}CH{-}CH_2{-}CH_2{-}C(Me){=}N{\sim}OH$ → (KOH/DMSO, 110°C, 200 psig) → (44) 80%, *E*:*Z* 7 : 3

(14.38)

(45) + NH_2OH → (46) *c.* 85%

(46) → (i, Bu^nLi/THF, -78°C; ii, Bu^nI) → $CH_3(CH_2)_4$-isoxazole-CH_3

→ (i, Bu^sLi/Et_2O, -78°C; ii, CH_3I) → $CH_3(CH_2)_4$-isoxazole-CH_2CH_3

→ (H_3O^+/EtOH) → $CH_3(CH_2)_4COCH_2COCH_2CH_3$ 75-90%

(14.39)

14.2.13 *Alkylation of β-diketones*

β-Diketones (**45**) may react with hydroxylamine to form isoxazoles (**46**). Subsequent alkylation of the alkyl substituents of the isoxazole (**46**) by reaction with a strong base and alkyl halides, followed by hydrolysis of the isoxazole, provides a path for the dialkylation of β-diketones is the least reactive positions (Scheme 14.39).[47]

14.2.14 *1,3-Dipolar addition of silyl oximes with alkenes and alkynes*

Silyl oximes (**47**) undergo 1,3-dipolar additions with alkenes and alkynes, reacting differently with each of these classes of compounds (Scheme 14.40).[48]

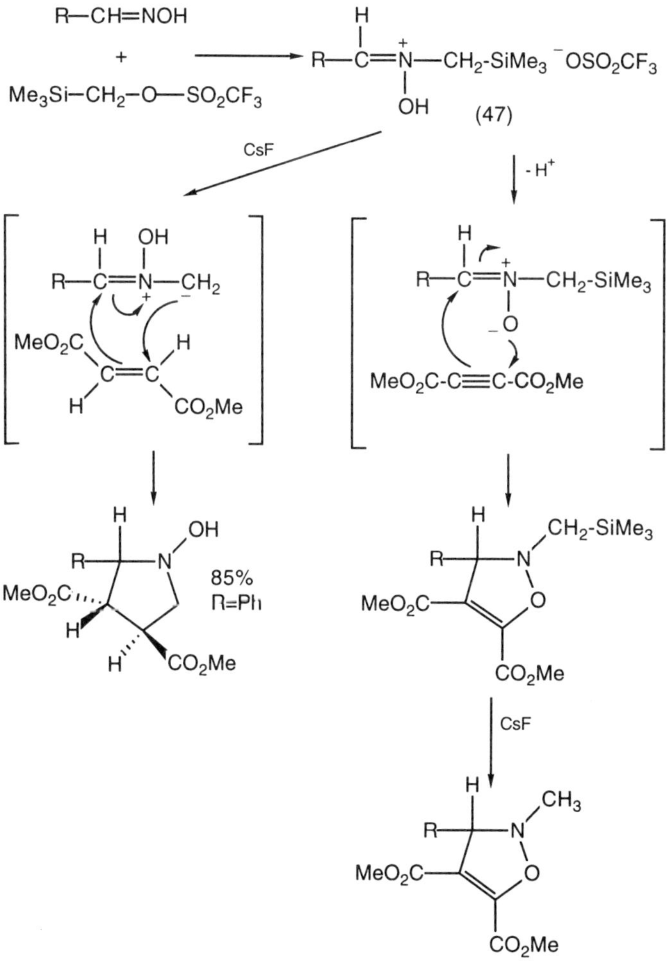

(14.40)

14.2.15 *Reaction of O-methylaldoximes with Grignard reagents*

Alkyllithiums or Grignard reagents (2 mole) react with *O*-methylaldoximes (**48**) to give ketones in good yield (Scheme 14.41).[49]

(48)

O-Alkyl oxime	Organometal	Product	Yield (%)
Ph-CH=N-OMe	EtLi	PhCOEt	84
Ph-CH=N-OMe	EtMgBr	PhCOEt	57
C_5H_{11}-CH=N-OMe	BuLi	C_5H_{11}COBu	54
C_6H_{11}-CH=N-OMe	PhMgBr	C_5H_{11}COPh	37

(14.41)

N-ALKYLHYDROXYLAMINES

14.3 Preparation of *N*-monoalkylhydroxylamines

14.3.1 *From nitroalkanes and nitroalkenes*

Preparation may be by reduction of nitroalkanes and nitroalkenes (Scheme 14.42, Table 14.1).

Nitroalkenes have been reduced to *N*-alkylhydroxylamines in low yields by the action of hydrogen over palladium charcoal in ethanol,[53] for example, Scheme 14.43.

By controlled reduction with lithium aluminium hydride, *N*-monoalkylhydroxylamines (**49**) have been prepared,[54] for example, Scheme 14.44.

The reduction of nitroalkenes to *N*-alkylhydroxlamines with boron trifluoride and tetrahydrofuran and a catalytic amount of sodium borohydride, or with sodium borohydride and boron trifluoride-etherate in tetrahydrofuran has been described,[55] for example, Scheme 14.45. With extended times of reaction, amines are produced.[56]

$$\text{R-NO}_2 \xrightarrow{\text{Reducing Agent}} \text{R-NHOH} \qquad (14.42)$$

Table 14.1 Preparation of *N*-monoalkylhydroxylamines from nitroalkanes and nitroalkenes

Reducing agent	Conditions	R	Yield (%)	Ref.
H_2, Pd/$BaSO_4$	Reduction of nitroalcohols in EtOH or AcOH		69–98	50
BH_3–THF	Reduction of	c-C_6H_{11}	56	51
>10 °C	Li, K, or NH_4 salt of nitroalkane	$H_3C-CH_2-CH(CH_3)$	49.4	
		c-$C_6H_{11}-CH_2$	62.4	
Al/Hg				52

$$\text{2,5-(MeO)}_2\text{-4-Me-C}_6\text{H}_2\text{-CH=C(Me)-NO}_2 \xrightarrow{\text{H}_2\text{/Pd-C, EtOH}} \text{2,5-(MeO)}_2\text{-4-Me-C}_6\text{H}_2\text{-CH}_2\text{CH(Me)-NHOH} \quad 12\% \qquad (14.43)$$

$$\text{Ph—CH=C(NO}_2\text{)—CH}_3 \xrightarrow[\text{ii, sodium potassium-tartrate (aq), 15°C}]{\text{i, LiAlH}_4\text{, -40°C}} \text{Ph—CH}_2\text{—CH(NHOH)—CH}_3 \ (49) \quad 23\% \qquad (14.44)$$

$$\text{Ar-CH=C(NO}_2\text{)-CH}_3 \xrightarrow{\text{BH}_3\text{, THF; NaBH}_4\text{, THF}} \text{Ar-CH}_2\text{-CH(NHOH)-CH}_3 \quad 70\text{-}85\ \% \qquad (14.45)$$

14.3.2 *From oximes*

A variety of reducing agents have been used to prepare *N*-monoalkylhydroxylamines from oximes (Scheme 14.46, Table 14.2).

$$\begin{matrix} R^1 \\ R^2 \end{matrix}\!\!>C=N-OH \xrightarrow{\text{Reducing agent}} \begin{matrix} R^1 \\ R^2 \end{matrix}\!\!>CH-NH-OH \qquad (14.46)$$

Table 14.2 Preparation of *N*-monoalkylhydroxylamines from oximes

Agent	Conditions	R^1, R^2	Yield (%)	Ref.
H_2/Pt	Aq.EtOH-HCl	*p*-MeO—C_6H_4—CH_2, Me	69	57
		m,p-$(MeO)_2$—C_6H_3—CH_2, Me	33	
		p-Cl—C_6H_4—CH_2, Me	78	
		p-Me—C_6H_4—CH_2, Me	74	
B_2H_6	i, B_2H_6–THF	H, H	50	58
	ii, Acid or base	Pr^n, H	71	
	hydrolysis, Acid, not	Et, Me	91	
	base, for aryl oximes	Ph, H	52.1	
		Ph, Me	54.8	
$NaBH_4$	On SiO_2 gel, in benzene	5-α-Cholestan-3-one oxime		59
$NaBH_3CN$	In MeOH	Et, Me	73	60
	pH 3, 1 h, aldoximes	Ph, Me	75	
	pH 4, 3 h, ketoximes	Ph, H	79	
		n-C_6H_{13}, H	53	

Reduction with sodium cyanoborohydride is the most suitable method for the reduction of oximes to *N*-monoalkylhydroxylamines. Ketoximes are reduced smoothly at pH 4 to monoalkylhydroxylamines, however reduction of aldoximes under these conditions forms mainly the corresponding *N,N*-dialkyl hydroxylamines. Aldoximes are reduced to monoalkylhydroxylamines if the pH is reduced to 3.

14.3.3 *From amines*

Careful oxidation of amines with hydrogen peroxide may yield *N*-monoalkylhydroxylamines. T. Kawaguchi *et al.*[61] have oxidized cyclohexylamine (**50**) in this manner (Scheme 14.47).

$$\text{C}_6\text{H}_{11}\text{NH}_2 \ (50) \xrightarrow[\text{silicotungstic acid or its salt acts as a catalyst}]{H_2O_2} \text{C}_6\text{H}_{11}\text{NHOH} \quad (14.47)$$

14.3.4 *From amides*

Iminoethers (**51**) can be prepared by the alkylation of amides with trimethyloxy fluoroborate (Meerweins Reagent).[62] Treatment of iminoethers with peracids gives oxiranes (**52**) which on reaction with aqueous acid yield esters and *N*-alkylhydroxylamines (Scheme 14.48).[63]

$$R^1\text{—NH—C(=O)—}R^2 \xrightarrow{Me_3O^+ BF_4^-} R^1\text{—N=C(OCH}_3\text{)—}R^2 \ (51)$$

$$(51) \xrightarrow[\text{in } CH_2Cl_2, -20^\circ C]{m\text{-}ClC_6H_4CO_3H} R^1\text{—N—C(OCH}_3\text{)—}R^2 \text{ (oxaziridine, N—O—C ring)} \ (52)$$

$$(52) \xrightarrow{H_3O^+} H_2O: \rightarrow R^1\text{—}\overset{+}{N}H\text{—C(OCH}_3\text{)—}R^2 \text{ (O-bridged)} \longrightarrow R^1\text{-NHOH} + R^2\text{-CO}_2CH_3$$

R^1=Bu^t, R_2=H, yield 86%; R^1=R_2=Bu^t, yield 88%. (14.48)

14.4 Reactivity

14.4.1 *Synthesis of imidazoles*

N-Alkylhydroxylamine (**54**) has been used for the synthesis of imidazoles (**55**), by means of a condensation reaction with aryl halides (Scheme 14.49).[64]

(53) + $HO{-}NH{-}CH_2R^2$ (54)

KCN, H_2O, EtOH

- H^+

(55)

$R^2 = H$,
40 - 90 %

(14.49)

14.4.2 *With allyl esters*

N-Alkylhydroxylamines undergo a catalysed reaction with allyl esters (**56**) to yield *N*-allylhydroxylamines (**57**) which can be reduced to yield *N*-allylamines (**58**) (Scheme 14.50).[65]

OAc; Ph–CH–CH=CH–CH$_3$ (56) + H$_3$C—NHOH; Pd(PPh$_3$)$_4$, THF; Ph–CH=CH–CH(CH$_3$)–N(OH)–CH$_3$ (57) *c.* 100%; Zn, HCl$_{(aq)}$; Ph–CH=CH–CH(CH$_3$)–NH-CH$_3$ (58) 93%

(14.50)

14.4.3 *Synthesis of a tricyclic system*

The reaction of *N*-alkylhydroxylamines with cyclohexanones which contain a suitable unsaturated side chain (**59**) has been applied to the synthesis of tricyclic system (**61**). The reaction occurs through a nitrone (**60**) which reacts in

(59); PhCH$_2$-NHOH; 80°C, 3hrs; azeotropic removal of water; PhCH$_2$; N; O$^-$; CH$_2$; CH; CH$_2$ (60); CH$_2$Ph; N; O; 100%; (61)

(14.51)

a 1,3-dipolar addition with a suitably placed carbon–carbon double bond (Scheme 14.51).[66]

N,N-DIALKYLHYDROXYLAMINES

14.5 Preparation

14.5.1 *By alkylation of hydroxylamine*

Unsatisfactorily low yields of *N*,*N*-dialkylhydroxylamine result from the alkylation of hydroxylamine with alkyl halides.[67] This method suffers from the same disadvantages as the reaction between ammonia and alkyl halides which yields a mixture of primary, secondary, and tertiary amines along with quaternary salts, the amounts of each depending upon the ratio of the two reactants used.

14.5.2 *By reduction*

Of N-oxides

N,*N*-Dialkylhydroxylamines (**62**) can be obtained by controlled catalytic reduction[68] or by reduction with lithium aluminium hydride of the corresponding *N*-oxide (**63**) (Scheme 14.52).[69]

$$R^1R^2C{=}\overset{+}{N}(O^-){-}R^3 \xrightarrow{LiAlH_4} R^1R^2CH{-}N(OH){-}R^3$$

(63) (62)

R^1	R^2	R^3	Yield (%)
H	C_6H_5	C_6H_5-CH_2	85
C_6H_5	C_6H_5	C_6H_5-CH_2	90
H	C_6H_5	CH_3	94
C_6H_5	C_6H_5	CH_3	91

(14.52)

Of oximes

The reduction of aldoximes with sodium cyanoborohydride in methanol at pH 4 gives *N*,*N*-dialkylhydroxylamines (**64**) by condensation of the first formed *N*-alkylhydroxylamine (**65**) with the original oxime, elimination of hydroxylamine, and further reduction (Scheme 14.53).

$$\text{R—CH=N—OH} \xrightarrow[\text{pH4}]{\text{NaBH}_3\text{CN}} \text{R-CH}_2\text{—N(H)—OH}$$

$$\text{MeOH (65)} \quad \text{R-CH=N-OH} \quad \text{H}^+$$

$$\downarrow$$

$$\text{R-CH}_2\text{—}\overset{+}{\text{N}}(\text{OH})\text{=CHR} + \text{NH}_2\text{OH} \longleftarrow \text{R-CH}_2\text{—N(OH)—CH(R)NHOH} \quad \text{H}^+$$

$$\downarrow \text{NaBH}_3\text{CN}$$

$$\underset{(64)}{\text{R-CH}_2\text{—N(OH)—CH}_2\text{-R}} \qquad (14.53)$$

The reduction of oximes, or of nitrones of imines, with sodium cyanoborohydride at pH 5–8, yields both *N*-substituted hydroxylamines (**66**) and *N*,*N*-dialkylhydroxylamines (**67**). It is not necessary to isolate the oximes or the nitrones (Scheme 14.54).[70]

14.5.3 *By oxidation of secondary amines*

Oxidation of secondary amines with benzoyl peroxide followed by hydrolysis of the resulting *N*-benzoyl ester with sodium ethoxide yields *N*,*N*dialkylhydroxylamines (Scheme 14.55).[71]

However, alkali can cause the decomposition of some products. A superior preparation therefore, from secondary amines, is one in which the oxidation is carried out with *m*-chloroperbenzoic acid followed by reduction with lithium aluminium hydride (Scheme 14.56).[72]

N,*N*-Dialkylhydroxylamines may be prepared, in increased yields, from amines by the action of benzoyl peroxide in the presence of sodium hydrogen phosphate, followed by hydrolysis with potassium methoxide in methanol (Scheme 14.57).[73]

$$R^1R^2C{=}O + NH_2OH \longrightarrow R^1R^2C{=}NOH \xrightarrow[\text{pH 6}]{NaBH_3CN} R^1R^2CH\text{-}NHOH \quad (66)$$

R^1 = Me R^2 = CH_2-3,4-dimethoxyphenyl (3,4-(OMe)$_2C_6H_3CH_2$) Yield = 49%

$$R^1R^2C{=}O + R^3\text{-}NHOH \longrightarrow R^1R^2C(OH)\text{-}\ddot{N}(OH)\text{-}R^3 \longrightarrow R^1R^2C{=}\overset{+}{N}(O^-)\text{-}R^3 \xrightarrow[\text{pH6}]{NaBH_3CN} R^1R^2CH\text{-}N(OH)\text{-}R^3 \quad (67)$$

R^1	R^2	R^3	Yield (%)
H	H	CH_2-3,4-(OMe)$_2C_6H_3$	51
Ph	H	CH_2-3,4-(OMe)$_2C_6H_3$	36

(14.54)

$$R_2NH \xrightarrow{(PhCO_2)_2} R_2N\text{-}O\text{-}C(=O)\text{-}Ph \xrightarrow{EtONa} R_2N\text{-}OH$$

R = $PhCH_2$ 57% 56.5% (14.55)

$$R^1R^2N\text{-}H \xrightarrow{3\text{-}ClC_6H_4CO_3H} R^1R^2N\text{-}O\text{-}C(=O)\text{-}C_6H_4\text{-}3\text{-}Cl \xrightarrow{LiAlH_4} R^1R^2N\text{-}OH$$

R^1	R^2	Yield (%)
$C_6H_5CH_2CH(Me)$	H	12
	Ph	11

(14.56)

$$R_2NH + (PhCO_2)_2 \xrightarrow[\text{THF}]{Na_2HPO_4} R_2N{-}O{-}C(=O){-}C_6H_5 \xrightarrow[\text{MeOH}]{\text{MeOK}} R_2N{-}OH$$

Amine	Yield (%)
R=Ph-CH_2	90
R=n-C_4H_9	69
piperidine (NH)	38
morpholine (O, NH)	50

(14.57)

14.5.4 *By pyrolysis of amino N-oxides*

N,N-dialkylhydroxylamines can also be prepared by the pyrolysis of amine *N*-oxides (**68**) (Scheme 14.58).[74]

$$R^1{-}CH(H){-}CH_2{-}\overset{+}{N}(R^2)(R^3){-}O^- \ \textbf{(68)} \xrightarrow{\text{heat}} R^2R^3N{-}OH + R^1{-}CH{=}CH_2$$

R^1	R^2	R^3	Yield (%)
Me	Et	Et	67
Me	Et	Me	82
Et	Pr^n	Pr^n	50–79

(14.58)

O-ALKYLHYDROXYLAMINES

14.6 Preparation

14.6.1 *By Gabriel type reaction*

O-Alkylhydroxylamines (**69**) are usually prepared by Gabriel-type alkylations using *N*-hydroxyphthalimide (**70**) (Scheme 14.59) or *N*-hydroxynorbornene-5-dicarboximide-2,3.[75]

O
N—OH
O
(70)
R-Hal
O
N—OR
O
98% H_2NNH_2
O
NH
NH
O
$R—ONH_2$ +
(69)

(14.59)

O
N-OH
O
+
$Cl—CH_2—CH=CH_2$
K_2CO_3
DMSO
O
$N-O-CH_2-CH=CH_2$
O
H_2N-NH_2
$H_2N-O-CH_2-CH=CH_2$
(71)
O
$CH_2=CH$
CH_2
O
N
heat, rearrangement
H_2C
CH
CH_2
+
N
O^-
- H_2O
H_2C
CH
CH_2
N
heat, O_2
O_2
N
(72)

(14.60)

14.7 Reactivity

14.7.1 *Synthesis of 2,3-disubstituted pyridines*

An *O*-allylhydroxylamine (**71**) synthesized by the Gabriel-type method, has been utilized in the synthesis of 2,3-disubstituted pyridines (**72**),[76] for example, Scheme 14.60.

14.7.2 *Synthesis of 1,3-amino alcohols from O-benzylhydroxylamine*

O-Benzylhydroxylamine reacts with β-hydroxy ketones (**73**) to yield mixtures of *syn*- and *anti*-benzyloximes (**74**), which are separable by chromatography (Scheme 14.61).[77] The *syn*-oximes are reduced, by lithium aluminium hydride, with high asymmetric induction, to *syn*-1,3 amino-alcohols (**75**). Reduction of the *anti*-oximes shows only moderate selectivity for the *syn*-amino alcohol.

$$R^1CH(OH)CH_2C(O)R^2 \ (73) \xrightarrow{BnO\text{-}NH_2} R^1CH(OH)CH_2C(=NOBn)R^2 + R^1CH(OH)CH_2C(=NOBn)R^2 \ (74) \quad 52:48$$

$$\text{first isomer} \xrightarrow{LiAlH_4,\ 87\%} R^1CH(OH)CH_2CH(NH_2)R^2 \ (75) \quad syn:anti = 95:5$$

$$\text{second isomer} \xrightarrow{LiAlH_4,\ 96\%} syn:anti = 77:23$$

$R^1 = R^2 = Bu^n$

(14.61)

N,*O*-DIALKYLHYDROXYLAMINES

14.8 Preparation

14.8.1 *By reduction of O-alkyloximes*

N,*O*-substituted hydroxylamines (**76**) can be prepared from *O*-substituted oximes (**77**) by reduction with dimethylphenylsilane (Scheme 14.62).[78]

$$\text{BnO-N=C(R}^1\text{)(R}^2\text{)} \ \textbf{(77)} \xrightarrow[\text{TFA}]{\text{PhMe}_2\text{SiH}} \text{BnO-NH-CH(R}^1\text{)(R}^2\text{)} \ \textbf{(76)} \quad c.\ 70\% \tag{14.62}$$

14.8.2 *By reduction of O-acyloximes*

The reduction of *O*-acyloximes (**78**) to *O*-acylhydroxylamines (**79**) with several different reagents has been investigated; with $NaBH_3CN$ in HOAc and Et_3SiH in F_3CCO_2H, the opposite stereochemistry is produced in the product (Scheme 14.63).[79]

(78) → (A) + (B) (79)

	Yield (%)	(A)	(B)
$NaBH_3CN$/HOAc	90	1	4
$Et_3SiH/F_3C.CO_2H$	85	5	1

(14.63)

Equatorial attack of the hydride is made favourable in the silicon hydride reduction because the benzoyl carbonyl participates in the transition state as an intramolecular nucleophile.

The reduction of *O*-acyloximes (**80**) to *O*-acylhydroxylamines (**81**) has also been carried out using pyridine–boranes in acid (Scheme 14.64).[80]

$$R^1R^2C{=}N{-}O{-}COR^3 \ (80) \xrightarrow{\text{py-}B_2H_6} R^1R^2CH{-}NH{-}O{-}COR^3 \ (81)$$

R^1	R^2	R^3	Yield (%)
Ph	H	Me	69
$-(CH_2)_5-$		Me	74
$-(CH_2)_5-$		Ph	87
Ph	Ph	Me	95

(14.64)

14.9 Reactivity

14.9.1 *Formation of aldehydes and ketones*

N,*O*-Dimethylhydroxylamine (**82**) reacts with α,β-unsaturated acid chlorides to give amides (**83**), which may be reduced by lithium aluminium hydride or DIBAH, forming aldehydes (**84**) almost exclusively. The amides also react with Grignard and alkyllithium reagents to form ketones (**85**) in generally high yield,[81] for example, Scheme 14.65.

$$PhHC{=}CH{-}COCl \xrightarrow[\text{MeOH}]{\text{MeO-NH-Me (82)}} PhHC{=}CH{-}C(O){-}N(OMe)Me \ (83) \xrightarrow{\text{DIBAH}} PhHC{=}CH{-}CHO \ (84)\ 76\% + PhHC{=}CH{-}CH_2OH \ 3\%$$

$$Ph{-}C(O){-}N(OMe)Me \ (83) \xrightarrow[\text{PhC}\equiv\text{CLi 20°C, 92\%}]{\text{PhC}\equiv\text{CMgBr 65°C, 92\%}} Ph{-}C(O){-}C{\equiv}C{-}Ph \ (85)$$

(14.65)

N-ACYLHYDROXYLAMINES (HYDROXAMIC ACIDS)

In the presence of polyphosphoric acid (**PPA**), hydroxamic acid (**86**)[82] attacks the *para*-position of a phenolic ether[83] (Scheme 14.66).

$CH_3CONHOH$ (86) $\xrightarrow{H^+}$ $H_2\overset{+}{O}$—N($\overset{+}{}$)(H)—$COCH_3$ + :OR–C$_6$H$_5$ $\xrightarrow{PPA}$ cyclohexadienyl cation ($\overset{+}{O}R$, H, $NHCOCH_3$) $\xrightarrow{-H^+}$ p-RO–C$_6$H$_4$–NH–C(=O)–CH_3

Yields are *c.* 50%, R = Me, Et (14.66)

References

1. *F. &F.*, **1**, 1097.
2. *F. &F.*, **1**, 1098; L. Claisen and O. Manasse, *Ber.*, 1889, **22**, 526.
3. D. H. R. Barton, I. Fernandez, C. S. Richard, and S. Z. Zard, *Tetrahedron*, 1987, **43**, 551.
4. G. N. Schrauzer, *Inorg. Synth.*, 1968, **11**, 61; *F. & F.*, **15**, 156; T. Okamoto, K. Kobayashi, S. Oka, and S. Tanimoto, *J. Org. Chem.*, 1988, **53**, 5089.
5. M. Friedfelder, W. D. Smart, and G. R. Stone, *J. Org. Chem.*, 1962, **27**, 2209.
6. *F. &F.*, **12**, 191; S. Satatani, T. Miyazaki, K. Marouka, and H. Yamamoto, *Tetrahedron Lett.*, 1983, **23**, 4711.
7. *F. &F.*, **3**, 76; H. Feuer, B. F. Vincent, Jr, and R. S. Bartlett, *J. Org. Chem.*, 1965, **30**, 2877.

8. H. Feuer and D. M. Braunstein, *J. Org. Chem.*, 1969, **34**, 1817.
9. *F. &F.*, **4**, 506; G. H. Timms and E. Wildsmith, *Tetrahedron Lett.*, 1971, 195.
10. *F. &F.*, **12**, 514; D. H. R. Barton, W. B. Motherwell, E. S. Simon, and S. Z. Zard, *J. Chem. Soc., Chem. Commun.*, 1984, 337.
11. E. J. Corey and J. E. Richman, *J. Am. Chem. Soc.*, 1970, **92**, 5276.
12. *F. &F.*, **4**, 495; A. McKillop, J. D. Hunt, R. D. Naylor, and E. C. Taylor, *J. Am. Chem. Soc.*, 1971, **93**, 4918.
13. *F. &F.*, **5**, 14; J. K. Sugden, *Chem. &Ind.*, 1972, 680.
14. G. Wilke, H. Schott, and P. Heimbach, *Angew. Chem., Int. Ed. Engl.*, 1967, **6**, 92.
15. *F. &F.*, **5**, 510; K. Maeda, I. Moritani, T. Hosokawa, and S.-I. Murahashi, *Tetrahedron Lett.*, 1974, 797.
16. E. J. Corey and J. E. Richman, *J. Am. Chem. Soc.*, 1970, **92**, 5267.
17. *F. &F.*, **6**, 9; S. R. Maynez, L. Pelavin, and G. Erker, *J. Org. Chem.*, 1975, **40**, 3302.
18. *F. &F.*, **3**, 45; J. W. Bird and D. G. M. Diaper, *Can. J. Chem.*, 1969, **47**, 145.
19. *F. &F.*, **15**, 178; R. Ballini and M. Petrini, *J. Chem. Soc., Perkin Trans. I*, 1988, 2563.
20. *F. &F.*, **11**, 61.
21. *F. &F.*, **12**, 62; H. Firouzabadi and A. Sardarian, *Synth. Commun.*, 1983, **13**, 863.
22. *F. &F.*, **12**, 526; M. Nitta, I. Sasaki, H. Miyano, and T. Kobayashi, *Bull. Soc. Chem. Japan*, 1984, **57**, 3357.
23. *F. &F.*, **11**, 453; S. Satish and N. Kalyanam, *Chem. Ind.*, 1981, 809.
24. *F. &F.*, **12**, 422; M. Gaudemar, *Tetrahedron Lett.*, 1983, **24**, 2749.
25. B. Jones, *Chem. Rev.*, 1944, **35**, 335; J. March, *Advanced organic chemistry* 3rd edn, Wiley-Interscience, Chichester, 1985, p. 987.
26. *F. &F.*, **1**, 900; E. C. Horning and V. L. Stromberg, *J. Am. Chem. Soc.*, 1952, **74**, 2680.
27. R. K. Hill and O. T. Chortyk, *J. Am. Chem. Soc.*, 1962, **84**, 1064.
28. *F. & F.*, **11**, 539; K. Hattori, Y. Matsumura, T. Miyazaki, K. Maruoka, and H. Yamamoto, *J. Am. Chem. Soc.*, 1981, **103**, 7368.
29. *F. &F.*, **12**, 7; Y. Ishida, S. Sasatani, K. Maruoka, and H. Yamamoto, *Tetrahedron Lett.*, 1983, **24**, 3255.
30. Y. M. Matsumura, J. Fuiwara, K. Maruoka, and H. Yamamoto, *J. Am. Chem. Soc.*, 1983, **105**, 6312.
31. *F. &F.*, **11**, 245; K. Hattori, M. Maruoka, and H. Yamamoto, *Tetrahedron Lett.*, 1982, **23**, 3395.
32. *F. &F.*, **11**, 466; A. Costa, R. Mestres, and J. M. Riego, *Synth. Commun.*, 1982, **12**, 1003.
33. *F. &F.*, **5**, 727; R. M. Walters, N. Wakabayashi, and E. S. Fields, *Org. Prep. Proc. Int.*, 1974, **6**, 53.
34. *F. &F.*, **14**, 28; E. J. Corey and S. G. Payne, *Tetrahedron Lett.*, 1983, **24**, 2821.
35. D. G. Hart and F. L. Seeley, *J. Am. Chem. Soc.*, 1988, **110**, 1631.
36. P. A. Bartlett, K. L. McLaren, and P. C. Ting, *J. Am. Chem. Soc.*, 1988, **110**, 1663.
37. *F. &F.*, **9**, 66; G. A. Olah, Y. D. Vankard, and G. K. S. Prakash, *Synthesis*, 1979, 113.
38. *F. &F.*, **10**, 364; P. M. Pojer, *Aust. J. Chem.*, 1979, **32**, 201.
39. *F. &F.*, **1**, 685; W. von Doering and P. M. LaFlamme, *Tetrahedron*, 1958, **2**, 75.
40. P. M. Walters and S. M. McElvain, *J. Am. Chem. Soc.*, 1940, **62**, 1482.
41. T. Mukaiyama, K. Tonooka, and K. Inoue, *J. Org. Chem.*, 1961, **26**, 2202.
42. *F. &F.*, **5**, 608; C. Bernhart and C. G. Wermuth, *Tetrahedron Lett.*, 1974, 2493.

43. *F. & F.*, **7**, 1; M. V. Bhatt, C. G. Rao, and S. Rengaraju, *J. Chem. Soc., Chem. Commun.*, 1976, 103.
44. *F. & F.*, **6**, 587; R. B. Boar, J. F. McGhie, M. Robinson, D. H. R. Barton, D. C. Horwell, and R. U. Stick, *J. Chem. Soc., Perkin Trans. I*, 1975, 1237.
45. *F. &F.*, **13**, 22; A. Alemagna, C. Baldoli, P. D. Buttero, E. Licandro, and S. Maiorana, *J. Chem. Soc., Chem. Commun.*, 1985, 417.
46. *F. & F.*, **11**, 439–440; B. A. Trofimov, *Russian Chem. Rev.*, 1981, **50**, 138; A. I. Mikhaleva, M. V. Sigalov, and G. A. Kalabin, *Tetrahedron Lett.*, 1982, **23**, 5063.
47. *F. &F.*, **11**, 257; D. J. Brunelle, *Tetrahedron Lett.*, 1981, **22**, 3699.
48. *F. &F.*, **14**, 332; A. Padwa, W. Dent, and P. E. Yeskee, *J. Org. Chem.*, 1987, **52**, 3944.
49. *F. &F.*, **13**, 177; S. Itsuno, K. Miyazaki, and K. Ito, *Tetrahedron Lett.*, 1986, **27**, 3033.
50. E. Schmidt, A. Ascherl, and L. Meyer, *Ber.*, 1925, **58B**, 2430.
51. H. Feuer, R. S. Bartlett, B. F. Vincent, and R. F. Anderson, *J. Org. Chem.*, 1965, **30**, 2880.
52. B. Lindeke, A. K. Cho, T. L. Thomas, and L. Michelson, *Acta Pharm. Suecica*, 1973, **10**, 493.
53. R. T. Coutts and J. L. Malicky, *Can. J. Chem.*, 1974, **52**, 395.
54. R. T. Gilsdorf and F. F. Nord, *J. Am. Chem. Soc.*, 1952, **74**, 1837.
55. *F. &F.*, **13**, 42; M. S. Mourad, R. S. Varma, and G. W. Kabalka, *J. Org. Chem.*, 1985, **50**, 133; R. S. Varma and G. W. Kabalka, *Org. Prep. Proc. India*, 1985, *17*, 243.
56. R. S. Varma and G. W. Kabalka, *Synth. Commun.*, 1985, **15**, 843.
57. F. Bennington, R. D. Morin, and L. C. Clark, Jr, *J. Med. Chem.*, 1965, **8**, 100; G. Vavon and M. Crajeinovic, *Bull. Soc. Chim. Belg.*, 1928, **4**, 231.
58. H. Feuer, B. F. Vincent, and R. S. Bartlett, *J. Org. Chem.*, 1965, **30**, 2877.
59. F. Hodesan and V. Ciurdaru, *Tetrahedron Lett.*, 1971, 1997.
60. R. F. Borch, M. D. Bernstein, and H. D. Durst, *J. Am. Chem. Soc.*, 1971, **93**, 2897; *F. &F.*, **4**, 450.
61. T. Kawaguchi, T. Matsubara, and H. Kato, *Jap. Patent*, 1966, 1949S; *Chem. Abstr.*, 1967, **66**, 85529q.
62. R. Rodger and D. G. Neilson, *Chem. Rev.*, 1961, **61**, 179.
63. *F. &F.*, **5**, 123; D. Thomas and D. H. Aue, *Tetrahedron Lett.*, 1973, 1807.
64. *F. &F.*, **10**, 324; E. Cawkill and N. G. Clark, *J. Chem. Soc., Perkin Trans. I*, 1980, 244.
65. *F. & F.*, **14**, 298; S.-I. Murahashi, Y. Imada, Y. Taniguchi, and Y. Kodera, *Tetrahedron Lett.*, 1988, **29**, 2973.
66. *F. &F.*, **12**, 13; R. L. Fink, L. M. H. Horeher II, J. U. Daggett, and M. M. Hansen, *J. Org. Chem.*, 1983, **48**, 2632.
67. W. R. Dunstan and E. Goulding, *J. Chem. Soc.*, 1899, **75**, 792.
68. C. Vavon and K. Krajeinovic, *C. R. Acad. Sci. Paris*, 1928, **187**, 420.
69. W. D. Emmons, *J. Am. Chem. Soc.*, 1957, **79**, 5739.
70. *F. &F.*, **6**, 538; P. H. Morgan and A. H. Beckett, *Tetrahedron*, 1975, **31**, 2595.
71. D. B. Denny and D. Z. Denny, *J. Am. Chem. Soc.*, 1960, **82**, 1389.
72. A. H. Beckett, R. T. Coutts, and F. A. Ogunbona, *Tetrahedron*, 1973, **29**, 4189.
73. *F. &F.*, **12**, 157; A. J. Boloski and B. Ganem, *Synthesis*, 1983, 537.
74. M. A. T. Rodgers, *J. Chem. Soc.*, 1955, 769.
75. *F. &F.*, **7**, 177; A. Rougny and M. Daudon, *Bull. Soc. Chim.*, 1976, 833.
76. *F. &F.*, **10**, 5; H. Irie, I. Katayama, and Y. Mizuno, *Heterocycles*, 1979, **12**, 771.
77. *F. &F.*, **12**, 272; N. Narasaka and Y. Ukaji, *Chem. Lett.*, 1984, 147.

78. *F. &F.*, **13**, 123; M. Fujita, H. Oishi, and T. Hiyama, *Chem. Lett.*, 1986, 837.
79. D. D. Sternbach and W. C. L. Jamison, *Tetrahedron Lett.*, 1981, **22**, 3331.
80. M. Kawase and Y. Kikugawa, *J. Chem. Soc., Perkin Trans. I*, 1979, 643.
81. *F. &F.*, **11**, 201; S. Nahm and S. M. Weinreb, *Tetrahedron Lett.*, 1981, **22**, 3815.
82. W. W. Fishbein, J. Daly, and C. Streeter, *Anal. Biochem.*, 1969, **28**, 13.
83. *F. &F.*, **13**, 2; J. March and J. S. Engenito, *J. Org. Chem.*, 1981, **46**, 4304.

15

HYDRAZINES

$R{-}NH{-}NH_2$ (1) $R_2N{\cdot}NR_2$ (2) $H_2N{\cdot}NH_2$ (3)

$R{-}C({=}O){-}NH{-}NH_2$ (4) $R^1R^2C{=}N{-}NH_2$ (5)

Alkyl hydrazines [e.g. (**1**) and (**2**)] are derived from hydrazine (**3**) by substituting alkyl groups for hydrogen atoms. Other important derivatives of hydrazine are the acyl hydrazines (**4**) and the hydrazones (**5**). Typical bond lengths of the hydrazine group are as shown below.

Bond	Length
a	1.48Å
b	1.40Å
c	1.37Å
d	1.40Å

ALKYL HYDRAZINES (1) AND (2)

15.1 Preparation

15.1.1 *From alkyl halides*

Alkyl hydrazines result from the reaction of alkyl halides with hydrazine, but, as with alkyl halides and ammonia, polyalkylation occurs and the reaction yields mixtures which are difficult to separate.

15.1.2 *From monoalkyl ureas*

A more useful method for the preparation of monoalkyl hydrazines (**6**) depends upon the action of bromine and alkali on a monoalkyl urea (**7**) which by a Hofmann degradation, analogous to that with primary amides, yields an alkyl hydrazine (Scheme 15.1).[1]

$$\underset{(7)}{R{-}NH{-}\overset{\displaystyle O}{\overset{\|}{C}}{-}NH_2} \xrightarrow[\text{aq. } HO^-]{Br_2} \underset{(6)}{R{-}NH{-}NH_2} + CO_2 \qquad (15.1)$$

15.1.3 *From hydrazones*

The most useful preparation of alkyl hydrazines is by the reduction of the readily available hydrazones and related compounds. Thus monoalkyl hydrazines, or their salts (**10**), result in high yield by the reaction of aldehydes and ketones with the *tert*-butyl carbamate of hydrazine (**8**) (*t*-butylcarbazate)[2] followed by reduction of the hydrazone (**9**) with diborane (Scheme 15.2).[3]

$$R^1R^2C{=}O + \underset{(8)}{H_2N{-}NH{-}\overset{\displaystyle O}{\overset{\|}{C}}{-}OBu^t} \xrightarrow[\text{hexane}]{\text{Heat in}} \underset{(9)}{R^1R^2C{=}N{-}NH{-}\overset{\displaystyle O}{\overset{\|}{C}}{-}OBu^t} \xrightarrow[\text{ii, aq. HCl}]{\text{i, } BH_3/THF} \underset{(10)}{R^1R^2CH{-}NH{-}\overset{+}{N}H_3\ Cl^-}$$

Carbonyl compound	Yield (**9**) (%)	Yield (**10**) (%)
n-$C_5H_{11}CHO$	96	92
n-$C_6H_{13}CHO$	95	93
PhCHO	96	95
Cyclopentanone	96	90
Cyclohexanone	96	93
Cycloheptanone	97	90

(15.2)

By reduction of the hydrazones (**11**), formed from 1-acetyl-1-methylhydrazine (**12**) and aldehydes or ketones, with sodium borohydride and hydrolysis of the acetyl group, 1-alkyl-2-methylhydrazines (**13**) can be prepared (Scheme 15.3).[4]

$$CH_3{-}NH{-}NH_2 \xrightarrow[\text{pyridine}]{Ac_2O} CH_3{-}N(COCH_3){-}NH_2 \; (12)\; 76\% \xrightarrow{R^1R^2C{=}O} CH_3{-}N(COCH_3){-}N{=}CR^1R^2 \; (11)$$

$$(11) \xrightarrow{NaBH_4} CH_3{-}N(COCH_3){-}NH{-}CHR^1R^2 \xrightarrow[\text{ii, NaOH}]{\text{i, } H_2O, HCl} CH_3{-}NH{-}NH{-}CHR^1R^2 \; (13)$$

R^1	R^2	Yield (%)
H	H	82
H	Me	65
H	Et	42
H	Pr	60
H	Pr^i	56
Me	Me	38
Me	Et	61

(15.3)

15.2 Reactivity

15.2.1 *Alkylation*

Alkyl hydrazines (**14**) can themselves be alkylated by treatment with an aldehyde and ensuing reduction with sodium borohydride in acetonitrile as solvent,[5] for example, Scheme 15.4.

15.2.2 *N,N′-Diisopropylhydrazine for protection of carboxylic acids*

N,N′-Diisopropylhydrazine (**15**) can be prepared by the catalytic reduction of acetone azine.[6] It can be used for the protection of carboxylic acids since its derivatives of acids are stable both to acids and to alkalis. The acid can be regenerated by the action of lead tetraacetate (Scheme 15.5).[7]

(14) $\xrightarrow{2\,CH_2O,\ NaBH_4\ \text{in}\ CH_3CN}$ piperidine–N–NMe_2

Piperidine–N–NH_2 (14) $\xrightarrow{CH_2O}$ piperidine–N–$N{=}CH_2$ $\xrightarrow{NaBH_4}$ piperidine–N–NH–CH_3 $\xrightarrow{CH_2O}$ piperidine–N–$\overset{+}{N}(CH_3)(H)$–H_2C–O^- $\rightarrow$ piperidine–N–$\overset{+}{N}(CH_3){=}CH_2$ HO^- $\xrightarrow{NaBH_4}$ piperidine–N–NMe_2

(15.4)

$Me_2C{=}N\text{-}N\text{-}C{=}CMe_2$ $\xrightarrow{H_2\,/\,\text{catalyst}}$ $Me_2CH\text{-}NH\text{-}NH\text{-}CHMe_2$ (15)

R^1—CO_2H ⟶ R^1—$COCl$ $\xrightarrow[\text{protection}]{Me_2CH\text{-}NH\text{-}NH\text{-}CHMe_2\ (15)}$ R^1—C(=O)—N($CHMe_2$)—NH—$CHMe_2$

$\xrightarrow{\text{Modify } R^1 \text{ to } R^2}$ R^2—C(=O)—N($CHMe_2$)—NH—$CHMe_2$

$\xrightarrow[-H^+]{Pb(OAc)_4}$ [R^2—C(=O)—$\overset{+}{N}$($CHMe_2$)=N—$CHMe_2$; H_2O:]

$\xrightarrow[\text{deprotection}]{H_2O}$ [R^2—C(O^-)(OH)—$\overset{+}{N}$($CHMe_2$)=N—$CHMe_2$] ⟶ R^2-CO_2H + $Me_2CH\text{-}N{=}N\text{-}CHMe_2$

(15.5)

15.2.3 *N,N′-Dimethylhydrazine*

Protection of ketones

N,N-Dimethylhydrazine (**16**) reacts with ketones (**17**) in mildly alkaline conditions to form *N,N*-dimethylhydrazones (**18**). The ketones may be reproduced under neutral conditions with methyl iodide in refluxing 95 per cent ethanol (Scheme 15.6). This is a useful method for the protection of ketones, since it can be carried out under acidic conditions.

$$\underset{(16)}{H_2N{-}NMe_2} + \underset{(17)}{R^1R^2C{=}O} \longrightarrow \underset{(18)}{R^1R^2C{=}N{-}NMe_2} \xrightarrow{MeI} \left[R^1R^2C{=}N{-}\overset{+}{N}Me_3\right] \longrightarrow \underset{(17)}{R^1R^2C{=}O} \quad (15.6)$$

Use in nitrile synthesis

N,N-Dimethylhydrazine may be used in a synthesis of nitriles (**19**) from aldehydes (**20**) by initially forming the *N,N*-dimethylhydrazone, followed by reaction with methyl iodide to produce an *N,N,N*-trimethyl hydrazonium salt. β-Elimination with methanolic sodium hydroxide is the final step in the production of the nitrile (**19**) (Scheme 15.7).[9]

$$\underset{(20)}{R{-}C(=O)H} + H_2NNMe_2 \longrightarrow R{-}CH{=}N{-}NMe_2 \xrightarrow{MeI} R{-}C(H){=}N{-}\overset{+}{N}Me_3I^- \xrightarrow[MeOH]{NaOH,\ HO^-} \underset{(19)}{R{-}C{\equiv}N} + Me_3N + H_2O + I^-$$

c. 70% overall yields (15.7)

ACYL HYDRAZINES (**3**)

15.3 Preparation

15.3.1 *From acyl halides and hydrazine*

Acyl hydrazines may be prepared from acyl halides (**21**) and hydrazine (**22**) (Scheme 15.8).

$$\mathrm{R{-}C({=}O){-}Cl} + \mathrm{H_2N{-}NH_2} \longrightarrow \mathrm{R{-}C({=}O){-}NH{-}NH_2}$$
(21) (22) (15.8)

15.3.2 *From carboxylic acids and tosyl hydrazine*

The reaction of carboxylic acids (**23**) with *p*-tosylhydrazine (**24**) forms a useful synthesis of acyl hydrazides (**25**) (Scheme 15.9).[10]

$$\mathrm{R{-}CO_2H} + \mathrm{NH_2NHTs} \longrightarrow \mathrm{R{-}C({=}O){-}NHNHTs}$$
(23) (24) (25) (15.9)

15.3.3 *From carboxylic acids and alkyl hydrazines*

Acyl hydrazines (**26**) may be prepared from carboxylic acids (**27**) and alkylhydrazines (**28**) using phosphonitrilic chloride as an activator (Scheme 15.10)[11] (compare the preparation of amides similarly, Chapter 8).

$$(\mathrm{PNCl_2})_3 + \mathrm{R^1{-}CO_2Na} \longrightarrow \mathrm{(PNCl_2)_2(PNCl){-}O{-}C({=}O){-}R^1} \xrightarrow{\mathrm{R^2{-}NH{-}NH_2}} \mathrm{R^1{-}C({=}O){-}NH{-}NHR^2}$$
(27) (28) (26) (15.10)

15.4 Reactivity

15.4.1 *Reduction to hydrocarbons*

The acyl hydrazines (**29**), formed from carboxylic acid chlorides and *p*-tosylhydrazine (see Section 15.3.2), may then be reduced by diborane in THF to tosylhydrazines (**30**) which react on heating in methanolic potassium hydroxide to give the hydrocarbon (**31**) (Scheme 15.11).[12]

$$R{-}COCl + H_2N{-}NHSO_2Ar \longrightarrow R{-}CO{-}NH{-}NH{-}SO_2{-}Ar \ (29)$$

$$(29) \xrightarrow{B_2H_6\,/\,THF\,/25^{o}C} [R{-}CH_2N{-}N(H){-}SO_2{-}Ar \ (30),\ HO^-] \xrightarrow[\text{heat}]{KOH\,/\,MeOH} [R{-}CH_2{-}N{=}N{-}H,\ ^-O_2S{-}Ar,\ H{-}O{-}H]$$

$$\longrightarrow R{-}CH_3 \ (31) + N_2 + HO{-}S(O){-}Ar \qquad Ar = {-}C_6H_4{-}CH_3$$

R	Yield (%)
n-$C_{17}H_{35}$	64
n-$C_{15}H_{31}$	62
n-$C_{13}H_{27}$	64
$PhCH_2CH_2$	59

(15.11)

Acyl tosylhydrazides (**32**) are also reduced to hydrocarbons by the action of lithium aluminium hydride. Initially the carbonyl group is reduced to CH_2, and the intermediate then decomposes on heating, resulting in the hydrocarbon (Scheme 15.12).[13]

$$R{-}CO{-}NH{-}NHTs \ (32) \xrightarrow{LiAlH_4} [R{-}CH_2{-}NH{-}NHTs] \xrightarrow{\text{heat}} TsH + N_2 + R{-}CH_3 \qquad (15.12)$$

15.4.2 *Oxidation*

Oxidation of acyl hydrazines (**33**) by oxygen in the presence of copper salts as catalysts yields derivatives of carboxylic acids (**34**); free acids in the presence of

$$\underset{(33)}{R{-}C({=}O){-}NH{-}NH_2} \xrightarrow{\text{Oxidation}} R{-}C({=}O){-}\overset{+}{N}{\equiv}N \longrightarrow \underset{(35)}{R{-}\overset{+}{C}{=}O} + N_2 \xrightarrow{Nu^-} \underset{(34)}{R{-}C({=}O){-}Nu} \quad (15.13)$$

$$R{-}C({=}O){-}NH{-}NH_2 \xrightarrow[\text{O}_2\text{ bubbled through water, THF or MeOH}]{Cu(OAc)_2} R{-}CO_2H$$

$$R^1{-}C({=}O){-}NH{-}NH_2 \xrightarrow[O_2,\ THF,\ 25^{\circ}C]{CuCl_2,\ NaOR^2} R^1{-}C({=}O){-}OR^2$$

R^2	Yield (%)
Me	86
Bu^n	80
Bu^t	77
c-C_6H_{11}	77

$$R^1{-}C({=}O){-}NH{-}NH_2 \xrightarrow[THF,\ under\ N_2]{CuCl_2,\ HNR^2R^3} R^1{-}C({=}O){-}NR^2R^3$$

Amine	Yield (%)
pyrrolidine	99
morpholine	96
NH_3	93

Using $CuCl_2$-DBU and $R^1{=}C_7H_{15}$

Amine	Yield (%)
Bu^iNH_2	92

(15.14)

water, esters with alkoxides, and amides with amines. The authors suggest a mechanism through an acyl cation (**35**) which captures a nucleophile as shown in Scheme 15.13.[14] Some specific examples of this general reaction are shown in Scheme 15.14.

15.4.3 *Oxidative cleavage with lead tetraacetate*

Acyl hydrazines (**36**) undergo oxidative cleavage on reaction with lead tetraacetate via an intramolecular transfer of an acetate group to produce a *cis*-azo compound (**37**) by virtue of the cyclic intermediate (Scheme 15.15).[15] However, if an external nucleophile is present (e.g. pyridine), the reaction forms a *trans*-azo compound (**38**) (Scheme 15.16).

$$R^2\text{—C(=O)—NR}^1\text{NHR}^1 \ (36) \xrightarrow{Ph(OAc)_4} [\text{cyclic intermediate: } O{=}C(R^2)\text{—N(R}^1)\text{—N(R}^1)\text{—Pb(OAc)}_2\text{—OAc}] \longrightarrow R^1\text{—N=N—R}^1 \ (37) + R^1\text{—C(=O)OAc} \qquad (15.15)$$

$$R^2\text{—C(=O)—NR}^1\text{NHR}^1 \longrightarrow [\text{Nu:} \rightarrow R^2\text{—C(=O)—N(R}^1)\text{—N(R}^1)\text{—Pb(OAc)}_2\text{—OAc}] \longrightarrow R^1\text{—N=N—R}^1 \ (38) + R^2\text{—C(=O)—Nu} \qquad (15.16)$$

Ceric ammonium nitrate also oxidatively cleaves acyl hydrazides (**39**) to give the acid in good yields (**40**) (Scheme 15.17).[16]

$$R{-}C({=}O)NH{-}NH_2 \ \textbf{(39)} \xrightarrow{(NH_4)_2Ce(NO_3)_6} R{-}C({=}O)OH \ \textbf{(40)}$$

R	Yield (%)
Ph	83
p-MePh	65
p-ClPh	80
n-C_6H_{13}	90
$PhCH_2$	85
PhCH=CH	70

(15.17)

HYDRAZINIUM SALTS

15.5 Preparation

15.5.1 *From o-mesitylenesulphonylhydroxylamine*

o-Mesitylenesulphonylhydroxylamine (**41**)[17] (Scheme 15.18) reacts with tertiary amines (**42**) to give good (75–100 per cent) yields of hydrazinium salts (**43**) (Scheme 15.19).[18]

$$2,4,6\text{-}(CH_3)_3C_6H_2SO_2Cl + HONH{-}CO_2Bu^t \longrightarrow 2,4,6\text{-}(CH_3)_3C_6H_2SO_2{-}O{-}NH_2 \ \textbf{(41)}$$

(15.18)

$$R^1R^2R^3N{:} \ \textbf{(42)} + H_2N{-}O{-}SO_2Ar \ \textbf{(41)} \longrightarrow R^1R^2R\overset{+}{N}{-}NH_2 \ ArSO_3^- \ \textbf{(43)}$$

(15.19)

HYDRAZONES (**5**)

15.6 Preparation

15.6.1 *From aldehydes or ketones*

Hydrazones may be prepared by the well-documented condensation of hydrazines (**44**) with aldehydes or ketones (**45**) e.g. (Scheme 15.20).[19]

$$CH_3-\overset{\overset{O}{\|}}{C}-CH_3 \;(45) + Ph-NH-NH_2 \;(44) \longrightarrow CH_3-\overset{\overset{N-NHPh}{\|}}{C}-CH_3 \qquad (15.20)$$

15.6.2 *From diazonium salts*

Hydrazones may be prepared by the reaction of β-dicarbonyl compounds (**46**) or nitroalkanes (**47**) with diazonium salts in alkaline solution.[20] Tautomerism of the less stable azo compound yields the hydrazone (Scheme 15.21).

$$CH_3-\overset{\overset{O}{\|}}{C}-CH_2-CO_2Me \;(46) \xrightarrow{\text{base}} CH_3-\overset{\overset{O}{\|}}{C}-\underset{-}{C}H-CO_2Me \xrightarrow{ArN_2^+} \left[CH_3-\overset{\overset{O}{\|}}{C}-\underset{\underset{N=N-Ar}{|}}{C}H-CO_2Me\right] \longrightarrow CH_3-\overset{\overset{O}{\|}}{C}-\underset{\underset{N-NH-Ar}{\|}}{C}-CO_2Me$$

$$R-CH_2-NO_2 \;(47) \xrightarrow{\text{base}} R-\underset{-}{C}H-NO_2 \xrightarrow{N\equiv\overset{+}{N}-Ar} \left[R-\underset{\underset{N=N-Ar}{|}}{C}H-NO_2\right] \longrightarrow R-\underset{\underset{N-NH-Ar}{\|}}{C}-NO_2 \qquad (15.21)$$

15.6.3 *The Japp–Klingemann reaction*

When no α-hydrogen is present in the intermediate azo compound, and therefore no tautomerism is possible, an acyl group or a carboxyl group is lost. This is known as the Japp–Klingemann reaction[21] (Scheme 15.22).

$$R^1\text{—C(=O)-CHR}^2\text{-CO}_2\text{Et} \xrightarrow[\text{Base}]{\text{Ar-N}_2^+} R^1\text{—C(=O)—CR}^2(\text{N=N—Ar})\text{—C(=O)—OEt} \;\; (^-\text{OH}) \longrightarrow R^2\text{—C(=N—NHAr)—CO}_2\text{Et} + R^1\text{-CO}_2^-$$

Yields > 90% (15.22)

A variation of the Japp–Klingemann reaction occurs when a β-ketoester (**48**) reacts with benzenediazonium tetrafluoroborate (**49**) in aqueous pyridine followed by cleavage of the azo compound thus formed with sodium borohydride (Scheme 15.23).[22]

$$\underset{(48)}{R^1\text{—C(=O)-CHR}^2\text{-CO}_2\text{Et}} \rightleftharpoons R^1\text{—C(:OH)=CR}^2\text{-CO}_2\text{Et} \;\; + \;\; \underset{(49)}{\text{N}\equiv\overset{+}{\text{N}}\text{—Ph BF}_4^-} \xrightarrow{\text{aq. pyridine}} R^1\text{—C(=O)—CR}^2(\text{N=N—Ph})\text{—C(=O)—OEt} \;\; (\text{H—}\overset{-}{\text{B}}\text{H}_3\ \text{Na}^+)$$

$$\longrightarrow \left[R^1\text{—CH=O} + R^2\text{—C(N=N—Ph)=C(O}^-\text{)—OEt} \right]$$

$$R^1\text{—CHO} \xrightarrow[\text{ii, aq. H}_3\text{O}^+]{\text{i, NaBH}_4} R^1\text{—CH}_2\text{OH} \qquad R^2\text{—C(N=N—Ph)=C(O}^-\text{)—OEt} \xrightarrow{\text{aq. H}_3\text{O}^+} R^2\text{—C(=N-NHPh)—CO}_2\text{Et}$$

(15.23)

15.7 Reactivity

15.7.1 *With iodine or PhSeBr and base*

Ketone hydrazones (**50**) react with iodine and a penta-alkylguanidine (as a base) to yield vinyl iodides (**51**) (Scheme 15.24).[23]

(50) —I_2, base→ [intermediate] —I_2→ (51) 54-91% (15.24)

Likewise if PhSeBr is used in place of iodine as the electrophile, then phenyl vinyl selenides (**52**) are produced (70–90 per cent).[24]

SePh (52)

15.7.2 *Cleavage of N,N′-dimethylhydrazones*

N,N′-Dimethylhydrazones (**53**) can be converted into ketones (**54**) by treatment with boron trifluoride in ether followed by work-up with water (Scheme 15.25).[25]

$$R^1-C(=N-NMe_2)-R^2 \xrightarrow[\text{ii, } H_2O]{\text{i, } BF_3\text{ / ether}} R^1-C(=O)-R^2$$

(53) (54) 65-75% (15.25)

Cleavage of *N,N′*-dimethylhydrazones (**53**) to give ketones (**54**) can also be effected by addition of sodium periodate,[26] but it has been found that MCPBA is superior (Scheme 15.26).[27]

$$R^1R^2C{=}N-NMe_2 \xrightarrow[\text{DMF, -63}^{\circ}\text{C}]{m\text{-}ClC_6H_4CO_3H} R^1R^2C{=}O$$

(53) (54) (15.26)

Conversion of N,N'-dimethylhydrazones (**55**) to ketones (**56**) occurs by the action of ferric nitrate on a clay support (K10 Bentonite [Clayfen]) giving 70–90 per cent yields (Scheme 15.27).[28]

$$R^1R^2C{=}N{-}NMe_2 \ (55) \longrightarrow R^1R^2C{=}O \ (56)$$

Carbonyl compound	Yield (%)
3-Hexanone	78
4-Heptanone	78
2-Octanone	85
Cyclopentanone	69
Cyclohexanone	87

(15.27)

15.7.3 *The Diels–Alder reaction*

N,N'-Dimethylhydrazones of α,β-unsaturated aldehydes (**57**) and ketones undergo Diels–Alder reactions with electrophilic dienophiles,[29] for example, with acrylonitrile (**58**) (Scheme 15.28). The reaction is regioselective as shown.

$$CH_3{-}C({=}CH_2){-}CHO + H_2N{-}NMe_2 \longrightarrow CH_3{-}C({=}CH_2){-}CH{=}N{-}NMe_2 \ (57) + CH_2{=}CH{-}C{\equiv}N \ (58)$$

$\downarrow$

CH$_3$-substituted tetrahydropyridine with N–NMe$_2$ and CN, 53%

(15.28)

15.7.4 *Oxidation to alkynes*

Dihydrazones of α-diketones (**59**) can be oxidized by oxygen in the presence of cuprous chloride as catalyst in pyridine to give high yields of alkynes (**60**) (Scheme 15.29).[30] HgO or Ag_2O can also be used as the oxidizing agents, but they are less efficient at effecting the conversion.[31]

$$R^1—C(=N—NH_2)—C(=N—NH_2)—R^2 \text{ (59)} \xrightarrow[\text{pyridine}]{O_2 / Cu_2Cl_2} R^1—C(—N{=}N—H)=C(—N{=}N—H)—R^2 \xrightarrow[\text{pyridine}]{O_2 / Cu_2Cl_2} R^1—C{\equiv}C—R^2 \text{ (60)} + 2\,N_2 + 2\,H_2O \qquad (15.29)$$

15.7.5 *Oxidation to azoalkenes*

Aldehyde hydrazones (**61**) can be oxidized to azoalkenes (**62**) by treatment with iodine and pyridine followed by an elimination or similarly through tosyl hydrazones (**63**).[32] Yields are moderate (20–70 per cent), for example, Scheme 15.30.

$$\text{Ph—NH—N=CH—(CH}_2)_2\text{CH}_3 \text{ (61)} \longrightarrow \text{Ph—N(SO}_2\text{Ts)—N=CH—(CH}_2)_2\text{CH}_3 \text{ (63)}$$

$$\text{(61)} \xrightarrow{I_2,\ \text{pyridine}} \text{Ph—N(N}^+C_5H_5)\text{—N=CH—(CH}_2)_2\text{CH}_3\ I^- \longrightarrow \text{Ph—N=N—CH=CH—CH}_2\text{CH}_3 \text{ (62)}\quad EE : EZ = 73{:}27$$

$$\text{(63)} \xrightarrow{\text{Bu}^t\text{OK}} \text{Ph—N=N—CH=CH—CH}_2\text{CH}_3 \text{ (62)}\quad EZ{:}EE = 95{:}5 \qquad (15.30)$$

15.7.6 *Regeneration of carbonyl compounds from 2,4-dinitrophenylhydrazones*

Carbonyl compounds (**64**) can be regenerated from 2,4-dinitrophenylhydrazones, tosyl hydrazones (**65**), or *N*-methyl-*N*-tosyl hydrazones by oxidation with sodium nitrite in TFA or acetic acid (Scheme 15.31).[33]

$$R^1R^2C{=}NNHTs\ (65) \xrightarrow{NaNO_2,\ TFA} R^1R^2C{=}O\ (64) + TsN_3 \qquad (15.31)$$

Another method for the regeneration of aldehydes and ketones from 2,4-dinitrophenylhydrazones is by reduction with vanadium dichloride (VCl_2) in aqueous tetrahydrofuran. The reaction depends upon the reduction of the nitro groups to amino groups and the hydrolysis of the resulting amines through their imino tautomers. The mixture is boiled under reflux for 1 h and the yields are in the range 65–95 per cent.[34]

15.7.7 *With iodine*

Hydrazones (**66**) react with iodine to yield *gem*-diiodides (**67**) (Scheme 15.32).[35]

$$R^1R^2C{=}N{-}NH_2\ (66) \xrightarrow[Et_3N]{I_2\,/\,ether} R^1R^2CI_2\ (67)\quad 40\text{-}60\% \qquad (15.32)$$

15.7.8 *With epoxide and base*

The dimethylhydrazone of acetone (**68**) reacts with *n*-butyl lithium and epoxides to yield spiroacetals (**69**) (Scheme 15.33).[36]

15.7.9 *Osazones*

Phenylhydrazines may be used to cleave sugars but the reaction ceases after phenylosazone is formed due to chelate ring formation (Scheme 15.34).[37]

$$\text{Me}_2\text{C}{=}\text{N}{-}\text{NMe}_2 \ \textbf{(68)} \xrightarrow[\text{ii, 1,2-epoxybutane}]{\text{i, BuLi}} \left[\text{Me}(\text{C}{=}\text{N}{-}\text{NMe}_2)\text{CH}_2{\cdot}\text{CH}_2{-}\text{CH(OH)}{-}\text{Et}\right] \xrightarrow[\text{ii, oxirane}]{\text{i, Bu}^n\text{Li}} \left[\text{CH}_2(\text{OH}){\cdot}\text{CH}_2{\cdot}\text{CH}_2{\cdot}\text{C}({=}\text{N}{-}\text{NMe}_2){-}\text{CH}_2{\cdot}\text{CH}_2{-}\text{CH(OH)}{-}\text{Et}\right] \xrightarrow{\text{Amberlite IR120}} \textbf{(69)} \qquad (15.33)$$

$$\text{HOCH}_2{-}\text{CO}{-}\text{CH}_2\text{OH} \xrightarrow[\text{3 times}]{\text{PhN(CH}_3\text{)NH}_2} \text{Ph(CH}_3\text{)N{-}N{=}CH{-}C(=N{-}N(CH}_3\text{)Ph){-}CH{=}N{-}N(CH}_3\text{)Ph} \qquad (15.34)$$

15.7.10 *Wolff–Kishner and Huang-Minlon reactions*

Hydrazones (**70**) may be reduced to the alkane (**71**) in the presence of base, in the Wolff–Kishner reaction. The mechanism is as shown in Scheme 15.35.[38]

$$\underset{(70)}{R_2C{=}N{-}NH_2} \xrightleftharpoons{OH^-} R_2CH{-}N{=}NH + OH^-$$

$$R_2CH{-}N{=}NH + OH^- \longrightarrow R_2CH{-}N{=}N^- + H_2O \xrightarrow{-N_2} R_2\bar{C}H \xrightarrow{H_2O} \underset{(71)}{R_2CH_2} + OH^- \qquad (15.35)$$

The Huang-Minlon modification of the Wolff–Kishner reaction involves carrying the reaction out in refluxing diethylene glycol and has completely replaced the original procedure,[39] for example, Scheme 15.36.

$$p\text{-}C_6H_5OC_6H_4COCH_2CH_2CO_2H \xrightarrow[\text{3 NaOH 195-200}^\circ\text{C}]{\text{diethyleneglycol, 85\% } N_2H_4.H_2O} p\text{-}C_6H_5OC_6H_4CH_2CH_2CH_2CO_2H$$

Yield >90% (15.36)

In another modified version of the Wolff–Kishner reaction, using dimethyl sulphoxide as solvent, the conversion of benzophenone hydrazone (**72**) to diphenylmethane (**73**) was effected at room temperature (Scheme 15.37).[40]

$$\underset{(72)}{Ph_2C{=}N{-}NH_2} \xrightarrow[\text{DMSO } 20^\circ\text{C}]{Bu^tOK} \underset{(73)}{Ph_2CH_2} \qquad (15.37)$$

Ketone hydrazones (**74**) can also be converted into hydrocarbons (**75**) by reaction with potassium *t*-butoxide in refluxing toluene (Scheme 15.38).[41]

$$\underset{(74)}{R_2C{=}N{-}NH_2} \xrightarrow[\text{reflux, toluene}]{Bu^tOK} \underset{(75)}{R_2CH_2} \qquad (15.38)$$

TOSYL HYDRAZONES

The hydrazones produced from tosyl hydrazine (**76**) can be used for a variety of purposes. Tosyl hydrazine can be made in high yield by the action of hydrazine on *p*-toluenesulphonyl (tosyl) chloride (**77**) in tetrahydrofuran (Scheme 15.39).

$$\underset{(77)}{p\text{-}CH_3C_6H_4SO_2Cl} + H_2N\text{-}NH_2 \xrightarrow[\text{THF}]{\text{10-15\%}} \underset{(76)}{p\text{-}CH_3C_6H_4SO_2NHNH_2} \qquad (15.39)$$

15.8 Reactivity

15.8.1 *With monosaccharides*

Monosaccharides (e.g. β-D-glucose, **78**) react with tosyl hydrazine to yield crystalline tosyl hydrazones (**79**) which can be used for characterization,[42] (Scheme 15.40).

$$(78) + Ts\text{-}NHNH_2 \longrightarrow CH{=}N{-}NHTs\text{-}(CHOH)_4CH_2OH\ (79) \xrightarrow{KBH_4} CH_3\text{-}(CHOH)_4CH_2OH \quad (15.40)$$

15.8.2 *Reduction*

The reduction of a tosyl hydrazone (**80**) with sodium borohydride constitutes a method for the reduction of an aldehydic or ketonic carbonyl group (>C=O) to a methylene group (>CH_2).[43] Under mild reaction conditions, side reactions are insignificant and good yields are obtained (Scheme 15.41). D-Glucose can also be reduced in this way.[44]

$$R^1R^2C{=}O + H_2N\text{-}NHTs \xrightarrow{MeOH} R^1R^2C{=}N\text{-}NHTs\ (80) \xrightarrow{NaBH_4} R^1R^2CH_2 + N_2 + TsH \quad (15.41)$$

15.8.3 *Synthesis of alkenes*

The action of a strong base on the tosyl hydrazone of an aldehyde or a ketone (**81**) which contains an α-hydrogen atom constitutes a good synthesis of alkenes.[45] A variety of bases have been used and several mechanisms have been suggested.[46] Bamford and Stevens[47] used sodium and ethylene glycol or sodium methoxide (Scheme 15.42). The thermodynamically more stable alkene is produced and the diazoalkane (**82**) is an intermediate which can decompose either in aprotic or protic conditions to yield the alkene (Scheme 15.43).

$$\mathrm{Ph{-}CH_2{-}C({=}O){-}CH_3 + H_2N{-}NHTs \longrightarrow Ph{-}CH_2{-}C(CH_3){=}N{-}NHTs\ (81) \xrightarrow{Na\,/\,HOCH_2{-}CH_2OH} Ph{-}CH{=}CHCH_3 + N_2 + TsH} \quad (15.42)$$

$$\mathrm{Ph{-}CH_2{-}C(CH_3){=}N{-}N(H){-}SO_2Ar \xrightarrow{^{-}OR} Ph{-}CH_2{-}C(CH_3){=}\ddot{N}{-}\bar{N}{-}SO_2Ar \longrightarrow Ph{-}CH_2{-}C(CH_3){=}\overset{+}{N}{=}\overset{-}{N}\ (82) + ArSO_2^-}$$

$$\mathrm{(82) \xrightarrow{Aprotic} Ph{-}CH(H){-}\bar{C}(CH_3){-}\overset{+}{N}{\equiv}N \longrightarrow PhCH{=}CHCH_3}$$

$$\mathrm{(82) \xrightarrow{Protic} Ph{-}CH(H){-}C(H)(CH_3){-}\overset{+}{N}{\equiv}N \longrightarrow PhCH{=}CHCH_3} \quad (15.43)$$

A better synthetic procedure involves the use of two equivalents of an alkyllithium which gives better yields, due to there being fewer side reactions. However this method forms the less thermodynamically stable (i.e. the less substituted) alkene (**83**),[48] for example, Scheme 15.44.

Tosyl hydrazones (**84**) react with *n*-butyllithium and a variety of reagents in TMEDA as solvent to produce substituted alkenes (**85**) in a modification of the

$Me_3C(H_2C-H)C{=}N{-}NH{-}SO_2Ar$ + Li–Bu^n ⟶ $Me_3C(H_2C{=})C{-}N{=}N{-}H$ + Bu^n–Li

⟶ $Me_3C(H_2C{=})C{-}N{=}\bar{N}$ + H–OH

⟶ $Me_3C(H_2C{=})C{-}H + N_2 + HO^-$

(83) (15.44)

Shapiro–Heath alkene synthesis. The reactions proceed via an intermediate vinyl carbanion (**86**) (Scheme 15.45).[49]

t-BUTYL HYDRAZONES

15.9 Synthesis of ketones or α-hydroxyketones

Tert-butyl hydrazones of aldehydes (**87**) can be deprotonated by alkyl lithiums to yield anions which react with aldehydes or ketones, eventually yielding α-hydroxyketones (**88**) or with alkyl halides to produce ketones (**89**) (Scheme 15.46).[50]

SULPHONYL HYDRAZONES

15.10 Generation of vinyl lithium reagents

Hydrazones as their 2,4,6-triisopropylbenzenesulphonyl derivatives (**90**) can be converted by *n*-butyl-lithium or *s*-butyl-lithium into anions of alkenes which can be trapped with chlorodiphenylphosphine to yield vinyl phosphines (**91**) (Scheme 15.47).[51]

$$\mathrm{Me\text{-}(CH_2)_5\text{-}C(=N\text{-}NHTs)\text{-}Me}\ (84) \xrightarrow[\text{TMEDA}]{\text{3 BuLi}} \left[\mathrm{Me\text{-}(CH_2)_5\text{-}\bar{C}{=}CH_2}\ \ \mathrm{Li^+}\right] (86) \xrightarrow{\text{PhCHO}} \mathrm{Ph{-}CH(OH){-}C({=}CH_2)(CH_2)_5Me}\ (85)$$

$$\mathrm{Me\text{-}(CH_2)_5\text{-}C(=N\text{-}NHTs)\text{-}Me} \xrightarrow{\text{i, BuLi, TMEDA; ii, cyclohex-2-enone}} \text{1-[C(=CH}_2\text{)(CH}_2\text{)}_5\text{Me]-cyclohex-2-en-1-ol (C–OH)}$$

$$\mathrm{Me\text{-}(CH_2)_5\text{-}C(=N\text{-}NHTs)\text{-}Me} \xrightarrow[\text{ii, CO}_2]{\text{i, BuLi, TMEDA}} \mathrm{CH_2{=}C(CO_2H)(CH_2)_5Me}$$

$$\mathrm{Me\text{-}(CH_2)_5\text{-}C(=N\text{-}NHTs)\text{-}Me} \xrightarrow{\text{i, BuLi, TMEDA; ii, MeI}} \mathrm{CH_2{=}C(Me)(CH_2)_5Me} \qquad (15.45)$$

Many other examples of reactions of these sulphonyl hydrazones are described with particular reference to the Shapiro reaction.[52]

Other examples of vinyl-lithium reagents generated from 2,4,6-triisopropylbenzene-sulphonyl hydrazines have been described.[53]

DIIMIDE

Diimide has been isolated by Wilberg and his co-workers.[54] Another preparation of diimide which is often advantageous makes use of potassium azodicarbonate ($K^{+-}O_2C{-}N{=}N{-}CO_2^-K^+$).[55] The preparation is usually performed *in situ* and diimide acts as a selective reducing agent. Diimide selectively reduces the less sterically hindered multiple bond in conjugated systems.

$Bu^t{-}NH{-}N{=}CH{-}R^1$ (87) $\xrightarrow{Bu^nLi}$ $[Bu^t{-}N{=}N{-}\bar{C}H{-}R^1]$

$[Bu^t{-}N{=}N{-}\bar{C}H{-}R^1] \xrightarrow{R^2\text{-}CO\text{-}R^3} [Bu^t{-}N{=}N{-}CH(R^1){-}C(R^2)(R^3){-}O^-\ Li^+] \xrightarrow{\text{i, } Bu^nLi;\ \text{ii, } H_2O} Bu^t{-}NH{-}N{=}C(R^1){-}C(R^2)(R^3){-}OH \xrightarrow{\text{aq. } H_3O^+} R^2R^3C(OH){-}C(=O){-}R^1$

(88) 40-95%

$[Bu^t{-}N{=}N{-}\bar{C}H{-}R^1] \xrightarrow{R^2Hal} [Bu^t{-}N{=}N{-}CH(R^2){-}R^1] \xrightarrow{F_3C.CO_2H} Bu^t{-}NH{-}N{=}C(R^2){-}R^1 \xrightarrow{\text{aq. } H_3O^+} R^1{-}C(=O){-}R^2$

(89) 15-83%

(15.46)

$R^1{-}C({=}N{-}NH{-}SO_2Ar){-}CH_2{-}R$ (90) $\xrightarrow{Bu^nLi}$ $[R^1{-}\bar{C}{=}CHR^2\ Li^+] \xrightarrow{Cl\text{-}PPh_2} R^1{-}C(PPh_2){=}CHR^2 \xrightarrow{30\%\ H_2O \text{ or } 40\%\ CH_3CO_3H} R^1{-}C(\overset{+}{P}Ph_2{-}O^-){=}CHR^2$

(91)

65-78% overall

(15.47)

For instance the dienol (**92**) was reduced to (**93**) (Scheme 15.48).[56] The alkyne (**94**) also produces (**93**) by diimide reduction (Scheme 15.49).

(92) —HN=NH→ (93) 74% (15.48)

(94) —HN=NH→ (93) 71% (15.49)

Diimide (**95**) has been prepared *in situ* by the oxidation of hydrazine (**96**) with phenyliodine diacetate (**97**) and again it has been used to reduce an alkyne (**98**) to an alkene by *cis*-reduction (Scheme 15.50).[57]

$$\underset{(96)}{H_2N{-}NH_2} + \underset{(97)}{PhI(OAc)_2} \longrightarrow \underset{(95)}{[HN{=}NH]} + PhI + 2\,AcOH$$

$$\underset{(98)}{Ph{-}C{\equiv}C{-}Ph} \xrightarrow{[HN=NH]} \text{(H)(Ph)C=C(H)(Ph)} \quad 80\% \qquad (15.50)$$

Diimide, prepared *in situ* by oxidation of hydrazine with a variety of oxidizing agents catalysed by copper(II) ions,[58] may be used to reduce alkenes, alkynes, and azo compounds. The reduction is more easily carried out on symmetrical multiple bonds (C=C, C≡C) than polar bonds (C≡N, C=O, C=N, $O{=}N^{+}{-}O^{-}$, $O{=}S^{+}{-}O^{-}$),[59] for example, Scheme 15.51.

As shown by the mechanism, the reaction proceeds by *cis* addition and therefore steric effects control the product formed. However, on occasion, electronic effects may override steric hindrance. For instance the expected product from the reduction of 7-oxygenated norbornadienes (**99**) is the corresponding *syn*-norbornene (**100**). However the main product is the *anti*-norbornene due to co-ordination of the partially positive N—N bond with the oxygen atom (Scheme 15.52).[60]

$$H_2N{-}NH_2 \longrightarrow H{-}N{=}N{-}H + R{-}CH{=}CH{-}R \longrightarrow \left[\begin{matrix} N \cdots H \cdots \overset{H}{C}{-}R \\ \| \qquad\quad \vdots \\ N \cdots H \cdots \underset{H}{C}{-}R \end{matrix}\right] \longrightarrow N{\equiv}N + \begin{matrix} H{-}CH{-}R \\ | \\ H{-}CH{-}R \end{matrix} \tag{15.51}$$

(99) $\xrightarrow{HN{=}NH}$ *anti* (major) + *syn* (minor) (100)

$X = H, \text{-}COCH_3, \text{-}Bu^t$ (15.52)

Allenes are reduced stereospecifically by diimide to *cis*-alkenes:[61]

1,2-cyclononadiene → *cis*-cyclononene
1,2-cyclodecadiene → *cis*-cyclodecene
1,2-nonadiene → *cis*-2-nonene

References

1. D. W. Lum and I. L. Mador, *Chem. Abstr.*, 1960, **54**, 6549.
2. *F. &F.*, **1**, 85; **2**, 46.
3. *F. &F.*, **11**, 88; N. I. Ghali, D. L. Venton, S. C. Hung, and G. C. LeBreton, *J. Org. Chem.*, 1981, **46**, 5413.
4. *F. &F.*, **4**, 7; F. E. Condon, *J. Org. Chem.*, 1972, **37**, 3608; 3615.
5. *F. &F.*, **5**, 609; S. F. Nelson and G. R. Weisman, *Tetrahedron Lett.*, 1973, 231.
6. *F. &F.*, **4**, 162; H. R. Lochte, J. R. Bailey, and W. A. Noyes, *J. Am. Chem. Soc.*, 1921, **43**, 2597.
7. D. H. R. Barton, M. Girijavallabhan, and P. G. Sammes, *J. Chem. Soc., Perkin Trans. I*, 1972, 929; see also T.-L. Ho, H. L. Ho, and C. M. Wong, *Synthesis*, 1972, 562.
8. *F. &F.*, **3**, 117; M. Avaro, J. Levisalles, and H. Rudler, *Chem. Commun.*, 1969, 445.

9. *F. &F.*, **1**, 289; R. F. Smith and L. E. Walker, *J. Org. Chem.*, 1962, **27**, 4372.
10. *F. &F.*, **6**, 161.
11. L. Caglioti, M. Poloni, and G. Rosini, *J. Org. Chem.*, 1968, **33**, 2979.
12. O. Attanasi, L. Caglioti, F. Gasparrini, and D. Misiti, *Tetrahedron*, 1975, **31**, 341.
13. L. Caglioti, *Tetrahedon*, 1966, **22**, 487.
14. *F. &F.*, **10**, 103; J. Tsugi, S. Hayakawa, and H. Takayanagi, *Chem. Lett.*, 1975, 1437; J. Tsugi, H. Takahashi, and Y. Toshida, *Chem. Lett.*, 1976, 147; definitive paper: J. Tsugi, T. Nagashima, N. T. Qui, and H. Takayanagi, *Tetrahedron*, 1980, **36**, 1311.
15. J. B. Aylward and R. O. C. Norman, *J. Chem. Soc. (C)*, 1968, 2399; D. H. R. Barton, M. Girijavaliabhan, and P. G. Sammes, *J. Chem. Soc., Perkin Trans. I*, 1972, 929.
16. T. L. Ho, H. C. Ho, and C. M. Wong, *Synthesis*, 1972, 562.
17. J. G. Kraus, *Synthesis*, 1972, 140; L. A. Carpino, C. A. Giza, and B. A. Carpino, *J. Am. Chem. Soc.*, 1959, **81**, 955; *F. &F.*, **5**, 431.
18. Y. Tamura, J. Minamikawa, Y. Kita, J. H. Kim, and M. Ikeda, *J. Org. Chem.*, 1973, **38**, 1239.
19. W. P. Jencks, *Prog. Phys. Org. Chem.*, 1964, **2**, 63; S. C. Johnson, *Adv. Phys. Org. Chem.*, 1967, **5**, 237.
20. Review: S. M. Parmerter, *Org. React.*, 1959, **10**, 1.
21. Review: R. R. Phillips, *Org. React.*, 159, **10**, 143; see R. P. Linstead and A. B. Wang, *J. Chem. Soc.*, 1937, 807.
22. *F. &F.*, **8**, 22; A. P. Kozikowski and W. C. Floyd, *Tetrahedron Lett.*, 1978, 19.
23. *F. &F.*, **12**, 241; D. H. R. Barton, G. Bashiardes, and J.-L. Fourrey, *Tetrahedron Lett.*, 1983, **24**, 1605.
24. D. H. R. Barton, G. Bashiardes, and J.-I. Fourrey, *Tetrahedron Lett.*, 1984, **25**, 1287.
25. *F. &F.*, **11**, 75; R. E. Gawley and E. J. Termine, *Synth. Commun.*, 1982, **12**, 15.
26. *F. &F.*, **7**, 126.
27. *F. &F.*, **12**, 119; M. Duraisamy and H. M. Walborsky, *J. Org. Chem.*, 1984, **49**, 3411.
28. P. Laszlo and E. Polla, *Tetrahedron Lett.*, 1984, **25**, 3309; *F. &F.*, **12**, 231.
29. *F. &F.*, **11**, 200; B. Serckx-Ponein, A.-M. Hesbain-Frisque, and I. Ghosez, *Tetrahedron Lett.*, 1982, **23**, 3261.
30. *F. &F.*, **5**, 165; J. Tsugi, H. Takahashi, and T. Kajimoto, *Tetrahedron Lett.*, 1973, 4573; J. Tsugi, H. Takahashi, and Y. Toshida, *Chem. Lett.*, 1976, 147.
31. J. March, *Advanced organic chemistry*, 3rd edn, Wiley-Interscience, Chichester, 1985, p. 943; A. T. Blomquist and C. Y. Lui, *J. Am. Chem. Soc.*, 1953, **93**, 761; A. Krebs and H. Kimling, *Tetrahedron Lett.*, 1970, 761.
32. J. C. Shantl and P. Hebeisen, *Tetrahedron*, 1990, **46**, 395.
33. *F. &F.*, **9**, 432; L. Caglioti, F. Gasparrini, D. Misiti, and G. Palmieri; *Synthesis*, 1979, 207.
34. *F. &F.*, **11**, 593; G. A. Ohla, Y.-L. Chao, M. Arvanaghi, and G. K. S. Prakash, *Synthesis*, 1981, 476.
35. *F. &F.*, **4**, 260; A. Pross and S. Sternhell, *Aust. J. Chem.*, 1970, **23**, 989.
36. *F. &F.*, **12**, 205; D. Enders, W. Dahmen, E. Dedericks, and P. Weuster, *Synth. Commun.*, 1983, **13**, 1235.
37. *F. &F.*, **1**, 694; L. F. Fieser and M. Fieser, *Organic chemistry*, 3rd edn, Reinhold, New York, 1956, p. 351; O. L. Chapman, W. J. Welstaed Jr, T. J. Murphy, and R. W. King, *J. Am. Chem. Soc.*, 1964, **86**, 732; 4968.
38. H. H. Szmant, *Angew. Chem., Int. Ed. Engl.*, 1968, **7**, 120.
39. Huang-Minlon, *J. Am. Chem. Soc.*, 1946, **68**, 2487.

40. *F. &F.*, **1**, 926; D. J. Cram, M. R. V. Sahyun, and G. R. Knox, *J. Am. Chem. Soc.*, 1962, **84**, 1734.
41. M. F. Grundon, H. B. Henbest, and M. D. Scott, *J. Chem. Soc.*, 1963, 1835; M. F. Grundon and M. D. Scott, *J. Chem. Soc.*, 1964, 5674.
42. K. Freudenberg and F. Blümmel, *Annalen*, 1924, **440**, 45; D. G. Easterby, L. Hough, and J. K. N. Jones, *J. Chem. Soc.*, 1951, 3416.
43. L. Caglioti, *Tetrahedron*, 1966, **22**, 487.
44. A. N. de Belder and H. Weigel, *Chem. &Ind.*, 1964, 1689.
45. *F. &F.*, **1**, 1185; *F. &F.*, **2**, 417.
46. Review: J. Casanova and B. Waegell, *Bull. Soc. Chim. Fr.*, 1975, 922.
47. W. R. Bamford and T. S. Stevens, *J. Chem. Soc.*, 1952, 4735; J. W. Powell and M. C. Whiting, *Tetrahedron*, 1959, **7**, 305.
48. *F. &F.*, **2**, 417; G. Kaufman, F. Cook, H. Shechter, J. Bayless, and L. Friedman, *J. Am. Chem. Soc.*, 1967, **89**, 5736.
49. *F. &F.*, **7**, 48; J. E. Stemke, A. R. Chamberlin, and F. T. Bond, *Tetrahedron Lett.*, 1976, 2947.
50. *F. &F.*, **12**, 87; R. M. Adlington, J. E. Baldwin, and M. W. D. Perry, *J. Chem. Soc., Chem. Commun.*, 1983, 1040.
51. *F. &F.*, **11**, 563; D. G. Mislanker, B. Mugrage, and S. D. Darling, *Tetrahedron Lett.*, 1981, **22**, 4619.
52. *F. &F.*, **11**, 564–6; J. E. Baldwin and J. C. Bottaro, *J. Chem. Soc., Chem. Commun.*, 1981, 1121; S. H. Bertz and G. Dabbagh, *J. Am. Chem. Soc.*, 1981, **103**, 5932.
53. *F. &F.*, **9**, 488; A. R. Chamberlain, J. E. Stemke, and F. T. Bond, *J. Org. Chem.*, 1978, **43**, 147; A. R. Chaimberlain and F. T. Bond, *Org. Synth.*, 1979, 44; R. M. Adlington and A. G. M. Barrett, 1978, 1071; *F. &F.*, **14**, 327; A. R. Chaimberlain, S. H. Blom, L. A. Cervini, and C. H. Fotsch, *J. Am. Chem. Soc.*, 1988, **110**, 4788.
54. N. Wiberg, H. Bachhuber, and G. Fischer, *Angew. Chem., Int. Ed. Engl.*, 1972, **11**, 829.
55. R. G. Powell, C. C. Smith, Jr, and I. A. Wolff, *J. Org. Chem.*, 1967, **32**, 1442.
56. *F. &F.*, **4**, 154; K. Mori, M. Ohki, A. Sato, and M. Matsui, *Tetrahedron*, 1972, **28**, 3739.
57. *F. &F.*, **14**, 258; R. M. Moriarty, R. K. Vaid, and M. P. Duncan, *Synth. Commun.*, 1987, **17**, 703.
58. E. J. Corey, W. L. Mock, and D. J. Pasto, *Tetrahedron Lett.*, 1961, 347.
59. *F. &F.*, **1**, 257; E. E. van Tamelen, M. Davis, and M. F. Deem, *Chem. Commun.*, 1965, 71.
60. *F. &F.*, **2**, 139; W. C. Baird, Jr, B. Franzos, and J. H. Surridge, *J. Am. Chem. Soc.*, 1967, **89**, 410.
61. *F. &F.*, **3**, 99; G. Nagendrappa and D. Devaprabhakara, *Tetrahedron Lett.*, 1970, 4243.

16

AZO AND AZOXY COMPOUNDS[1]

AZO COMPOUNDS

$$R{-}N{=}N{-}R$$

(1)

Aliphatic azo compounds (**1**) have not been extensively studied due to their ready decomposition to nitrogen and hydrocarbons.

The N=N double bond gives rise to the possibility of geometric isomerism, but most azo compounds are planar with a preference for the *trans* form.[2] For aliphatic azo compounds the N=N bond distance has been found to be between 1.22–1.25 Å, with most falling in the range 1.22–1.24 Å.[3] For example, the molecular dimensions of diimide (**2**)[4] and azomethane (**3**) are shown below.

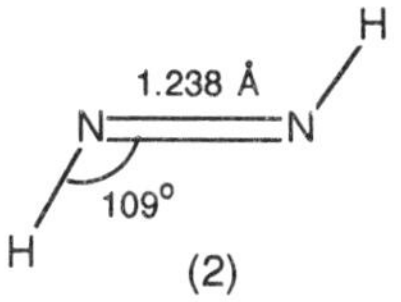

1.238 Å; 109°

(2)

1.254 Å; 1.474 Å; 111.9°

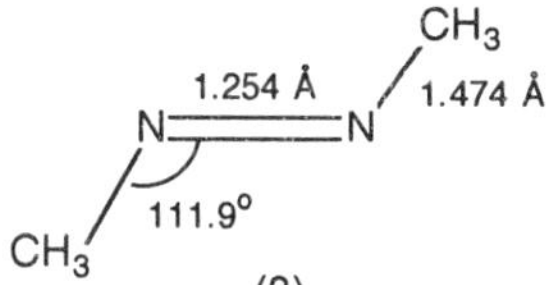

(3)

In compounds which have carbonyl groups associated with the azo linkage the N=N bond length is shown to increase to within the range 1.24–1.26 Å.[5] The effects of substitution on the bond lengths of azo compounds have been more comprehensively studied by Chang and his co-workers.[3]

NMR studies of *trans*-azomethane have shown the protons to resonate at 3.68δ, showing the protons to be more shielded than those in alkenes.[6]

16.1 Preparation[1,7,8]

16.1.1 *Oxidation of 1,2-dialkylhydrazines*

Any synthesis of a 1,2-dialkylhydrazine (**4**) (see Chapter 15) may potentially be extended by oxidation to provide an azo alkane (**5**) (Scheme 16.1). Thus a general approach to synthesizing azo alkanes with primary or secondary alkyl groups attached in low yields (<50 per cent) is by a condensation–reduction–oxidation approach.

$$\mathrm{(R)(R^1)CH{-}NH{-}NH{-}CH(R^2)(R^3)}\ \textbf{(4)} \xrightarrow{[O]} \mathrm{(R)(R^1)CH{-}N{=}N{-}CH(R^2)(R^3)}\ \textbf{(5)} \qquad (16.1)$$

For symmetrical azo compounds

This has been applied to symmetrical compounds using yellow mercuric oxide for the oxidation step (Scheme 16.2).[9]

$$\mathrm{(R^1)(R^2)C{=}O} \xrightarrow[\text{Condensation}]{NH_2NH_2} \mathrm{(R^1)(R^2)C{=}N{-}N{=}C(R^1)(R^2)} \xrightarrow[\text{Reduction}]{H_2/Pd\text{-}C} \mathrm{(R^1)(R^2)CH{-}NH{-}NH{-}CH(R^1)(R^2)} \xrightarrow[\text{Oxidation}]{HgO} \mathrm{(R^1)(R^2)CH{-}N{=}N{-}CH(R^1)(R^2)} \qquad (16.2)$$

For asymmetric azo compounds

The same reagents (reduction by catalytic hydrogenation and oxidation by yellow mercuric oxide) have been applied to the synthesis of unsymmetrical substituted azo alkanes when the hydrazines are readily available.[10] Although the oxidation step takes place in high (60 per cent) yield low overall yields predominate (Scheme 16.3).

$$\mathrm{CH_3NHNH_2 + O{=}C(Ph)(CH_3)} \longrightarrow \mathrm{CH_3NHN{=}C(Ph)(CH_3)} \xrightarrow[Pd/C]{H_2} \mathrm{CH_3{-}NH{-}NH{-}CH(Ph)(CH_3)} \xrightarrow{HgO} \mathrm{CH_3{-}N{=}N{-}CH(Ph)(CH_3)} \qquad (16.3)$$

Low Overall Yields

Azo compounds (**6**, **7**, **8**, **9**) derived from less readily available hydrazines have been prepared by modifications of the synthetic procedures developed by Overberger and Di Guilio,[11] for example, Scheme 16.4.

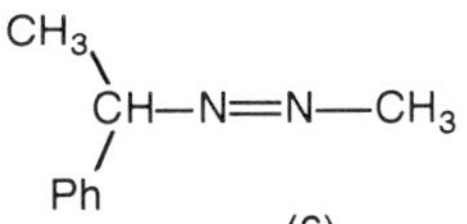

$$CH_3(Ph)CH{-}N{=}N{-}CH_3 \quad (6)$$

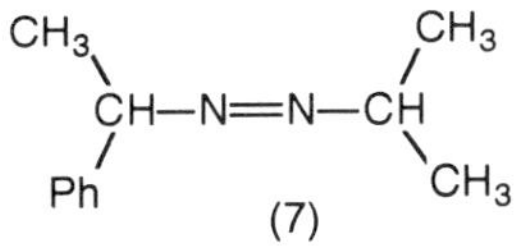

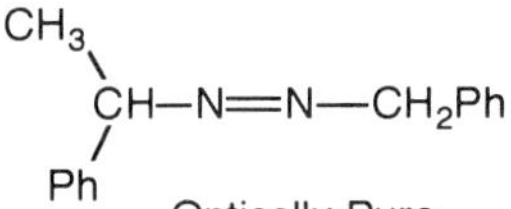

$$CH_3(Ph)CH{-}N{=}N{-}CH(CH_3)_2 \quad (7)$$

$$CH_3(Ph)CH{-}N{=}N{-}CH_2Ph \quad \text{Optically Pure} \quad (8)$$

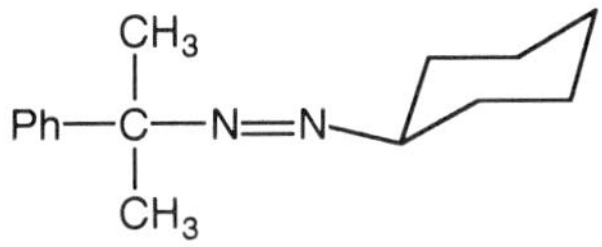

$$Ph{-}C(CH_3)_2{-}N{=}N{-}C_6H_{11} \quad (9)$$

$$RR^1C{=}N{-}N{=}CRR^1 \xrightarrow{H_2/Pd\text{-}C} RR^1CH{-}NH{-}N{=}CRR^1 \xrightarrow{(COOH)_2.2H_2O} RR^1CH{-}NH{-}NH_3^+ \left(\begin{array}{c}COO \\ COOH\end{array}\right)^-_2$$

$$\xrightarrow{NaOH} RR^1CH{-}NH{-}NH_2 \xrightarrow{R^2R^3C{=}O} RR^1CH{-}NH{-}N{=}CR^2R^3 \xrightarrow[\text{Oxidation}]{\text{Redn.}} RR^1CH{-}N{=}N{-}CHR^2R^3 \quad (16.4)$$

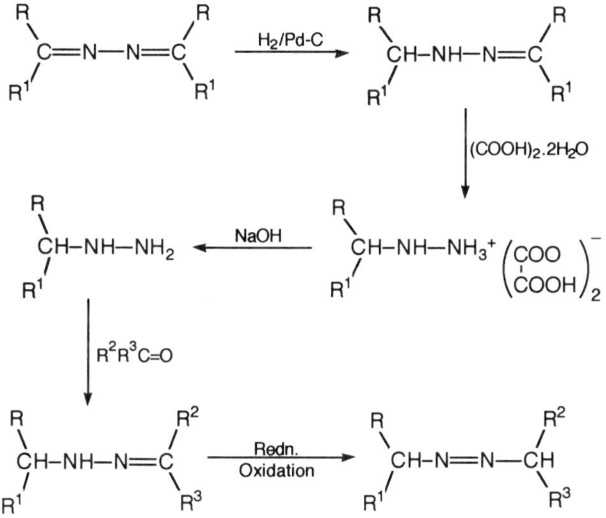

Asymmetric azo alkanes (**10**) have been prepared in low to medium yields (23–59 per cent) by reaction of an ester (**11**) with hydrazine followed by condensation with a ketone and ensuing reduction and oxidation, as shown in Scheme 16.5.[12]

From acyl hydrazines

Lead tetraacetate has been used to prepare both *cis*- and *trans*-alkyl azo compounds from acyl hydrazines (see Chapter 15, p. 596).

For acyl azo compounds

A variety of aliphatic hydrazines have been oxidized by *N*-bromosuccinimide,[13] for example, in the preparation of *t*-butyl azodiformate (**12**) (Scheme 16.6). Diacylazo compounds may be prepared in the same way.[14]

$$\underset{(11)}{R{-}\overset{O}{\overset{\|}{C}}{-}OR^1} \xrightarrow{NH_2NH_2} R{-}\overset{O}{\overset{\|}{C}}{-}NH{-}NH_2 \xrightarrow{R^1R^2C{=}O} R{-}\overset{O}{\overset{\|}{C}}{-}NH{-}N{=}C(R^2)(R^1)$$

$$\xrightarrow{LiAlH_4} R{-}CH_2{-}NH{-}NH{-}CH(R^2)(R^1) \xrightarrow{HgO} \underset{(10)}{R{-}CH_2{-}N{=}N{-}CH(R^2)(R^1)} \qquad (16.5)$$

$$Bu^tO{-}\overset{O}{\overset{\|}{C}}{-}NH{-}NH{-}\overset{O}{\overset{\|}{C}}{-}OBu^t \xrightarrow{NBS} \underset{\underset{(12)}{90\%}}{Bu^tO{-}\overset{O}{\overset{\|}{C}}{-}N{=}N{-}\overset{O}{\overset{\|}{C}}{-}OBu^t} \qquad (16.6)$$

For bicyclic azo compounds

Hydrazines may be oxidized to azo alkanes (**13**) in high yields (80–95 per cent) by bubbling oxygen through the compound dissolved in methanol with a small amount of palladium black suspended in solution,[15] for example, Scheme 16.7.

$$\text{bicyclic hydrazine } [(CH_2)_n \text{ bridge}, NH{-}NH]\ (n = 1, 2, 3) \xrightarrow{O_2/Pd\text{-}C} \text{bicyclic azo compound } [(CH_2)_n \text{ bridge}, N{=}N]\ (\mathbf{13}) \qquad (16.7)$$

For preparation of α-substituted azo compounds

Azo bisnitriles (**14**), prepared by addition of HCN to azines followed by oxidation,[16] may be used as intermediates for transformation to other α-substituted azo alkanes, to esters (**15**, **17**),[17] and to alcohols (**16**)[18] (Scheme 16.8).

For preparation of diazines

Diazines of the type R—N=NH may be prepared by condensation of carbonyl compounds with methylcarbazate (**18**) to yield carbomethoxy hydrazones (**19**) followed by catalytic reduction, oxidation with a peracid, and finally decarboxylation (Scheme 16.9).[19]

$$RR^1C{=}N{-}N{=}CRR^1 \xrightarrow{HCN} R^1{-}C(R)(CN){-}NH{-}NH{-}C(R)(CN){-}R^1 \xrightarrow{Br_2} R^1{-}C(R)(CN){-}N{=}N{-}C(R)(CN){-}R^1\ (14)$$

$$(14) \xrightarrow[74\%]{HCl,\ CH_3OH} R^1{-}C(R)(CO_2CH_3){-}N{=}N{-}C(R)(CO_2CH_3){-}R^1\ (15)$$

$$(15) \xrightarrow[67\%]{LiAlH_4} R^1{-}C(R)(CH_2OH){-}N{=}N{-}C(R)(CH_2OH){-}R^1\ (16) \xrightarrow[77\%]{Ac_2O} R^1{-}C(R)(CH_2OCOCH_3){-}N{=}N{-}C(R)(CH_2OCOCH_3){-}R^1\ (17) \quad (16.8)$$

$$R^1R^2C{=}O + H_2N{-}NH{-}C(O){-}OCH_3\ (18) \longrightarrow R^1R^2C{=}N{-}NH{-}C(O){-}OCH_3\ (19)$$

$$(19) \xrightarrow{H_2,\ \text{Raney Ni}} R^1R^2CH{-}NH{-}NH{-}C(O){-}OCH_3 \xrightarrow[\text{Oxidation, 80-90\%}]{CH_3COOOH,\ \text{peracetic acid}} R^1R^2CH{-}N{=}N{-}C(O){-}OCH_3\ \text{(Diazine ester)}$$

$$\text{Diazine ester} \xrightarrow{\text{Xs NaOH or KOH}} R^1R^2CH{-}N{=}N{-}CO_2^- \xrightarrow{\text{decarboxylation}} R^1R^2CH{-}N{=}N{-}H \quad (16.9)$$

16.1.2 *Oxidative coupling of amines*

Iodine pentafluoride, a powerful dehydrogenating reagent, has been used to achieve oxidative coupling of tertiary alkyl primary amines, for example *t*-butylamine (**20**),[20] providing a synthesis of symmetrical tertiary substituted azo alkanes (Scheme 16.10). This reagent has subsequently been applied to the synthesis of many other tertiary substituted alkyl azo alkanes.[21]

$$(CH_3)_3C{-}NH_2 \xrightarrow[\text{CHCl}_3,\ 0^{\circ}\text{C, pyridine}]{IF_5} (CH_3)_3C{-}N{=}N{-}C(CH_3)_3 \quad \text{(20)} \rightarrow 48\% \qquad (16.10)$$

Oxidative coupling using sodium hypobromite (NaOBr) followed by treatment with base provides a useful route to azo compounds. For example 1,1′-azoadamantane (**21**) has been prepared in high yields by this method (Scheme 16.11).[22]

$$\text{Ad}{-}NH_2 \xrightarrow{\text{NaOBr}} \text{Ad}{-}NHBr \xrightarrow{\text{Bu}^t\text{OK / Bu}^t\text{OH}} \text{Ad}{-}N{=}N{-}\text{Ad} \quad 81\% \ \text{(21)} \qquad (16.11)$$

Similarly, oxidative coupling has been performed using sodium hypochlorite (NaOCl) (Scheme 16.12).[23]

$$CH_3{-}C(CH_3)(CN){-}NH_2 \xrightarrow{\text{NaOCl}} CH_3{-}C(CH_3)(CN){-}N{=}N{-}C(CH_3)(CN){-}CH_3 \quad 86\% \qquad (16.12)$$

16.1.3 *Oxidation of ureas*

Alkyl azo compounds may be prepared from ureas (**22**) by initial conversion to diaziridones (**23**) by reaction with *t*-butylhypochlorite and a base. These cyclic ureas are readily hydrolysed to hydrazines (**24**), which may subsequently be oxidized to azo alkanes (**25**) (Scheme 16.13).[24] This method of synthesis may be applied to symmetrical or unsymmetrical azo compounds but nevertheless remains dependent on the availability of the substituted urea (**22**) (see Chapter 9, p. 406).

$$R{-}NHC(=O){-}NHR^1 \xrightarrow[\text{ii, } Bu^tO^- K^+]{\text{i, } Bu^tOCl} \underset{(23)}{R{-}N{-}N{-}R^1 \text{ (with } C{=}O)} \xrightarrow{HCl} \left[RNH{-}N(CO_2H){-}R^1 \right]$$
(22)

$$\left[RNH{-}N(CO_2H){-}R^1 \right] \longrightarrow \underset{(24)}{R{-}NH{-}NHR^1} \xrightarrow{[O]} \underset{(25)}{R{-}N{=}N{-}R^1} \qquad (16.13)$$

16.1.4 *From hydrazodicarboxylate esters*

A multistep procedure for synthesizing allylic azo compounds has been developed by Crawford and his co-workers in which hydrazocarboxylate esters (**26**) are mono/dialkylated, hydrolysed, and finally oxidized with red mercuric oxide,[25] for example, 3,3-azo-1-propene (Scheme 16.14).

$$\underset{(26)}{C_2H_5CO_2NH{-}NHCO_2C_2H_5} \xrightarrow[\text{ii, } CH_2{=}CHCH_2OSO_2C_6H_5]{\text{i, NaH}} C_2H_5O_2CN(CH_2{-}CH{=}CH_2){-}N(CH_2{-}CH{=}CH_2)CO_2C_2H_5$$

$$\xrightarrow[\text{ii, HCl}]{\text{i, KOH / MeOH}} CH_2{=}CHCH_2NH{-}NHCH_2CH{=}CH_2 \xrightarrow[Na_2SO_4]{\text{HgO (red)}} \underset{21\%}{CH_2{=}CHCH_2{-}N{=}N{-}CH_2CH{=}CH_2} \qquad (16.14)$$

This method has been extended to mixed azo alkanes, but only very low yields have been obtained.[26]

16.1.5 *From azines*

Synthesis of α-substituted azo compounds, by 1,4 addition of chlorine to azines (**27**)[27] accompanied by nucleophilic displacement of the chlorine by use of tri-alkyl aluminium (AlR_3), provides a powerful synthesis of tertiary azo alkanes (**28**) (Scheme 16.15).[28]

Similarly lithium aluminium hydride may be used to reduce the chlorinated compound providing a synthesis of secondary azo compounds (**29**) (Scheme 16.16).[29]

$$\underset{(27)}{R^1R^2C{=}N{-}N{=}CR^1R^2} + Cl_2 \longrightarrow R^2{-}\overset{R^1}{\underset{Cl}{C}}{-}N{=}N{-}\overset{R^1}{\underset{Cl}{C}}{-}R^2 \xrightarrow{AlR^3_3} \underset{\text{moderate to high yields (28)}}{R^2{-}\overset{R^1}{\underset{R^3}{C}}{-}N{=}N{-}\overset{R^1}{\underset{R^3}{C}}{-}R^2} \quad (16.15)$$

$$R^1R^2C{=}N{-}N{=}CR^1R^2 + Cl_2 \longrightarrow R^2{-}\overset{R^1}{\underset{Cl}{C}}{-}N{=}N{-}\overset{R^1}{\underset{Cl}{C}}{-}R^2 \xrightarrow{LiAlH_4} \underset{\text{40-60\% yield (29)}}{R^2{-}\overset{R^1}{\underset{H}{C}}{-}N{=}N{-}\overset{R^1}{\underset{H}{C}}{-}R^2} \quad (16.16)$$

Ready displacement of the chlorine by the appropriate nucleophile has led to the synthesis of other interesting α-substituted azo compounds (**30**)[30] with $R^3 =$ —SCN, CH_3CO_2, CH_3COS, PhS, CN, N_3;[31] $R^3 = CH_3$—;[32] $R^3 = CH_3O$—;[33] and $R^3 = CH_3S$—.[34]

$$\underset{(30)}{R^2{-}\overset{R^1}{\underset{R^3}{C}}{-}N{=}N{-}\overset{R^1}{\underset{R^3}{C}}{-}R^2}$$

16.1.6 *Oxidation of N,N′-dialkylsulphamides*

Probably the most widely applicable synthesis of azo alkanes, useful for both symmetrical and unsymmetrical, alkyl or aryl azo compounds, has been pioneered by Ohme, Schmitz, and Preuschhoff.[35] In this procedure azo alkanes are prepared by the oxidation of *N,N*-dialkylsulphamides (**31**) using sodium hypochlorite. The reaction proceeds by chlorination of nitrogen, formation of an anion with ensuing ring closure to form the thiadiaziridine-1,1-dioxide (**32**), which may be hydrolysed to the 1,2 dialkylhydrazine (**33**) which is in turn oxidized to an azo compound (Scheme 16.17). This method has been extensively used to prepare azoethane (**34**) (Scheme 16.18).[36]

R–NH–SO_2–NH–R (31) —NaOH, Excess NaOCl→ [R—N(Cl)—SO_2—N^-—R] → R—N<SO_2>N—R (32) —hydrolysis→ R—NH-NH—R (33) —NaOCl→ R—N=N—R

R	Yield (%)
Et	54
$CH_3(CH_2)_2$-	54
$CH_3(CH_2)_3$-	54
c-C_6H_{11}-	80

(16.17)

Timberlake *et al.*[37] showed that azo alkanes (**35**) may be synthesized from *N,N*-dialkylsulphamides (**36**)[38] using *t*-butyl hypochlorite as the chlorinating reagent in a modification of the Ohme–Schmitz–Preuschoff process (Scheme 16.19).

Similarly treatment of 1,1-di-*t*-butylsulphonamine (**37**) with sodium hydride in pentane (0 °C) produces a quantitative yield of the sodo anion (**38**), and with addition of *t*-butyl hypochlorite followed by refluxing leads to the formation of the azo compound (**39**) (Scheme 16.20).[39]

$$2\,C_2H_5NH_2 + SO_2Cl_2 \xrightarrow[\text{ether}]{\text{Pyridine, Pet.}} (C_2H_5NH)_2SO_2 \xrightarrow[H_2O]{\text{NaOCl, NaOH}} [NaO_3S{-}N(C_2H_5){-}NH{-}C_2H_5 \rightarrow C_2H_5NH{-}NHC_2H_5] \xrightarrow[H_2O,\ 25^{\circ}C]{\text{NaOCl, NaOH}} C_2H_5N{=}NC_2H_5$$

(34)

(16.18)

$$R{-}NHSO_2NHR \xrightarrow[Bu^tOCl]{\text{Na / ether}} R{-}NHSO_2\bar{N}R\ Na^+ \xrightarrow[\text{pentane}]{Bu^tOCl} \text{RN–NR (cyclic, } SO_2) \rightarrow R{-}N{=}N{-}R$$

(36) → 35-85% (35)

R	Yield (%)
Bu^n	37
Bu^s	64
Bu^t	85

(16.19)

Mixed solvent systems have also been successfully employed in a variation of the Ohme–Schmitz–Preuschhoff preparation.[40]

The procedure has been applied to the synthesis of mixed azo alkanes.[41]

16.1.7 *Thermal isomerization*

Good yields of allylic azo alkanes (**40**) have been prepared by thermal isomerism of suitably substituted precursors for example (Scheme 16.21).[42]

16.2 Reactivity

Azo compounds are on the whole non-reactive substances except for azo compounds with strong electron withdrawing groups which are treated separately.

Bu^t—NH—SO_2—NH—Bu^t (37) $\xrightarrow{NaH}$ Bu^t—N^-(Na^+)—SO_2—NH—Bu^t (38)

$\downarrow Bu^tOCl$

Na^+ Bu^t—N^-—SO_2—N(Bu^t)—Cl $\longrightarrow$ Bu^t—N—SO_2—N—Bu^t (three-membered ring)

$\downarrow \Delta$, $-SO_2$

Bu^t—N=N—Bu^t (39)

(16.20)

R(CH$_3$)C=N—N(CH$_3$)—CH$_2$—CH=CH$_2$ $\xrightarrow{140\text{-}200°C}$ CH_2=CH—CH_2—C(R)(CH_3)—N=N—CH_3 (40)

R	Yield (%)
Me	94
Et	80

(16.21)

16.2.1 *Tautomerism and basicity*

The pK_a values of azobenzenes have been investigated at some length and show them to be weaker bases than water.[43] There is some controversy as to the site of protonation with earlier proposals of a symmetrically bound proton in a π-complex[44] giving way to NMR studies suggesting that the proton is singly σ-bound to one nitrogen atom.[45]

However, unlike those of their aromatic analogues the acid/base properties of aliphatic azo compounds have not been greatly studied due to their rapid isomerization to hydrazones (**41**) under strongly acidic conditions (Scheme 16.22).[46]

$$CH_3{-}N{=}N{-}R \rightleftharpoons CH_2{=}N{-}NHR \quad (41) \qquad (16.22)$$

This isomerization is also displayed in the Japp–Klingemann reaction of enolic forms of β-diketones in which an azo compound is formed as an intermediate before rapid tautomerism (Scheme 16.23).[47]

$$R{-}C(OH){=}CH{-}C(=O){-}R^1 \longrightarrow R{-}C(O{-}N{=}N{-}Ar){=}CH{-}C(=O){-}R^1 \longrightarrow R{-}C(=O){-}CH(N{=}N{-}Ar){-}C(=O){-}R^1 \xrightarrow{\text{Tautomerism}} R{-}C(=O){-}C(=N{-}NH{-}Ar){-}C(=O){-}R^1 \qquad (16.23)$$

Bellamy and Gutherie showed that in non-polar solvents the equilibrium shown in Scheme 16.22 lies almost exclusively to the right.[48] This has been confirmed by many other workers.[49] However, in polar solvents and in the presence of strongly basic or acidic catalysts a few per cent of the azo compound may well be present.[50]

16.2.2 *Oxidation*

Azo compounds are readily oxidized providing a good synthesis of azoxy compounds (see Section 16.4.1), but reaction conditions must be carefully controlled to prevent isomerism to hydrazones. Peracids and hydrogen peroxide in dichloromethane have been the preferred reagents for oxidation of azo compounds.[51] With unsymmetric azo compounds it had been thought that oxidation occurred on the least hindered nitrogen atom,[52] but since then it has been shown that the oxidation may occur at either atom.[53] The reaction with peracids proceeds by electrophilic attack on the azo linkage with the stereochemistry of the azo compond being retained in the azoxy product.[54]

16.2.3 *Reduction*

Earlier reduction work is covered thoroughly in reviews.[7,55]

To hydrazo derivatives

Reduction of acylazo compounds (**42**) to acylhydrazo compounds may be performed using hydrazine hydrate,[56] for example, Scheme 16.24, or by catalytic hydrogenation,[57] for example, Scheme 16.25.

$$R{-}\overset{\overset{O}{\|}}{C}{-}N{=}N{-}\overset{\overset{O}{\|}}{C}{-}R \xrightarrow{N_2H_4.H_2O} R{-}\overset{\overset{O}{\|}}{C}{-}NH{-}NH{-}\overset{\overset{O}{\|}}{C}{-}R$$

(42) (16.24)

$$Ph_3CCH_2{-}\overset{\overset{O}{\|}}{C}{-}N{=}N{-}\overset{\overset{O}{\|}}{C}{-}CCH_2CPh_3 \xrightarrow[25^{\circ}C]{H_2\,/\,PtO_2} Ph_3CCH_2{-}\overset{\overset{O}{\|}}{C}{-}NH{-}NH{-}\overset{\overset{O}{\|}}{C}{-}CCH_2CPh_3$$

70% (16.25)

Reductive cleavage to amines

Reductive cleavage has been performed with a wide variety of reagents including lithium aluminium hydride and metal–acid combinations. The amines which are produced in the reduction procedure have been used in structural determination studies.

Hydrogenolysis of *S*,*S*-(−)-1,1′-diphenylazoethane (**43**) to *S*-(−)-α-phenylethylamine (**44**) has been performed using zinc dust in aqueous ethanol and glacial acetic acid (Scheme 16.26).[58]

$$Ph{-}\underset{\underset{Me}{|}}{\overset{\overset{H}{|}}{C}}{-}N{=}N{-}\underset{\underset{H}{|}}{\overset{\overset{Me}{|}}{C}}{-}Ph \xrightarrow[HOAc\,/\,H_2O,\ 50^{\circ}C]{Zn\,/\,EtOH} 2\ Ph{-}\underset{\underset{Me}{|}}{\overset{\overset{H}{|}}{C}}{-}NH_2$$

(43) (44) (16.26)

Iron(II) chloride/acetic acid[59] and tin(II) chloride/tin/conc. HCl[60] are two alternative metal/acid combinations which have been used in structural studies for similar reductive cleavages.

16.2.4 *Synthesis of three-membered rings*

Five-membered cyclic azo compounds (**45**) are frequently used for the synthesis of three-membered rings (**46**), for example, Scheme 16.27,[61] and similarly Scheme 16.28.[62]

hυ, - 40°C; (45) → (46); Cope Rearrangement → Cycloheptadienes (16.27)

hυ, C_6D_6; $N-CH_2Bu^t$ → $N-CH_2Bu^t$ + N_2 (16.28)

16.2.5 *Formation of epoxides from azo acetates*

Azo acetates (**47**) are particularly reactive and cyclize at room temperatures and, on losing nitrogen, epoxides (**48**) are formed,[63] for example, Scheme 16.29.

OAc, O, R^1—N=N—C—R^3, R^2 (47) → R^1, N=N, OAc, R^2, O, R^3 → R^1, OAc, R^2, R^3, O (48) (16.29)

16.2.6 *Use of azodicarbonyl compounds*

For oxidation

Alcohols (**49**) have been selectively oxidized to their corresponding aldehydes or ketones (**50**) under mild neutral conditions using:

(1) 2,6-di-*t*-butyl-4-nitrophenol, diethyl azodicarboxylate, and triphenylphosphine (Scheme 16.30);[64]

$$R^1CH(OH)R^2 \ (49) \xrightarrow[\text{ii, } Ph_3P \quad \text{iii, } EtO_2CCH_2NO_2]{\text{i, } EtO_2C\text{-}N{=}N\text{-}CO_2Et} R^1C(=O)R^2 \ (50) \ 52\text{-}85\% \qquad (16.30)$$

(2) *N*,*N*-(azodicarbonyl) dipiperidine (Scheme 16.31).[65]

$$R^1CH(OH)R^2 \ (49) \xrightarrow{C_5H_{10}N\text{-}C(=O)\text{-}N{=}N\text{-}C(=O)\text{-}NC_5H_{10}} R^1C(=O)R^2 \ (50) \qquad (16.31)$$

In 'oxidative–reductive' dehydrative coupling

'Oxidative–reductive' dehydrative coupling occurs with certain compounds in the presence of triphenylphosphine (TPP) and diethyl azodiacetate (DEAD). These reagents have been applied to the synthesis of oxiranes and other heterocycles (Scheme 16.32).[66]

$$\text{HO}\frown\text{OH} \xrightarrow[\text{ii, } Ph_3P]{\text{i, } EtO_2C\text{-}N{=}N\text{-}CO_2Et} \text{cyclic ether (O), 4-7 membered rings} \quad {>}94\% \qquad (16.32)$$

Amination

Azodicarboxylic esters have been used to aminate chiral substrates, providing a convenient synthesis of optically active α-hydrazino and α-amino acids (**51**). The diastereoselectivity increases with the increase in size of the ester group. This ensures that the di-*t*-butyl ester is particularly useful. Lithium enolates of chiral *N*-acyloxazolidones have been used as chiral precursors for reaction with di-*t*-butyl azodicarboxylate and then α-hydrazino acids, obtained by trans-esterification, acid hydrolysis, and catalytic hydrogenation of the product. Chiral α-amino acids prevail on hydrogenation of the α-hydrazino acids with Raney nickel, for example, Scheme 16.33.[67]

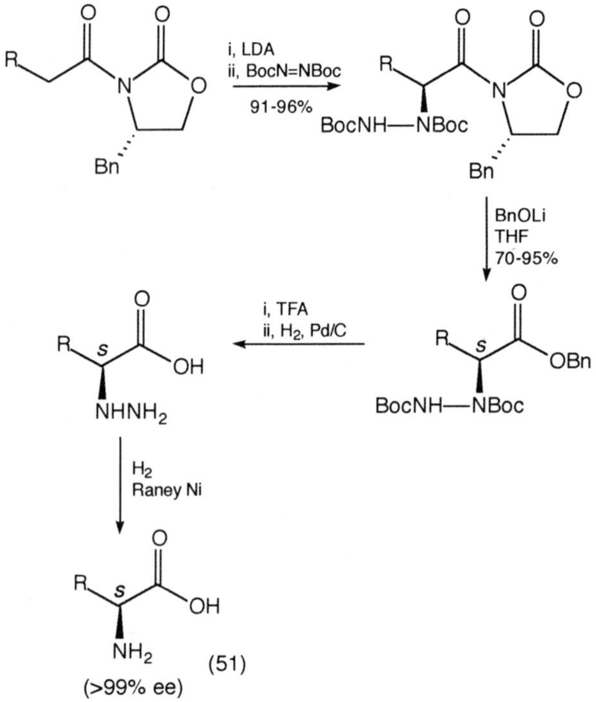

i, LDA
ii, BocN=NBoc
91-96%

BnOLi
THF
70-95%

i, TFA
ii, H_2, Pd/C

H_2
Raney Ni

(51)
(>99% ee)

(16.33)

AZOBISISOBUTYRONITRILE (AIBN)

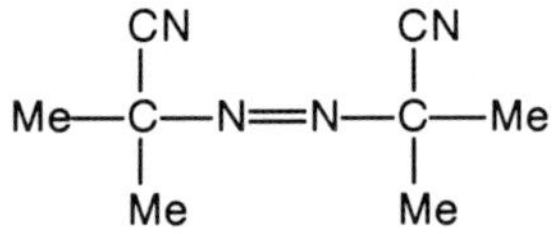

$$\mathrm{Me{-}C(CN)(Me){-}N{=}N{-}C(CN)(Me){-}Me}$$

(52)

AIBN (**52**) may be purified by recrystallization from methanol and dried over phosphorus pentoxide.[68]

16.3 Reactivity

AIBN (**52**) decomposes under mild conditions (40 °C) into cyanopropyl radicals (**53**) (Scheme 16.34), and hence finds use as an initiator in a variety of radical reactions.[69]

$$\mathrm{Me{-}C(CN)(Me){-}N{=}N{-}C(CN)(Me){-}Me}\ \textbf{(52)} \xrightarrow{-N_2} 2\ \mathrm{Me{-}\dot{C}(CN)(Me)}\ \textbf{(53)} \qquad (16.34)$$

16.3.1 *Halogenation of hydrocarbons*

Reaction of *t*-butyl hypochlorite with toluene in the presence of AIBN with toluene affords benzyl chloride (**54**) by a radical-induced chain chlorination reaction (Scheme 16.35).[70] Similarly alkenes react to form allylic chlorides in good yield.[71] Side chain bromination may also be effected.[72]

$$\mathrm{AIBN} \xrightarrow[-N_2]{40^\circ C} (CH_3)_2\dot{C}(CN) \xrightarrow{Bu^tOCl} Bu^tO^{\cdot} \xrightarrow{C_6H_5CH_3} C_6H_5CH_2^{\cdot} \xrightarrow{Bu^tOCl} C_6H_5CH_2Cl\ \ 84\%\ \textbf{(54)} \qquad (16.35)$$

16.3.2 *Dehalogenation*[73]

Alkyl halides may be reduced under radical conditions using combinations of organotin hydrides and AIBN.[74] In these reactions the following reactivity sequence is observed: $Ph_2SnH_2 = C_4H_9SnH_3 > Ph_3SnH = (Bu^n)_2SnH > (Bu^n)_3SnH$; for example Scheme 16.36.[75]

If the alkyl halide undergoing dehalogenation has a β-hydrogen then elimination of HHal may be observed, for example, Scheme 16.37.[76]

(16.36)

$$\mathrm{CH_3{-}CH(Br){-}CH(Br){-}CH_3} \xrightarrow[AIBN]{Bu_3SnH} \mathrm{CH_3{-}CH{=}CH{-}CH_3}\ \ 99\% \qquad (16.37)$$

16.3.3 *Reduction*

Organotin halides in combination with AIBN have been extensively reviewed in reduction.[77]

ELECTROPHILIC AZO COMPOUNDS[78]

16.4 Reactivity

16.4.1 *Electrophilic substitution reactions*

Azo compounds may react as electrophiles, displaying high reactivity towards nucleophilic reagents, when strong electron withdrawing groups (e.g. —CN, $—CO_2R$, —COR) are attached to the azo linkage. This is demonstrated by an electrophilic substitution reaction at an activated aromatic nucleus which may be catalysed by a Lewis acid,[79] for example, Scheme 16.38.

$RO_2C—N{=}N—CO_2R$ + Me-benzene $\xrightarrow[\text{or } BF_3]{H^+}$ Me-C$_6$H$_4$-N(CO$_2$R)—NHCO$_2$R (16.38)

Similar substitution takes place on activated alkenes (**55**),[80] for example, Scheme 16.39.

Ph—C(OMe)=CH$_2$ (55) + $RO_2C—N{=}N—CO_2R$ → Ph—C(OMe)=CH-N(CO$_2$R)—NHCO$_2$R (16.39)

Substituted hydrazines (**56**) may be formed by the alkylation of activated methylene compounds in the form of Grignard reagents or more usually as sodium salts (**57**) (Scheme 16.40).[81]

$$RO_2C—N{=}N—CO_2R \; + \; Me—\overset{O}{\overset{\|}{C}}—\overset{-}{C}H—\overset{O}{\overset{\|}{C}}—Me \;\; Na^+ \; (57) \longrightarrow Me—\overset{O}{\overset{\|}{C}}—\underset{\underset{NH—CO_2R}{|}}{\underset{N—CO_2R}{|}}{CH}—\overset{O}{\overset{\|}{C}}—Me \; (56) \qquad (16.40)$$

16.4.2 *Synthesis of carbodiimides*

The reaction of diethylazo-dicarboxylate with thioureas (**58**) provides a good synthesis of carbodiimides (**59**) (Scheme 16.41).[82]

$$R^1—NH—\overset{S}{\overset{\|}{C}}—NH—R^1 \; (58) \; + \; EtO_2C—N{=}N—CO_2Et \longrightarrow EtO_2C—N(—S—C(=NR^1)NHR^1)—NHCO_2Et \xrightarrow{PPh_3} EtO_2C—NH—NH—CO_2Et + R^1—N{=}C{=}N—R^1 \; (59) \qquad (16.41)$$

16.4.3 *With silanes*

Addition of triarylsilanes to azo compounds occurs via dipolar intermediates, for example, Scheme 16.42.[83]

$$RO_2C—N{=}N—CO_2R \; + \; Ph_3SiH \longrightarrow Ph_3Si\underset{}{\overset{CO_2R}{\overset{|}{N}}}—NHCO_2R \qquad (16.42)$$

16.4.4 *Hydroboration*

Hydroboration products are formed with organoboranes in their reaction with electrophilic azo compounds (Scheme 16.43).[84]

$$RO_2C—N{=}N—CO_2R + R^1{}_3B \longrightarrow (RO_2C)(R^1{}_2B)N–NH(H)(CO_2R)\ \ \ \text{i.e.}\ \ R^1{}_2B(RO_2C)N–N(H)CO_2R + R^1{}_2B(RO_2C)N–N(R^1)CO_2R \quad (16.43)$$

AZOXY COMPOUNDS

$$R—N{=}\overset{\overset{O}{\uparrow}}{N}—R$$

(60)

Azoxy compounds (**60**) exist as both *cis* and *trans* diastereoisomers and may be prepared with the oxygen attached to either nitrogen atom.

16.5 Preparation[7,85]

16.5.1 *Oxidation of azo compounds*

Oxidation of azo compounds has been performed using peracetic acid, *m*-chloroperbenzoic acid, or hydrogen peroxide to provide a general synthesis of aliphatic azoxy compounds (Scheme 16.44).[86] For example, Scheme 16.45 shows the use of *m*-chloroperbenzoic acid.[87] See also Section 16.1.2.

$$RN{=}NR + R'CO_3H \longrightarrow R(O\leftarrow)N{=}NR \quad (16.44)$$

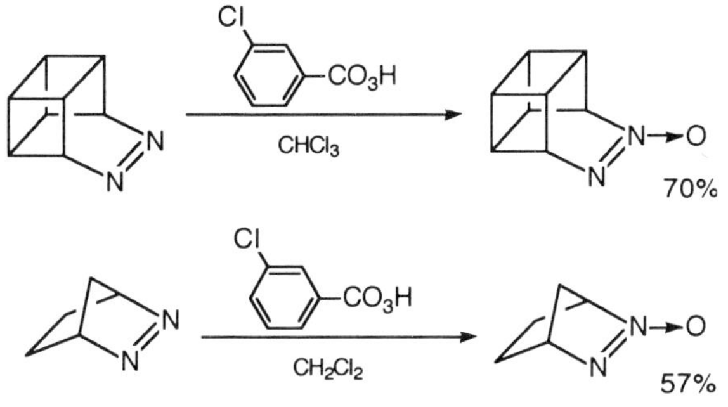

(16.45)

16.5.2 *Reduction of azodioxy compounds*

The conversion of these nitroso dimers (**61**) may be effected either by catalytic hydrogenation or using stannous chloride,[88] for example, Scheme 16.46.

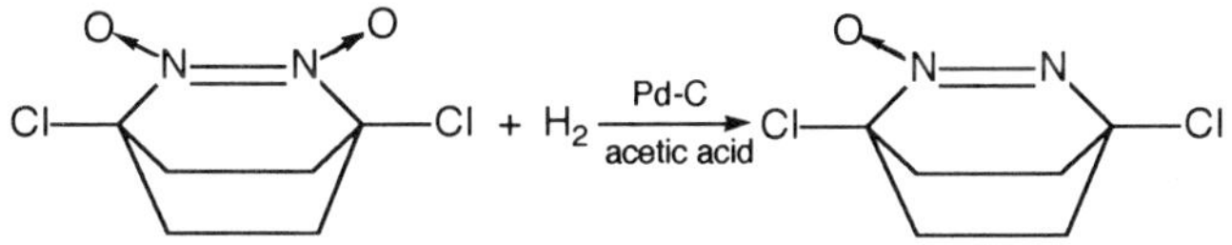

(61)

(16.46)

16.5.3 *Condensation of nitroso compounds*

With hydroxylamines

Azoxy compounds may be prepared by heating equimolar amounts of hydroxylamines (**62**) and nitroso compounds (**63**),[89] for example, Scheme 16.47.

R-NO (dimer) + RNHOH ⟶ RN(→O)=NR

(63) (62)

R	Yield (%)
c-C_6H_{11}-	47
1-adamantyl-	70–75

(16.47)

Asymmetric azoxy compounds have been prepared in this manner from methylhydroxylamine (**64**) and nitroso compounds with the oxygen ending up attached to the nitrogen originating from the nitroso compound,[90] for example, Scheme 16.48.

$$CH_3NHOH + Bu^tNO \longrightarrow CH_3N{=}\overset{O\uparrow}{N}{-}Bu^t \qquad (16.48)$$

(64)

With N,N-dichloramines

Azoxy compounds may be prepared by the condensation of *N*,*N*-dichloro compounds (**65**) with nitroso compounds in the presence of cuprous chloride (Scheme 16.49).[91]

$$R{-}NCl_2 + O{=}N{-}C_6H_5 \xrightarrow{Cu_2Cl_2} R{-}N{=}\underset{\downarrow O}{N}{-}C_6H_5 \quad 50\text{-}80\% \qquad (16.49)$$

(65)

16.5.4 *Oxidation of hydroxylamines*

A similar condensation takes place when hydroxylamines are partially oxidized to azoxy compounds using either air and a cobalt catalyst or lead tetraacetate,[92] for example, Scheme 16.50.

$$C_6H_{11}{-}NHOH + Pb(OAc)_4 \xrightarrow[0\text{-}10^\circ C]{CH_3OH} C_6H_{11}{-}\overset{O\uparrow}{N}{=}N{-}C_6H_{11} \quad 86\% \qquad (16.50)$$

16.5.5 *Alkylation of diazotates*

Reaction of alkane diazotates (**66**) prepared from amines[93] with a suitable alkylating reagent constitutes a good synthesis of azoxyalkanes.

Triethyloxonium fluoroborate (Meerwein's reagent) in dichloromethane may be used as the alkylating reagent producing good yields of azoxyalkanes (50–60 per cent) (Scheme 16.51).[94] Or similarly reaction of alkane diazotates (**66**) with alkyl halides, preferably iodides, in hexamethylphosphoric triamide (HMPT) may be the final step (Scheme 16.52).

$$\text{R-N=N-O}^-\text{K}^+ \ (66) + (C_2H_5)_3O^+BF_4^- \xrightarrow{CH_2Cl_2} \text{R-N=N(}\rightarrow\text{O)-}C_2H_5 + KBF_4 \quad 50\text{-}60\% \tag{16.51}$$

$$\text{R-N=N-O}^-\text{K}^+ \ (66) + R^1I \xrightarrow{HMPT} \text{R-N=N(O}^-\text{)-R}^1 + KI \quad 50\text{-}60\% \tag{16.52}$$

Either of these procedures may be applied to the synthesis of an optically pure azoxyalkane (**67**) with alkylation taking place by S_N2 attack. Partial isomerization may be induced by photochemical conversion to yield the isomeric azo alkane (**68**) (Scheme 16.53).[95] In the above reactions the oxygen is initially attached to the nitrogen which bears the group from the alkylating reagent.

$$\text{(R-)} \ H_3C(R)(H)C\text{-N=N-O}^-K^+ \xrightarrow[\text{in CHCl}_2 \text{ or EtI in HMPT}]{Et_3O^+BF_4^-} \text{(R-) (67)} \ H_3C(R)(H)C\text{N=N}^+(O^-)\text{Et} \xrightarrow{h\upsilon \ + \text{heat}} \text{(R-) (68)} \ H_3C(R)(H)C\text{-N}^+(O^-)\text{=NEt} \tag{16.53}$$

16.6 Reactivity

16.6.1 *Basicity*

Azoxy compounds are weaker bases than azo compounds with protonation occurring on the oxygen atom only in strong acidic media.[96]

16.6.2 *Rearrangement*

In acid aliphatic azoxy compounds undergo a rearrangement similar to the Wallach rearrangement. The product hydrazines are dependent upon the nature of the substituents:[97]

(1) When both substituents are primary alkyl the oxygen migrates to the nearest carbon atom (Scheme 16.54);

$$R{-}CH_2{-}\overset{\overset{O}{\uparrow}}{N}{=}NCH_2R^1 \xrightarrow{H^+} R{-}\overset{\overset{O}{\|}}{C}{-}NH{-}NH{-}CH_2R^1 \qquad (16.54)$$

(2) when the substituents are secondary alkyl groups the products are ketones, nitrogen, and hydrazine (Scheme 16.55).

$$R^1{-}\overset{H}{\underset{R^2}{C}}{-}\overset{\overset{O}{\uparrow}}{N}{=}N{-}\overset{H}{\underset{R^2}{C}}{-}R^1 \longrightarrow R^1R^2C{=}O + N_2 + NH_2NH_2 \qquad (16.55)$$

16.6.3 *Reduction to azo compounds*

Magnesium turnings in dry methanol have been used to reduce azoxy systems to azo compounds,[98] for example, Scheme 16.56. However, there has been more interest in reduction of alicyclic azoxy systems rather than open chain,[99] for example, Scheme 16.57.

$$Cl{-}C_6H_4{-}CH_2{-}\overset{\overset{O}{\uparrow}}{N}{=}N{-}CH_2{-}C_6H_4{-}Cl \xrightarrow{Mg,\ MeOH} Cl{-}C_6H_4{-}CH_2{-}N{=}N{-}CH_2{-}C_6H_4{-}Cl \qquad (16.56)$$

$$\text{bicyclic } N{=}N{\rightarrow}O \xrightarrow{LiAlH_4} \text{bicyclic } N{=}N + \text{bicyclic } HN{-}NH \qquad (16.57)$$

16.6.4 *Reduction to hydrazo compounds*

Azoxy compounds may be reduced either by catalytic hydrogenation (MeOH over PtO_2) or by using stannous chloride,[100] for example, Scheme 16.58. Diimide has been used for similar such reductions (see p. 609).

$$\mathrm{Me{-}N{=}\overset{\overset{O}{\uparrow}}{N}{-}Me} \xrightarrow[\mathrm{HCl}]{\mathrm{SnCl_2}} \mathrm{Me{-}NH{-}NH{-}Me} \qquad (16.58)$$

16.6.5 *Reductive cleavage*

Azoxy compounds may be cleaved by the action of metal/acid combinations, for example (Scheme 16.59).[101]

$$\mathrm{R^1{-}N{=}\overset{\overset{O}{\uparrow}}{N}{-}R^2} \xrightarrow[\mathrm{KOH,\ H_2O,\ MeOH}\ \mathrm{r.t.,\ 20h}]{\mathrm{Ni\text{-}Al\ Alloy}} \mathrm{R^1{-}NH_2} \qquad (16.59)$$

16.6.6 *Photochemical isomerism*

New aliphatic azoxy compounds may be synthesized by photolytic *trans* to *cis* isomerizations.[102]

References

1. Extensive reviews of preparations and reactions of aliphatic and aromatic azo compounds to 1973: S. Patai (ed.), *The chemistry of the hydrazo, azo and azoxy groups*, Parts 1 & 2, Wiley Interscience, London, 1975.
2. J. P. Freeman, *J. Org. Chem.*, 1963, **28**, 2508; G. H. Hartley, *Nature*, 1937, **140**, 281; G. H. Hartley, *J. Chem. Soc.*, 1938, 663.
3. C. H. Chang, R. F. Porter, and S. H. Bauer, *J. Am. Chem. Soc.*, 1970, **92**, 5313.
4. A. Trombetti, *J. Can. Phys.*, 1968, **46**, 1005.
5. A. Mostad and C. Rømming, *Acta Chem. Scand.*, 1971, **25**, 3561; H. Hope and D. Victor, *Acta Crystallogr.*, 1969, **B25**, 1849.
6. J. P. Freeman, *J. Org. Chem.*, 1963, **28**, 2508.
7. P. A. S. Smith, *The chemistry of open chain nitrogen compounds*, W. A. Benjamen, New York, 1966, Vol. 2.
8. E. Müller, *Methoden der organischen Chemie* (ed. Houben-Weyl), Georg. Thieme Verlag, Stuttgart, 1965, Vol. 10/3; H. Zollinger, *Azo and diazo chemistry*, Interscience Publishers, New York, 1961.

9. S. G. Cohen, S. J. Groszos, and D. B. Sparrow, *J. Am. Chem. Soc.*, 1950, **72**, 3947; J. R. Shelton and C. K. Liang, *Synthesis*, 1971, 204; S. E. Scheppele and S. Seltzer, *J. Am. Chem. Soc.*, 1968, **90**, 358.
10. S. Seltzer and F. T. Dunne, *J. Am. Chem. Soc.*, 1965, **87**, 2628.
11. C. G. Overberger and A. V. Di Guilio, *J. Am. Chem. Soc.*, 1958, **80**, 6562; (**6**): A. Tosolis, S. G. Mylonakis, M. T. Nieh, and S. Seltzer, *J. Am. Chem. Soc.*, 1972, **94**, 829; (**7**): S. Seltzer, *J. Am. Chem. Soc.*, 1963, **85**, 14; (**8**) K. R. Kopecky and T. Gillian, *Can. J. Chem.*, 1969, **47**, 2371; (**9**): R. C. Neuman, E. S. Neuman, and E. S. Alhardfef, *J. Org. Chem.*, 1970, **35**, 3401.
12. L. Spialter, D. H. O'Brien, G. L. Untereiner, and W. A. Reusch, *J. Org. Chem.*, 1965, **30**, 3278.
13. L. A. Carpino, P. H. Terry, and P. J. Crowley, *J. Org. Chem.*, 1961, **26**, 4336; H. Bock, *Angew. Chem., Int. Ed. Engl.*, 1965, **4**, 457; H. Bock and J. Kroner, *Chem. Ber.*, 1966, **99**, 2039.
14. J. A. Campbell, D. Mackay and T. D. Sauer, *Can. J. Chem.*, 1972, **50**, 371.
15. *F. &F.*, **5**, 491; M. L. Heyman and J. P. Snyder, *Tetrahedron Lett.*, 1973, 2859.
16. C. G. Overberger, J. P. Anselme, and J. G. Lombardino, *Organic compounds with nitrogen–nitrogen bonds*, Ronald Press, New York, 1966.
17. G. A. Mortimer, *J. Org. Chem.*, 1965, **30**, 1632.
18. G. A. Mortimer, *J. Org. Chem.*, 1965, **30**, 1632.
19. T. Tsuji and E. M. Kosower, *J. Am. Chem. Soc.*, 1971, **93**, 1992.
20. T. E. Stevens, *J. Org. Chem.*, 1961, **26**, 2531 or 3451.
21. S. F. Nelsen and P. D. Bartlett, *J. Am. Chem. Soc.*, 1966, **88**, 137; J. R. Shelton, J. F. Gormish, C. K. Liang, P. L. Samuel, and P. Kovacic, *Can. J. Chem.*, 1968, **46**, 1149; P. D. Bartlett and J. M. McBride, *Pure Appl. Chem.*, 1967, **15**, 89; J. W. Timberlake and J. C. Martin, *J. Org. Chem.*, 1968, **33**, 4054; M. Procazka, O. Ryba, and D. Lim, *Coll. Czech. Chem. Commun.*, 1968, **33**, 3387; D. J. Severn and E. M. Kossower, *J. Am. Chem. Soc.*, 1969, **91**, 1710; J. B. Levy and E. J. Lehmann, *J. Am. Chem. Soc.*, 1971, **93**, 5790.
22. H. Stetter and E. Smulders, *Chem. Ber.*, 1971, **104**, 917.
23. M. A. Berwick and R. E. Rondeau, *J. Org. Chem.*, 1972, **37**, 2409.
24. F. D. Greene, J. C. Stowell and W. R. Bergmark, *J. Org. Chem.*, 1969, **34**, 2254.
25. B. H. Al-Sader and R. J. Crawford, *Can. J. Chem.*, 1970, **48**, 2745; R. J. Crawford and K. Taguki, *J. Am. Chem. Soc.*, 1972, **94**, 7406.
26. B. H. Al-Sader and R. J. Crawford, *Can. J. Chem.*, 1970, **48**, 2745; R. J. Crawford and K. Takugi, *J. Am. Chem. Soc.*, 1972, **94**, 7406.
27. Radical preparations: E. Benzing, *Ann. Chem.*, 1960, **631**, 1; 10; S. Goldschmidt and B. Acksteiner, *Annalen*, 1958, **618**, 173; ionic preparation: D. S. Malament and J. M. McBride, *J. Am. Chem. Soc.*, 1970, **92**, 4586; 4593.
28. W. Duismann, H. D. Beckhaus, and C. Rüchard, *Tetrahedron Lett.*, 1974, 265; 1348.
29. P. L. Grizzle, D. W. Miller, and S. E. Scheppele, *J. Org. Chem.*, 1975, **40**, 1902.
30. S. Goldschmidt and B. Acksteiner, *Justus Liebigs Ann. Chem.*, 1958, **618**, 173.
31. E. Benzing, *Justus Liebigs Ann. Chem.*, 1960, **631**, 1.
32. J. W. Timberlake and J. C. Martin, *J. Org. Chem.*, 1968, **33**, 4054.
33. J. W. Timberlake and M. L. Hodges, *Tetrahedron Lett.*, 1970, 4147.
34. J. W. Timberlake, A. W. Garner, and M. L. Hodges, *Tetrahedron Lett.*, 1973, 301.
35. R. Ohme, H. Preuschhoff, and H.-U. Heyne, *Justus Liebigs Ann. Chem.*, 1968, **713**, 74.
36. *F. &F.*, **4**, 456; R. Ohme, H. Preuschhoff, and H.-U. Heyne, *Org. Synth.*, 1972, **52**,

11; R. Ohme, H. Preuschhoff, and H.-U. Heyne, *Justus Liebigs Ann. Chem.*, 1968, **713**, 74.
37. *F. &F.*, **4**, 58; J. W. Timberlake, M. L. Hodges, and K. Betterton, *Synthesis*, 1972, 632; J. W. Timberlake and M. L. Hodges, *J. Am. Chem. Soc.*, 1973, **95**, 634.
38. Preparation: R. Ohme and H. Preuschhoff, *Annalen*, 1968, **713**, 74.
39. *F. &F.*, **5**, 77; H.-H. Chang and B. Weinstein, *J. Chem. Soc., Chem. Commun.*, 1973, 397.
40. J. C. Stowell, *J. Org. Chem.*, 1967, **32**, 2360; P. S. Engel and D. J. Bishop, *J. Am. Chem. Soc.*, 1972, **94**, 2148; C. C. Wamser and P. L. Chang, *J. Am. Chem. Soc.*, 1973, **95**, 2044; F. D. Greene and S. S. Hecht, *J. Org. Chem.*, 1970, **35**, 2482.
41. J. W. Timberlake, M. L. Hodges, and A. W. Garner, *Tetrahedron Lett.*, 1973, 3843.
42. A. Isidorov, B. V. Ioffe, and I. G. Zenkevich, *Dokl. Chem. (Engl. Trans.)*, 1976, **230**, 584.
43. M. Isaks and H. H. Jaffe, *J. Am. Chem. Soc.*, 1964, **86**, 2209; S.-J. Yeh and H. H. Jaffe, *J. Am. Chem. Soc.*, 1959, **81**, 3279; 3283.
44. H. H. Jaffe and R. W. Gardner, *J. Am. Chem. Soc.*, 1958, **80**, 319.
45. F. Gerson and E. Heilbronner, *Helv. Chim. Acta*, 1962, **45**, 51.
46. F. B. Culp, A. Nabeya, and J. A. Moore, *J. Org. Chem.*, 1973, **38**, 2949.
47. O. Dimroth and M. Hartmann, *Chem. Ber.*, 1908, **41**, 4012; B. Eistert and M. Regitz, *Chem. Ber.*, 1963, **96**, 2290.
48. A. J. Bellamy, R. D. Gutherie, and G. J. F. Chittenden, *J. Chem. Soc. (C)*, 1966, 1989; A. J. Bellamy and R. D. Gutherie, *J. Chem. Soc.*, 1965, 2788; 3528.
49. G. J. Karabatsos and R. A. Taller, *J. Am. Chem. Soc.*, 1963, **85**, 3624; G. J. Karabatsos, F. M. Vane, R. A. Taller, and N. Hsi, *J. Am. Chem. Soc.*, 1964, **86**, 3351; M. C. Yao and P. Resnick, *J. Am. Chem. Soc.*, 1962, **84**, 3524.
50. B. V. Ioffe and S. V. Stopsky, *Tetrahedron Lett.*, 1968, 1333.
51. J. P. Freeman, *J. Org. Chem.*, 1963, **28**, 2508; B. W. Langley, B. Lythgoe, and L. S. Rayner, *J. Chem. Soc.*, 1952, 4191.
52. S. R. Sandler and W. Karo, *Organic functional group preparations*, (ed. T. A. Blomquist), Academic Press, NY, 1971, Vol. II.
53. B. T. Gillis and J. T. Hagarty, *J. Org. Chem.*, 1967, **32**, 95.
54. G. M. Badger, R. G. Buttery, and E. G. Lewis, *J. Chem. Soc.*, 1953, 2143.
55. H. Zollinger, *Azo and diazo chemistry—aliphatic and aromatic compounds*, Interscience Publishers, New York, 1961; K. L. Rinehart, *Oxidation and reduction of organic compounds*, Prentice-Hall, Englewood Cliffs, 1973.
56. L. A. Carpino, P. H. Terry, and P. J. Crowley, *J. Org. Chem.*, 1961, **26**, 4336.
57. D. Y. Curtin and T. C. Miller, *J. Org. Chem.*, 1960, **25**, 885; H. Bock, *Angew. Chem., Int. Ed. Engl.*, 1965, **4**, 457; M. L. Heyman and J. P. Snyder, *Tetrahedron Lett.*, 1975, 2859.
58. F. D. Greene, M. A. Berwick, and J. C. Stowell, *J. Am. Chem. Soc.*, 1970, **92**, 867.
59. G. E. Lewis and R. J. Mayfield, *Aust. J. Chem.*, 1966, **19**, 1445.
60. B. J. Newbold, *J. Chem. Soc.*, 1965, 6972.
61. M. P. Schneider and A. Rau, *J. Am. Chem. Soc.*, 1979, **101**, 4426.
62. M. Quast, A. Fuss, and A. Meublein, *Angew. Chem., Int. Ed. Engl.*, 1980, **19**, 49; J. Martelli and A. Gree, *Chem. Commun.*, 1980, 355; R. K. Huff and E. G. Savins, *Chem. Commun.*, 1980, 742.
63. R. W. Hoffmann and H. J. Luthardt, *Chem. Ber.*, 1968, **101**, 3851; 3861.
64. O. Mitsunoubu and N. Yoskida, *Tetrahedron Lett.*, 1981, **22**, 2295.
65. K. Narasaka, A. Movikawa, K. Saigo, and T. Mukaiyama, *Bull. Chem. Soc. Japan*, 1977, **50**, 2773.

66. J. T. Carlock and M. P. Mack, *Tetrahedron Lett.*, 1978, 5153.
67. D. A. Evans, T. C. Britton, R. L. Dorrow, and J. F. Delamina, *J. Am. Chem. Soc.*, 1986, **108**, 6395; L. A. Trimble and J. C. Vederas, *J. Am. Chem. Soc.*, 1986, **108**, 6397; C. Gennani, L. Colombo, and G. Bertolini, *J. Am. Chem. Soc.*, 1986, **108**, 6394.
68. G. S. Hammond, L. R. Mahoney, and U. S. Nandi, *J. Am. Chem. Soc.*, 1963, **85**, 740.
69. C. Walling and E. S. Huyser, *Org. React.*, 1963, **13**, 115.
70. C. Walling and B. B. Jacknow, *J. Am. Chem. Soc.*, 1960, **82**, 6108, 6113.
71. C. Walling and W. Thaler, *J. Am. Chem. Soc.*, 1961, **83**, 3877; C. A. Grob, H. Kny, and A. Gagneux, *Helv. Chim. Acta.*, 1957, **40**, 130.
72. J. A. Clark and O. Meth-Cohn, *Tetrahedron Lett.*, 1975, 4705.
73. *F. &F.*, **2**, 118.
74. M. G. Kuivila, *Acc. Chem. Res.*, 1968, **1**, 299; H. G. Kuivila and L. W. Menapace, *J. Org. Chem.*, 1963, **28**, 2165.
75. A. P. Marchand and W. R. Weimar, Jr, *J. Org. Chem.*, 1969, **34**, 1109.
76. R. J. Strunk, P. M. DiGiacomo, K. Also, and H. G. Kuivila, *J. Am. Chem. Soc.*, 1970, **92**, 2849.
77. *F. &F.*, **3**, 294; H. G. Kuivila, *Synthesis*, 1970, 499.
78. Review: E. Fahr and H. Lind, *Angew. Chem., Int. Ed. Engl.*, 1966, **5**, 372.
79. R. Huisgen, F. Jakob, W. Siegel, and A. Cadus, *Ann. Chem.*, 1954, **590**, 1; R. B. Carlin and W. S. Moores, *J. Am. Chem. Soc.*, 1962, **84**, 4107.
80. E. E. Smissman and A. Makriyannis, *J. Org. Chem.*, 1973, **38**, 1652.
81. O. Diels and H. Behnike, *Chem. Ber.*, 1924, **57**, 653.
82. O. Mitsunobu, K. Kato, and M. Tomari, *Tetrahedron*, 1970, **26**, 5731.
83. K. H. Linke and H. J. Gohausen, *Chem. Ber.*, 1973, **103**, 3438.
84. A. G. Davis, B. R. Roberts, and J. C. Scaiano, *J. Chem. Soc., Perkin Trans. I*, 1972, 803.
85. Reviews for the synthesis of aromatic and aliphatic azoxy compounds: E. Müller, *Methoden der organischen Chemie* (Houben-Weyl), Georg Thieme Verlag, Stuttgart, 1967, Vol. 10/2; E. Müller, *Methoden der organischen Chemie* (Houben-Weyl), Georg Thieme Verlag, Stuttgart, 1965, Vol. 10/3; S. R. Sandler and W. Karo, *Organic functional group preparations*, Academic Press, New York, 1971, Vol. 2.
86. E. Müller, *Methoden der organischen Chemie* (ed. Houben-Weyl), Georg Thieme Verlag, Stuttgart, 1967, Vol. 10/2; J. P. Freeman, *J. Org. Chem.*, 1963, **28**, 2508; F. D. Greene and S. S. Hecht, *J. Org. Chem.*, 1970, **35**, 2482.
87. J. P. Snyder, L. Lee, and F. D. Farnum, *J. Am. Chem. Soc.*, 1971, **93**, 3816; F. D. Greene and S. S. Hecht, *J. Org. Chem.*, 1970, **35**, 2482.
88. E. Müller, *Methoden der organischen Chemie* (ed. Houben-Weyl), Georg. Thieme Verlag, Stuttgart, 1965, Vol. 10/3; W. Lüttke and V. Schubacker, *Justus Liebigs Ann. Chem.*, 1965, **687**, 236.
89. E. Müller, *Methoden der organischen Chemie* (ed. Houben-Weyl), Georg Thieme Verlag, Stuttgart, 1967, Vol. 10/2; 1965, Vol. 10/3; H. Meister, *Justus Liebigs Ann. Chem.*, 1964, **679**, 83; H. Stetter and E. Smulders, *Chem. Ber.*, 1971, **104**, 917.
90. J. P. Freeman, *J. Org. Chem.*, 1963, **28**, 2508.
91. *F. &F.*, **6**, 146; V. Nelson and P. Kovacic, *J. Chem. Soc., Chem. Commun.*, 1975, 312.
92. H. Meister, *Justus Liebigs Ann. Chem.*, 1964, **679**, 83.
93. R. A. Moss, *J. Org. Chem.*, 1966, **31**, 1082.

94. *F. &F.*, **4**, 528; R. A. Moss, M. J. Landon, K. M. Luchter, and A. Mamantov, *J. Am. Chem. Soc.*, 1972, **94**, 4392.
95. *F. &F.*, **5**, 692; R. A. Moss and G. M. Love, *J. Am. Chem. Soc.*, 1973, **95**, 3070.
96. C. S. Hahn and H. H. Jaffe, *J. Am. Chem. Soc.*, 1962, **84**, 949.
97. J. N. Brough, B. Lythgoe, and P. Waterhouse, *J. Chem. Soc.*, 1954, 4069.
98. B. W. Langley, B. Lythgoe, and L. S. Rayner, *J. Chem. Soc.*, 1952, 4191.
99. F. D. Greene and S. S. Hecht, *Tetrahedron Lett.*, 1969, 2818.
100. B. W. Langley, B. Lythgoe, and L. S. Rayner, *J. Chem. Soc.*, 1952, 4191.
101. G. Lunn, E. G. Sansone, and L. K. Keefer, *Synthesis*, 1985, 1104.
102. K. G. Taylor and T. Riehl, *J. Am. Chem. Soc.*, 1972, **94**, 250.

17

AZIDES

$R\text{-}N_3$ (1) (alkyl) $\qquad$ $R\text{-}CO\text{-}N_3$ (2) (acyl)

$$R{-}N{=}\overset{+}{N}{=}\overset{-}{N} \longleftrightarrow R{-}\overset{-}{N}{-}\overset{+}{N}{\equiv}N$$

(A) $\qquad$ (B)

(3)

Alkyl (**1**) and acyl (**2**) azides contain three contiguous nitrogen atoms, comprising the azide group, $-N_3$, which is stabilized by delocalization (**3**). Methyl azide has the bond lengths and bond angles shown, which fit the values

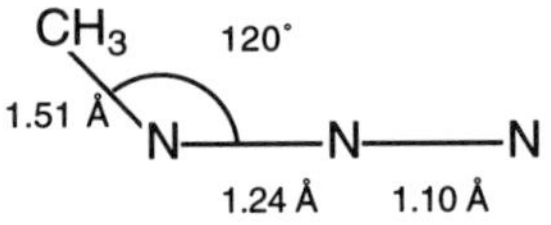

CH_3 120° 1.51 Å N—N—N 1.24 Å 1.10 Å

calculated for equal contributions from the hybrids of (**3**), viz. (A) and (B). The small dipole moment of 1.5 D is also in accord with this interpretation of the structure. The anti-symmetric stretching of N=N appears in the infra-red as a strong band in the region 2070–2160 cm^{-1}. The Raman spectrum has a strong anti-symmetric N—N—N stretch at 2080–2170 cm^{-1} and a symmetric N—N—N stretch at 1175–1345 cm^{-1}.

ALKYL AZIDES

17.1 Preparation

17.1.1 *By the use of the azide ion in S_N2 reactions*[1]

From alkyl halides or sulphates

Using sodium azide Alkyl azides can be prepared from alkyl halides (Scheme 17.1) or by the reaction between alkyl sulphates ($R^1{-}OSO_3R^2$) (**4**) and the azide ion, for example, the reaction between dimethylsulphate (**4**) and sodium azide, Scheme 17.2.[6]

$$R\text{-}Hal + N_3^- \longrightarrow R\text{-}N_3 + Hal^-$$

R-Hal	Yield (%)	Ref.
1-iodopropane	64	2
1-bromobutane	90	3
1-bromocyclopentane	82	2
1-iodocyclohexane	75	4
1,3-dichloropentane	68	5

(17.1)

$$\underset{(4)}{(CH_3)_2SO_4} + N_3^- \longrightarrow CH_3N_3 + CH_3SO_4^- \qquad (17.2)$$

Tertiary, allylic, and benzylic azides (**5**) can be prepared from the corresponding halides (**6**) by the action of sodium azide (**7**) catalysed by zinc chloride (Scheme 17.3).[7]

$$\underset{(6)}{R\text{-}Cl} + \underset{(7)}{NaN_3} \xrightarrow[\text{10 to 100 hours}]{ZnCl_2 / CS_2 / 20^\circ C} \underset{(5)}{R\text{-}N_3} + NaCl$$

$R\text{-}N_3$	Yield (%)
Bu^tN_3	96
Am^tN_3	89
1-methylcyclohexyl azide	86
1-adamantyl azide	72

(17.3)

Phase transfer greatly improves the reaction of sodium azide with alkyl halides.[8] Other authors used methyl tricapryl-ammonium chloride[9] (the capryl group consists of mixed C_8–C_{12} chains) or methyl tri-octylammonium chloride as the phase transfer catalysts with aqueous sodium azide. Alkyl bromides and iodides gave 74–93 per cent yields.

Later sodium azide with Amberlite IR-400 was found to convert alkyl bromides, iodides, and tosylates into azides at room temperature in yields of 100 per cent.[10] The solvents used were acetonitrile, chloroform, ether, or dimethylformamide.

Use of the tetramethylguanidinium cation The tetramethylguanidinium cation (**8**) can be used as a source of azide ion to prepare azides (**9**) from alkyl halides or by the action of nitriles to produce tetrazoles (**10**) (Scheme 17.4).[11] This gives high yields with short reaction times of 1–3 h.

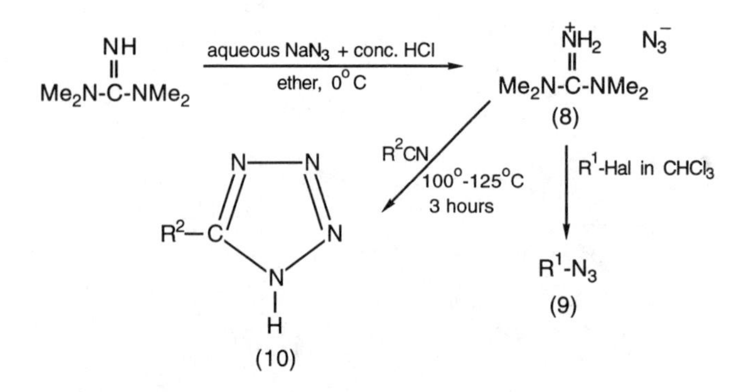

R^1Hal	Yield (%) azide
β-bromoethylbenzene	100
ethyl α-chloroacetate	89
1-chlorohexane	60

R^2	Yield (%) tetrazole
Ph	100
$PhCH_2$	94
Me_2N	83

(17.4)

Using trimethylsilyl azide Trimethylsilyl azide (Me_3SiN_3) (**11**) can be prepared as shown in Scheme 17.5,[12] and can be used to obtain good yields (65–85 per cent) of alkyl azides (**12**) from alkyl halides in hexamethylphosphoric triamide, $(Me_2N)_3P{=}O$, at 60 °C (Scheme 17.6).[13]

$$Me_3Si\text{-}Cl + NaN_3 \xrightarrow{THF} Me_3Si\text{-}N{=}\overset{+}{N}{=}\overset{-}{N} + NaCl \quad \text{(11) 87 \%} \tag{17.5}$$

$$R\text{-}Hal \xrightarrow[\text{HMPT, }60^\circ C]{Me_3SiN_3} R\text{-}N{=}\overset{+}{N}{=}\overset{-}{N} \quad \text{(12)} \tag{17.6}$$

From epoxides

Using sodium azide[14] As expected from the S_N2 mechanism of the ring opening of epoxides (**13**) by nucleophiles, the azide ion acts both regiospecifically and stereospecifically to produce β-hydroxyazides (**14**) (Scheme 17.7).

Using an azido titanium compound Alternatively the azido titanium compound (**15**) can be used to open epoxides (**16**) to give β-hydroxyazides (**17**) (Scheme 17.8).[15]

Using alkylsilyl azides Trimethylsilylmethyl azide ($Me_3SiCH_2N_3$) (**18**) can be prepared in high yield as shown in Scheme 17.9.[16]

$$CH_3\text{-}CH(O)CH_2 \ (13) \xrightarrow[\text{dioxan, reflux}]{\text{satd. aqueous } NaN_3} CH_3\text{-}CH(OH)\text{-}CH_2N_3 \ (14)$$

$$\text{cyclopentene oxide (13)} \xrightarrow[\text{dioxan, reflux}]{\text{satd. aqueous } NaN_3} \text{trans-2-azidocyclopentanol (14)}$$

Epoxide	Yield (%)
2,3-epoxybutane	52.5
cyclopentene oxide	64.0
cyclohexene oxide	61.4

(17.7)

$$CH_3(CH_2)_2\text{—}CH(O)CH\text{—}CH_2\text{-}OH \ (16) \xrightarrow[\text{heat}]{(15),\ PhH} CH_3(CH_2)_2\text{—}CH(N_3)\text{-}CH(OH)\text{—}CH_2\text{-}OH \ (17)$$

$(N_3)_2Ti(OPr^i)_2$ (15)

+ a small amount of

$$CH_3(CH_2)_2\text{—}CH(OH)\text{-}CH(N_3)\text{—}CH_2\text{-}OH \ (17)$$

(17.8)

$$Me_3SiCH_2Cl + NaN_3 \xrightarrow[\text{or DMF}]{\text{HMPT}} Me_3SiCH_2N{=}\overset{+}{N}{=}\overset{-}{N} \ (18)$$

(17.9)

Azides (**19**) can be prepared by the opening of epoxides (**20**) with trimethylsilyl azide and aluminium isopropoxide in methylene dichloride (Scheme 17.10).[17]

17.1.2 *From diazoalkanes*

Azides can be prepared from a diazoalkane (**21**) by reaction with hydrazoic acid (see Chapter 13, p. 524) (Scheme 17.11).[18]

$$AcO\text{-}CH_2\text{-}\underset{(20)}{\text{epoxide}} \xrightarrow[Al(OPr^i)_3,\ CH_2Cl_2]{Me_3SiN_3} \underset{(19)}{AcO\text{-}CH_2CH(OSiMe_3)\text{-}CH_2N_3}\quad 77\ \%$$

$$\underset{(20)}{\text{cyclohexene oxide}} \xrightarrow[Al\ (OPr^i)_3,\ CH_2Cl_2]{Me_3SiN_3} \text{2-}(N_3)\text{-1-}(OSiMe_3)\text{cyclohexane}\quad 74\ \% \qquad (17.10)$$

$$\underset{(21)}{CH_2N_2} + HN_3 \longrightarrow CH_3\text{-}\overset{+}{N}\equiv N + N_3^- ;\quad CH_3\text{-}\overset{+}{N}\equiv N \rightarrow N^2 + \overset{+}{C}H_3;\quad \overset{+}{C}H_3 + N_3^- \longrightarrow CH_3N_3 \qquad (17.11)$$

17.1.3 *From aldehydes and ketones*

Trimethylsilyl azide has been used in two procedures to convert aldehydes and ketones (**22**) into *gem*-diazides (**23**) when catalysed with $ZnCl_2$ or $SnCl_2$ or into α-silyloxy azides (**24**) when catalysed solely by sodium azide and 5-crown-15 (Scheme 17.12).[19]

$$\underset{(22)}{R^1R^2C{=}O} \xrightarrow[ZnCl_2\ \text{or}\ SnCl_2]{Me_3SiN_3} \underset{(23)}{R^1R^2C(N{=}\overset{+}{N}{=}\overset{-}{N})_2}$$

$$\underset{(22)}{R^1R^2C{=}O} \xrightarrow[\text{5-crown-15}]{NaN_3} \underset{(24)}{R^1R^2C(O\text{-}SiMe_3)(N{=}\overset{+}{N}{=}\overset{-}{N})}$$

$$R^2 = H \text{ or alkyl} \qquad (17.12)$$

17.1.4 *From Grignard reagents or lithium derivatives of primary amines by diazo transfer*

Tosyl azide (**25**) reacts with Grignard reagent (**26**) derived from an amine (**27**) by diazo transfer to yield an alkyl azide (**28**),[20] for example, Scheme 17.13. Similarly, the lithium derivative of a primary amine (**29**) reacts with tosyl azide to form an azide by diazo transfer (Scheme 17.14).

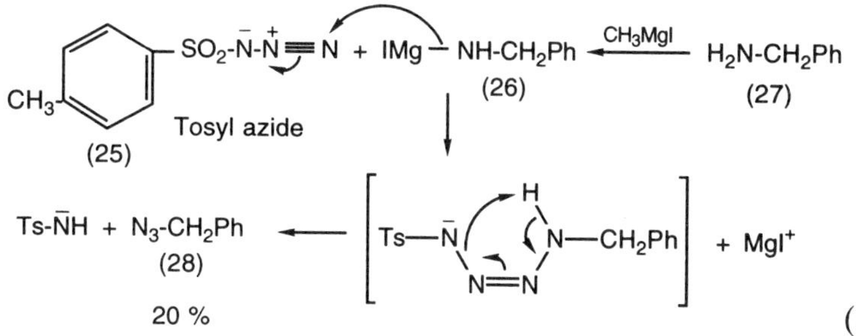

(17.13)

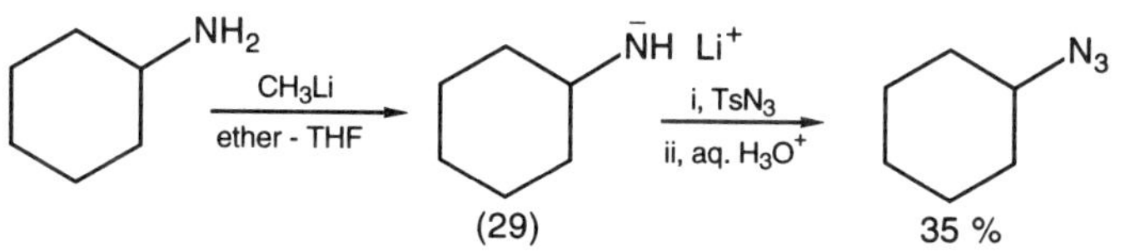

(17.14)

For diazo transfer as above, *p*-dodecylbenzene sulphonyl azide or naphthalene-2-sulphonyl azide are less explosive than tosyl azide which is usually used.[21]

The compound *p*-$AcNHC_6H_4SO_2N_3$ (**30**), which may be prepared by reaction of the sulphonyl chloride (**31**) with sodium azide as shown in Scheme 17.15, has been used as a safer azide than tosyl azide for diazo transfer reactions and for the synthesis of vinyl diazo compounds (**32**) (Scheme 17.15).[22]

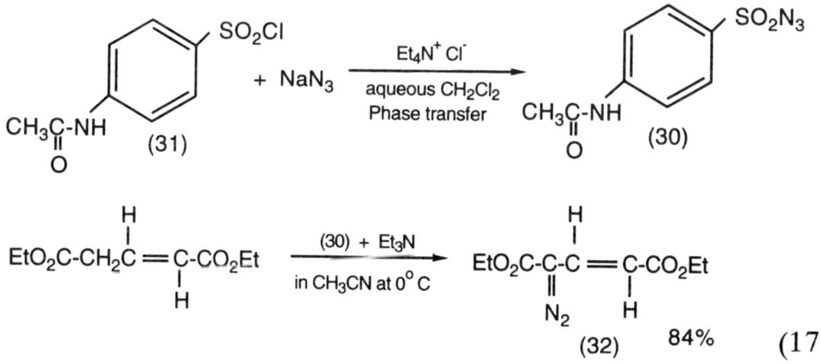

(17.15)

17.1.5 *By addition to alkenes*

Using bromine azide

As expected, bromine azide, BrN_3 (**33**), adds by an ionic mechanism stereospecifically *trans* across alkenes (**34**).[23] Bromine azide is prepared as shown in Scheme 17.16, but is not isolated.

$$NaN_3 \text{ in } CH_2Cl_2 \text{ at } 0^\circ C \xrightarrow[\text{ii, } Br_2]{\text{i, 30 \% aq. HCl}} BrN_3 \text{ (used } \textit{in situ}\text{)} \; (33)$$

$$\text{cis-}CH_3CH{=}CHCH_3 \; (34) \text{ and trans-}CH_3CH{=}CHCH_3 \; (34) \xrightarrow{BrN_3 \text{ in } CH_3NO_2 - CH_2Cl_2} CH_3CH(Br)\text{-}CH(N_3)CH_3 \quad (17.16)$$

Using hydrazoic acid

1,1-Disubstituted or trisubstituted alkenes (**35**) react with hydrazoic acid in the presence of a Lewis acid ($TiCl_4$ or $AlCl_3$) as catalyst to yield azides (**36**) (Scheme 17.17).[24]

$$>C{=}C< \; (35) \xrightarrow[AlCl_3 \text{ or } TiCl_4]{HN_3} >CH\text{-}C(N_3)< \; (36) \quad (17.17)$$

Via organoboranes

Azides can be prepared in good yield (*c.* 90 per cent) from organoboranes (**37**) by treatment with hydrogen peroxide and ferric azide which is prepared *in situ* from sodium azide and ferric sulphate (Scheme 17.18).[25]

$$3\ R\text{-}CH{=}CH_2 \xrightarrow{B_2H_6} (R\text{-}CH_2CH_2)_3B \; (37) \xrightarrow[MeOH]{Fe(N_3)_3 + H_2O_2} R\text{-}CH_2CH_2N_3$$

Alkene	Yield (%)
cyclopentene	100
$Bu^nCH{=}CH_2$	85

(17.18)

Using mercuric azide

Mercuric azide (**38**), $Hg(N_3)_2$, generated *in situ* from mercuric acetate and sodium azide in 50 per cent aqueous tetrahydrofuran (THF),[26] will add across a terminal double bond in an alkene (**39**) to produce a mercuro-organic intermediate (**40**), which can be reduced *in situ* with sodium borohydride to give azides (Scheme 17.19). The reaction is regiospecific as shown.

$$(39)\ \text{Terminal only} + Hg(N_3)_2\ (38) \longrightarrow f(40) \xrightarrow{NaBH_4}$$

Alkene	Yield (%)
$C_5H_{11}CH{=}CH_2$	83
methylenecyclohexane ($=CH_2$)	60

(17.19)

17.1.6 *Alkylaryl azides*

Alkylaryl azides (**41**) can be prepared by the action of trimethylsilyl azide and dichlorodicyanobenzoquinone (DDQ) on benzylic compounds (**42**) (Scheme 17.20).[27]

$$\underset{(42)}{Ar\text{-}CH(R^1)\text{-}R^2} \xrightarrow[CHCl_3,\ 25^\circ C]{DDQ,\ Me_3SiN_3} \underset{(41)}{Ar\text{-}C(R^1)(N{=}\overset{+}{N}{=}\overset{-}{N})\text{-}R^2} \quad 50\text{ - }95\ \%$$

(17.20)

17.1.7 *By azide transfer*

To enols

The electron-rich, hindered azide, 2,4,6-triisopropylbenzenesulphonyl azide (**43**)[28] reacts with enolates (**44**) to produce overall azide transfer,[29] for example, Scheme 17.21.

When R = CMe_3 a better final hydrolysis can be achieved by the use of hydrogen peroxide (4 mole) and lithium hydroxide (3 mole) as hydrolysing agent.[30] Less electron-rich azides, for example *p*-nitrobenzenesulphonyl azide (**45**) cause an azide transfer with the sodium enolate (Scheme 17.22).

Pr^i SO_2Cl Pr^i Pr^i + NaN_3 → Pr^i SO_2N_3 Pr^i Pr^i

(43) 80%

$R{-}CH_2{-}CO{-}N$ (oxazolidinone, Bn) (44)

i, $K^+\ N^-(SiMe_3)_2$

ii, $ArSO_2N_3$ / THF / $PhCH_3$

iii, HOAc

→ $R{-}C(H)(N_3){-}CO{-}N$ (oxazolidinone, Bn) 74 - 91 %

$LiOH / H_2O / THF$

ee > 91 % $R{-}CH(N_3){-}C(=O){-}OH$

Mechanism :

(43)

$R{-}CH(N_3){-}C(=O){-}NR^1R^2$

(17.21)

$NO_2{-}C_6H_4{-}SO_2N_3$

(45)

$R{-}CH{=}C(O^-Na^+){-}N$ (oxazolidinone, Bn)

i, reagents as in Scheme 17.21

ii, phosphate buffer, pH 7.0

→ $R{-}C(H)(N_3){-}CO{-}N$ (oxazolidinone, Bn) 85 %

(17.22)

17.1.8 *α-Azidoethers*

Hydrazoic acid adds to enol ethers (**46**) to yield α-azidoethers (**47**) (Scheme 17.23).[31]

$$R^1\text{-CH=C(OR}^3\text{)-R}^2 \ (46) \xrightarrow{HN_3} R^1\text{-CH(N}_3\text{)-C(H)(OR}^3\text{)-R}^2 \ (47) \qquad (17.23)$$

17.2 Reactivity

Consideration of the canonical form (B) indicates that one can expect electrophiles to attack at the nitrogen atom adjacent to the alkyl group (1) and nucleophiles at the terminal nitrogen atom (3) (Scheme 17.24). In addition, azides being 1,3-dipoles (C) undergo 1,3-dipolar additions with unsaturated compounds (Scheme 17.25). Azides also readily lose nitrogen to yield nitrenes (Scheme 17.26).

$$R\text{—}\overset{-}{\underset{1}{N}}\text{—}\overset{+}{\underset{2}{N}}\equiv\underset{3}{N} \quad (B)$$

Electrophiles : $E^+ \curvearrowright R\text{—}\overset{-}{N}\text{—}\overset{+}{N}\equiv N$

Nucleophiles : $R\text{—}\overset{-}{N}\text{—}\overset{+}{N}\equiv N \curvearrowleft Nu$ (17.24)

$$R\text{—}\overset{-}{N}\text{—}\overset{+}{N}\equiv N \ (B) \longleftrightarrow R\text{—}\overset{-}{N}\text{—}N\text{=}\overset{+}{N} \ (C) \qquad (17.25)$$

$$R\text{—}\overset{-}{N}\text{—}\overset{+}{N}\equiv N \xrightarrow[\text{or heat}]{h\nu} R\text{—}\ddot{N}: + N_2 \qquad (17.26)$$

17.2.1 *With nucleophiles*

Basic hydrolysis

Alkyl azides behave with the hydroxyl ion in aqueous media like alkyl halides, i.e. they either undergo S_N2 reactions or eliminations (Scheme 17.27).[32]

$$\gt C-N=\overset{+}{N}=\overset{-}{N} + HO^- \longrightarrow \gt C\text{-}OH + \overset{-}{N}=\overset{+}{N}=\overset{-}{N} \tag{17.27}$$

Elimination occurs if there is an easily ionizable hydrogen atom β to the azide group, for example, Scheme 17.28.

$$N_3-CH_2-CH(H)-CO_2Et + {}^-OH \longrightarrow N_3^- + CH_2{=}CH\text{-}CO_2Et + H_2O \tag{17.28}$$

Reduction to amines

For a review of reducing agents for azides see Ref. 33.

Stannous chloride (**48**), thiophenol (**49**), and ethylamine (**50**) react in benzene to yield a solution which rapidly and almost quantitatively reduces azides to spectroscopically pure amines (Scheme 17.29).[34] The mechanism proposed is as shown in Scheme 17.30.

$$\underset{(48)}{SnCl_2} + \underset{(49)}{PhSH} + \underset{(50)}{Et_3N} \xrightarrow[20^\circ C]{Ph\,H} Et_3NH^+\ Sn(SPh)_3^- \tag{17.29}$$

Mechanism

$$R\text{-}\overset{-}{N}\text{-}\overset{+}{N}\equiv N + \overset{-}{Sn}(SPh)_3\ Et_3\overset{+}{N}H \longrightarrow R\text{-}\overset{-}{N}\text{-}N{=}N\text{-}Sn(SPh)_3\ Et_3\overset{+}{N}H$$

$$\rightleftharpoons R\text{-}NH\text{-}N{=}N\text{-}Sn(SPh)_3 + Et_3N \longrightarrow N_2 + R\text{-}NH\text{-}Sn(SPh)_3 \xrightarrow{H_2O} R\text{-}NH_2$$

$$\text{2-methyl-1,3-dioxolan-2-yl-}CH_2\text{-}C(Me)_2N_3 \xrightarrow{Sn(II),PhSH,Et_3N} \text{2-methyl-1,3-dioxolan-2-yl-}CH_2\text{-}C(Me)_2NH_2 \quad 95\text{-}98\% \tag{17.30}$$

Catalytic hydrogenation takes place under mild conditions,[35] for example, Scheme 17.31. The same reduction can be carried out with lithium aluminium hydride in an ethereal solution (Scheme 17.32).[36] Reduction with lithium

$$\begin{array}{l} H_2C-N_3 \\ N_3-CH \\ HC-N_3 \\ H_2C-N_3 \end{array} \xrightarrow[\text{MeOH, 1 h, 20°C}]{H_2,\text{Pd-C}} \begin{array}{l} H_2C-NH_2 \\ H_2N-CH \\ HC-NH_2 \\ H_2C-NH_2 \end{array} \quad >80\% \qquad (17.31)$$

$$4\ RN_3 + LiAlH_4 \longrightarrow (RNH)_4AlLi + N_2 \longrightarrow 4\ RNH_2$$

Mechanism:

$$LiAlH_3\text{-}H + \overset{}{N}\equiv\overset{+}{N}-\overset{-}{N}\text{-}R \longrightarrow H\text{-}N{=}N\text{-}\overset{-}{N}\text{-}R \rightleftharpoons \overset{-}{N}{=}N\text{-}NH\text{-}R \xrightarrow{H_3O^+} N_2 + H_2N\text{-}R$$

for example:

$$Ph\text{-}CH_2\text{-}CH_2\text{-}N_3 \longrightarrow Ph\text{-}CH_2\text{-}CH_2\text{-}NH_2 \quad 89\ \% \qquad (17.32)$$

aluminium hydride has the great advantage that optical activity is retained in the resulting amine (Scheme 17.33).[37]

$$(CH_3\text{-}CH_2)(CH_3)(H)C\text{-}N_3\ (+) \longrightarrow (CH_3\text{-}CH_2)(CH_3)(H)C\text{-}NH_2\ (+) \quad 70\ \% \quad \text{No racemization} \qquad (17.33)$$

Reduction of an α-azidoketone (**51**) with lithium aluminium hydride furnishes a route to an α,β-amino alcohol (**52**), for example, Scheme 17.34.

$$\underset{N_3\ (51)}{CH_2\text{-}CO\text{-}CH_3} \xrightarrow{LiAlH_4} \underset{NH_2\ (52)}{CH_2\text{-}CHOH\text{-}CH_3} \quad 48\ \% \qquad (17.34)$$

Reduction to a primary amine (**53**) can also be achieved with diborane,[38] for example, Scheme 17.35.

Another way for reducing azides to amines utilizes palladium charcoal, hydrogen, and sodium hypophosphate,[39] for example, Scheme 17.36.

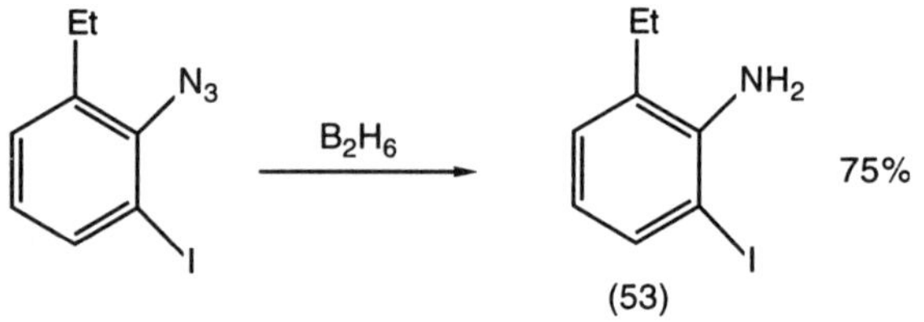

Et, N_3, I $\xrightarrow{B_2H_6}$ Et, NH_2, I (53) 75% (17.35)

N_3, N–H, O $\xrightarrow[Na_2CO_3, THF]{NaH_2PO_2, Pd-C,}$ NH_2, N–H, O 73% (17.36)

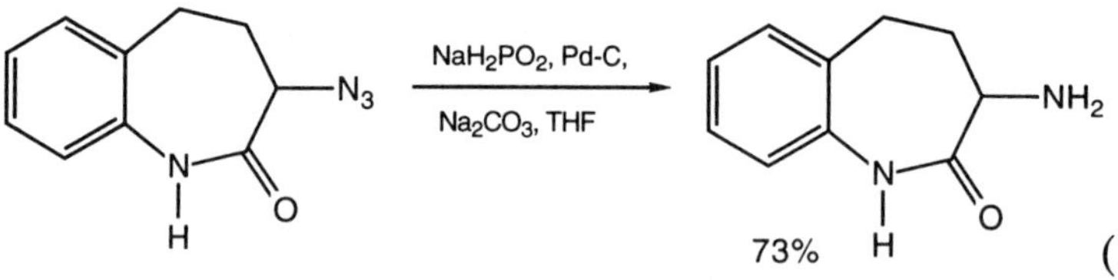

Synthesis of triazoles

With the enolate anions from acetoacetic ester, malonic ester, and cyanoacetic ester triazoles are produced, for example, from cyanoacetic ester (**54**) (Scheme 17.37).[40]

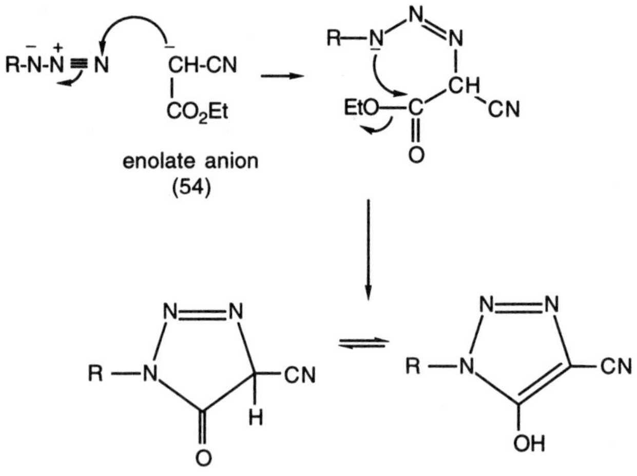

R-N-N≡N + CH-CN, CO_2Et (enolate anion (54)) → R–N, N, N, CH, CN, EtO–C=O → R–N, N=N, CN, H, O ⇌ R–N, N=N, CN, OH (17.37)

The Staudinger reaction[41]

The reaction involves the attack of a substituted phosphine (**55**), acting as a nucleophile, on an azide (**56**). Nitrogen is expelled and a phosphazo compound (**57**) is the product (Scheme 17.38). The accepted mechanism of the reaction is as shown in Scheme 17.39.

The intermediate phosphazide (**58**) is thought to be linear, not branched, i.e. the phosphine attacks the terminal nitrogen atom, and the disposition of the

$$\underset{(55)}{R^1{}_3P} + \underset{(56)}{N_3\text{-}R^1} \longrightarrow \underset{\text{Phosphazide}}{R^1{}_3P\text{=}N\text{-}N\text{=}N\text{-}R^2} \xrightarrow{-N_2} \underset{\text{Phosphazo compound}}{\underset{(57)}{R^1{}_3P\text{=}N\text{-}R^2}} \qquad (17.38)$$

$$R^1_3P: + {}^{15}N\equiv\overset{+}{N}-\overset{-}{N}-R^2 \rightleftharpoons \left[R^1_3\overset{+}{P}-{}^{15}N\text{=}N-\overset{-}{N}-R^2 \longleftrightarrow R^1_3P\text{=}{}^{15}N-N\text{=}N-R^2 \right] (58)$$

$$\longrightarrow \left[R_3P\cdots\overset{-}{N}\text{-}R^2 \;;\; {}^{15}N\equiv N \right] (59) \longrightarrow R^1_3P\text{=}N\text{-}R^2 + {}^{15}N\equiv N \qquad (17.39)$$

isotopically labelled nitrogen atom (done in an experiment where $R^1 = Ph$ and $R^2 = Ts$) shows that the $N—R^2$ bond does not break and is incorporated into the phosphazo product. Furthermore, there is evidence for the four-membered ring intramolecular transition state (**59**).

There are a large number of applications of the Staudinger reaction in synthesis. Reaction of a phosphazo compound (**60**) with an acid chloride (**61**) yields a halogenated imine (an imidoyl halide) (**62**) (Scheme 17.40).

$$\underset{(60)}{R^1{}_3P\text{=}N\text{-}R^2} + \underset{(61)}{R^3\text{-COCl}} \longrightarrow R^1{}_3P\text{=}\overset{+}{N}(—R^2)—C(\text{=}O)—R^3 \;\; Cl^- \longleftrightarrow R^1{}_3\overset{+}{P}—N(—R^2)—C(\text{=}O)—R^3 + Cl^-$$

$$\rightleftharpoons R^1{}_3\overset{+}{P}—N(—R^2)—C(—{}^-O)(—R^3)(—Cl) \longrightarrow R^1{}_3P\text{=}O + \underset{(62)}{R^3—C(\text{=}N\text{-}R^2)—Cl} \qquad (17.40)$$

With Grignard reagents

As expected, a Grignard reagent (**63**), acting as a nucleophile, attacks an azide at N-3 to yield unstable compounds which contain three contiguous nitrogen atoms (**64**) (Scheme 17.41). Yields are generally poor (22 per cent from methyl azide and methylmagnesium iodide),[42] and the reaction is of little value though the products are more stable if the R groups are aromatic (**65**),[43] for example, Scheme 17.42.

$$R^1\text{-}\bar{N}\text{-}\overset{+}{N}\equiv N \quad R^2\text{-}MgX\ (63) \longrightarrow R^1\text{-}\bar{N}\text{-}N{=}N\text{-}R^2 + MgBr^+ \xrightarrow{H_3O^+} R^1\text{-}NH\text{-}N{=}N\text{-}R^2 \rightleftharpoons R^1\text{-}N{=}N\text{-}NH\text{-}R^2\ (64) \qquad (17.41)$$

$$PhN_3 + MeMgI \longrightarrow Ph\text{-}NH\text{-}N{=}N\text{-}Me \xrightleftharpoons{\text{tautomerism}} Ph\text{-}N{=}N\text{-}NH\text{-}Me\ \ 75\% \qquad (17.42)$$

However, utilizing this reaction, certain azides can convert Grignard reagents (**66**) into amides (**67**), and one of the most useful for this purpose is azidomethyl phenyl sulphide (**68**) (Scheme 17.43).[44] An example is shown in Scheme 17.44.[45]

$$Ph\text{-}S\text{-}CH_3 \xrightarrow{SO_2Cl_2} Ph\text{-}S\text{-}CH_2Cl \xrightarrow{KI / KN_3} Ph\text{-}S\text{-}CH_2N_3\ \ (68)\ \ 93\ \%\ \text{overall} \qquad (17.43)$$

$$PhMgBr\ (66) + Ph\text{-}S\text{-}CH_2N_3\ (68) \xrightarrow{-78^\circ C} [Ph\text{-}N{=}N\text{-}N(MgBr)\text{-}CH_2\text{-}S\text{-}Ph] \xrightarrow{Ac_2O,\ -78^\circ C} Ph\text{-}N(Ac)\text{-}N{=}N\text{-}CH_2\text{-}S\text{-}Ph \xrightarrow{Bu^n_4N^+\ HCO_3^-,\ DMF,\ 45^\circ C} PhNHAc\ (67)\ \ 95\ \% \qquad (17.44)$$

17.2.2 *With electrophiles*

Acid hydrolysis

The hydrolysis of azides with aqueous acids occurs easily but a complicated set of products results.[46] Possibly the complication is caused by the formation of a nitrene (**69**) (Scheme 17.45).

$$R-\overset{-}{N}-\overset{+}{N}\equiv N \xrightarrow{H^+} R-NH-\overset{+}{N}\equiv N \longrightarrow R\text{-}\overset{+}{N}H + N_2$$

$$R\text{-}\overset{+}{N}H \xrightarrow{-H^+} R\text{-}\ddot{N}: \ (\mathbf{69}) \qquad R\text{-}\overset{+}{N}H \xrightarrow{H_2O} R\text{-}NHOH \qquad (17.45)$$

With isonitriles

The reaction with isonitriles (**70**) catalysed by iron pentacarbonyl yields carbodiimides (**71**) (Scheme 17.46).[47]

$$R^1\text{-}N_3 + \underset{(\mathbf{70})}{R^2\text{-}N{=}C} \xrightarrow{Fe(CO)_5} \underset{(\mathbf{71})}{R^1\text{-}N{=}C{=}N\text{-}R^2} + N_2$$

R^1	R^2	Yield (%)
c-C_6H_{11}	Bu^t	60
c-C_6H_{11}	c-C_6H_{11}	48
c-C_6H_{11}	Ph	51

(17.46)

With alkenes

Azides (**72**), for example, phenyl azide, methyl azide, or benzyl azide, add to trisubstituted alkenes (**73**) containing electron-attracting substitutents in two ways corresponding to two possible Michael additions (Scheme 17.47).[48]

With alkylboranes

Trialkylboranes, for example triethylborane (**74**) react with alkyl azides (**75**) to yield secondary amines (**76**) (Scheme 17.48),[49] but more convenient are alkyl-dichloroboranes (**77**) or dialkyl-chloroboranes (**78**) (Scheme 17.49).[50]

α-Azidostyrene[51] *with organoboranes*

α-Azidostyrene (**79**) is prepared as shown in Scheme 17.50.[52] The more reactive benzylic bromine atom is substituted. To ensure that the substitution

OMe
O=C
$R\text{-}\overset{-}{N}\text{-}\overset{+}{N}{\equiv}N$
(72)
CH=C-CN
CO_2Me
(73)
and
$R\text{-}N\text{-}\overset{+}{N}{\equiv}N$
CO_2Me
CH=C-C≡N
CO_2Me
(73)

OMe
O^--C
N
N^+
N-R
CH-C-CN
CO_2Me

CO_2Me
CH-C=C=N^-
CO_2Me
R-N-$\overset{+}{N}$≡N

CO_2Me
H
N
N
NC
N
MeO_2C
R H^+

CO_2Me
CN
H
CO_2Me
N
R
N
N

$MeO_2C\text{-}C{=}\overset{+}{N}{=}\overset{-}{N}$
NC-C-NHR
CO_2Me

(17.47)

$$RN_3 + Et_3B \xrightarrow{\text{heat}} REtN\text{-}BEt_2 \xrightarrow{\text{MeOH}} REtNH + Et_2B\text{-}OMe$$

(75) (74) (76)

R	Yield (%)
Bu^n	72
Bu^i	78
sec-Bu	80
$c\text{-}C_5H_9$	77
$c\text{-}C_6H_{11}$	73
Ph	78

(17.48)

$$R^1\text{—}BCl_2 + R^2\text{—}\overset{-}{N}\text{—}\overset{+}{N}\equiv N \longrightarrow \begin{matrix}R^1\\ \\R^2\end{matrix}\!\!>\!N\text{—}BCl_2 \xrightarrow{\text{aq. }HO^-} \begin{matrix}R^1\\ \\R^2\end{matrix}\!\!>\!NH$$

(77)

R^1	R^2	Yield (%)
Bu^n	Bu^n	71
Bu^n	Ph	72
Bu^n	$c\text{-}C_6H_{11}$	72
Bu^i	$c\text{-}C_6H_{11}$	73
$c\text{-}C_6H_{11}$	$c\text{-}C_6H_{11}$	76

$$\begin{matrix}R^1\\ \\R^1\end{matrix}\!\!>\!BCl + R^2\text{—}\overset{-}{N}\text{—}\overset{+}{N}\equiv N \xrightarrow[\text{ii, }H_2O]{\text{i, ether}} \begin{matrix}R^1\\ \\R^2\end{matrix}\!\!>\!NH$$

(78)

R^1	R^2	Yield (%)
Bu^n	Bu^n	84
Bu^n	Ph	89
Bu^n	$c\text{-}C_6H_{11}$	92
$c\text{-}C_6H_{11}$	$c\text{-}C_6H_{11}$	92
Ph	Ph	92

(17.49)

$$X\text{—}C_6H_4\text{—}CH{=}CH_2 \xrightarrow[CCl_4]{Br_2} X\text{—}C_6H_4\text{—}\underset{|}{\overset{Br}{C}}H\text{—}\overset{Br}{C}H_2$$

$$\xrightarrow[15\text{-}20\,^\circ C]{NaN_3\,/\,DMSO} X\text{—}C_6H_4\text{—}\overset{N_3}{C}H\text{—}\overset{Br}{C}H_2 \xrightarrow[DMSO]{NaOH} X\text{—}C_6H_4\text{—}\overset{N_3}{C}{=}CH_2$$

(79)

$$\xrightarrow{\text{reflux, toluene}} X\text{—}C_6H_4\text{—}C_2H_2N \text{ (azirine)}$$

R	Yield (%)
H	63
OMe	54
Me	57
F	63
Cl	50
Br	54

(17.50)

goes to completion, an excess of sodium azide should be used. The reaction of α-azidostyrene (**79**) with trialkyl boranes (**80**) yields, after hydrolysis, alkyl aryl ketones (**81**),[53] for example, Scheme 17.51. The rearrangement is caused by the loss of nitrogen in conjugation with the double bond and the migration of the ethyl group to the electron deficient carbon atom thus generated. The final stage is the hydrolysis of an imine.

$$3\,CH_2{=}CH_2 + BH_3 \longrightarrow (CH_3CH_2)_3B\ (80)$$

$$(CH_3CH_2)_3B + \text{Ph-C}(N^-{-}N^+{\equiv}N){=}CH_2\ (79) \longrightarrow [\,CH_3CH_2{-}B^-(CH_2CH_3)_2{-}N^-{-}N^+{\equiv}N,\ CH_2{=}C(Ph)\,] \xrightarrow{-N_2} [\,(CH_3CH_2)_2B{-}N{=}C(Ph){-}CH_2CH_2CH_3\,]$$

$$\xrightarrow{H_2O} [\,CH_3CH_2CH_2{-}C({=}NH){-}Ph\,] \longrightarrow CH_3CH_2CH_2{-}C({=}O){-}Ph\ (81)$$

Alkene	Yield (%)
ethylene	95
cyclopentene	80

(17.51)

With carbon monoxide—formation of isocyanates

Isocyanates (**82**) have been prepared by the reaction of azides (**83**) with carbon monoxide,[54] for example, Scheme 17.52.

$$PhN_3 + CO \xrightarrow{200\text{-}300\ atm.} PhN{=}C{=}O + N_2$$

(83) → (82) 47%

$$EtO{-}CH_2{-}N_3 + CO \xrightarrow{200\text{-}300\ atm.} EtO{-}CH_2{-}N{=}C{=}O$$

(83) → (82) 32% (17.52)

Tetramethyltin azide in a synthesis of tetrahydropyrrole

An azide of tetramethyltin (**84**) has been used in a reaction with an aldehyde and an alkene to synthesize a tetrahydropyrrole (**85**) (Scheme 17.53).[55]

$$Me_3SnCH_2I \xrightarrow{NaN_3} Me_3SnCH_2N_3\ (84) \xrightarrow[Ph_3P,\ -N_2]{CH_3CH_2CH_2CHO} CH_3(CH_2)_2\text{-}CH{=}N\text{-}CH_2\text{-}SnMe_3$$

$$\xrightarrow{Bu^nLi} CH_3(CH_2)_2\text{-}CH{=}N\text{-}\bar{C}H_2\ Li^+ \longleftrightarrow Li^+\ CH_3(CH_2)_2\text{-}\bar{C}H\text{-}N{=}CH_2$$

$$\xrightarrow{PhCH{=}CHPh}$$ 3,4-diphenyl-2-propylpyrrolidine (85) (17.53)

17.2.3 *1,3-Dipolar additions*

Azides (**86**) add to alkenes (**87**) and to alkynes (**88**) to yield syntheses of heterocyclic compounds (**89**) (Scheme 17.54).[56]

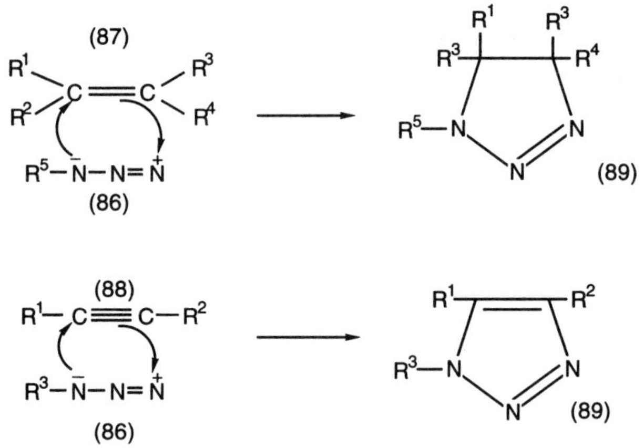

(17.54)

17.2.4 *Nitrenes*[57]

Ethyl azidoformate

Ethyl azidoformate (N_3COOEt) (**90**)[58] reacts on irradiation to form a nitrene which may react with a variety of unsaturated compounds: cyclohexene reacts to give an aziridine (**91**)[59] and similarly benzene yields a ring-enlarged carbamate (**92**)[60] (Scheme 17.55). With a nitrile (**93**) a heterocyclic compound (**94**) is obtained (Scheme 17.56).[61] Ethyl azidoformate reacts with enolic compounds (**95**) (Scheme 17.57).[62] It also reacts with a variety of double bonds eventually yielding heterocyclic compounds, for example, with phosphorus ylides (**96**)[63] and with acetylenes (**97**) and ynamines (**98**)[64] (Scheme 17.58).

$$Cl\text{-}CO_2Et + NaN_3\,(aq.) \longrightarrow N_3\text{-}CO_2Et + NaCl$$

(90)

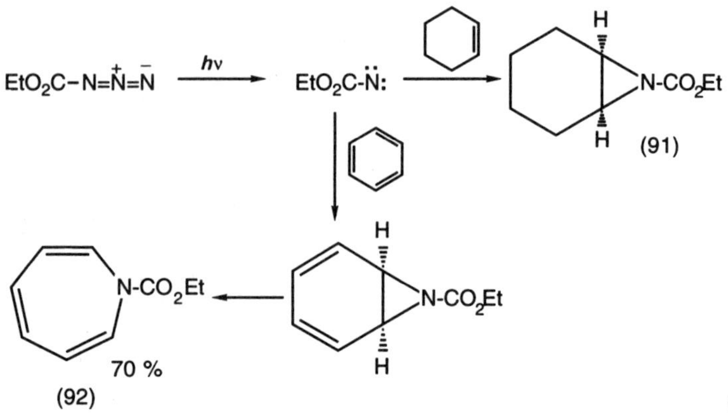

(17.55)

R—C≡N + N_3-CO_2Et —hν→ R, O, N, N, OEt (94)

(93)

(17.56)

(95) —EtO_2C-N_3, hν→ [O, N-CO_2Et] —H_2O→ NHCO$_2$Et, O, OH

(17.57)

CH_3C-CH=PPh_3 (O) (96) + EtO_2CN_3 → [Ph_3P^+, O^-, H-C—C-CH_3, N, N, N-CO_2Et] → H-C=C-CH_3, N, N, N-CO_2Et

A triazole 65 %

H-C≡C-OEt + EtO_2C-N_3 —20° C, 35 days→ H-C=C-OEt, N, N, N-CO_2Et + H-C=C-OEt, EtO_2C-N, N, N

(97)

C. 1 : 1

DABCO — DABCO

H-C—C-OEt, N, N, N, CO_2Et

CH_3C≡C-NEt_2 + EtO_2C-N_3 —CH_2Cl_2, 20°C→ CH_3-C=C-NEt_2, N, N, N-CO_2Et + CH_3-C—C-NEt_2, N, N, N, CO_2Et

(98)

95 : 5

CH_3C≡ $CNEt_2$

(17.58)

17.2.5 *Cyclization with boron trichloride*

Suitable boronic esters which contain an ω-azide group (**99**) can be caused to cyclize by boron trichloride. Alcoholysis then yields a pyrrolidine or a piperidine (Scheme 17.59).[65]

$$(CH_2)_n(CH_2\text{-}CH_2\text{-}B(OEt)_2)(CH_2\text{-}N_3) \xrightarrow[-78^\circ C]{BCl_3} \left[(CH_2)_n(CH_2\text{-}CH_2\text{-}BCl_2)(CH_2\text{-}N_3) \xrightarrow{25^\circ C} (CH_2)_n(CH_2\text{-}CH_2)(CH_2)\text{N-}BCl_2 \right] \xrightarrow{MeOH} (CH_2)_n(CH_2\text{-}CH_2)(CH_2)\text{N-H} \quad (17.59)$$

n = 1 or 2

(99)

n = 1, yield 89 %
n = 2, yield 79 %

17.2.6 *Aza-Wittig reaction*[66]

Azido ketones (**100**) react with triphenylphosphine to give a cyclic compound (**101**) by an intra-molecular aza-Wittig reaction (Scheme 17.60).[67]

$$R^1\text{-}CH((CH_2)_nN_3)\text{-}CHR^2\text{-}C(=O)\text{-}CH_3 \xrightarrow[\text{ether}]{PPh_3,\ 20^\circ C} \left[R^1\text{-}CH((CH_2)_nN{=}PPh_3)\text{-}CHR^2\text{-}C(=O)\text{-}CH_3 \right] \xrightarrow{-Ph_3P{=}O} \text{cyclic } R^1\text{-}CH\text{-}CH(R^2)\text{-}C(CH_3){=}N\text{-}(CH_2)_n \quad (17.60)$$

n = 1,2 or 3

(100)

60 - 90 %

(101)

High yields of heterocycles (**102**) have been obtained by ring closure of azides (**103**) [prepared from amides (**104**)], under the influence of tripheny-phosphine (a cyclic aza-Wittig reaction) (Scheme 17.61).[68]

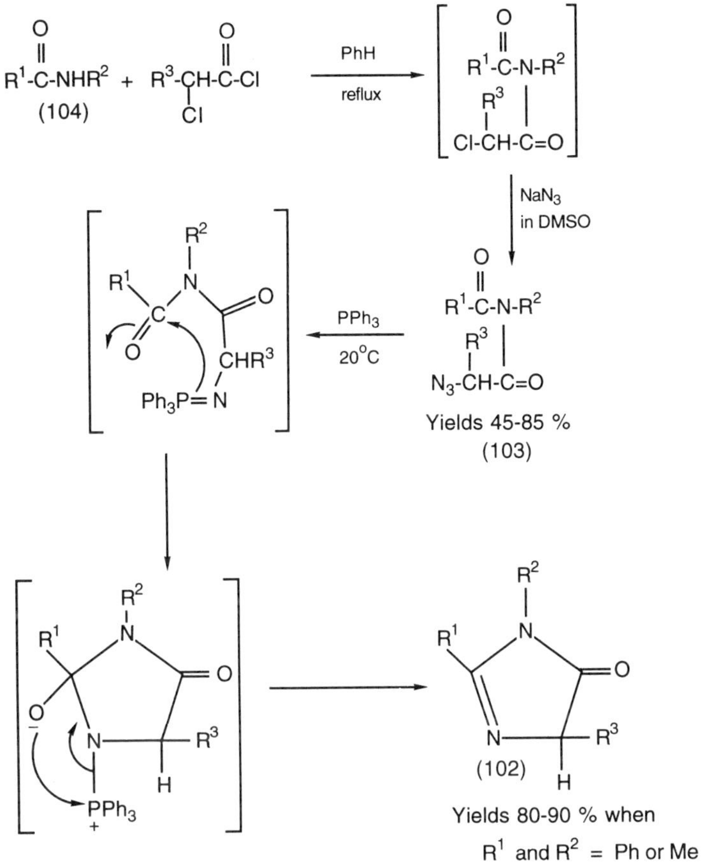

(17.61)

17.2.7 *Intramolecular cyclization of vinyl azides*

Vinyl azides (**105**) undergo intramolecular cyclization to yield 1-azirines (**106**) e.g Schemes 17.62[69] and 17.63.[70]

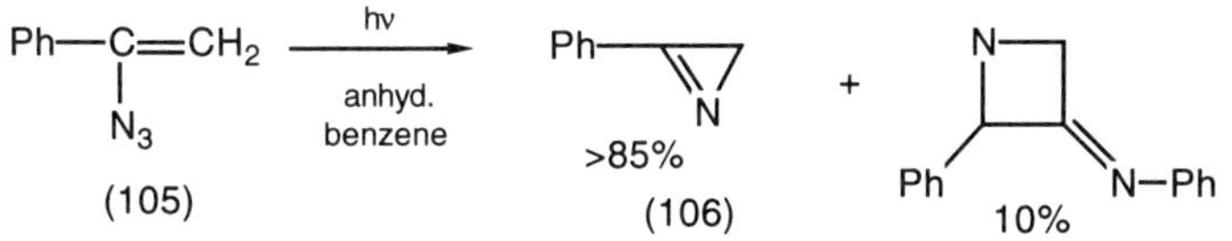

(17.62)

17.2.8 *Rearrangement to imines*

Alkyl azides (**107**) on being heated undergo rearrangement analogous to the Curtius reaction to yield imines (**108**) (Scheme 17.64).[71]

$$R^1(N_3)C{=}CH(R^2) \xrightarrow{\text{heat, catalyst}} \text{2H-azirine}$$

(105) → (106)

R^1	R^2	Catalyst	Yield (%)
Ph	H	NEt_3	85
Ph	H	Dabco*	95
Ph	Me	Dabco	92
n-C_6H_{13}	H	Dabco	44

* Dabco=1,4-diazabicyclo[2.2.2]octane (17.63)

$$R^1R^2R^3C\text{-}N{=}\overset{+}{N}{=}\overset{-}{N} \xrightarrow{\text{heat}} R^1R^2C{=}N\text{-}R^3 \qquad (17.64)$$

(107) → (108)

ACYL AZIDES

17.3 Preparation

17.3.1 *By nitrosation of acyl hydrazines*

Acyl azides (**109**) result from nitrosation of acyl hydrazines (**110**) (Scheme 17.65).[72]

$$\text{R-CO-NH-NH}_2 \xrightarrow{HNO_2} \text{R-CO-N}_3 + H_2O \qquad (17.65)$$

(110) → (109)

Putrescine dihydrochloride (**111**) may be prepared by Curtius degradation of adipyl azide (**112**), obtained from adipylhydrazine (**113**) by nitrosation (Scheme 17.66).[73] The use of nitrosonium tetrafluoroborate (**114**) facilitates the nitrosation reaction (Scheme 17.67).[74]

17.3.2 *From acyl chlorides*

Substitution of an azide ion for the halogen on an acyl halide (**115**) using alkali metal azides gives acyl azides (**116**) in variable yields (Scheme 17.68).[75]

$$EtO_2C\text{-}(CH_2)_4\text{-}CO_2Et \xrightarrow[\text{EtOH, boiled}]{H_2NNH_2(aq)} H_2NNH\text{-}CO\text{-}(CH_2)_4\text{-}CO\text{-}NHNH_2 \; (113) \; 91\text{-}95\%$$

$$\xrightarrow{HNO_2} N_3\text{-}CO\text{-}(CH_2)_4\text{-}CO\text{-}N_3 \; (112) \xrightarrow{\text{heat}} OCN\text{-}(CH_2)_4\text{-}NCO \xrightarrow{\text{heat, } HCl, H_2O} HCl.NH_2\text{-}(CH_2)_4\text{-}NH_2.HCl \; (111) \; 73\text{-}77\% \quad (17.66)$$

$$R\text{-}C(=O)\text{-}NHNH_2 \xrightarrow[CH_2Cl_2, -40^{\circ}C]{NO^+ BF_4^- \, (114)} R\text{-}C(=O)\text{-}N{=}\overset{+}{N}{=}\overset{-}{N}$$

$$R\text{-}C(=O)\text{-}NHNH_2 \xrightarrow{NO^+, \, -H^+} R\text{-}C(=O)\text{-}NH\text{-}NH\text{-}N{=}O \rightleftharpoons R\text{-}C(=O)\text{-}N(H)\text{-}N{=}N\text{-}OH \xrightarrow{-H_2O} R\text{-}C(=O)\text{-}N{=}\overset{+}{N}{=}\overset{-}{N}$$

R	Yield (%)
C_6H_5	58
p-NO_2-C_6H_4	84
o,p-$(NO_2)_2$-C_6H_3	86
-CO-$(CH_2)_4$-CO-	74

(17.67)

$$\underset{(115)}{R\text{-}COHal} + NaN_3 \longrightarrow \underset{(116)}{R\text{-}CON_3} + NaHal \quad (17.68)$$

However, the use of pyridinium azide (**117**) instead of sodium azide is often advantageous (Scheme 17.69).[76]

17.3.3 *From carboxylic acids*

Good yields (75–95 per cent) of acyl azides (**118**) can be obtained from a reaction between a carboxylic acid, the iminium ion (**119**) formed from

R-COCl + (pyridinium, $\overset{+}{N}$-H) N_3^- $\xrightarrow{0^\circ C}$ R-CON_3

(117)

R^1	Yield (%)
Ph-CH_2	71
PhCH=CH_2	50
n-$C_{11}H_{23}$	92
-$(CH_2)_4$-	77

(17.69)

dimethylformamide (**120**) and thionyl chloride (**121**) (see Chapter 8), sodium azide, and pyridine catalysed by tetra-n-butylammonium bromide.[77] The reaction proceeds through the formation of an activated carboxylic acid (**122**) (Scheme 17.70).

$$(Me)_2N{-}CHO\ (120) + SOCl_2\ (121) \longrightarrow (Me)_2\overset{+}{N}{=}CH{-}O{-}S(=O)Cl\ \ Cl^-\ (119)$$

$$(119) + Ph{-}CH_2{-}COOH \xrightarrow{-\,HCl} (Me)_2\overset{+}{N}{=}CH{-}O{-}S(=O){-}O{-}C(=O){-}CH_2{-}Ph\ (122)$$

$$(122) \xrightarrow[-HCl,\ -DMF,\ -SO_2]{NaN_3} Ph{-}CH_2{-}C(=O)N_3\ (118)$$

yield=(75%)

(17.70)

17.4 Reactivity

17.4.1 *Rearrangements (Curtius and Schmidt)*

The main use of acyl azides is in the Curtius and Schmidt reactions for the preparation of isocyanates (**123**) by rearrangement (see Chapter 7, p. 306) (Scheme 17.71).

$$\text{R-C(=O)-N=}\overset{+}{\text{N}}\text{=}\overset{-}{\text{N}} \xrightarrow[\text{Curtius}]{\text{heat}} N_2 + \text{R-C(=O)-}\ddot{\text{N}}\text{:} \longrightarrow \text{R-N=C=O} \quad (123) \qquad (17.71)$$

17.4.2 *Decomposition to an aldehyde or ketone and a nitrile*

Epoxidation of vinyl azides, to give 2-azido oxiranes (**124**), followed by thermolysis of the epoxide, leads to the production of ketones (**125**) and nitriles (**126**) (Scheme 17.72).[78] *N*-Benzoylperoxycarbamic acid has been used for the epoxidation.[79] Likewise, MCPBA has also been used for the epoxidation step (Scheme 17.73).[80]

$$R^1R^2C(-O-)C(R^3)N_3 \ (124) \xrightarrow{\text{heat}} R^1R^2C(O^-)\text{-}C(R^3)\text{=}\overset{+}{N}\text{-}N_2 \longrightarrow R^1R^2C\text{=O} \ (125) + R^3\text{-C}\equiv\text{N} \ (126)$$

R^1	R^2	R^3	
Bu^t	H	H	Yields are quantitative
Bu^t	Me	H	
Ph	Ph	H	

(17.72)

$$\text{1-azidocyclohexene} \xrightarrow{\text{MCPBA}} \text{1-azido-7-oxabicyclo[4.1.0]heptane} \xrightarrow{\text{heat}} N\equiv C\text{-}(CH_2)_4\text{-}CHO \qquad (17.73)$$

References

1. M. E. C. Biffen, J. Miller, and D. B. Paul, in *Chemistry in the azido group*, (ed. S. Patai), Interscience, New York, 1976, p. 57.
2. E. Lieber, T. S. Chao, and C. N. R. Rao, *J. Org. Chem.*, 1957, **22**, 238.
3. J. H. Boyer and J. Hamer, *J. Am. Chem. Soc.*, 1955, **77**, 951.
4. J. H. Boyer, F. C. Canter, J. Hamer, and R. K. Putney, *J. Am. Chem. Soc.*, 1956, **78**, 325.

5. J. A. Durden, Jr, H. A. Stansbury, and W. H. Catlette, *J. Chem. Eng. Data*, 1964, **9**, 228.
6. O. Dimroth and W. Wislicenius, *Ber.*, 1905, **38**, 1573.
7. *F. &F.*, **6**, 676; J. A. Miller, *Tetrahedron Lett.*, 1975, 2959.
8. *F. &F.*, **7**, 246; W. P. Reeves and M. L. Bahr, *Synthesis*, 1976, 823.
9. *F. &F.*, **4**, 28; C. M. Starks, *J. Am. Chem. Soc.*, 1971, **93**, 195.
10. *F. &F.*, **13**, 240; A. Hassner and M. Stern, *Angew. Chem., Int. Ed. Engl.*, 1986, **25**, 478.
11. *F. &F.*, **2**, 403; A. J. Papa, *J. Org. Chem.*, 1966, **31**, 1426.
12. *F. &F.*, **1**, 1236; L. Birkofer and A. Ritter, *Angew. Chem., Int. Ed. Engl.*, 1965, **4**, 417.
13. *F. &F.*, **11**, 32; K. Nishiyama and H. Karigomi, *Chem. Lett.*, 1982, 1477.
14. *F. &F.*, **1**, 1043; C. A. Vanderwerf, R. Y. Heisler, and W. E. McEwen, *J. Am. Chem. Soc.*, 1954, **76**, 1231.
15. *F. &F.*, **13**, 217; C. Blandy, R. Choukroun, and D. Gervais, *Tetrahedron Lett.*, 1983, **24**, 4189; M. Caron and K. B. Sharpless, *J. Org. Chem.*, 1985, 1557, Ref. 17.
16. *F. &F.*, **12**, 538; O. Tsuge, S. Kanemasa, and K. Matsuda, *Chem. Lett.*, 1983, 1131; N. Nishiyama and N. Tanaka, *J. Chem. Soc., Chem. Commun.*, 1983, 1322.
17. *F. &F.*, **15**, 342; M. Emziane, P. Lhoste, and D. Simon, *Synthesis*, 1988, 541.
18. E. Oliveri-Mandalà, *Gazz. Chim. Ital.*, 1932, **62**, 716.
19. *F. &F.*, **14**, 25; K. Nishiyama and Y. Yamaguchi, *Synthesis*, 1988, 106.
20. *F. &F.*, **2**, 415; W. Fischer and J.-P. Anselme, *J. Am. Chem. Soc.*, 1967, **89**, 5284; J.-P. Anselme and W. Fischer, *Tetrahedron*, 1969, **25**, 855.
21. *F. &F.*, **11**, 535; G. H. Hazen, L. M. Weistock, R. Connell, and F. W. Bollinger, *Synth. Commun.*, 1981, **11**, 947; *F. &F.*, **11**, 947; *F. &F.*, **12**, 220.
22. *F. &F.*, **14**, 1; J. S. Baum, D. A. Shook, H. M. L. Davies, and H. D. Smith, *Synth. Commun.*, 1987, **17**, 1709.
23. *F. &F.*, **2**, 37; A. Hassner and F. Boerwinkle, *J. Am. Chem. Soc.*, 1968, **90**, 216.
24. *F. &F.*, **12**, 242; A. Hassner, R. Fibiger, and D. Andisik, *J. Org. Chem.*, 1984, **49**, 4237.
25. *F. &F.*, **7**, 153; A. Suzuki, M. Ishidoya, and M. Tabata, *Synthesis*, 1976, 687.
26. *F. &F.*, **3**, 196; C. H. Heathcock, *Angew. Chem., Int. Ed. Engl.*, 1969, **8**, 134.
27. *F. &F.*, **15**, 126; A. Guy, A. Lemor, J. Doussot, and M. Lemaire, *Synthesis*, 1988, 900.
28. R. E. Harmon, G. Wellman, and S.-K. Gupta, *J. Org. Chem.*, 1973, **38**, 11.
29. *F. &F.*, **14**, 327; D. A. Evans and T. C. Britton, *J. Am. Chem. Soc.*, 1987, **109**, 6881.
30. D. A. Evans, T. C. Britton, and J. A. Ellman, *Tetrahedron Lett.*, 1987, **28**, 6141.
31. *F. &F.*, **12**, 242; A. Hassner, R. Fibiger, and D. Andisik, *J. Org. Chem.*, 1984, **49**, 4237.
32. T. Curtius, *Ber.*, 1912, **45**, 1057.
33. E. F. V. Scriven and K. Turnbull, *Chem. Rev.*, 1988, **88**, 297.
34. M. Bartra, P. Romea, F. Urpí, and J. Vilarrasa, *Tetrahedron*, 1990, **46**, 587.
35. R. L. Willer, *J. Org. Chem.*, 1984, **49**, 5150.
36. M. Pankova, J. Sicher, and J. Zavada, *J. Chem. Soc., Chem. Commun.*, 1967, 394; A. K. Bose, J. F. Kistner, and L. Faber, *J. Org. Chem.*, 1962, **27**, 2925; L. J. Dolby and D. L. Booth, *J. Org. Chem.*, 1965, **30**, 1550.
37. A. Streitweiser, Jr, and W. D. Schaeffer, *J. Am. Chem. Soc.*, 1956, **78**, 5597.
38. A. Hassner and L. A. Levy, *J. Am. Chem. Soc.*, 1965, **87**, 4203.
39. *F. &F.*, **13**, 230; S. K. Boyer, J. Bach, J. McKenna, and E. Jagdmann, Jr, *J. Org. Chem.*, 1985, **50**, 3408.

40. O. Dimroth, *Ber.*, 1905, **38**, 670.
41. Y. G. Gololobov, I. N. Zhmurova, and L. F. Kasukhin, *Tetrahedron*, 1981, **37**, 437.
42. M. S. Kharasch and O. Reinmuth, *Grignard reactions of non-metallic substances*, Constable, London, Chapter 8, p. 1227; O. Dimroth, *Ber.*, 1905, **39**, 3905.
43. O. Dimroth, *Ber.*, 1905, **38**, 670.
44. *F. &F.*, **10**, 14; B. M. Trost and W. H. Pearson, *J. Am. Chem. Soc.*, 1981, **103**, 2483.
45. *F. &F.*, **12**, 37; B. M. Trost and W. H. Pearson, *J. Am. Chem. Soc.*, 1983, **103**, 1054.
46. E. Bamberger, *Annalen*, 1921, **424**, 233; 1925, **443**, 192.
47. *F. &F.*, **3**, 167; T. Saegusa, Y. Ito, and T. Shimizu, *J. Org. Chem.*, 1970, **35**, 3995.
48. M. S. Ouali, M. Vaultier, and R. Carrié, *Tetrahedron*, 1980, **36**, 1821.
49. A. Suzuki, S. Sono, M. Itoh, H. C. Brown, and M. M. Midland, *J. Am. Chem. Soc.*, 1971, **93**, 4329.
50. H. C. Brown, M. M. Midland, and A. S. Levy, *J. Am. Chem. Soc.*, 1972, **94**, 2114; H. C. Brown, M. M. Midland, and A. S. Levy, *J. Am. Chem. Soc.*, 1973, **95**, 2394.
51. *F. &F.*, **6**, 24.
52. A. G. Hortmann, D. A. Robertson, and B. K. Gillard, *J. Org. Chem.*, 1972, **37**, 322.
53. A. Suzuki, M. Tabata, and M. Ueda, *Tetrahedron Lett.*, 1975, 2195.
54. R. P. Bennett and W. B. Hardy, *J. Am. Chem. Soc.*, 1968, **90**, 3295.
55. *F. &F.*, **14**, 16; W. H. Pearson, D. P. Szura, and W. G. Harter, *Tetrahedron Lett.*, 1988, **29**, 1336; see also *F. &F.*, **13**, 163.
56. R. Huisgen, *Angew. Chem., Int. Ed. Engl.*, 1963, **2**, 565; 633; C. G. Stuckwisch, *Synthesis*, 1973, 469; D. S. Black, R. F. Crozier, and V. C. Davis, *Synthesis*, 1975, 205.
57. G. L'Abbé, *Chem. Rev.*, 1969, **69**, 345.
58. *F. &F.*, **1**, 363; M. O. Forster and H. E. Fierz, *J. Chem. Soc.*, 1908, **93**, 81.
59. R. S. Berry, D. Cornell, and W. Lwowski, *J. Am. Chem. Soc.*, 1963, **85**, 1199; W. Lwowski, T. J. Maricich, and T. W. Mattingly, Jr, *J. Am. Chem. Soc.*, 1963, **85**, 1200.
60. K. Hafner and C. König, *Angew. Chem., Int. Ed. Engl.*, 1963, **2**, 96; K. Hafner, D. Zinser, and K.-I. Moritz, *Tetrahedron Lett.*, 1964, 1733.
61. W. Lwowski, A. Hartenstein, C. DeVita, and R. L. Smick, *Tetrahedron Lett.*, 1964, 2497; R. Huisgen and H. Blaschke, *Annalen*, 1965, **686**, 145.
62. I. Brown and O. E. Edwards, *Can. J. Chem.*, 1965, **43**, 1266; J. F. W. Keana, S. B. Keana, and D. Beetham, *J. Org. Chem.*, 1967, **32**, 3057.
63. G. L'Abbé and H. J. Bestmann, *Tetrahedron Lett.*, 1969, 63.
64. P. Ykman, G. L'Abbé, and G. Smets, *Chem. &Ind.*, 1972, 886.
65. *F. &F.*, **15**, 44; J. M. Jego, B. Carboni, M. Vaultier, and R. Carrie, *J. Chem. Soc., Chem. Commun.*, 1989, 142.
66. H. Staudinger and I. Meyer, *Helv. Chim. Acta*, 1919, **2**, 635.
67. *F. &F.*, **11**, 588; P. H. Lambert, M. Vaultier and R. Carrié, *J. Chem. Soc., Chem. Commun.*, 1982, 1224.
68. H. Takeuchi, S. Hagiwara, and S. Eguchi, *Tetrahedron*, 1989, **45**, 6375.
69. F. P. Woerner, H. Reimlinger, and D. R. Arnold, *Angew. Chem., Int. Ed. Engl.*, 1968, **7**, 130.
70. K. Isomura, S. Kobayashi, and H. Taniguchi, *J. Org. Chem.*, 1973, **38**, 4341.
71. R. A. Abramovitch and E. P. Kyba, *J. Am. Chem. Soc.*, 1974, **96**, 480.
72. P. A. Smith, *Org. React.*, 1946, **3**, 337.
73. *F. &F.*, **1**, 1097; P. A. Smith, *Org. Synth., Coll. Vol. 4*, 1963, 819.
74. *F. &F.*, **14**, 215; V. Pozsgay and H. J. Jennings, *Tetrahedron Lett.*, 1987, **28**, 5091.
75. P. A. Smith, *Org. React.*, 1946, **3**, 374.

76. *F. &F.*, **5**, 330; J. W. van Rejendam and F. Baardman, *Synthesis*, 1973, 413.
77. *F. &F.*, **12**, 204; A. Arrieta, J. M. Aizpura, and C. Palomo, *Tetrahedron Lett.*, 1984, **25**, 3365.
78. H. W. Moore and D. M. Goldish in Supplement D, *The chemistry of halides, pseudo-halides, and azides* (ed. S. Patai), Part 1, Wiley Interscience, Chichester, 1983, p. 321.
79. C. P. Kyba and D. C. Alexander, *Tetrahedron Lett.*, 1976, 4563.
80. E. Zbiral, *Synthesis*, 1972, 285.

18

ALKYL NITRITES AND NITRATES

Alkyl nitrites (**1**) and alkyl nitrates (**2**) are esters of nitrous and nitric acids, respectively. Both classes of compounds are useful for the preparation of other compounds.

R-O-N=O (1) R-O-N^{+}(-O^{-})=O (2)

ALKYL NITRITES

18.1 Preparation

18.1.1 *From alcohols*

Alkyl nitrites are easily formed by treating alcohols with sodium nitrite and mineral acid at 0 °C (Scheme 18.1). However, the reaction is reversible and they are easily hydrolysed by mineral acid. This may happen, for example, if the reaction mixture from their preparation is allowed to warm to room temperature.

$$R{-}OH \xrightarrow[\text{dil. HCl at } 0^{\circ}C]{NaNO_2} R{-}O{-}N{=}O \qquad (18.1)$$

18.2 Reactivity

18.2.1 *As a substitute for nitrous acid*

The ease with which alkyl nitrites decompose, yielding nitrous acid, has led to their being used in reactions as substitutes for nitrous acid. For example, they have been used with advantage in an azide synthesis of peptides (**3**) which originally utilized nitrous acid (Scheme 18.2).[1]

$$\text{Cb–NH–CHR}^1\text{–}\overset{\overset{\large O}{\|}}{C}\text{—NHNH}_2 \xrightarrow{HNO_2} \text{Cb–NH–CHR}^1\text{–}\overset{\overset{\large O}{\|}}{C}\text{—N}{=}\overset{+}{N}{=}\overset{-}{N}$$

$$\downarrow H_2N\text{-CHR}^2CO_2Me$$

$$\text{Cb–NH–CHR}^1\text{–}\overset{\overset{\large O}{\|}}{C}\text{—NH—}\underset{\underset{\large CO_2Me}{|}}{CHR^2}$$

(3) (18.2)

An improvement results from the use of *n*-butyl nitrite ($Bu^n—O—N{=}O$) in tetrahydrofuran instead of nitrous acid,[2] and even better is *tert*-butyl nitrite ($Bu^t—O—N{=}O$) and hydrogen chloride in dimethylformamide.[3] Methyl nitrite has also been used in amino acid synthesis.[4]

18.2.2 *α-Nitrosation of ketones*

α-Nitrosation of ketones is usually carried out by the action of alkyl nitrites, for example methyl nitrite (Scheme 18.3)[5] and *tert*-butyl nitrite (Scheme 18.4).[6]

$$\text{Ph—}\overset{\overset{\large O}{\|}}{C}\text{–CH}_2\text{–CH}_3 \xrightarrow[\text{aqueous HCl}]{\text{MeO-N=O}} \text{Ph—}\overset{\overset{\large O}{\|}}{C}\text{—}\underset{\underset{\large NOH}{\|}}{C}\text{—CH}_3$$

66% (18.3)

$$\text{Ph—}\overset{\overset{\large O}{\|}}{C}\text{–CH}_2\text{–Cl} \xrightarrow[\text{Reflux in ether}]{Bu^tO\text{-N=O / HCl}} \text{Ph—}\overset{\overset{\large O}{\|}}{C}\text{—}\underset{\underset{\large NOH}{\|}}{C}\text{—Cl}$$

85% (18.4)

18.2.3 *Use in the protection of amino acids (as BOC)*

The reagent (**4**), used for the protection of amino acids as *t*-butoxycarbonyloxy (BOC) derivatives, is prepared from benzyl cyanide and methyl nitrite (Scheme 18.5).[7]

$$Ph-CH_2-C\equiv N + CH_3O-N=O \xrightarrow{NaOH} Ph-C(=NOH)-C\equiv N \quad c.\ 80\ \%$$

$$\xrightarrow[\text{ii, Bu}^t\text{OH, pyridine}]{\text{i, COCl}_2\text{, PhNMe}_2} Ph-C(=N-O-C(=O)-OBu^t)-C\equiv N \quad (4)\ c.\ 60\ \% \qquad (18.5)$$

18.2.4 *Oxidation of disubstituted phenols*

Isoamyl nitrite has been used in methylene dichloride as an oxidizing agent to convert 2,6-disubstituted phenols (**5**) into 3,3′,5,5′-tetrasubstituted diphenoquinones (**6**) (Scheme 18.6).[8]

$$2\ (5) \xrightarrow[CH_2Cl_2]{i\text{-}C_5H_{11}O\text{-}N=O} (6) \qquad (18.6)$$

R^1	Me	Ph	Me	OMe	Bu^t	Cl
R_2	Me	Ph	Ph	OMe	Bu^t	Ph
Yield (%)	53	58	51	65	16	10

ALKYL NITRATES

The molecular geometry is as shown below.

$$\left\{ >C \overset{1.44\,\text{Å}}{—} O — \overset{+}{N} \begin{smallmatrix} \cdots O \\ \cdots O \end{smallmatrix} \right\}^{-}$$

18.3 Preparation

18.3.1 *From alcohols*

The direct esterification of alcohols with nitric acid alone, or with concentrated sulphuric acid or with acetic acid or anhydride gives good yields of the nitrate esters. However, the reaction must be carefully controlled and steps taken to remove small quantities of nitrous acid by using, for example, urea or hydrazine.[9] The esterification of methanol is a typical example (Scheme 18.7).[10]

$$CH_3OH \xrightarrow[\text{conc. } H_2SO_4,\ \text{below } 10^{\circ}C]{\text{conc. } HNO_3} CH_3\text{-}ONO_2 \tag{18.7}$$

18.3.2 *From amines*

Alkyl nitrates also result from amines by deamination with dinitrogen tetraoxide (N_2O_4) in tetrahydrofuran, which as a Lewis acid forms a complex with the nitrogen oxide. The mechanism is ionic with a high degree of retention of configuration (Scheme 18.8).[11]

18.3.3 *From alkyl halides*

Alkyl nitrates can be prepared from alkyl halides by the action of silver nitrate in, for preference, acetonitrile (Scheme 18.9).[12]

A better procedure utilizes mercurous nitrate in diglyme, $(MeOCH_2)_2$, (Scheme 18.10).[13] For a wide range of alkyl halides the yields are between 74 and 99 per cent. Tertiary bromides do not react.

$$R\text{-}NH_2 + N_2O_4 \xrightarrow[-60^\circ C]{THF} RO\text{-}NO_2 + N_2 + H_2O$$

Mechanism:

$$R{-}NH_2 + N_2O_4.(THF) \longrightarrow [RNH{-}N{=}O]\ HONO_2(THF)$$

$$\downarrow$$

$$\left[\begin{array}{c} R{-}\ddot{N}{=}N{-}O{-}H(THF) \\ O_2N{-}O{-}H \end{array}\right] \longrightarrow [R{-}N{\equiv}N\ \ \bar{O}{-}NO_2].(THF) + H_2O$$

$$\downarrow$$

$$[R^+\ {}^-ONO_2].(THF) + N_2 \longrightarrow R{-}O{-}NO_2$$

R:	$CH_3(CH)_2)_5$	$CH_3(CH_2)_3CH(C_2H_5)CH_2$	$C_6H_5CH(CH_3)$	$(CH_3)_2CHCH_2$
Yield (%):	20	65	50	20

(18.8)

$$R{-}Hal \xrightarrow[CH_3CN]{AgNO_3} R{-}O{-}NO_2 \qquad (18.9)$$

$$R{-}Br \xrightarrow[\text{in }(MeOCH_2)_2]{HgNO_3} R{-}O{-}\overset{+}{N}(=O){-}O^- \qquad (18.10)$$

18.4 Reactivity

18.4.1 *With nitriles*

Methyl nitrate reacts with nitriles to yield nitroalkanes, for example, Scheme 18.11.[14]

$$Ph{-}CH_2{-}C{\equiv}N \xrightarrow[EtO^- \text{ in EtOH}]{MeO{-}NO_2} Ph{-}C(=NO_2^-){-}C{\equiv}N \; Na^+ \xrightarrow{aq.\ NaOH} Ph{-}C(=NO_2^-){-}CO_2^- \; 2Na^+ \xrightarrow[0^\circ - 10^\circ C]{conc.\ HCl} Ph{-}CH_2{-}NO_2 \quad 50\text{-}55\% \qquad (18.11)$$

18.4.2 *Vasodilation*

Alkyl nitrates, particularly glyceryl trinitrate and octyl nitrate, have been used clinically for many years to treat angina and other circulatory disorders. Recent work has shown that *in vivo* nitric oxide is formed (probably enzymatically) which then effects vasodilation.[15]

References

1. N. A. Smart, G. T. Young, and M. W. Williams, *J. Chem. Soc.*, 1960, 9202; T. Wieland and H. Determann, *Angew. Chem., Int. Ed. Engl.*, 1963, **2**, 358.
2. J. Honzyl and J. Rudinger, *Coll. Czech.*, 1961, **26**, 2333.
3. K. Hofmann, W. Naas, M. J. Smithers, R. D. Wells, Y. Wolman, W. Yaihara, and G. Zaretti, *J. Am. Chem. Soc.*, 1965, **87**, 620.
4. *F. &F.*, **1**, 692.
5. W. H. Harting and F. Crossley, *Org. Synth. Coll. Vol. 2*, 1943, 363.
6. N. Levin and W. H. Harting, *Org. Synth. Coll. Vol. 3*, 1955, **3**, 191.
7. *F. &F.*, **10**, 61; M. Itoh, D. Hagiwera, and T. Kamiya, *Bull. Chem. Soc. Japan*, 1977, **50**, 718; *Org. Synth.*, 1979, **59**, 95.
8. *F. &F.*, **4**, 270; R. A. Jerussi, *J. Org. Chem.*, 1970, **35**, 2105.
9. R. Boscham, R. Merrow, and R. W. Van Dolah, *Chem. Rev.*, 1955, **55**, 485; D. R. Goddard, E. D. Hughes, and C. K. Ingold, *Nature*, 1958, **158**, 480; *J. Chem. Soc.*, 1950, 2559; C. K. Ingold, D. J. Millen, and H. G. Poole, *J. Chem. Soc.*, 1950, 2576; E. L. Blackall, E. D. Hughes, C. K. Ingold, and R. B. Pearson, *J. Chem. Soc.*, 1958, 4366; A. Topchiev, *Nitration of hydrocarbons and other organic compounds*, Pergamon Press, London, 1959; F. Kaufman, H. J. Cook, and S. M. David, *J. Am. Chem. Soc.*, 1952, **74**, 4997; J. E. Lufkin, *US Patent* 2 396 330, 1946; *Chem. Abstr.*, 1946, **40**, 3462; N. W. Connon, *Eastman Org. Bull.*, 1970, 42.
10. *F. &F.*, **1**, 691; W. H. Hartung and F. Crossley, *Org. Synth. Coll. Vol. 2*, 1943, 363.
11. *F. &F.*, **4**, 202; F. Wudl and T. B. K. Lee, *J. Am. Chem. Soc.*, 1971, **93**, 271.
12. N. Kornblum, N. N. Lichtin, J. T. Patton, and D. C. Iffland, *J. Am. Chem. Soc.*, 1947, **69**, 307; J. W. Baker and D. M. Easty, *J. Chem. Soc.*, 1952, 1193; E. Gand, *Bull. Chim. Soc. Fr.*, 1950, 120; L. F. Fieser and W. Von E. Doering, *J. Am. Chem. Soc.*,

1946, **68**, 2252; A. F. Ferris, K. W. Mclean, I. G. Marks, and W. D. Emmons, *J. Am. Chem. Soc.*, 1953, **75**, 4078.
13. *F. &F.*, **5**, 430; A. Mckillop and M. E. Ford, *Tetrahedron*, 1974, **30**, 2467.
14. *F. &F.*, **1**, 691; A. P. Black and F. H. Babers, *Org. Synth. Coll. Vol. 2*, 1043, 512.
15. A. Burger (ed.), *Medicinal chemistry*, 2nd edn., Interscience Publishers, New York, p. 635; M. M. Rath and J. C. Krantz, Jr, *J. Pharmacol. Exptl. Therap.*, 1942, **76**, 33; J. C. Krantz, Jr, C. J. Carr, and H. H. Bryant, *J. Pharmacol. Exptl. Therap.*, 1951, **102**, 16; F. E. Hunter, Jr, and L. Ford, *J. Pharmacol. Exptl. Therap.*, 1955, **113**, 186.

19

NITRAMINES

$$\underset{(1)}{\mathrm{R\text{-}NH\text{-}\overset{+}{N}(=O)\text{-}O^{-}}} \qquad \underset{(2)}{\mathrm{R\text{-}N{=}\overset{+}{N}(\text{-}OH)\text{-}O^{-}}}$$

Nitramines (**1**) contain a nitro group attached to a nitrogen atom and, like the nitroalkanes, they give alkali salts dervied from the aci form (**2**). Their known molecular dimensions are as shown.

$$\mathrm{{>}C \overset{1.45\ \text{Å}}{—} N(H) — \overset{+}{N}(\cdots O)_2^{\,-}}$$

These compounds are not well known nor have they been used to any great extent for synthetic purposes, however, some nitramines have attracted attention in connection with explosives.

19.1 Preparation

19.1.1 *From carbamates*

The earliest preparation was from carbamates (urethanes) (**3**) by nitration, treatment with ammonia, and careful acidification of the resulting ammonium salts (Scheme 19.1).

$$\underset{(3)}{R^1\text{—NH—C(=O)—OR}^2} \xrightarrow[\mathrm{HNO_3}]{\text{conc.}} R^1\text{—N(NO}_2\text{)—C(=O)—OR}^2 \xrightarrow{\mathrm{NH_3}} R^1\text{—N=N—O}^-\mathrm{NH_4^+} \xrightarrow{\text{aq. }\mathrm{H_3O^+}} R^1\text{—NH—NO}_2 \qquad (19.1)$$

19.1.2 *By nitration of acyl derivatives of secondary amines*

In a more recent preparation a mixture of nitric acid and trifluoroacetic anhydride generates the nitronium ion NO_2^+ and can be used to prepare nitramines from acyl derivatives of secondary amines (**4**), i.e. from amides (Scheme 19.2).[1]

$$\overset{+}{N}O_2 + R^1R^2N{-}C(=O){-}CH_3 \;(4) \longrightarrow R^1R^2N{-}NO_2 + CH_3\overset{+}{C}O \qquad (19.2)$$

19.1.3 *Nitration of secondary amines*

Another preparation from secondary amines (**5**) involves their nitration with the nitrate of acetone cyanhydrin (**6**) (Scheme 19.3).[2]

$$2\,R_2NH\;(5) \xrightarrow{(CH_3)_2C(ONO_2)CN\;(6)} R_2N.NO_2 + R_2N{-}C(CH_3)_2{-}CN$$

$$(CH_3)_2C(OH)CN \xrightarrow{HNO_3,\;(CH_3CO_2)_2O} (CH_3)_2C(ONO_2)CN\;(6) \qquad (19.3)$$

19.1.4 *Oxidation of nitrosamines*

The oxidation of nitrosamines (**7**) with a peroxide (**8**) derived from an alkene gives good yields (*c*.80 per cent) of nitramines (Scheme 19.4).[3]

$$R_2N{-}NO\;(7) \xrightarrow{CH_3CH_2{-}CMe_2{-}O_2H\;(8)} R_2N{-}NO_2\;\;80\%$$

$$CH_3CH_2{-}CMe_2{-}HSO_4 \xrightarrow{H_2O_2,\;0^\circ C} CH_3CH_2{-}CMe_2{-}O_2H\;(8) \qquad (19.4)$$

19.1.5 *Polynitration of hexamethylene tetramine*

The polynitration of hexamethylene tetramine (**9**) has been studied (Scheme 19.5).[4] The mechanism of this reaction has been investigated with ^{15}N NMR.[5]

Hexamethylene tetramine (9)

Nitrating mixture NH_4NO_3 HNO_3 Ac_2O

Acetylation

Nitration

Nitration

(19.5)

19.1.6 *Preparation of N-nitroamides*

N-Nitroamides (**10**) can be prepared by treating an amide with ammonium nitrate in trifluoroacetic anhydride (Scheme 19.6).[6]

NH_4NO_3 $(F_3C.CO)_2O$ in CH_3NO_2 $0^{\circ}C$

(10)

(19.6)

19.2 Reactivity

19.2.1 *Acidity*

In aqueous solvents primary nitramines are acidic, which is a result of equilibrium with the aci form (Scheme 19.7). In solvents such as benzene the eqilibrium moves more to the left.

$$R\text{-}NH\text{-}NO_2 \rightleftharpoons R\text{-}N{=}N\text{-}OH \qquad (19.7)$$

19.2.2 *Reduction*

The reduction of nitramines (**11**) with lithium aluminium hydride yields hydrazines (Scheme 19.8).[7]

$$R^1R^2N\text{-}NO_2 \; (11) \xrightarrow{LiAlH_4} R^1R^2N\text{-}NH_2 \qquad (19.8)$$

References

1. J. H. Robson and J. Reinhart, *J. Am. Chem. Soc.*, 1955, **77**, 2453; M. B. Frankel, C. H. Tieman, C. R. Vanneman, and M. H. Gold, *J. Org. Chem.*, 1960, **25**, 744.
2. W. D. Emmons and J. P. Freeman, *J. Am. Chem. Soc.*, 1955, **77**, 4387.
3. *F. &F.*, **4**, 20; G. A. Tolslikev, V. M. Jemilev, V. P. Jujev, F. B. Gerchanov, and S. R. Pafikov, *Tetrahedron Lett.*, 1971, 2807; N. A. Milas and D. M. Surgenor, *J. Am. Chem. Soc.*, 1946, **68**, 643.
4. W. E. Bachmann and J. C. Sheehan, *J. Am. Chem. Soc.*, 1949, **71**, 1812; W. E. Bachmann, W. J. Horton, E. L. Jenner, N. N. MacNaughton, and L. B. Scott, *J. Am. Chem. Soc.*, 1951, **73**, 2769; W. E. Bachmann and E. L. Jenner, *J. Am. Chem. Soc.*, 1951, **73**, 2773; V. I. Seele, M. Warman, J. Leccacorvi, R. N. Hutchinson, R. Motto, E. E. Gilbert, T. M. Coburn, R. K. Rohwer, and R. K. Davey, *Propellants & Explosives*, 1981, **6**, 67.
5. M. R. Crampton, M. Jones, J. C. Scranage, and P. Golding, *Tetrahedron*, 1988, **44**, 1679.
6. *F. &F.*, **15**, 14; S. C. Suri and R. D. Chapman, *Synthesis*, 1988, 743.
7. A. I. Vogel, W. T. Cresswell, G. H. Jeffery, and J. Leicester, *J. Chem. Soc.*, 1952, 514; F. W. Schueler and C. Hanna, *J. Am. Chem. Soc.*, 11951, **73**, 4996.

20

β-AMINO ALCOHOLS

$$\underset{\text{HO}}{\overset{|}{\text{R}^1\text{—CH}}}\text{-}\underset{\text{NH}_2}{\overset{|}{\text{CH}}}\text{—R}^2$$

(1)

β-Amino alcohols (**1**) have achieved importance in synthesis because of the versatility of their reactions and the useful properties of some of the products formed.

20.1 Preparation

20.1.1 *From epoxides*

Epoxides (**2**) are opened by the action of ammonia to yield β-amino alcohols, (Scheme 20.1).[1] Primary and secondary amines can also be used to give *N*-substituted products (**4**) (Scheme 20.2).

$$R_2C\overset{O}{—}CH_2 \ (2) \xrightarrow{NH_3} R_2C(OH)\text{-}CH_2\text{—}NH_2 \ (3) \qquad (20.1)$$

$$R_2C\overset{O}{—}CH_2 \xrightarrow{R_2NH} R_2C(OH)\text{-}CH_2\text{—}NHR_2 \ (4) \qquad (20.2)$$

Good yields of β-amino alcohols can be obtained from epoxides by using tetraphenylantimony trifluoromethanesulphonate to catalyse their reaction with amines. Again both primary and secondary amines can be used, for example (Scheme 20.3).[2]

β-Amino alcohols may be synthesized by the reaction of trialkylsilyl azides (**5**) and aluminium isopropoxide, $Al(OPr^i)_3$, with epoxides (**6**). Nucleophilic attack at the least substituted carbon gives a trialkyl siloxy azide (**7**), a precursor of β-amino alcohols (see Chapter 17, p. 655) (Scheme 20.4).[3]

$$\text{cyclohexene oxide} + HNEt_2 \xrightarrow[CH_2Cl_2,\ 40^\circ C]{Ph_4SbOTf} \text{trans-2-(diethylamino)cyclohexanol}\ (70\text{-}75\%)$$

$$AcO\text{-}CH_2\text{-epoxide} + HNEt_2 \xrightarrow[CH_2Cl_2,\ 40^\circ C]{Ph_4SbOTf} AcOCH_2CH(OH)CH_2NEt_2\ (89\%) \quad (20.3)$$

$$R^1CH\text{-}CHR^2\ (\text{epoxide, } \mathbf{6}) + Me_3SiN_3\ (\mathbf{5}) \xrightarrow{Al(OPr^i)_3} R^1CH(OSiMe_3)\text{-}CH(N_3)R^2\ (\mathbf{7}) \quad (20.4)$$

20.1.2 *From silyl enol ethers*

Silyl enol ethers (**8**), produced from aldehydes or ketones,[4] react with hydrogen cyanide and a trace of concentrated sulphuric acid to give silyl ethers of cyanohydrins (**9**), which can be reduced by lithium aluminium hydride and then hydrolysed to give β-amino alcohols (**10**) (Scheme 20.5).[5]

$$R^1(H)C{=}C(R^2)OSiMe_3\ (\mathbf{8}) \xrightarrow[HCN]{H_2SO_4} R^1CH_2C(R^2)(OSiMe_3)CN\ (\mathbf{9}) \xrightarrow{\text{i, } LiAlH_4;\ \text{ii, } H_2O} R^1CH_2C(R^2)(OH)CH_2NH_2\ (\mathbf{10}) \quad (20.5)$$

20.1.3 *Using organometallics*

O-Trimethylsilylcyanohydrins (**11**) react with alkyllithium reagents by double addition to form β-trimethylsilyloxy amines. Removal of the trimethylsilyl group with aqueous acetic acid then yields a β-amino alcohol (**12**) (Scheme 20.6).[6]

$$R^1-\underset{\underset{OSiMe_3}{|}}{\overset{R^2}{\overset{|}{C}}}-CN \xrightarrow[\text{ii, } H_2O]{\text{i, } 2R^3Li} R^1-\underset{\underset{Me_3SiO}{|}}{\overset{R^2}{\overset{|}{C}}}-\underset{\underset{R^3}{|}}{\overset{R^3}{\overset{|}{C}}}-NH_2 \xrightarrow{\text{aq. HOAc}} R^1-\underset{\underset{HO}{|}}{\overset{R^2}{\overset{|}{C}}}-\underset{\underset{R^3}{|}}{\overset{R^3}{\overset{|}{C}}}-NH_2 \quad (12)$$

Prepared from cyanohydrins (11); (12) *c.* 80%

(20.6)

Anti-β-amino alcohols (**13**) have been prepared in a highly diastereoselective reaction by hydrogenolysis of the product of the reaction between organolithium compounds and dimethyl hydrazones of α-benzyloxy aldehydes (**14**), obtained from the appropriate aldehyde,[7] for example, Scheme 20.7.

$$C_5H_9CH(OBn)CH{=}N-NMe_2 \ (14) \xrightarrow{\text{MeLi}} C_5H_9CH(OBn)CH(Me)NH-NMe_2 \quad anti : syn = 97 : 3$$

$$\xrightarrow{H_2,\ PtO_2,\ \text{HOAc, MeOH}} C_5H_9CH(OH)CH(Me)NH_2 \ (13)$$

(20.7)

β-Amino alcohols (**15**) can be synthesized by metallating *N*-(diphenylmethylene)methylamine (**16**), with lithium diisopropylamide and reacting the product with a ketone. After work-up in aqueous acid, the β-amino alcohol is formed, for example, Scheme 20.8.

Chiral β-amino alcohols (**17**) are synthesized by the reaction of L-*N*,*N*-dibenzoyl-α-amino aldehyde (**18**) with alkyllithium and Grignard reagents.

$$Me-N{=}CPh_2 \ (16) \xrightarrow[\text{iii, } H_2O]{\text{i, } LiNPr^i_2;\ \text{ii, } Ph_2CO} Ph_2\overset{OH}{\overset{|}{C}}-CH_2-N{=}CPh_2 \xrightarrow{HCl} Ph_2\overset{OH}{\overset{|}{C}}-CH_2-NH_2 \ (15)\ 76\ \%$$

(20.8)

Addition of Lewis acids, for instance CH_3TiCl_3 or $SnCl_4$, can favour the chelation controlled product (**17A**). Hydrolysis removes the protecting groups to reveal the β-amino alcohol (Scheme 20.9).[8]

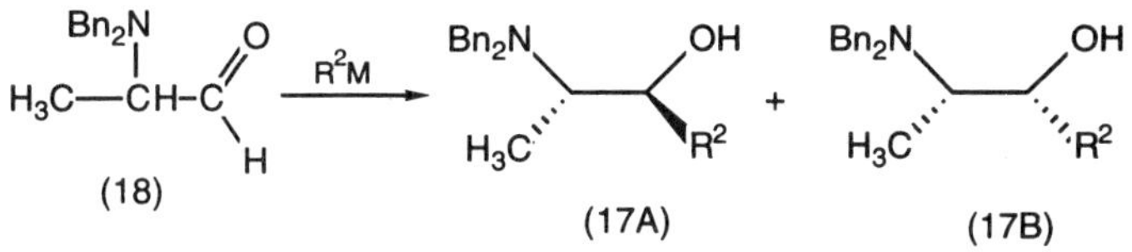

R^2M	Yield (%)	Ratio A:B
CH_3MgI	87	5:95
CH_3Li	91	9:91
CH_3TiCl_3	82	94:6

(20.9)

20.1.4 *From nitroalkanes by reaction with aldehydes or ketones*

β-Amino alcohols (**19**) have been prepared from nitroalkanes (**20**) by reaction of their *O*-trimethylsilyl derivatives (**21**) with aldehydes yielding β-nitro alcohols (**22**), which are easily reduced by lithium aluminium hydride (Scheme 20.10).[9]

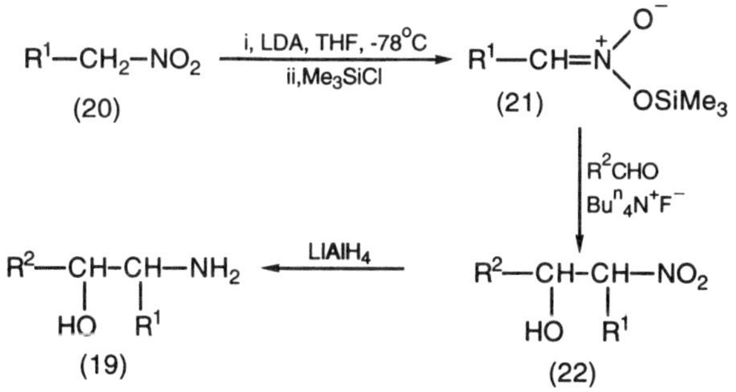

(20.10)

Ketones do not react in the same way as aldehydes above, but a modified synthesis has been developed which allows ketones to be used. In this case the nitroalkane (**23**) is doubly deprotonated at the α-position with a strong base, followed by nucleophilic attack of the anion (**24**) so produced at the carbonyl carbon of a ketone, to give (**29**).[5] The nitro group may then be reduced and the protection removed from the alcohol function by hydrolysis. (Scheme 20.11).

$$R^1\text{—}CH_2\text{—}NO_2 \ (23) \xrightarrow[\text{THF-HMPT}]{2Bu^nLi} R^1\text{—}\bar{C}\text{=}\overset{+}{N}(O^-)_2 \ 2Li^+ \ (24)$$

$$(24) \xrightarrow{\text{i, } R^2COR^3;\ \text{ii, } Bu^tMe_2SiCl} R^2\text{—}C(R^3)(OSiMe_2Bu^t)\text{—}CH(R^1)\text{-}NO_2 \ (25) \xrightarrow{\text{i, LiAlH}_4;\ \text{ii, H}_2\text{O}} R^2\text{—}C(R^3)(OH)\text{—}CH(R^1)\text{—}NH_2 \qquad (20.11)$$

20.1.5 *From nitroalkenes*

Erythro-β-amino alcohols (**26**) are produced by conjugate addition of sodium benzyloxide to 1-nitro-1-butenes (**27**) followed by catalytic hydrogenation (Scheme 20.12).[10]

$$R^1\text{—}CH\text{=}C(R^2)\text{—}NO_2 \ (27) \xrightarrow[\text{ii, HOAc, -78°C}]{\text{i, PhCH}_2\text{OH / NaH / THF / 25°C}} R^1\text{—}CH(OCH_2Ph)\text{-}CH(R^2)\text{—}NO_2 \xrightarrow{H_2\text{ / Pd-C, EtOH / HCl}} R^1\text{—}CH(OH)\text{-}CH(R^2)\text{—}NH_2 \ (26)$$

(26) 50-70%

ratio *erythro* : *threo*
80-90 % : 20-10 %

(20.12)

A route to cyclic β-amino alcohols (**28**) is via a Diels–Alder reaction of the nitroalkene derived from a nitroalkane (**29**) with an acetal (**30**), with butadiene followed by reduction of the resulting β-nitro alcohol (**31**), with $LiAlH_4$ (Scheme 20.13).[11]

CH_3NO_2 + $Me_2N-CH(OMe)_2$ ⟶ $O_2N-CH=CH-NMe_2$
(29) (30)

i, KOH / EtOH / 0°C
ii, PhCOCl

$O_2N-CH=CH-C(=O)-OPh$ (trans) 34 % + butadiene ⟶ (31) (OH, NO_2)

$LiAlH_4$

(28) (OH, NH_2)

(20.13)

20.1.6 *From α-amino acids*

Optically pure β-amino alcohols (**33**) can be prepared from optically pure α-amino acids (**32**) by reduction, using trimethylsilyl chloride, Me_3SiCl, to enhance reactivity (Scheme 20.14).[12]

$R^1R^2CH-CH(NH_2)-CO_2H$ (32) $\xrightarrow[\text{in THF}]{NaBH_4\text{, excess } Me_3SiCl}$ $R^1R^2CH-CH(NH_2)-CH_2OH$ (33) (20.14)

If no Me_3SiCl is present, the β-amino alcohol can be alkylated *in situ* yielding a β-amino methyl ether (**34**),[13] for example, Scheme 20.15.

(S)-$Ph-CH_2-CH(CO_2^-)-\overset{+}{N}H_3$ $\xrightarrow[\text{ii, methylation}]{\text{i, } NaBH_4}$ $H_2N-C(H)(CH_2OMe)(CH_2Ph)$ (34) (20.15)

Scheme 20.16 is an example of the synthesis of a chiral β-amino alcohol (**35**) from a derivative of an α-amino acid (**36**) by a Friedel–Crafts reaction.[14]

(36) COCl, NHCO$_2$Me, H — Phenylalanine → $AlCl_3 / CH_2Cl_2$, 25°C → NHCO$_2$Me, H, (S) 98%ee → $NaBH_4$ → HO H, NH$_2$, H (35)

(20.16)

20.1.7 *From ethyl esters of α-amino acids*

By the reduction of hydrochlorides of ethyl esters of α-amino acids (**37**) with borane and dimethyl sulphide, β-amino alcohols (**38**) can be prepared in yields of about 50 per cent (Scheme 20.17). Optical activity is retained in the reduction.[15]

$$\underset{(37)}{R-CH(CO_2Et)(NH_3^+Cl^-)} \xrightarrow[Me_2S]{B_2H_6} \underset{(38)}{R-CH(CH_2OH)(NH_2)} \qquad (20.17)$$

20.1.8 *From β-hydroxy amides*

On addition of lead tetraacetate, β-hydroxyamides (**39**) undergo a Hofmann-type rearrangement rapidly followed by ring closure to give a 2-oxazolidinone (**40**). Base hydrolysis of this compound leads after decarboxylation to the formation of β-amino alcohols (**41**) (Scheme 20.18).[16] Overall retention of configuration occurs in this reaction, for example, Scheme 20.19.

(39) $R^1CH(OH)-CH(R^2)CONH_2$ → $Pb(OAc)_4$, pyridine, 20°C → (40) oxazolidinone (R^1, R^2, O, NH, C=O) *c.* 95% → aqueous OH^- → (41) $R^1CH(OH)-CH(R^2)NH_2$

(20.18)

$\longrightarrow$ 80 % (20.19)

20.1.9 *From alkenes*

A method has been developed for the preparation of β-amino alcohols (**42**) from alkenes (**43**), by reaction with the trihydrate of chloramine-T (**44**) using osmium tetroxide as a catalyst (Scheme 20.20).[17] The intermediate reagent is thought to be R—N=OsO_3 (**45**) and this reagent can be pre-formed and used (Scheme 20.21).[18]

(43) + $TsNClNa.3H_2O$ (44) $\xrightarrow[\text{in Bu}^t\text{OH}]{1\%\ OsO_4}$ HO–C–C–NHTs $\xrightarrow{\text{hydrolysis}}$ HO–C–C–NH_2 (42) (20.20)

C=C $\xrightarrow[\text{ii, LiAlH}_4]{\text{i, Bu}^t\text{-N=OsO}_3\ (45)}$ HO–CH–CH–NHBu^t (20.21)

Another synthesis from alkenes (**46**) has been reported, employing Pd co-ordination (Scheme 20.22).[19]

(46) $\xrightarrow{PdCl_2}$ (Pd-alkene complex) $\xrightarrow[\text{ii, Pb(OAc)}_4]{\text{i, R}_2\text{NH}}$ HO–C–C–NR_2 (20.22)

20.1.10 *From oximes*

β-Amino alcohols (**47**) have been prepared from oximes (**48**) by reaction with Grignard reagents (**49**) to produce an aziridine (**50**) which opens on hydrolysis to form the β-amino alcohol (Scheme 20.23). An α-hydrogen atom must be present in the oxime for the reaction to go ahead.[20,21]

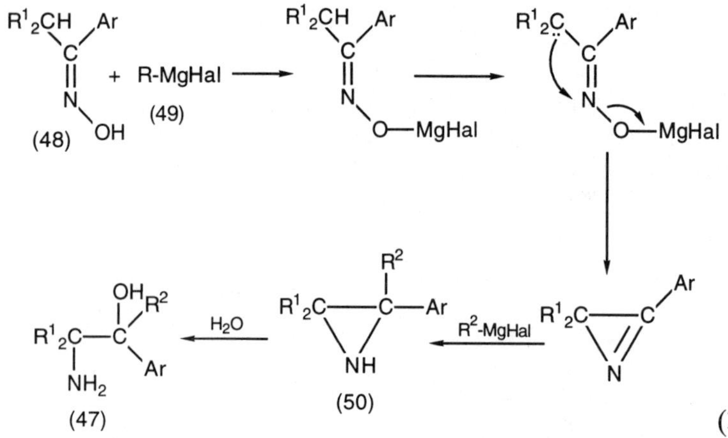

(20.23)

20.1.11 *From trans-1,2-bromohydrins*

The consecutive reaction of a *trans*-1,2-bromohydrin (**51**) with phosgene and benzylamine yields a bromo carbamate (**52**) which, followed by treatment with NaH cyclises to form a *cis*-oxazolidinone (**53**). N-Debenzylation of the

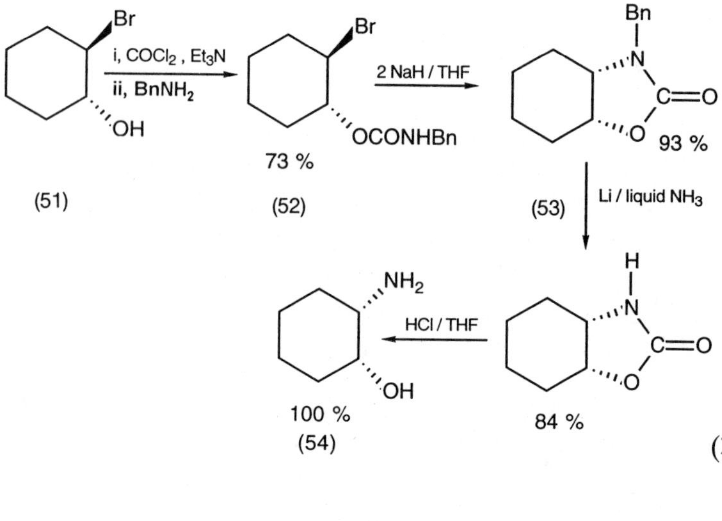

(20.24)

product, and finally hydrolysis produces a *cis*-cyclic β-amino alcohol (**54**) (Scheme 20.24).[22]

20.1.12 *From allylic alcohols*

β-Amino diols (**55**) are prepared from allylic alcohols (**56**) by iodoamination followed by hydrolysis (Scheme 20.25).[23] A closely related reaction also occurs with homoallylic alcohols (**57**) (Scheme 20.26).[23]

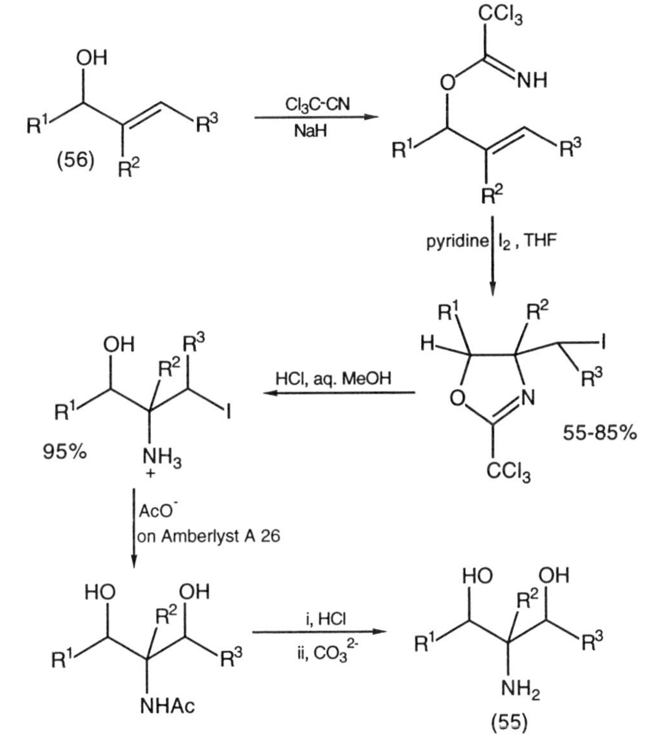

(20.25)

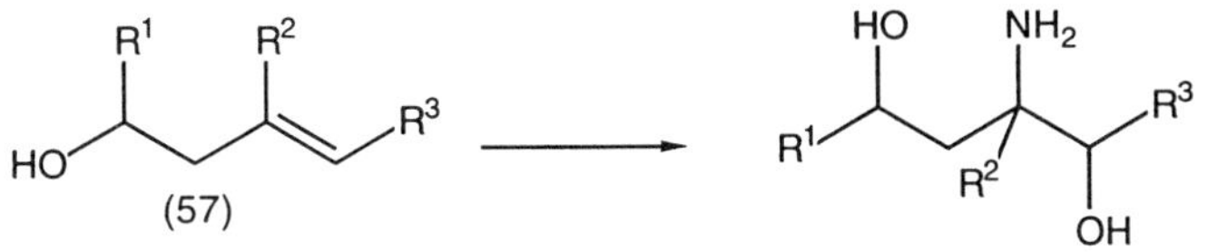

(20.26)

20.1.13 *From imines*

The trivalent niobium product (**58**), from the reduction shown in Scheme 20.27, reacts with imines (**59**), and the cyclic intermediate (**60**) can be

$$NbCl_5 + CH_2OMe\text{–}CH_2OMe \xrightarrow{Bu^n_3SnH} NbCl_3.CH_2OMe\text{–}CH_2OMe \ \textbf{(58)} \quad (20.27)$$

hydrolysed to yield an amine (**61**). However, it also reacts with aldehydes or ketones to yield a β-amino alcohol (**62**),[24] for example, Scheme 20.28.

$$Bn\text{–}N{=}CH\text{–}Ph \ \textbf{(59)} + NbCl_3.(CH_2OMe)_2 \ \textbf{(58)} \xrightarrow{THF} [(CH_2OMe)_2Cl_3.Nb(\text{–}N(Bn)\text{–}CH(Ph)\text{–})] \ \textbf{(60)}$$

$$\textbf{(60)} \xrightarrow{Et_2C=O} Bn\text{–}NH\text{–}CH(Ph)\text{–}C(Et)_2\text{–}OH \ \textbf{(62)} \ 97\%$$

$$\textbf{(60)} \xrightarrow{\text{hydrolysis}} Bn\text{–}NH\text{–}CH_2\text{-}Ph \ \textbf{(61)} \quad (20.28)$$

A further synthesis of β-amino alcohols involves the use of a niobium(III) reagent, (niobium(III) chloride–dimethoxyethane complex) (**63**). This produces β-amino alcohols via the cross-coupling of an imine (**65**) with an aldehyde or ketone (**66**) (Scheme 20.29).[25]

$$R\text{–}CH{=}N\text{–}R^1 \ \textbf{(65)} \xrightarrow[\text{ii, } R^2COR^3 \ \textbf{(66)}]{\text{i, } NbCl_3(DME) \ \textbf{(63)}} R^1\text{–}NH\text{–}CH(R)\text{–}C(R^3)(OH)\text{–}R^2 \ \textbf{(64)}$$

R	R¹	R²	R³	Yield (%)
Ph	Bn	Et	Et	97
Ph	Bn	H	Bu^t	79
Ph	Bn	Me	CO_2Et	82
Ph	Bn	H	n-C_6H_{13}	69
Ph	C_3H_5	H	Pr^i	60

(20.29)

20.1.14 *Stereoselective syntheses*

A highly stereoselective synthesis of β-amino alcohols (**67**) has been detailed. It consists of the fluoride ion catalysed condensation of a silyl ester of a nitro-alkane (**68**) with an aldehyde (**69**) followed by reduction with high *erythro* selectivity (Scheme 20.30).[26] Some stereoselectivity can also be achieved in the preparation of complex β-amino alcohols (**70**) related to ephedrine (Scheme 20.31).[27]

(69) (68) F⁻, THF, -78 °C; *erythro* > 95 %; LiAlH$_4$, ether; H$_2$, Raney Ni; (67)

(20.30)

i, LDA; ii, PhCHO; iii, NaBH$_4$, CH$_3$OH; (70) *erythro* : *threo* 3.3 : 1

(20.31)

20.2 Reactivity

20.2.1 *With carbonyl compounds*

With aldehydes and ketones

β-Amino alcohols (**71**) add to carbonyl groups (**72**) with dehydration to yield five membered heterocyclic compounds (**73**) (Scheme 20.32).[28]

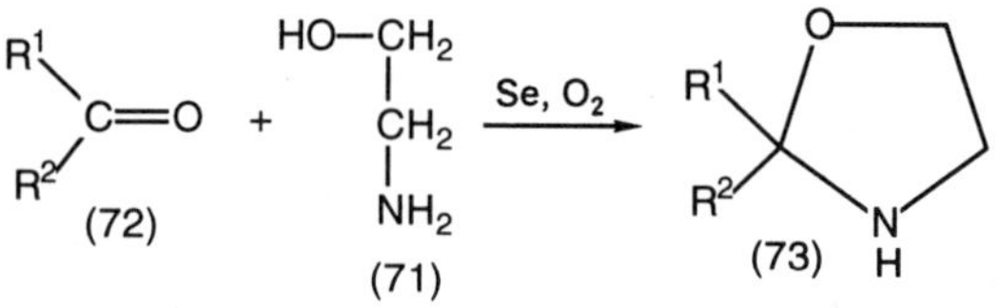

(20.32)

With carboxylic acids

Similarly, carboxylic acids (**74**) and β-amino alcohols (**75**) react to give 2-oxazolines (**76**) when treated with triphenylphosphine in carbon tetrachloride (Scheme 20.33).[29]

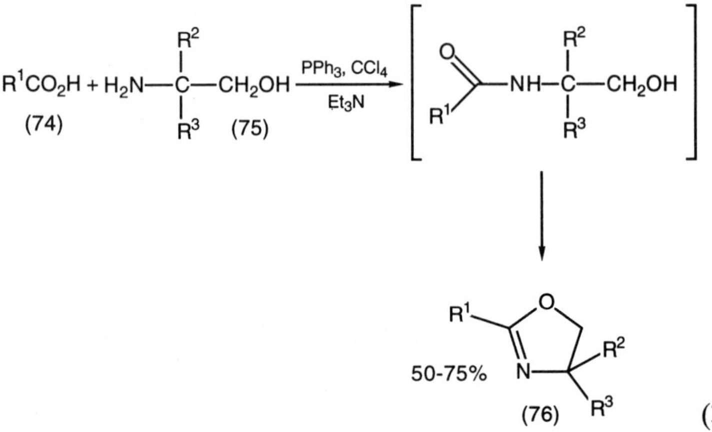

(20.33)

With acyl chlorides

2-Oxazolines (**77**) can also be prepared from β-amino alcohols by acylation with acid chlorides and subsequent dehydration in a very similar reaction to that shown in Scheme 20.33 (Scheme 20.34).[30]

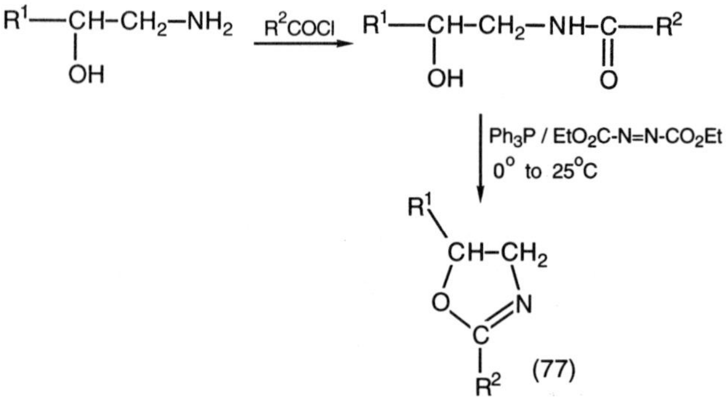

(20.34)

Protection of the carbonyl function

This reaction is often used to protect carboxyl groups because oxazolines do not react with alkyl metal reagents, as demonstrated in Scheme 20.35.[31]

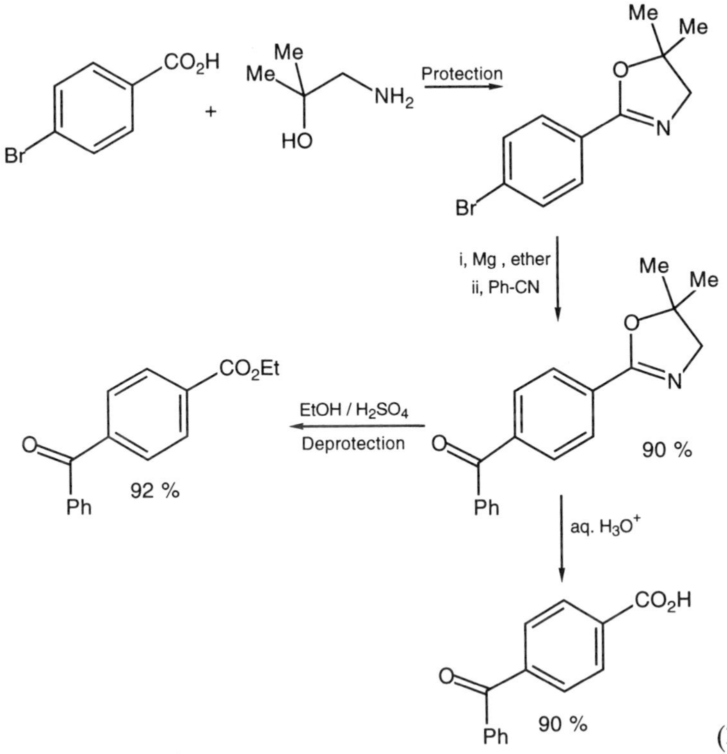

(20.35)

Advantage can also be taken of the acidity of the α-hydrogen atom of a suitably substituted oxazoline (**78**),[32] thus enabling esters (**79**) and β-hydroxyesters (**80**) to be prepared (Scheme 20.36).[33]

Michael reaction with acetylenic esters and β-keto esters

The reaction of β-amino alcohols (**81**) with acetylenic esters (**82**) or β-keto esters (**83**) by a Michael reaction yields compounds (**84**) and (**85**) (Scheme 20.37). On pyrolysis (**85**) gave moderate yields (*c.* 30 per cent maximum) of pyrroles (**86**).[34] An example of the result of a typical pyrolysis of, for example, (**85**) is shown in Scheme 20.38.

(78) → i, Bu^n-Li; ii, R^1-Hal → (R^1CH_2-oxazoline) → i, Bu^n-Li; ii, R^2-Hal → (R^1R^2CH-oxazoline)

(78) → i, Bu^n-Li; ii, R-CHO → (HO–CH(R)–CH_2-oxazoline) → EtOH, H_2SO_4 → R–CH(OH)–CH_2CO_2Et (80)

(R^1CH_2-oxazoline) → EtOH, H_2SO_4 → R^1–CH_2–CO_2Et (79)

(R^1R^2CH-oxazoline) → EtOH, H_2SO_4 → R^1–CH(R^2)–CO_2Et (79)

(20.36)

R–CH(OH)–CH_2–NH_2 (81) + H–C≡C–CO_2Et (82) → Michael reaction → R–CH(OH)–CH_2–NH–CH=CH–CO_2Et (84)

R^2–C(=O)–CH(R^3)–CO_2Et (83) + R^1–CH(OH)–CH_2–NH_2 (81) → R^1–CH(OH)–CH_2–NH–C(R^2)=C(R^3)–CO_2Et (85)

(20.37)

(85) → 400°C, 15 torr → 2,5-dimethylpyrrole-3-carbaldehyde (N–H) 34 % + (86) 1-formyl-2,5-dimethylpyrrole (N–CHO) a trace

(20.38)

20.2.2 *Alkaline hydrolysis of N-chloro-β-amino alcohols*

The alkaline decomposition of *N*-chloro-β-amino alcohols is fast compared to that of *N*-chloramines. Kinetic measurements of the decomposition of *N*-chlorodiethanolamine (**87**) by aqueous alkali at various pH values compared to that of *N*-chloramines (**88**) indicate the mechanism shown in Scheme 20.39.[35]

$$\text{Cl—N(CH}_2\text{–CH}_2\text{OH)}_2\ \textbf{(87)} \underset{}{\overset{HO^-}{\rightleftharpoons}} \text{Cl—N(CH}_2\text{–CH}_2\text{O}^-\text{)(CH}_2\text{–CH}_2\text{OH)} \rightarrow \text{CH}_2\text{=N–CH}_2\text{–CH}_2\text{OH} + \text{Cl}^- + \text{CH}_2\text{O} \xrightarrow{H_2O} \text{CH}_2\text{=O} + \text{H}_2\text{N–CH}_2\text{–CH}_2\text{OH} \quad (20.39)$$

This mechanism differs from the much slower one which takes place in the alkaline decomposition of *N*-chloramines (**88**) (Scheme 20.40).

$$\text{R}^1\text{–CH}_2\text{–N(R}^2\text{)–Cl}\ \textbf{(88)} \overset{HO^-}{\rightleftharpoons} \text{R}^1\text{–}\ddot{\text{C}}\text{H}^-\text{–N(R}^2\text{)–Cl} \rightarrow \text{R}^1\text{–CH=N–R}^2 + \text{Cl}^- \xrightarrow{H_2O} \text{H}_2\text{N–R}^2 + \text{R}^1\text{–C(=O)H} \quad (20.40)$$

20.2.3 *Interconversion of N- and O-acyl derivatives by acid and alkali*

N- and *O*-acyl derivatives of β-amino alcohols are interconvertible as shown in Scheme 20.41. The acyl group migrates from nitrogen to oxygen in the presence of acid (amine trapped as cation) and from oxygen to nitrogen in the presence of alkali. The amide (**89**) is more stable than the ester (**90**), and hence at equilibrium the amide predominates.

$$\text{(89)} \underset{HO^-}{\rightleftharpoons} \text{cyclic intermediate (R}^1\text{–N, O, R}^2\text{, OH)} \underset{H^+}{\rightleftharpoons} \text{(90)}\ R^1NH\text{–C–C–O–C(=O)R}^2 \qquad (20.41)$$

20.2.4 *Selective N-benzoylation*

Benzoyl cyanides (**91**), prepared from cyanohydrins (**92**) by oxidation, are good reagents for selective *N*-benzoylation of *β*-amino alcohols (**93**) (Scheme 20.42).

$$\underset{(92)}{Ar\text{—}CH(OH)\text{—}CN} \xrightarrow[RuCl_2(PPh_3)_3 \text{ in } PhH]{Me_3C\text{-}O\text{-}O\text{-}H} \underset{(91)}{Ar\text{—}C(=O)\text{—}CN} \quad 75\text{-}99\ \%$$

$$\underset{(93)}{R^1\text{—}CH(OH)\text{-}CH(NH_2)\text{—}R^2} \xrightarrow{Ar\text{-}COCN} R^1\text{—}CH(OH)\text{-}CH(NH\text{—}COAr)\text{—}R^2 \qquad (20.42)$$

20.2.5 *Cleavage by periodate/lead tetra-acetate*

β-Amino alcohols (**94**) are cleaved by periodate or by lead tetra-acetate in the same fashion as diols (Scheme 20.43). Periodate cleaves *N*-primary and *N*-secondary *β*-amino alcohols fastest at pH 7–8. The resultant imine undergoing further hydrolysis.[36]

$$\underset{(94)}{R^1{}_2C(NHR^2)\text{—}CH_2OH} \xrightarrow{IO_4^-} R^2\text{—}NH_2 + H_2C{=}O + R^1{}_2C{=}O \qquad (20.43)$$

20.2.6 *The Tiffeneau–Demyanov rearrangement*

The Tiffeneu–Demyanov rearrangement occurs when β-amino alcohols (**95**) are treated with nitrous acid, resulting in the production of a ketone (**96**). Cyclic β-amino alcohols undergo ring expansion (Scheme 20.44).[37]

(95) $\xrightarrow{HNO_2}$ $CH_2-\overset{+}{N}\equiv N$ / OH → (96) + N_2

(20.44)

20.2.7 *Arylation*

By the use of bismuth compounds β-amino alcohols (**97**) can be arylated. A mixture of products is formed. (Scheme 20.45).[38]

H_2N–CH₂CH₂–OH (97) $\xrightarrow{Ph_3Bi(OAc)_2}$ PhNH–CH₂CH₂–OH + Ph_2N–CH₂CH₂–OH + PhNH–CH₂CH₂–OPh

(20.45)

20.2.8 *Chiral synthesis of α-amino acids*

Chiral α-amino acids (**98**) have been prepared from chiral β-amino alcohols (**99**). The β-amino alcohol undergoes a condensation reaction with a bromoester (**100**) followed by reaction with Boc_2O. The product of this reaction (**101**) is deprotonated and the enolate reacts with alkyl halides at −78 °C with high diastereoselectivity. The alkylated product (**102**) is easily reduced to an α-amino ester (**103**) (Scheme 20.46).[39]

(99) + (100) → i, $(Pr^i)_2NEt$; ii, Boc_2O → (101)

(101) → i, $NaN(SiMe_3)_2$; ii, MeI, -78°C → (102) 3R / 3S > 200 / 1

(102) → HCl, EtOH → (103)

(103) → H_2, Pd / C → (98)

(20.46)

20.2.9 *Use in asymmetric reductions*

Appropriate β-amino alcohols have been used as chiral ligands for preparation of modified lithium aluminium hydrides for use in asymmetric reductions. For example, (**104**) is useful for reduction of ketones, and is prepared from (*S*)-propylene oxide and (*S*)-α-methylbenzylamine.[40]

(104)

References

1. J. March, *Advanced organic chemistry*, Wiley Interscience, New York, 1985, p. 368; S. P. McManus, C. A. Larson, and R. A. Hearn, *Synth. Commun.*, 1973, **3**, 177.
2. *F. & F.*, **15**, 306; M. Fujiwara, M. Imada, A. Baba, and H. Matsuda, *Tetrahedron Lett.*, 1989, **30**, 739.

3. *F. &F.*, **15**, 17; M. Emziane, P. Lhoste, and D. Simon, *Synthesis*, 1988, 541.
4. G. Stork and P. F. Hudrlik, *J. Am. Chem. Soc.*, 1968, **90**, 4462; *F. &F.*, **2**, 436; H. O. House, L. J. Czuba, M. Gall, and H. D. Olmstead, *J. Org. Chem.*, 1969, **34**, 2324; *F. &F.*, **3**, 310.
5. *F. &F.*, **4**, 538; W. E. Parnham and C. S. Roosevelt, *Tetrahedron Lett.*, 1971, 923.
6. *F. &F.*, **10**, 3; R. Amouroux and G. P. Axiotis, *Synthesis*, 1981, 270.
7. *F. &F.*, **14**, 149; D. A. Claremon, P. K. Lumma, and B. T. Phillips, *J. Am. Chem. Soc.*, 1986, **108**, 8215.
8. *F. &F.*, **15**, 111; M. T. Reetz, M. W. Drewes, and A. Schmitz, *Angew. Chem., Int. Ed. Engl.*, 1987, **26**, 1141.
9. *F. &F.*, **9**, 444; L. A. Carpino and A. C. Saw, *J. Chem. Soc., Chem. Commun.*, 1979, 514; D. Seebach and F. Lehr, *Angew. Chem., Int. Ed. Engl.*, 1976, **15**, 505; D. Seebach, A. K. Lehr, T. Weller, and E. Colvin, *Angew. Chem., Int. Ed. Engl.*, 1981, **20** 397.
10. *F. &F.*, **15**, 289; A. Kamimura and N. Ono, *Tetrahedron Lett.*, 1989, **30**, 731.
11. *F. &F.*, **14**, 28; G. A. Kraus, J. Thurston, P. J. Thomas, R. A. Jacobson, and Y. Su, *Tetrahedron Lett.*, 1988, **29**, 1879.
12. *F. &F.*, **15**, 186; A. Giannis and K. Sandhoff, *Angew. Chem., Int. Ed. Engl.*, 1989, **28**, 218.
13. *F. & F.*, **10**, 11; A. J. Meyers, D. R. Williams, G. W. Erickson, S. White, and M. Druelinger, *J. Am. Chem. Soc.*, 1981, **103**, 3081.
14. *F. &F.*, **10**, 10; D. E. McClure, B. H. Arison, J. H. Jones, and J. J. Baldwin, *J. Org. Chem.*, 1981, **46**, 2431.
15. *F. &F.*, **12**, 64; G. A. Smith and R. E. Gawley, *Org. Synth.*, 1983, **63**, 136.
16. *F. &F.*, **4**, 280; S. Simons, Jr, *J. Org. Chem.*, 1973, **38**, 414.
17. J. March, *Advanced organic chemistry*, Wiley Interscience, New York, 1985, p. 754; K. B. Sharpless, A. O. Chong, and K. Oshima, *J. Org. Chem.*, 1976, **41**, 177.
18. K. B. Sharpless, D. W. Patrick, L. K. Truesdale, and S. A. Biller, *J. Am. Chem. Soc.*, 1975, **97**, 2305.
19. J. E. Bäckvall, *Tetrahedron Lett.*, 1975, 2225.
20. J. March, *Advanced organic chemistry*, Wiley Interscience, New York, 1985, p. 846; K. N. Campbell, B. K. Campbell, J. F. McKenna, and E. P. Chaput, *J. Org. Chem.*, 1943, **8**, 103.
21. J. P. Freeman, *Chem. Rev.*, 1973, **73**, 283; G. Alveruhe, S. Arsenyiades, R. Chanbouni, and A. Larent, *Tetrahedron Lett.*, 1975, 355.
22. *F. &F.*, **15**, 265; J. Das, *Synth. Commun.*, 1988, **18**, 907.
23. *F. &F.*, **11**, 265; G. Cardillo, M. Orena, G. Porzi, and S. Sandri, *J. Chem. Soc., Chem. Commun.*, 1982, 1308; 1309.
24. *F. &F.*, **14**, 213; E. J. Roskamp and S. F. Pedersen, *J. Am. Chem. Soc.*, 1987, **109**, 655.
25. S. F. Pederson and E. J. Roskamp, *J. Am. Chem. Soc.*, 1987, **109**, 3152.
26. *F. &F.*, **11**, 499; D. Seebach, A. K. Beck, T. Mukhopadhyay, and E. Thomas, *Helv. Chim. Acta*, 1982, **65**, 1101; E. W. Kolvin, A. K. Beck, B. Bastini, D. Seebach, Y. Kai, and J. D. Dunitz, *Helv. Chim. Acta*, 1980, **63**, 697.
27. *F. &F.*, **9**, 420.
28. *F. &F.*, **6**, 508.
29. *F. &F.*, **11**, 588; H. Vorbrüggen and K. Krolikiewicz, *Tetrahedron Lett.*, 1981, 4471.
30. *F. &F.*, **13**, 332; D. M. Roush and M. M. Patel, *Synth. Commun.*, 1985, **15**, 675.
31. *F. &F.*, **3**, 14; A. I. Meyers and D. L. Temple, *J. Am. Chem. Soc.*, 1970, **92**, 6644; 6646; J. W. Cornforth, *Heterocyclic compounds*, 1957, **5**, 386.

32. Preparation: H. L. Wehrmeister, *J. Org. Chem.*, 1962, **27**, 4418.
33. *F. &F.*, **3**, 313; A. I. Meyers and D. L. Temple, *J. Am. Chem. Soc.*, 1970, **92**, 6644.
34. C. Pale-Grosdemange and J. Chuche, *Tetrahedron*, 1989, **45**, 3397.
35. J. M. Antelo, F. Arce, D. Casal, P. Rodriguez, and A. Varela, *Tetrahedron*, 1989, **45**, 3955.
36. B. H. Nicholet and L. A. Shinn, *J. Am. Chem. Soc.*, 1939, **61**, 1615; 1941, **63**, 1456; 1486; P. F. Fleury, J. E. Courtois, and M. Grandchamp, *Bull. Soc. Chim. Fr.*, 1949, 88.
37. A. J. Sisti, *Tetrahedron Lett.*, 1967, 5327; A. J. Sisti, *J. Org. Chem.*, 1968, **33**, 453; A. J. Sisti and A. C. Vitale, 1972, **37**, 4090.
38. R. A. Abramovitch, D. H. R. Barton, and J. P. Finet, *Tetrahedron*, 1988, **44**, 3039; D. H. R. Barton, N. Y. Bhatnagar, J. C. Blazejeweki, B. Charpiot, J. P. Finet, D. J. Lester, W. B. Motherwell, M. T. B. Papoula, and S. P. Stanforth, *J. Chem. Soc., Perkin Trans I*, 1985, 2657; D. H. R. Barton, J. P. Finet, and C. Pichon, *J. Chem. Soc., Chem. Commun.*, 1986, 65.
39. *F. &F.*, **15**, 256; J. F. Dellaria and B. D. Santarsiero, *Tetrahedron Lett.*, 1988, **29**, 6079.
40. *F. &F.*, **11**, 153; J. D. Morrison, E. R. Grandbois, S. I. Howard, and G. R. Weisman, *Tetrahedron Lett.*, 1981, **22**, 2619.

21

AZIRIDINES

Aziridines (**1**) are the nitrogen analogues of epoxides and, like epoxides, their strained three-membered heterocyclic ring makes them more reactive than the corresponding open chain compounds (secondary or tertiary amines). As a result of this they have achieved usefulness in organic synthesis by virtue of their ring-opening reactions.

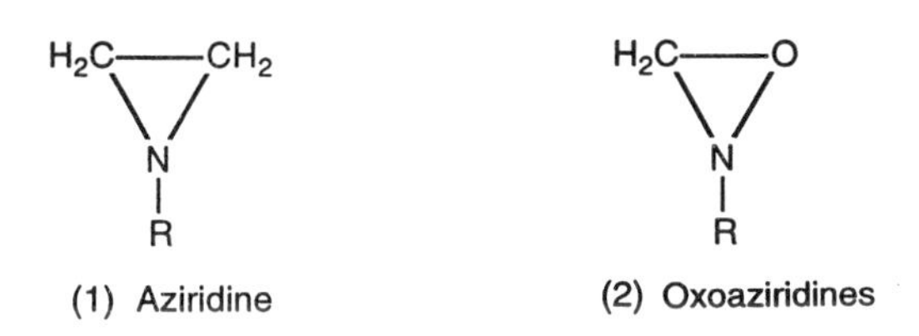

(1) Aziridine (2) Oxoaziridines

Aziridines may contain chiral nitrogen atoms. In some cases the interconversion has been shown to be slow enough to allow separation of the enantiomers.[1] This is the case with 1-chloro-2-methylaziridine. The same chirality is displayed in oxaziridines (**2**).

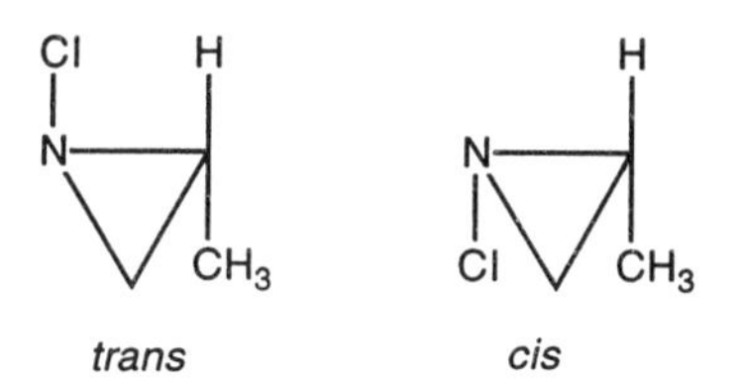

trans *cis*

21.1 Preparation

21.1.1 *From alkenes*

By reaction with iodine azide

Iodine azide (**3**), which may be prepared either by the reaction of silver azide and iodine,[2] or more conveniently using sodium azide and iodine chloride, as shown in Scheme 21.1,[3] reacts with an alkene (**4**) to form a β-iodoazide (**5**). Reduction of the product with lithium aluminium hydride constitutes a good synthesis of aziridines (**6**) (Scheme 21.2).[4]

$$NaN_3 + ICl \xrightarrow[\text{or } CH_3CN]{DMF} \underset{(3)}{IN_3} \qquad (21.1)$$

$$\underset{(4)}{>C=C<} + \underset{(3)}{IN_3} \longrightarrow \underset{(5)}{>C(I)-C(N_3)<} \xrightarrow{LiAlH_4} \underset{(6)}{\text{aziridine (N-H)}}$$

Alkene	Yield (%)
cyclohexene	80
styrene	70
cis-stilbene	63
trans-stilbene	80

(21.2)

$$AgCNO + I_2 \longrightarrow \underset{(7)}{I\text{-}N{=}C{=}O} + AgI$$

$$>C=C< + \underset{(7)}{I-N=C=O} \xrightarrow{MeOH} \text{iodonium ion} + O=C=N^- \longrightarrow \underset{(8)}{>C(I)-C(N{=}C{=}O)<} \xrightarrow{MeOH / LiOMe} \underset{(9)}{>C(I)-C(NHC(=O)OMe)<} \xrightarrow{KOH\text{-}CH_3OH} \underset{(10)}{\text{aziridine (N-H)}}$$

(21.3)

By reaction with iodine isocyanate

Iodine isocyanate (**7**) (prepared as shown in Scheme 21.3[5]) adds to alkenes to produce an iodine isocyanate (**8**). Conversion to the carbonate (**9**) is catalysed by lithium or sodium methoxide. The aziridine (**10**) is produced by reaction of base with the carbonate (Scheme 21.3).[6]

21.1.2 *From β-amino alcohols*

Reaction of β-amino alcohols (**11**) with triphenylphosphine dibromide (**12**) in triethylamine produces aziridines in moderate yields (Scheme 21.4).[7]

R²—CH(OH)-CH₂-NH-R¹ (11) + Ph_3PBr_2 (12) —Et_3N→ R²—CH—CH₂ (N-R¹ aziridine) *c.* 50 % + $Ph_3P{=}O$ (21.4)

21.1.3 *The Staudinger reaction (from epoxides)*

The Staudinger reaction (see Chapter 17, p. 658) has been applied to a synthesis of aziridines (**13**).[8] In this way epoxides may be converted into aziridines with overall retention of configuration (Scheme 21.5).

R—C(H)—C(R)—H (epoxide, O) —aq. NaN_3→ HO(R)HC—CH(R)N_3 —PPh_3, Staudinger reaction→ HO(R)HC—CH(R)N=PPh_3 ⇌ R(O⁻)HC—CH(R)NH—$\overset{+}{P}Ph_3$ → HO(R)HC—CH(R):NH_2 → R—C(H)—C(R)—H (N—H aziridine) (13) (21.5)

21.1.4 *From 1,1-disubstituted hydrazones*

A further synthesis of aziridines has been developed from certain 1,1-disubstituted hydrazones (**14**) by reaction with $Pb(OAc)_4$ in the presence of nucleophilic alkenes (**15**), which act as nitrene traps. Aziridines are produced in good to fair yield (Scheme 21.6).[9]

$$R_2N{-}NH_2 \ (14) \xrightarrow{Pb(OAc)_4} R_2N{-}\ddot{N}: \xrightarrow{R^3R^4C{=}CR^1R^2 \ (15)} R_2N{-}N\langle C(R^2)(R^1){-}C(R^3)(R^4) \rangle \qquad (21.6)$$

21.1.5 *From sulphimides*

The Michael addition of free sulphimides, in particular diphenyl free sulphimide (**16**), to alkenes in which the double bond is conjugated with an electron-withdrawing group (**17**) yields roughly equal amounts of *trans*-2-acylaziridines (**18**) and *trans*-enaminoketones (**19**) which can be separated (Scheme 21.7).[10] Optically active aziridines have been obtained by this method by the use of chiral diaryl free sulphimides.

$$Ph{-}S(=N{-}Ts){-}Ph \longrightarrow Ph{-}S(=NH){-}Ph \ (16) + PhCO{-}CH{=}CH{-}COPh \ (17)$$

$$\xrightarrow{\text{benzene, } 30^{\circ}C, \text{ 1 hour}} PhCO{-}C(NH_2){=}CH{-}COPh \ (19) + \textit{trans}\text{-2,3-dibenzoylaziridine (N–H)} \ (18) \qquad (21.7)$$

Mechanism:

$$Ph-\overset{\displaystyle NH}{\underset{}{S}}-Ph \;+\; PhCO-CH=CH-COPh \;\longrightarrow\; \left[Ph_2\overset{+}{S}-NH-\underset{\displaystyle C(=O)Ph}{CH}-\overset{-}{C}H-C(=O)-Ph \right] \longrightarrow \text{2,3-dibenzoylaziridine}$$

$$\rightleftharpoons \left[Ph_2\overset{+}{S}-NH-\underset{\displaystyle C(=O)Ph}{\overset{-}{C}}-CH=C(OH)-Ph \longleftrightarrow Ph_2\overset{+}{S}-NH-\underset{\displaystyle \|\;{}^{-}O-C-Ph}{C}-CH=C(OH)-Ph \right]$$

$$\longrightarrow PhC(=O)-C(=NH)-CH=C(OH)-Ph \longrightarrow PhC(=O)-C(NH_2)=CH-C(=O)-Ph \qquad (21.7)$$

Alkene	Aziridine Yield (%)	Enamine Yield (%)
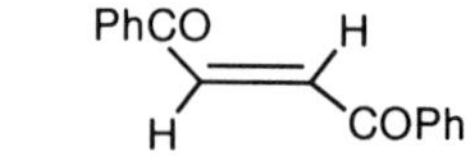 PhCO(H)C=C(H)COPh (PhCO and COPh trans)	50	50
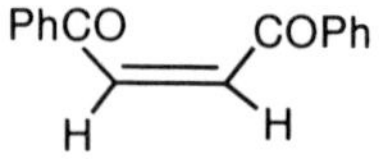 PhCO(H)C=C(H)COPh (PhCO and COPh cis)	43	55
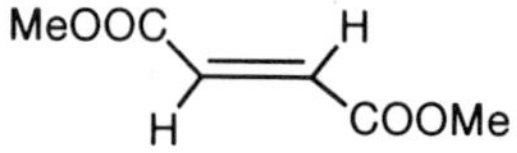 MeOOC(H)C=C(H)COOMe (MeOOC and COOMe trans)	46	23
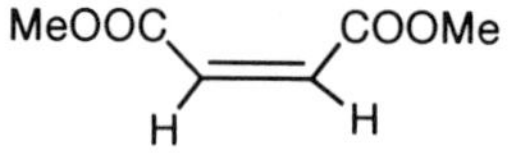 MeOOC(H)C=C(H)COOMe (MeOOC and COOMe cis)	40	26
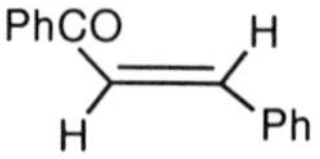 PhCO(H)C=C(H)Ph (PhCO and Ph trans)	73	0
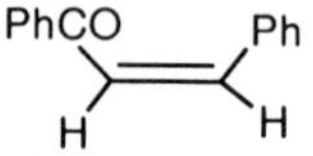 PhCO(H)C=C(H)Ph (PhCO and Ph cis)	67	0

(21.7)

21.1.6 *From 1,2-diols*

In a version of Paulsen's method dimesylates of 1,2-diols (**20**) react with hydrazine at room temperature to yield *N*-aminoaziridines (**21**), (Scheme 21.8).[11]

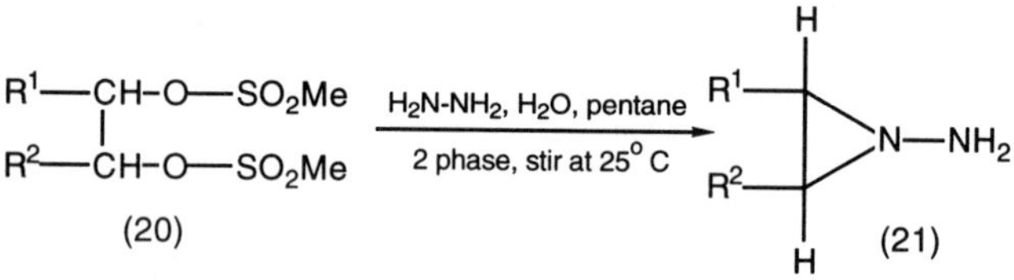

$$R^1\text{–CH–O–}SO_2Me \,/\, R^2\text{–CH–O–}SO_2Me \ (20) \xrightarrow[\text{2 phase, stir at }25^\circ\text{ C}]{H_2N\text{-}NH_2,\ H_2O,\ \text{pentane}} \text{aziridine }(R^1, R^2, H, H)\text{–N–}NH_2 \ (21)$$

R^1	R^2	Yield (%)
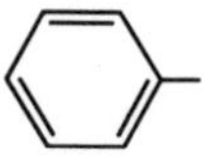 Phenyl (C_6H_5–)	H	90

(21.8)

21.1.7 *From hydroxamic acid*

A hydroxamic acid (**22**) reacts with triflic anhydride and triethylamine in CH_2Cl_2 at $-70\,°C$ to yield an aziridone (**23**) in very good yield (Scheme 21.9).[12]

Bu^t—N(OH)—C(=O)—CH_2·Ph (22) $\xrightarrow[CH_2Cl_2,\ -70°C]{(F_3C\text{-}SO_2)_2O,\ Et_3N}$ (23)

(21.9)

21.1.8 *From an enamine and phenyl azide*

A 1,3-dipolar cycloaddition of phenyl azide to an enamine (**24**) yields a triazoline (**25**) which on photolysis gives a 2-aminoaziridine (**26**) (Scheme 21.10).[13]

(24) $\xrightarrow[CHCl_3]{PhN_3\ in}$ (25) $\xrightarrow{h\nu}$ (26) + Me_2CH—C(=N—Ph)—NMe_2

3 : 2

(21.10)

21.1.9 *By dehydration of β-amino alcohols*

Aziridines can be prepared by the dehydration of β-amino alcohols (**27**). Phosphorus compounds have been used with success for this purpose (Scheme 21.11).[14]

Likewise diethoxytriphenylphosphorane ($Ph_3P(OEt)_2$) effects cyclodehydration of β-amino alcohols (**28**) to aziridines (**29**) with yields ranging between 85 and 95 per cent,[15] for example, Scheme 21.12.

$$\mathrm{BnNH\text{-}C(CH_3)_2\text{-}CH_2OH}\ (27) \xrightarrow[\mathrm{EtO_2C\text{-}N{=}N\text{-}CO_2Et}]{\mathrm{PPh_3}} \mathrm{Bn{-}N}\langle\mathrm{C(CH_3)_2{-}CH_2}\rangle \qquad (21.11)$$

$$\mathrm{H_2N\text{-}CH(Bn)\text{-}CH_2OH}\ (S)\ (28) \xrightarrow[\mathrm{Ph_3P(OEt)_2}]{\text{Toluene, } 60\ ^{\circ}\mathrm{C}} \text{2-benzylaziridine (NH)}\ (S)\ 85\,\%\ (29) \qquad (21.12)$$

Reaction of β-amino alcohols (**30**) with diphosphorus tetraiodide also effects cyclodehydration and conversion into aziridines (**31**) (Scheme 21.13).[16]

$$\mathrm{R^1\text{-}CH(OH)\text{-}CH_2\text{-}NHR^2}\ (30) \xrightarrow[\mathrm{PhH},\ 25^{\circ}\mathrm{C}]{\mathrm{P_2I_4}} \text{1-}\mathrm{R^2}\text{-2-}\mathrm{R^1}\text{-aziridine}\ \ 50\text{-}75\%\ (31) \qquad (21.13)$$

21.1.10 *From a primary amine and chloroacetyl chloride*

Aziridines (**32**) can be synthesized from a primary amine and chloroacetyl chloride (**33**), followed by reduction with aluminium hydride (Scheme 21.14). Applied to steroidal amines and to benzylamines the yields were in the range 20–85 per cent.[17]

$$\mathrm{R\text{-}NH_2} + \mathrm{Cl{-}C({=}O){-}CH_2Cl}\ (33) \longrightarrow \mathrm{RNH{-}C({=}O){-}CH_2Cl} \xrightarrow{\mathrm{AlH_3}} \mathrm{RNH{-}CH_2{-}CH_2Cl} \longrightarrow \mathrm{R{-}N}\langle\mathrm{CH_2{-}CH_2}\rangle\ (32) \qquad (21.14)$$

21.1.11 *α-Cyanoaziridines from α-chloroketimines*

α-Chloroketimines (**34**) react with KCN in methanol in a synthesis of α-cyanoaziridines (**35**) as shown in Scheme 21.15.[18]

$$R^4-C(R^3)(Cl)-C(=N-R^1)-R^2 \xrightarrow[\text{Heat}]{\text{KCN, MeOH}} \left[R^4-C(R^3)(Cl)-C(NH-R^1)(CN)-R^2 \right] \longrightarrow (35)$$

(34) → (35): aziridine with N–R^1, ring carbons bearing R^3, R^4 and R^2, CN

R^1	R^2	R^3	R^4	Yield (%)
Pr^i	Me	Me	H	88
Bu^t	Me	Me	H	74
Pr^i	Me	Me	Me	73

(21.15)

21.1.12 *Formation of tetrachloroaziridines*

Dichlorocarbene generated from phenyl(bromodichloromethyl) mercury (**36**) with carbonimidoyl dichlorides (**37**) to give tetrachloroaziridines (**38**) in fair yield (Scheme 21.16).[19]

$$\underset{(37)}{RN{=}CCl_2} + \underset{(36)}{C_6H_5HgCCl_2Br} \longrightarrow \underset{\substack{(30\text{-}55\%)\\(38)}}{R-N\langle CCl_2CCl_2 \rangle} + C_6H_5HgBr \qquad (21.16)$$

21.1.13 *Preparation of oxaziridines by oxidation of aziridines*

N-H and *N*-alkyl oxaziridines (**39**) may be prepared by the oxidation of imines (**40**) using peracids (Scheme 21.17).[20]

$$R^1{-}N{=}C(R^3)(R^2)\ \textbf{(40)}\ \xrightarrow{R^4CO_3H}\ R^1{-}N{-}C(R^3){-}R^2 \text{ (N–O–C ring) } \textbf{(39)}$$

R^1, R^2, R^3 = alkyl, aryl. (21.17)

21.2 Reactivity

21.2.1 *Ring-opening reactions*

Conversion to azomethine ylides

Aziridines (**41**) may be converted into azomethine ylides (**42**)[21] thermally (Scheme 21.18),[22] by the action of acid,[23] and by iodine ion assistance.[24]

$$R^1{-}CH{-}CH{-}C({=}O){-}R^2 \text{ (CH–CH bridged by N–H) } \textbf{(41)}\ \xrightarrow{\text{heat}}\ R^1{-}CH{=}\overset{+}{N}H{-}C(H){=}\overset{-}{C}(O){-}NHR^2\ \textbf{(42)}$$

(21.18)

ω-Aldehydoacetylenes from N-aminoaziridines

The *N*-aminoaziridines (**43**) produced above are synthetically useful since they react with cyclic epoxy ketones (**44**) to yield *ω*-aldehydoacetylenes (**45**) (Scheme 21.19).[25]

Reaction of N-protected aziridines with lithium dialkyl cuprates to give protected primary amines

N-Protected aziridines (**46**) (alkyl, benzyl, silyl, Boc) react with lithium dialkyl cuprates in the presence of excess boron trifluoride etherate. The ring opens and protected primary and secondary amines (**47**) result,[26] for example, Scheme 21.20.

(44) (43) 25° C PhH 150° C c. 60% (45)

(21.19)

Me_2CuLi + (46) $\xrightarrow[\text{THF}]{BF_3 \cdot Et_2O}$ $CH_3CH_2CH_2\text{-}NHBn$ 80 % (47)

(21.20)

Ring fission yielding fluoro compounds

Aziridines (**48**) undergo ring fission to yield fluoro compounds (**49**) on treatment with amino–hydrogen fluoride reagents,[27] for example, Scheme 21.21.

(48) $\xrightarrow{Et_3N,\ 2.5M\ HF}$ 63 % (49)

(21.21)

Synthesis of primary allylic amines

Aziridines (**50**) may be converted into allyl carbamates (**51**) by reaction with ethylchloroformate followed by thermolysis in benzene. Base hydrolysis of the resulting allyl carbamate provides a convenient synthesis of allylic amines (**52**) for example (Scheme 21.22).[28]

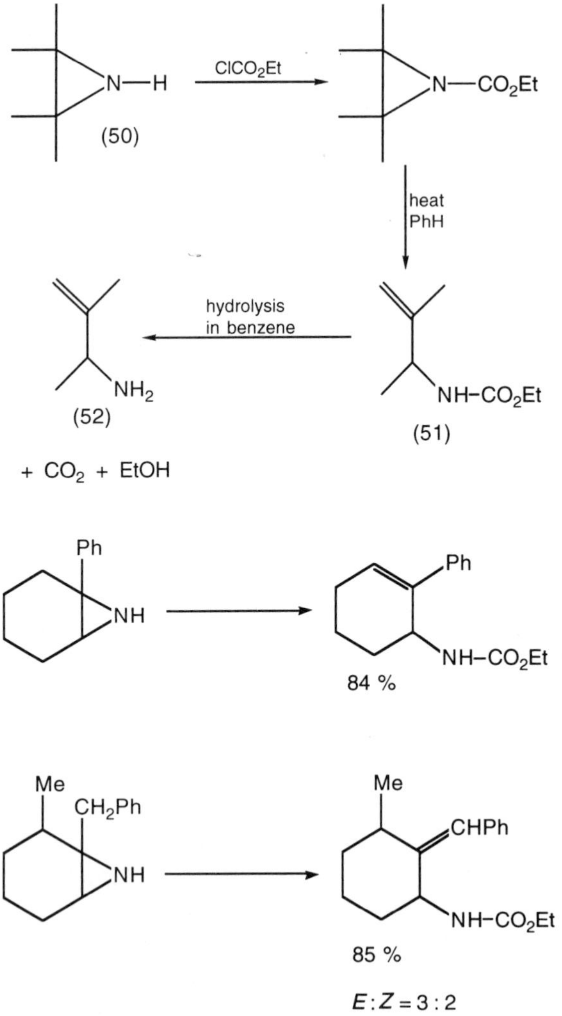

N—H (50) $\xrightarrow{ClCO_2Et}$ N—CO_2Et

heat, PhH

NH_2 (52) + CO_2 + EtOH $\xleftarrow{\text{hydrolysis in benzene}}$ NH–CO_2Et (51)

Ph, NH ⟶ Ph, NH–CO_2Et 84 %

Me, CH_2Ph, NH ⟶ Me, CHPh, NH–CO_2Et 85 % $E:Z = 3:2$ (21.22)

Formation of ethene

Aziridine (**53**) reacts with difluoroamine (**54**) in a ring opening reaction releasing N_2 and forming ethene (**55**) (Scheme 21.23).[29]

(53) $\xrightarrow[(54)]{HNF_2}$ $H_2C{=}CH_2$ + N_2 + 2 HF

80 %

(55)

(21.23)

The acylaziridine–2-oxazoline rearrangement

Reaction in the presence of a carboxylic acid results in the well known acylaziridine (**56**)–2-oxazoline (**57**) rearrangement (Scheme 21.24).[30]

R—C(=O)OH + H—N(aziridine, Me, Me) ⟶ R—C(=O)—N(aziridine, Me, Me) (56)

H^+ / Et_2O

(57)

R	Yield (%)
Me—C(=O)CH$_2$CH$_2$-	80
MeO$_2$C—(CH$_2$)$_4$-	50

(21.24)

21.2.2 *With chloroform*

Aziridine (**58**) reacts with chloroform as shown in Scheme 21.25, in a reaction in which CH_3O— and aziridine groups compete to replace Cl atoms.[31]

(58) NH $\xrightarrow[\text{Benzene}]{CHCl_3,\ CH_3ONa}$ N-C-H + N-C-H + N—C-H

OCH_3 OCH_3 OCH_3 OCH_3

(21.25)

References

1. Review: S. J. Brois, *Trans. N.Y. Acad. Sci.*, 1969, **31**, 931.
2. A. Hantzch, *Ber.*, 1900, **33**, 524.
3. A. Hassner and L. A. Levy, *J. Am. Chem. Soc.*, 1965, **87**, 4203; review: A. Hassner, *Acc. Chem. Res.*, 1971, **4**, 9; F. W. Fowler, A. Hassner, and L. A. Levy, *J. Am. Chem. Soc.*, 1967; **89**, 2077; A. Hassner and F. W. Fowler, *Tetrahedron Lett.*, 1967, 1545; A. Hassner and F. W. Fowler, *J. Org. Chem.*, **33**, 2686.
4. *F. &F.*, **3**, 160; A. Hassner, G. J. Matthews, and F. W. Fowler, *J. Am. Chem. Soc.*, 1969, **91**, 5046.
5. L. Birkenbach and M. Lindhard, *Ber.*, 1931, **64**, 961; A. Hassner and C. C. Heathcock, *Tetrahedron*, 1964, **20**, 1037.
6. *F. &F.*, **2**, 223; A. Hassner, M. E. Lorber, and C. Heathcock, *J. Org. Chem.*, 1967, **32**, 540.
7. *F. &F.*, **3**, 322; I. Okada, K. Ichimura, and R. Sudo, *Bull. Chem. Soc. Japan*, 1970, **43**, 1185.
8. *F. &F.*, **9**, 418; Y. I. Hah, Y. Sasson, I. Shahak, S. Tsaroom, and J. Blum, *J. Org. Chem.*, 1978, **43**, 4271.
9. D. J. Anderson, T. L. Gilchrist, D. C. Horwell, and C. W. Rees, *J. Chem. Soc.*, 1970, 576; *F. &F.*, **3**, 170.
10. N. Furukawa, T. Yoshimura, M. Ohtsu, T. Akasaka, and S. Oae, *Tetrahedron*, 1980, **36**, 73.
11. D. Felix, R. K. Müller, U. Horn, R. Joos, J. Schreiber, and A. Eschenmoser, *Helv. Chim. Acta.*, 1972, **55**, 1276; R. K. Müller, R. Joos, D. Felix, J. Schreiber, and C. Winter, *Org. Synth.*, 1976, **55**, 114; *F. &F.*, **5**, 328.
12. *F. &F.*, **11**, 562; C. M. Bladon and G. W. Kirby, *J. Chem. Soc., Chem. Commun.*, 1982, 1402.
13. *F. &F.*, **5**, 513; M. De Poortere and F. C. De Schryver, *Tetrahedron Lett.*, 1970, 3949.
14. *F. &F.*, **13**, 332; J. R. Pfister, *Synthesis*, 1984, 969; I. Okada, K. Ichimura, and R. Sudo, *Bull. Chem. Soc. Japan*, 1970, **43**, 1185.
15. *F. &F.*, **13**, 110; J. W. Kelly, N. L. Eskew, and S. A. Evans, Jr, *J. Org. Chem.*, 1986, **51**, 95.
16. *F. &F.*, **13**, 127; H. Suzuki and H. Tani, *Chem. Lett.*, 1984, 2129.
17. *F. &F.*, **3**, 46; Y. Langlois, H. P. Husson, and P. Potier, *Tetrahedron Lett.*, 1969, 2085.
18. *F. &F.*, **11**, 434; N. De Kimpe, L. Moëns, R. Verhé, L. De Buyck, and N. Schamp, *J. Chem. Soc., Chem. Commun.*, 1982, 19.
19. *F. &F.*, **5**, 514; D. Seyferth, W. Tronich, and H. Shih, *J. Org. Chem.*, 1974, **39**, 158.
20. F. A. Davis and A. C. Sheppard, *Tetrahedron*, 1989, **45**, 5703.
21. I. D. Blackburne, M. J. Cook, and C. D. Johnson, *C.S. Annual Reports B*, 1972, 428.
22. L. P. Kuhn, *J. Am. Chem. Soc.*, 1952, **74**, 2492.
23. L. P. Kuhn, P. V. R. Schleyer, W. F. Baitinger, and L. Eberson, *Tetrahedron Lett.*, 1964, **86**, 650.
24. F. V. Brutcher, Jr and W. Bruce, Jr, *Tetrahedron Lett.*, 1962, **84**, 2236.
25. *F. &F.*, **5**, 328; D. Felix, R. K. Müller, U. Horn, R. Joos, J. Schreiber, and A. Eschenmoser, *Helv. Chim. Acta*, 1972, **55**, 1276; R. K. Müller, R. Joos, D. Felix, J. Schreiber, and C. Winter, *Org. Synth.*, 1976, **55**, 114.

26. *F. &F.*, **13**, 208; M. J. Eis and B. Ganem, *Tetrahedron Lett.*, 1985, **26**, 1153.
27. *F. &F.*, **11**, 453–4; G. M. Alvernhe, C. M. Ennakova, S. M. Lambcombe, and A. J. Laurent, *J. Org. Chem.*, 1981, 4938; T. N. Wade, *J. Org. Chem.*, 1980, **45**, 5328.
28. *F. &F.*, **12**, 223; A. Laurent, P. Mison, A. Nafti, R. B. Cheich, and R. Chaabouni, *J. Chem. Res. (M)*, 1984, 354.
29. *F. &F.*, **1**, 254; C. L. Bumgardner, K. L. Martin, and J. P. Freeman, *J. Am. Chem. Soc.*, 1963, **85**, 99; C. L. Bumgardner and J. P. Freeman, *J. Am. Chem. Soc.*, 1964, **86**, 2233.
30. D. Haidukewych and A. I. Meyers, *Tetrahedron Lett.*, 1972, 3031.
31. *F. &F.*, **4**, 22; W. Funke, *Annalen*, 1969, **725**, 15.

22

β-LACTAMS (AZETIDINONES)

(1)

m.p. 73-75°C

(2)

(3)

β-Lactams (**1**) are cyclic amides derived from β-amino acids and they have achieved importance as one of the basic heterocyclic structures present in the antibiotics including the penicillins (**2**) (R = Bn, *n*-pentyl etc.) and the cephalosporins (**3**) (cephalosporin C 3, R = $HOOC.CH(NH_2)(CH_2)_3$). From X-ray analysis the following molecular dimensions have been derived.[1]

Infra-red data carried out on various substituted β-lactams gives the carbonyl stretching frequency between 1730 and 1760 cm^{-1}.[2]

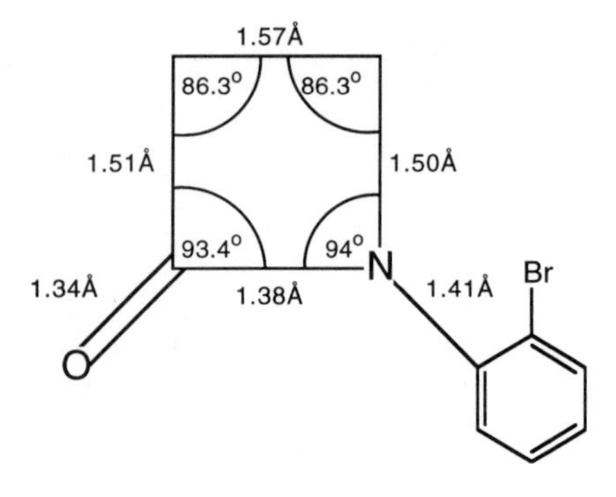

22.1 Preparation

22.1.1 *From β-amino acids*

Methods with special reagents have been developed for the preparation of β-lactams from the most obvious starting materials, viz. the β-amino acids which by the action of heat alone do not yield β-lactams. For example, a preparation from β-amino acids using a diphenylphosphoryl chloride has been described[3] (Scheme 22.1).

$HO_2CCH_2-CH(Ph)-NH_2$ (4) $\xrightarrow[Et_3N \text{ in } CH_3CN,\ 80^\circ C]{Ph_2P(O)-Cl}$ 4-phenyl-2-azetidinone (O, NH, Ph)

(22.1)

22.1.2 *From imines*

Several preparations of β-lactams from imines (**5**) have been described. Substituted acetic acids (**6**) have been caused to react with imines when added to a suspension of triphenylphosphene dibromide to give a synthesis of β-lactams (7) to moderate yields (40–60 per cent) (Scheme 22.2).[4]

$R^1-CH_2-CO_2H$ (6) + $H(R^2)C{=}N-R^3$ (5) $\xrightarrow[Et_3N,\ CH_2Cl_2]{Ph_3PBr_2}$ β-lactam (7) (R^1, R^2, $N-R^3$, O)

R^1	R^2	R^3	Yield (%)	Stereochemistry
N-phthalimido	Ph	Ph	50	*trans*
p-$MeOC_6H_4$	Ph	Ph	46	*trans*
PhO	Ph	Ph	55	*cis*
CH_3O	Ph	Ph	40	*cis*

(22.2)

β-Lactams can be prepared from imines (**8**) and chromium carbene complexes (**9**),[5] for example, Scheme 22.3.

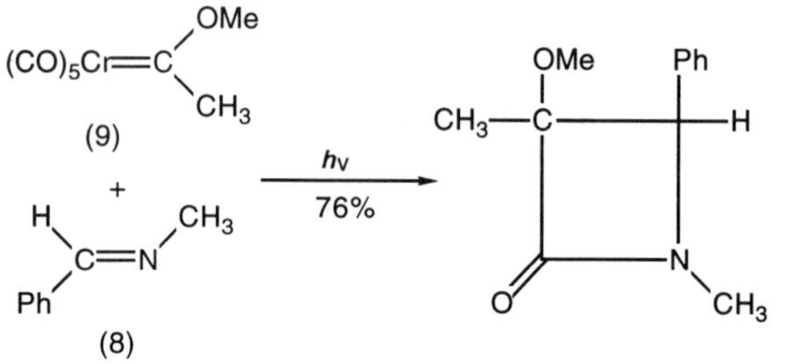

(22.3)

α-Phthalimido-β-lactams (**10**) can be synthesized by using the phenyl dichlorophosphate–dimethylformamide complex (**11**) to phosphonate a phthalimidoacetic acid (**12**) to form an intermediate which then reacts with imines to produce the β-lactams (**10**) (Scheme 22.4).[6]

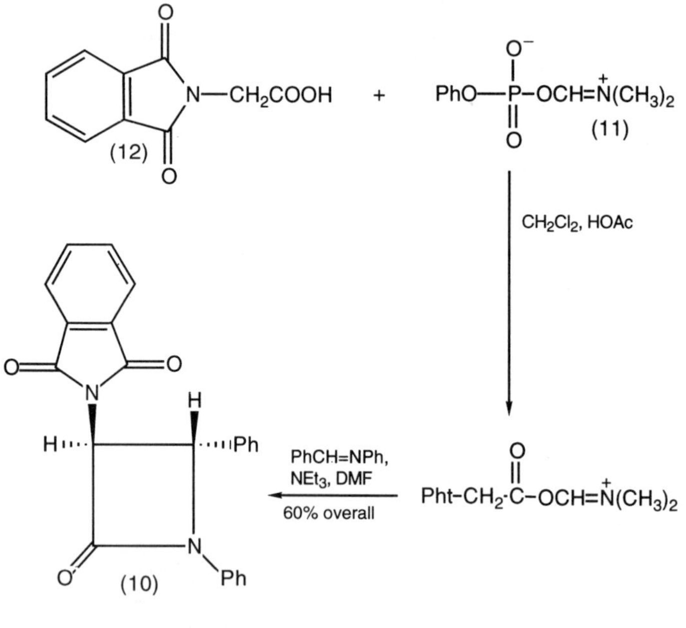

(22.4)

β-Lactams (**13**) have been prepared from carboxylic acids (**14**) and imines (**15**) in the presence of dimethylformamide and thionyl chloride, viz. with $Me_2N{=}CH{-}O{-}SOCl^+Cl^-$ (1 mole), and an excess of triethylamine (Scheme 22.5).[7]

An earlier preparation of β-lactams from α-bromoesters (**16**) and Schiff's bases (**17**) in a Reformatsky-type reaction[8] has been improved by the use of ultrasonic irradiation[9] with dioxan as solvent[10] (Scheme 22.6).

(22.5)

(22.6)

Another preparation of β-lactams from imines utilizes an iron cyclopentadienyl–carbon monoxide complex of triphenylphosphine (**18**) which can be prepared as shown in Scheme 22.7.[11] The complex (**18**) reacts with imines as shown in Scheme 22.8.[12]

(22.7)

(22.8)

N-Trimethylsilyl imines (**19**) react with silyl ketene acetals (**20**) in the presence of zinc iodide and *t*-butyl alcohol to form β-lactams (Scheme 22.9).[13]

$$R^1\text{-CHO} \xrightarrow[\text{ii, } Me_3SiCl]{\text{i, } LiN(SiMe_3)_2 / THF / 0^\circ C} R^1\text{-CH=N-SiMe}_3 \;(19)$$

$$R^1\text{-CH=N-SiMe}_3\;(19) + (R^2)(R^3)C{=}C(OCH_3)(OSiMe_3)\;(20) \xrightarrow[\text{ii, } CH_3MgBr]{\text{i, } ZnI_2 / Bu^tOH} \beta\text{-lactam } (R^2, R^3; R^1, H; \text{N-H})$$

R^1	R^2	R^3	Yield (%)	*cis*: *trans* ratio
Ph	Me	Me	75	—
Ph	H	H	27	—
Ph	Et	H	61	1:9
$Me_3SiC{\equiv}C$	Me	H	62	1:1.5
$Me_3SiC{\equiv}C$	Ph	H	53	1:12

(22.9)

Another similar preparation of *trans*-β-lactams from imines by reaction with a silyl ketene acetal (**21**) is as shown in Scheme 22.10.[14]

$$RCH{=}NR + (CH_3)(H)C{=}C(OCH_3)(OSiMe_3)\;(21) \xrightarrow[\text{(catalyst)}]{Me_3Si\text{-O-}SO_2CF_3} R\text{—CH(NHR)—CH(CH}_3\text{)—CO}_2Me \xrightarrow{LiNMe_2,\ THF} \beta\text{-lactam (CH}_3\text{, R; N-R)}$$

R = Ph
c. 60%

(22.10)

S-Tritylsulphinimine (**22**), prepared from acetaldehyde as shown in Scheme 22.11,[15] has been used for the preparation of β-lactams.[16] Reaction of *S*-tritylsulphenimine (**22**) with lithium enolates of esters (**23**) forms *N*-tritylsulphenyl β-lactams (**24**) in which the *cis* isomer predominates (Scheme 22.12).

CH_3-CHO
+
Ph_3C-S-NH_2

(pyridinium, N–H) Me–C_6H_4–SO_3^- →

CH_3-CH=N-S-CPh_3
(22)

(22.11)

$(CH_3)_2CHCH_2COOEt$
(23)

1) LDA, THF
2) $CH_3CH{=}NSPH_3$ →

H, CH_3
$(CH_3)_2CH$ ····· ····· H
O, N, $SCPh_3$
2%

+

H, H
$(CH_3)_2CH$ ····· ····· CH_3
O, N, $SCPh_3$
69%

Li-NH_3
85%
←

H, CH_3
$(CH_3)_2CH$ ····· ····· H
O, N, H
(24)

(22.12)

Another preparation from an imine utilizes (*S*)-phenylglycine (**25**) as the starting material and enables chiral β-lactams (**26**) to be prepared (Scheme 22.13).[17]

(*S*)-Ph—CH-NH_2
CO_2H
(25)

→

O, O, N-CH_2-COCl +, Ph

R, H
C
N
Bn

Et_3N, CH_2Cl_2
↓

O, O, N, Ph
H, H, R
O, N, Bn
92 - 98%

Li / liq. NH_3
THF
←

H, H
H_2N, R
O, NH
(26)
80 - 90%

(22.13)

22.1.3 *From (S)-lactic aldehyde*

Another chiral starting material which has been used is (*S*)-lactic aldehyde (**27**),[18] which after reaction with lithium hexamethylsilane produces the optically active silylimine (**28**). This reacts with the enolate of *t*-butyl butanoate to produce the optically active β-lactam (**29**) (Scheme 22.14).[19]

O
C
H
H
HO
CH3
(27)
LiN[Si(CH3)3]2
OSiR3
CH3
N
(CH3)3Si
(28)
i, EtCH=C(OBu^t)(OLi) (61%)
ii, HF, CH3CN (97%)
OH
Et
CH3
NH
O
(29)
(22.14)

22.1.4 *From methyl-(R)-3-hydroxybutyrate*

Methyl-(*R*)-3-hydroxybutyrate (**30**) is also used to produce chiral β-lactams (**31**) using diethyl zinc (Scheme 22.15).[20]

OH
CH3
COOCH3
(30)
Et2Zn
THF
OZnEt
CH3
COOCH3
LiN[Si(CH3)3]2
Et
Zn
O
OLi
CH3
OCH3
(CH3)3SiN:CHC:CPh
85%
HO
H
H
CH3
C≡CPh
NH
O
(31)
(22.15)

22.1.5 *From zinc enolates of α-amino acids*

Zinc enolates of *N*,*N*-disubstituted α-amino acids (**32**) react with imines to form β-lactams (**33**) in 60–90 per cent yield. Depending on the substituent on the imine and the solvent the reaction can be *trans* selective (Scheme 22.16).[21]

$(CH_3)_2NCH_2COOEt$ (32) —[i) LDA; ii) $ZnCl_2$; iii) $(CH_3)_3SiN{:}CHC{:}CSi(CH_3)_3$]→ (33): β-lactam bearing H, $(CH_3)_2N$, $C{\equiv}CSi(CH_3)_3$ and N–$Si(CH_3)_3$; 90%; *trans*:*cis* = >99:1

(22.16)

22.1.6 *From α-diazoketoamide*

Preparation of β-lactams from a diazoketoamide (**34**) catalysed by dirhodium tetraacetate is also possible via the insertion of the carbene into the β C—H bond. The use of *tert*-butyl and the rhodium catalyst ensures high yields of the four-membered ring (Scheme 22.17).[22]

$(CH_3)_3C$–N($PhCH_2$)–C(=O)–C(=N_2)–$COCH_3$ (34) —[$Rh_2(OAc)_4$, C_6H_6; 98%]→ β-lactam bearing CH_3–C(=O), Ph and N–$C(CH_3)_3$

(22.17)

22.1.7 *From an oxime from an α-alkyl-β-ester*

Cis-3,4-disubstituted β-lactams (**35**) can be produced from the relevant oxime of an α-alkyl-β-ketoester (**36**). The high selectivity of the sodium cyanoborohydride reduction leading to the *syn*-β-hydroxyamino ester allows cyclization to the *cis*-3,4-disubstituted β-lactam after reduction and hydrolysis (Scheme 22.18).[23]

(36) $\xrightarrow{NaBH_3CN,\ HOAc}$ RO–CO–CH(CH_3)–CH(CH_3)–NHOH $\xrightarrow{1)\ H_2,\ Ni;\ 2)\ H_3O^+}$ RO–CO–CH(CH_3)–CH(CH_3)–NH_2 $\xrightarrow{DCC}$ (35) 62% overall

(22.18)

22.1.8 *From N-(cyanomethyl)amines*

Benzylamine hydrochloride (**37**) reacts with potassium cyanide and aqueous formaldehyde (**38**) to give benzylaminoacetonitrile (**39**) (Scheme 22.19).[24] The reaction of *N*-(cyanomethyl)amines (**40**) of this type with lithium ester enolates (**41**) provides a synthesis of *β*-lactams (Scheme 22.20).[25]

$$\underset{(37)}{Ph\text{-}CH_2\text{-}\overset{+}{N}H_3\ Cl^-} + KCN + \underset{(38)}{CH_2O} \xrightarrow{\text{in } H_2O} \underset{(39)\ 65\ \%}{Ph\text{-}CH_2NH\text{-}CH_2\text{-}CN} \qquad (22.19)$$

22.1.9 *From aziridines*

Several methods have been described for the conversion of aziridines (**42**) into *β*-lactams by carbonylation. This may be catalysed by chlorodicarbonyl-rhodium(I) dimer (Scheme 22.21).[26] The substituent on the aziridine ring must be aromatic and the substituent on the nitrogen atom must be *tert*-butyl. When Ar = Ph the yield is 93 per cent and the product is optically pure.

Use of nickel tetracarbonyl to effect carbonylation after the aziridine (**43**) has been treated with lithium iodide leads to the production of the *β*-lactam (**44**) in 51 per cent yield (Scheme 22.22).[27]

R^1—CH(R^2)—CO_2Et —(LDA, THF, -70°C)→ R^1—$\bar{C}$(R^2)—CO_2Et (41) + NC-CH_2-NH-CH_2Ph (40)

R^1—CH(R^2)-C(O-Et)=O + NC—CH_2—$\bar{N}$—CH_2Ph

EtO⁻ + R^1—CH(R^2)-C=O, NC-CH_2—NCH_2Ph

EtOH + R^1—$\bar{C}$(R^2)—C=O, NC-CH_2-NCH_2Ph

R^1, R^2, O, N, CH_2Ph

R^1	R^2	Yield (%)
Me	Me	66
Me	SPh	62
-$(CH_2)_5$	-$(CH_2)_5$	65

(22.20)

Ar, N, Bu^t (42) —($[(CO)_2RhCl]_2$ catalyst, 90°C, 20 atom)→ Ar, O, N, Bu^t

(22.21)

Bn, N, Me (43) —(LiI, THF, Reflux)→ [Bn, N—Li, Me—CH—CH_2-I] —($Ni(CO)_4$)→ Bn, N, O, Me, 51% (44)

(22.22)

22.1.10 *From amides*

From α,β-unsaturated amides

β-Lactams can be prepared from reaction of α,β-unsaturated amides (**45**) and benzenesulphenyl chloride, PhSCl, followed by ring closure induced by base (Scheme 22.23).[28,29]

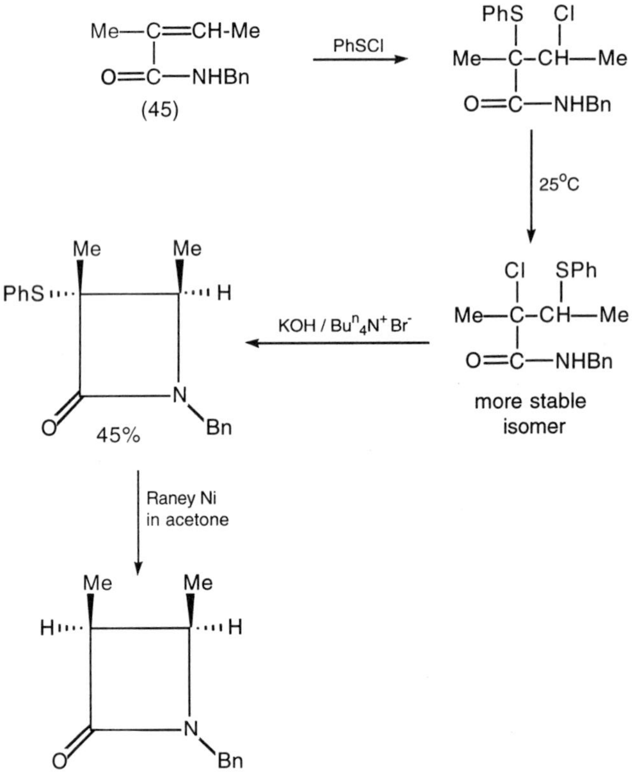

(22.23)

β-Lactams can also be prepared from *N*-tosylates of unsaturated amides (**46**) as shown in Scheme 22.24,[30] for example.

From β-bromoamides

Phase transfer catalysis has been used to prepare β-lactams from β-bromoamides (**47**) (Scheme 22.25).[31]

From α-ketoamides

β-Lactams can be synthesized from α-ketoamides by first converting them into arylsulphonylhydrazides (**48**) and then removing an acidic proton with butyl

CH_2-CH=CH_2 / O=C—NHTs (46) $\xrightarrow[NaHCO_3]{Br_2 / CH_2Cl_2}$ 4-(CH_2Br)-1-Ts-azetidin-2-one, 67%

$\downarrow$ $Bu^n{}_3SnH$ / THF

4-CH_3-1-Ts-azetidin-2-one, 92% (22.24)

R-NH_2 + Br-CH_2-CH_2-COCl $\xrightarrow[CH_2Cl_2]{PhNMe_2}$ R-NH—C(=O)—$(CH_2)_2$Br (47) 54 - 93%

$\downarrow$ KOH, $Bu^n{}_4N^+ Br^-$, CH_2Cl_2, CH_3CN, in high dilution

1-R-azetidin-2-one, 70 - 95% (22.25)

lithium, causing the anion (**49**) so formed to react with an aldehyde, followed by tosylation of the resulting hydroxy group and reductive ring closure with sodium hydride (Scheme 22.26).[32]

22.1.11 *From γ-amino acids*

High yields of β-lactams result from the treatment of γ-amino acids (**50**) with 2-chloro-*N*-methylpyridinium chloride and triethylamine,[33] for example, Scheme 22.27.

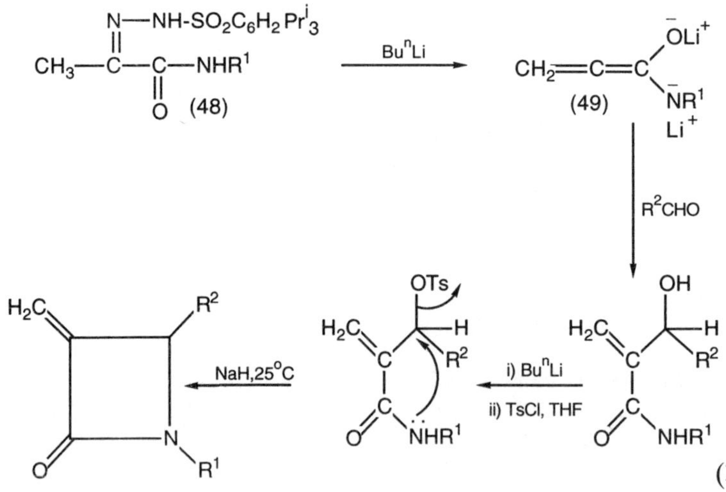

(22.26)

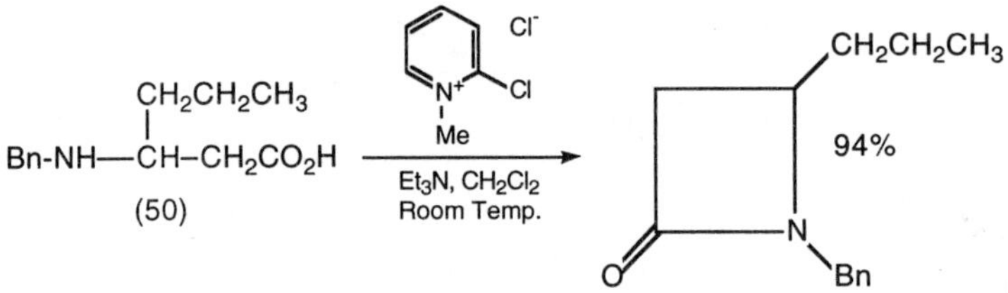

(22.27)

22.1.12 *From allyl iodide*

A preparation in low yield from a $[2+2]$ cycloaddition has been described from an allyl iodide (**51**) (Scheme 22.28).[34]

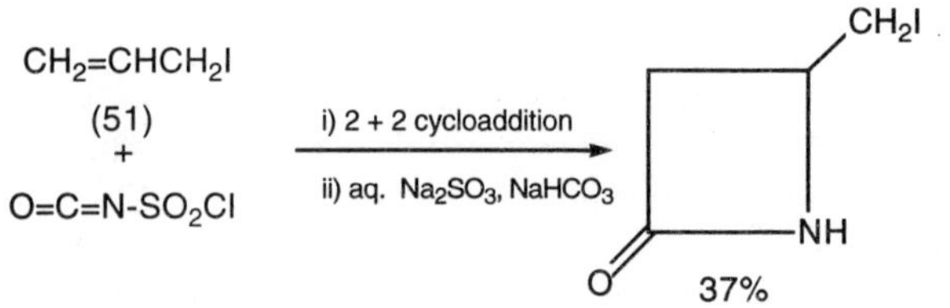

(22.28)

22.1.13 *From phenylserine*

A β-lactam can be prepared from protected phenylserine (**52**) by dehydration as shown in Scheme 22.29.[35]

(52) → (22.29)

22.1.14 *Using N,N-bis(trimethylsilyl)methoxymethylamine*

Reaction of *N*,*N*-bis(trimethylsilyl)methoxymethylamine (**53**) with ketene silyl acetals (**54**) affords *N*,*N*-bis(trimethylsilyl)-β-amino esters (**55**), which are readily converted to β-lactams via the β-amino acids (**56**), after desilylation and treatment with base (Scheme 22.30).[36]

(22.30)

22.1.15 *From keteniminium salts*

Keteniminium salts (**57**) can react with imines to yield β-lactams (**58**) (Scheme 22.31). The starting material is an enamine (**59**) which can be isolated. For the preparation see Ref. 37.

Ketenimine ions (**60**) react with imines to form β-lactams (Scheme 22.32).[38]

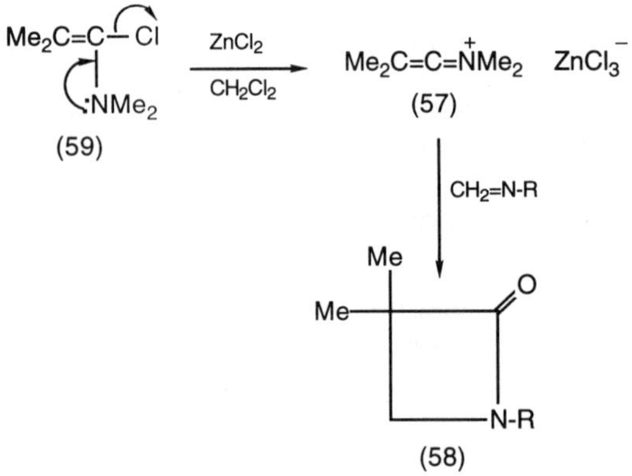

(22.31)

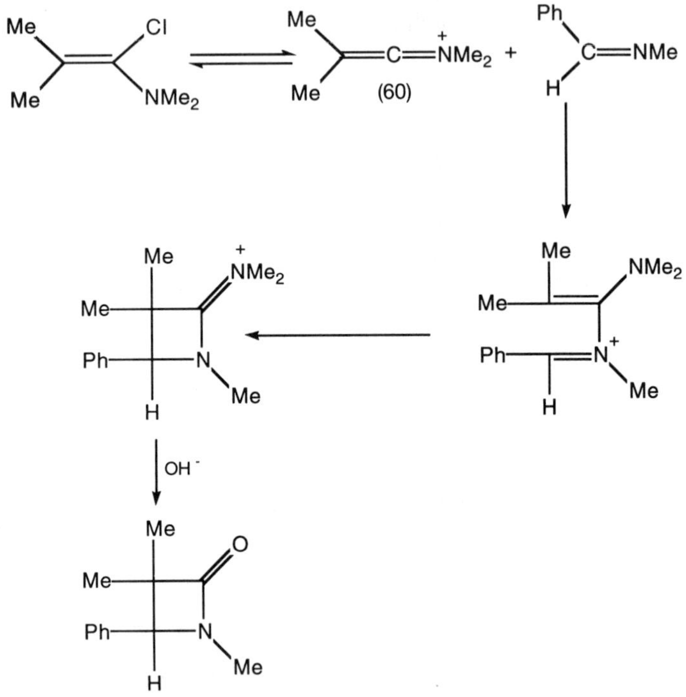

(22.32)

22.2 Reactivity

As is to be expected from the presence of a strained four-membered ring, β-lactams are more easily hydrolysed by acids and alkalis than are non-cyclic aliphatic amides. The reactivity can be accounted for by the relief of strain in the C—C—N bond angle as the geometry round the carbonyl carbon atom changes from planar sp^2 to tetrahedral sp^3.[39]

22.2.1 *Reduction*

The reduction of β-lactams with diborane in tetrahydrofuran yields 1,3-amino alcohols (**61**),[40] for example, Scheme 22.33.

$(CH_2)_3CH_3$ N Me O $\xrightarrow[-10^{\circ}C]{B_2H_6,\ THF}$ HO-CH_2CH_2—CH—$(CH_2)_3CH_3$ (NHMe) 78% (61) (22.33)

N-Substituted β-lactams (**62**) can be reduced without ring fission by chloroaluminium hydride or with dichloroaluminium hydride (for preparation see Ref. 41) (Scheme 22.34).[42]

BnO R^1 N R^2 O (62) $\xrightarrow{AlH_2Cl}$ BnO R^1 N R^2 85-100% (22.34)

22.2.2 *Benzoylation*

β-Lactams are benzoylated α to the nitrogen atom by *t*-butyl perbenzoate (**63**) in the presence of cupric octanoate. However, the substituent on the nitrogen atom must not contain an α-hydrogen atom (Scheme 22.35).[43]

N Bu^t O + Ph-CO_3Bu^t (63) $\xrightarrow{\text{cupric octanoate}}$ O-CO-Ph N Bu^t O 59% (22.35)

22.2.3 *Alkylation*

Acyloxy-β-lactams (**64**) can by alkylated by allyl trimethylsilane (**65**) in the presence of a Lewis acid (BF_3 or $SnCl_4$) (Scheme 22.36).[44]

OAc

NH

(64)

+

$Me_3Si\text{-}CH_2\text{-}CH{=}CH_2$

(65)

$\xrightarrow[\text{CH}_2\text{Cl}_2,\ 25^{\circ}\text{C}]{\text{BF}_3.\text{Et}_2\text{O}}$

$CH_2CH{=}CH_2$

NH

63%

(22.36)

References

1. H. Fujiwara, R. L. Varley, and J. M. van der Veen, *J. Chem. Soc., Perkin Trans. II*, 1977, 547.
2. A. R. Katrinsky (ed.), *Comprehensive heterocyclic chemistry*, Pergamon Press, Oxford, 1984, p. 248.
3. S. Kim, P. H. Lee, and T. A. Lee, *J. Chem. Soc., Chem. Commun.*, 1988, 1242.
4. *F. &F.*, **13**, 333; F. P. Cossio, I. Ganboa, and C. Palomo, *Tetrahedon Lett.*, 1985, **26**, 3041.
5. *F. &F.*, **11**, 400; M. A. Mcguire and L. S. Hagadus, *J. Am. Chem. Soc.*, 1982, **104**, 5538.
6. *F. &F.*, **11**, 410; A. Arrieta, J. M. Aizpurua, and C. Palomo, *Synth. Commun.*, 1982, **12**, 967.
7. *F. &F.*, **12**, 204; A. Arrieta, J. M. Aizpura, and C. Palomo, *Tetrahedron Lett.*, 1984, **25**, 3365.
8. H. Gilman and M. Speeter, *J. Am. Chem. Soc.*, 1943, **65**, 2255.
9. *F. &F.*, **11**, 599.
10. *F. &F.*, **12**, 567; A. K. Bose, K. Gupta, and M. S. Manhas, *J. Chem. Soc., Chem. Commun.*, 1984, 86.
11. J. P. Bibler and A. Wojcicki, *Inorg. Chem.*, 1966, **5**, 889.
12. L. S. Liebeskind, M. E. Welker, and V. Goedken, *J. Am. Chem. Soc.*, 1984, **106**, 441.
13. *F. &F.*, **13**, 350; E. W. Colvin and D. G. McGarry, *J. Chem. Soc., Chem. Commun.*, 1985, 539.
14. *F. &F.*, **14**, 333; J. R. Hwu and J. M. Wetzel, *J. Org. Chem.*, 1985, **50**, 3946.
15. E. P. Branchaud, *J. Org. Chem.*, 1983, **48**, 3531.
16. *F. &F.*, **13**, 1; D. A. Burnett, D. J. Hart, and J. Liu, *J. Org. Chem.*, 1986, **51**, 1929.
17. *F. &F.*, **13**, 225; D. A. Evans and E. B. Sjogren, *Tetrahedron Lett.*, 1985, **16**, 3783.
18. M. Hirana, I. Nishizaki, T. Shigemoto, and S. I. Oto, *J. Chem. Soc., Chem., Commun.*, 1986, 393.
19. *F. &F.*, **15**, 181; G. Cainelli, M. Panuncio, D. Giaconini, G. Martello, and G. Spunta, *J. Am. Chem. Soc.*, 1988, **110**, 6879.
20. *F. &F.*, **15**, 207; N. Oguni and Y. Ohkawa, *J. Chem. Soc., Chem. Commun.*, 1988, 1376.

21. *F. &F.*, **15**, 370; F. H. van der Steen, H. Kleijn, J. T. B. H. Jastrzebski, and G. van Koten, *Tetrahedron Lett.*, 1989, **30**, 765.
22. *F. &F.*, **15**, 279; M. P. Doyle, M. S. Shanklin, S.-M. Oon, H. O. Pho, F. R. van der Heide, and W. R. Veal, *J. Org. Chem.*, 1988, **53**, 3384.
23. *F. &F.*, **14**, 287; T. Chiba, T. Ishizawa, J. Sakaki, and C. Kaneko, *Chem. Pharm. Bull.*, 1987, **35**, 4672.
24. W. Baker, W. D. Ollis, and V. D. Poole, *J. Chem. Soc.*, 1949, 307.
25. *F. &F.*, **13**, 29; L. E. Overman and T. Osawa, *J. Am. Chem. Soc.*, 1985, **107**, 1698.
26. *F. &F.*, **15**, 82; S. Calet, F. Urso, and H. Alper, *J. Am. Chem. Soc.*, 1989, **111**, 931.
27. *F. &F.*, **15**, 216; W. Chamchaang and A. R. Pinhas, *J. Chem. Soc., Chem. Commun.*, 1988, 710.
28. *F. &F.*, **11**, 40; **12**, 42; M. Ihara and K. Fukumoto, *Heterocycles*, 1982, **19**, 1435.
29. M. Ihara, Y. Haga, M. Yonekura, T. Ohsawa, K. Fukumoto, and T. Kametani, *J. Am. Chem. Soc.*, 1983, **105**, 7345.
30. *F. &F.*, **11**, 76; A. J. Biloski, R. D. Wood, and B. Ganem, *J. Am. Chem. Soc.*, 1982, **104**, 3233.
31. *F. & F.*, **11**, 406; H. Takahata, Y. Ohnishi, H. Takehara, K. Tsuitana, and Y. Yamazaki, *Chem. Pharm. Bull. Japan*, 1981, **29**, 1063.
32. *F. & F.*, **11**, 536; R. M. Adlington and A. G. M. Barrett, *J. Chem. Soc., Chem. Commun.*, 1981, 65; R. M. Adlington, A. G. M. Barrett, P. Quayle, and M. J. Betts, *J. Chem. Soc., Chem. Commun.*, 1981, 404.
33. H. Huang, N. Iwasawa, and T. Mukaiyama, *Chem. Lett.*, 1984, 1465.
34. *F. &F.*, **11**, 125; T. Tanaka and T. Miyadera, *Heterocycles*, 1982, 1497.
35. *F. &F.*, **12**, 553; A. K. Bose, M. S. Manhas, D. P. Sahu, and V. R. Hegde, *Can. J. Chem.*, 1984, **62**, 2498.
36. *F. &F.*, **12**, 62; K. Okano, T. Morimoto, and M. Sekiya, *J. Chem. Soc., Chem. Commun.*, 1984, 883.
37. L. Ghosez, B. Haveaux, and H. G. Viehe, *Angew. Chem., Int. Ed. Engl.*, 1969, **6**, 454.
38. M. De Poortere, J. Manchand-Brynaert, and L. Ghowez, *Angew. Chem., Int. Ed. Engl.*, 1974, **13**, 267.
39. G. M. Blackburn and J. D. Plackett, *J. Chem. Soc., Perkin Trans. II*, 1972, 1366.
40. *F. &F.*, **11**, 156; P. G. Sammes and S. Smith, *J. Chem. Soc., Chem. Commun.*, 1982, 1143.
41. *F. &F.*, **1**, 595.
42. *F. &F.*, **12**, 333; M. Yamashita and I. Ojima, *J. Am. Chem. Soc.*, 1983, **105**, 6339.
43. *F. &F.*, **13**, 58–9; C. J. Easton and S. G. Love, *Tetrahedron Lett.*, 1986, **27**, 2315.
44. *F. &F.*, **11**, 16; G. A. Kraus and K. Neuenschwander, *J. Chem. Soc., Chem. Commun.*, 1982, 134.

SUBJECT INDEX